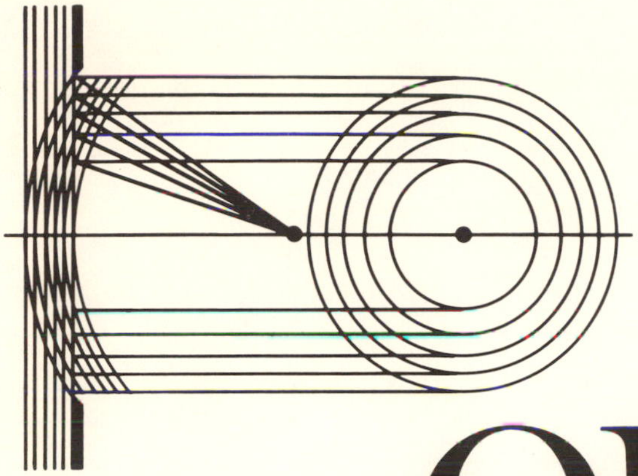

OPTICS

MILES V. KLEIN
PROFESSOR OF PHYSICS, UNIVERSITY OF ILLINOIS AT URBANA—CHAMPAIGN

JOHN WILEY & SONS, INC. New York, London, Sydney, Toronto

Library of Congress Catalogue Card Number: 73-107584

SBN 471 49080 6

Printed in the United States of America

10 9 8 7 6 5 4 3 2 1

To Bobbi, Cindy, and Gail

Preface

This book has evolved from lecture notes used for several years in a one-semester course at the University of Illinois. The students were juniors, seniors, and first-year graduate students in physics, primarily, but there were also some in electrical engineering, mechanical engineering, and chemistry. The subject matter of the course varied from year to year, and the notes were accordingly revised and expanded. Additional material has also been included from a senior-level course on electromagnetic waves that I taught. This text is intended to serve in a one-semester to one-year course at the advanced undergraduate, first-year graduate level.

Within the last decade the new laser technology has revitalized and extended the subject of optics. The high degree of coherence of laser light has been exploited in areas such as holography and spatial filtering. The high intensities possible with some lasers have made nonlinear optics a vigorous new field. Such developments have had their impact on the more traditional branches of optics because of technical demands for better and more varied components and sometimes through changes in viewpoint.

Given these innovations, the instructor in an intermediate optics course may have difficulty deciding what to include and what to leave out. In many institutions, such as my own, such a course is a terminal one. A further complication results from the varied backgrounds of the students. This can be met by keeping the list of formal prerequisites low.

Subject matter of this book has been chosen with two general aims in mind. The first is to give the reader enough of the principles behind practical optical components and systems so that he can do effective laboratory work. The second aim is for the reader to learn a sufficient amount of the theory of physical optics to understand well the important phenomena, old and new. This includes being able to take the next step, namely, to gain at least an entry into the advanced treatments and the optics literature. I have placed some limitations on this second aim because of lack of space and because of inadequate background on the part of the students. The most significant omission is probably the quantum theory of the emission and absorption of light. Consequently there is not a detailed discussion of laser action, although an elementary treatment is given in an appendix along with a discussion of the optics of laser cavities.

I have also tried to cover as much optics as I can without using the full electromagnetic theory. This simplifies the presentation and does not burden the student who has not yet been exposed to an intermediate course in electromagnetic theory. Indeed, the only formal prerequisites are sophomore physics and advanced calculus. Those readers and instructors who prefer to cover the electromagnetic nature of light earlier may wish to include parts of Chapters 10 and 11, for example Sections 10.1 and 11.1, immediately after the elementary discussion of wave motion in Chapter 1 or after the discussion of interference in Chapter 5.

The reader will find material of varying difficulty in this book. Wherever possible the discussion of a new major subject has been started at essentially the sophomore physics level, and the level of difficulty has increased as the subject matter becomes more complex. For example, the first two chapters build toward the rather detailed geometrical optics in Chapters 3 and 4. In Chapter 3 itself the foundation of the matrix theory of Gaussian optics is gradually laid, but the matrices themselves are not used until after principal planes and their uses have been explored.

Chapter 5 begins slowly with a discussion of the addition of two or more simple harmonic motions. The classical interference phenomena are then discussed in terms of this superposition principle.

Chapter 6, which covers various aspects of coherence, starts with a discussion of how to determine spectral distributions with an interferometer. The result is then reformulated in terms of the autocorrelation function of the light signal. A similar analysis is then given of the role of a spatial distribution of light flux in a two-slit interference experiment. These two devices are used to discuss coherence time and coherence length. Intensity fluctuations are then discussed. This material is considerably more difficult than that in Chapter 5, and it involves topics such as the Fourier integral and correlation functions which students often find strange. I feel, however, that these are important subjects of great usefulness, and students should be acquainted with them as early as possible.

Diffraction is introduced at the start of Chapter 7 as a simple extension of the superposition principle, and the more rigorous proofs are postponed until Chapter 8. Fraunhofer diffraction takes up most of Chapter 7. The Fourier-transform properties of the diffraction patterns are noted, and most of the standard topics are treated.

In Chapter 8 the discussion of Fresnel diffraction leads to zone plates and on to holography—a topic growing so rapidly that only the essential ideas can be covered. This is done at an elementary level so that about 80 per cent of the discussion can be understood without reference to the previous material on diffraction.

Chapter 9 covers some of the recent developments of modern optics that depend on the diffraction theory of image formation. Included are spatial filtering and the optical transfer function; these are concepts brought into optics from electrical engineering. Extensive use is made of Fourier integrals, and the material, as in Chapter 6, is more difficult than that in the rest of the book.

In Chapter 10 the transverse nature of light oscillations is discussed from an elementary point of view. The basic properties of polarized light are covered and, finally, a more sophisticated treatment of partially polarized light is given using the coherency matrix.

Chapter 11 is on the electromagnetic theory of light. Familiarity with Maxwell's equations is assumed. The classical theory of dispersion is discussed in some detail. There is also a complete treatment of the reflection and transmission of light at an interface and through a layer. The chapter closes with an introduction to crystal optics. Enough material is presented to enable the reader to follow advanced treatises or to begin a study of nonlinear optics.

There are numerous problems. These are to be regarded as an integral part of the text. They often provide examples to illustrate general principles, and in some instances they lead the student through an extension of the textual material.

I am indebted to many colleagues and students for helpful comments and remarks, especially to Prof. R. D. Sard and to Mr. H. Macksey, who made countless useful suggestions. I am also grateful to Miss Nila Meredith for her careful typing of the various versions of the lecture notes and to Mr. Marc de Peo for making some of the diffraction photographs.

Miles V. Klein

Urbana, Illinois
May, 1969

Contents

OPTICS

Visible light consists of electromagnetic radiation with wavelengths restricted to lie in a narrow region of the spectrum from about 4×10^{-5} cm *to* 7.2×10^{-5} cm. *Many of its physical properties are determined by the nature of electromagnetic waves and their interactions with matter and are thus shared with nonvisible forms of electromagnetic radiation, particularly infrared and ultraviolet radiation. Indeed, many important optical phenomena can be understood as manifestations of wave phenomena in general; it is not necessary to invoke the peculiar properties of electromagnetic fields to explain them. It is therefore quite appropriate to begin our discussion of optics with a treatment of wave motion and propagation of disturbances in simple systems.*

We start with a case having only one independent spatial variable which represents the distance along the direction of propagation of a wave or wavelike disturbance. The dependent variable, which represents the quantity doing the "waving," will be transverse and thus represent motion at right angles to the direction of propagation. Light waves share this property of being transverse.

CHAPTER 1

WAVELIKE PROPAGATION

CHAPTER 1 CONTENTS

1.1 WAVES ON A STRING

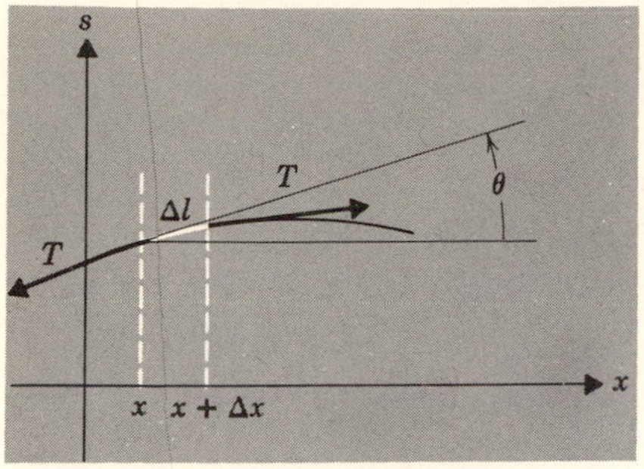

Fig. 1.1. Transverse displacement of a string. The segment between x and $x+\Delta x$ is acted upon by the tensions T along the tangent to the string at both ends.

A. WAVE EQUATION

The propagation of a transverse disturbance, or a transverse wave, on a string is easy to treat mathematically. Consider a uniform string stretched along the x axis under a tension T. It will be assumed to lie always in a plane, the plane of the paper in Fig. 1.1. (Bowing of the string because of its weight is irrelevant and will be neglected.) We intend to show that small transverse displacements, s, of the string obey the one-dimensional wave equation

$$\frac{\partial^2 s}{\partial x^2} - \frac{1}{v^2}\frac{\partial^2 s}{\partial t^2} = 0 \tag{1.1}$$

Here v represents the speed of the waves. This equation is of quite general applicability. Its importance transcends that of any particular system obeying it.

The displacement of the string from equilibrium $s(x, t)$ is plotted as a function of x for a particular value of the time t in Fig. 1.1. To derive Eq. (1.1) for the string, we begin by applying Newton's second law of motion to a small section of the string of length Δl located between x and $x+\Delta x$, as shown in Fig. 1.1. At any given point the tension always acts along the tangent to the string. The inclination angle θ can be obtained from

$$\tan\theta = \frac{\partial s}{\partial x} \tag{1.2}$$

We now make the necessary but reasonable assumption that the angle θ measured in radians is small compared with unity, that is, that

$$|\theta| \ll 1 \tag{1.3}$$

This allows us to make these approximations:

$$\sin\theta \approx \tan\theta \approx \theta \tag{1.4}$$

and

$$\cos\theta \approx 1 \tag{1.5}$$

The horizontal and vertical components of the tension at a point where the inclination is θ then have magnitudes

$$T_{\text{hor}} = T\cos\theta \approx T, \quad \text{a constant} \tag{1.6}$$

and

$$T_{\text{vert}} = T\sin\theta \approx T\tan\theta = T\frac{\partial s}{\partial x} \tag{1.7}$$

The last equality follows from Eq. (1.2).

The net horizontal force on our segment of string is then zero, implying no acceleration in the x direction, whereas there is a net vertical force equal to

$$F_{\text{vert}} = T\frac{\partial s}{\partial x}(x+\Delta x, t) - T\frac{\partial s}{\partial x}(x, t) \tag{1.8}$$

This must be equated to $(\mu \Delta x a_c/\cos\theta) \approx \mu \Delta x a_c$, where μ is the mass per unit length of the string and a_c is the vertical or transverse component of the acceleration of the center of mass of the segment. The latter will be located near the center of the segment, and we can write approximately

$$a_c = \frac{\partial^2 s}{\partial t^2}\left(x+\frac{\Delta x}{2}, t\right) \tag{1.9}$$

Putting Eqs. (1.8) and (1.9) into the equation of motion gives us

$$T\left[\frac{\partial s}{\partial x}(x+\Delta x, t) - \frac{\partial s}{\partial x}(x, t)\right] = \mu \Delta x \frac{\partial^2 s}{\partial t^2}\left(x+\frac{1}{2}\Delta x, t\right) \tag{1.10}$$

We next divide this equation by Δx and take the limit as Δx tends to zero, which allows us to apply the definition of a derivative and obtain

$$T\frac{\partial^2 s}{\partial x^2} = \mu \frac{\partial^2 s}{\partial t^2} \tag{1.11}$$

Equation (1.11) is of the form of (1.1) provided that

$$v^2 = \frac{T}{\mu} \tag{1.12}$$

The reader should check for himself that the units of T/μ are indeed those of a velocity squared.

B. GENERAL SOLUTION—SUPERPOSITION PRINCIPLE

The general solution of Eq. (1.1) takes the form

$$s(x, t) = f\left(t - \frac{x}{v}\right) + g\left(t + \frac{x}{v}\right) \tag{1.13}$$

where $f(u)$ and $g(u)$ are arbitrary functions of the variable u having well-defined second derivatives. By setting $u = t - x/v$ or $t + x/v$, we obtain a function of two variables. Before proving that Eq. (1.13) is a solution, we make a very important comment about the solutions of the wave equation (1.1) and almost all other wave equations, namely, that *the sum of two solutions is also a solution.* More generally, if s_1 and s_2 are solutions, so is

$$s = as_1 + bs_2$$

where a and b are arbitrary constants. The proof is straightforward. We write

$$\frac{\partial^2 s}{\partial x^2}-\frac{1}{v^2}\frac{\partial^2 s}{\partial t^2}=\frac{\partial^2}{\partial x^2}(as_1+bs_2)-\frac{1}{v^2}\frac{\partial^2}{\partial t^2}(as_1+bs_2)$$

$$=a\frac{\partial^2 s_1}{\partial x^2}+b\frac{\partial^2 s_2}{\partial x^2}-\frac{a}{v^2}\frac{\partial^2 s_1}{\partial t^2}-\frac{b}{v^2}\frac{\partial^2 s_2}{\partial t^2}$$

$$=a\left(\frac{\partial^2 s_1}{\partial x^2}-\frac{1}{v^2}\frac{\partial^2 s_1}{\partial t^2}\right)+b\left(\frac{\partial^2 s_2}{\partial x^2}-\frac{1}{v^2}\frac{\partial^2 s_2}{\partial t^2}\right)=0+0=0$$

This property of the solutions is often called the *superposition principle*. Using it, we can content ourselves with showing that the functions f and g in Eq. (1.13) separately satisfy the wave equation.

We give only the proof for f, which goes as follows. We differentiate f once with respect to x and t,

$$\frac{\partial f}{\partial t}=f'\left(t-\frac{x}{v}\right); \qquad \frac{\partial f}{\partial x}=-\frac{1}{v}f'\left(t-\frac{x}{v}\right)$$

where $f'(u)$ is an abbreviation for df/du and u represents the argument of f: $u=(t-x/v)$. We differentiate again and obtain

$$\frac{\partial^2 f}{\partial t^2}=f''\left(t-\frac{x}{v}\right); \qquad \frac{\partial^2 f}{\partial x^2}=\frac{1}{v^2}f''\left(t-\frac{x}{v}\right)$$

where

$$f''(u)=\frac{d^2 f}{du^2}$$

These expressions can be directly combined to give Eq. (1.1). The proof for g proceeds in a similar way.

The two terms in Eq. (1.13) have the following simple interpretation. The term $f(t-x/v)$ describes a disturbance moving with velocity of magnitude v in the $+x$ direction with no change in size or shape. And $g(t+x/v)$ describes a similar undistorted disturbance moving with velocity v in the $-x$ direction.

To prove the statement about f, consider the displacement at time $t+\Delta t$ and compare it with the displacement at time t,

$$f\left(t+\Delta t-\frac{x}{v}\right)=f\left(t-\frac{x-v\Delta t}{v}\right)=f\left(t-\frac{x'}{v}\right)$$

where $x'=x-v\Delta t$. This means that the value of the function f at the point x and at the later time $t+\Delta t$ is the same as it was at time t not at the point x but at the point x', which lies to the left of x at a distance $v\Delta t$. Hence the entire curve moves to the right with velocity v. This behavior is illustrated in Fig. 1.2. A similar argument applies to $g(t+x/v)$.

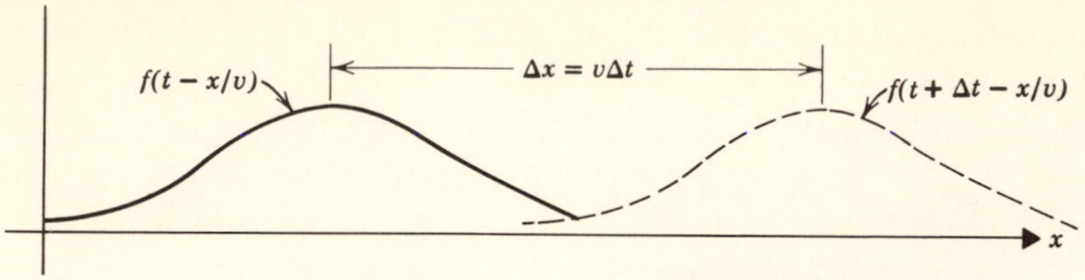

Fig. 1.2. Illustration of a disturbance moving in the $+x$ direction with velocity v. Solid curve: the displacement at time t. Dotted curve: The displacement at time $t+\Delta t$. The dotted curve is obtained from the solid curve by a rigid shift of amount $\Delta x = v\Delta t$.

C. ENERGY IN WAVES

1. Kinetic and Potential Energy Densities. The two oppositely moving components of the general solution of the wave equation represented by Eq. (1.13) carry energy. We can associate an energy density $U(x, t)$ with the total disturbance defined to be the total mechanical energy per unit length. The kinetic energy of a mass dm having a transverse velocity $\partial s/\partial t$ will be

$$\tfrac{1}{2}\,dm\left(\frac{\partial s}{\partial t}\right)^2$$

But since $dm = \mu dx$, we obtain for the kinetic energy per unit length

$$U_k = \tfrac{1}{2}\mu\left[\frac{\partial s(x, t)}{\partial t}\right]^2 \tag{1.14}$$

The potential energy density can be calculated by computing the work required to deform the string the required amount. We can see that the extra work needed to stretch the portion of the string lying between x and $x+\Delta x$ in Fig. 1.1 is

$$\Delta W = T\Delta l - T\Delta x = T\sqrt{(\Delta x)^2 + (\Delta s)^2} - T\Delta x$$

It is consistent with our earlier assumption about the smallness of θ to put $|\Delta s/\Delta x| \ll 1$ and to expand the square root as follows:

$$\sqrt{(\Delta x)^2 + (\Delta s)^2} = \left\{(\Delta x)^2\left[1 + \frac{(\Delta s)^2}{(\Delta x)^2}\right]\right\}^{1/2} \approx \Delta x\left[1 + \frac{1}{2}\left(\frac{\Delta s}{\Delta x}\right)^2 + \cdots\right]$$

so that

$$\Delta W \approx \frac{1}{2}T\Delta x\left(\frac{\Delta s}{\Delta x}\right)^2$$

In the limit as $\Delta x \to 0$, the work per unit length $\Delta w/\Delta x$ becomes the potential energy density U_p given by

$$U_p(x, t) = \frac{T}{2}\left[\frac{\partial s(x, t)}{\partial x}\right]^2 \tag{1.15}$$

If only one component in Eq. (1.13) is present, the kinetic and potential energy densities are equal, since $v = \sqrt{T/\mu}$, giving

$$\left[\frac{\partial s(x, t)}{\partial x}\right]^2 = \frac{1}{v^2}\left(\frac{\partial s}{\partial t}\right)^2$$

and we can write

$$U = U_p + U_k = \mu \left(\frac{\partial s}{\partial t}\right)^2 \tag{1.16}$$

This equation, which involves the square of the first derivative, is not to be confused with Eq. (1.11), which involves the second derivative. In the general case, we must be content with

$$U = \frac{\mu}{2}\left(\frac{\partial s}{\partial t}\right)^2 + \frac{T}{2}\left(\frac{\partial s}{\partial x}\right)^2 \tag{1.17}$$

The important property possessed by Eqs. (1.16) and (1.17) is the "quadratic dependence" of the energy density on the amplitude of the disturbance. This simply means that if $s(x, t)$ is increased by a factor α, then $U(x, t)$ is increased by the factor α^2. Other types of wave motion share this quadratic dependence.

2. Energy Flux. The energy flux $S(x, t)$ is defined to be the net energy transferred past the point x per unit time. If only one of the components of Eq. (1.13) is present, there is an amount of energy between x and $x + \delta x$ given by

$$\delta E = U(x, t)\, \delta x$$

In time $\delta t = \delta x/v$ this has all moved past the point $x + \delta x$. Thus

$$S(x, t) = \lim_{\delta t \to 0} \frac{\delta E}{\delta t} = vU(x, t) \tag{1.18}$$

When both components are present, the net flux in the $+x$ direction is given by

$$S(x, t) = v[U^+(x, t) - U^-(x, t)] \tag{1.19}$$

where $U^{\pm}(x, t)$ are given by Eq. (1.16):

$$U^+(x, t) = \mu\left[f'\left(t - \frac{x}{v}\right)\right]^2 \tag{1.16a}$$

and

$$U^-(x, t) = \mu\left[g'\left(t + \frac{x}{v}\right)\right]^2 \tag{1.16b}$$

Equations (1.18) and (1.19) have quite general applicability.

The net inward flow of energy into a segment of the string must equal the rate of increase of the total energy in that segment of the string, since no outside agents are doing work on it. Thus, in integral form we have

$$S(x_1, t) - S(x_2, t) = \frac{\partial}{\partial t}\int_{x_1}^{x_2} U(x, t)\, dx \tag{1.20}$$

The differential form of this law is called the *equation of energy continuity*. We get it by differentiating Eq. (1.20) with respect to x_2 (and then replacing x_2 by x):

$$\frac{\partial S(x,t)}{\partial x} = -\frac{\partial U(x,t)}{\partial t} \tag{1.21}$$

It is a straightforward matter to show that this equation follows from Eqs. (1.16*a*, *b*), (1.17), and (1.19).

D. REFLECTION AND TRANSMISSION AT A DISCONTINUITY

When the propagating medium abruptly changes character, an incident disturbance is reflected at the discontinuity, and only part is transmitted. Both the reflected and the transmitted disturbances can be determined from the incident disturbance and are in a sense proportional to it. Consider the string to have a massless joint located at $x = 0$ so that $\mu = \mu_1$ for $x < 0$, and $\mu = \mu_2 \neq \mu_1$ for $x \geqslant 0$, as indicated in Fig. 1.3.

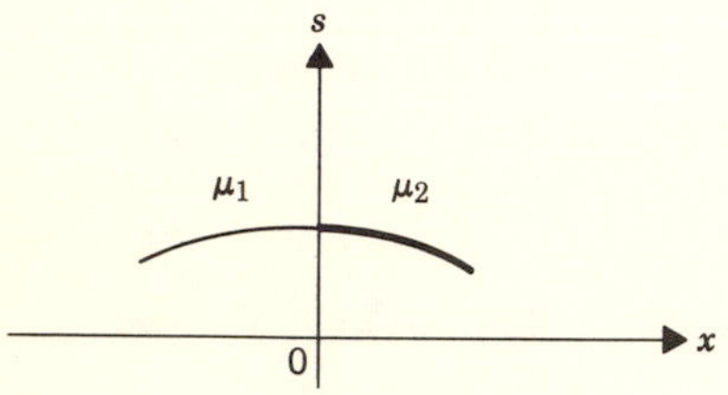

Fig. 1.3. A joint in the string at $x = 0$.

On the two sides of the joint these wave equations must hold:

$$\frac{\partial^2 s}{\partial x^2} - \frac{1}{v_1^2}\frac{\partial^2 s}{\partial t^2} = 0 \qquad \text{for } x < 0, \text{ where } \quad v_1^2 = \frac{T}{\mu_1} \tag{1.22}$$

and

$$\frac{\partial^2 s}{\partial x^2} - \frac{1}{v_2^2}\frac{\partial^2 s}{\partial t^2} = 0 \qquad \text{for } x \geqslant 0, \text{ where } \quad v_2^2 = \frac{T}{\mu_2} \tag{1.23}$$

The solutions of these equations must satisfy the following boundary conditions. (1) The displacement s must be continuous at $x = 0$, for if not, there would be a break in the string. (2) The derivative $\partial s/\partial x$ must also be continuous at $x = 0$, for if not, there would be an unbalanced vertical force on the massless joint and hence an infinite acceleration (Fig. 1.4).

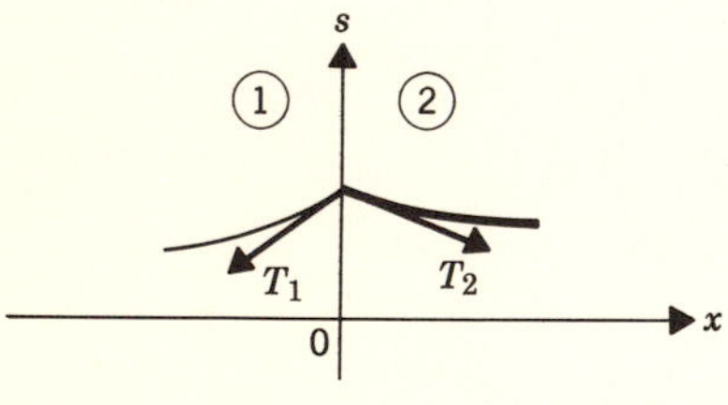

Fig. 1.4. A discontinuity in slope at $x = 0$ would imply the existence of a net unbalanced vertical force on the massless joint at $x = 0$.

Equations (1.22) and (1.23) admit of the general solutions

$$s = f_1\left(t - \frac{x}{v_1}\right) + g_1\left(t + \frac{x}{v_1}\right), \qquad x < 0 \tag{1.24a}$$

and

$$s = f_2\left(t - \frac{x}{v_2}\right) + g_2\left(t + \frac{x}{v_2}\right), \qquad x \geqslant 0 \tag{1.24b}$$

The two boundary conditions give two relations between these four functions: f_1, g_1, f_2, and g_2. Two of the functions may still be chosen arbitrarily. An ideal case that is usually of interest involves a known wave incident onto the discontinuity from the left in medium 1 (that is, f_1 is assumed to be known) and no wave propagating to the left in medium 2 (that is, $g_2 \equiv 0$). The required continuity of s at $x = 0$ then gives

$$f_1(t) + g_1(t) = f_2(t) \tag{1.25}$$

The required continuity of $\partial s/\partial x$ at $x = 0$ then gives

$$-\frac{1}{v_1} f_1'(t) + \frac{1}{v_1} g_1'(t) = -\frac{1}{v_2} f_2'(t) \tag{1.26a}$$

These functional relationships must hold for all values of t.

Equation (1.26a) can be integrated if it is assumed that $f_1(-\infty) = g_1(-\infty) = f_2(-\infty) = 0$. Such an assumption means that the disturbances in question originated at a finite time in the past. Integration of Eq. (1.26a) from $t = -\infty$ to $t = t$ then gives simply

$$f_1(t) - g_1(t) = \frac{v_1}{v_2} f_2(t) \tag{1.26b}$$

Equations (1.25) and (1.26b) can be solved for $g_1(t)$ and $f_2(t)$ with the results

$$f_2(t) = \tau f_1(t) \qquad \text{with} \quad \tau = \frac{2v_2}{v_1 + v_2} \tag{1.27}$$

and

$$g_1(t) = \rho f_1(t) \qquad \text{with} \quad \rho = \frac{v_2 - v_1}{v_2 + v_1} \tag{1.28}$$

The quantities τ and ρ are called the *amplitude transmission and reflection coefficients*, respectively. Note that for $\mu_2 > \mu_1$, it follows that $v_2 < v_1$ and hence $\rho < 0$. This change in sign upon reflection from an interface with a more dense medium is quite a general property and is also shared by light waves. Equations (1.27) and (1.28) also apply to light incident normally onto a dielectric interface, where v_1 and v_2 represent the velocities of light in the two media. The boundary conditions are the same in this case also. (See Section 11.3.)

When Eq. (1.28) is inserted into Eq. (1.24a), we obtain the incident disturbance and the reflected disturbance in medium 1,

$$s(x, t) = f_1\left(t - \frac{x}{v_1}\right) + \rho f_1\left(t + \frac{x}{v_1}\right), \qquad x < 0 \tag{1.29}$$

From Eq. (1.24b) with Eq. (1.27) and the condition $g_2 = 0$ we obtain the transmitted disturbance in medium 2,

$$s(x, t) = \tau f_1\left(t - \frac{x}{v_2}\right), \qquad x \geqslant 0 \tag{1.30}$$

From Eq. (1.30) it is apparent in what sense the transmitted disturbance is proportional to the incident disturbance. As a function of x it is reduced by the factor τ and stretched out by the ratio v_2/v_1 (Fig. 1.5).

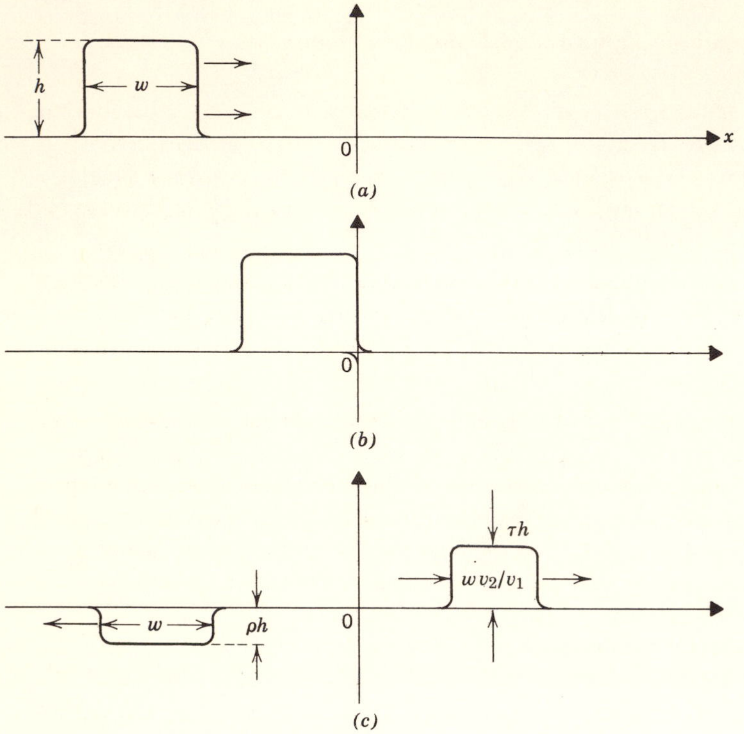

Fig. 1.5. A pulse of height h and width w is both transmitted and reflected. (*a*) Before reaching joint. (*b*) Upon reaching joint. (*c*) Reflected and transmitted pulses.

PROBLEMS

1.1.1 Consider a transverse wave on a string of known density μ of the form

$$s(x, t) = A \sin\left[2\pi\nu\left(t - \frac{x}{v}\right)\right]$$

Find analytical expressions as functions of x and t for the transverse velocity, the vertical force acting on the end of the string just to the left of the point x across an imaginary "cut" at x, and the energy density in the string.

1.1.2 Prove that Eq. (1.21) is consistent with Eqs. (1.16*a*, *b*), (1.17), and (1.19).

1.1.3 A Gaussian displacement

$$s(x, t) = A \exp - \left[\frac{(t - x/v_1)^2}{t_1^2}\right]$$

is incident from the left onto a junction at $x = 0$ where the mass density changes from μ_1 to μ_2 and the velocity changes from v_1 to v_2. Write down expressions for $s_{\text{trans}}(x, t)$ and $s_{\text{refl}}(x, t)$.

1.1.4 An infinitely long string is under tension T and has uniform mass per unit length μ. It is given an initial displacement at $t = 0$, $s(x, 0) = Ae^{-(x/x_0)^2}$, and an initial velocity $\partial s/\partial t\ (x, 0) = 0$.

a. Find the displacement $s(x, t)$ for all x and for $t > 0$.
b. Interpret your answer to (a).

1.1.5 A wave is incident on a junction where the mass density μ suddenly increases by a factor of 2. What fraction of the energy density in the incident wave is transmitted? Does the incident energy density equal the sum of the reflected plus transmitted energy densities? Would you expect it to? Is the total energy *flux* conserved? Explain.

1.1.6 Consider a joint between two pieces of string that is not massless, but very, very small in size. Let the mass of the knot be m, let it be at the origin, and let it separate the regions $x < 0$ where $\mu = \mu_1$ and $x > 0$ where $\mu = \mu_2$.

a. The condition that $\partial s/\partial x$ must be continuous now becomes modified. How?
b. *Find functional expressions for the transmitted and reflected disturbances for part (a) when the incident disturbance has the general form $f_1(t - x/v_1)$. (You will have to solve a differential equation.)
c. *Do part (b) with $f_1(t - x/v_1) = A \cos 2\pi\nu\,(t - x/v_1)$ and discuss appropriate high and low frequency limits.
d. *Discuss the limit $m \to 0$.
e. *Discuss the limit $\mu_1 \to \mu_2$.

1.1.7 *Consider two massless joints in a string located at $x = 0$ and $x = a$ such that the mass density μ equals μ_1 for $x < 0$, $\mu = \mu_3$ for $0 < x < a$, and $\mu = \mu_2$ for $x > a$. Let there be an incident disturbance of the form $f_1(t - x/v_1)$. You wish to calculate the reflected disturbance for $x < 0$ and the final transmitted disturbance for $x > a$. You will have to introduce disturbances moving in both directions in the intermediate medium where $0 < x < a$ and match boundary conditions at $x = 0$ and $x = a$.

a. The resulting equations can be solved relatively easily for a sine wave: $f_1 = A \sin 2\pi\nu(t - x/v_1)$. Find the overall amplitude transmission and reflection coefficients and discuss their behavior as a function of a.
b. Solve the equations for the general case. You cannot do this in closed form, but will obtain an infinite series for the overall reflected and transmitted disturbances that will have a simple physical interpretation in terms of multiple reflections. Supply this interpretation.†
c. Consider the results of (b) in the limit as $a \to 0$, $\mu_3 \to \infty$, but where $a\mu_3 = m$ (a constant), and compare your expressions with those obtained in 1.16(*b*).

*These parts present unusual difficulty.

†The reader is invited to make a comparison of the problem stated here and the approach to multiple reflection taken in Section 5.6.

1.2 PROPAGATION OF DISTURBANCES AND WAVES IN THREE DIMENSIONS

We have just discussed some of the important properties of wave motion in general that are already revealed in the one-dimensional case. Other important properties do not show up until we consider three spatial coordinates. We thus turn to an introductory discussion of waves and wavelike disturbances in three dimensions.

A. THE THREE-DIMENSIONAL WAVE EQUATION

In three dimensions the wave equation (1.1)

$$\frac{\partial^2 s}{\partial x^2} - \frac{1}{v^2}\frac{\partial^2 s}{\partial t^2} = 0$$

generalizes to

$$\frac{\partial^2 p}{\partial x^2} + \frac{\partial^2 p}{\partial y^2} + \frac{\partial^2 p}{\partial z^2} - \frac{1}{v^2}\frac{\partial^2 p}{\partial t^2} = 0 \qquad \text{or} \qquad \nabla^2 p - \frac{1}{v^2}\frac{\partial^2 p}{\partial t^2} = 0 \tag{1.31}$$

Here ∇ stands for the vector differential operator "dell" or "nabla," whose coefficients in a Cartesian coordinate system are given by $(\partial/\partial x, \partial/\partial y, \partial/\partial z)$, and ∇^2 is an abbreviation for the inner product or dot product of ∇ with itself: $\nabla \cdot \nabla = \nabla^2 = \partial^2/\partial x^2 + \partial^2/\partial y^2 + \partial/\partial z^2$. The function $p(x, y, z, t)$ could represent one of several physical quantities that can obey an equation like (1.31). For instance, it could represent the pressure in an isotropic fluid, in which case the velocity v would be given by

$$v^2 = \frac{\beta}{\rho} \tag{1.32}$$

Here ρ is the fluid mass per unit volume and β is the bulk modulus (the reciprocal of the compressibility), defined as the ratio

$$\beta = \rho\,\frac{dp}{d\rho} \tag{1.33}$$

Light consists of electromagnetic waves. In free space each spatial component of the electric field **E** and the magnetic flux density **B** obeys an equation of the form of (1.31), with v replaced by the velocity of light c. In addition to satisfying the wave equation, **E** and **B** must satisfy the equations of electromagnetic theory known as *Maxwell's equations*, which will be discussed in Chapter 11. These extra equations imply that for the individual elementary sinusoidal components of the waves, the vectors **E** and **B** must be mutually perpendicular and also perpendicular to the direction of propagation. This *transverse* nature of light waves gives rise to the various phenomena associated with the term *polarization*. Many other optical phenomena, however, do not depend on it. They can be described most simply by using a *scalar* wave theory such as that appropriate to sound waves in a fluid. In such a theory the vector nature of the optical disturbance, that is, the vector nature of **E** and **B**, is overlooked, and it is treated as a scalar. We shall, therefore, use the scalar theory for much of our discussion and postpone the discussion of polarized light until later.

In the following few sections we shall denote our dependent scalar variable by p. The reader may think of p as the pressure in a sound wave or as the electric field in a light wave. In the latter case the results may not be literally true all of the time, because we have supressed the vector nature of light. At any rate, our aim here is to discuss a few simple solutions of the three-dimensional wave equation (1.31).

We first remark that the superposition principle applies to the solutions of Eq. (1.31) for the same reason that it applies to the solutions of Eq. (1.1). Thus the sum of two solutions is also a solution, and complicated solutions can be obtained by the superposition of simple solutions.

B. PLANE DISTURBANCES AND WAVES

A simple situation occurs under the assumption that p is a function only of x and t. Then Eq. (1.31) reduces to Eq. (1.1) and has the same general solution, namely,

$$p(x,t) = f\left(t-\frac{x}{v}\right) + g\left(t+\frac{x}{v}\right) \tag{1.34}$$

Here the disturbance is constant over each yz plane and consists of two parts, one propagating along the $+x$ axis with velocity of magnitude v (the f term) and one propagating along the $-x$ axis with velocity of magnitude v (the g term).

A *sinusoidal traveling plane wave* is a special case of Eq. (1.34) with, say, $g=0$ and

$$p(x,t) = f\left(t-\frac{x}{v}\right) = A\cos\left[2\pi\nu\left(t-\frac{x}{v}\right)+\phi\right] \tag{1.35}$$

Here A is called the *amplitude* of the wave, ν the *frequency* (in cycles per second), and ϕ the *constant phase angle*. We often use the angular frequency given by

$$\omega = 2\pi\nu \tag{1.36}$$

measured in radians per second. Equation (1.35) can also be written

$$p(x,t) = A\cos\left(\omega t - \frac{2\pi x}{\lambda} + \phi\right) = A\cos\left(\omega t - kx + \phi\right) \tag{1.37}$$

where the *wavelength* λ is given by

$$\lambda = \frac{v}{\nu} \tag{1.38}$$

and represents the crest-to-crest or trough-to-trough spatial distance between waves at a given time. The parameter $k = 2\pi/\lambda$ is called the *wave number*. The total argument of the cosine function

$$\left[2\pi\nu\left(t - \frac{x}{v}\right) + \phi\right] = \left(\omega t - \frac{2\pi x}{\lambda} + \phi\right) = (\omega t - kx + \phi)$$

represents the *total phase* at x and t. At a given position the total phase increases linearly with time; at a given time the phase decreases linearly with x (for the case of a wave propagating in the $+x$ direction).

Expressions exactly analogous to Eq. (1.34) could be written to describe uniform disturbances over the xz plane propagating along the $+y$ or $-y$ axis or uniform disturbances over the xy plane propagating along the $+z$ or $-z$ axis.

What if the normal to the plane of the disturbance is not along a coordinate axis, but along an arbitrary direction having unit vector $\mathbf{n} = (n_x, n_y, n_z)$? It is not difficult to see that a general plane disturbance propagating in the $+\mathbf{n}$ direction would be of the form

$$p(r, t) = f\left(t - \frac{\mathbf{r} \cdot \mathbf{n}}{v}\right) = f\left(t - \frac{xn_x + yn_y + zn_z}{v}\right) \tag{1.39}$$

First, we note that this expression is a solution of Eq. (1.31). We only have to notice that

$$\frac{\partial f}{\partial x} = -\frac{n_x}{v} f'$$

and that

$$\frac{\partial^2 f}{\partial x^2} = \frac{n_x{}^2}{v^2} f''$$

to obtain

$$\frac{\partial^2 f}{\partial x^2} + \frac{\partial^2 f}{\partial y^2} + \frac{\partial^2 f}{\partial z^2} = \frac{n_x{}^2 + n_y{}^2 + n_z{}^2}{v^2} f'' = \frac{1}{v^2} f''$$

$$= \frac{1}{v^2} \frac{\partial^2 f}{\partial t^2}$$

as required. As to the interpretation of Eq. (1.39), it gives a constant value for the disturbance p at a given value of time for all values of the position vector $\mathbf{r}$ obeying $\mathbf{r} \cdot \mathbf{n} =$ const. Such an equation defines

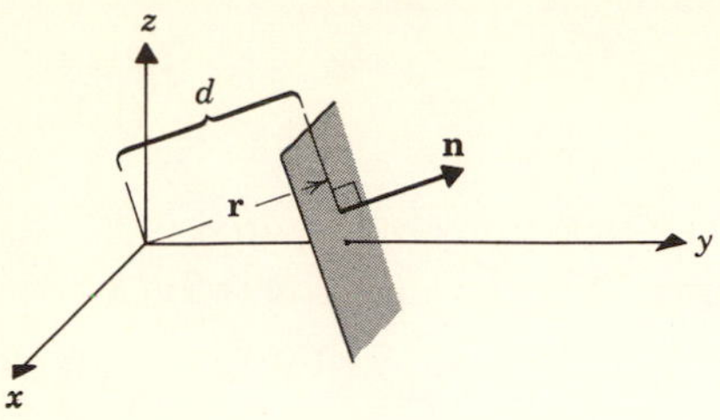

Fig. 1.6. The equation $\mathbf{r} \cdot \mathbf{n} = d$ defines a plane perpendicular to the unit vector $\mathbf{n}$ and a distance d from the origin.

a plane perpendicular to $\mathbf{n}$. As the value of $\mathbf{r} \cdot \mathbf{n}$ increases algebraically, the plane moves in the $+\mathbf{n}$ direction (Fig. 1.6).

The general expressions for a disturbance moving in the $-\mathbf{n}$ direction is

$$p(r, t) = g\left(t + \frac{\mathbf{r} \cdot \mathbf{n}}{v}\right) \tag{1.40}$$

C. SPHERICAL DISTURBANCES AND WAVES

Another important simple solution of the three-dimensional wave equation is obtained when we make the assumption that the function $p(\mathbf{r}, t)$ has spherical symmetry about the origin, that is, we assume that

$$p(\mathbf{r}, t) = p(r, t) \tag{1.41}$$

only, where

$$r = \sqrt{x^2 + y^2 + z^2}$$

To calculate $\nabla^2 p$ in this case we begin with

$$\frac{\partial p}{\partial x} = \frac{\partial p(r, t)}{\partial r} \frac{\partial r}{\partial x} = \left(\frac{\partial p}{\partial r}\right) \frac{x}{r}$$

and differentiate once more to obtain

$$\frac{\partial^2 p}{\partial x^2} = \frac{\partial}{\partial x}\left(\frac{x}{r}\frac{\partial p}{\partial r}\right) = \frac{1}{r}\frac{\partial p}{\partial r} + x\frac{\partial}{\partial x}\left(\frac{1}{r}\frac{\partial p}{\partial r}\right)$$

$$= \frac{1}{r}\frac{\partial p}{\partial r} + x\frac{\partial}{\partial r}\left(\frac{1}{r}\frac{\partial p}{\partial r}\right)\frac{\partial r}{\partial x}$$

$$= \frac{1}{r}\frac{\partial p}{\partial r} + \frac{x^2}{r}\left(-\frac{1}{r^2}\frac{\partial p}{\partial r} + \frac{1}{r}\frac{\partial^2 p}{\partial r^2}\right)$$

$$\frac{\partial^2 p}{\partial x^2} = \frac{1}{r}\frac{\partial p}{\partial r} - \frac{x^2}{r^3}\frac{\partial p}{\partial r} + \frac{x^2}{r^2}\frac{\partial^2 p}{\partial r^2} \tag{1.42}$$

In the same way we obtain the other derivatives:

$$\frac{\partial^2 p}{\partial y^2} = \frac{1}{r}\frac{\partial p}{\partial r} - \frac{y^2}{r^3}\frac{\partial p}{\partial r} + \frac{y^2}{r^2}\frac{\partial^2 p}{\partial r^2} \tag{1.43}$$

$$\frac{\partial^2 p}{\partial z^2} = \frac{1}{r}\frac{\partial p}{\partial r} - \frac{z^2}{r^3}\frac{\partial p}{\partial r} + \frac{z^2}{r^2}\frac{\partial^2 p}{\partial r^2} \tag{1.44}$$

The sum of these three equations yields

$$\nabla^2 p = \frac{3}{r}\frac{\partial p}{\partial r} - \frac{(x^2 + y^2 + z^2)}{r^3}\frac{\partial p}{\partial r} + \frac{(x^2 + y^2 + z^2)}{r^2}\frac{\partial^2 p}{\partial r^2}$$

$$= \frac{2}{r}\frac{\partial p}{\partial r} + \frac{\partial^2 p}{\partial r^2} = \frac{1}{r}\frac{\partial^2}{\partial r^2}(rp)$$

The wave equation then becomes

$$\frac{1}{r}\frac{\partial^2}{\partial r^2}[rp(r,t)] - \frac{1}{v^2}\frac{\partial^2}{\partial t^2}p(r,t) = 0$$

or

$$\frac{\partial^2}{\partial r^2}[rp(r,t)] - \frac{1}{v^2}\frac{\partial^2}{\partial t^2}[rp(r,t)] = 0 \qquad (1.45)$$

This differential equation for the function $[rp(r, t)]$ has mathematically the same form as the one-dimensional wave equation discussed earlier for the function $s(x, t)$. Thus the general mathematical solution can be written in the form

$$rp(r,t) = f\left(t - \frac{r}{v}\right) + g\left(t + \frac{r}{v}\right)$$

The actual disturbance p then takes the form

$$p(r,t) = \frac{1}{r}f\left(t - \frac{r}{v}\right) + \frac{1}{r}g\left(t + \frac{r}{v}\right) \qquad (1.46)$$

The first term in Eq. (1.46) can be interpreted as a spherically symmetric disturbance that originates at the origin and propagates outward with velocity of magnitude v. The amplitude of the disturbance falls off as $1/r$. In the case of a sound wave, such a disturbance could be caused by a small round source at the origin. The second term in Eq. (1.46) represents an incoming spherical disturbance that would be hard to realize experimentally. An outgoing spherical sine wave is obtained by letting f equal an expression similar to Eq. (1.35) or (1.36):

$$p(r,t) = \frac{A}{r}\cos\left[2\pi\nu\left(t - \frac{r}{v}\right) + \phi\right]$$

Associated with a disturbance in three dimensions there is always an energy density for which we use the same symbol U as in one dimension. But now $U(\mathbf{r}, t)$ represents the energy per unit volume at the point $\mathbf{r}$ and at time t. It will be quadratic in p in the sense that if we replace p by αp, we must replace U by $\alpha^2 U$. The expression for U in terms of p will not be exactly like Eq. (1.17), but if only one disturbance is present with a well-defined propagation direction at $\mathbf{r}$, then the energy flux $S(\mathbf{r}, t)$ in that direction is given by

$$S(\mathbf{r}, t) = vU(\mathbf{r}, t) \qquad (1.47)$$

an equation analogous to (1.18). Here S has the units of power/area.

For an outgoing spherical disturbance we can combine the quadratic dependence of U on p with Eq. (1.47) to obtain for the outward flux an expression of the form

$$S(\mathbf{r}, t) = S(r, t) = \frac{\text{const}}{r^2} F\left(t - \frac{r}{v}\right) \tag{1.48}$$

where F is a function related quadratically to f. Its exact form depends on the detailed physical nature of the disturbance p. The total power flowing through a sphere of radius r_0 is then

$$P(r_0, t) = 4\pi r_0^2 S(r_0, t) = \text{const}\, 4\pi\, F\left(t - \frac{r_0}{v}\right) \tag{1.49}$$

If the disturbance lasts a finite time, there is a total amount of energy flow through the sphere given by

$$E_{\text{tot}}(r_0) = \int_{-\infty}^{+\infty} P(r_0, t)\, dt = \text{const}\, 4\pi \int_{-\infty}^{\infty} F\left(t - \frac{r_0}{v}\right) dt$$

$$= \text{const}\, 4\pi \int_{-\infty}^{\infty} F(u)\, du$$

The total outward flux is thus independent of the radius r_0 of the sphere through which it passes. The energy passing through a sphere of radius r_1 ultimately will pass through a sphere of radius $r_2 > r_1$. All the energy in the disturbance radiates away; none is dissipated as heat and none remains behind. This conservation of energy holds only because of the $1/r$ factors in Eq. (1.46).

PROBLEMS

1.2.1 Write down an equation describing a sinusoidal plane wave in three dimensions with wavelength λ, velocity v, propagating in the following directions:

a. $+z$ axis.
b. Along the line $x = y$, $z = 0$.
c. Perpendicular to the planes $x + y + z = \text{const}$.

1.2.2 a. A spherical disturbance propagating away from a source at the origin with velocity v is found to have the value $\psi(t)$ at a radius $r = a$. Find the disturbance $p(r, t)$ for $r \neq a$ and for all t.

b. Redo part (a) with the same source now moved to $\mathbf{r}_0 \neq 0$, so that on a sphere of radius a about $\mathbf{r}_0$ the disturbance has the value $\psi(t)$. Find $p(\mathbf{r}, t)$ for all $\mathbf{r}$, t. (*Hint.* Change the origin of the coordinate system so that the result of part (a) may be used.)

1.3 HUYGENS' PRINCIPLE

Wave propagation in three dimensions is generally quite a complicated process when considered in detail, especially when matter is present to absorb, reflect, refract, or scatter the radiation. Inside matter the waves propagate with different velocities. At discontinuous interfaces between media there are boundary conditions to be imposed on the disturbance and its derivatives. We met these effects in the case of waves on a string, where they could be handled exactly. But in three-dimensions only the most elementary geometries can be treated exactly. We are thus forced to look for approximate descriptions of wave propagation which contain some physical truth without the full mathematical complexity of the exact solutions of the wave equation. One such approximation is Huygens' principle.

A. INTUITIVE JUSTIFICATION

We have just seen how a spherical wave or disturbance propagates in three dimensions, but we have not yet discussed the methods of exciting such disturbances. They clearly depend on the specific problem at hand and will vary with the nature of the waves. In a mechanical medium such as a fluid, forces between one part of the medium and a nearby part transmit the disturbance. The physical *source* of the disturbance will act locally through similar forces on the medium near it. (Think of a loudspeaker emitting sound waves in air, for instance.) You might suspect that it would make little difference whether a given region of the medium was excited by an external source or by a nearby region of the medium. In fact, it is reasonable to consider each small excited region of the medium as a source of spherical waves of its own. The suggested procedure then is to imagine the disturbance at a given time as composed of the sum of many separate disturbances, each of which acts like a point source that radiates a spherically symmetric wavelike disturbance, or wavelet. The superposition of these wavelets then gives the disturbance at a later time.

This scheme for thinking about the nature of wave propagation is due to Huygens. The method tells how to calculate the position in space of a disturbance at a later time if it is known at an earlier time. It does not give quantitative information about the resulting amplitude, although refinements of the procedure made by Fresnel and later by Kirchoff allow one to do this. Kirchoff's mathematical formulation of the problem is almost equivalent to the wave equation itself, as we shall see below in Chapter 8.

B. STATEMENT OF METHOD

Huygens' principle contains two important rules about the way the secondary wavelets should be combined. We shall state them as they apply to the situation shown in Fig. 1.7, where a source is sending a spherical disturbance onto a screen with an aperture in it. Suppose that at time t the disturbance can be localized to lie

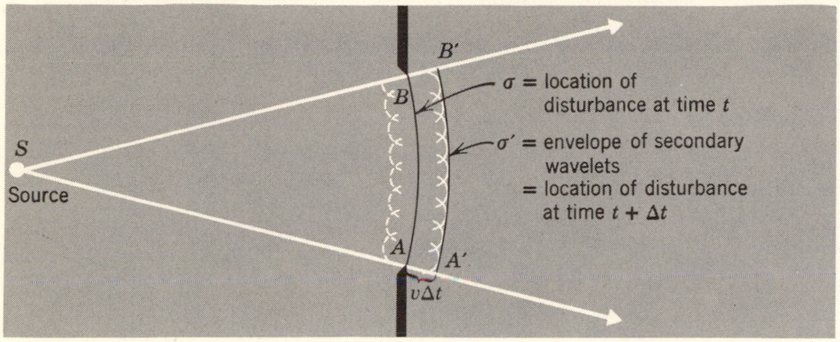

Fig. 1.7. Huygens' construction and the law of rectilinear propagation.

just inside the aperture on the surface σ and we wish to know where the disturbance is at time $t+\Delta t$. Then Part 1 of Huygens' principle says to construct secondary spherical wavelets of radius $v\Delta t$ originating at each point on σ.

We must now form the superposition of these secondary wavelets to produce the disturbance at time $t+\Delta t$. Note that the wavelets are most dense in the region included between the lines SA and SB in Fig. 1.7. This region is called the *region of geometrical brightness* because if light is assumed to travel outward from the source S in straight lines or *rays*, this region will be illuminated, whereas the regions outside, the *regions of geometrical shadow*, will not be illuminated.

Note also that the regions in the geometrical shadow are reached by only a few of the secondary wavelets originating from σ. We assume in Part 2 of Huygens' prescription that we can neglect these wavelets, and we determine the new disturbance σ' as the *envelope* of all the outgoing secondary wavelets. Each point on the envelope is tangent to one and only one Huygens wavelet. The envelope ends abruptly at the edge of the geometrical shadow. The backward-moving wavelets shown dotted also have an envelope which would be propagating toward the source. They are the analogue of the $g(t+x/v)$ terms in the one-dimensional case, whereas the forward-moving wavelets are the analogue of the $f(t-x/v)$ terms. These are neglected in Huygens' prescription, without any justification beyond an appeal to the experimental fact that the back wave envelope does not occur in nature. The later versions of the theory are able to explain this nonoccurrence. The later versions also take account of the wavelets in the shadow region, which give rise to diffraction effects.

A more detailed analysis* will show that the entity that propagates in the way just described is not the optical disturbance itself, but rather those portions of it that are discontinuous as functions of time and space. These discontinuities are confined to surfaces and propagate according to Huygens' principle. The propagation of a continuous disturbance will be approximately described by this principle if the disturbance changes rapidly over distances that are small compared with the sizes of relevant physical objects.

Our ability to neglect safely those parts of the Huygens' wavelets that do not form the envelope, that is, those in the geometrical shadow region, is closely connected with the localization of the changing part of the disturbance. This will be made clearer later when we discuss diffraction and consider explicitly these neglected parts of the wavelets.

In order to follow the propagation of the optical disturbance through space or through optical media, we have, in theory, only to make repeated applications of Huygens' principle. In the example of Fig. 1.7, to get the location of the disturbance at time $t + 2\Delta t$ we would draw the envelope of all forward-moving wavelets of radius $v\Delta t$ originating on the surface σ'. This process can be repeated indefinitely.

The notion of *rays* follows quite simply from our idealized picture of a localized optical disturbance. *Light rays are directed lines that are always perpendicular to the surface occupied by the disturbance at a given time and point along the direction of its motion.*

C. THE LAWS OF GEOMETRICAL OPTICS

The three laws of geometrical optics are usually stated in terms of light rays.

1. *The law of rectilinear propagation.* Light rays in homogeneous media propagate in straight lines.

2. *The law of reflection.* At an interface between two different homogeneous, isotropic† optical media, an incident disturbance is (partially) reflected, and the reflected ray is in the plane of incidence (the plane determined by the incident ray and the normal to the surface). The angle it makes with the normal (the angle of reflection θ_r) equals the angle made by the incident ray with the normal (the angle of incidence θ).

*See Rossi, *Optics*, or for more rigor see Kline and Kay, *Electromagnetic Theory and Geometrical Optics*.

†In nonisotropic media (that is, those whose optical properties are direction-dependent) more complicated laws of reflection and refraction than those stated here are necessary.

3. *The law of refraction.* At an interface between dielectric media, there is also a refracted ray in the second medium, lying in the plane of incidence, making an angle θ_t with the normal, and obeying Snell's law:

$$\frac{\sin\theta}{\sin\theta_t} = \frac{n_2}{n_1} = \frac{v_1}{v_2} \tag{1.50}$$

where v_1 and v_2 are the velocities of propagation in the two media, and $n_1 = (c/v_1)$, $n_2 = (c/v_2)$ are the indices of refraction. Here c is the velocity of light in vacuum, a universal constant: $c = 3.0 \times 10^8$ m/sec. For sound such a unique reference velocity does not exist.

D. PROOF OF THE LAWS OF GEOMETRICAL OPTICS BY USE OF HUYGENS' PRINCIPLE

Huygens' principle can account for the three laws of geometrical optics, as we now show.

1. Rectilinear Propagation. A single light ray has no meaning in the context of Huygens' principle, since the wave surface, however small, will have a finite area, and infinitely many normals can be drawn through it. If the surface is limited, for example by an aperture, there will be limiting rays such as SB in Fig. 1.7 that are normal to it at its edge. We now argue that these rays obey the law of rectilinear propagation.

The new surface σ' will also be sharply limited, since it is defined to be the outgoing envelope of all secondary Huygens' wavelets originating on σ. A wavelet centered at a point on the edge of σ, such as B in Fig. 1.7, will be tangent to σ' at its edge (point B' in the figure). There is no other wavelet beyond that centered at B; thus the envelope ends sharply at B'. Here σ and σ' are readily seen to be normal to the radius vector SB. If we repeat the Huygens construction to determine yet another surface σ'', the extension of the line SB will just pass through its edge. Thus a line through S and B defines a limiting ray that obeys the law of rectilinear propagation.

It is the envelope prescription in Huygens' principle that gives well-defined edges to wave surfaces and allows us to speak of these limiting rays. When we give up the envelope prescription in favor of a quantitative treatment of the secondary wavelets, we should not be surprised to find that nongeometrical behavior results.

2. Reflection and Refraction. To prove these two laws using Huygens' principle, we find it useful to modify the principle to cover cases where the wavelets are centered on a surface Σ that does not coincide with a wave surface. The total optical disturbance at time t consists of the incident disturbance σ, wherever it has not yet reached Σ, and the envelope of secondary wavelets *centered*

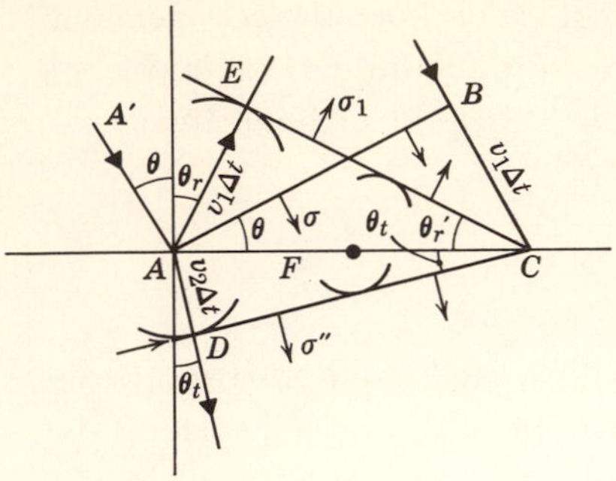

Fig. 1.8. Huygens' construction for proving the laws of reflection and refraction.

on Σ to replace the part of σ that has passed through Σ. The wavelets have radius $v\Delta t$, where $t-\Delta t$ is the local time of arrival of the disturbance on Σ.

This occurs in Fig. 1.8. Assume that the primary disturbance is on σ at time t. The limiting rays are in the direction of $A'A$ and BC. The reference surface is represented by AFC. The secondary wavelet from point A is a hemisphere of radius $v_1\Delta t$ above the interface and another hemisphere of radius $v_2\Delta t$ below the interface. If Δt is selected so that $v_1\Delta t = BC$, then the primary wave is just arriving at point C at time $t+\Delta t$, and there is no secondary wavelet there. At an intermediate point such as F, the upper secondary wavelet is tangent to the line EC, and the lower wavelet is tangent to the line DC, as shown. The envelopes at time $t+\Delta t$ are the lines EC and DC, and the respective rays are along AE and AD. The three right triangles AEC, ABC, and ADC have a common hypotenuse AC. In addition, $AE = BC = v_1\Delta t$ and $AD = v_2\Delta t$. Thus, $\theta = \theta_r$ and

$$\frac{\sin\theta}{\sin\theta_t} = \frac{v_1\Delta t/AC}{v_2\Delta t/AC} = \frac{v_1}{v_2}$$

Note also that the points of tangency E and D occur in the plane of incidence, so that both the reflected ray (AE) and the refracted ray (AD) are in this same plane. This completes the proof.

Some elementary applications of these laws will be found in the problems that follow.

PROBLEMS

1.3.1 Show that the rays from S that are reflected by the plane mirror MM' appear to be coming from the image point S'. Locate S'.

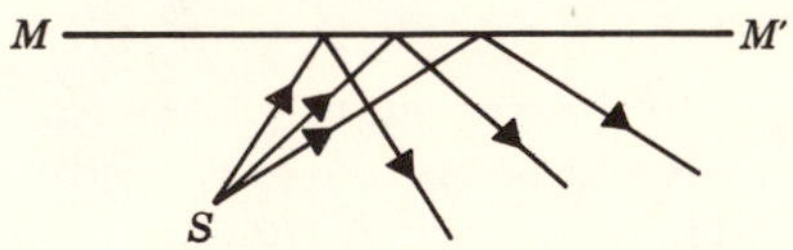

1.3.2 a. Prove that a ray reflected from both faces of a corner mirror is sent back parallel to its original direction.

b. Locate the image point S' from which the doubly reflected rays appear to be coming if S is a point source.

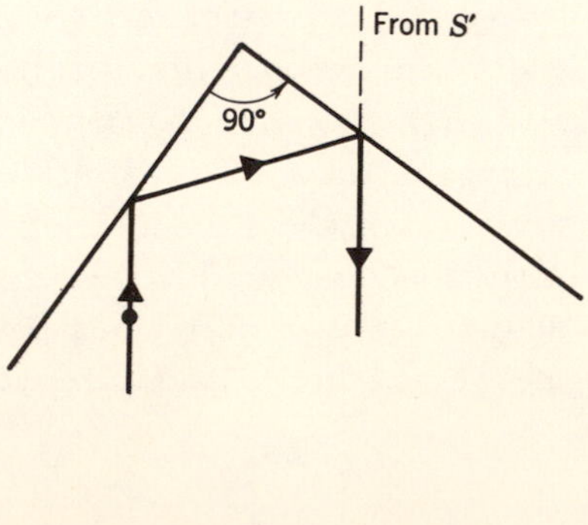

1.3.3 A source S embedded in a medium with refractive index n_1 emits a narrow bundle of rays with an infinitesimal angular spread $\delta\theta$ about a center ray having an angle of incidence θ onto a plane interface with a medium with refractive index n_2. Find the position of the image point S' from which the rays appear to be diverging and calculate the angular spread $\delta\phi$.

a. Consider only rays in the plane of the paper.
b. Consider only a fan of rays about the center ray in a plane perpendicular to the plane of the paper.

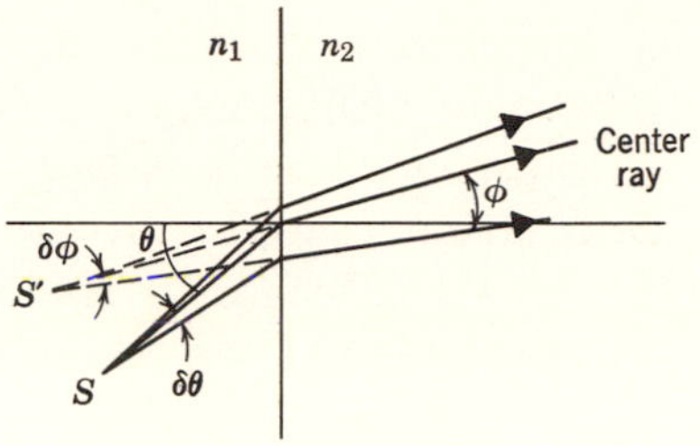

1.3.4 Calculate the deviation Δ produced by the plane parallel slab as a function of n_1, n_2, t, θ.

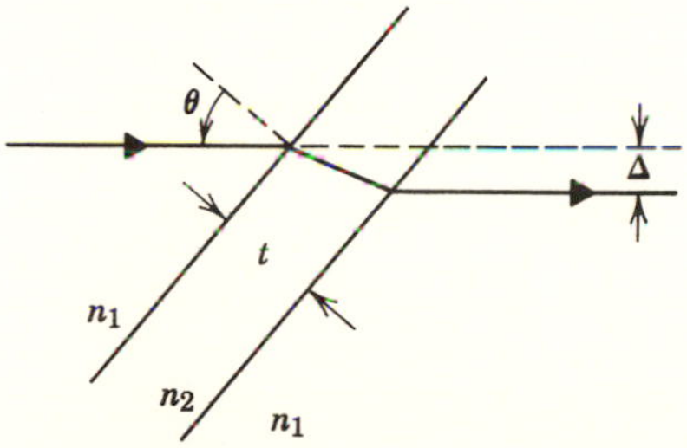

1.3.5 A ray with glancing incidence onto an Abbe prism from a medium of refractive index n_x will emerge at angle i.

a. If n_g is the refractive index of the glass and α the prism angle, show that

$$n_x = \sin\alpha\,(n_g^2 - \sin^2 i)^{1/2} + \cos\alpha \sin i$$

(The above effect is used to measure indices of refraction.)

b. Find n_x for $n_g = 1.600$ and $\alpha = 45°$ if $i = +15°$ and $i = -15°$.

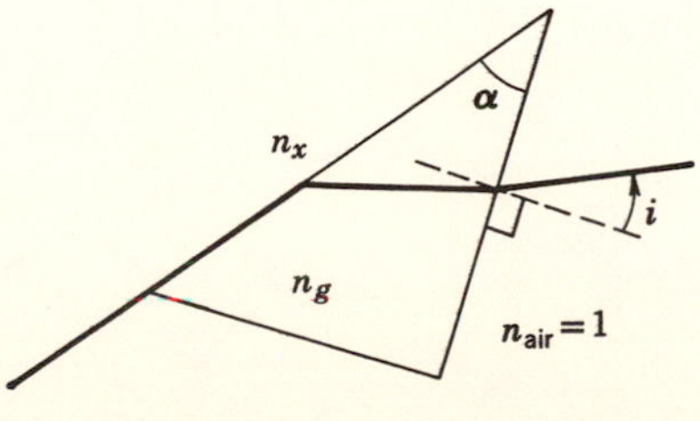

1.3.6 The deviation angle δ gives the total deviation of a ray incident onto a prism. It is given by $\delta = \delta_1 + \delta_2$. Minimum deviation occurs when $\delta_1 = \delta_2$.

a. Show that in this case δ_m, the value of δ, obeys the equation

$$\frac{n_2}{n_1} = \frac{\sin \frac{1}{2}(\alpha + \delta_m)}{\sin \frac{1}{2}\alpha}$$

b. Find δ_m for $\alpha = 60°$ and $n_2/n_1 = 1.69$.

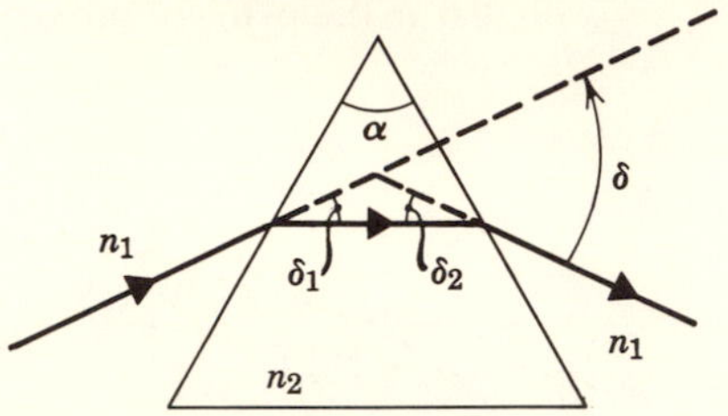

REFERENCES FOR CHAPTER 1

R. W. Ditchburn, *Light*, second edition. Parts of Chapter 1.
J. M. Pearson, *A Theory of Waves*. Most of the book.
B. Rossi, *Optics*. Chapters 1 and 2 and the appendices.

Geometrical optics is that part of optics involving image formation and related phenomena that can be discussed within the framework of the three laws of geometrical optics. In this chapter we discuss some aspects of this broad subject. We give first a rather abstract discussion of the general properties of rays and wave surfaces, which will include a reformulation of the laws of geometrical optics known as Fermat's principle. Then image formation will be discussed, and eventually the more practical aspects of the subject will be introduced.

CHAPTER 2

THE FOUNDATIONS OF GEOMETRICAL OPTICS

CHAPTER 2 CONTENTS

2.1 RAYS AND WAVE SURFACES

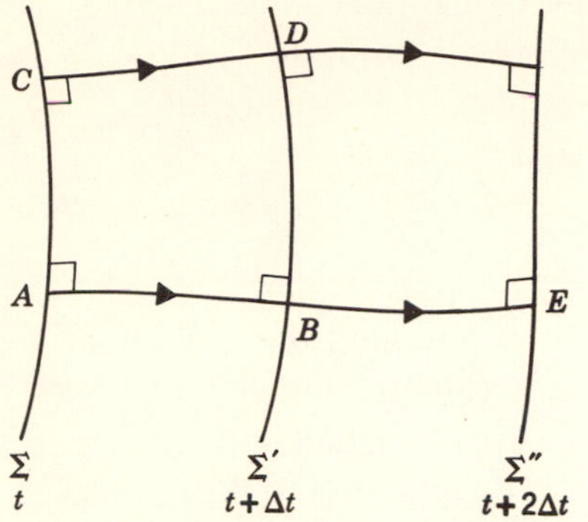

Fig. 2.1. Development of a wave surface Σ at time t to Σ' at time $t+\Delta t$ to Σ'' at time $t+2\Delta t$.

We mentioned earlier that the use of Huygens' principle in its original form implies that the optical disturbance is of short duration (or is made up of a series of short pulses). When this notion is carried to the logical limit, we obtain the so-called *wave surface*, which is the infinitely thin surface on which the optical disturbance is assumed to be located at a particular instant. As time progresses, the disturbance propagates, and we obtain a family of surfaces. *Light rays* are defined to be directed lines everywhere perpendicular to the wave surfaces. This relationship is shown in Fig. 2.1.

The discussion that follows in Sections 2.1A–E contains a somewhat abstract description of the behavior of these rays and wave surfaces, with emphasis on media with a continuously varying index of refraction. Not all of this material will be used subsequently in this book, but the notion of optical path length and an acquaintance with Fermat's principle (Sections 2.1D,E) will be necessary for understanding later subject matter.

A. OPTICAL PATH LENGTH

Let Σ be the wave surface at time t and Σ' be the wave surface at time $t+\Delta t$. Let A be a point on Σ and B be on Σ', both on the same ray (Fig. 2.1). The *optical path length* from A to B is defined to be the line integral

$$\Delta l \equiv \int_{(\text{ray})A}^{B} \frac{c}{v}\,ds = \int_{(\text{ray})A}^{B} n\,ds \tag{2.1}$$

Here c is the velocity of light in vacuum, n is the index of refraction, ds is a line element, and the line integral goes along the ray from A to B. The velocity of light in the medium is $v = ds/dt$, so that along the ray one can write $c\,ds/v = c\,dt$ and hence

$$\Delta l = \int_{A}^{B} c\,dt = c\Delta t \tag{2.2}$$

Thus the optical path length is proportional to the travel time for light along the path. One sees from this result that the optical path length from another point C on Σ along a ray to D on Σ' (Fig. 2.1) is also $\Delta l = c\Delta t$. *This means that the optical path length between two wave surfaces is a constant independent of the ray used to calculate it.*

If the index of refraction is a function of position, $n = n(x,y,z)$, Δl is still given by Eq. (2.1). Explicitly we would have

$$\Delta l = \int_{A(\text{ray})}^{B} n(x,y,z)\,ds$$

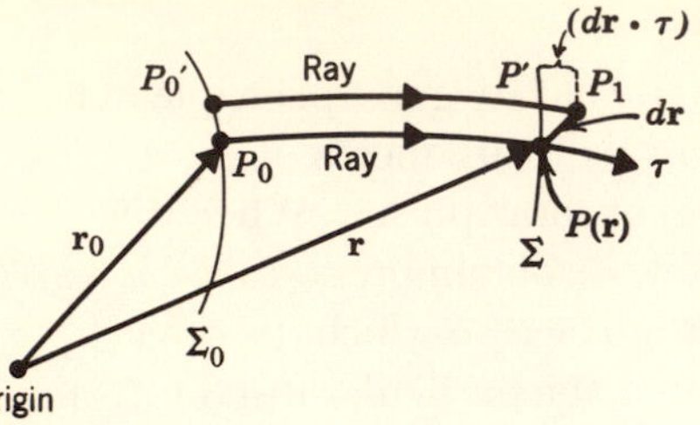

Fig. 2.2. Illustration pertinent to the definition of the optical path length function and its gradient.

B. OPTICAL PATH FUNCTION

An optical path function $L(\mathbf{r})$ can be defined as the value of the optical path length from a convenient reference wave surface Σ_0 to an arbitrary point P with position vector $\mathbf{r}$ that can be reached by a ray from Σ_0. In Fig. 2.2 let us denote the point on Σ_0 at the foot of the ray by P_0 with position vector $\mathbf{r}_0$. Then we obtain

$$L(\mathbf{r}) = \int_{\substack{\mathbf{r}_0 \\ (\text{ray})}}^{\mathbf{r}} n\,ds \tag{2.3}$$

Suppose that the optical disturbance is on Σ_0 at time t_0. Then at another time t the disturbance will be on the two-dimensional surface Σ consisting of all points that can be reached in a time interval $(t - t_0)$ by rays starting on Σ_0. The equation defining this surface is thus

$$L(\mathbf{r}) = c(t - t_0) \tag{2.4}$$

What is the change in the path function $L(\mathbf{r})$ as $\mathbf{r}$ moves a distance $d\mathbf{r}$ away from P to P_1? Construct the ray $P_0'P'P_1$ from Σ_0 to P_1. Then the change in L is $dL = n|P'P_1| = n(\mathbf{dr} \cdot \boldsymbol{\tau})$, where $\boldsymbol{\tau}$ is a unit vector at P tangent to the original ray P_0P. Since in general dL will be given by

$$dL = \frac{\partial L}{\partial x}dx + \frac{\partial L}{\partial y}dy + \frac{\partial L}{\partial z}dz = \boldsymbol{\nabla} L \cdot \mathbf{dr}$$

we find that the gradient of the optical path function is given by

$$\nabla L = n\boldsymbol{\tau} \tag{2.5}$$

The absolute value gives a result known as the *eikonal equation*

$$|\nabla L| = n \qquad \text{or} \qquad \left(\frac{\partial L}{\partial x}\right)^2 + \left(\frac{\partial L}{\partial y}\right)^2 + \left(\frac{\partial L}{\partial z}\right)^2 = n^2 \tag{2.6}$$

We now show that graphical integration of the eikonal equation is equivalent to application of Huygens' construction. Suppose that we know that the disturbance is on Σ_1 at time t_1. Then from Eq. (2.4) the surface Σ_1 must obey the equation

$$L(\mathbf{r}) = c(t_1 - t_0)$$

Where is the surface at $t_1 + \Delta t$? Let Δt be small. Then the new surface can be obtained from the old one at time t_1 by starting on Σ_1 and then by advancing in the direction of the normal to Σ_1, that is, in the direction of the ray, a distance Δs. Over this distance, L will change an amount $\Delta L = |\nabla L|\Delta s = n\Delta s$; the last equality follows from Eq. (2.6). From Eq. (2.4) we know that ΔL must also be given by

$$\Delta L = c\Delta t$$

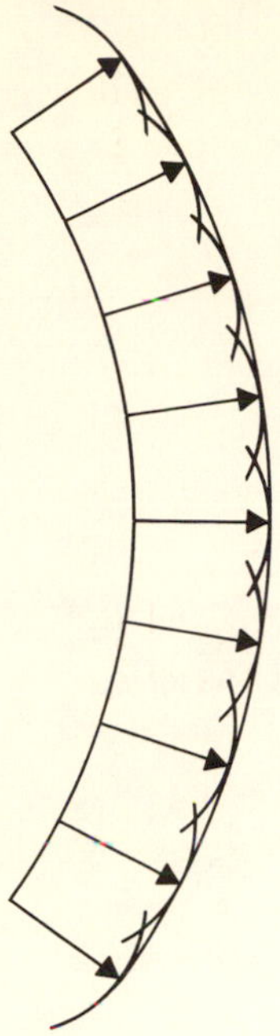

Fig. 2.3

Thus, Δs must satisfy

$$n\Delta s = c\Delta t \qquad \text{or} \qquad \Delta s = \frac{c}{n}\Delta t = v\Delta t$$

The new surface obtained by advancing up the gradient ∇L (that is, along the ray) by an amount $v\Delta t$ is equivalent to the surface obtained from the envelope of all forward-moving wavelets of radius $v\Delta t$ originating on the old surface. In fact, each forward-moving wavelet touches the envelope at the same place the ray does (Fig. 2.3).

C. THE EQUATION OF A RAY

The light ray at a given point is along the gradient of the optical path function. Let $\boldsymbol{\tau}$ be a unit vector along the ray. The ray will in general follow a curve, and $\boldsymbol{\tau}$ will be along the tangent to that curve. We shall now derive an equation that tells us how $\boldsymbol{\tau}$ changes along this curve.

Consider first how a general function of position $f(\mathbf{r})$ will change along a curve. Let $\mathbf{r}$ be on the curve. Then for a general displacement $d\mathbf{r}$ we have

$$df = \frac{\partial f}{\partial x}dx + \frac{\partial f}{\partial y}dy + \frac{\partial f}{\partial z}dz = \nabla f \cdot d\mathbf{r}$$

Now if $d\mathbf{r}$ is along the curve and has length ds, we may write $d\mathbf{r} = \boldsymbol{\tau} ds$, where $\boldsymbol{\tau}$ is a unit vector tangent to the curve, and obtain

$$df = (\nabla f \cdot \boldsymbol{\tau})\, ds$$

Thus the directional derivative of f along the curve is

$$\frac{df}{ds} = (\boldsymbol{\tau} \cdot \nabla f)$$

Since the curve in question for our problem is a ray for which $n\boldsymbol{\tau} = \nabla L$,

$$\frac{df}{ds} = \frac{1}{n}(\nabla L \cdot \nabla) f \tag{2.7}$$

We now make a special choice of f, namely, that $f = n\tau_x$ or, by Eq. (2.5),

$$f = n\tau_x = \frac{\partial L}{\partial x}$$

Then Eq. (2.7) gives

$$\frac{d}{ds}(n\tau_x) = \frac{d}{ds}\frac{\partial L}{\partial x} = \frac{1}{n}(\nabla L \cdot \nabla)\frac{\partial L}{\partial x} \tag{2.8}$$

We write out the scalar product $(\nabla L \cdot \nabla \partial L/\partial x)$ to obtain

$$\frac{d}{ds}(n\tau_x) = \frac{1}{n}\left(\frac{\partial L}{\partial x}\frac{\partial}{\partial x}\frac{\partial L}{\partial x} + \frac{\partial L}{\partial y}\frac{\partial}{\partial y}\frac{\partial L}{\partial x} + \frac{\partial L}{\partial z}\frac{\partial}{\partial z}\frac{\partial L}{\partial x}\right)$$

$$= \frac{1}{n}\left(\frac{\partial L}{\partial x}\frac{\partial^2 L}{\partial x^2} + \frac{\partial L}{\partial y}\frac{\partial}{\partial x}\frac{\partial L}{\partial y} + \frac{\partial L}{\partial z}\frac{\partial}{\partial x}\frac{\partial L}{\partial z}\right)$$

$$= \frac{1}{2n}\frac{\partial}{\partial x}\left[\left(\frac{\partial L}{\partial x}\right)^2 + \left(\frac{\partial L}{\partial y}\right)^2 + \left(\frac{\partial L}{\partial z}\right)^2\right]$$

$$= \frac{1}{2n}\frac{\partial}{\partial x}|\nabla L|^2 = \frac{1}{2n}\frac{\partial}{\partial x}n^2 = \frac{\partial n}{\partial x} \tag{2.9}$$

where we again have made use of Eq. (2.6). Equations similar to Eq. (2.9) hold for the y and z components. We are thus led to the vector equation

$$\frac{d}{ds}(n\boldsymbol{\tau}) = \nabla n \tag{2.10}$$

Equation (2.10) is one form of our desired result and has some important direct consequences (see Problems 2.1.2, 2.1.3). It enables us to follow the ray throughout its course, and describes how the ray direction changes. When $n =$ constant, Eq. (2.10) reads $d\boldsymbol{\tau}/ds = 0$ and has the general solution

$$\boldsymbol{\tau} = \mathbf{a}, \qquad \text{a constant} \tag{2.11}$$

which corresponds to an unchanging direction for the ray. We have thus shown that this approach gives the expected result that light travels in straight lines in a homogeneous medium. If $n \neq$ constant, the rays generally are curved.

The ray can be characterized by a vector equation of the form

$$\mathbf{r} = \mathbf{r}(s) \tag{2.12}$$

where $\mathbf{r}$ lies on the ray, and s represents the distance along the ray from a fixed reference point on it (Fig. 2.4). The unit vector $\boldsymbol{\tau}$ is then given simply by

$$\boldsymbol{\tau} = \frac{d\mathbf{r}}{ds} \tag{2.13}$$

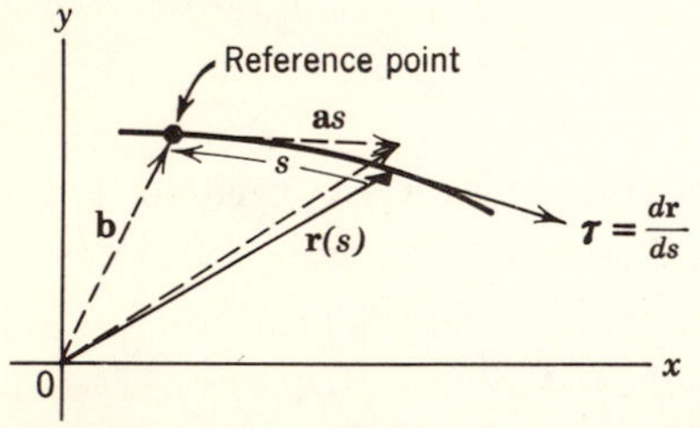

Fig. 2.4

and Eq. (2.10) becomes

$$\frac{d}{ds}\left(n\frac{d\mathbf{r}}{ds}\right) = \nabla n \tag{2.14}$$

Equations (2.11) and (2.13) then give for a straight ray the equation

$$\mathbf{r}(s) = \mathbf{a}s + \mathbf{b}, \qquad \mathbf{a}, \mathbf{b}, \text{ constant vectors} \tag{2.15}$$

as represented by the dotted lines in Fig. 2.4.

In the general case with a spatially varying $n = n(\mathbf{r})$ we can always solve Eq. (2.14) numerically or graphically. The procedure is analogous to that sometimes employed in the mechanics of a particle moving in a three-dimensional potential where the acceleration is a function of position.

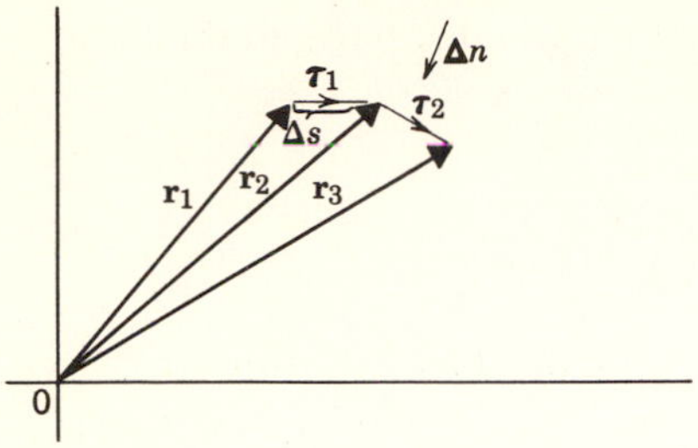

Fig. 2.5. Graphical integration of the equation of a ray. The index is increasing in the direction shown.

Assume that we know the function $n(\mathbf{r})$ and that we know the ray direction τ, at a point $\mathbf{r}_1$ (Fig. 2.5). If we choose a small but finite change in path length Δs, the change $\Delta\mathbf{r} = \mathbf{r}_2 - \mathbf{r}_1$ will be given approximately by

$$\Delta\mathbf{r} = \boldsymbol{\tau}_1 \Delta s$$

or

$$\mathbf{r}_2 = \mathbf{r}_1 + \boldsymbol{\tau}_1 \Delta s$$

This is shown in Fig. 2.5. At the new position $\mathbf{r}_2$, $\boldsymbol{\tau}_2$ may be slightly different from $\boldsymbol{\tau}_1$. From Eq. (2.10), $d(n\boldsymbol{\tau}) = (\nabla n)\, ds$, so that we have approximately

$$\Delta(n\boldsymbol{\tau}) \equiv n(\mathbf{r}_2)\boldsymbol{\tau}_2 - n(\mathbf{r}_1)\boldsymbol{\tau}_1 = \nabla n \Delta s$$

where for good accuracy ∇n should be evaluated by differentiating the known function $n(\mathbf{r})$ halfway between $\mathbf{r}_1$ and $\mathbf{r}_2$. We then obtain

$$\boldsymbol{\tau}_2 = \frac{n(\mathbf{r}_1)}{n(\mathbf{r}_2)}\boldsymbol{\tau}_1 + \frac{\Delta s}{n(\mathbf{r}_2)}\nabla n$$

Note how a nonzero gradient of the refractive index causes the ray direction $\boldsymbol{\tau}_2$ to be nonparallel to $\boldsymbol{\tau}_1$, except for the special case where $\boldsymbol{\tau}_1$ and ∇n are parallel. Knowing the ray direction at $\mathbf{r}_2$ enables us to locate approximately $\mathbf{r}_3 = \mathbf{r}_2 + \boldsymbol{\tau}_2 \Delta s$. And so on.

D. FERMAT'S PRINCIPLE

This principle is a close analogue of the action principle in classical mechanics. As is true of similar principles, Fermat's principle is a concise statement that contains all the pertinent physical laws, in this case the three laws of geometrical optics. Suppose that you have two points, P_1 and P_2, and you wish to know the path of the light ray or rays that goes through both of them. *Fermat's principle states that any deviation of the path from that taken by the true ray that is of first order in small quantities will produce a deviation in optical path length that is at least of second order in small quantities.* More precisely, the difference

$$\int_{P_1(\gamma)}^{P_2} n\, ds - \int_{P_1(\mathrm{ray})}^{P_2} n\, ds$$

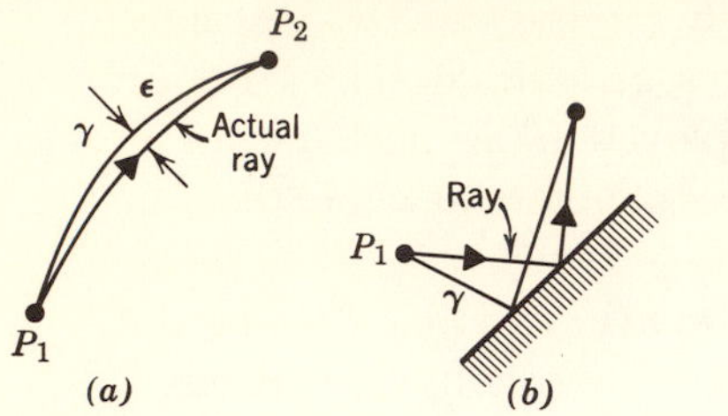

Fig. 2.6. Illustration of Fermat's principle. (*a*) Direct propagation. (*b*) Single reflection.

must be at least of second order in the parameters that measure the deviation of γ from the true ray (Fig. 2.6). This means, for instance, that if the parameter ϵ is a measure of the maximum separation of γ from the true path, then we must have, in the limit as $\epsilon \to 0$,

$$I(\epsilon) = \int_{P_1(\gamma)}^{P_2} n\,ds - \int_{P_1(\text{ray})}^{P_2} n\,ds = \text{const} \cdot \epsilon^2 \tag{2.16}$$

or a higher power of ϵ. The parameter ϵ must be a continuous "function" of the curve γ (such a "function" is called a *functional*) that equals zero when γ coincides with the true ray. A precise formulation of Fermat's principle requires the use of the calculus of variations and will not be presented here. The main point to be emphasized is that one can always define a quantitative measure of how much the distorted ray γ deviates from the true ray.

One simple way of defining the curve γ is illustrated for two dimensions as follows. Suppose that the function $y = f(x)$ represents the curve in the xy plane followed by the true ray (Fig. 2.7*a*). Then if $h(x)$ has the form shown in Fig. 2.7*b*, with $h(x_1) = h(x_2) = 0$, a family of possible paths γ is given by functions of the form

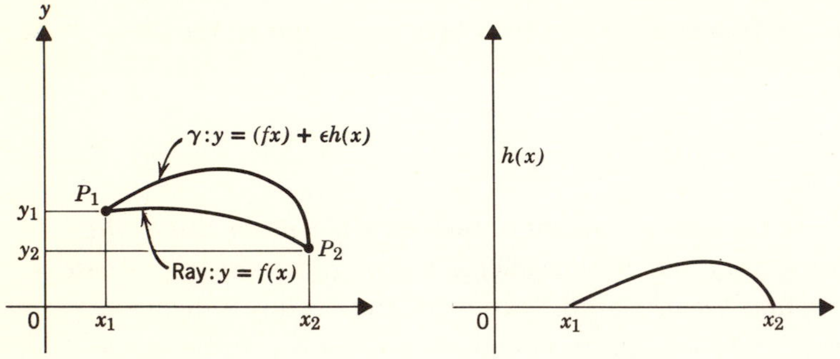

Fig. 2.7

$$y = f(x) + \epsilon h(x)$$

Then as $\epsilon \to 0$, $\gamma \to$ ray, and we must have for the optical path lengths

$$\text{O.P.L. } (\gamma) - \text{O.P.L. (ray)} \sim \epsilon^2 \quad \text{(or higher power)}$$

The path γ is knows as a *virtual path*, since the light does not actually propagate along it. This will become clearer below. There is a mathematical equivalence between Huygens' construction, the eikonal equation, and Fermat's principle; that is, any one of the three can be derived from one of the others. Thus each formulation has the same physical content, and each may be derived from the wave equation in the limit as the wavelength or length of the changing part of the disturbance goes to zero.

E. DERIVATION OF THE LAWS OF GEOMETRICAL OPTICS

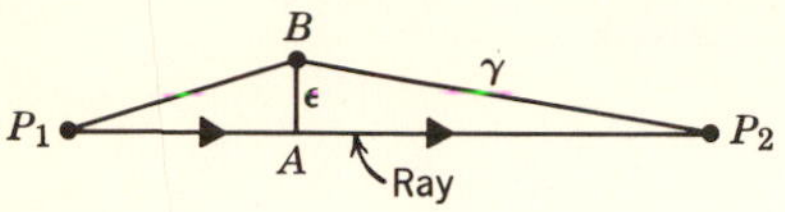

Fig. 2.8. Fermat's principle yields the law of rectilinear propagation.

1. Rectilinear Propagation in a Uniform, Homogeneous Medium. In this case n is a constant, and Eq. (2.1) gives $\Delta l = n\Delta s$, where Δs equals the actual distance from P_1 to P_2 (Fig. 2.8). The straight line P_1AP_2 will be the shortest distance between P_1 and P_2. Any other possible path γ such as P_1BP_2 will have a $\Delta s > P_1P_2$ for we can write

$$\Delta s(\gamma) = \sqrt{|P_1A|^2 + \epsilon^2} + \sqrt{|AP_2|^2 + \epsilon^2}$$

which for small $\epsilon = |AB|$ reduces to (expand the radicals)

$$\Delta s(\gamma) = |P_1A|\left[1 + \frac{\epsilon^2}{|P_1A|^2}\right] + |AP_2|\left[1 + \frac{\epsilon^2}{2(|AP_2|)^2}\right] + 0(\epsilon^4)$$

where $0(\epsilon^4)$ represents a function proportional to ϵ^4 for small ϵ. Thus

$$\Delta s(\gamma) = |P_1A| + |AP_2| + \frac{\epsilon^2}{2}\left(\frac{1}{|P_1A|} + \frac{1}{|AP_2|}\right) + 0(\epsilon^4)$$

Thus the left-hand side of Eq. (2.16) becomes

$$I(\epsilon) = n\Delta s(\gamma) - n\Delta s(\text{ray}) = n\frac{\epsilon^2}{2}\left(\frac{1}{|P_1A|} + \frac{1}{|AP_2|}\right) + 0(\epsilon^4)$$
$$= \text{const}\,\epsilon^2 \text{ as required}$$

In this case the optical path length for the actual ray is a minimum—other virtual rays, such as γ, possess a greater optical path length. This derivation is not mathematically rigorous, because only a limited class of virtual paths has been considered.

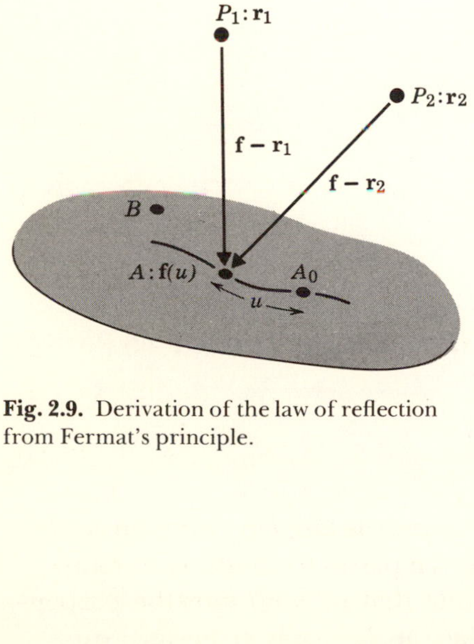

Fig. 2.9. Derivation of the law of reflection from Fermat's principle.

2. Law of Reflection. We shall treat the case of reflection from a smooth surface that is not necessarily a plane (Fig. 2.9). By the previous result we may assume the incident and reflected rays to be straight lines. Let O be an arbitrary origin of coordinates, let A_0 be the point where the true ray from P_1 will be reflected to P_2, and let A be a point on the reflecting surface. Imagine $\overrightarrow{OA}$ to move on the surface along a curve through A_0 that obeys the equation $\overrightarrow{OA} = \mathbf{f}(u)$. The parameter u is defined to be equal to the length of this curve from the point A_0 where the real ray meets the reflecting surface. Points P_1 and P_2 are the fixed end points. Consider a virtual path γ consisting of the straight lines P_1A and AP_2. The optical path length will be

$$l(\gamma) = n(|P_1A| + |AP_2|) = n[|\mathbf{f}(u) - \mathbf{r}_1| + |\mathbf{f}(u) - \mathbf{r}_2|] \tag{2.17}$$

As u varies, a whole family of paths is generated. When $u = 0$ we have $A = A_0$, and the real path is obtained having length $l(0) = n(|P_1A_0| + |A_0P_2|)$. We can take u as a measure of the deviation of γ from the true path. Then Fermat's principle requires

$$l(u) - l(0) = \text{const} \cdot u^2 \quad \text{(or higher power of } u) \quad \text{as } u \to 0$$

which, if we make a series expansion,

$$l(u) = l(0) + \left(\frac{dl}{du}\right)_{u=0} u + \frac{1}{2}\left(\frac{d^2l}{du^2}\right)_{u=0} u^2 + \cdots$$

is seen to be equivalent to the statement

$$\left(\frac{dl(u)}{du}\right)_{u=0} = 0 \tag{2.18}$$

We can write

$$\begin{aligned} |\mathbf{f}(u) - \mathbf{r}_1|^2 &= (\mathbf{f} - \mathbf{r}_1) \cdot (\mathbf{f} - \mathbf{r}_1) \\ &= \mathbf{f} \cdot \mathbf{f} + \mathbf{r}_1 \cdot \mathbf{r}_1 - 2\mathbf{f} \cdot \mathbf{r}_1 \end{aligned}$$

or

$$|\mathbf{f}(u) - \mathbf{r}_1| = \sqrt{\mathbf{f} \cdot \mathbf{f} + \mathbf{r}_1 \cdot \mathbf{r}_1 - 2\mathbf{f} \cdot \mathbf{r}_1}$$

and obtain

$$\frac{d}{du}|\mathbf{f}(u) - \mathbf{r}_1| = \frac{\left[(\mathbf{f}(u) - \mathbf{r}_1) \cdot \dfrac{d\mathbf{f}}{du}\right]}{|\mathbf{f}(u) - \mathbf{r}_1|} \tag{2.19}$$

and similarly

$$\frac{d}{du}|\mathbf{f}(u) - \mathbf{r}_2| = \left[\frac{(\mathbf{f}(u) - \mathbf{r}_2)}{|\mathbf{f}(u) - \mathbf{r}_2|}\right] \cdot \frac{d\mathbf{f}}{du} \tag{2.20}$$

Define

$$\mathbf{e}_1 = \frac{\mathbf{f} - \mathbf{r}_1}{|\mathbf{f} - \mathbf{r}_1|} \tag{2.21}$$

a unit vector in the direction of $\overrightarrow{P_1A}$, and

$$\mathbf{e}_2 = \frac{\mathbf{r}_2 - \mathbf{f}}{|\mathbf{r}_2 - \mathbf{f}|} \tag{2.22}$$

a unit vector in the direction of $\overrightarrow{AP_2}$. From Eqs. (2.18)–(2.22) we obtain as our condition

$$0 = \left[\frac{dl(u)}{du}\right]_{u=0} = n\left[(\mathbf{e}_1 - \mathbf{e}_2) \cdot \frac{d\mathbf{f}}{du}\right]_{u=0} \tag{2.23}$$

Note that the vector $d\mathbf{f}/du$ points along the tangent to the curve at A_0. The curve is arbitrary as long as it passes through A_0 and stays on the surface. We thus conclude that $(\mathbf{e}_1 - \mathbf{e}_2)$ must be perpendicular to all tangents to the surface at A_0, that is to say, $(\mathbf{e}_1 - \mathbf{e}_2)$

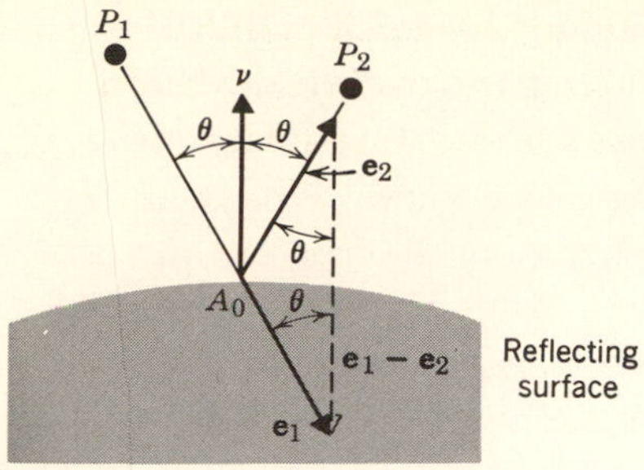

Fig. 2.10. View in the $P_1A_0P_2$ plane of Fig. 2.9.

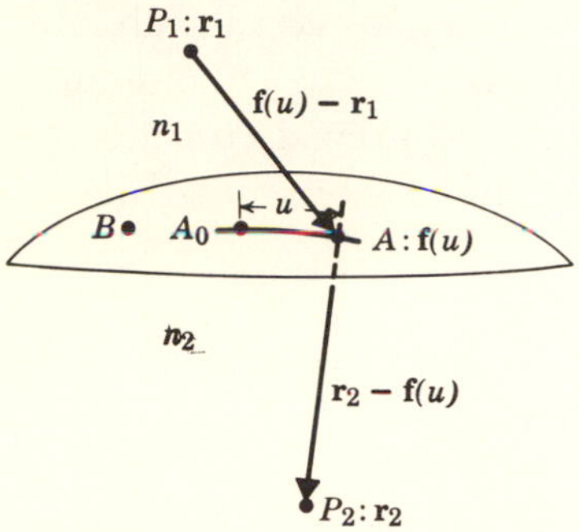

Fig. 2.11. Derivation of the law of refraction from Fermat's principle.

must be along the normal $\boldsymbol{\nu}$ to the tangent plane at A_0 (Fig. 2.10). Thus we can write $\mathbf{e}_1 - \mathbf{e}_2 = a\boldsymbol{\nu}$, where a is a constant. Thus, $\mathbf{e}_2$ (in the direction of the reflected ray) must lie in the plane of incidence (the plane determined by $\mathbf{e}_1$ and $\boldsymbol{\nu}$). Since $\mathbf{e}_1$ and $\mathbf{e}_2$ have equal lengths, they intersect $(\mathbf{e}_1 - \mathbf{e}_2)$ and hence $\boldsymbol{\nu}$ at the same angle θ. Thus, angle of incidence = angle of reflection = θ. This is the desired proof.

3. ***Law of Refraction.*** The argument here is quite similar to that just advanced for reflection. Now P_1 and P_2 are on opposite sides of the refracting surface (Fig. 2.11) in media having indices n_1 and n_2. The real ray is $P_1A_0P_2$; P_1AP_2 represents a virtual ray. A curve runs on the surface from A_0 to A with length u such that $\overrightarrow{OA} = \mathbf{f}(u)$ and $\overrightarrow{OA_0} = \mathbf{f}(0)$. Equations (2.18)–(2.22) are still valid, but the optical path length for the virtual path P_1AP_2 is given by

$$l(u) = n_1|\mathbf{f}(u) - \mathbf{r}_1| + n_2|\mathbf{r}_2 - \mathbf{f}(u)| \tag{2.24}$$

Equation (2.18) then becomes

$$0 = \left(\frac{dl(u)}{du}\right)_0 = \left[(n_1\mathbf{e}_1 - n_2\mathbf{e}_2) \cdot \frac{d\mathbf{f}}{du}\right] \tag{2.25}$$

Again $d\mathbf{f}/du$ is tangent to the surface at A_0, but otherwise arbitrary. This implies that $n_1\mathbf{e}_1 - n_2\mathbf{e}_2$ must be parallel to the normal $\boldsymbol{\nu}$, and we can write

$$n_1\mathbf{e}_1 - n_2\mathbf{e}_2 = b\boldsymbol{\nu}$$

where b is a constant, or

$$\mathbf{e}_2 = \frac{n_1\mathbf{e}_1 - b\boldsymbol{\nu}}{n_2} \tag{2.26}$$

Since $\mathbf{e}_2$ is a linear combination of $\mathbf{e}_1$ and $\boldsymbol{\nu}$, it must lie in the common plane, the plane of incidence. Let $d\mathbf{f}/du$ also be in the plane of incidence. Pick a coordinate system so that $d\mathbf{f}/du$ is along the x axis and $\boldsymbol{\nu}$ is along the y axis (Fig. 2.12). Equation (2.25) can then be written

$$0 = (n_1e_{1x} - n_2e_{2x})\left|\frac{df}{du}\right| \tag{2.27}$$

But $e_{1x} = \sin\theta$; $e_{2x} = \sin\varphi$, so that we obtain Snell's law:

$$n_1 \sin\theta = n_2 \sin\varphi \tag{2.28}$$

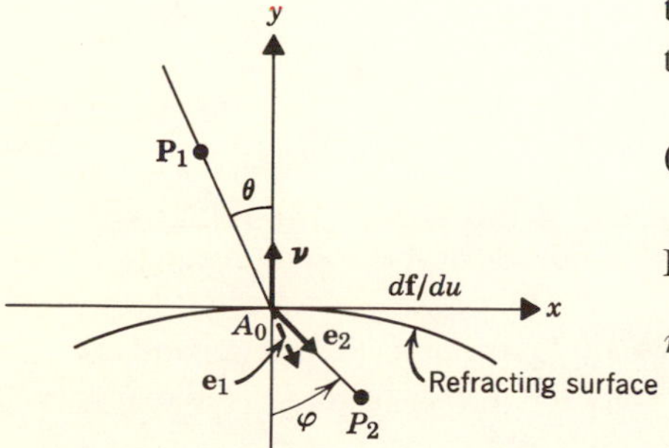

Fig. 2.12. View in the $P_1A_0P_2$ plane of Fig. 2.11.

4. ***Further Discussion of Fermat's Principle.*** It would take us too far afield to study Fermat's principle and its implications in any depth, but a few additional comments seem in order. First, if no reflecting or refracting surfaces are involved in the propagation of the disturbance so that the index of refraction varies continuously, then the actual path length will be a *minimum*, and any nearby optical paths will be longer. This happens because the actual ray is everywhere normal to the wave surfaces. Any deviation of the path must somewhere be skew to the wave surfaces and hence travel a greater distance than does the true ray, just as P_1B and $\widehat{BP_2}$ do in Fig. 2.8.

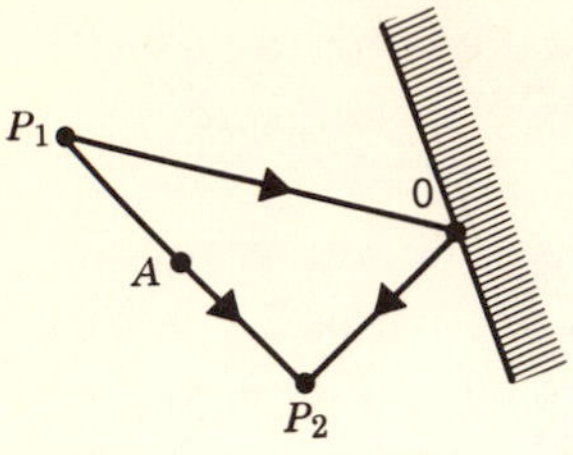

Fig. 2.13. The path $\widehat{P_1OP_2}$ is a local minimum; $\widehat{P_1AP_2}$ is an absolute minimum.

There may be more than one allowed path from a point P_1 to P_2. An example is shown in Fig. 2.13. Each of the paths shown is an extremum with respect to *small* deviations in optical path. This is all that is required by Fermat's principle. It does not demand an absolute maximum or minimum.

When reflections or refractions are involved, the actual path followed by the ray may be either a local maximum or a minimum—the key point is that it be *stationary* with respect to small changes in path. There will be more discussion of this property in the next section.

PROBLEMS

2.1.1 Consider a case of light propagation in two dimensions where n is a function of x only: $n = n(x)$. Starting from Eq. (2.10),

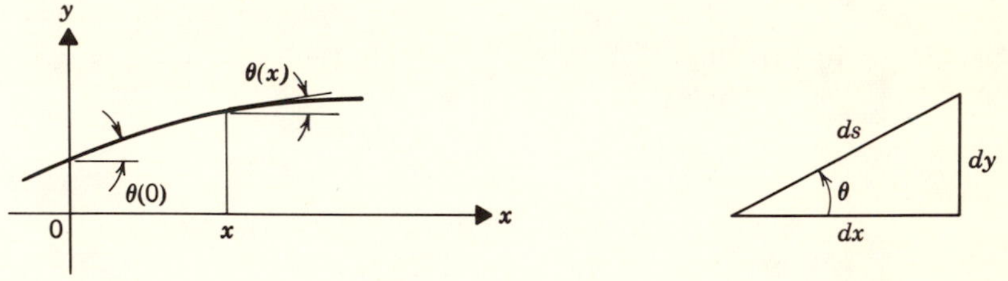

a. Prove that

$$n(x) \sin\theta\,(x) = n(0) \sin\theta\,(0)$$

Note that

$$\sin^2\theta = \left(\frac{dy}{ds}\right)^2 = \frac{(dy)^2}{(dx)^2 + (dy)^2}$$

b. Show that

$$\left(\frac{dx}{dy}\right)^2 + 1 = \frac{n^2(x)}{n^2(0)\sin^2\theta\,(0)}$$

What happens when the right-hand side of this equation equals unity?

This equation gives the slope of the ray in terms of $n(x)$ and the initial angle of incidence $\theta(0)$ at $x = 0$. It may be integrated to obtain $x = x(y)$ or $y = y(x)$ as the equation of the ray.

c. Find the equation of the ray when

$n(x) = a + bx$

if the ray passes through the origin at $x = 0$ at an angle of inclination

θ_0 having magnitude less than $\pi/2$. Caution: y will not be a single-valued function of x when b is negative.

2.1.2 Show that $\oint n\boldsymbol{\tau} \cdot dl = 0$ for a line integral around a closed path. Apply this equation to the case of a discontinuity in n across a plane surface, say the $x = 0$ plane, so that $n = n_1$ for $x < 0$ and $n = n_2$ for $x > 0$. Use the path in the figure in order to derive Snell's law for this case as $\epsilon \to 0$.

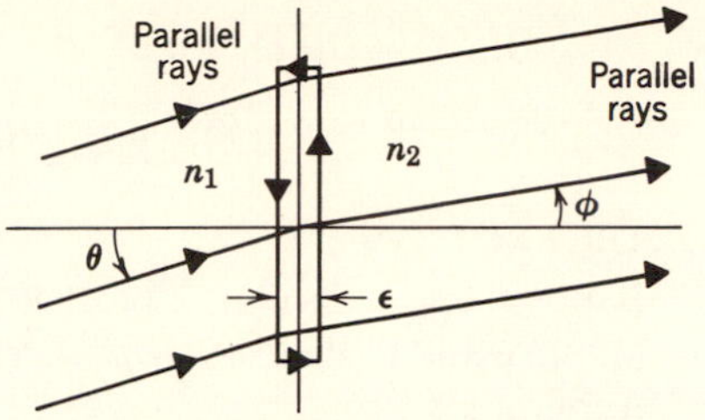

2.1.3 Consider a series of plane interfaces all parallel. At the first, the index changes from n_0 to n_1; at the second, from n_1 to n_2; at the mth, from n_{m-1} to n_m; etc. Let θ_m be the angle of refraction at the mth interface and θ_{m-1} be the angle of incidence there.

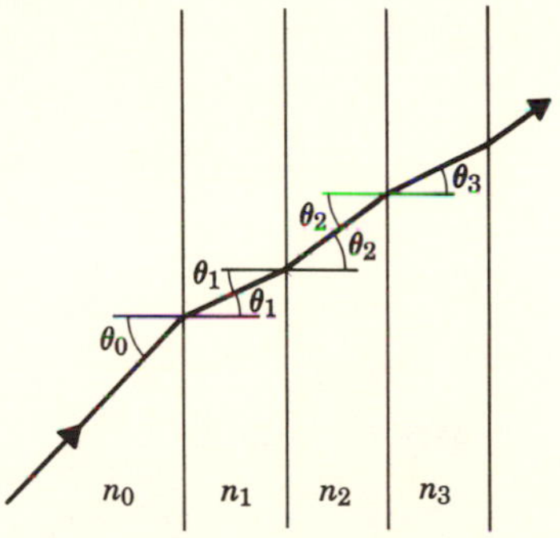

a. Show that repeated application of Snell's law gives

$$n_0 \sin \theta_0 = n_m \sin \theta_m$$

b. Derive this result using the method of Problem 2.1.2.

2.1.4 Point C is at the center of a reflecting sphere of radius R.

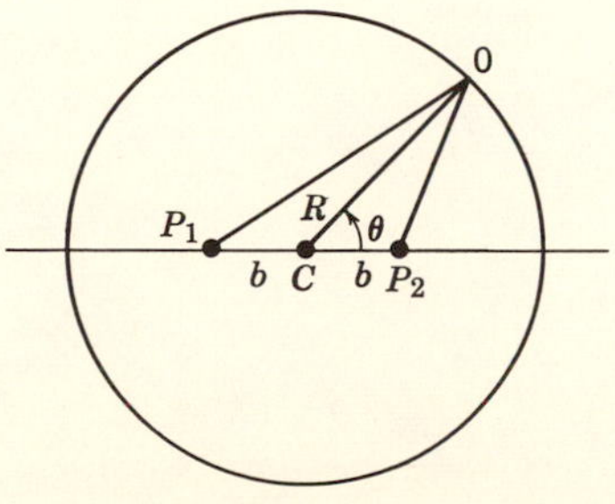

a. Find values of θ for which the line segments $P_1O + OP_2$ will be actual paths of light rays from P_1 to P_2 via reflection at O.
b. Find an analytic expression for the square of the optical path length OPL as a function of θ.
c. Plot $(OPL)^2$ as a function of θ, showing quantitatively the extremal values of $(OPL)^2$ and of the corresponding values of θ.

Answer to Part (b):

$$(OPL)^2 = (|\overrightarrow{P_1O}| + |\overrightarrow{OP_2}|)^2 = 2A^2(1 + \sqrt{1 - d^2 \cos^2 \theta})$$

where $A^2 = b^2 + R^2$, $\quad d = 2bR/(b^2 + R^2)$.

REFERENCES FOR SECTION 2.1

B. Rossi, *Optics*. Sections 2.1 to 2.4.

M. Born and E. Wolf, *Principles of Optics*. Section 3.2 and 3.3 and Appendix I.

2.2 IMAGE FORMATION

In the two derivations of the laws of reflection and refraction from Fermat's principle just completed, note that the single ray $P_1A_0P_2$ would be reflected the same way by all surfaces tangent to the given surface at A_0 in Fig. 2.9. Similarly the single ray $P_1A_0P_2$ in Fig. 2.11 would be refracted the same way by all surfaces tangent to the given surface at A_0. If we go a finite distance away on these surfaces to a point B, the ray P_1B will generally not be reflected or refracted to P_2. Of all the infinitely many surfaces with common tangents at A_0 that send P_1A_0 to P_2, only one, the so-called *Cartesian surface*, will send P_1B to P_2 for an arbitrary point B a finite distance from A_0. If this occurs, we say that P_2 is an *image* of P_1.

A. BEHAVIOR OF RAYS NEAR AN IMAGE

An image can also be formed by a combination of several reflections and refractions and/or by the continuous bending of light rays in a medium where the index of refraction is a smoothly varying function of position. We say that P_2 is a *real image* of P_1 if a bundle of rays leaving P_1 and subtending a *finite* solid angle comes together at P_2 (Fig. 2.14). If the medium near P_1 is uniform and

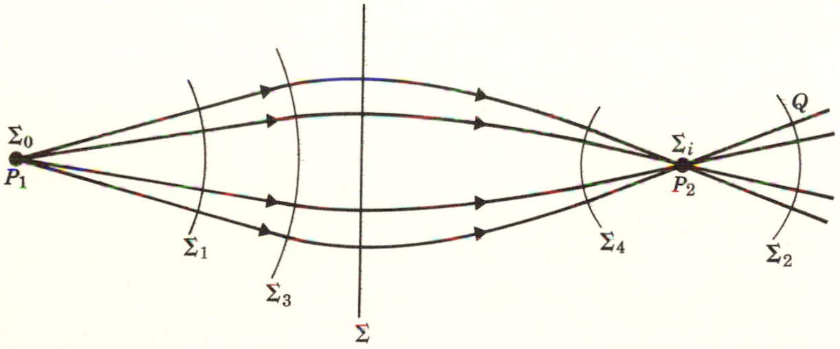

Fig. 2.14. P_2 is a real image of P_1.

isotropic, the wave surface Σ_1 near P_1 will be spherical. Similarly the wave surface Σ_2 near P_2 will be spherical, if the medium there is uniform and isotropic. (The intermediate wave surfaces Σ_3 may be distorted, however.) At P_1 and P_2 themselves the wave surfaces Σ_0 and Σ_i degenerate to points. Since the optical path length along all rays between two wave surfaces is the same, it follows that *the optical path length along all rays from a point P_1 to its image P_2 is the same.**

The rays from P_1 may not actually cross at P_2; they may diverge *as though they met there*, forming what is known as a *virtual image* at

*This will have the very important consequence for sinusoidal waves that the relative phases of all waves leaving the object point will be the same at the image point.

P_2 (Fig. 2.15). The region of divergence must occur in uniform media so that the wave surface Σ_2 is spherical; otherwise there is no way to be sure that the rays near Σ_2 appear to be coming from P_2.

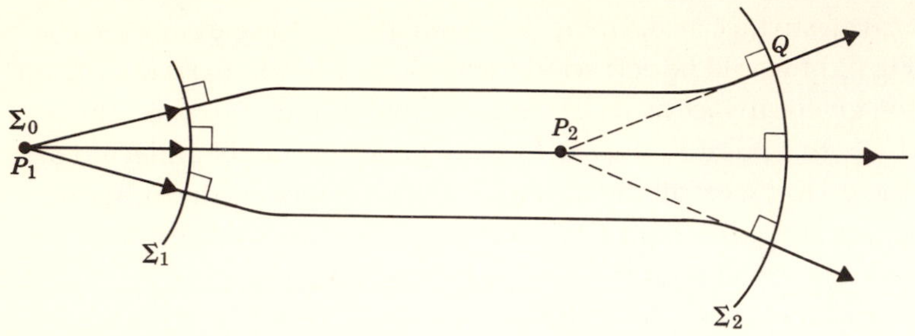

Fig. 2.15. P_2 is a virtual image of P_1.

The optical path from P_1 to Σ_2 is the same for all rays. The apparent (dotted) path from P_2 to Σ_2 is the same for all rays, and so is the apparent optical path from P_1 to P_2, which must evidently be defined by the relation

$$\begin{aligned} \text{opt. path } (P_1P_2) &\equiv \text{opt. path } (P_1Q) - \text{opt. path } (P_2Q) \\ &= \text{opt. path } (P_1Q) - n'|P_2Q| \end{aligned} \tag{2.29}$$

The minus sign occurs because the rays at Q point away from P_2. This equation also holds for the real image illustrated in Fig. 2.14. In both cases n' is the index of refraction near Σ_2.

B. EXAMPLES OF CARTESIAN SURFACES

1. The Cartesian surfaces for reflection are conic sections. We illustrate them in Fig. 2.16, but leave to the reader the proofs that they are indeed what we claim.

2. The Cartesian surfaces for refraction are generally not simple curves. In Fig. 2.17*a* the case of a real image is shown. In this case $n_1|P_1O| + n_2|P_2O| =$ constant, and the curve is a Cartesian ovoid of revolution. Figure 2.17*b* illustrates virtual image formation for which the equation, based on (2.29),

$$\text{opt. path } (P_1P_2) = n_1|P_1O| - n_2|P_2O| = \text{const} \tag{2.30}$$

holds. The refracting surface can sometimes be spherical in this case. See Problem 2.2.2.

If the object or image is at infinity, then the Cartesian surfaces may be conic sections, as indicated in Fig. 2.17*c* and 2.17*d*.

C. APPROXIMATE IMAGE OF A POINT

Perfect image formation often does not occur in real optical systems, in part because of the difficulty of preparing good Cartesian surfaces, and in part because Cartesian surfaces often give good imaging for only one object point. *Approximate image forma-*

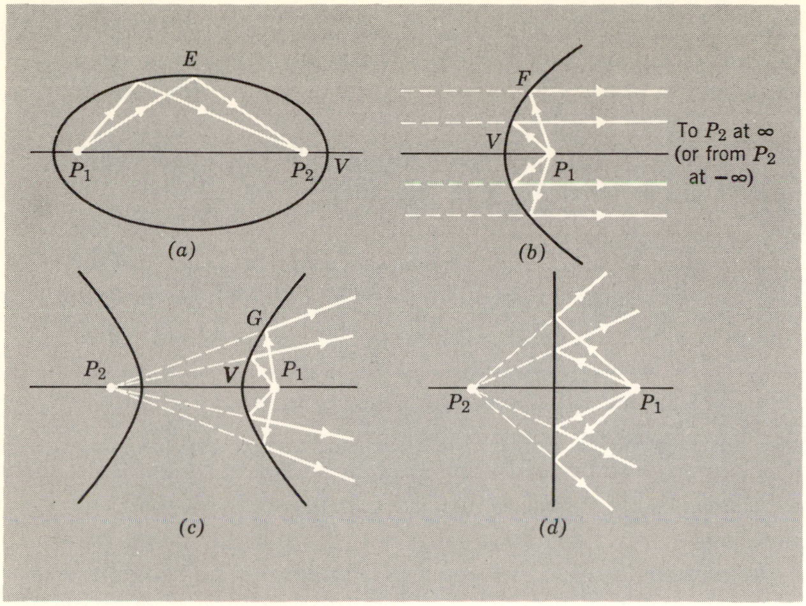

Fig. 2.16. Cartesian surfaces for reflection. (*a*) Ellipsoid—P_1 and P_2 at Foci. (*b*) Paraboloid—P_1 at focus. (*c*) Hyperboloid—P_1 and P_2 at foci. (*d*) Plane.

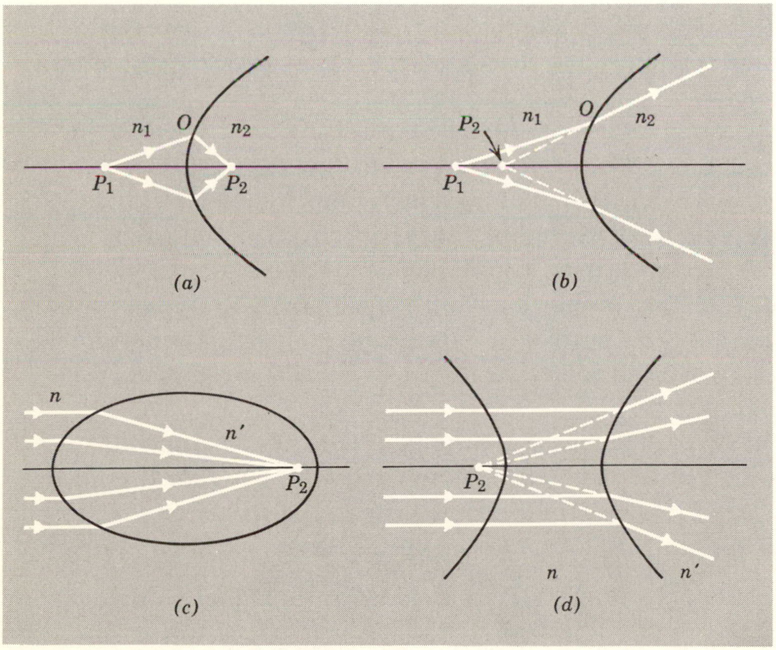

Fig. 2.17. Cartesian surfaces for refraction: (*a*) Real image. (*b*) Virtual image. (*c*) Real image, object at infinity. (Ellipsoid—Eccentricity $e = n/n' < 1$.) (*d*) Virtual image, object at infinity. (Hyperboloid.)

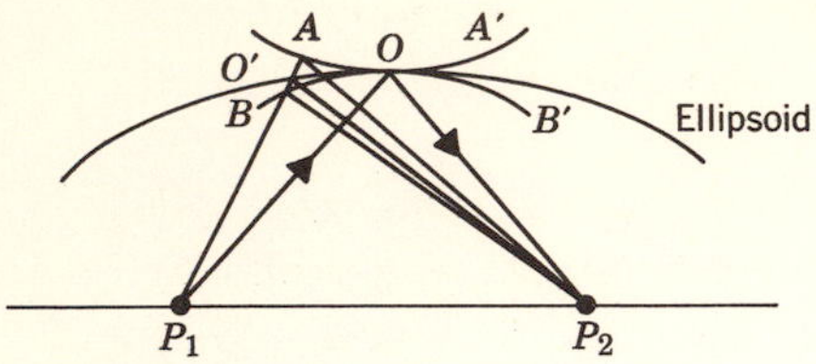

Fig. 2.18. Illustration of the stationary property of the optical path length P_1OP_2. It is a maximum for BOB', a minimum for AOA' and constant for the Cartesean surface OO'.

tion, although not satisfying the condition for exact image formation, is often much easier to obtain, and it can be made quite satisfactory in practice by minimizing the extent to which it is an approximation.

To introduce the subject, let us consider the Cartesian ellipsoid of revolution shown in Fig. 2.18. Any surface tangent to the Cartesian surface at O will reflect the ray P_1O to P_2. The Cartesian surface is the one surface that reflects all other rays such as P_1O′ to P_2. There are other types of surface tangent at O: (1) those for which the optical path P_1AP_2 is greater than or equal to P_1OP_2, and (2) those for which the optical path P_1BP_2 is less than or equal to P_1OP_2. In a first approximation, these two cases are distinguished by the curvature of the surface at O; BOB' has more curvature than the Cartesian surface, and AOA' has less. In the special case when the curvature matches that of the Cartesian surface at O, the ray P_1A or P_1B will in a first approximation be reflected to P_2. Thus we may say that the requirement for a *single ray* P_1O to be reflected to P_2 is that the reflecting surface be tangent to the Cartesian surface at O; that is, the intersection curve of the two surfaces with any plane through O must have the same slope there. If, in addition, *nearby rays* are to be reflected close to P_2, the reflecting surface must have the same curvature at O as the Cartesian surface; that is, the intersection curves of these two surfaces with any plane through O must have the same second derivative.

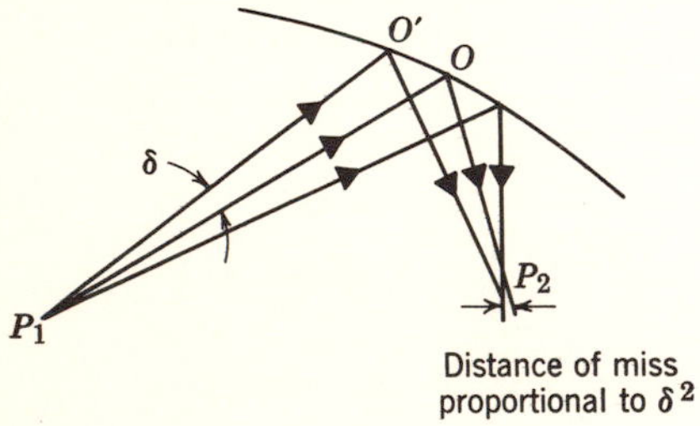

Fig. 2.19. Imaging of a narrow pencil.

The condition for *approximate imaging* can also be stated in the following way. (Consider Fig. 2.19 as an illustration.) Consider a *pencil*, that is, a narrow bundle of light rays originating from a single point. If all the rays of a pencil having a small angular spread 2δ when leaving P_1 miss passing through P_2 by no more than an amount proportional to δ^2, or a higher power, then in the limit as $\delta \to 0$ we say that we have focusing of an infinitesimal pencil. When this occurs, we have approximate imaging of P_1 onto P_2.

The rays just discussed are the exact light rays. We may also consider virtual paths γ from P_1 to P_2 via O′. The angle δ between P_1O′ and P_1O can serve to characterize the path γ. Fermat's principle says that if $\widehat{P_1OP_2}$ is a true ray, then

$$|\widehat{P_1O'} + \widehat{O'P_2}| - |\widehat{P_1O} + \widehat{OP_2}| = \text{const} \cdot \delta^2 \quad \text{(or higher power)}$$

Approximate imaging can be shown to occur if the stricter requirement

$$|\widehat{P_1O'} + \widehat{O'P_1}| - |\widehat{P_1O} + \widehat{OP_2}| = \text{const} \cdot \delta^3 \quad \text{(or higher power)} \tag{2.31}$$

is obeyed.

A picture like Fig. 2.19 could also be drawn to illustrate approximate imaging by a refracting surface. Again the requirement is that the surface be tangent to the Cartesian surface and have the same curvature.

We can test for the presence of approximate imaging in reflection from a single surface by first determining the appropriate Cartesian surface passing through the point of tangency. We then choose the sphere or other surface that matches the curvature of the Cartesian surface. Looking back at the Cartesian surfaces for reflection in Fig. 2.16, we see that at the vertices V there is rotational symmetry; hence approximate imaging can be obtained by placing a spherical surface there with the correct radius of curvature. Off axis, such as at the points E, F, or G in Fig. 2.16*a*, 2.16*b*, 2.16*c*, respectively, we must match the two different principal radii of curvature there. A relatively inexpensively manufactured surface that can match two different curvatures can be cut from the outer surface of a toroid ("donut").

If the condition for approximate imaging is not satisfied, the best that we can obtain is two mutually perpendicular line "images" of an object point, even for a very narrow pencil. This phenomenon is called *astigmatism.*

We shall now discuss some examples of approximate image formation to illustrate these statements. These examples will serve as an introduction to the subject of Gaussian optics in the next chapter.

D. APPROXIMATE IMAGE FORMATION BY A SINGLE REFRACTING SURFACE

1. Path Length Calculation. The geometry of the refracting surface is illustrated in Fig. 2.20. The spherical surface inter-

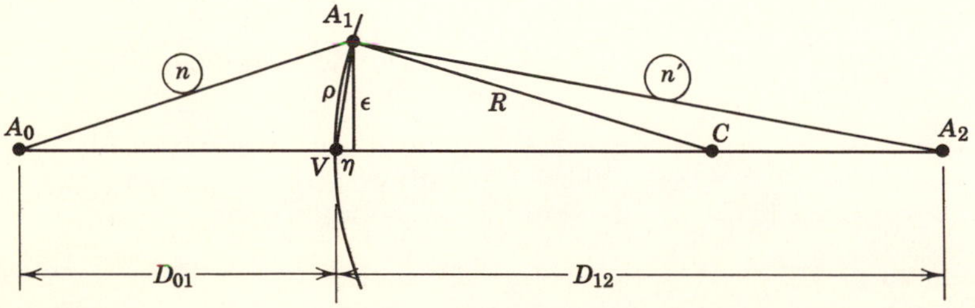

Fig. 2.20

sects the plane of the paper along the arc A_1V and has its center of curvature at C. The radius of curvature is R. Then A_0 is the object point, and A_0VA_2 an exact ray to the approximate image point A_2. Because of the rotational symmetry about this ray, the proper Cartesian surface to be placed at V to image A_0 on A_2 may be approximated by our spherical surface if they both have the same radius of curvature R. Then a narrow cone about the ray A_0V will be approximately imaged at A_2.

We first take the approach of Eq. (2.31). We consider the virtual path $A_0A_1A_2$ and let the distance $\rho = |VA_1|$ be the parameter that characterizes it. We then seek the difference in optical paths denoted by $I(\rho)$:

$$\begin{aligned} I(\rho) &= \text{O.P.L.}(A_0A_1A_2) - \text{O.P.L.}(A_0VA_2) \\ &= [n|A_0A_1| + n'|A_1A_2|] - [nD_{01} + n'D_{12}] \\ &= n\,[|A_0A_1| - D_{01}] + n'\,[|A_1A_2| - D_{12}] \end{aligned} \tag{2.32}$$

We have these exact relations:

$$|A_0A_1| = \sqrt{(D_{01}+\eta)^2+\epsilon^2} \tag{2.33a}$$

$$|A_1A_2| = \sqrt{(D_{12}-\eta)^2+\epsilon^2} \tag{2.33b}$$

$$R^2 = (R-\eta)^2+\epsilon^2 \tag{2.33c}$$

From Eq. (2.33c) we obtain

$$0 = -2R\eta + \epsilon^2 + \eta^2$$

or, since

$$\epsilon^2 + \eta^2 = \rho^2,$$

$$\eta = \frac{\rho^2}{2R} \tag{2.34}$$

Note that

$$\epsilon = \rho\sqrt{1-\frac{\rho^2}{4R^2}} \tag{2.34'}$$

so that for $\rho \ll R$, $\epsilon \approx \rho$. We then obtain, from Eq. (2.33a),

$$\begin{aligned} |A_0A_1| &= (D_{01}{}^2 + 2D_{01}\eta + \eta^2 + \epsilon^2)^{1/2} = D_{01}\left[1 + \frac{2\eta}{D_{01}} + \frac{\eta^2+\epsilon^2}{(D_{01})^2}\right]^{1/2} \\ &= D_{01}\left[1 + \frac{\rho^2}{D_{01}}\left(\frac{1}{D_{01}} + \frac{1}{R}\right)\right]^{1/2} \end{aligned}$$

where we have used (2.34) in the last equality.

We approximate the square root by the binomial theorem using

$$(1+x)^{1/2} = 1 + \tfrac{1}{2}x - \tfrac{1}{8}x^2 + \cdots$$

to obtain

$$|A_0A_1| = D_{01} + \frac{\rho^2}{2}\left(\frac{1}{D_{01}} + \frac{1}{R}\right) - \frac{\rho^4}{8D_{01}}\left(\frac{1}{D_{01}} + \frac{1}{R}\right)^2 + \text{O}(\rho^6) \tag{2.35a}$$

where the notation $\text{O}(\rho^6)$ means "terms that contain a factor which is at least sixth power in ρ."

In a similar fashion we obtain

$$|A_1A_2| = D_{12} + \frac{\rho^2}{2}\left(\frac{1}{D_{12}} - \frac{1}{R}\right) - \frac{\rho^4}{8D_{12}}\left(\frac{1}{D_{12}} - \frac{1}{R}\right)^2 + O(\rho^6) \tag{2.35b}$$

Thus Eq. (2.32) becomes

$$I(\rho) = \frac{\rho^2}{2}\left[\frac{n}{D_{01}} + \frac{n'}{D_{12}} - \frac{(n'-n)}{R}\right] - \frac{\rho^4}{8}\left\{\left[n\left(\frac{1}{D_{01}} + \frac{1}{R}\right)\right]^2 \frac{1}{nD_{01}} + \left[n'\left(\frac{1}{D_{12}} - \frac{1}{R}\right)\right]^2 \frac{1}{n'D_{12}}\right\} + O(\rho^6) \tag{2.36}$$

We see from Eq. (2.36) that for a given value of D_{01} and D_{12} and a general value of R the optical path difference is of second order in ρ. We expect this because of Fermat's principle, since A_0VA_2 is a true ray, and $A_0A_1A_2$ a virtual ray. For approximate image formation to occur, according to Eq. (2.31), the power must be higher, and the coefficient of ρ^2 in Eq. (2.36) must vanish. This gives the condition

$$\frac{n}{D_{01}} + \frac{n'}{D_{12}} = \frac{n'-n}{R} \tag{2.37}$$

Then

$$I(\rho) = \frac{-C\rho^4}{4} + O(\rho^6)$$

$$C = \frac{1}{2}\left\{\left[n\left(\frac{1}{D_{01}} + \frac{1}{R}\right)\right]^2 \frac{1}{nD_{01}} + \left[n'\left(\frac{1}{D_{12}} - \frac{1}{R}\right)\right]^2 \frac{1}{n'D_{12}}\right\} \tag{2.38}$$

The reason this is an even function of ρ has to do with the rotational symmetry of the system about the line A_0V. When A_1 is a distance ρ below V, the optical path difference will be the same as when it is a distance ρ above V, giving $I(-\rho) = I(+\rho)$. Equation (2.38) also implies that the violation of the condition for perfect image formation, namely,

$$I(\rho) = 0$$

is very small as long as ρ is small.

*2. **Ray Aberrations.*** We now change our point of view slightly. Note that $A_0A_1A_2$ is a virtual path. When Eq. (2.37) is obeyed, the true ray from A_0 to A_1 will be refracted and actually cross the axis at a point A_2', which is not exactly at A_2 (Fig. 2.21). We now want to find the separation of A_2' from A_2. From this we can then calculate the analogue of the "miss distance" in Fig. 2.19.

Since CA_1 is the normal to the refracting surface at A_1, we can apply Snell's law at A_1 to obtain

$$n \sin\theta = n' \sin\theta' \tag{2.39a}$$

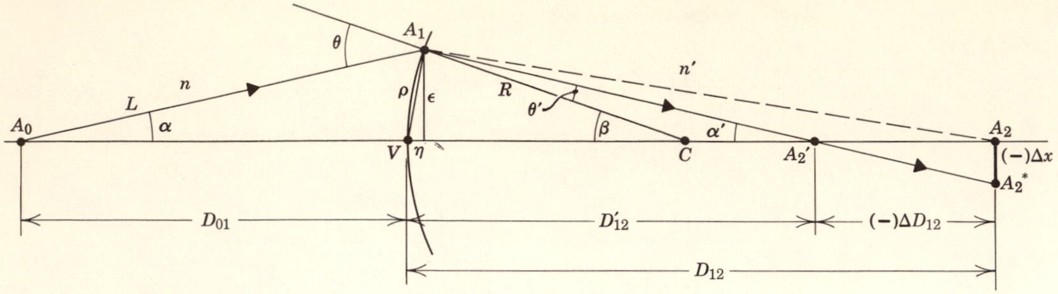

Fig. 2.21

From triangle A_0CA_1 we obtain

$$\theta = \alpha + \beta \tag{2.39b}$$

From triangle A_1CA_2' we obtain

$$\beta = \theta' + \alpha' \tag{2.39c}$$

Thus we have

$$n \sin(\alpha+\beta) = n' \sin(\beta - \alpha')$$

or

$$n(\sin\alpha \cos\beta + \cos\alpha \sin\beta) = n'(\sin\beta \cos\alpha' - \cos\beta \sin\alpha') \tag{2.40}$$

We now expand the sines and cosines appearing in Eq. (2.40) in a power series in ϵ and take the lowest few terms. Later we shall switch to ρ as the independent parameter, since, according to Eq. (2.34′) they are almost equal. For example,

$$\sin\alpha = \frac{\epsilon}{L} = \epsilon\,(D_{01}{}^2 + 2D_{01}\eta + \epsilon^2 + \eta^2)^{-1/2} \approx \epsilon\left[D_{01}{}^2 + \epsilon^2\left(1 + \frac{D_{01}}{R}\right)\right]^{-1/2}$$

$$= \frac{\epsilon}{D_{01}}\left[1 - \frac{\epsilon^2}{2D_{01}{}^2}\left(1 + \frac{D_{01}}{R}\right)\right] + O(\epsilon^5) \tag{2.41a}$$

In the first equality we dropped $\eta^2 = \rho^4/(2R)^2 \sim \epsilon^4$ since it will also lead to corrections of order ϵ^5.

$$\cos\alpha = 1 - \frac{\epsilon^2}{2D_{01}{}^2} + O(\epsilon^4) \tag{2.41b}$$

In addition,

$$\sin\beta = \frac{\epsilon}{R} \qquad \text{(this is exact)} \tag{2.42a}$$

and

$$\cos\beta = 1 - \frac{\epsilon^2}{2R^2} + O(\epsilon^4) \tag{2.42b}$$

Similarly

$$\sin\alpha' = \frac{\epsilon}{D'_{12}}\left[1 - \frac{\epsilon^2}{2(D'_{12})^2}\left(1 - \frac{D'_{12}}{R}\right)\right] + O(\epsilon^5) \tag{2.43a}$$

and

$$\cos\alpha' = 1 - \frac{\epsilon^2}{2(D'_{12})^2} + O(\epsilon^4) \tag{2.43b}$$

In Eqs. (2.43) the distance $D'_{12} = |VA'_2|$ is not exactly equal to $D_{12} = |VA_2|$. When we insert Eqs. (2.41)–(2.43) into (2.40), we obtain

$$n\left[\frac{\epsilon}{D_{01}}\left(1 - \frac{\epsilon^2}{2D_{01}{}^2} - \frac{\epsilon^2}{2D_{01}R} - \frac{\epsilon^2}{2R^2}\right) + \frac{\epsilon}{R}\left(1 - \frac{\epsilon^2}{2D_{01}{}^2}\right)\right]$$

$$= n'\left[\frac{\epsilon}{R}\left(1 - \frac{\epsilon^2}{2(D'_{12})^2}\right) - \frac{\epsilon}{D'_{12}}\left(1 - \frac{\epsilon^2}{2(D'_{12})^2} + \frac{\epsilon^2}{2D'_{12}R} - \frac{\epsilon^2}{2R^2}\right)\right] + O(\epsilon^5)$$

This can be rearranged, and a factor of ϵ canceled, to give

$$\frac{n}{D_{01}} + \frac{n'}{D'_{12}} - \frac{(n'-n)}{R} = C\epsilon^2 + O(\epsilon^4) \tag{2.44a}$$

where C is given by Eq. (2.38) with D_{12} replaced by D'_{12}.

To switch from the parameter ϵ to the parameter ρ, we obtain from Eq. (2.34),

$$\epsilon^2 = \rho^2 - \eta^2 = \left(\rho^2 - \frac{\rho^4}{4R^2}\right)$$

so that Eq. (2.44*a*) gives

$$\frac{n}{D_{01}} + \frac{n'}{D'_{12}} - \frac{(n'-n)}{R} = C\rho^2 + O(\rho^4) \tag{2.44b}$$

From Eq. (2.44*b*) we can see what happens to D'_{12} for fixed D_{01} and R as ρ decreases toward zero. The left side of the equation must tend toward zero. Hence D'_{12} must tend toward D_{12}, which obeys Eq. (2.37) and gives the distance to the approximate image point. We now keep ρ finite in Eq. (2.44*b*) and find an approximate expression for

$$\Delta D_{12} = D'_{12} - D_{12}$$

Let the small dimensionless parameter a be defined by

$$a = \frac{-\Delta D_{12}}{D_{12}} \tag{2.45}$$

Then we can write

$$\frac{1}{D'_{12}} = \frac{1}{D_{12}(1-a)} = \frac{1}{D_{12}}(1+a) + O(a^2)$$

When this is inserted in the left-hand side of Eq. (2.44*b*), it becomes, because of Eq. (2.37), simply

$$\frac{n'a}{D_{12}} + O(a^2)$$

The quantity C appearing on the right side of Eq. (2.44*b*) can be approximated well enough by replacing D'_{12} by D_{12}. We then obtain

$$\frac{n'a}{D_{12}} = C\rho^2 + O(\rho^4)$$

or

$$-\Delta D_{12} = D_{12}a = \frac{CD_{12}{}^2}{n'}\rho^2 + O(\rho^4) = \frac{\rho^2 D_{12}{}^2}{n'}\frac{1}{2}\left\{\left[n\left(\frac{1}{D_{01}}+\frac{1}{R}\right)\right]^2\frac{1}{nD_{01}} + \left[n'\left(\frac{1}{D_{12}}-\frac{1}{R}\right)\right]^2\frac{1}{n'D_{12}}\right\} + O(\rho^4) \tag{2.46}$$

3. ***Spherical Aberration.*** This difference ($-\Delta D_{12}$) is called the *longitudinal spherical aberration.* For the case illustrated it will be positive. Note the proportionality to ρ^2.

The vertical distance $|A_2A_2^*| = \Delta x$ in Fig. 2.21 is called the *lateral spherical aberration.* It will be given by

$$\Delta x = \Delta D_{12} \tan \alpha' = \Delta D_{12}\left[\frac{\rho}{D_{12}} + O(\rho^3)\right]$$

or

$$-\Delta x = \frac{CD_{12}}{n'}\rho^3 + O(\rho^5) = \frac{\rho^3 D_{12}}{n'}\frac{1}{2}\left\{\left[n\left(\frac{1}{D_{01}}+\frac{1}{R}\right)\right]^2\frac{1}{nD_{01}} + \left[n'\left(\frac{1}{D_{12}}-\frac{1}{R}\right)\right]^2\frac{1}{n'D_{12}}\right\} + O(\rho^5) \tag{2.47}$$

We stated in connection with Fig. 2.19 that the "miss distance" at the approximate image point must be proportional to at least the second power of the angle of deviation of the approximate ray from the true ray at the source point. In the present case α represents that angle, and when it is small it can be approximated by ρ/D_{01}. Thus, in this case the "miss distance" $|\Delta x|$ is proportional to the third power of the deviation angle:

$$|\Delta x| \approx \frac{D_{01}{}^3 D_{12}\alpha^3}{n'}\{\cdots\}$$

4. ***Use of Reference Surface.*** Another way of looking at this system is to examine the behavior of the wave surfaces. Consider Fig. 2.22. Here Σ_1 and Σ_2 represent spherical wave surfaces leaving the point source A_0. Let Σ_e be a wave surface emerging after refraction. It cuts the axis at the point P selected to be on a sphere with radius $|A_1A_2|$ centered at A_2. Then perfect image formation would be obtained if Σ_e coincided with this sphere, because the emerging wave surface must be a sphere to give a perfect point image. The distance $|A_1Q|$ represents the deviation of Σ_e from this "reference sphere" about A_2.

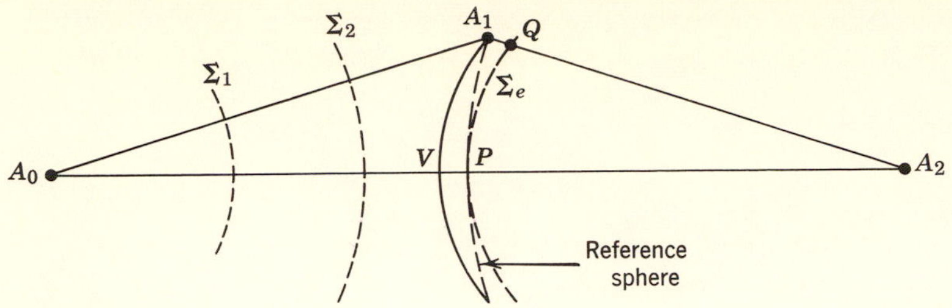

Fig. 2.22

Since Q and P both lie on the wave surface Σ_e, we can write

$$\text{O.P.L.}(\widehat{A_0A_1Q}) = \text{O.P.L.}(\widehat{A_0VP}) \tag{2.48}$$

Since A_1 and P lie on the reference sphere, we have

$$\text{O.P.L.}(\widehat{A_1A_2}) = \text{O.P.L.}(\widehat{\text{PA}_2}) \tag{2.49}$$

Then we can write

$$I(\rho) = \text{O.P.L.}(\widehat{A_0A_1QA_2}) - \text{O.P.L.}(\widehat{A_0VPA_2})$$

$$I(\rho) = \text{O.P.L.}(\widehat{A_0A_1Q}) + \text{O.P.L.}(\widehat{QA_2}) - \text{O.P.L.}(\widehat{A_0VP}) - \text{O.P.L.}(\widehat{PA_2})$$

Now write $\text{O.P.L.}(\widehat{QA_2}) = \text{O.P.L.}(\widehat{A_1QA_2}) - \text{O.P.L.}(\widehat{A_1Q})$ and rearrange:

$$\begin{aligned} I(\rho) = {} & [\text{O.P.L.}(\widehat{A_0A_1Q}) - \text{O.P.L.}(\widehat{A_0VP})] + [\text{O.P.L.}(\widehat{A_1QA_2}) \\ & - \text{O.P.L.}(\widehat{PA_2})] - \text{O.P.L.}(\widehat{A_1Q}) \end{aligned} \tag{2.50}$$

Use of Eqs. (2.48) and (2.49) in Eq. (2.50) gives

$$I(\rho) = -\text{O.P.L.}(\widehat{A_1Q}) = -n'|\widehat{A_1Q}|$$

Thus, from Eq. (2.38) we obtain

$$n'|A_1Q| = \frac{C\rho^4}{4} + O(\rho^6) \tag{2.51}$$

The directed distance *$l(Q) = |\overrightarrow{A_1Q}| = -I(\rho)/n'$ from A_1 to Q is an extremely useful quantity, since the aberration properties of the system may be derived from it. For example, we can easily derive Eq. (2.47) for the lateral spherical aberration from $l(Q)$. Consider Fig. 2.23, which is greatly exaggerated for clarity. The true ray at Q is QA_2', which is normal to the wave surface $\Sigma_e = PQQ'$. On the other hand, the line A_1QA_2 is normal to the reference sphere PA_1A_1'. We have, approximately, the result

$$-\Delta x = \gamma D_{12}$$

A_1' is slightly displaced from A_1 by a distance $\Delta\rho \ll \rho$, and Q' is the corresponding point along the line $A_1'A_2$.

*By "directed" we mean that the sign of $l(Q)$ is positive if A_1 is to the left of Q and negative if A_1 is to the right of Q in Fig. 2.22.

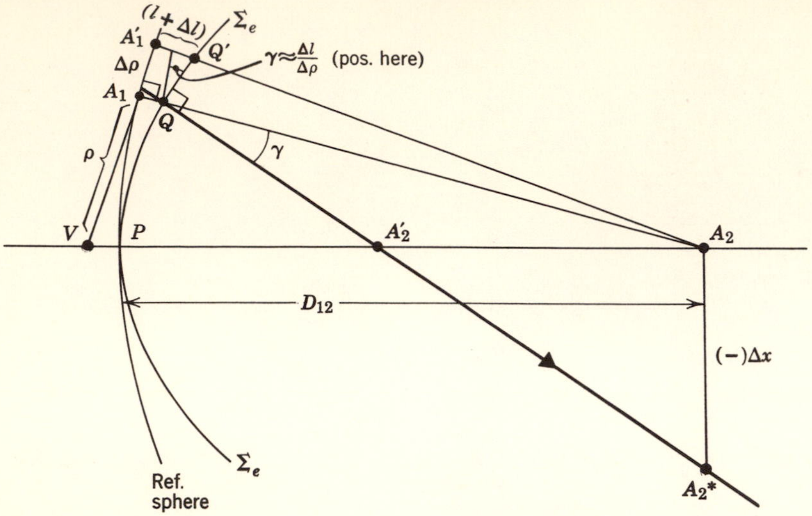

Fig. 2.23

If we consider the approximately straight-sided figure $A_1QQ'A_1'$, we obtain

$$\gamma \approx \frac{\Delta l}{\Delta\rho} \approx \frac{dl}{d\rho} = -\frac{1}{n'}\frac{dI}{d\rho}$$

Hence

$$\Delta x \cong -D_{12}\gamma$$

$$\Delta x \cong \frac{D_{12}}{n'}\frac{dI}{d\rho} = -\frac{D_{12}}{n'}C\rho^3 + O(\rho^5) \qquad (2.52)$$

This is a much simpler derivation of Eq. (2.47) than the earlier one. The quantity $W = n'|\overrightarrow{A_2Q}|$, called the *aberration function* for this refraction problem, will appear in subsequent discussions of aberrations.

E. ADDITIONAL EXAMPLES

1. Reflection off a Spherical Surface. The situation here is quite analogous to the case of refraction just considered, and we can omit most of the details. Figure 2.24 defines the geometry; A_0VA_2 is the true ray, $A_0A_1A_2$ a virtual path. The optical path difference $I(\rho) = n[|A_0A_1| + |A_1A_2| - D_{01} - D_{12}]$ can be readily shown to be given by

$$\frac{I(\rho)}{n} = \frac{\rho^2}{2}\left(\frac{1}{D_{01}} + \frac{1}{D_{12}} - \frac{2}{R}\right) - \frac{\rho^4}{8}\left[\left(\frac{1}{D_{01}} - \frac{1}{R}\right)^2\frac{1}{D_{01}} + \left(\frac{1}{D_{12}} - \frac{1}{R}\right)^2\frac{1}{D_{12}}\right] + O(\rho^6)$$

This is analogous to Eq. (2.36). When the approximate image-forming condition

$$\frac{1}{D_{01}} + \frac{1}{D_{12}} = \frac{2}{R} \qquad (2.53)$$

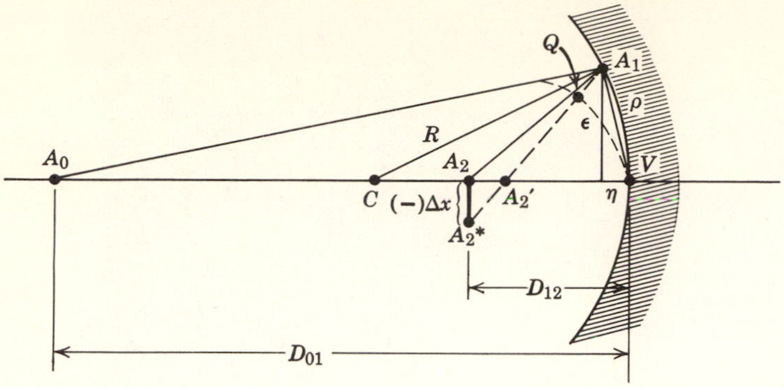

Fig. 2.24

is satisfied, this becomes

$$I(\rho) = -\frac{C\rho^4}{4} + O(\rho^6) \tag{2.54}$$

$$C = \frac{n}{2}\left(\frac{1}{D_{01}} - \frac{1}{R}\right)^2 \left(\frac{1}{D_{01}} + \frac{1}{D_{12}}\right)$$

If we consider the true reflected ray $A_1A_2^*$ and the true wave surface VQ, then for the separation between the true wave surface and the reference sphere we obtain

$$|\overrightarrow{A_1Q}| = -\frac{I(\rho)}{n} \tag{2.55}$$

The quantity $W = n|\overrightarrow{A_1Q}| = -I(\rho)$, called the *aberration function* for this reflection problem, will appear in any discussion of aberrations.

The lateral spherical aberration is given by

$$\Delta x \cong \frac{D_{12}}{n}\frac{dI}{d\rho} = -\frac{D_{12}}{n}C\rho^3 + O(\rho^5) \tag{2.56}$$

2. Sequences of Approximate Images. Real optical devices almost always have more than one refracting or reflecting surface. The image produced by the first surface acts as an object for the second surface. The image it produces in turn acts as an object for the next surface. And so on. It is therefore important to know how the aberration properties of the entire system depend on those of each surface.

If a point A_0 is imaged exactly onto a point A_2 by an optical system and if A_2 is in turn imaged exactly onto A_4, then we may say that A_0 is imaged exactly onto A_4. The analogous statement holds for approximate image formation. This is illustrated in Fig. 2.25 for a system with two refracting surfaces and a symmetry axis, which is along the true ray $\widehat{A_0V_1A_2V_2A_4}$. The system illustrated is, of course, a thick lens and might form part of a still larger system. Another true ray is $\widehat{A_0A_1A_2^*A_3A_4^*}$; $\widehat{A_0A_1A_2A_3A_4}$ is a virtual

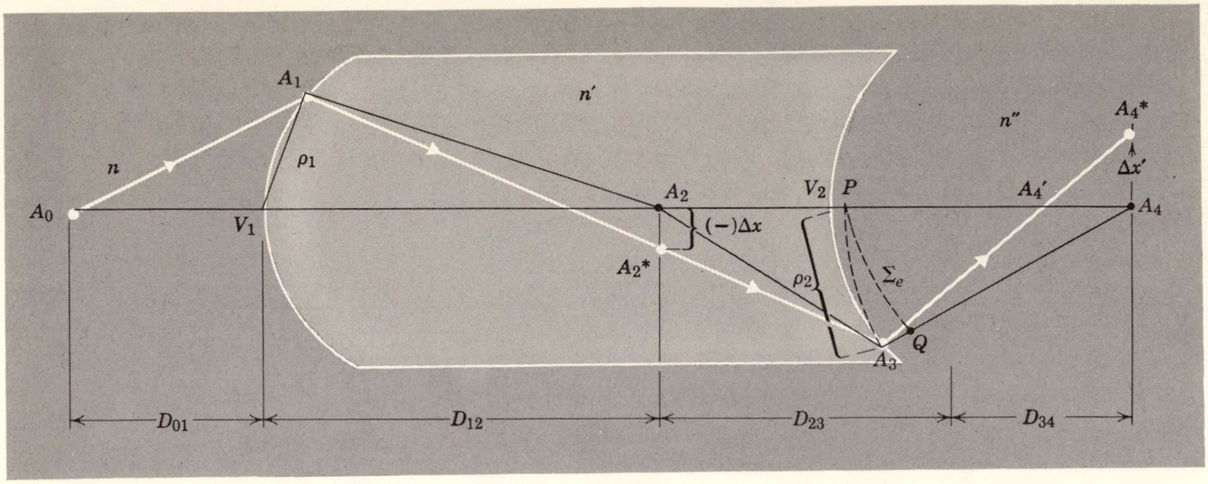

Fig. 2.25

ray; A_2 is the approximate image of A_0 and A_4 is the approximate image of A_2. We now show that A_4 is the approximate image of A_0 by use of optical path differences.

Consider $I_t(\rho_1) = \text{O.P.L.}(\widehat{A_0A_1A_2A_3A_4}) - \text{O.P.L.}(\widehat{A_0V_1A_2V_2A_4})$ Then

$$I_t(\rho_1) = I_1(\rho_1) + I_2(\rho_2) \tag{2.57}$$

where

$$I_1(\rho_1) = \text{O.P.L.}(\widehat{A_0A_1A_2}) - \text{O.P.L.}(\widehat{A_0V_1A_2}) = -\frac{C_1\rho_1^4}{4} + O(\rho_1^6)$$

and

$$I_2(\rho_2) = \text{O.P.L.}(\widehat{A_2A_3A_4}) - \text{O.P.L.}(\widehat{A_2V_2A_4}) = -\frac{C_2\rho_2^4}{4} + O(\rho_2^6)$$

represent optical path differences for the two separate image-forming processes. We have approximately

$$\rho_2 = \frac{\rho_1 D_{23}}{D_{12}}$$

Thus

$$I_t(\rho_1) = -\frac{C_t\rho_1^4}{4} + O(\rho_1^6)$$

where

$$C_t = C_1 + C_2\left(\frac{D_{23}}{D_{12}}\right)^4 \tag{2.58}$$

This result establishes that A_4 is an approximate image of A_0. It can also be expressed as a function of ρ_2;

$$I_t(\rho_2) = -\frac{C_t^*\rho_2^4}{4} + O(\rho_2^6), \qquad C_t^* = C_2 + C_1\left(\frac{D_{12}}{D_{23}}\right)^4 \tag{2.58′}$$

Note in Fig. 2.25 that the length of $\widehat{A_1A_2A_3}$ differs from that of the straight line $\widehat{A_1A_2^*A_3}$ by an amount proportional to $(\Delta x)^2 = O(\rho_1{}^6)$. This gives

$$\text{O.P.L.}(A_0A_1A_2A_3) = \text{O.P.L.}(A_0A_1A_2^*A_3) + O(\rho^6)$$

or

$$I_t(\rho_1) = \text{O.P.L.}(A_0A_1A_2^*A_3A_4) - \text{O.P.L.}(A_0V_1A_2V_2A_4) + O(\rho^6) \tag{2.59}$$

If we take a reference sphere of radius $|A_3A_4| = |PA_4|$ with A_4 at its center and consider the emerging wavefront $\Sigma_e = PQ$ that describes the light that left the original point source A_0, we have, in analogy with Eq. (2.59),

$$\text{O.P.L.}(\widehat{A_0A_1A_2^*A_3Q}) = \text{O.P.L.}(\widehat{A_0V_1A_2P})$$

and, in analogy with Eq. (2.49),

$$\text{O.P.L.}(\widehat{A_3A_4}) = \text{O.P.L.}(\widehat{PA_4})$$

Then, with the help of Eq. (2.59) we easily derive

$$I_t(\rho_1) = -\text{O.P.L.}(\widehat{A_3Q}) + O(\rho_1{}^6)$$

Thus

$$\text{O.P.L.}(\widehat{A_3Q}) = n''|\overrightarrow{A_3Q}| = -I_1(\rho_1) - I_2(\rho_2) \tag{2.60}$$

The *additive property* of Eq. (2.60) implies that the wavefront aberration function $W_t = n''|\overrightarrow{A_3Q}|$ is given in terms of the sum of the aberration functions W_1 and W_2 produced by the two refracting surfaces. By use of Eq. (2.58′) we may express W_t in terms of ρ_2 and can then obtain the lateral spherical aberration Δx in a manner similar to that employed above for a single refracting surface.

These results can be immediately generalized to a system with an arbitrary number of refracting or reflecting surfaces. The approximate image formed by the last surface will be the approximate image for the entire system, which will have an aberration function W that is the sum of the W_i's from each surface.

3. ***Behavior for an Off-Axis Object Point.*** So far we have considered systems with axial symmetry, with the object point A_0 located on the axis. The situation becomes much more complicated as soon as A_0 moves off axis, because in general we no longer have approximate image formation in the technical sense considered above where terms in ρ^2 do not appear in the expression for the optical path difference. This subject will be discussed quantitatively in Chapter 4. Here we merely mention that in lowest order the aberration function W is no longer simply proportional to ρ^4. It contains additional terms varying as ρ^2x^2, $\rho^2x^2\cos^2\theta$, and $\rho^3x\cos\theta$, where x is the off-axis distance and θ the azimuthal angle (Fig. 2.26). Thus, for fixed x and θ the optical path difference

$$\text{O.P.L.}(A_0A_1'A_2) - \text{O.P.L.}(A_0VA_2)$$

is not proportional to ρ^3 or a higher power, but contains terms with a lower power in ρ. Strictly speaking, then, approximate image formation is not obtained at A_2.

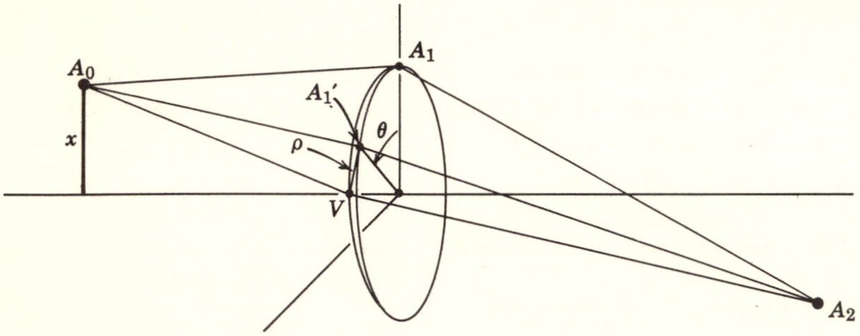

Fig. 2.26. Optical paths when the object point A_0 is off-axis. $\widehat{A_0A_1A_2}$ lies in the plane of the paper (meridian plane). $\widehat{A_0A_1'A_2}$ is out of the plane of the paper in a plane making an angle θ with that of the paper.

Note, however, that the combined power of ρ and x in each of the above terms is 4; thus as long as x and ρ are both small, the optical path difference can be made quite small. In the limit as x and ρ are both very small, we obtain a kind of approximate image formation referred to as *Gaussian or paraxial image formation.* This is the subject of Chapter 3.

PROBLEMS

2.2.1 Prove that the conic surfaces of revolution in Fig. 2.16 are Cartesian surfaces for reflection.

2.2.2 Prove that the spherical surface shown is Cartesian for refraction with a virtual image formed at P_2 provided that s_1 and s_2 satisfy

$$n_1{}^2 s_1 = n_2{}^2 s_2$$
$$= n_1 n_1 R$$

(Use Eq. 2.29 and establish equality of the appropriate optical paths.)

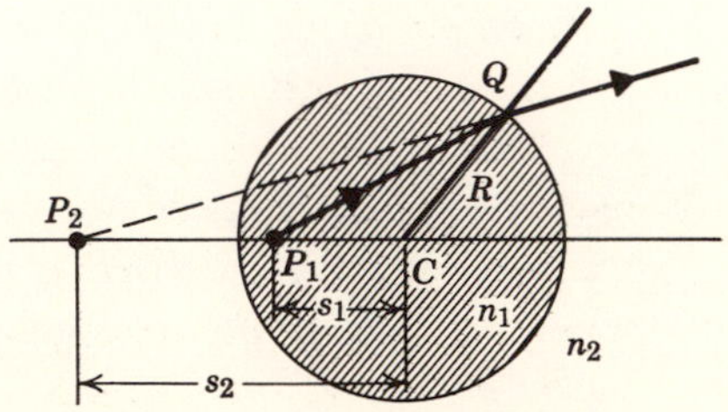

2.2.3 a. What is the nature of the Cartesian reflecting surface to be inserted at O so that a virtual image of P is formed at P'?

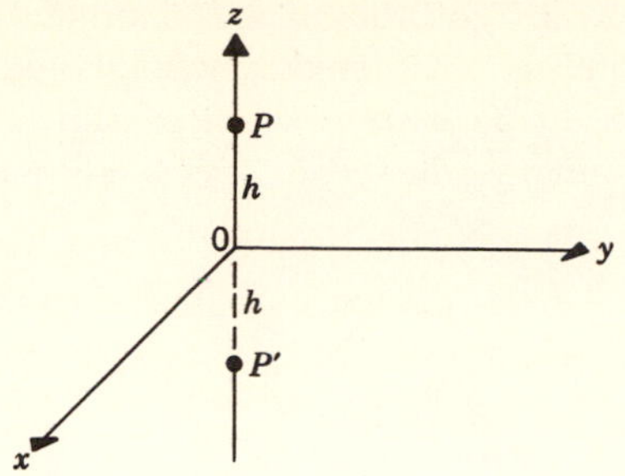

b. Which of the following surfaces will give approximate virtual imaging of P at P' in the above illustration? Here are the equations for the surfaces:

(i) $z = a(x^2 + y^2)$
(ii) $z = ax^2 + by^2, \qquad a \neq b$
(iii) $z = ax^3 + by^3$
(iv) $z = a(x^3 + y^3) + cy^4$
(v) $z = a(x^4 + y^4)$

Explain.

2.2.4 Show that the ellipsoidal surface of revolution in the illustration will give perfect image formation with refraction for the parallel beam at the focus F' of the ellipse provided that the eccentricity $e = c/a$ satisfies

$$e = \frac{n}{n'} \tag{1}$$

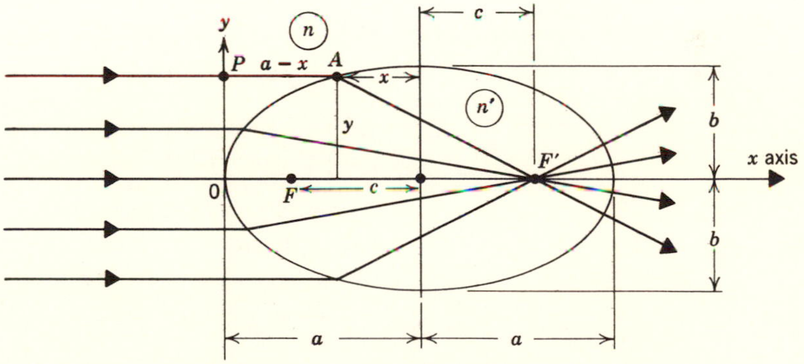

Hint. Take an origin at O. Then the point A will have coordinates $[(a - x), y]$. Show that the condition

$$\text{O.P.L.}(\widehat{PAF'}) = \text{O.P.L.}(\widehat{OFF'})$$

can lead to the equation of the ellipse

$$\frac{x^2}{a^2} + \frac{y^2}{b^2} = 1$$

Provided that Eq. (1) is satisfied.

2.2.5 You want to place a toroidal mirror at O to obtain approximate imaging of P_1 onto P_2. What should be the values of the two principal radii of curvature? (One principal curve lies in the plane of the figure, and the other lies in a perpendicular plane.)

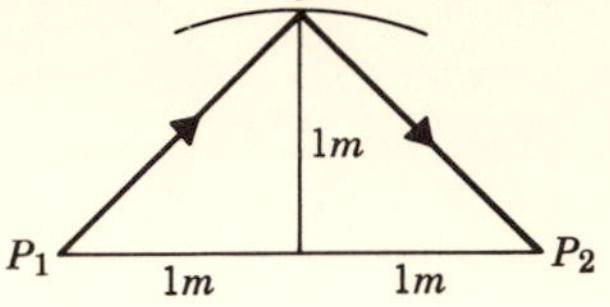

2.2.6 Consider reflection from a spherical surface with center of curvature at C, radius of curvature R. The surface cuts the plane of the paper in the arc A_0V. A_1VA_2 is an exact ray, and $A_1A_0A_2$ is a virtual ray γ.

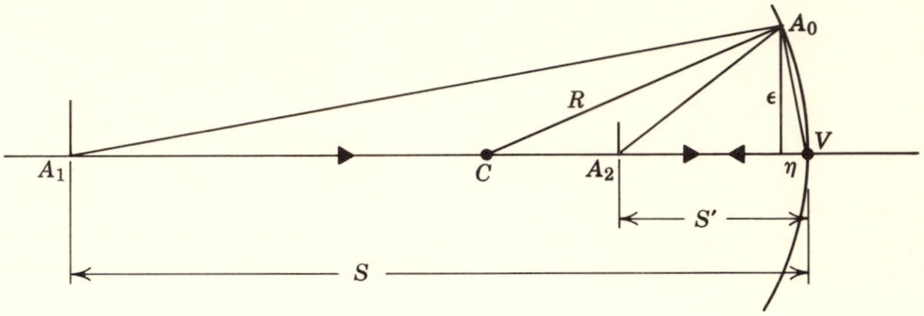

a. Show that the condition for approximate image formation, namely,

$$\text{O.P.L.}(\gamma) - \text{O.P.L.}(\text{ray}) \sim \epsilon^3 \text{ (or higher power)} \tag{1}$$

is satisfied provided that

$$\frac{1}{S} + \frac{1}{S'} = \frac{2}{R} \tag{2}$$

and find the expression for the coefficient of the leading term on the *RHS* of Eq. (1).

b. When $|A_2V| < S'$, with S' satisfying Eq. (2), is the optical path length of the ray A_1VA_2 a maximum or a minimum?

2.3 RAY TRACING

Most optical systems consist of a series of refracting or reflecting surfaces, usually plane or spherical, to provide the required deviation of the light rays together with appropriate stops and apertures to limit the angular and spatial extent of the rays. The optical designer must often trace the path of rays through the system. This is done by repeated applications of the laws of geometrical optics—the law of reflection or refraction at the deviating surfaces, and the law of rectilinear propagation between the surfaces. This is simple in principle, but in practice it can become a formidable task if high accuracy is required. Approximate calculations can be carried out relatively easily, as we shall see in the next chapter. In this section we indicate the type of calculation necessary to perform accurate ray tracing through a series of coaxial spherical refracting surfaces. In subsequent work we shall need only approximate versions of these formulas, which will be derived as we go along.

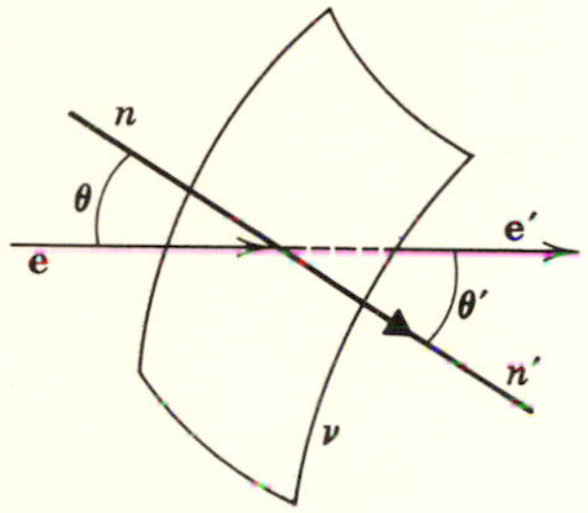

Fig. 2.27

A. REFRACTION

1. General Equations. We begin by considering the refraction of a ray with unit vector **e** by a surface having a normal $\boldsymbol{\nu}$. The refracted ray has unit vector **e**′; the angles of incidence and refraction are θ and θ', respectively, as shown in Fig. 2.27. With a slight change in notation, this is exactly the situation illustrated earlier in Fig. 2.11. The law of refraction can be written in the compact form of Eq. (2.26), which in our new notation becomes

$$n'\mathbf{e}' = n\mathbf{e} + b\boldsymbol{\nu} \tag{2.61}$$

The constant b can be determined by taking the scalar product of both sides of (2.61) with the unit vector $\boldsymbol{\nu}$. By noting the equalities

$$(\boldsymbol{\nu}\cdot\boldsymbol{\nu}) = 1; \qquad (\mathbf{e}\cdot\boldsymbol{\nu}) = \cos\theta; \qquad (\mathbf{e}'\cdot\boldsymbol{\nu}) = \cos\theta' \tag{2.62}$$

we obtain the result

$$b = n'\cos\theta' - n\cos\theta \tag{2.63}$$

We can determine $\cos\theta'$ in terms of $\cos\theta$ by Snell's law, which, incidentally, results from taking the cross product of (2.61) with $\boldsymbol{\nu}$ (since $|\boldsymbol{\nu}\times\mathbf{e}| = \sin\theta$, $|\boldsymbol{\nu}\times\mathbf{e}'| = \sin\theta'$, and $\boldsymbol{\nu}\times\boldsymbol{\nu} = 0$):

$$n'\cos\theta' = \sqrt{n'^2 - n'^2\sin^2\theta'} = \sqrt{n'^2 - n^2\sin^2\theta}$$
$$= \sqrt{n'^2 - n^2 + n^2\cos^2\theta} \tag{2.64}$$

Thus, if we are initially given only the vectors **e**, $\boldsymbol{\nu}$, (that is, their components) and the indices n, n', we first determine $\cos\theta$ by

(2.62) and then cos θ' by (2.64). This allows us to calculate b by Eq. (2.63), which is then used in Eq. (2.61) to obtain $\mathbf{e}'$.

*2. **Refraction at a Spherical Surface.*** Let us apply the above results to Fig. 2.28. The spherical refracting surface has radius of curvature R and center of curvature C, with coordinates

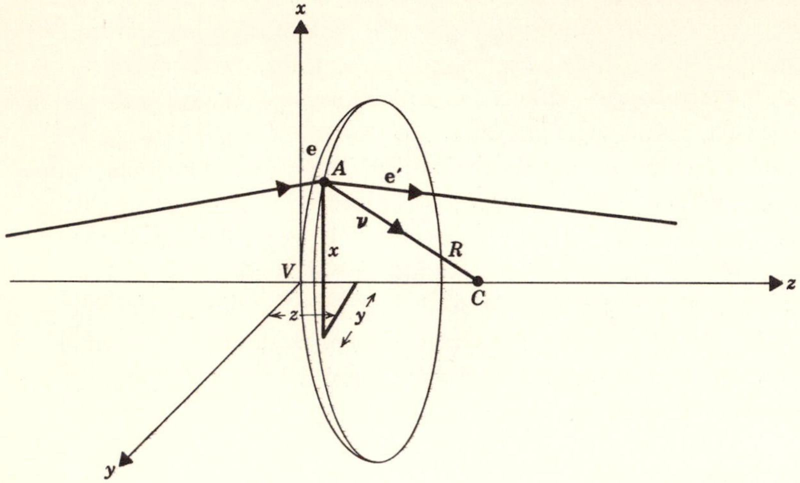

Fig. 2.28

$$\mathbf{C}\text{:}\quad (0, 0, R)$$

The vertex V where the surface intersects the z axis is at the origin. The incident ray meets the surface at the point A having coordinates

$$\mathbf{A}\text{: } (x, y, z) = (x, y, R - \sqrt{R^2 - x^2 - y^2})$$

The z coordinate of $\mathbf{A}$ is determined by the requirement that the distance $|AC| = R$.

The normal to the surface is a unit vector pointing from A toward C. It is given by

$$\boldsymbol{\nu} = \frac{\mathbf{C}-\mathbf{A}}{|\mathbf{C}-\mathbf{A}|} = \frac{\mathbf{C}-\mathbf{A}}{R}$$

with components

$$\boldsymbol{\nu}\text{: } \left(-\frac{x}{R}, -\frac{y}{R}, \frac{\sqrt{R^2 - x^2 - y^2}}{R}\right) \tag{2.65}$$

with

$$\nu_x^2 + \nu_y^2 + \nu_z^2 = 1$$

We now introduce the components (called direction cosines) of the unit vectors $\mathbf{e}$ and $\mathbf{e}'$:

$$\begin{aligned}&\mathbf{e} \text{ has components } (\alpha, \beta, \gamma) \text{ with length } \alpha^2 + \beta^2 + \gamma^2 = 1\\&\mathbf{e}' \text{ has components } (\alpha', \beta', \gamma') \text{ with length } \alpha'^2 + \beta'^2 + \gamma'^2 = 1\end{aligned} \tag{2.66}$$

With the help of Eq. (2.63) we can write the x and y components of Eq. (2.61) as

$$
\begin{aligned}
n'\alpha' &= n\alpha - \frac{(n'\cos\theta' - n\cos\theta)x}{R} \\
n'\beta' &= n\beta - \frac{(n'\cos\theta' - n\cos\theta)y}{R} \\
n'\gamma' &= n\gamma + \frac{(n'\cos\theta' - n\cos\theta)\,(R^2 - x^2 - y^2)^{1/2}}{R}
\end{aligned}
\tag{2.67a}
$$

[Note that γ and γ' can always be determined from the normalization condition, Eq. (2.66).]

The *approximate* versions of the first two equations of (2.67*a*) will be needed later. They result from the assumption that all light rays and normals to surfaces at the point of incidence make small angles with the z axis. Then $|\alpha|$, $|\beta|$, $|\alpha'|$, $|\beta'|$, $|x/R|$, $|y/R|$, $|\theta|$, and $|\theta'|$ are all small, and we may approximate the first two equations of (2.67*a*) by

$$
\begin{aligned}
n'\alpha' &= n\alpha - \frac{(n' - n)x}{R} \\
n'\beta' &= n\beta - \frac{(n' - n)y}{R}
\end{aligned}
\tag{2.68}
$$

We return to the general case. Since $\cos\theta = (\mathbf{e}\cdot\boldsymbol{\nu})$, we may use Eqs. (2.65) and (2.66) to determine $\cos\theta$ in terms of the direction cosines α, β of $\mathbf{e}$ and the transverse coordinates x, y of A:

$$
\cos\theta = \left\{\left[1 - \left(\frac{x}{R}\right)^2 - \left(\frac{y}{R}\right)^2\right](1 - \alpha^2 - \beta^2)\right\}^{1/2} - \frac{x\alpha}{R} - \frac{y\beta}{R}
\tag{2.69}
$$

We then determine $\cos\theta'$ by Eq. (2.64) and α' and β' by use of (2.67*a*). Note that α' and β' in Eq. (2.67*a*) would depend linearly* on (α, x) and (β, y), respectively, if it were not for the (x, y) and (α, β) dependence of $\cos\theta$ and $\cos\theta'$. The approximate version in Eq. (2.68) has this linear dependence.

We also have the obvious relation

$$
\begin{aligned}
x' &= x \\
y' &= y
\end{aligned}
\tag{2.67b}
$$

between the transverse coordinates of the ray just after refraction and its coordinates just before refraction.

Regarding sign conventions, we note that the implicit assumption has been made that the refracting surface is convex towards the left, that is, that V is to the left of C. Equations (2.64), (2.67*a*), (2.68), and (2.69) hold also for a concave surface, that is, one with V

*By linear dependence we mean, for example, that α' would be given by a constant times α plus another constant times x.

to the right of C provided that R is defined to be negative, and provided that $\cos\theta$ and $\cos\theta'$ are defined to remain positive. These conventions require some rewriting of intermediate steps of the derivation; the results remain valid.

Equations (2.67) can be considered as the equations of a transformation, denoted symbolically by $\mathscr{R}$, that the direction cosines α, β and coordinates x, y of a ray undergo during a single refraction.

3. Translation. Before and after refraction by the surface in Fig. 2.28 the rays will travel undeviated in straight lines until they meet other surfaces. Let us describe this linear propagation mathematically. To follow the process, it is convenient to adopt a subscript notation. We add the subscript 1 to all symbols in the preceding discussion. Thus the refraction takes place at A_1 with coordinates $(x_1, y_1, z_1) = (x_1, y_1, R_1 - \sqrt{R_1^2 - x_1^2 - y_1^2})$. The equations governing the refraction transformation $\mathscr{R}_1$ are

$$\mathscr{R}_1 \begin{cases} n_1'\alpha_1' = n_1\alpha_1 - \dfrac{(n_1'\cos\theta_1' - n_1\cos\theta_1)x_1}{R_1} \\ n_1'\beta_1' = n_1\beta_1 - \dfrac{(n_1'\cos\theta_1' - n_1\cos\theta_1)y_1}{R_1} \end{cases} \tag{2.70a}$$

$$\begin{cases} x_1' = x_1 \\ y_1' = y_1 \end{cases} \tag{2.70b}$$

together with the auxiliary relations

$$\cos\theta_1 = \left\{\left[1 - \left(\frac{x_1}{R_1}\right)^2 - \left(\frac{y_1}{R_1}\right)^2\right](1 - \alpha_1^2 - \beta_1^2)\right\}^{1/2} - \frac{x_1\alpha_1}{R_1} - \frac{y_1\beta_1}{R_1}$$

$$n_1'\cos\theta_1' = \sqrt{n_1'^2 - n_1^2 + n_1^2\cos^2\theta_1} \tag{2.70c}$$

As shown in Fig. (2.29), we now pick a point A_0 on the incident

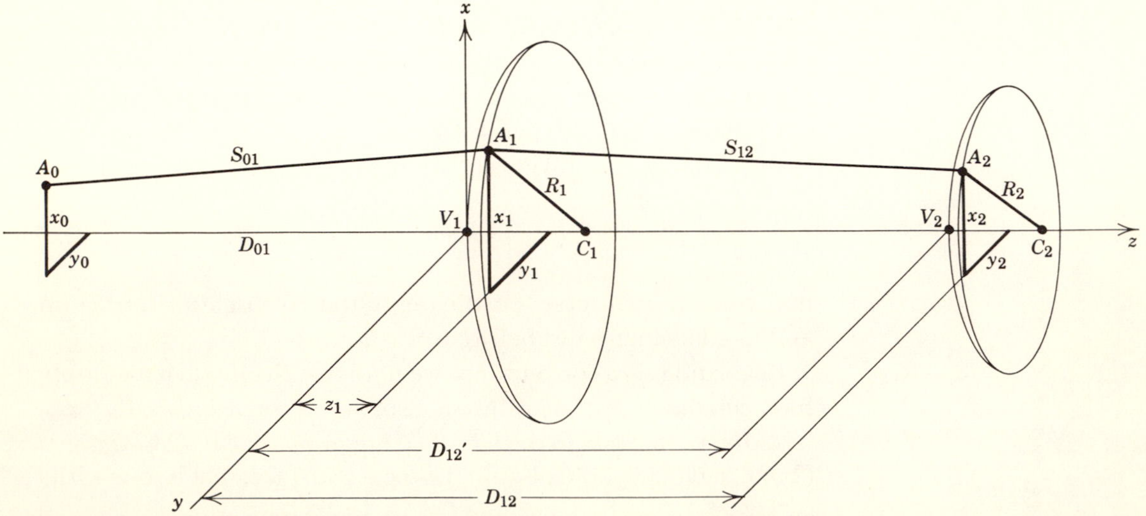

Fig. 2.29

ray with a projection on the z axis that is a distance D_{01} to the left of V_1. Thus, A_0 has coordinates $(x_0, y_0, -D_{01})$. We let $\mathbf{e}_0$ be the unit vector along the ray at A_0 with direction cosines $(\alpha_0, \beta_0, \gamma_0)$. Of course we have $\mathbf{e}_1 = \mathbf{e}_0$, or

$$\begin{aligned} \alpha_1 &= \alpha_0 \\ \beta_1 &= \beta_0 \end{aligned} \tag{2.71a}$$

(from which $\gamma_1 = \gamma_0$ follows). The x, y coordinates of A_1 are given in terms of those of A_0 by the equation

$$\mathbf{A}_1 = \mathbf{A}_0 + S_{01}\mathbf{e}_0 \qquad \text{or} \quad x_1 = x_0 + \alpha_0 S_{01}, \qquad y_1 = y_0 + \beta_0 S_{01} \tag{2.71b}$$

where S_{01} is the directed distance between A_0 and A_1 with magnitude given by

$$S_{01}{}^2 = (x_1 - x_0)^2 + (y_1 - y_0)^2 + (D_{01} + R_1 - \sqrt{R_1{}^2 - x_1{}^2 - y_1{}^2})^2 \tag{2.71c}$$

The sign of S_{01} will be the same as that of $(z_1 + D_{01})$. Equations (2.71) give us the equations of a transformation, denoted schematically by $\mathscr{T}_{01}$, that describes the change in position and direction of a ray during the translation from A_0 to A_1. Equations of exactly the same form will describe the transformation undergone during the translation from A_1 to another point A_2 with a projection on the z axis that is a distance D_{12} to the right of V_1 (Fig. 2.29). The equations of this transformation, denoted by $\mathscr{T}_{12}$, are

$$\mathscr{T}_{12}\begin{cases} \alpha_2 = \alpha'_1 \\ \beta_2 = \beta'_1 & (2.72a) \\ x_2 = x'_1 + \alpha'_1 S_{12} \\ y_2 = y'_1 + \beta'_1 S_{12} & (2.72b) \\ S_{12}{}^2 = (x_2 - x'_1)^2 + (y_2 - y'_1)^2 + [R_1 - D_{12} - \sqrt{R_1{}^2 - (x'_1)^2 - (y'_1)^2}] \end{cases}$$

(2.72c)

[The sign of S_{12} is the same as that of $(D_{12} - z_1)$.]

The overall transformation from A_0 to A_2 is then given by applying first $\mathscr{T}_{01}$, then $\mathscr{R}_1$, and finally $\mathscr{T}_{12}$.

The *approximate version* of Eq. (2.72b) is obtained by using D'_{12} for S_{12}. The other equations remain unchanged.

4. Multiple Refractions and Translations. The above mathematical formalism may be applied repeatedly as the ray makes its way through the optical system. Suppose that at A_2 (Fig. 2.29) there is another coaxial refracting surface where the index changes from $n_2 = n'_1$ to n'_2, having its vertex at V_2, radius of curvature R_2, and so forth. The refraction transformation $\mathscr{R}_2$ from $\alpha_2, \beta_2, x_2, y_2$ to $\alpha'_2, \beta'_2, x'_2, y'_2$ will be given by a set of equations exactly like (2.70), with the subscript 2 replacing the subscript 1.

Before we can actually perform the second refraction calculation, we must express the translation $\mathscr{T}_{12}$ of Eqs. (2.72) in terms of the vertex separation D'_{12}. We can express D_{12} in terms of D'_{12} by

$$D_{12} = D'_{12} + R_2 - \sqrt{R_2^2 - x_2^2 - y_2^2} \tag{2.72d}$$

This equation used in (2.72*c*) will give S_{12} in terms of D'_{12}, x_2, and y_2. But x_2 and y_2 are in turn given by (2.72*b*) in terms of S_{12}. These coupled equations must be solved together to determine x_2 and y_2. This is equivalent to determining the exact position of the point A_2 on the second refracting surface. Then the equations for the refracting $\mathscr{R}_2$ can be applied without difficulty.

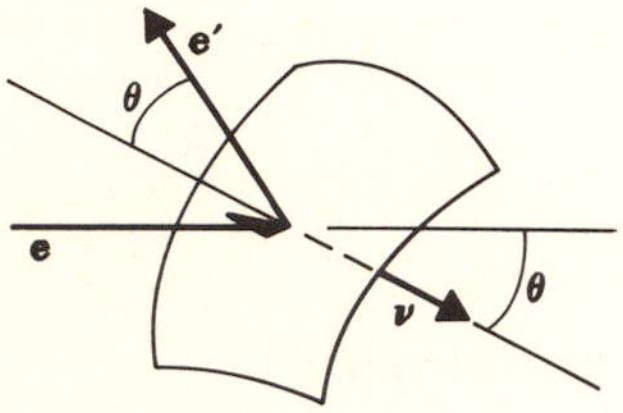

Fig. 2.30

B. REFLECTION

Ray tracing when there is a single reflecting surface differs little from the case of a single refraction. Consider Fig. 2.30. The unit vector along the incident ray, **e**, makes an angle θ with the normal $\boldsymbol{\nu}$ to the reflecting surface. By the law of reflection, the unit vector along the reflected ray will make an angle θ with $-\boldsymbol{\nu}$ In fact, we can write the entire law of reflection in vector form:

$$\mathbf{e}' = \mathbf{e} - 2\boldsymbol{\nu}\cos\theta \tag{2.73}$$

Now we consider reflection off a spherical reflecting surface with vertex V_1 and center of curvature C_1, both on the z axis, as shown in Fig. 2.31. The radius of curvature R_1 will be considered

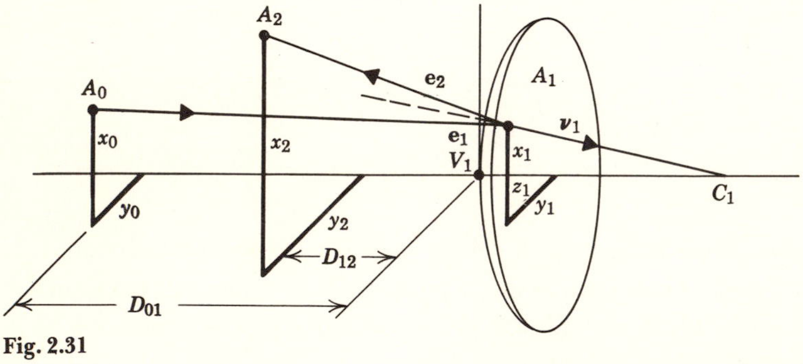

Fig. 2.31

positive if C_1 is to the right of V_1. (This is the same sign convention we used for refraction.) The incident ray meets the surface at the point A_1 with coordinates

$$(x_1, y_1, z_1) = (x_1, y_1, R_1 - \sqrt{R_1^2 - x_1^2 - y_1^2})$$

As before, the surface normal $\boldsymbol{\nu}_1$ has components given by Eq. (2.65) with subscripts 1 added, and the unit vectors $\mathbf{e}_1$, $\mathbf{e}'_1$ have components given by (2.66) with subscripts 1 added.

Equation (2.73) then leads us to the following equations for the reflection transformation

$$\mathscr{R}_1 \begin{cases} \alpha_1' = \alpha_1 + 2\cos\theta_1 \dfrac{x_1}{R_1} \\ \beta_1' = \beta_1 + 2\cos\theta_1 \dfrac{y_1}{R_1} & (2.74a) \\ \gamma_1' = \gamma_1 - 2\cos\theta_1 \dfrac{(R_1^2 - x_1^2 - y_1^2)^{1/2}}{R_1} \\ x_1' = x_1 & (2.74b) \\ y_1' = y_1 \end{cases}$$

together with the auxiliary relation to determine $\cos\theta_1$ from Eq. (2.69),

$$\cos\theta_1 = \left\{\left[1 - \left(\frac{x_1}{R_1}\right)^2 - \left(\frac{y_1}{R_1}\right)^2\right]^2 (1 - \alpha_1^2 - \beta_1^2)\right\}^{1/2} - \frac{x_1\alpha_1}{R_1} - \frac{y_1\beta_1}{R_1} \tag{2.74c}$$

We have given the equation for γ_1' also, so that the sign of γ_1' may be correctly determined. (Its magnitude follows from $\gamma_1'^2 = 1 - \alpha_1'^2 - \beta_1'^2$.)

We now translate back along the incident ray a distance S_{01} to the point A_0 having coordinates $(x_0, y_0, -D_{01})$, as shown in Fig. 2.31. This is identical to the earlier situation shown in Fig. 2.29, for which Eqs. (2.71) applied. Thus the translation transformation $\mathscr{T}_{01}$ is still given by these equations.

If we translate along the reflected ray a distance S_{12} to the point A_2 with coordinates $(x_2, y_2, -D_{12})$, the transformation $\mathscr{T}'_{12}$ will be given by Eqs. (2.72*a*, *b*), but with S_{12} given by

$$S_{12}^2 = (x_2 - x_1'^2) - (y_2 - y_1')^2 + (D_{12} + R_1 - \sqrt{R_1^2 - (x_1')^2 - (y_1')^2}$$
(2.72*c'*)

The sign of S_{12} is the same as that of $D_{12} + z_1 = D_{12} + R_1 - \sqrt{R_1^2 - x_1^2 - y_1^2}$.

Similar treatments may be given for ray tracing through non-coaxial surfaces or nonspherical surfaces. For more information the reader should consult the references, particularly the book by Herzberger.

There have been extensive applications of similar tracing techniques to study the trajectories of charged particles in electric and magnetic fields. These are used in the design of all sorts of beam transport systems: electron microscopes, mass spectrometers, particle accelerators, and so on.

REFERENCES FOR SECTION 2.3

M. Herzberger, *Modern Geometrical Optics*.
M. Born and E. Wolf, *Principles of Optics*.
E. L. O'Neill, *Introduction to Statistical Optics*.
Willem Brouwer, *Matrix Methods in Optical Instrument Design*.

This branch of geometrical optics, named after the great nineteenth-century mathematician, Carl Frederick Gauss, is based on an extremely useful approximation to the correct theory. The resulting formulas form a starting point for more accurate calculations, but they are often good enough to be used for many applications as they stand.

The equations of Gaussian optics are derived by taking the first order approximations to the equations of the exact theory. We assume that an optical axis can be defined for our system and that all light rays and all normals to refracting or reflecting surfaces make small angles with the axis. Such light rays are called paraxial rays.

CHAPTER 3

GAUSSIAN OPTICS

CHAPTER 3 CONTENTS

3.1 REFRACTION OF PARAXIAL RAYS BY A SPHERICAL SURFACE

A. DERIVATION

We now develop the first order approximations to the exact ray tracing formulas of Section 2.3. In the paraxial ray approximation the direction cosines α, β are treated as small and $\gamma = \sqrt{1-\alpha^2-\beta^2}$ is approximated by unity. Similarly all transverse dimensions x, y are treated as small compared with the radii of curvature R. This means that the exact coordinates $(x_1, y_1, R_1 - \sqrt{R_1^2 - x_1^2 - y_1^2})$ of the point A_1 in Fig. 3.1 are replaced by $(x_1, y_1, 0)$, and the components of the unit vector $\boldsymbol{\nu}_1$ are approximated by $(-x_1/R_1, -y_1/R_1, 1)$.

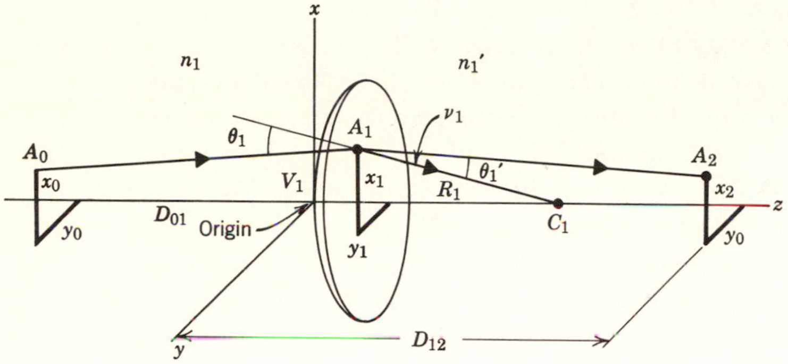

Fig. 3.1

The angles θ_1 and θ'_1 are also considered to be small, so that Snell's law becomes

$$n_1\theta_1 = n_1'\theta_1' \tag{3.1}$$

and

$$\cos\theta_1 \approx 1, \qquad \cos\theta_1' \approx 1 \tag{3.2}$$

The equations of the refraction transformation $\mathscr{R}_1$ given by Eqs. (2.70a,b) then become the following linear equations:

$$\mathscr{R}_1 \begin{cases} n_1'\alpha_1' = n_1\alpha_1 + \dfrac{(n_1 - n_1')x_1}{R_1} \\ n_1'\beta_1' = n_1\beta_1 + \dfrac{(n_1 - n_1')y_1}{R_1} & (3.3a) \\ x_1' = x_1 \\ y_1' = y_1 & (3.3b) \end{cases}$$

Note that these equations remain valid in the limit $R_1 \to \infty$.

The translation equations also simplify under the above approximations. Equation (2.71c) gives simply $S_{01} = D_{01}$, and Eqs. (2.71a,b)

become

$$\mathscr{T}_{01}\begin{cases}\alpha_2 = \alpha_1' & \\ \beta_1 = \beta_0 & (3.4a) \\ x_1 = x_0 + D_{01}\alpha_0 & \\ y_1 = y_0 + D_{01}\beta_0 & (3.4b)\end{cases}$$

These are also stricly linear. In a similar way the equations (2.72) of the translation $\mathscr{T}_{12}$ become

$$\mathscr{T}_{12}\begin{cases}\alpha_2 = \alpha_1' & \\ \beta_2 = \beta_1' & (3.5a) \\ x_2 = x_1' + D_{12}\alpha_1' & \\ y_2 = y_1' + D_{12}\beta_1' & (3.5b)\end{cases}$$

Note that in Eqs. (3.3), (3.4), and (3.5), x and α at a given point depend only on x and α at other points, not at all on y and β. In other words, the pairs of variables (x, α) and (y, β) are decoupled from one another and may be treated independently. Still another way of saying this is that the projections of the rays onto the xz or yz planes behave independently. (This is not true in the exact theory.) Because of this independence, it is no longer necessary to perform calculations of both projections simultaneously; we choose here to concentrate on the projection in the xz plane. The projections behave as though the rays were actually lying in the xz plane, that is, as though y and β were zero. Such rays that lie in a single plane containing the z axis are called *meridional rays*.

When we make the assumption of meridional rays lying in the xz plane, we are led to the two-dimensional situation shown in Fig. 3.2. The direction cosine α is in general defined to be the

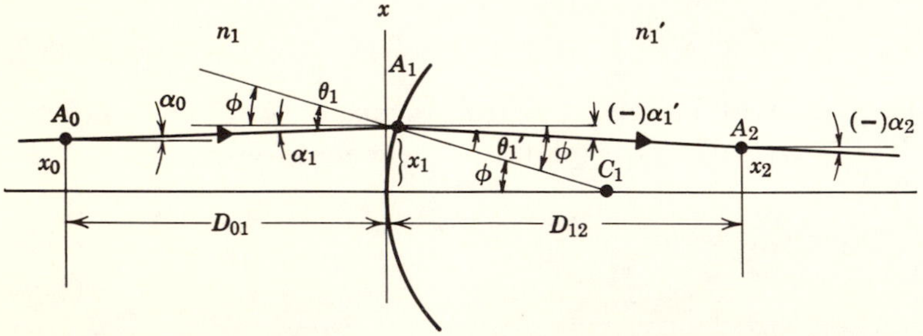

Fig. 3.2

cosine of the angle the ray makes with the x axis. If the rays lie in the xz plane, α also becomes the sine of the angle the ray makes with the z axis. In the paraxial ray approximation the sine of this angle equals the angle itself. Thus we are justified in using α_0, α_1, and α_2 to label the angles of inclination of the rays in Fig. 3.2. The relevant parts of the transformations (3.3) to (3.5) are given by

$$\mathscr{T}_{01}\begin{cases}x_1 = x_0 + D_{01}\alpha_0 \\ \alpha_1 = \alpha_0\end{cases} \tag{3.6}$$

$$\mathscr{R}_1 \begin{cases} x_1' = x_1 \\ \alpha_1' = \dfrac{n_1\alpha_1}{n_1'} + \left[\dfrac{(n_1 - n_1')}{(R_1 n_1')}\right] x_1 \end{cases} \tag{3.7}$$

$$\mathscr{T}_{12} \begin{cases} x_2 = x_1' + D_{12}\alpha_1' \\ \alpha_2 = \alpha_1' \end{cases} \tag{3.8}$$

Once the restriction to meridional rays and small angles is accepted, we can easily derive these approximate relations directly from Fig. 3.2. Equations (3.6) and (3.8) follow by inspection (for small angles). The second of Eqs. (3.7) comes from applying Snell's law at A_1,

$$n_1\theta_1 = n_1'\theta_1'$$

and using the auxiliary angle $\phi = x_1/R_1$ to write

$$\theta_1 = \phi + \alpha_1$$
$$\theta_1' = \phi + \alpha_1'$$

One must be careful about the signs of the angles. Our sign convention is stated below. Those angles that are negative in Fig. 3.2 are so marked. The use of the auxiliary relations involving ϕ brings Snell's law into the form

$$n_1'(\phi + \alpha_1') = n_1(\phi + \alpha_1)$$

or

$$n_1'\alpha_1' = n_1\alpha_1 - (n_1' - n_1)\phi = n_1\alpha_1 - \frac{(n_1' - n_1)x_1}{R_1}$$

as required, to give the second of Eqs. (3.7).

Equations (3.6), (3.7), and (3.8) can be used to obtain the overall transformation that gives the pair (x_2, α_2) in terms of the pair (x_0, α_0):

$$x_2 = \left[\frac{(n_1 - n_1')D_{12}}{n_1'R_1} + 1\right] x_0 + \left[D_{01} + \frac{n_1 D_{12}}{n_1'} + \frac{(n_1 - n_1')D_{01}D_{12}}{n_1'R_1}\right]\alpha_0 \tag{3.9a}$$

$$\alpha_2 = \left[\frac{n_1 - n_1'}{n_1'R_1}\right] x_0 + \left[\frac{n_1}{n_1'} + \frac{(n_1 - n_1')D_{01}}{n_1'R_1}\right]\alpha_0 \tag{3.9b}$$

Note the linear dependence of x_2 and α_2 on x_0 and α_0. This follows from the linear dependence of each individual transformation and will be exploited when we introduce the matrix formalism a bit later.

B. IMAGE FORMATION

1. Location of Image. If A_2 is to be an image of A_0, then any ray leaving A_0, regardless of the value of α_0 (as long as it is small), should arrive at A_2. This means that x_2 must be independent of α_0, that is, the coefficient of α_0 in Eq. (3.9*a*) must vanish if A_2 is to be an image of A_0. The imaging condition thus becomes

$$D_{01} + \frac{n_1 D_{12}}{n_1'} + \frac{(n_1 - n_1')D_{01}D_{12}}{n_1' R_1} = 0$$

This can be rearranged to yield the well-known equation already encountered as Eq. (2.37):

$$\frac{n_1'}{D_{12}} + \frac{n_1}{D_{01}} = \frac{n_1' - n_1}{R_1} \tag{3.10}$$

The *sign convention* we have been using is as follows.

1. Light rays travel from left to right (in the $+z$ direction).
2. R_1 is positive if V_1 is to the left of C_1, that is, if the refracting surface is convex towards the left; otherwise R_1 is negative.
3. D_{01} is positive if A_0 is to the left of V_1; otherwise D_{01} is negative.
4. D_{12} is positive if A_2 is to the right of V_1; otherwise D_{12} is negative.
5. The angles α_0, α_1, and so on, are positive if the ray direction is obtained by rotating the $+z$ axis counterclockwise through an acute angle; otherwise the angles are negative.
6. The distances x_0, x_1, and so on, are positive if up, negative if down.

It is useful to note that when $D_{01} = \infty$, the image will be located at the point

$$D_{12} = f_1' \equiv \frac{n_1' R_1}{n_1' - n_1} \tag{3.11a}$$

and if $D_{12} = +\infty$, the object must have been located at the point

$$D_{01} = f_1 \equiv \frac{n_1 R_1}{n_1' - n_1} \tag{3.11b}$$

These distances are called the *right* and *left focal lengths*, respectively. Using them, we can write Eq. (3.10) in the form

$$\frac{n_1'}{D_{12}} + \frac{n_1}{D_{01}} = \frac{n_1' - n_1}{R_1} = \frac{n_1'}{f_1'} = \frac{n_1}{f_1} \tag{3.12}$$

2. Lateral Magnification. Now that we have made x_2 in Eq. (3.9*a*) independent of α_0, we are left with an equation of the form $x_2 = m_x x_0$, where the lateral magnification m_x is given by

$$m_x = \frac{n_1 - n_1'}{R_1} \cdot \frac{D_{12}}{n_1'} + 1 = \frac{D_{12}}{n_1'}\left[\frac{n_1 - n_1'}{R_1} + \frac{n_1'}{D_{12}}\right]$$

This is the usual magnification that gives image size in terms of object size. We use Eq. (3.10) to express D_{12} in terms of D_{01} and obtain

$$m_x = \frac{x_2}{x_0} = \frac{D_{12}}{n_1'}\left[\frac{n_1 - n_1'}{R_1} + \frac{n_1' - n_1}{R_1} - \frac{n_1}{D_{01}}\right] = -\frac{n_1 D_{12}}{n_1' D_{01}} \tag{3.13}$$

Hence, if D_{01} and D_{12} are both positive, m_x must be negative; this means that with x_0 positive, x_2 will be negative, or A_2 will be below the z axis if A_0 is above it. This is shown in Fig. 3.3.

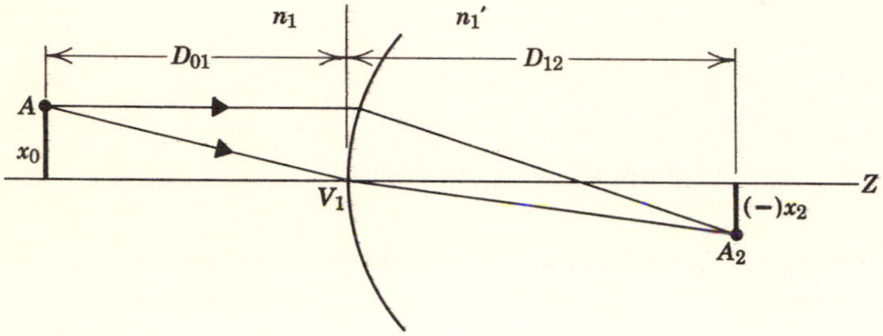

Fig. 3.3

3. ***Ray Angle Magnification.*** Consider two rays leaving A_0 with angular separation $\Delta\alpha_0$ and arriving at A_2 with angular separation $\Delta\alpha_2$ as shown in Fig. 3.4*a*. One of the rays could equivalently cross the z axis, and the other could be the axis itself, as shown in Fig. 3.4*b*. The ray angle magnification m_α is given by

$$m_\alpha = \frac{\Delta\alpha_2}{\Delta\alpha_0} = -\frac{D_{01}}{D_{12}} \tag{3.14}$$

as can be seen from the illustrations. This result can also be obtained from Eq. (3.9*b*) by taking differences for fixed x_0. The ray angle

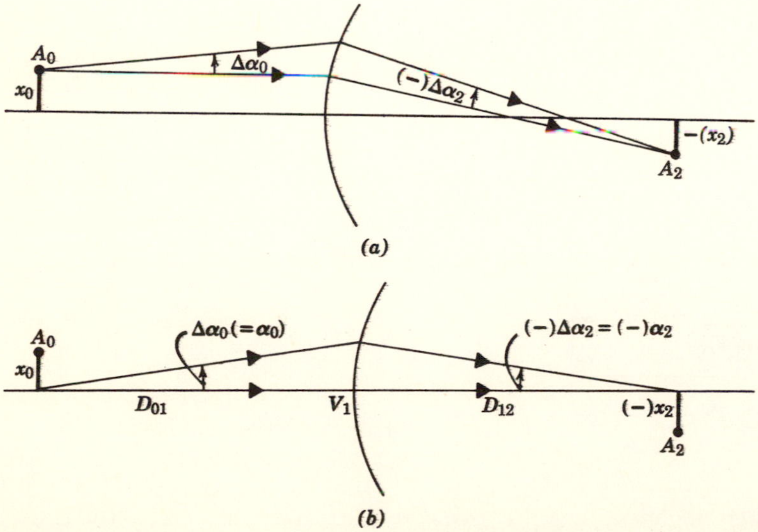

Fig. 3.4

magnification is a useful concept when the angular spread of rays forming the image must be considered. From Eqs. (3.13) and (3.14) we obtain a useful relation called the *Lagrange or Smith-Helmholtz equation*:

$$\frac{x_2}{x_0}\cdot\frac{\Delta\alpha_2}{\Delta\alpha_0}=\frac{n_1}{n_1'},\qquad \text{or}\qquad n_1'x_2\Delta\alpha_2=n_1x_0\Delta\alpha_0,\qquad \text{or}\qquad n_1'x_2\alpha_2=n_1x_0\alpha_0 \tag{3.15}$$

Another version of this is

$$m_x m_\alpha=\frac{n_1}{n_1'} \tag{3.15$'$}$$

which says that large linear magnification can be obtained only at the expense of a large demagnification of the ray angles, that is, only by having much smaller angular divergence at the image than at the object.

When the image formation condition (3.10) is satisfied, we may use Eqs. (3.11*a*), (3.13), and (3.14) in Eqs. (3.9) to obtain for the ray-tracing transformation from object point A_0 to Gaussian image point A_2 the following simple result:

$$\begin{aligned} x_2 &= m_x x_0 \\ \alpha_2 &= \frac{-x_0}{f_1'}+m_\alpha\alpha_0 \end{aligned} \tag{3.16}$$

*4. **Virtual Object and Image.*** The above formulas hold when D_{01} and/or D_{12} is negative. A negative value for D_{12} means that the image is to the left of V_1; the light rays in the second medium are diverging as though they were coming from an image A_2 to the left of the vertex V_1, as shown in Fig. 3.5*a*. This is called a *virtual image*.

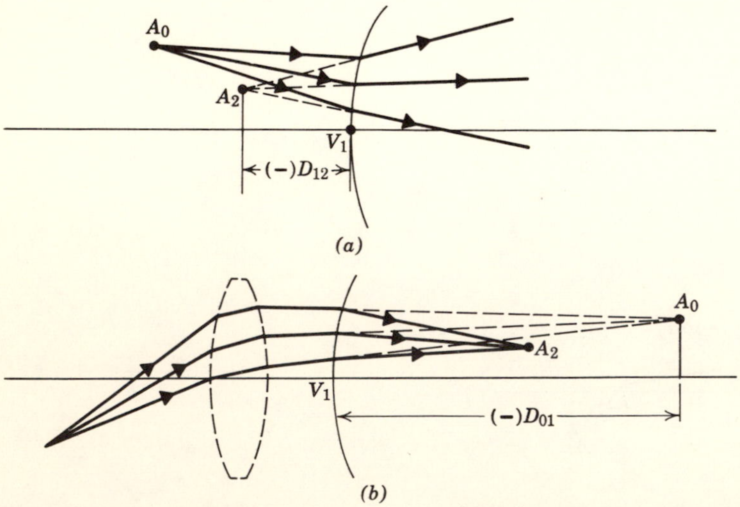

Fig. 3.5. (*a*) A_2 is a virtual image. (*b*) A_0 is a virtual object.

If D_{01} is negative, the rays appear to be converging at A_0 to the right of the vertex. This is called a *virtual object* and is shown in Fig. 3.5*b*.

An auxiliary lens (shown dotted) is necessary to prepare the converging rays for such an object.

PROBLEMS

3.1.1 Derive the lateral magnification relation (3.13) by means of the ray A_0V_1 in Fig. 3.3.

3.1.2 Derive the ray angle magnification relation (3.14) from Eq. (3.9*b*).

3.1.3 a. Find the left and right focal lengths of the system in the illustration.

b. Find the image position and size for a 1 cm high erect object located
50 cm to the left of V,
30 cm to the left of V,
20 cm to the left of V,
20 cm to the right of V (virtual object).
(If the image is located at $\pm\infty$, give, instead of image size, a value for the angle the parallel exit rays make with the z axis.)

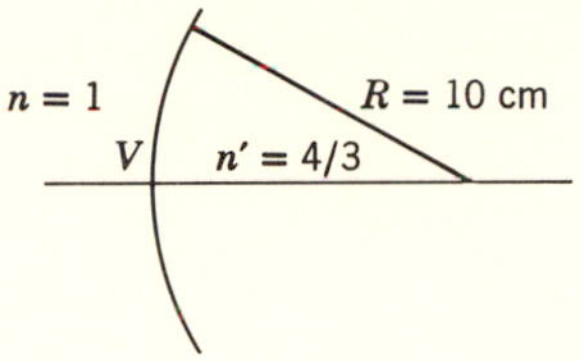

3.1.4 Locate the images of P_1, P_2, P_3. Find the lateral magnification $\widehat{P_2'P_1'}/\widehat{P_2P_1}$ and the longitudinal magnification $\widehat{P_3'P_1'}/\widehat{P_3P_1}$. Assume

$|P_2-P_1| \ll 10$ cm, $|P_3-P_1| \ll 10$ cm.

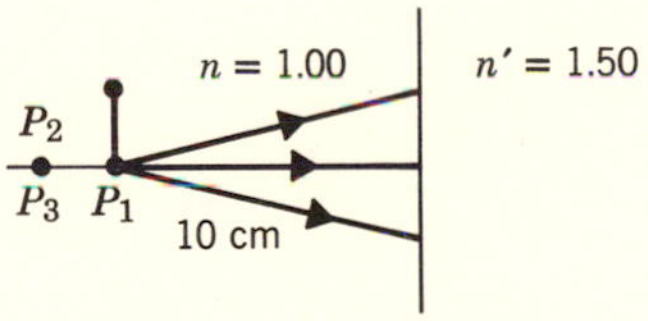

3.1.5 Find the location and size of the final images of the objects at O_1 and O_2.

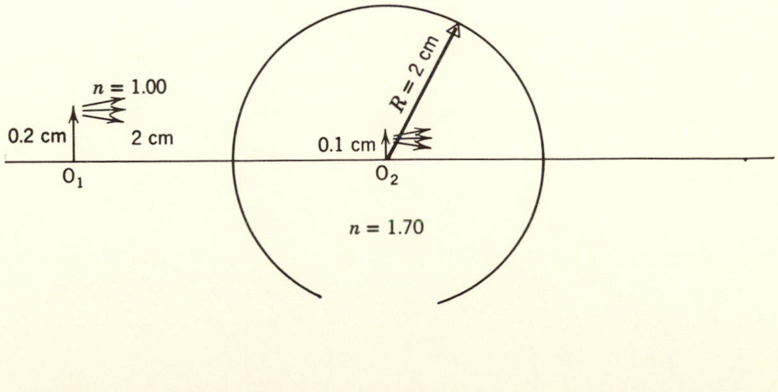

3.1.6 a. Calculate where a parallel beam of light incident from the left would be focused.

b. Turn the lens around and repeat the calculation.

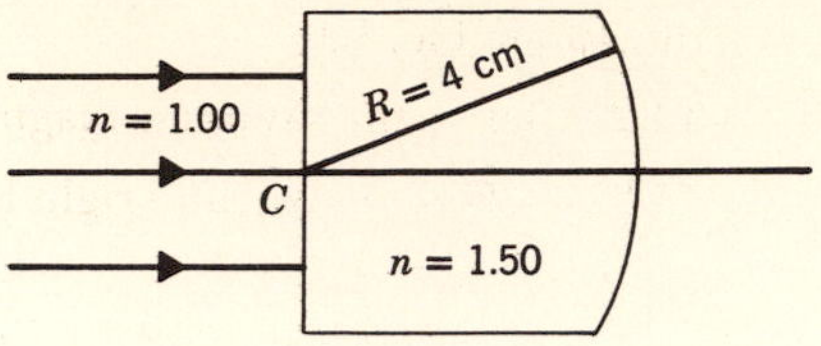

3.1.7 Consider the refracting surface used in Problem 3.1.3. An incident ray passes a point 15 cm to the left of the vertex at a height of 1 cm and angle of inclination −0.030 radians. Find the height and angle of inclination of the refracted ray 20 cm to the right of the vertex. Use the ray-tracing equations.

3.2 REFLECTION OF PARAXIAL RAYS BY A SPHERICAL MIRROR

A. DERIVATION

We now make the same paraxial ray assumptions that we made in Section 3.1 and apply them to the reflection equations of Section 2.3B. The result is that Eqs. (2.74*a*,*b*) become

$$\mathscr{R}_1' \begin{cases} \alpha_1' = \alpha_1 + \dfrac{2x_1}{R_1} \\ \beta_1' = \beta_1 + \dfrac{2y_1}{R_1} \qquad\qquad (3.17a) \\ \gamma_1' = \gamma_1 - \dfrac{2\sqrt{R_1{}^2 - x_1{}^2 - y_1{}^2}}{R_1} \approx -1 \\ x_1' = x_1 \\ y_1' = y_1 \qquad\qquad (3.17b) \end{cases}$$

The translation equations (2.71*a*,*b*) and 2.72*a*,*b*) are unaltered, but now $S_{01} = D_{01}$ and $S_{12} = D_{12}$. As was the case for refraction in the paraxial ray approximation, the pairs of variables (x,α) and (y,β) are uncoupled from one another and, as before, we choose to work with the pair (x,α). This is equivalent to working with meridional rays lying in the xz plane. The translation transformations $\mathscr{T}_{01}$ and $\mathscr{T}_{12}$, and the reflection transformation $\mathscr{R}_1'$ are given by

$$\mathscr{T}_{01} \begin{bmatrix} x_1 = x_0 + D_{01}\alpha_0 \\ \alpha_1 = \alpha_0 \end{bmatrix} \tag{3.18}$$

$$\mathscr{R}_1' \begin{cases} x_1' = x_1 \\ \alpha_1' = \dfrac{2x_1}{R_1} + \alpha_1 \\ (\gamma_1' = -1) \end{cases} \tag{3.19}$$

$$\mathscr{T}_{12} \begin{bmatrix} x_2 = x_1' + D_{12}\alpha_1' \\ \alpha_2 = \alpha_1' \end{bmatrix} \tag{3.20}$$

The situation is illustrated in Fig. 3.6 (for a negative value of the radius of curvature R_1). We may derive the second of Eqs. (3.19) directly from an application of the law of reflection to the rays A_0A_1 and A_1A_2 in the figure:

$$\alpha_1 - \phi = \theta = \phi + \alpha_1'$$

But

$$\phi = \frac{-x_1}{R_1}$$

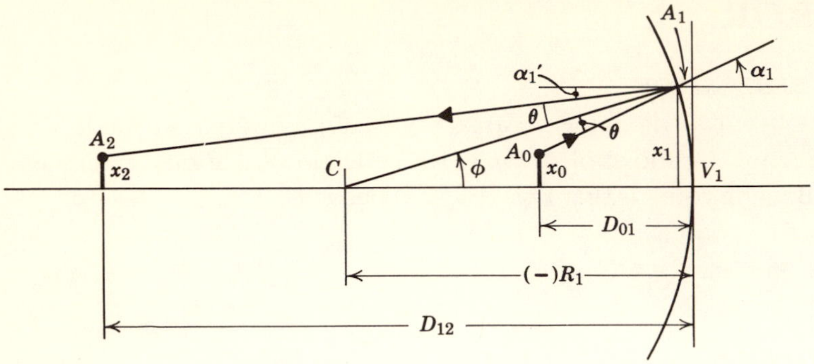

Fig. 3.6. [The angles α_1' and θ are negative by the sign convention discussed below Eq. (3.22*b*).]

Thus

$$\alpha_1' = \alpha_1 - 2\phi = \frac{2x_1}{R_1} + \alpha_1$$

as required.

Equations (3.18), (3.19), and (3.20) may be combined to give the overall transformation from the pair of variables (x_0,α_0) to the pair (x_2,α_2):

$$x_2 = \left(1 + \frac{2D_{12}}{R_1}\right)x_0 + \left(D_{01} + \frac{2D_{01}D_{12}}{R_1 + D_{12}}\right)\alpha_0$$

$$\alpha_2 = \frac{2x_0}{R_1} + \left(\frac{2D_{01}}{R_1} + 1\right)\alpha_0 \tag{3.21}$$

B. IMAGE FORMATION

1. Location of Image. We continue to use the line of reasoning employed in the refraction case. The condition for image formation is that x_2 be independent of α_0 or

$$\frac{1}{D_{12}} + \frac{1}{D_{01}} = -\frac{2}{R_1} = \frac{1}{f} \tag{3.22a}$$

(which differs from Eq. (2.53) because of our sign convention for R_1), where the focal length f is defined by

$$f = -\frac{R_1}{2} \tag{3.22b}$$

If $D_{12} = \pm\infty$, then $D_{01} = f$, and vice versa.

The *sign convention* for this case of *reflection* is as follows.

1. Positive ray direction is from left to right before reflection and from right to left after reflection.
2. R_1 is positive if V_1 is to the left of C_1.
3. D_{01} is positive if A_0 is to the left of V_1.
4. D_{12} is positive if A_2 is to the *left* of V_1.

5. The angles α_0, α_1 are positive if the ray direction is obtained by rotating the $+z$ *axis counterclockwise* through an acute angle.
6. The angles α_1', α_2 are positive if the ray direction is obtained by rotating the $-z$ *axis clockwise* through an acute angle.

(The sign conventions in (5) and (6) are designed to preserve the interpretation of α_0, α_1, α_1', α_2 as x components of unit vectors along the rays.)

*2. **Lateral Magnification.*** Once the imaging condition (3.22*a*) is satisfied, we may compute the lateral magnification by

$$m_x = \frac{x_2}{x_0} = -\frac{D_{12}}{D_{01}} \tag{3.23}$$

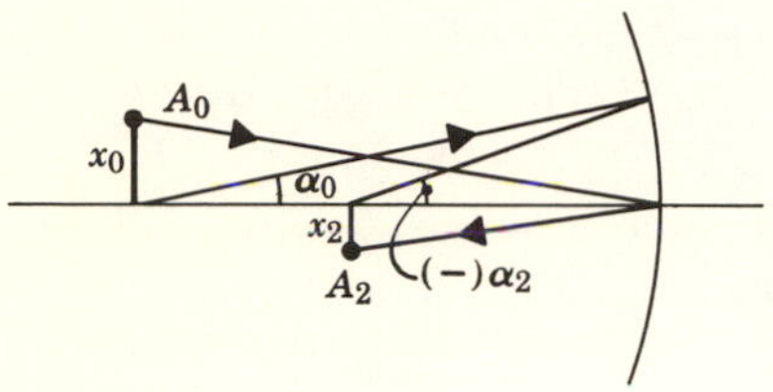

Fig. 3.7

Again, for both a real image and a real object (D_{12} and D_{01} both positive), they cannot both be erect or both inverted (Fig. 3.7).

*3. **Ray Angle Magnification.*** We may derive an expression for the ray angle magnification from the second of Eqs. (3.21) if we put $x_0 = 0$ (Fig. 3.7):

$$m_\alpha = \frac{\alpha_2}{\alpha_0} = -\frac{D_{01}}{D_{12}} \tag{3.24}$$

The Lagrange or Smith-Helmholtz equation then becomes

$$\frac{x_2\alpha_2}{x_0\alpha_0} = 1 \tag{3.25}$$

*4. **Virtual Object and Image.*** Our sign convention is still such that the object is real when D_{01} is positive, and the image is real when D_{12} is positive. The object is virtual when D_{01} is negative, and the image is virtual when D_{12} is negative.

PROBLEMS

3.2.1 Locate the position and size of the image of a 1 cm high object formed by a spherical mirror under the following conditions.

a. $R = -20$ cm, $D_{01} = 30$ cm
b. $R = -20$ cm, $D_{01} = 15$ cm
c. $R = -20$ cm, $D_{01} = 8$ cm
d. $R = -20$ cm, $D_{01} = -15$ cm
e. $R = +20$ cm, $D_{01} = -8$ cm

Illustrate each case with a rough sketch.

3.2.2 An incident ray passes a point 20 cm to the left of a reflecting spherical surface (of radius -30 cm) at a height of -2 cm below the z axis (which passes through the center of curvature) and with an inclination angle with the z axis of 1/20 radian. Find the height and inclination angle of the reflected ray 10 cm to the left of the surface.

3.3 IMAGE FORMATION PROPERTIES OF THIN LENSES AND THEIR GENERALIZATION

The transformation approach of the last section has given us the basic equations for Gaussian ray tracing through an optical system. Before applying it to a general system, we shall adopt a simpler approach for the thin lens and then indicate how its properties may be generalized for a general image-forming system. Mathematical proofs of these and other properties will be given in Section 3.4.

A. THE THIN LENS

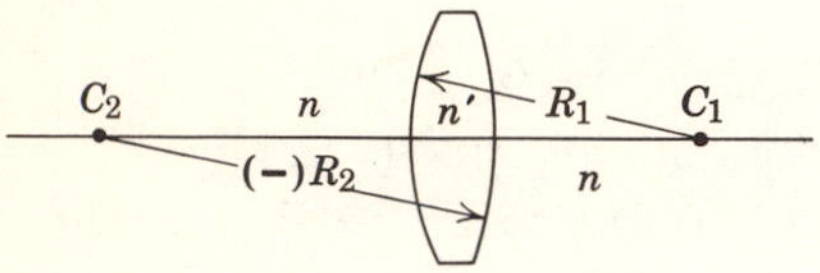

Fig. 3.8. A thin lens.

A thin lens consists of two refracting surfaces close enough for their separation to be negligible with respect to the object and image distances and the focal length. The refractive index of the lens material will be n', that of the surrounding medium will be n, as shown in Fig. 3.8. The sign convention is that adopted in Section 3.1B.

1. ***Derivation of Lens Maker's Equation.*** For the first refraction the refracting surface has radius R_1, the object distance is S_1, and the image distance is S'_1. The first medium has index n, the second has index n'. Equation (3.10) then applies

$$\frac{n}{S_1}+\frac{n'}{S'_1}=\frac{1}{R_1}(n'-n) \tag{3.26}$$

For the next refraction the corresponding quantities are R_2, S_2, and S'_2. The first medium has index n' and the second has index n. This equation therefore holds:

$$\frac{n'}{S_2}+\frac{n}{S'_2}=-\frac{1}{R_2}(n'-n) \tag{3.27}$$

The image for the first surface is the object for the second. Because the lens is thin and because of our sign convention, we have the relation

$$S_2=-S'_1$$

This equation is inserted into Eq. (3.27) to give

$$-\frac{n'}{S'_1}+\frac{n}{S'_2}=-\frac{1}{R_2}(n'-n) \tag{3.28}$$

We add Eqs. (3.26) and (3.28) and cancel a factor of n to yield

$$\frac{1}{S}+\frac{1}{S'}=\left(\frac{n'}{n}-1\right)\left(\frac{1}{R_1}-\frac{1}{R_2}\right) \tag{3.29}$$

with $S_1=S$, $S'_2=S'$. The right side of this equation defines the reciprocal of the focal length f:

$$\frac{1}{f}=\left(\frac{n'}{n}-1\right)\left(\frac{1}{R_1}-\frac{1}{R_2}\right) \tag{3.30}$$

This is the *lens maker's* equation. Equation (3.29) then becomes

$$\frac{1}{S}+\frac{1}{S'}=\frac{1}{f} \tag{3.31}$$

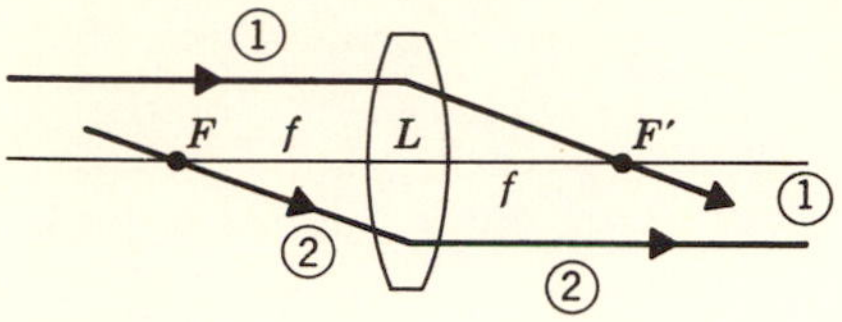

Fig. 3.9. Focusing of parallel rays—focal length.

The significance of the focal length can be seen from this equation and Fig. 3.9. If $S=\infty$, then $S'=f$, and an image is formed at F' (ray 1). Hence $|LF'|=f$. On the other hand, if the source is at F, where $|FL|=f$, then $S'=\infty$ (ray 2).

2. Determination of Image Position and Size for a Thin Lens. The image of an off-axis point A can be located by the intersection of any two of the following three rays (Fig. 3.10).

1. A ray parallel to the axis that is refracted through F'.
2. A ray through F that is refracted parallel to the axis.
3. A ray through the center of the lens that remains undeviated and undisplaced (for a thin lens).

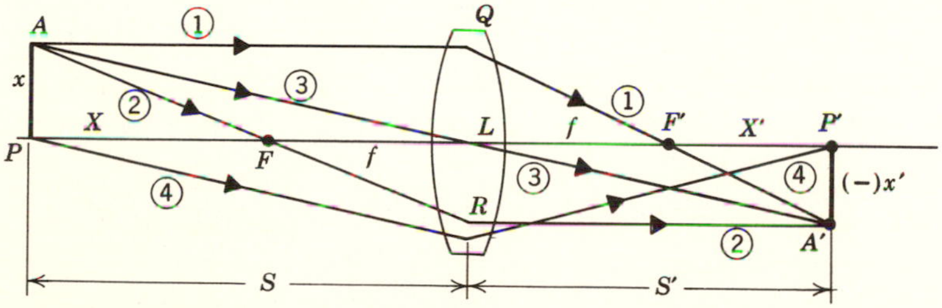

Fig. 3.10. Graphical location of the image formed by a thin lens.

Extension of some of these rays may be necessary in order for them to intersect. Several useful relations can be obtained from Fig. 3.10. The acute angles of the right triangles APL and $A'P'L$ are equal. Hence the magnification is given by

$$m_x=\frac{x'}{x}=-\frac{S'}{S} \tag{3.32}$$

From the similar triangles APF and RLF we obtain the relation

$$\frac{X}{f}=-\frac{x}{x'} \tag{3.33}$$

From the similar right triangles QLF' and $A'P'F'$ we obtain the relation

$$\frac{X'}{f}=\frac{-x'}{x} \tag{3.34}$$

Multiplying Eq. (3.33) by Eq. (3.34) then gives

$$\frac{XX'}{f^2} = \frac{x}{-x'} \cdot \frac{-x'}{x} = 1$$

or

$$XX' = f^2 \tag{3.35}$$

This is the *Newtonian form of the lens equation.* The quantities $X = S - f$ and $X' = S' - f$ are algebraic, and change sign as P moves past F or as P' moves past F'.

B. THICK LENS, COMPOUND LENS

A thick lens consists of two refracting surfaces that are not necessarily close together. A compound lens is made up of several thick or thin lenses, and again the refracting surfaces may be far apart. Thick lenses are necessary to obtain short focal lengths. Compound lenses are necessary if the aberrations, or deviations from paraxial Gaussian behavior, of single thin lenses are to be corrected. One can always treat the focusing of paraxial rays by repeated applications of the image formation equations. We can show generally that a complicated lens system produces a net focusing action in many respects like that for a thin lens. This will be done in Section 3.4. It is necessary to know only the location of four of the six cardinal points of the lens, its two principal points, and its two focal points.

*1. **Principal Points and Focal Points.*** Figures 3.11 and 3.12 show how the principal points and focal points can be defined. Parallel rays coming from the left will be focused at the *second focal point, F'*. The plane perpendicular to the axis of the system at F' is called the *second focal plane.* The rays near F', when extended backwards, intersect the extension of the original parallel rays in the *second principal plane,* which intersects the axis at H'. These extensions, of course, do not represent the true paths of the rays, but the rays do behave *as though* all the deviation were taking place in the principal plane.

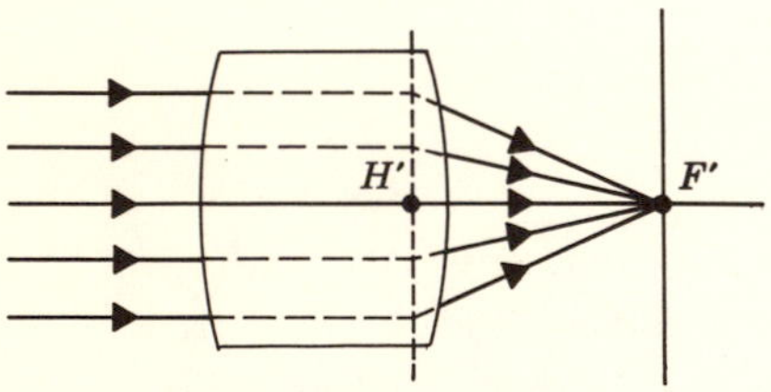

Fig. 3.11. Illustration of the definition of the second focal point F', the second principal point H', and the related focal plane and principal plane.

Figure 3.12 shows the defining properties of the first focal point, F, the first focal plane, and the first principal point H and plane. Rays leaving F form a collimated beam parallel to the axis. Each ray in the collimated beam, when extended backward, appears to intersect the original ray from F in the plane through H. Again, one notices that the system behaves *as though* all the deviation of light rays coming to or from F takes place in the first principal plane. *It is not necessary that the relative position of F, H, H', F' be as shown in the figures.*

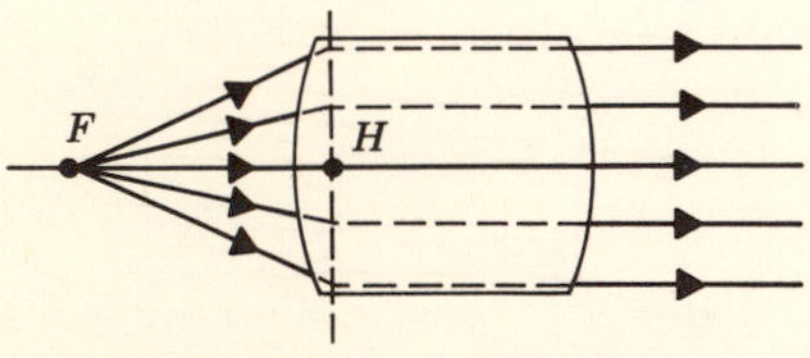

Fig. 3.12. First focal point F and principal point H with their respective planes.

*2. **Construction of Image.*** The location of these four points completely characterizes the image-forming properties of a system

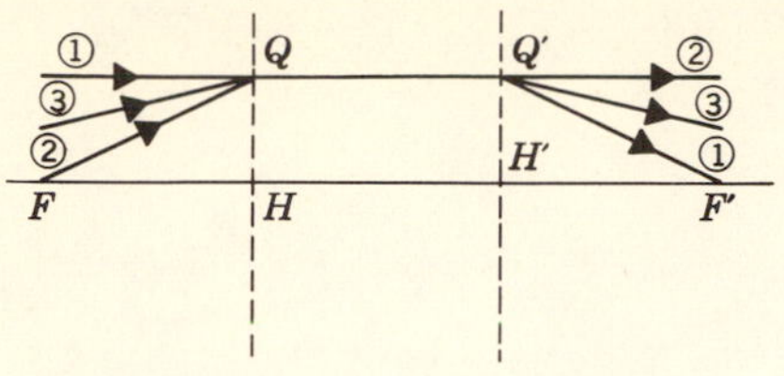

Fig. 3.13. Illustration of the unit lateral magnification property of the principal planes.

for paraxial rays. Before we discuss the procedure for locating the position and size of the image of a given object, we make the following remark.

The principal planes are planes of unit lateral magnification. This property can be seen from Fig. 3.13. Here the apparent paths of two incoming and outgoing rays are traced. Rays 1 and 2 appear to intersect first at Q and then at Q'; hence Q' is an image of Q. Since $|HQ| = |H'Q'|$, the lateral magnification is unity. Because they possess this property, *principal planes are often called unit planes*. Any ray, such as 3, that appears to cross the H plane at height h will appear to be crossing the H' plane also at height h.

Pairs of points which are images of each other are called *conjugate points*. The principal points are conjugate, but the focal points are not.

The general case of image formation is illustrated in Fig. 3.14. Because of the conjugate property, ray 3, which is aimed at H,

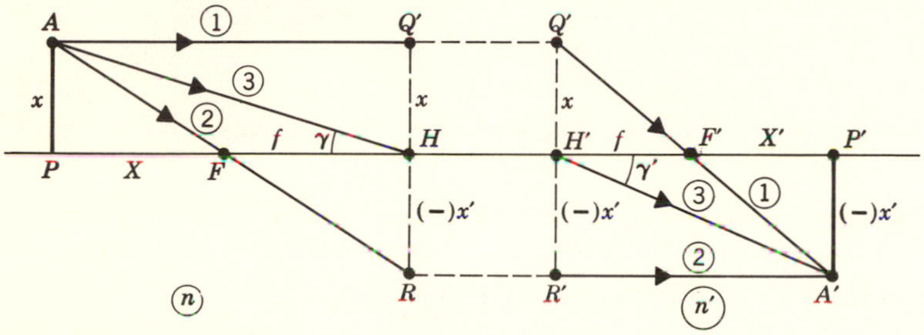

Fig. 3.14. Graphical location of the image formed by a thick lens or compound lens system.

must appear to be coming from H'. From the similar right triangles FPA and FHR we deduce

$$\frac{-x'}{x} = \frac{f}{X} \tag{3.36}$$

and from the similar triangles $Q'H'F'$ and $A'P'F'$ we deduce

$$\frac{x}{-x'} = \frac{f'}{A} \tag{3.37}$$

Multiplication of the two equations yields the Newtonian form

$$ff' = XX' \tag{3.38}$$

The focal lengths f and f' will now be related to one another for this small-angle case. Consider ray 3 in Fig. 3.14. Use will be made of the Lagrange equation (3.15), which can be shown to generalize to a general image-forming system. See Eq. (3.85) in Section 3.4D. We apply it in the context of Fig. 3.14 to ray 3 at H and H'. The lateral magnification between the conjugate planes at H and H' is

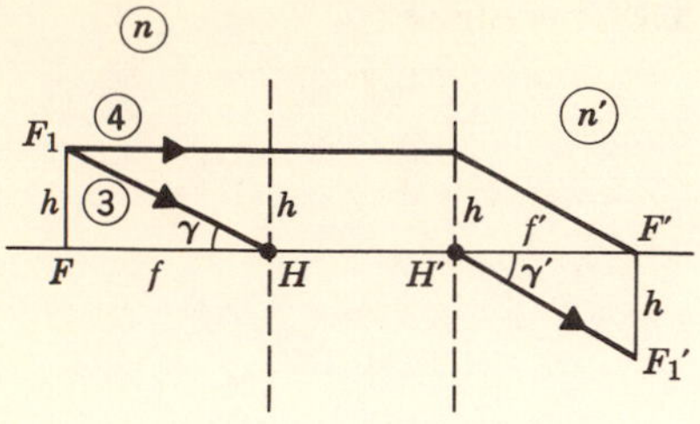

Fig. 3.15. Part of Fig. 3.14

unity. We identify γ with α_0 in (3.15) and γ' with α_2. The condition (3.15'),

$$m_x m_\alpha = \frac{n_1}{n_1'}$$

with unit magnification m_x and with $m_\alpha = \gamma'/\gamma$ reduces to

$$n\gamma = n'\gamma' \tag{3.39}$$

Figure 3.15 shows part of Fig. 3.14 with another ray, ray 4, added. Since rays 3 and 4 leave the F plane from the same point, they form an image at $+\infty$; that is, they must be parallel after refraction. Thus the acute angle at F' is also γ'. From the triangles F_1FH and $F_1'F'H'$ we have

$$\gamma = \frac{h}{f}; \qquad \gamma' = \frac{h}{f'}$$

Hence, from (3.39) we get

$$\frac{n}{f} = \frac{n'}{f'} \tag{3.40}$$

Another relation can be obtained from Fig. 3.14. We can write

$$n\frac{x}{X+f} = n\frac{x}{S} = n\gamma = n'\gamma' = \frac{-n'x'}{S'} = \frac{-n'x'}{X'+f'}$$

Thus the lateral magnification is given by

$$m_x = \frac{x'}{x} = \frac{-S'n}{Sn'} \tag{3.41}$$

which we have seen for a single surface in Eq. (3.13). Equation (3.36) can be rewritten

$$\frac{x}{x'} = -\frac{X}{f} = \frac{-(S+f)}{f} \tag{3.42}$$

From (3.41) we have

$$\frac{x}{x'} = \frac{-Sn'}{S'n}$$

which with (3.42) gives

$$\frac{n'S}{nS'} = \frac{S-f}{f}$$

which can be rewritten using (3.40) to yield the Gaussian form

$$\frac{n}{S} + \frac{n'}{S'} = \frac{n}{f} = \frac{n'}{f'} \tag{3.43}$$

When $n = n'$, f then equals f' and (3.38) becomes

$$XX' = f^2 \tag{3.44}$$

Then (3.43) becomes

$$\frac{1}{S} + \frac{1}{S'} = \frac{1}{f} \tag{3.45}$$

Equation (3.44) should be compared with (3.35); (3.43) with (3.31); (3.41) with (3.32). *The only formal change necessary to carry the thin-lens formulas over to the general image-forming case is the measurement of S and S' from H and H' instead of from the lens itself.*

REFERENCES FOR SECTION 3.3

The material in this section is covered in most optics books—for example, F. A. Jenkins and H. E. White, *Fundamentals of Optics*; A. C. Hardy and F. H. Perrin, *The Principles of Optics*; W. T. Welford, *Geometrical Optics*.

PROBLEMS

3.3.1 Locate the image graphically for a positive lens as in Fig. 3.10 when (a) $0 < S < f$, (b) $-f < S < 0$ (virtual object), and (c) $S < -f$.

3.3.2 For a negative lens we have $f < 0$.

a. Find the focal length of such a lens assumed thin when $R_1 = -40$ cm, $R_2 = 50$ cm, $n = 1{\cdot}75$.
b. Locate graphically and analytically the position and size of an image when $S = 2f$, $S = f$, $S = f/2$, $S = -f/2$. (These are algebraic equations.)

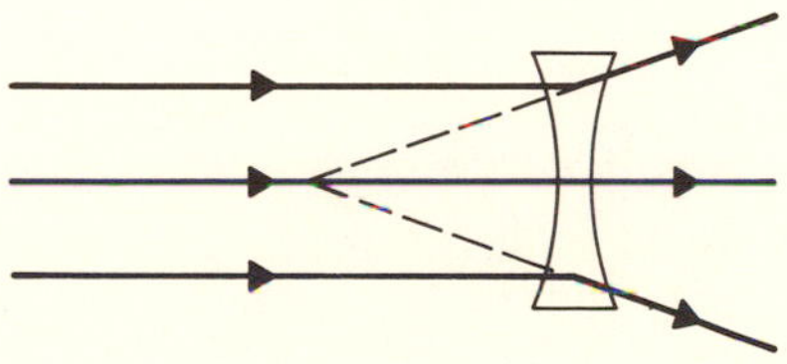

3.3.3 Find the principal points and focal points of the spherical lens in Problem 3.1.5.

3.3.4 a. Two thin lenses, one with focal length $+f$, the other with focal length $-f$, are mounted a distance f apart. Find the principal planes and focal planes of the system.
b. As above, but the focal lengths are $+f$ and $-f/6$, and the lens separation is $2/3f$.

You may use graphical or analytical means.

3.4 MATRIX METHODS IN GAUSSIAN OPTICS

The ray-tracing equations of Gaussian optics are linear in the variables x and α. The individual refraction and translation equations are linear, and the resulting overall transformation is also linear. The equations of a linear transformation are very conveniently written using matrices, and the results of several consecutive linear transformations can be compactly written in terms of the product of the individual matrices. In addition to providing computational convenience, the matrix method will enable us to prove some important theorems about Gaussian optical systems. The most important of these will validate the statements made in the previous section about principal planes, focal planes, and their role in a general image-forming system.

A. THE USE OF MATRICES TO DESCRIBE LINEAR TRANSFORMATIONS

In the general theory of linear transformations we deal with n independent variables $x_1, x_2, x_3, \ldots x_n$. In the present case of Gaussian optics, $n = 2$ and we have $x_1 = x$ and $x_2 = \alpha$. Thus, x_1 and x_2 do not have to be Cartesian coordinates; they can be much more general than that. In this subsection we shall treat them simply as mathematical variables.

A linear transformation from the variables x_1, x_2 to new variables y_1, y_2 can be written in the form

$$\begin{aligned} y_1 &= M_{11}x_1 + M_{12}x_2 \\ y_2 &= M_{21}x_1 + M_{22}x_2 \end{aligned} \tag{3.46}$$

We now introduce the *matrix* $\mathbf{M}$, which in this case is a 2×2 array of *elements*,

$$\mathbf{M} = \begin{pmatrix} M_{11} & M_{12} \\ M_{21} & M_{22} \end{pmatrix}$$

We may also form the pair of variables x_1, x_2 into a 2×1 array or *column matrix* (also called *column vector*),

$$\mathbf{x} = \begin{pmatrix} x_1 \\ x_2 \end{pmatrix}$$

And similarly for the pair y_1, y_2,

$$\mathbf{y} = \begin{pmatrix} y_1 \\ y_2 \end{pmatrix}$$

We then write Eqs. (3.46) as a product,

$$\mathbf{y} = \mathbf{Mx} \tag{3.47a}$$

or equivalently

$$\begin{pmatrix} y_1 \\ y_2 \end{pmatrix} = \begin{pmatrix} M_{11} & M_{12} \\ M_{21} & M_{22} \end{pmatrix} \begin{pmatrix} x_1 \\ x_2 \end{pmatrix} \tag{3.47b}$$

The multiplication implied in Eqs. (3.47) is defined by Eqs. (3.46), which also can be written using summation notation as

$$y_i = \sum_{j=1}^{2} M_{ij} x_j, \qquad \text{for } i = 1,2 \tag{3.48}$$

Now suppose that there are two new variables z_1, z_2, which are given in terms of y_1, y_2 by the linear equations

$$z_1 = N_{11} y_1 + N_{12} y_2$$
$$z_2 = N_{21} y_1 + N_{22} y_2$$

or equivalently

$$z_k = \sum_{i=1}^{2} N_{ki} y_i, \qquad k = 1,2 \tag{3.49a}$$

or equivalently

$$\mathbf{z} = \mathbf{N}\mathbf{y} \tag{3.49b}$$

with

$$\mathbf{z} = \begin{pmatrix} z_1 \\ z_2 \end{pmatrix}, \qquad \mathbf{N} = \begin{pmatrix} N_{11} & N_{12} \\ N_{21} & N_{22} \end{pmatrix} \tag{3.49c}$$

The equations giving z_1, z_2 in terms of x_1, x_2 are obtained by substituting Eq. (3.48) into (3.49*a*):

$$z_k = \sum_{i=1}^{2} N_{ki} y_i = \sum_{i=1}^{2} N_{ki} \left(\sum_{j=1}^{2} M_{ij} x_j \right)$$

Now interchange the order of summation:

$$z_k = \sum_{j=1}^{2} \left(\sum_{i=1}^{2} N_{ki} M_{ij} \right) x_j \tag{3.50}$$

We define a 2×2 matrix $\mathbf{Q}$ with elements Q_{kj} given by

$$Q_{kj} = \sum_{i=1}^{2} N_{ki} M_{ij} \tag{3.51}$$

Equation (3.51) defines the operation of *matrix multiplication*. Note that the summation here as in (3-48) is over adjacent indices. Thus

$$\mathbf{Q} = \mathbf{N}\mathbf{M} \tag{3.52}$$

Then Eq. (3.50) gives

$$z_k = \sum_{j=1}^{2} Q_{kj} x_j$$

or

$$\mathbf{z} = \mathbf{Q}\mathbf{x} \tag{3.53}$$

We may summarize the above operations by writing

$$\mathbf{z} = \mathbf{N}\mathbf{y} \text{ with } \mathbf{y} = \mathbf{M}\mathbf{x} \quad \text{gives} \quad \mathbf{z} = \mathbf{N}(\mathbf{M}\mathbf{x}) = (\mathbf{N}\mathbf{M})\mathbf{x} = \mathbf{Q}\mathbf{x} \tag{3.54}$$

In component form, Eq. (3.52) or (3.51) gives these four equations:

$$\begin{aligned} Q_{11} &= N_{11}M_{11} + N_{12}M_{21} \\ Q_{12} &= N_{11}M_{12} + N_{12}M_{22} \\ Q_{21} &= N_{21}M_{11} + N_{22}M_{21} \\ Q_{22} &= N_{21}M_{12} + N_{22}M_{22} \end{aligned} \tag{3.55}$$

We may continue this process of introducing new variables and new matrices. This will give us the overall transformation matrix in terms of the products of the individual transformation matrices *written in order from right to left*, as in Eq. (3.54). The order of matrix multiplication is important, for it is *not* generally true for two matrices $\mathbf{A}$, $\mathbf{B}$ that $\mathbf{AB} = \mathbf{BA}$. Thus, matrix multiplication is *not commutative*. It is *associative*, however, and this is a very important property. It is expressed by the equation

$$\mathbf{A}(\mathbf{BC}) = (\mathbf{AB})\mathbf{C} = \mathbf{ABC} \tag{3.56}$$

The proof can be given by introducing the elements of $\mathbf{A}$, $\mathbf{B}$, and $\mathbf{C}$, writing out the matrix multiplications as the sums of products of elements, and then interchanging the order of summation, in direct analogy with the steps that lead to Eq. (3.50).

The *determinant* of the 2×2 matrix $\mathbf{M}$ is defined by the equation

$$\det \mathbf{M} = M_{11}M_{22} - M_{21}M_{12} \tag{3.57}$$

We can easily show that if $\mathbf{Q} = \mathbf{NM}$,

$$\det \mathbf{Q} = (\det \mathbf{N})\,(\det \mathbf{M}) \tag{3.58}$$

This readily generalizes to the product of more than two matrices:

$$\det(\mathbf{M}_1\mathbf{M}_2\mathbf{M}_3 \cdots \mathbf{M}_n) = (\det \mathbf{M}_1)(\det \mathbf{M}_2)(\det \mathbf{M}_3) \cdots (\det \mathbf{M}_n) \tag{3.59}$$

Since the determinant is an ordinary number, we may commute the order of the products on the right-hand side of Eq. (3.59) anyway we like.

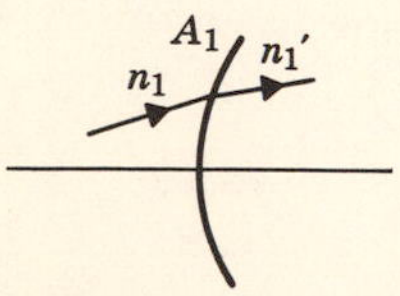

Fig. 3.16

B. THE MATRICES FOR REFRACTION AND TRANSLATION TRANSFORMATIONS

At a spherical refracting surface between media with refractive indices n_1 and n_1' with a radius of curvature R_1 (Fig. 3.16), the equations of the refraction transformation are (Eq. 3.7)

$$x_1' = x_1$$
$$\alpha_1' = \frac{n_1 - n_1'}{n_1' R_1} x_1 + \frac{n_1}{n_1'}\alpha_1 = -\frac{P_1}{n_1'} x_1 + \frac{n_1}{n_1'}\alpha_1$$

where P_1 is called the *power* of the surface:

$$P_1 \equiv \frac{n_1' - n_1}{R_1} \tag{3.60}$$

The transformation can be written in matrix form as

$$\begin{pmatrix} x_1' \\ \alpha_1' \end{pmatrix} = \mathbf{R}_1 \begin{pmatrix} x_1 \\ \alpha_1 \end{pmatrix}$$

where

$$\mathbf{R}_1 = \begin{pmatrix} 1 & 0 \\ -\dfrac{P_1}{n_1'} & \dfrac{n_1}{n_1'} \end{pmatrix} \tag{3.61}$$

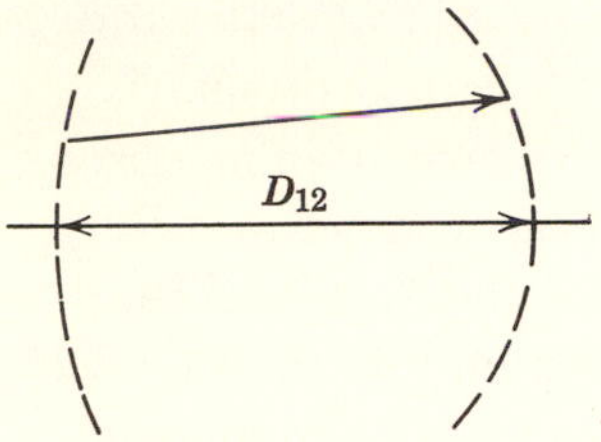

Fig. 3.17

represents the refraction matrix for the surface. The equations of a translation $\mathscr{T}_{12}$ by a distance D_{12} (Fig. 3.17) are

$$x_2 = x_1' + D_{12}\alpha_1'$$
$$\alpha_2 = \alpha_1'$$

They can be written in matrix form as

$$\begin{pmatrix} x_2 \\ \alpha_2 \end{pmatrix} = \mathbf{T}_{12} \begin{pmatrix} x_1' \\ \alpha_1' \end{pmatrix} \quad \text{with} \quad \mathbf{T}_{12} = \begin{pmatrix} 1 & D_{12} \\ 0 & 1 \end{pmatrix} \tag{3.62}$$

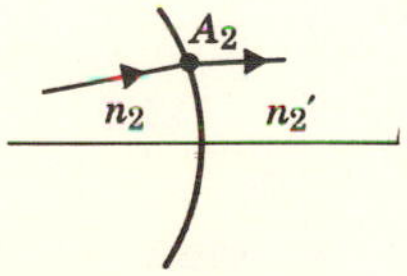

Fig. 3.18

A refraction at A_2 (Fig. 3.18) can be described by

$$\begin{pmatrix} x_2' \\ \alpha_2' \end{pmatrix} = \mathbf{R}_2 \begin{pmatrix} x_2 \\ \alpha_2 \end{pmatrix} \quad \text{with} \quad \mathbf{R}_2 = \begin{pmatrix} 1 & 0 \\ -\dfrac{P_2}{n_2'} & \dfrac{n_2}{n_2'} \end{pmatrix} \tag{3.63}$$

$$P_2 = \frac{n_2' - n_2}{R_2}$$

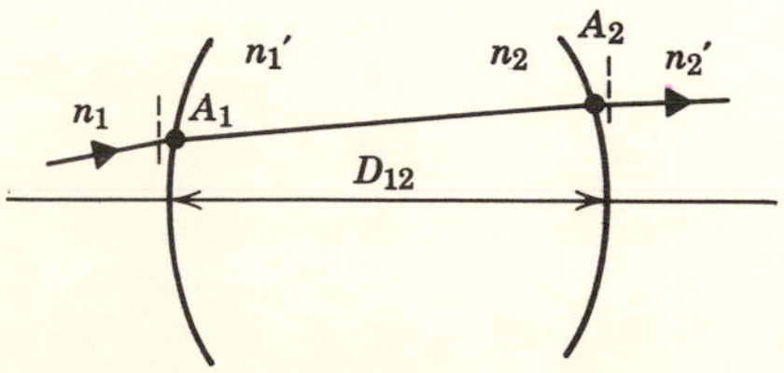

Fig. 3.19

These equations may be combined to give the overall transformation from a point just to the left of A_1 to a point just to the right of A_2 (Fig. 3.19):

$$\begin{pmatrix} x_2' \\ \alpha_2' \end{pmatrix} = \mathbf{R}_2\mathbf{T}_{12}\mathbf{R}_1 \begin{pmatrix} x_1 \\ \alpha_1 \end{pmatrix} = \mathbf{M}_{12} \begin{pmatrix} x_1 \\ \alpha_1 \end{pmatrix}, \qquad \mathbf{M}_{12} = \mathbf{R}_2\mathbf{T}_{12}\mathbf{R}_1 \tag{3.64}$$

This process can be repeated as often as necessary. The linear transformation between the initial position and angle x, α and the final position and angle x', α' can then be written in the matrix form

$$\begin{pmatrix} x' \\ \alpha' \end{pmatrix} = \mathbf{M} \begin{pmatrix} x \\ \alpha \end{pmatrix} \tag{3.65}$$

where **M** is the product of all the refraction and translation matrices, written in order, from right to left, in the same sequence followed by the light ray.

The determinant of **M** is the product of all the determinants of the refraction and translation matrices. We see from Eq. (3.62) that the determinant of a translation matrix is always unity, and from Eq. (3.61) or (3.63) that the determinant of a refraction matrix is given by the ratio of initial to final refractive indices. Thus the determinant of **M** is the product of the determinants of the refraction matrices and takes the form

$$\det \mathbf{M} = \left(\frac{n_1}{n_1'}\right)\left(\frac{n_2}{n_2'}\right)\cdots$$

But $n_1' = n_2$, $n_2' = n_3$, and so on. This produces a great deal of cancellation, leaving

$$\det \mathbf{M} = \frac{n}{n'} \tag{3.66}$$

where n is the index of the medium to the left of the first refracting surface, and n' is the index of the medium to the right of the last refracting surface.

C. THE TRANSFORMATION MATRICES FOR SIMPLE SYSTEMS

We give here some important transformation matrices using a common notation. The initial transverse coordinate, inclination angle, and refractive index will be denoted by x, α, and n, respectively; the final coordinate, angle, and index by x', α', and n'. The transformation carries from a point B to a point B' on the ray. In other words, the ray at point B has coordinate x and inclination angle α.

Fig. 3.20

1. Refraction at a Spherical Surface (Fig. 3.20). This has been discussed already with different notation. We have

$$\mathbf{M} = \begin{pmatrix} 1 & 0 \\ -\dfrac{P}{n'} & \dfrac{n}{n'} \end{pmatrix} \tag{3.67}$$

$$P = \frac{n' - n}{R}$$

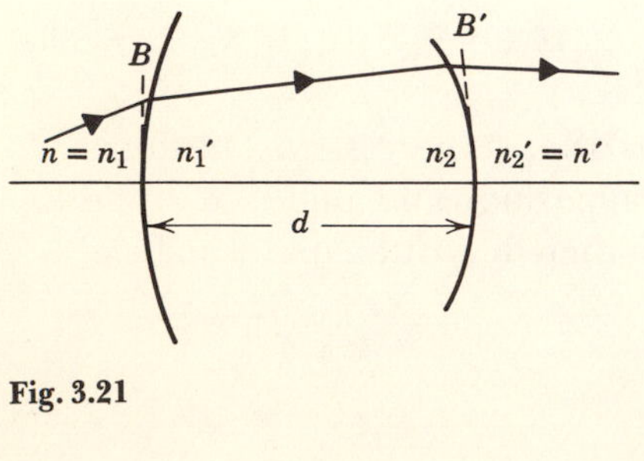

Fig. 3.21

2. Simple Lens (Fig. 3.21). The matrix **M** is the same as $\mathbf{M}_{12}$ in Eq. (3.64). We now perform the matrix multiplication (using $n_1' = n_2$, $n_2' = n'$):

$$\mathbf{M} = \mathbf{R}_2\mathbf{T}_{12}\mathbf{R}_1$$

$$\begin{pmatrix} 1 & 0 \\ -\dfrac{P_2}{n'} & \dfrac{n_2}{n'} \end{pmatrix}\begin{pmatrix} 1 & d \\ 0 & 1 \end{pmatrix}\begin{pmatrix} 1 & 0 \\ -\dfrac{P_1}{n_2} & \dfrac{n}{n_2} \end{pmatrix} = \begin{pmatrix} 1 & 0 \\ -\dfrac{P_2}{n'} & \dfrac{n_2}{n'} \end{pmatrix}\begin{pmatrix} 1 - \dfrac{P_1 d}{n_2} & \dfrac{nd}{n_2} \\ -\dfrac{P_1}{n_2} & \dfrac{n}{n_2} \end{pmatrix}$$

This gives

$$\mathbf{M} = \begin{pmatrix} 1-\dfrac{P_1 d}{n_2} & \dfrac{nd}{n_2} \\ -\dfrac{P_2}{n'}+\dfrac{P_1P_2d}{n_2n'}-\dfrac{P_1}{n'}, & \dfrac{n}{n'}\left(1-\dfrac{P_2d}{n_2}\right) \end{pmatrix} \tag{3.68}$$

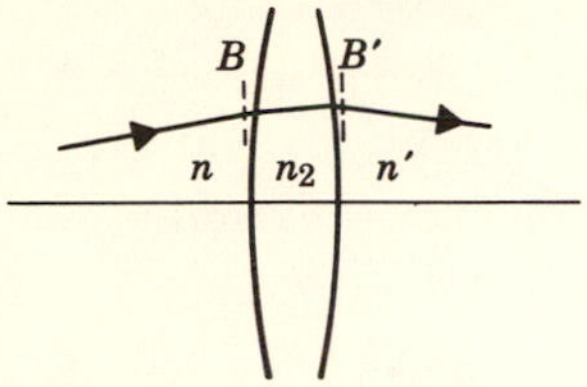

Fig. 3.22

3. ***Thin Lens*** (Fig. 3.22). A thin lens is a simple lens with a negligible thickness d. We let $d \to 0$ in Eq. (3.68) and obtain

$$\mathbf{M} = \begin{pmatrix} 1 & 0 \\ -\dfrac{P}{n'} & \dfrac{n}{n'} \end{pmatrix} \tag{3.69a}$$

where the total power P is given by

$$P = P_1 + P_2 = \frac{n_2 - n}{R_1} + \frac{n' - n_2}{R_2} \tag{3.69b}$$

Note that $\mathbf{M}$ has the same form for a thin lens as for a single refracting surface. Note also that for both we have $M_{11} = 1$ and $M_{12} = 0$. This means that we have $x' = x$, independently of the value of α. Thus, B' is an image of B formed with unit magnification.

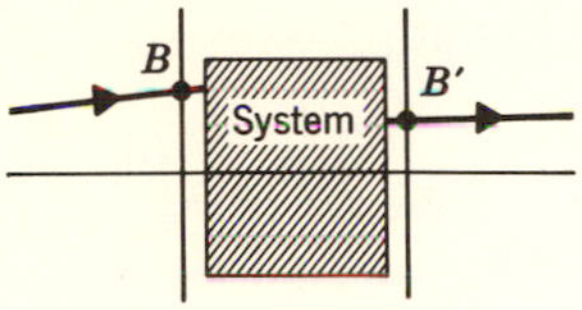

Fig. 3.23

4. ***The General Image-Forming System.*** In the general case, indicated schematically in Fig. 3.23, we have a general transformation matrix from B to B':

$$\mathbf{M} = \begin{pmatrix} M_{11} & M_{12} \\ M_{21} & M_{22} \end{pmatrix}$$

and the only requirement so far is

$$\det \mathbf{M} = \frac{n}{n'} \tag{3.66}$$

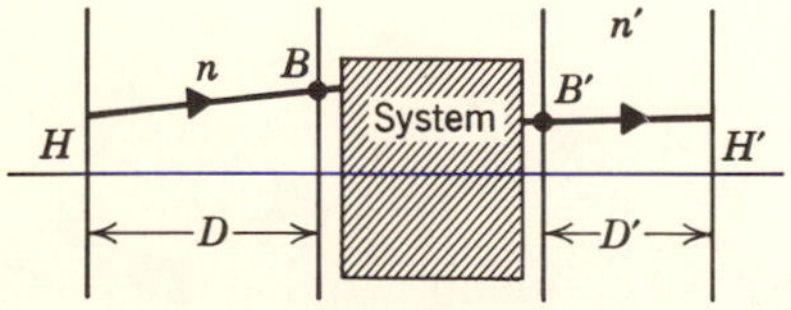

Fig. 3.24

We now ask if it is possible to find new reference planes instead of B and B' for which this general matrix $\mathbf{M}$ will take the form of that for a thin lens. These, of course, will turn out to be the principal planes that intersect the axis at H and H' in Fig. 3.24. The transformation matrix from the H plane to the H' plane can be written in terms of $\mathbf{M}$ by adding translations $\mathbf{T}$ and $\mathbf{T}'$:

$$\begin{aligned} \mathbf{M}_{HH'} = \mathbf{T}'\mathbf{M}\mathbf{T} &= \begin{pmatrix} 1 & D' \\ 0 & 1 \end{pmatrix}\begin{pmatrix} M_{11} & M_{12} \\ M_{21} & M_{22} \end{pmatrix}\begin{pmatrix} 1 & D \\ 0 & 1 \end{pmatrix} \\ &= \begin{pmatrix} 1 & D' \\ 0 & 1 \end{pmatrix}\begin{pmatrix} M_{11}, & M_{11}D + M_{12} \\ M_{21}, & M_{21}D + M_{22} \end{pmatrix} \\ &= \begin{pmatrix} M_{11} + D'M_{21}, & M_{11}D + M_{12} + D'(M_{21}D + M_{22}) \\ M_{21}, & M_{21}D + M_{22} \end{pmatrix} \end{aligned} \tag{3.70}$$

So far this is completely general. We now require that the H' plane be conjugate to the H plane with unit magnification; thus the 11 element of $\mathbf{M}_{HH'}$ must be unity and the 12 element must be zero. This gives

$$\mathbf{M}_{HH'} = \begin{pmatrix} 1 & 0 \\ M_{21} & \dfrac{n}{n'} \end{pmatrix} \tag{3.71}$$

We have written the 22 element as n/n' as required by the condition

$$\det \mathbf{M}_{HH'} = \frac{n}{n'}$$

We now equate the elements of the matrices in Eqs. (3.70) and (3.71):

$$11\colon\ M_{11} + D'M_{21} = 1, \qquad \text{or} \quad D' = \frac{1 - M_{11}}{M_{21}} \tag{3.72}$$

$$22\colon\ M_{21}D + M_{22} = \frac{n}{n'}, \qquad \text{or} \quad D = \frac{(n/n') - M_{22}}{M_{21}} \tag{3.73}$$

These equations are meaningful only if the condition

$$M_{21} \neq 0 \tag{3.74}$$

is satisfied. This then becomes the requirement that our general Gaussian system be image-forming. Returning to the elements of (3.70) and (3.71), we obtain

$$12\colon\ M_{11}D + M_{12} + D'(M_{21}D + M_{22}) = 0 \tag{3.75}$$

This equation follows readily from Eqs. (3.72) and (3.73).

To complete the formal equivalence between our general image-forming system and a thin lens, it is only necessary to make the identification

$$-\frac{P}{n'} = M_{21} \tag{3.76}$$

Thus the image formation condition, Eq. (3.74), guarantees that our system has a nonzero power P. We have thus shown that the Gaussian ray-tracing transformation between the H plane and the H' plane can be written

$$\mathbf{M}_{HH'} = \begin{pmatrix} 1 & 0 \\ -\dfrac{P}{n'} & \dfrac{n}{n'} \end{pmatrix} \tag{3.77}$$

This means that all image forming systems have the same formal behavior in Gaussian optics, as far as ray tracing is concerned.

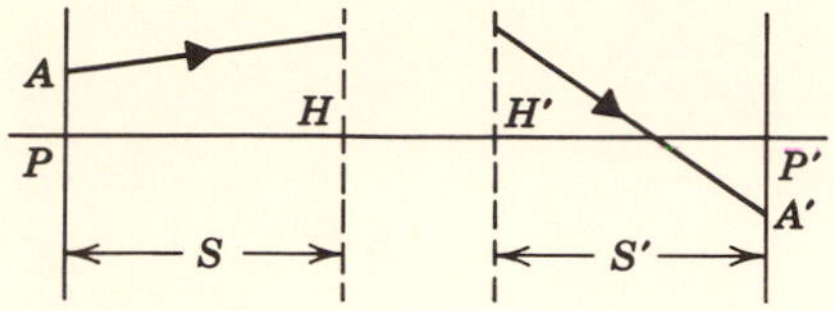

Fig. 3.25

D. THE IMAGE FORMATION PROCESS

We now want to move from the principal planes to other conjugate planes and determine the object-image relationships that result. This is done by translation transformations over the distances S and S' in Fig. 3.25. The overall transformation matrix from A to A' is given by

$$\mathbf{M}_{AA'} = \begin{pmatrix} 1 & S' \\ 0 & 1 \end{pmatrix} \begin{pmatrix} 1 & 0 \\ -\dfrac{P}{n'} & \dfrac{n}{n'} \end{pmatrix} \begin{pmatrix} 1 & S \\ 0 & 1 \end{pmatrix}$$

$$= \begin{pmatrix} 1 & S' \\ 0 & 1 \end{pmatrix} \begin{pmatrix} 1 & S \\ -\dfrac{P}{n'} & -\dfrac{SP}{n'} + \dfrac{n}{n'} \end{pmatrix}$$

$$= \begin{pmatrix} 1 - \dfrac{S'P}{n'} & S - \dfrac{S'SP}{n'} + \dfrac{nS'}{n'} \\ -\dfrac{P}{n'} & -\dfrac{SP}{n'} + \dfrac{n}{n'} \end{pmatrix} \tag{3.78}$$

This result holds for general values of S and S'.
The image-formation condition is that the 12 element of this matrix be zero, or

$$S - \frac{SS'P}{n'} + \frac{nS'}{n'} = 0 \tag{3.79}$$

This can be rearranged to give the Gaussian form

$$\frac{n}{S} + \frac{n'}{S'} = P \tag{3.80}$$

When the image is at $+\infty$, the object is in the first focal plane at a distance

$$S = \frac{n}{P} \equiv f \tag{3.81}$$

to the left of the first principal plane. When the object is at $+\infty$, the image is in the second focal plane at a distance

$$S' = \frac{n'}{P} \equiv f' \tag{3.82}$$

to the right of the second principal plane. Thus, Eq. (3.80) may also be written in the Gaussian form

$$\frac{n}{S} + \frac{n'}{S'} = \frac{n}{f} = \frac{n'}{f'}, \qquad \text{or in the Newtonian form } XX' = ff' \tag{3.83}$$

When we have image formation, our matrix equation can be written

$$\begin{pmatrix} x' \\ \alpha' \end{pmatrix} = \mathbf{M}_{AA'} \begin{pmatrix} x \\ \alpha \end{pmatrix}, \qquad \mathbf{M}_{AA'} = \begin{pmatrix} m_x & 0 \\ -\dfrac{P}{n'} & m_\alpha \end{pmatrix} \tag{3.84a}$$

where the lateral magnification is

$$m_x = 1 - \frac{S'P}{n'} = 1 - \frac{S'}{f'} = -\frac{X'}{f'} = -\frac{nS'}{n'S} \tag{3.84b}$$

and the ray angle magnification is

$$m_\alpha = -\frac{SP}{n'} + \frac{n}{n'} = -\frac{S}{S'} \tag{3.84c}$$

The component form of Eq. (3.84*a*) is of course

$$\begin{aligned} x' &= m_x x \\ \alpha' &= -\frac{P}{n'}x + m_\alpha \alpha \end{aligned} \tag{3.84d}$$

From the condition det $\mathbf{M}_{AA'} = n/n'$ we obtain the result

$$m_x m_\alpha = \frac{n}{n'}$$

If we then insert $m_x = x'/x$ and $m_\alpha = \Delta\alpha'/\Delta\alpha$, this becomes the LaGrange or Smith-Helmholtz equation for a general image-forming system

$$nx\,\Delta\alpha = n'x'\,\Delta\alpha' \tag{3.85}$$

The behavior of some key rays is shown in Fig. 3.26. They are consistent with Eq. (3.84).

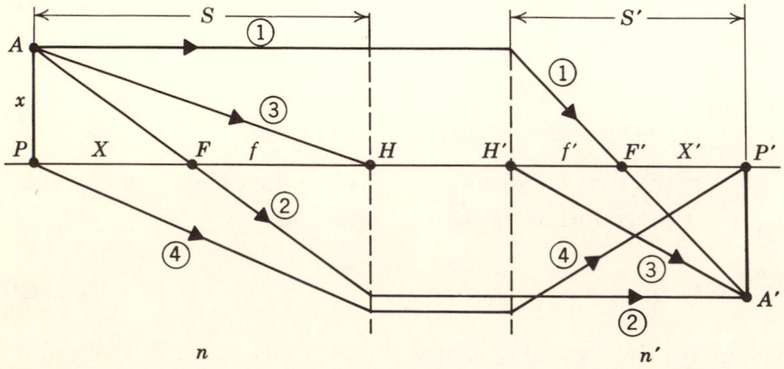

Fig. 3.26

It should be emphasized that the physical location of the principal planes could be inside one of the components of the image-forming system. Or they could be outside. The point to be made is that these are mathematical planes, and the rays behave as though they were deviated as shown in Fig. 3.26. There is no a priori reason for the order of the principal planes to be as shown in Fig. 3.26. The point H could be to the right of H', as shown in Fig. 3.27.

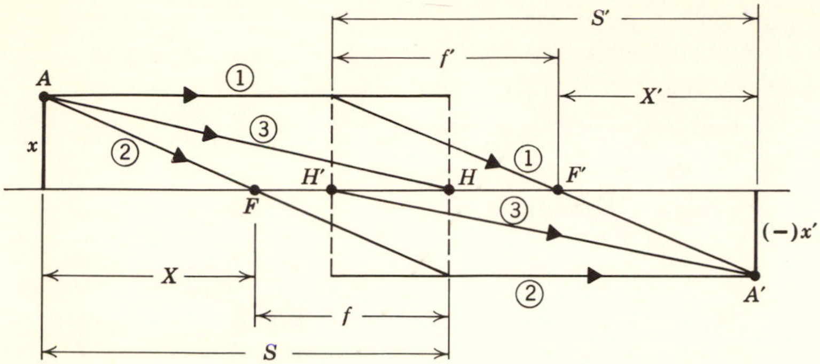

Fig. 3.27

The point H will be to the right of F and H' to the left of F' if f and f' are positive.

E. THE COMBINATION OF TWO IMAGE-FORMING SYSTEMS

1. Derivation of Formulas. We now develop formulas that will enable us to combine two image-forming systems into a single overall image-forming system. The matrix method makes this and similar problems relatively easy. Figure 3.28 defines the situation.

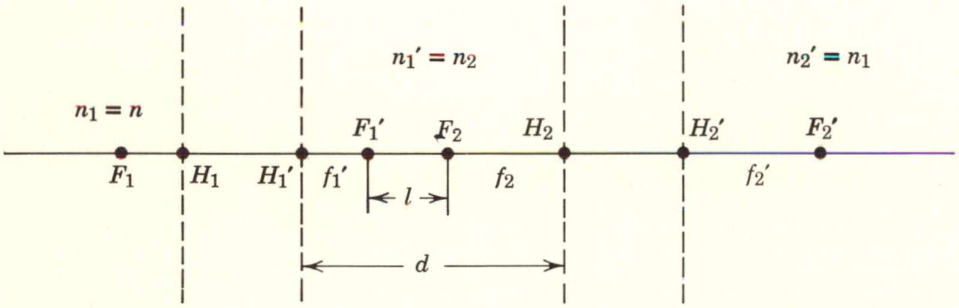

Fig. 3.28

The focal lengths of the first system are f_1, f'_1; those of the second, f_2, f'_2. The respective powers are

$$P_1 = \frac{n_1}{f_1} = \frac{n'_1}{f'_1} \quad \text{and} \quad P_2 = \frac{n_2}{f_2} = \frac{n'_2}{f'_2} \tag{3.86}$$

The transformation matrix between the principal planes of the first system is

$$\mathbf{M}_{H_1H'_1} = \begin{pmatrix} 1 & 0 \\ -\dfrac{P_1}{n'_1} & \dfrac{n_1}{n'_1} \end{pmatrix} \tag{3.87a}$$

Between the principal planes of the second system the transformation matrix is

$$\mathbf{M}_{H_2H_2'} = \begin{pmatrix} 1 & 0 \\ -\dfrac{P_2}{n_2'} & \dfrac{n_2}{n_2'} \end{pmatrix} \tag{3.87b}$$

The overall transformation matrix between the first principal plane of the first system and the second principal plane of the second system is identical in form to $\mathbf{M}$ in Eq. (3.68) and is given by

$$\mathbf{M}_{H_1H_2'} = \mathbf{M}_{H_2H_2'}\begin{pmatrix} 1 & d \\ 0 & 1 \end{pmatrix}\mathbf{M}_{H_1H_1'}$$

$$= \begin{pmatrix} 1-\dfrac{P_1 d}{n_2} & \dfrac{nd}{n_2} \\ -\dfrac{P_1}{n'}-\dfrac{P_2}{n'}+\dfrac{P_1P_2d}{n_2n'}, & \dfrac{n}{n'}\left(1-\dfrac{P_2d}{n_2}\right) \end{pmatrix} \tag{3.88}$$

This matrix now plays the role of the general matrix $\mathbf{M}$ of the last section,

$$\mathbf{M}_{H_1H_2'} = \mathbf{M} = \begin{pmatrix} M_{11} & M_{12} \\ M_{21} & M_{22} \end{pmatrix} \tag{3.89}$$

for which we know how to find the principal planes H, H'. We write

$$\mathbf{M}_{HH'} = \begin{pmatrix} 1 & D' \\ 0 & 1 \end{pmatrix}\mathbf{M}_{H_1H_2'}\begin{pmatrix} 1 & D \\ 0 & 1 \end{pmatrix}$$

$$= \begin{pmatrix} 1 & 0 \\ -\dfrac{P}{n'} & \dfrac{n}{n'} \end{pmatrix} \tag{3.90}$$

where from (3.76), (3.72), and (3.73) we have

$$P = -n'M_{21} = P_1 + P_2 - \frac{P_1P_2d}{n_2} \tag{3.91a}$$

$$D' = \frac{1-M_{11}}{M_{21}} = \frac{P_1n'd}{n_2P} \tag{3.91b}$$

$$D = \frac{(n/n') - M_{22}}{M_{21}} = \frac{-P_2nd}{n_2P} \tag{3.91c}$$

In Eq. (3.91*c*), D is the distance of the first principal plane of the overall system to the left of the first principal plane of the first component, and D' is the distance of the second principal plane of the overall system to the right of the second principal plane of the second component.

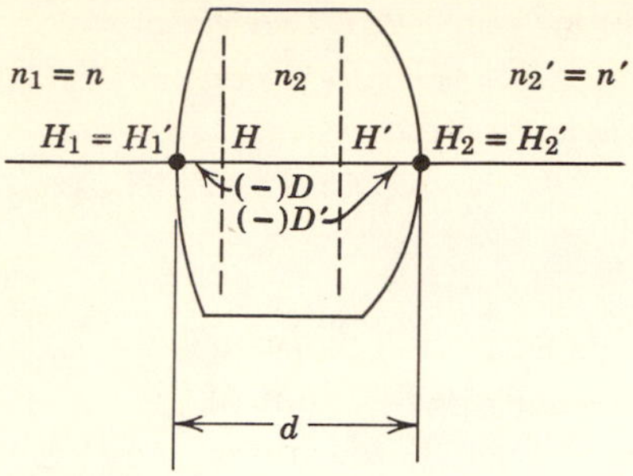

Fig. 3.29

2. ***Application to a Thick Lens.*** We may consider the thick lens shown schematically in Fig. 3.29 to be the simplest example of the general situation illustrated in Fig. 3.28. Here the principal planes of the first system coincide at $H_1 = H_1'$, and the principal planes of the second system coincide at $H_2 = H_2'$. We have

$$P_1 = \frac{n_2 - n}{R_1}, \qquad P_2 = \frac{n' - n_2}{R_2}$$

$$P = P_1 + P_2 - \frac{P_1 P_2 d}{n_2} = P_{\text{thin}} - \frac{P_1 P_2 d}{n_2} \tag{3.92}$$

where $P_{\text{thin}} = P_1 + P_2$ is the power of a thin lens utilizing the same two refracting surfaces. Equations (3.91*b*,*c*) may be used as they are.

The separation of the principal planes is given by

$$t = d + D + D' = d\left(1 - \frac{P_1 n'}{P n_2} - \frac{P_2 n}{P n_2}\right) = \frac{d}{P n_2}(P n_2 - P_1 n' - P_2 n) \tag{3.93}$$

In the usual case where $n = n'$, this becomes

$$t = d\left[1 - \frac{n}{n_2}\frac{(P_1 + P_2)}{P}\right] \tag{3.94}$$

There are times when we wish to treat even a thin lens as having a nonzero thickness d in order to locate its principal planes. When d is small compared with R_1 and R_2, but not entirely negligible, we have

$$P = P_{\text{thin}} = P_1 + P_2$$

and (for a thin lens in air)

$$t = d\left[1 - \frac{1}{n_2}\right]$$

$$D = \frac{-P_2}{P_1 + P_2}\frac{d}{n_2}$$

$$D' = \frac{-P_1}{P_1 + P_2}\frac{d}{n_2} \tag{3.95}$$

when $n_2 = 1.50$ (a typical value), we have $t = d/3$. Some examples are shown in Fig. 3.30.

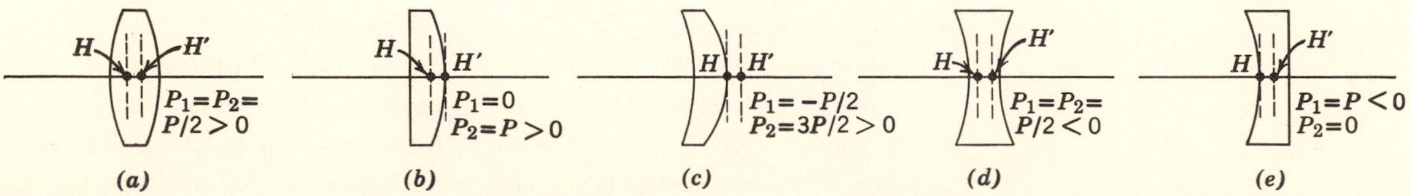

Fig. 3.30

3. Application to a Doublet. A doublet consists of two lenses with principal plane separation d. Figure 3.28 applies, as do Eqs. (3.90) and (3.91a,b,c). We set $n_2 = n' = 1$, but anticipate use of a doublet as an immersion microscope where $n \neq 1$. It is convenient to introduce the separation of focal planes l:

$$d = f_1' + f_2 + l \tag{3.96}$$

Then the power of the doublet can be written

$$P = -P_1 P_2 l \tag{3.97a}$$

and its second focal length is

$$f' = -\frac{f_1' f_2}{l} \tag{3.97b}$$

(The first focal length is $f = -f_1 f_2 / l$.)
Since by Eq. (3.84b) the lateral magnification is given by $m_x = -(S' - f')/f'$, we may obtain large magnifications for constant S' by combining two lenses of moderately small focal lengths and separating them by a distance much greater than either focal length, so that $|f'|$ is small. This is the essential feature of the compound microscope.

4. The Simple Magnifier. To appreciate the performance of the compound microscope, we first need to understand how the simple magnifier works. A one-lens version of a magnifier is shown in Fig. 3.31. Multilens versions will function the same way. A

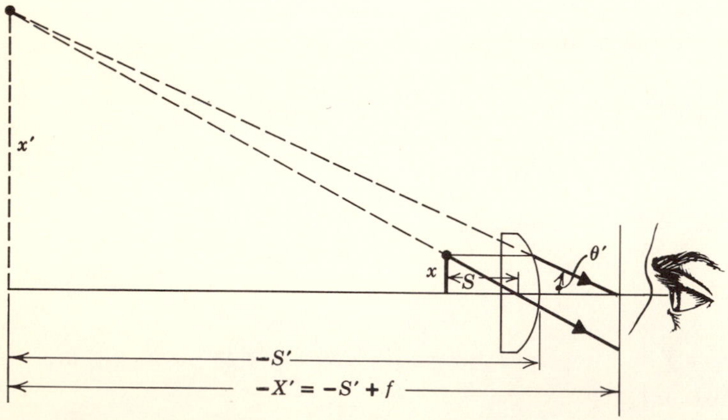

Fig. 3.31. The simple magnifier.

virtual image is formed at a distance $|S'|$ to the left of the lens. Ordinarily this distance is much greater than the focal length f, and hence the object distance is very close to f. The rays from a point in the object become almost parallel, and the angle θ' subtended by the image at the eye of the observer is a good measure of the apparent size of the image.

Typically the eye of the observer will be located at about the second focal plane, so that by Fig. 3.31, $\theta' \cong x'/(-X')$. Since the transverse magnification can be written $m_x = -X'/f$, we have $x' = m_x x$ and

$$\theta' \cong \frac{-X'}{f} \frac{1}{(-X')} x = \frac{x}{f} \tag{3.98}$$

A comfortable viewing distance for the eye is about 250 mm; at distances less than this the eye cannot readily accomodate. Thus we need

$$-X' \gtrsim 250 \text{ mm}$$

If the object were observed at a distance of 250 mm unaided, it would subtend an angle

$$\theta = \frac{x}{250} \qquad (x \text{ in mm})$$

The magnifier then multiplies this angle by the factor

$$m_m = \frac{\theta'}{\theta} = \frac{250}{f} \qquad (f \text{ in mm}) \tag{3.99}$$

which is conventionally taken as the definition of the magnification of a microscope, simple or compound.

*5. **The Compound Microscope.*** As shown earlier (Eq. 3.97*b*), for a doublet the focal length is given by $f' = -f_1' f_2 / l$. If l is large compared with f_1' and f_2, a doublet used as a microscope will have a large m_m.

As an example we consider the combination shown in Fig. 3.32 where $f_1 = f_1' = f_2 = 16$ mm and where the separation l is 160 mm, which is standard for many microscopes. We then find, from Eqs. (3.97*b*) and (3.91),

$$f = -1.6 \text{ mm}, \qquad D = D' = 19.2 \text{ mm}$$

For comfortable visual observation it is convenient to have $|X'| \geqslant 250$ mm.
We then obtain from $XX' = f^2$ the result

$$|X| \leqslant 0.01024 \text{ mm}$$

Thus the object must be located practically in the primary focal plane of the combination which passes through F in Fig. 3.32. The magnification is

$$m_m = \frac{250}{f} = \frac{250}{(-1.6)} = -156$$

The objective forms an inverted, real intermediate image which is then magnified to give a final virtual image. In performing this

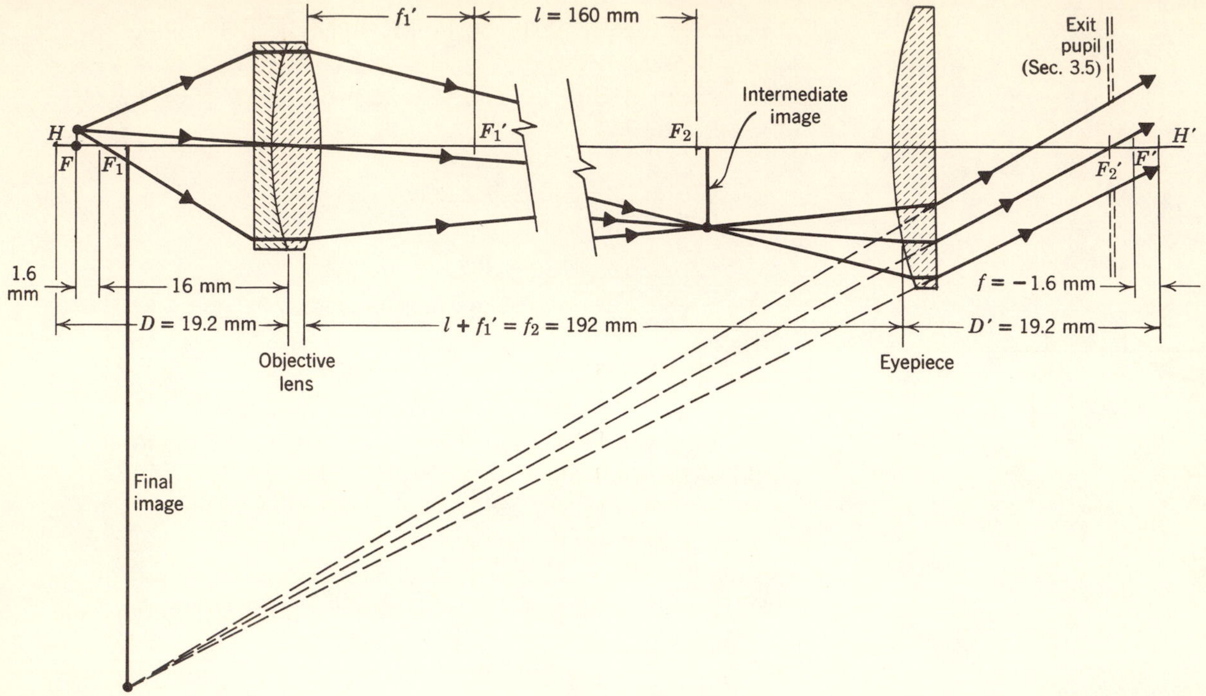

Fig. 3.32. The compound microscope.

magnification the eyepiece is acting like a simple magnifier having an object close to its first focal plane through F_2. The Newtonian distance $X_1' = S_1' - f_1'$ for the objective can be approximated by l, and the transverse magnification produced by the objective can be written

$$m_0 = \frac{-X'}{f_1'} = \frac{-l}{f_1'} \tag{3.100a}$$

In general, we have allowed the possibility of an immersion objective with $n \neq 1$ and $f_1 = nf_1' \neq f_1'$. The magnification of the eyepiece considered as a simple magnifier is, from (3.99),

$$m_e = \frac{250}{f_2} \tag{3.100b}$$

The magnification of the compound microscope can be obtained from Eqs. (3.99), (3.97*b*), and (3.100*a*, *b*):

$$m_m = \frac{250}{f} = \frac{250}{f_2}\left(\frac{-l}{f_1'}\right) = m_e m_0 \tag{3.101}$$

The magnification of a compound microscope is the product of the magnifications of eyepiece and objective.

F. NODAL POINTS

The transformation matrix between principal planes of an image-forming system becomes especially simple when the refractive indices in image and object space are the same. We obtain

$$\mathbf{M}_{HH'} = \begin{pmatrix} 1 & 0 \\ -\dfrac{1}{f} & 1 \end{pmatrix} \tag{3.102}$$

The transformation equations become

$$x' = x$$

$$\alpha' = -\frac{x}{f} + \alpha \tag{3.103}$$

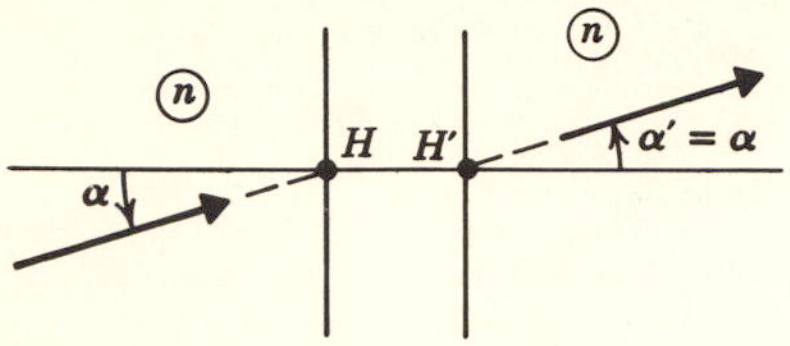

Fig. 3.33

The second of these equations implies that a ray headed for H (at $x = 0$) at angle α will leave H' at the same angle, that is, with unit angular magnification (Fig. 3.33).

In general, when $n \neq n'$ this property of unit angular magnification does not hold for the principal points. There is another pair of conjugate points lying on the axis of the system which has this property. They are called *nodal points*, and planes through them perpendicular to the axis are called *nodal planes*. Let us find these points.

The equation that applies is (3.78). We let the first nodal point N be a distance $S = l$ to the left of H and the second nodal point N' be a distance $S' = l'$ to the right of H'. Since the nodal points are to be conjugate, we must have

$$\frac{n'}{l'} + \frac{n}{l} = P \tag{3.104}$$

To obtain the unit angular magnification property, we set the (2, 2) matrix element equal to unity and obtain

$$n - lP = n'$$

or

$$\frac{n}{l} - P = \frac{n'}{l} \tag{3.105}$$

But, from (3.104),

$$\frac{n}{l} - P = -\frac{n'}{l'}$$

Thus

$$l = -l' \tag{3.106}$$

and, from (3.105),

$$\frac{n' - n}{l} = -P$$

$$l = \frac{n - n'}{P} = \frac{f(n - n')}{n} = \frac{f'(n - n')}{n'} \tag{3.107}$$

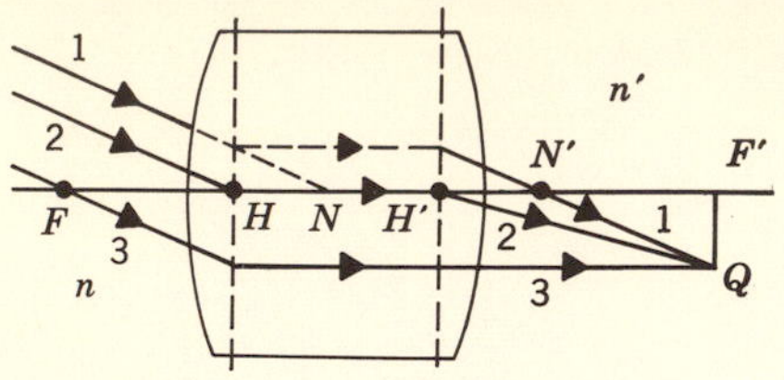

Fig. 3.34. Illustration of nodal points.

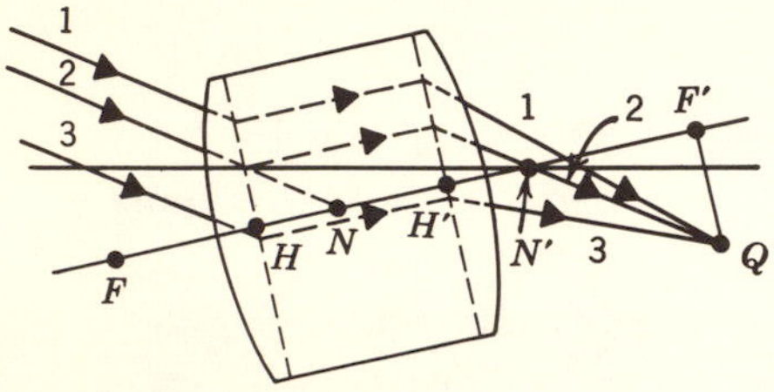

Fig. 3.35. The parallel beam in Fig. 3.34 remains focused at Q as the lens is rotated about its second nodal point N'.

This equation tells us that N is to the left of H when $n > n'$ and to the right of H when $n < n'$. Since $l = -l'$, the separation of the nodal points equals the separation of the principal points.

A case where $n < n'$ is shown in Fig. 3.34. Ray 1 is headed for N at an angle $(-)\alpha$. It behaves as shown and passes through the point N' also at angle $(-)\alpha$. If we construct two other rays parallel to ray 1, ray 2 aimed at H and ray 3 aimed at F, they will behave as shown, giving an image at Q in the secondary focal plane. Note that ray 1 travels parallel to its original direction when it leaves the lens. If the lens is rotated through a small angle about N', the same parallel incident beam will continue to be focused at Q. This will occur because there will always be a ray in the beam that is headed for N and which therefore must leave N' at the same inclination angle, as shown in Fig. 3.35. Thus the image formed with parallel, collimated light remains stationary as a lens is rotated about its nodal points. This property may be used to locate experimentally the nodal points of a lens using a device called a *nodal slide*.

G. TELESCOPIC SYSTEMS

1. Two-Lens Astronomical Telescope. A "telescopic system" may be defined as one for which $M_{21} = 0$. This makes $\alpha' = M_{22}\alpha$, independently of x. Now, in general, it is true that the (2, 1) matrix element is invariant under translations; that is, the (2, 1) matrix element of the matrix

$$\begin{pmatrix} 1 & D \\ 0 & 1 \end{pmatrix} \begin{pmatrix} M_{11} & M_{12} \\ M_{21} & M_{22} \end{pmatrix} \begin{pmatrix} 1 & D' \\ 0 & 1 \end{pmatrix}$$

is M_{21}. Thus, M_{21}, once zero, will remain zero (unless refracting surfaces are added to the system).

One of the simplest examples of a telescopic system is the astronomical telescope (Fig. 3.36), which consists of two positive

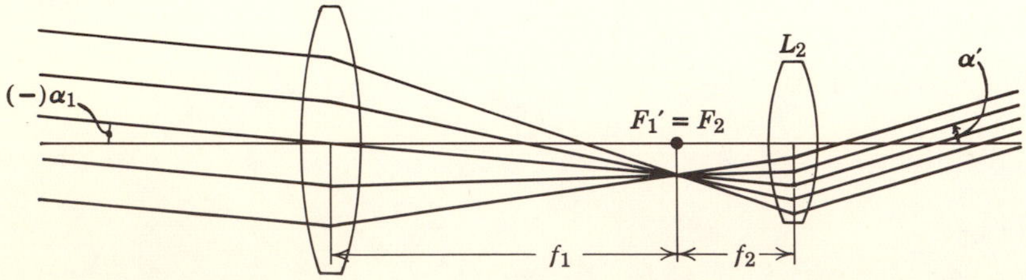

Fig. 3.36. Astronomical telescope.

lenses, the objective lens L_1 with focal length f_1 and the eye lens L_2 with focal length f_2 separated a distance $d = f_1 + f_2$. Then the total focusing power P is

$$P = \frac{1}{f_1} + \frac{1}{f_2} - \frac{d}{f_1 f_2} = 0$$

A parallel incident beam emerges from the telescope as another parallel beam making a different angle. The transformation matrix from L_1 to L_2, obtained from Eq. (3.88), is

$$\mathbf{M}_{L_1L_2} = \begin{pmatrix} -\dfrac{f_2}{f_1} & f_1+f_2 \\ 0 & -\dfrac{f_1}{f_2} \end{pmatrix} \tag{3.108a}$$

The transformation equations become

$$\begin{aligned} x' &= -\frac{f_2 x}{f_1} + (f_1+f_2)\,\alpha \\ \alpha' &= -\frac{f_1\alpha}{f_2} \end{aligned} \tag{3.108b}$$

We introduce the angular magnification or power p_α (not to be confused with the power P of an image-forming system) by

$$p_\alpha = -\frac{f_1}{f_2} \tag{3.109}$$

Then $\alpha' = p_\alpha \alpha$. *Here p_α gives the factor by which the telescope magnifies the angular separation of two distant objects.*

Then we have

$$\mathbf{M}_{L_1L_2} = \begin{pmatrix} \dfrac{1}{p_\alpha} & f_1+f_2 \\ 0 & p_\alpha \end{pmatrix} \tag{3.110}$$

We now look for an object-image relationship by asking for the transformation between a point A a distance Z to the left of L_1 and a point A' a distance Z' to the right of L_2. We obtain

$$\begin{aligned} \mathbf{M}_{AA'} &= \begin{pmatrix} 1 & Z' \\ 0 & 1 \end{pmatrix} \begin{pmatrix} \dfrac{1}{p_\alpha} & f_1+f_2 \\ 0 & p_\alpha \end{pmatrix} \begin{pmatrix} 1 & Z \\ 0 & 1 \end{pmatrix} \\ &= \begin{pmatrix} \dfrac{1}{p_\alpha} & f_1+f_2+Z'p_\alpha+\dfrac{Z}{p_\alpha} \\ 0 & p_\alpha \end{pmatrix} \end{aligned} \tag{3.111}$$

Thus, when

$$Z' = -\frac{Z}{p_\alpha^{\,2}} - \frac{f_1+f_2}{p_\alpha} \tag{3.112}$$

we obtain

$$\mathbf{M}_{AA'} = \begin{pmatrix} \dfrac{1}{p_\alpha} & 0 \\ 0 & p_\alpha \end{pmatrix} \tag{3.113a}$$

Thus the transformation equations are

$$x' = \frac{x}{p_\alpha}$$
$$\alpha' = p_\alpha \alpha \qquad (3.113b)$$

(so that the Lagrange equation $x'\alpha' = x\alpha$ remains satisfied). Then the transverse magnification is

$$m_x = \frac{1}{p_\alpha} = -\frac{f_2}{f_1} \qquad (3.114a)$$

and the angular magnification remains

$$m_\alpha = p_\alpha \qquad (3.114b)$$

The *longitudinal magnification* $\Delta Z'/\Delta Z$ can be obtained by differentiating (3.112):

$$\frac{\Delta Z'}{\Delta Z} = -\left(\frac{1}{p_\alpha}\right)^2 = -\frac{f_2^2}{f_1^2} \qquad (3.114c)$$

Note that these values for m_x, m_α, and $\Delta Z'/\Delta Z$ are independent of the object and image distances Z and Z'. Location of images by ray tracing is shown in Fig. 3.37.

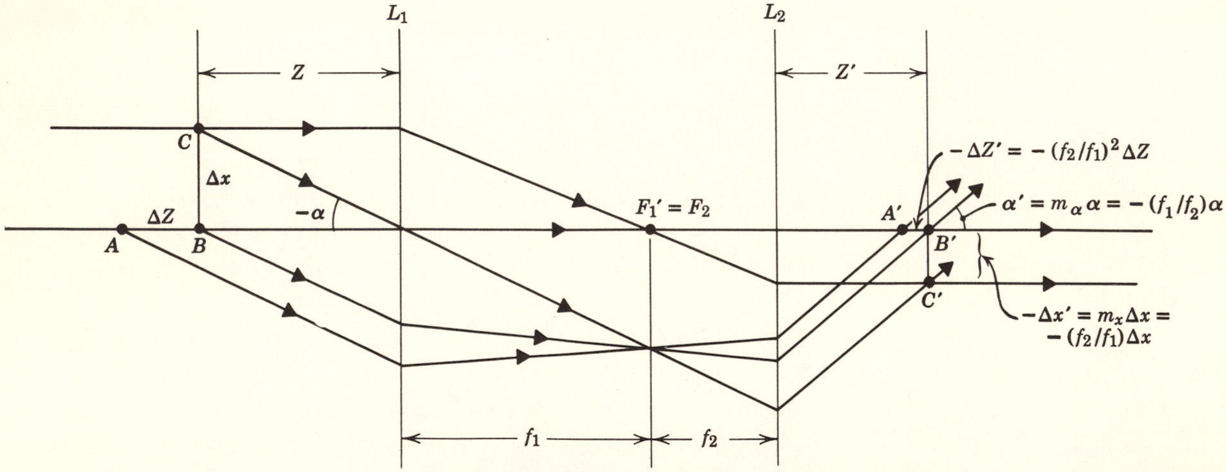

Fig. 3.37. Image formation with an astronomical telescope. (Here $f_1 = 2f_2$).

Note that these values for m_x, m_α, and $\Delta Z'/\Delta Z$ are independent of negative. In this case we have a *Galilean telescope* (Fig. 3.38). It gives a positive angular magnification $f_1/(-f_2)$.

*2. **General Telescopic System.*** The general telescopic system is quite similar to the two-lens system we have just discussed. The general form for the transformation matrix between two points A_1 and A_2 will then be

$$\mathbf{M}_{12} = \begin{pmatrix} M_{11} & M_{12} \\ 0 & M_{22} \end{pmatrix}$$

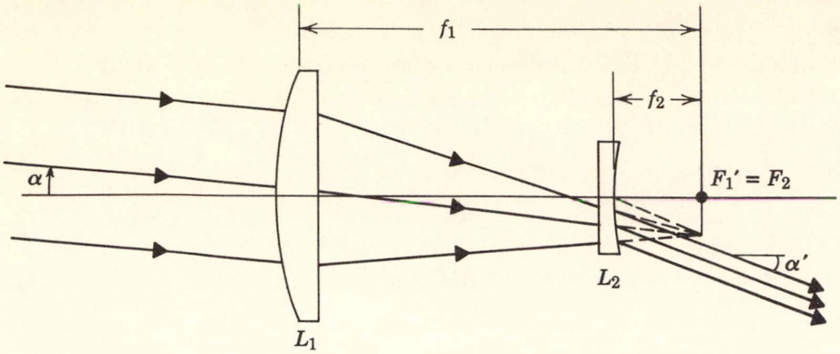

Fig. 3.38. Galilean telescope.

With the system in air, we must have

$$1 = \det \mathbf{M}_{12} = M_{11} \cdot M_{22}$$

We then define the power, or angular magnification p_α, by

$$p_\alpha = M_{22}$$

We then obtain

$$\mathbf{M}_{12} = \begin{pmatrix} \frac{1}{p_\alpha} & M_{12} \\ 0 & p_\alpha \end{pmatrix} \tag{3.115}$$

This has the same form as Eq. (3.110). An object-image relation is obtained between A and A' when

$$\mathbf{M}_{AA'} = \begin{pmatrix} 1 & D' \\ 0 & 1 \end{pmatrix} \begin{pmatrix} \frac{1}{p_\alpha} & M_{12} \\ 0 & p_\alpha \end{pmatrix} \begin{pmatrix} 1 & D \\ 0 & 1 \end{pmatrix} = \begin{pmatrix} \frac{1}{p_\alpha} & 0 \\ 0 & p_\alpha \end{pmatrix} \tag{3.116a}$$

provided that

$$D' = -\frac{D}{p_\alpha^{\,2}} - \frac{M_{12}}{p_\alpha} \tag{3.116b}$$

This is the condition that the (1, 2) element of the matrix $\mathbf{M}_{AA'}$ be zero.

A telescopic system is often used together with an image-forming system to form another image-forming system of longer focal length. An example is found in the telephoto lens (Fig. 3.39).

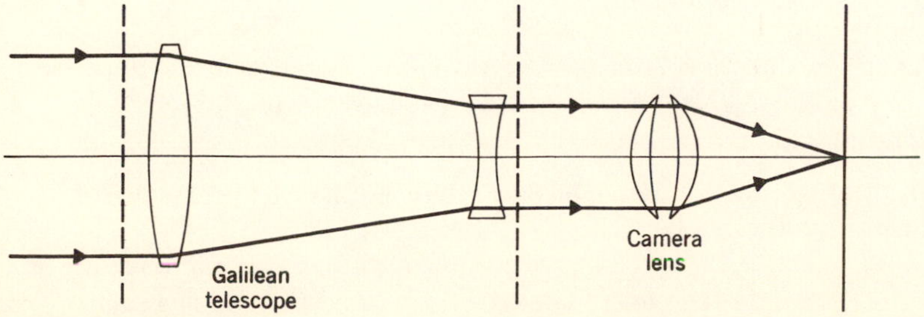

Fig. 3.39. Telephoto lens.

REFERENCES FOR SECTION 3.4

Much of the material of this section is based on the article by Klaus Halbach entitled "Matrix Representation of Gaussian Optics," *American Journal of Physics*, **32**, 90–108 (1964).

Other references on the matrix method:

W. Brouwer, *Matrix Methods in Optical Instrument Design.*
M. Herzberger, *Modern Geometrical Optics.*
E. L. O'Neill, *Introduction to Statistical Optics.*

PROBLEMS

3.4.1 Prove the associative law of matrix multiplication:

$$(\mathbf{AB})\mathbf{C} = \mathbf{A}(\mathbf{BC})$$

3.4.2 Find two 2×2 matrices **A** and **B** for which

$$\mathbf{AB} \neq \mathbf{BA}$$

3.4.3 Prove for two 2×2 matrices **A**, **B** that

$$\det(\mathbf{AB}) = (\det \mathbf{A})(\det \mathbf{B})$$

3.4.4 Verify directly by matrix methods that use of the matrix in Eq. (3.84*a*) will yield values of (x', α') for rays 1, 2, 3, 4 in Fig. 3.26 so that they behave as shown.

3.4.5 Verify the location of the principal planes shown in Fig. 3.30 for $n_2 = 1.50$.

3.4.6 Find the power and the locations of the principal planes for the lenses shown in Problems 3.1.5 and 3.1.6.

3.4.7 Find the power and the locations of the principal planes for a combination of two thin lenses each with the same focal length $f > 0$ separated by a distance d: (a) where $d = f$, (b) where $d = 3f/4$.

3.4.8 Consider a small glass sphere with index n, radius R. Find a value of n for which a virtual image is formed at infinity. What then is the resulting microscope magnification m_m?

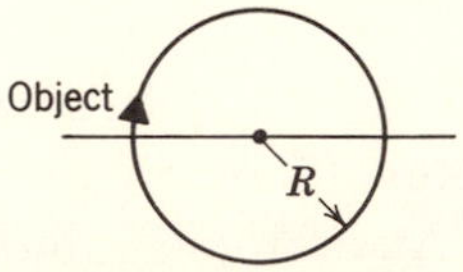

Van Leeuwenhoek microscope.

3.4.9 Show that the combination of two lenses having equal and opposite powers a finite, positive, distance d apart has a net positive power P, and find P as a function of d.

3.4.10 A microscope is assembled with elements having focal planes as shown.

a. Find the focal length and the location of the principal planes of the combination. This microscope is used to project a real image a distance 500 mm to the right of F_2'.

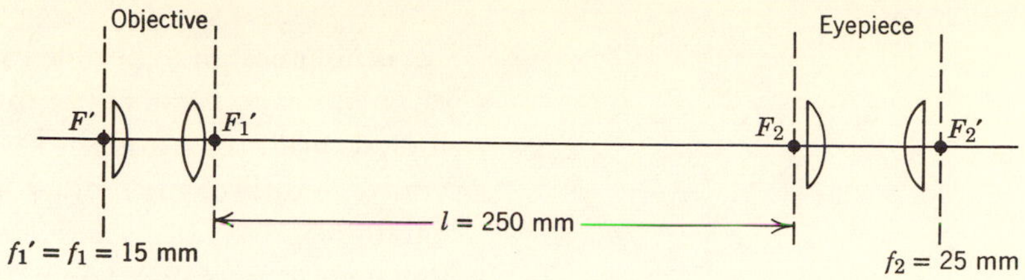

b. How far to the left of F_1 must the object be placed?
c. What is the overall lateral magnification?
d. Find the answers to (b) and (c) when the image is virtual and 250 mm to the left of F_2'.

3.4.11 An astronomical telescope (Fig. 3.36) is made of two thin positive lenses such that $p_\alpha = -10$ and $f_1 + f_2 = 1$ m. It is used with a lens separation $d = 1$ m. An observer looks through the telescope at L_2. The human eye can conveniently focus on virtual images located 250 mm or farther away.

a. What is the closest distance an object may be to L_1 and still be conveniently seen?
b. Find the answer to (a) when a Galilean telescope is used with $p_\alpha = +10$, $f_1 + f_2 = 0.8$ m, $d = 0.8$ m.

3.4.12 The telescope in Problem 3.4.11(a) is used as a doublet to observe an object 2 m to the left of L_1. The lens separation d is adjusted to give a virtual image located 0.3 m to the left of L_2. Find d and the overall magnification.

3.4.13 Consider a telephoto lens consisting of a Galilean telescope of angular power p_α together with a lens of focal length f. Show that the focal length of the combination is $p_\alpha f$ and locate the principal planes (see Fig. 3.39).

3.5 STOPS AND APERTURES

It is not enough to be able to predict the image position and size for an optical system. Other important properties of an image include its brightness and the size of the field of view. These topics require a study of the limitations of the spatial and angular extent of light beams by stops and apertures. To make such a study we shall have to introduce some new definitions and concepts. Even if the optical system under consideration is designed to accept light rays that are not paraxial, it is usually sufficient to treat stops and apertures using the methods of Gaussian optics.

A. APERTURE STOP AND PUPILS

Consider an object located on the axis of an optical system. Imagine the object to be emitting a cone of light of varying apex angle. For a small angle, all the light rays will pass through the system, but as the angle is increased, some of the rays will be stopped by the rim of a lens or a mechanical aperture somewhere in the system. The *aperture stop* (AS) is defined to be that stop or lens rim which physically limits the solid angle of rays passing through the system from an on-axis object point.

1. Simple System with the Aperture Stop in Object Space. A simple example of an aperture stop is shown in Fig. 3.40*a*. Here the hole in the screen limits the solid angle of rays from the object at P_0 that can pass through the system. It is the aperture stop.

The rays are cut off at A and B. The images of A and B are A' and B'. To an observer looking back through the lens from a position near P_0' it will appear as if A' and B' are cutting off the rays.

If we move the screen to the left of F, we have the situation shown in Fig. 3.40*b*. The screen still is the aperture stop, but the images A', B' of A and B are now to the right of the image P_0'. To an observer who moves sufficiently far to the right it still appears as if the rays are being cut off by A' and B'.

A "space" may be defined that contains all physical objects to the right of the lens plus all points conjugate to physical objects that are to the left of the lens. It is called *image space*. In Figs. 3.40*a*, and 3.40*b* all primed points are in image space.

The image of the aperture stop in image space is called the *exit pupil*. To an observer in image space it appears either as if the rays *converging* to an on-axis image P_0' are limited in solid angle by the exit pupil $A'B'$ as in Fig. 3.40*a*, or as if the rays *diverging* from P_0' are limited in solid angle by $A'B'$ as in Fig. 3.40*b*.

By analogy to image space, a space called *object space* may be defined that contains all physical objects to the left of the lens plus all points conjugate to any physical object that may be to the right

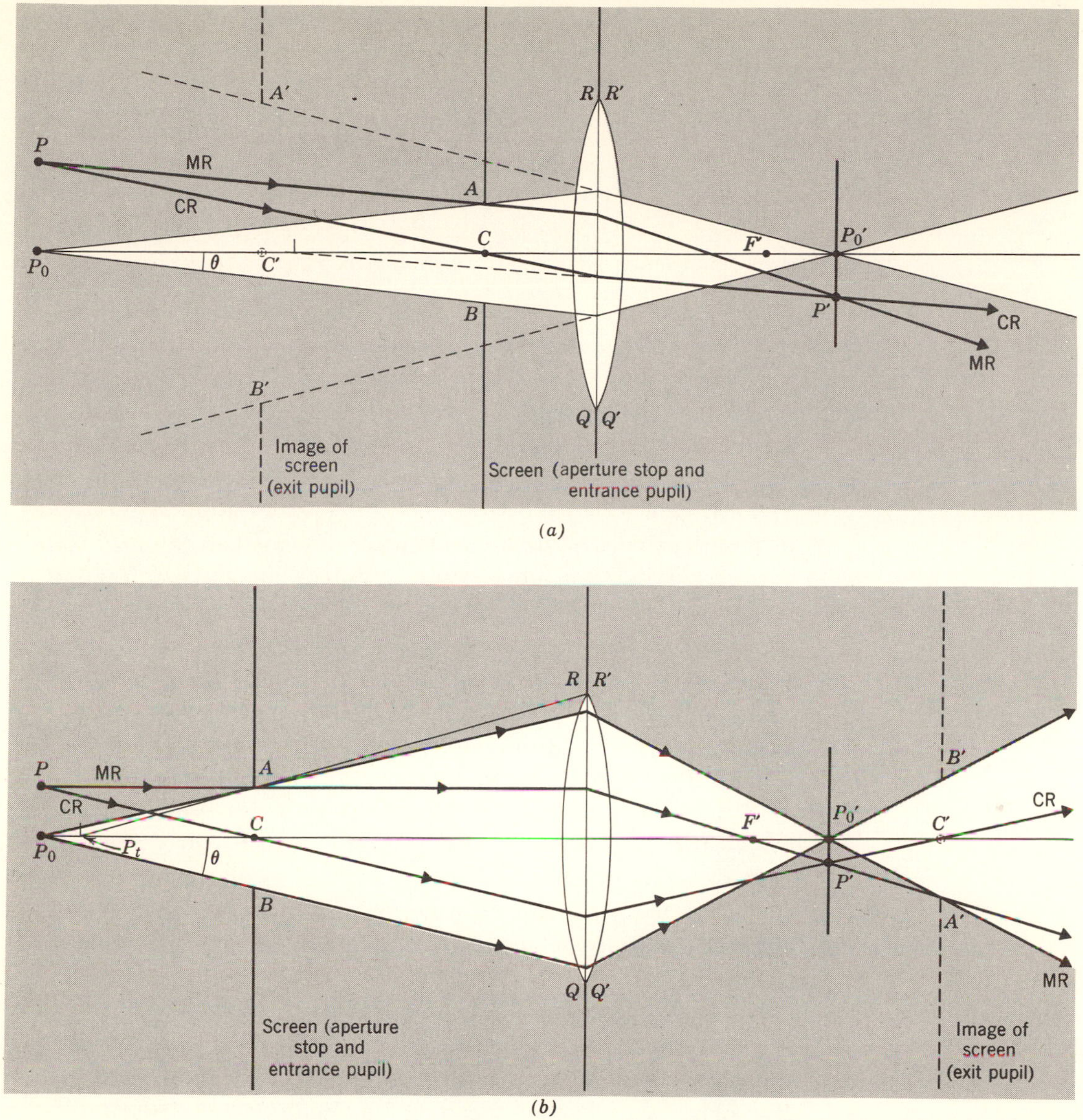

Fig. 3.40. *PA* will not generally be parallel to the axis.

of the lens. In Figs. 3.40*a* and 3.40*b* all unprimed objects are in object space. The first principal and focal planes of the lens are in object space; the second principal and focal planes are in image space.

The image of the aperture stop in object space is defined to be the *entrance pupil.* The aperture stop in Figs. 3.40*a* and 3.40*b* is already in object space; hence it itself is the entrance pupil.

The term "pupil" has to do with the fact that in a well-designed optical system intended for use with visual observation the exit pupil often coincides approximately with the pupil of the observer's eye. This could happen in Fig. 3.40*b* but not in Fig. 3.40*a*.

A ray from an off-axis object P through the center of C of the aperture stop is called a *chief ray* (CR). Its initial portion (extended if necessary) passes through the center of the entrance pupil. The conjugate ray $P'C'$ will (appear to) pass through the center C' of the exit pupil.

A ray such as PA from an off-axis object point through the edge of the aperture stop is called a *marginal ray* (MR). The conjugate ray $P'A'$ in image space appears to pass through the edge of the exit pupil.

Note that if the object plane in Fig. 3.40*b* were somewhat closer to the screen, the rays from P_0 would cease to be limited by AB but instead would be limited by the lens mount QR. Hence the latter would become the aperture stop. The transition point occurs for P_0 at P_t, where P_t, A, and R are colinear. For P_0 closer to the lens than this, its rim is the aperture stop.

Note that if AB is the aperture stop, the corresponding points $A'B'$ of the exit pupil subtend an angle at P_0' smaller than that from any other points in image space, such as $R'Q'$, that are on the edge of a potential aperture stop.

2. A Simple Example with the Aperture Stop in Image Space. Two simple cases where the aperture stop is in image space are shown in Figs. 3.41*a*, and 3.41*b*. It then is its own exit pupil. We continue to use the notation with primes for points in image space and no primes for points in object space. Thus the physical screen is at $A'B'$.

Note that rays from $A'B'$ to P_0' still subtend the smallest angle of all those subtended by rays from points in image space such as $R'Q'$ to P_0'. Similarly AB subtends a smaller angle at P_0 than RQ or any other aperture, if it exists in object space, or its image in object space.

3. The Aperture Stop in a Multilens System. In a multilens system some physical objects will be neither in object nor in image space. If a given point is imaged by all lens elements to its right, it will give an image in image space; if imaged by all elements to its left, it will give an image in object space.

An example of this more complicated situation is provided by the compound lens system of Fig. 3.42. The notation is the same as that used in Figs. 3.40 and 3.41 except that AB denotes the physical aperture stop, which is now neither the entrance pupil nor the exit pupil.

A systematic method of finding the entrance pupil is to image all stops and lens rims to the left through all intervening refracting elements of the system into object space. Find the solid angle subtended by each at P_0. The one with the smallest solid angle is the entrance pupil, and the physical object corresponding to it is

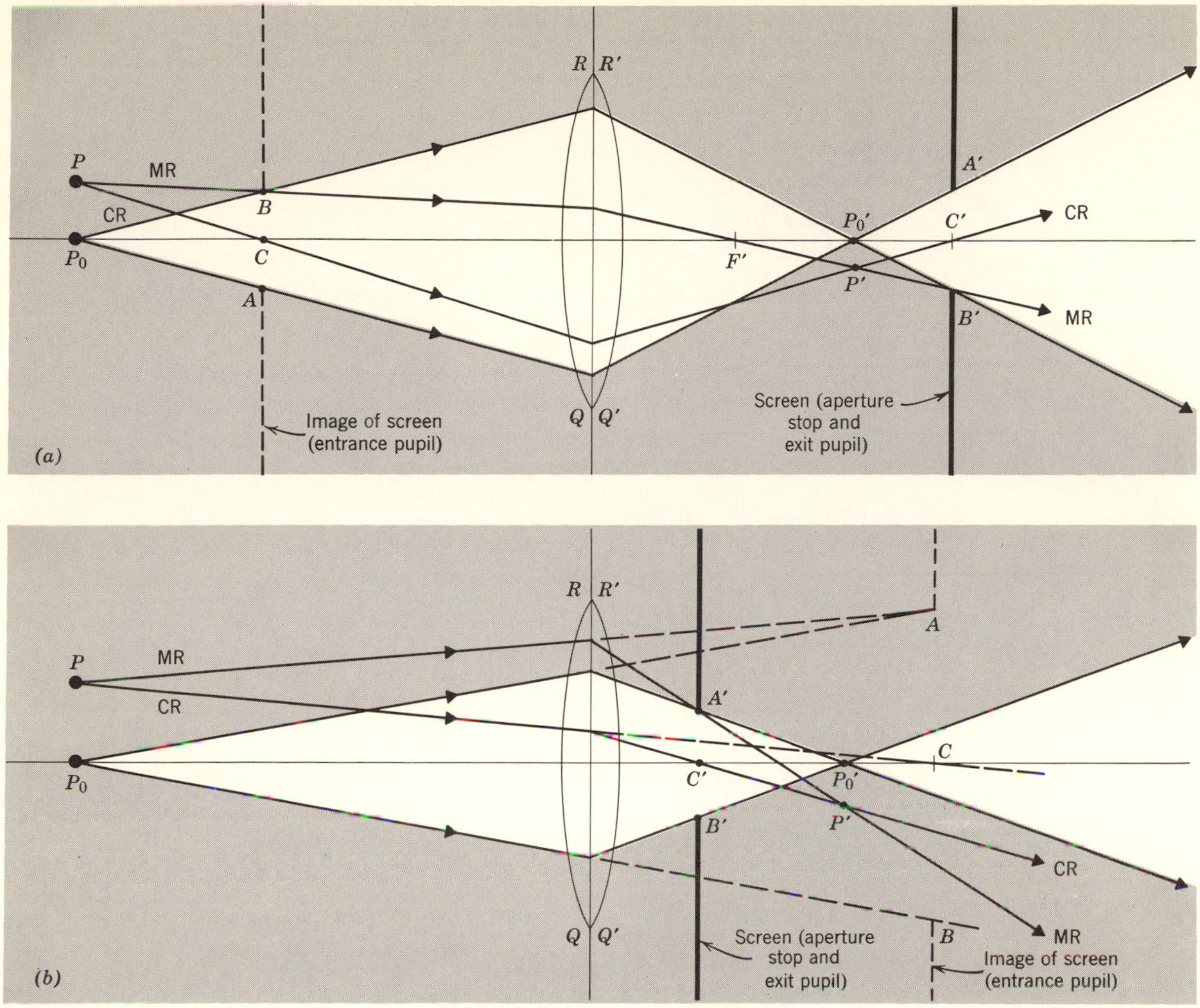

Fig. 3.41

the aperture stop. Alternatively we may just as correctly image all stops and lens rims to the right through all intervening refracting elements into image space and determine the solid angle subtended by each image at P_0'. The one with the smallest solid angle is the exit pupil, and the corresponding real physical object is the aperture stop.

*4. **Matrix Method of Locating the Aperture Stop.*** The matrix method may be useful in locating the aperture stop, especially when a one-dimensional use of angles rather than solid angles is sufficient. Let the object be at the plane that cuts the axis at A in Fig. 3.43. Then a ray from A that just misses the rim of the aperture stop will be the ray that just passes through the system. Let α_0 be the angle it makes with the axis. If $\mathbf{M}_{AB}$ represents the transformation matrix from the plane at A to the plane of the aperture stop,

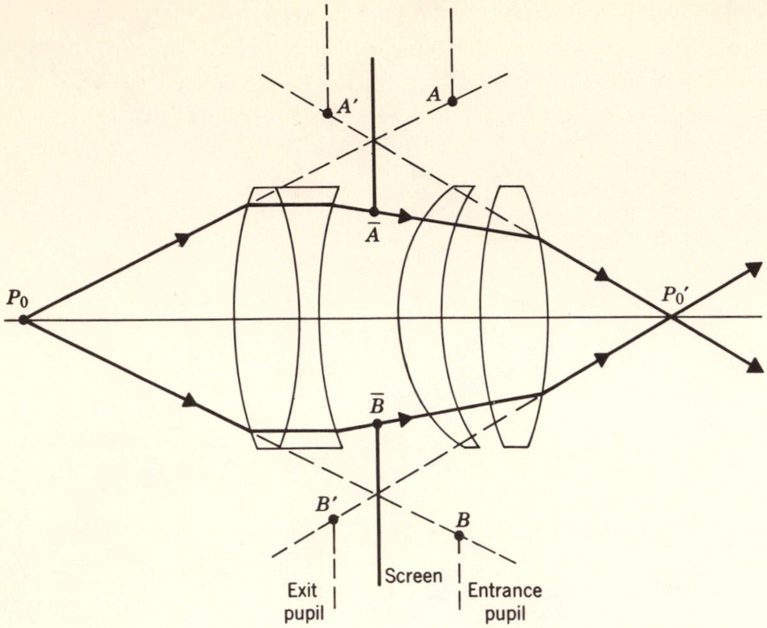

Fig. 3.42

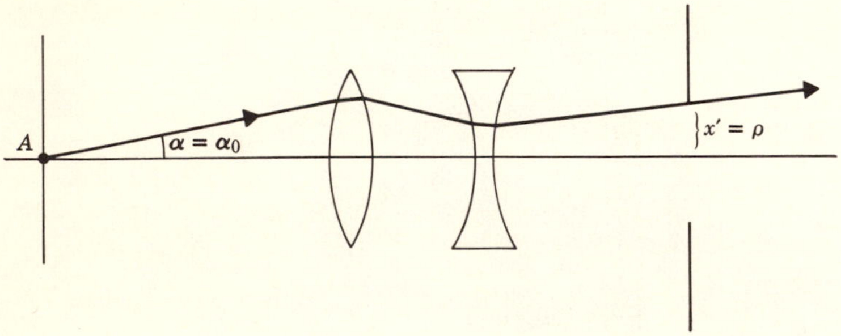

Fig. 3.43

then in general we have

$$\begin{pmatrix} x' \\ \alpha' \end{pmatrix} = \mathbf{M}_{AB} \begin{pmatrix} x \\ \alpha \end{pmatrix}$$

or

$$\begin{aligned} x' &= M_{11}x + M_{12}\alpha \\ \alpha' &= M_{21}x + M_{22}\alpha \end{aligned} \tag{3.117}$$

Now for the ray in question we have

$$x = 0, \qquad \alpha = \alpha_0, \qquad \text{and } x' = \rho$$

where ρ is the radius of the aperture stop. The first of Eqs. (3.117) then gives

$$\alpha_0 = \frac{\rho}{M_{12}} \tag{3.118}$$

Thus the aperture stop has the *smallest* ratio of radius ρ to M_{12} for all the stops or lens rims in the system.

*5. **An Example.*** As a simple example we consider the case illustrated in Fig. 3.44. The issue to be decided is whether L_1 or the

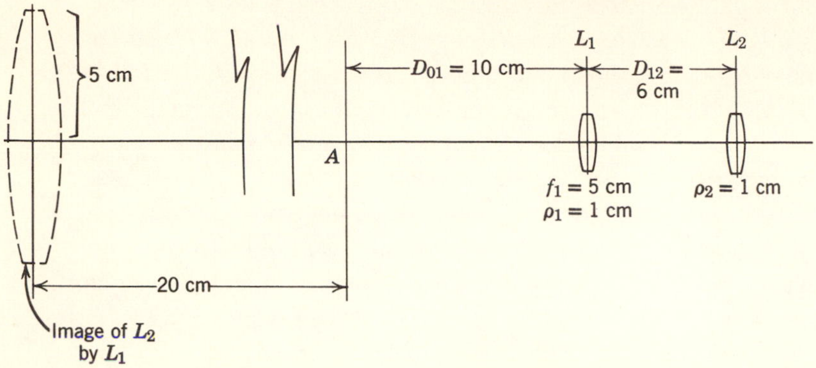

Fig. 3.44

image of L_2 subtends the smallest angle at A. Now L_1 subtends the half-angle

$$\alpha_1 = \frac{1}{10}\,\text{rad}$$

The *direct method* for L_2 is to find its image as formed by L_1. It is located a distance $S' = 30$ cm to the left of L_1 and is $1(S'/S) = 30/6 =$ 5 cm high. Hence the angle subtended *at A* is

$$\alpha_2 = \frac{5}{30-10} = \frac{1}{4}\,\text{rad}$$

Therefore, L_1 is the aperture stop and the entrance pupil.

The matrix method requires the calculation of the (1, 2) element of the matrix

$$\mathbf{M}_{AL_2} = \begin{pmatrix} 1 & D_{12} \\ 0 & 1 \end{pmatrix} \begin{pmatrix} 1 & 0 \\ -\dfrac{1}{f_1} & 1 \end{pmatrix} \begin{pmatrix} 1 & D_{01} \\ 0 & 1 \end{pmatrix} = \begin{pmatrix} -\frac{1}{5} & 4 \\ -\frac{1}{5} & -1 \end{pmatrix}$$

which is $M_{12} = 4$. Hence, by Eq. (3.118) we have

$$\alpha_2 = \frac{\rho_2}{M_{12}} = \frac{1}{4}\,\text{rad}$$

Thus, for the system of Fig. 3.44 the rim of L_1 is the aperture stop and the entrance pupil.

B. FIELD STOP AND WINDOWS

*1. **Definitions.*** It should be clear that the aperture stop determines the illumination of an on-axis point image, since the greater the solid angle of the transmitted light cone, the greater the transmitted light flux. The flux density in the off-axis parts of an image depends in part on the size and location of another stop, the field stop.

Chief rays (CR) are defined to be those passing through the center of the entrance pupil. *Marginal rays* (MR) are those passing through the edge of the entrance pupil (Figs. 3.40–3.42). Both kinds of ray will leave an off-axis object, but not all such rays may pass through to the image. The *field stop* is that stop or lens rim which limits the solid angle formed by chief rays. Its image in object space is called the *entrance window* (En.W.); its image in image space is called the *exit window* (Ex.W.).

If the entrance window is near the object plane, it tends to limit the *field of view*, that is, the lateral extent of the object that will be imaged by the system. To an observer in image space the exit window appears to limit the area of the image just as a window limits the area that an observer can see when he looks outdoors. The exit pupil, on the other hand, tends to limit the solid angle of rays that converge to each point in the image.

We have used the term "tends to" in the preceding paragraph, because in many optical systems the field of view is not sharply limited by the field stop; instead there is a gradual loss of light as we move farther off axis. In addition to this effect, and related to it, is another effect: a cone of light rays from an off-axis source to the entrance pupil will not necessarily be transmitted entirely. It can be *partially* cut off by the field stop or by other stops or lens rims in the system. This is called *vignetting.*

2. *An Example.* To illustrate these ideas, we consider an optical system consisting of two thin lenses of the same focal length f, where

$$PL_1 = 2f, \qquad L_1L_2 = 4f, \qquad L_2P' = 2f$$

There is an aperture stop right behind L_1 (Fig. 3.45). An on-axis object P is imaged first at P' and then at P''. The aperture stop and entrance pupil are at L_1. From the lens equation applied to the imaging of the aperture stop by L_2,

$$\frac{1}{S}+\frac{1}{S'}=\frac{1}{f}$$

with $S = 4f$, we find

$$\frac{S'}{S}=\frac{4}{3}\frac{f}{4f}=\frac{1}{3}$$

Thus the exit pupil will be one-third as large as the aperture stop.

Lens L_2 is the field stop and also the exit window in this case. Because of the symmetric placement of the lenses, the entrance window is one-third the size of L_2 and located a distance $(4/3)f$ to the left of L_1.

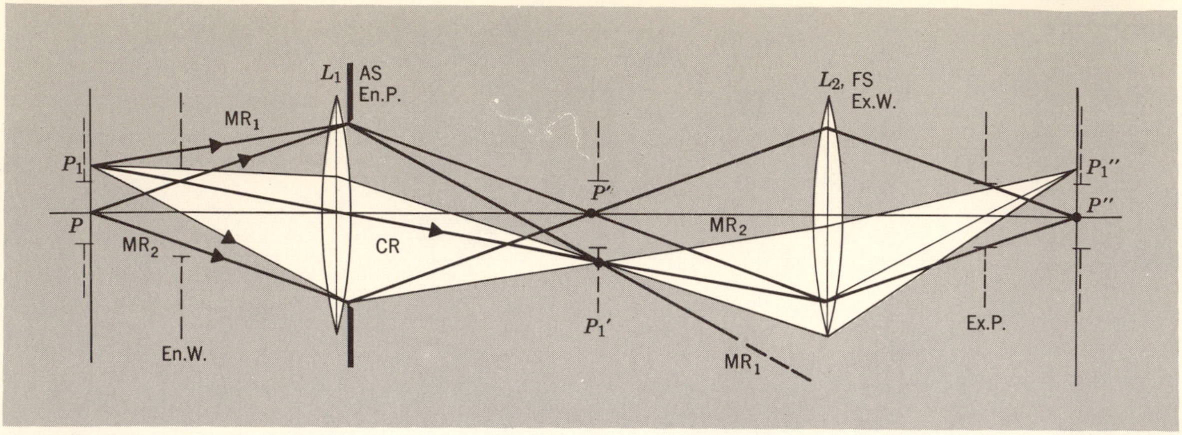

Fig. 3.45. Illustration of vignetting.

As P moves off axis, *vignetting* occurs. At P_1, only the bright bundle of rays passes through the system. The other rays such as MR_1 are lost by vignetting. If a uniformly illuminated screen were placed in the object plane at P, its image would be nonuniformly illuminated and would be roughly half as bright at P_1'' as at P''.

One can put a stop at the intermediate image point P' as shown by the dotted lines. If it subtends a smaller angle as seen from the center of the aperture stop L_1 than does L_2, it will become the field stop. The entrance window will now be in the object plane, and (in this example) it will be the same size as the field stop (both are shown dotted in Fig. 3.45). As P moves off axis toward P_1, the chief rays will now be cut off first by this new field stop, but some of the marginal rays will still miss L_2, and there will still be loss of intensity in the image plane. The *field of view* in the object plane is now sharply limited by the new entrance window, but we still have vignetting.

C. FIELD LENS

The vignetting in the above example can be completely eliminated by putting a *field lens* L_3 at P' (Fig. 3.46). Because of its location, it will not affect the imaging of light from the on-axis point P. It will deviate the whole light cone originating at an off-axis point P_1. This deviation will be most effective if L_3 focuses L_1 onto L_2, for a marginal ray leaving the bottom of L_1 will be deflected to the top of L_2, whereas without the field lens L_3, this marginal ray would completely miss L_2. If L_3 is stronger than necessary to focus L_1 onto L_2, then a marginal ray leaving the bottom of L_1 will be deviated so much that it misses the top of L_2.

In this optimum case, an entire cone of rays from an off-axis point in the object plane to the full entrance pupil is either completely successful in passing through the rest of the system or completely unsuccessful, depending on whether or not P_1 is within

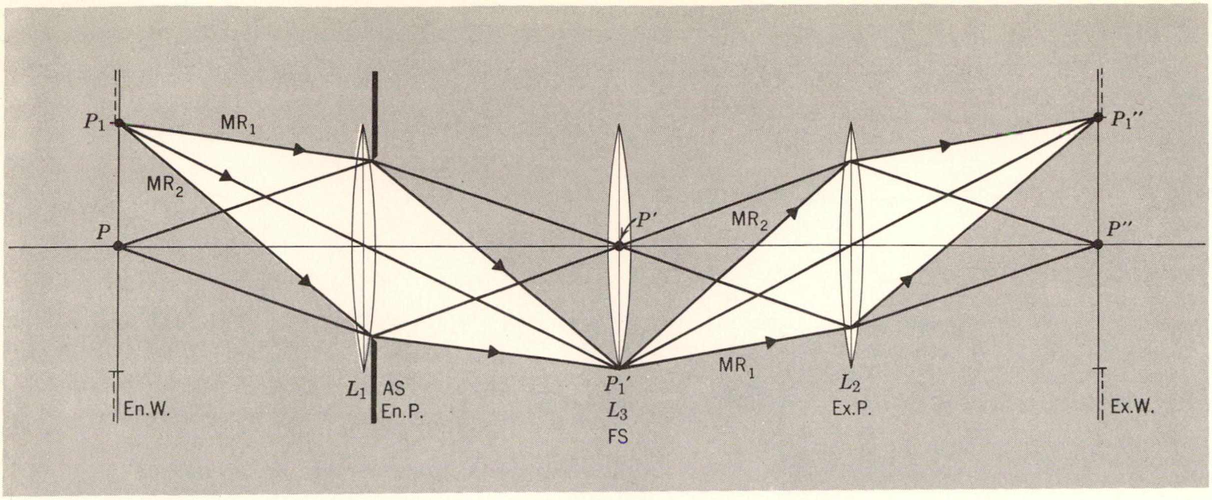

Fig. 3.46. A field lens L_3 eliminates all vignetting.

the entrance window. The field stop is at L_3, the entrance window is in the object plane, the field of view is sharply limited by the entrance window, and the brightness of the image matches, in a first approximation, that of the object. This will be discussed in detail in a later section.

The insertion of a field lens at key places in an optical system can make a marked difference in the ability of the system to transmit a useful amount of light. Of course, this can be done in a system made up of well-separated individual components, but not for a compound lens.

The concept of field stop is often not particularly useful in discussing a compound lens, such as that in Fig. 3.42. For such a lens the vignetting of a cone of rays from an off-axis point can be described by vignetting diagrams as shown in Fig. 3.47. We first image all ray-limiting apertures or lens rims into object space (or image space). Usually there are only three important limiting apertures, the aperture stop (AS), the rim of the first lens surface, and the rim of the last lens. The image in object space of the rim of the last lens is a circle we shall call the ILLR; the first lens rim (FLR) is already in object space. As the object point moves off the axis, the top marginal rays are cut off by the FLR, and the bottom marginal rays are cut off by the ILLR. The cone of light that passes through the system has a vertical section that is shown bright in the left half of each drawing.

To illustrate the limitations imposed on rays out of the plane of the paper, we can project the FLR and the ILLR from the object point onto the plane of the entrance pupil (En.P.). If the angles are not too large, these projections can be approximated by circles. (They really are ellipses.) We then find the area common to all three projections, shown bright on the right side of each drawing

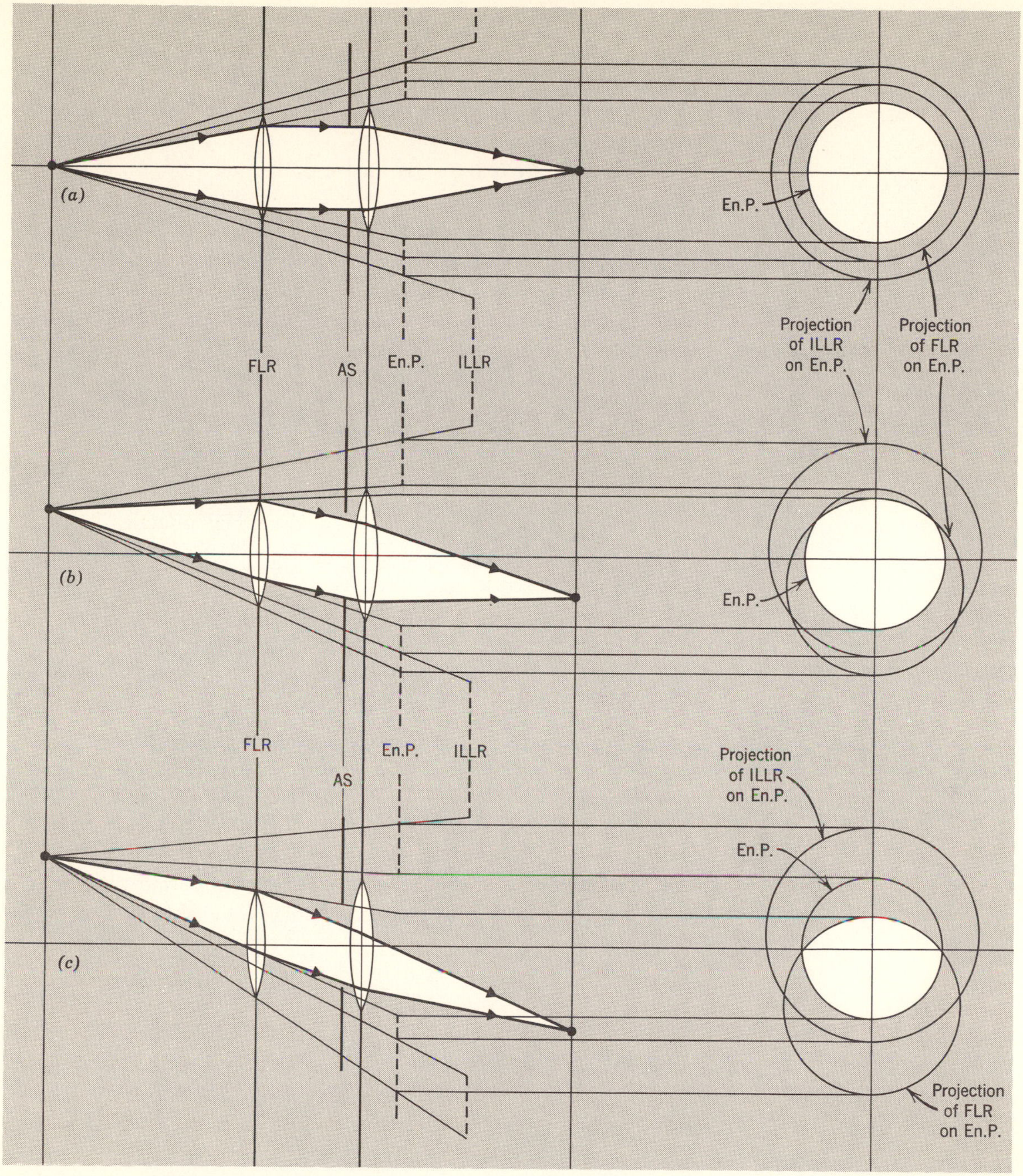

Fig. 3.47. Vignetting diagram for a doublet.

in Fig. 3.47. Then the light passing through their system consists of all rays from the object point that are headed for this common (bright) area in the plane of the entrance pupil.

The effect of stopping down the lens can be seen by decreasing the size of the circle that represents the entrance pupil. To obtain a uniform decrease in illumination (and, ordinarily, to minimize aberrations) it is desirable to have the center of the En.P. correspond to the center of the shaded area. This is not the case in Fig. 3.47, because the aperture stop is not centered between the two lenses, which in this example have equal power.

REFERENCES FOR SECTION 3.5

F. A. Jenkins and H. E. White, *Fundamentals of Optics*. Chapter 7.
J. Strong, *Concepts of Classical Optics*. Sections 15-7 through 15-11.
W. Brouwer, *Matrix Methods in Optical Instrument Design*. Chapter 5.
M. Herzberger, *Modern Geometrical Optics*. Pages 101–107.

PROBLEMS

3.5.1 A thick lens as shown is used in air. The first and second radii of curvature are $R_1 > 0$ and $R_2 < 0$, the index is $n > 1$, and the thickness $|V_1V_2|$ is d. What will be the aperture stop for this lens for an axial object at a general distance S_1 to the left of V_1? Is the aperture stop always the same? (No calculation is necessary to solve this problem.)

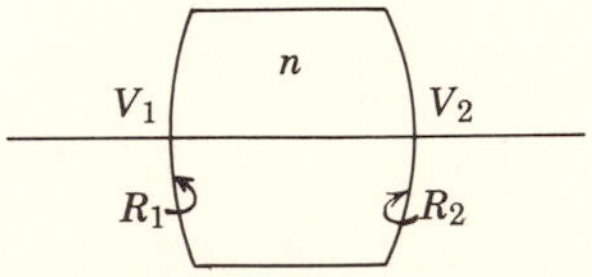

3.5.2 The Coddington magnifier is cut as shown from a single sphere.

a. Where must an object be held so that a virtual image is formed 25 cm in front of the lens?
b. Under the conditions of (a), find the entrance and exit pupils.

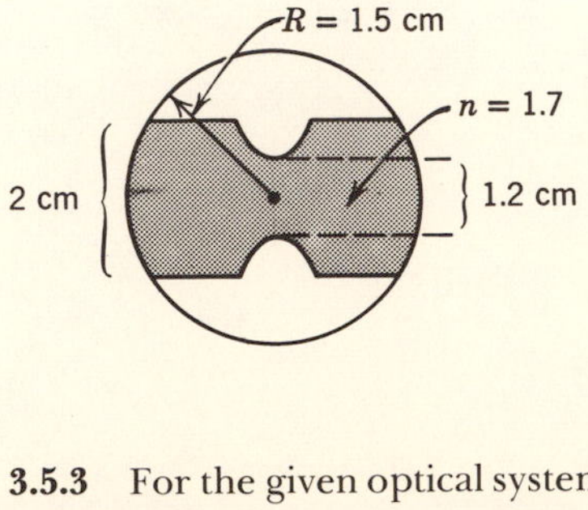

3.5.3 For the given optical system, located (a) the image, (b) the position and size of the entrance window.

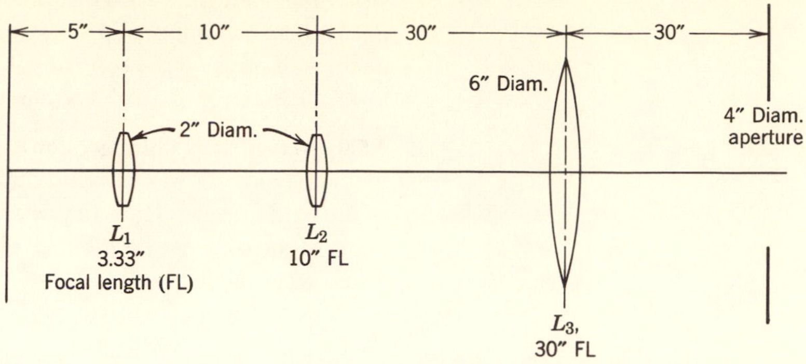

3.5.4 A thin lens L_1 with a 5.0-cm diameter aperture and focal length +4.0 cm is placed 4.0 cm to the left of another lens L_2 4.0 cm in diameter with a focal length of +10.0 cm. A 2.0-cm-high object is located with its center on the axis 5 cm in front of L_1. There is a 3.0-cm diameter stop centered halfway between L_1 and L_2. Find analytically the position and size of (a) the entrance pupil, (b) the exit pupil, (c) the image. Make a brief sketch to scale.

3.5.5

$$f_1 = 12 \text{ cm} \quad R = 0.9 \text{ cm} \quad f_2 = 12 \text{ cm}$$
$$R_1 = 1.0 \text{ cm} \qquad\qquad R_2 = 1.0 \text{ cm}$$

a. Locate the aperture stop for the given system for an object at A as a function of the distance S where S ranges from infinity to zero.
b. For each value of S in (a), find the field stop. The lenses are thin.

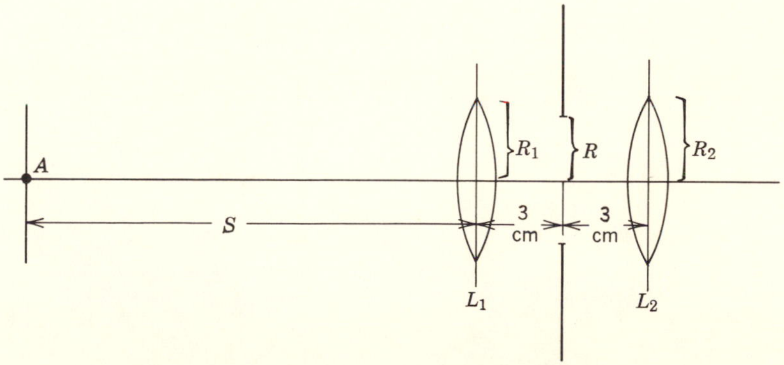

3.5.6 Standard "35-mm" cameras have a 24 × 37 mm aperture just in front of the film that acts as a field stop. Consider a camera with a 50-mm-focal-length lens. What is the size of the field of view for an object distance of 1 m? 30 m?

3.5.7 The "F-number" of a lens is defined as the ratio of the focal length to the diameter of the entrance pupil with the iris diaphram (if any) wide open and with a distant object. Consider an $F/2.8$ lens, that is, one with an F-number of 2.8, having a focal length of 45 mm. What will be the half-angle of the cone of light converging from the exit pupil toward the image of an object at a distance of 0.50 m? 5.0 m? 50 m?

3.5.8 A field stop 4 mm in diameter is placed at the position of the intermediate image in the compound microscope of Fig. 3.32. What is the size of the field of view? The exit pupil is 4 mm in diameter. What is the half-angle of the cone of rays from an object point, if the cone just fills the entrance pupil?

In this chapter we discuss two additional topics of importance for real optical systems. The first has to do with the intensity of radiation, particularly visible radiation in the system. The second deals with the departure of the performance of real optical systems from the predictions of Gaussian optics, that is, with aberrations.

CHAPTER 4

ADDITIONAL TOPICS RELATED TO PRACTICAL PROBLEMS IN GEOMETRICAL OPTICS

CHAPTER 4 CONTENTS

4.1 RADIOMETRY AND PHOTOMETRY

A real optical system must not only produce an image of the desired size at the right place, but the image must be bright enough to be useful. Image brightness will depend on object brightness and on the size and location of stops within the optical system. Before quantitative calculations can be performed, some definitions and other relationships must first be discussed.

A. PHYSICAL OR RADIOMETRIC NOMENCLATURE

1. Definitions. In optics there are two types of units for energy-related quantities, physical and psychophysical. In our notation the subscript e (for energy) will denote physical units, and v (for visual) psychophysical units. We shall discuss the physical system first, and shall use the m.k.s. system or a hybrid system involving watts and centimeters. *Radiant Energy* Q_e in joules refers to the total amount of energy emitted, transferred, or collected in a radiation process. The *radiant energy density* U_e in joules/m^3 is the radiant energy contained in a unit volume of space. Radiant power or *radiant flux* Φ_e in watts is the time rate of change, or rate of transfer, of radiant energy.

The above three quantities refer to total radiation of all wavelengths. A *spectral version* of each may be defined, for example, a *spectral flux* $\Phi_e(\lambda)\,d\lambda$ represents the flux in a wavelength range between λ and $\lambda + d\lambda$. Wavelengths are often measured in nanometers nm (also called millimicrons mμ): 1 mμ = 1 nm = 10^{-9} m = 10^{-7} cm = 10 angstroms. Another convenient unit is the micron = 10^{-6} m. Thus, $\Phi_e(\lambda)$ would be in watts per nanometer.

We obtain by integration

$$\Phi_e = \int_0^\infty \Phi_e(\lambda)\,d\lambda$$

$$Q_e = \int_0^\infty Q_e(\lambda)\,d\lambda \qquad (4.1)$$

$$U_e = \int_0^\infty Ue(\lambda)\,d\lambda$$

Similar spectral versions of the quantities to be discussed below may be defined.

The flux of radiant energy flowing across unit area of a real or imaginary surface is often of interest. Our nomenclature distinguishes source surfaces from other surfaces. The *radiant exitance* M_e (in watts/m^2) is the total emitted flux density, that is, the

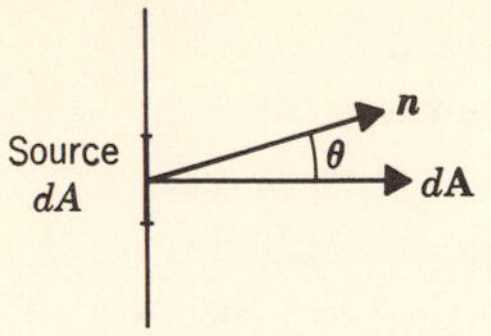

Fig. 4.1. The radiant exitance is the power per unit area leaving dA and going to the right. It includes all propagation directions θ from 0 to $\pi/2$.

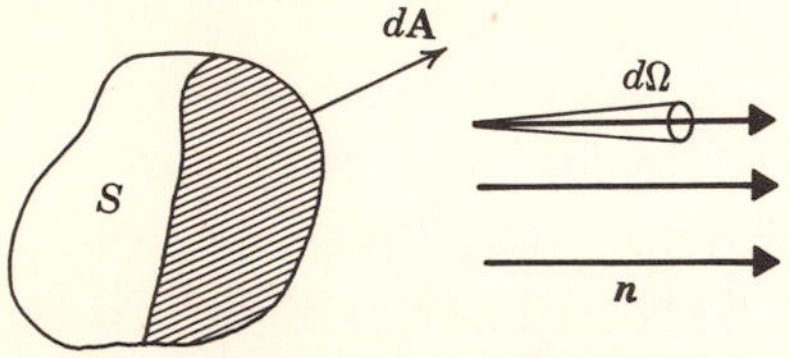

Fig. 4.2. The radiant intensity $I(\mathbf{n})$ represents the total power per unit solid angle radiating in direction **n** from a finite source S.

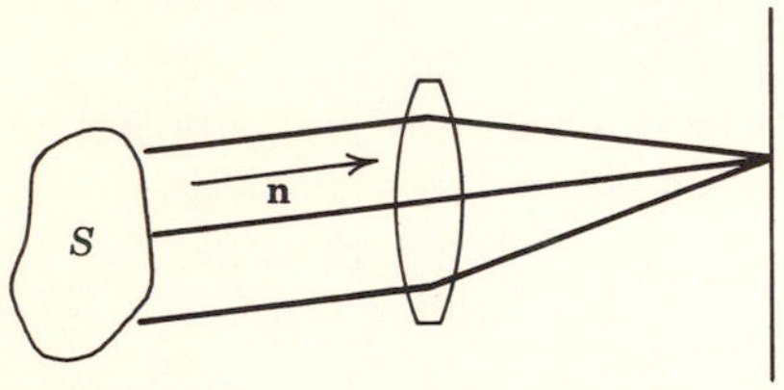

Fig. 4.3. All rays in the same direction **n** are focused at the same point by the lens.

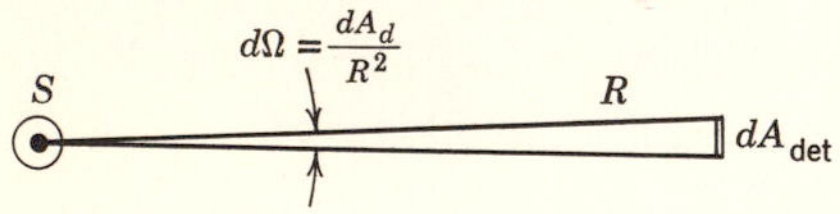

Fig. 4.4

flux emitted per unit area of a source of radiation (Fig. 4.1). The value M_e may vary with position on the surface of the source. Similarly the *radiant incidence* E_e (in watts/m^2) is the incident radiant flux density or the flux per unit area incident onto a real or imaginary surface.* This is proportional to the square of the electric field strength as given by Eqs. (7.38) and (7.38′) in Chapter 7. Thus we have for the flux to or from area dA

$$d\Phi_e = M_e\, dA \qquad \text{(out from } dA) \qquad\qquad (4.2)$$

$$d\Phi_e = E_e\, dA \qquad \text{(onto } dA)$$

The *radiant intensity* I_e in watts/steradian represents the flux per unit solid angle radiated by an entire source in a given direction (represented by unit vector **n**). This is shown in Fig. 4.2. The intensity is of interest if the source is far away and if one has a small detector so that all rays from any part of the source to any part of the detector are essentially parallel. It is also of interest if the light is studied in the focal plane of a lens, for then all rays from the source with a given **n** are focused at the same point (Fig. 4.3). If we consider a cone of rays making a solid angle $d\Omega$ about **n** (Fig. 4.2), then the flux is given by

$$d\Phi_e = I_e\, d\Omega \qquad (4.3)$$

The intensity I_e will, in general, depend on the direction of **n**.

For a uniform spherical source of radius R_0 emitting total flux $\Phi_{e,\text{tot}}$, we obtain, by integrating over the entire surface,

$$\Phi_{e,\text{tot}} = \iint M_e\, dA = 4\pi R_0^2 M_e \qquad (4.4a)$$

Then a small detector of area dA_{det} placed as shown in Fig. 4.4 will receive flux

$$d\Phi_e = \Phi_{e,\text{tot}} \frac{dA_{\text{det}}}{4\pi R^2} = \frac{\Phi_{e,\text{tot}}}{4\pi}\frac{dA_{\text{det}}}{R^2} = \frac{\Phi_{e,\text{tot}}}{4\pi}\, d\Omega \qquad (4.4b)$$

By comparison with Eq. (4.3) we obtain

$$I_e = \frac{\Phi_{e,\text{tot}}}{4\pi} \qquad (4.5)$$

for an isotropic source. For such a source, assumed spherical, we also obtain, using (4.4*a*),

$$I_e = R_0^2 M_e \qquad (4.6)$$

*This quantity is also called "irradiance". We shall use it or "flux density" in most of what follows. Physicists like to use "intensity," but this is technically incorrect in optics.

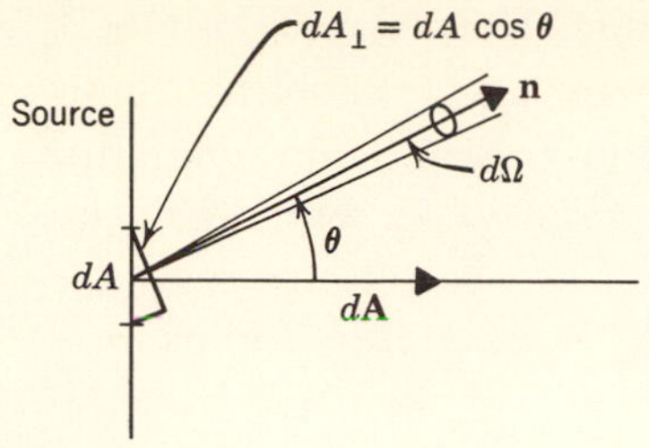

Fig. 4.5

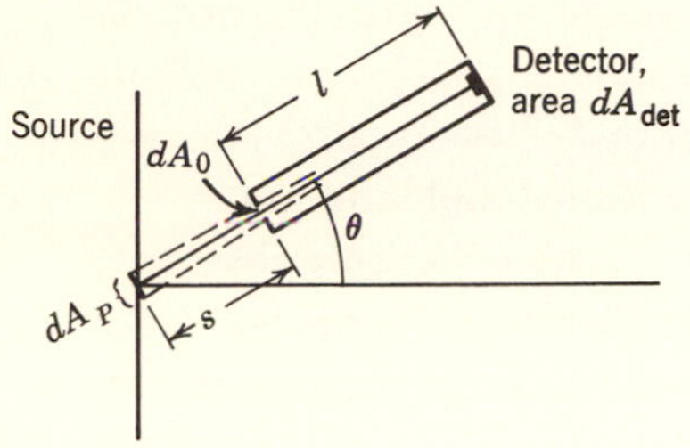

Fig. 4.6

The emission properties of real sources often vary over their surfaces. Thus the radiant intensity I_e, which describes the integrated dependence of the emission integrated over the surface, may be misleading. We therefore define a differential angular dependence, the *radiance* L_e in watts/(m^2-ster), defined as the flux per unit solid angle per unit *projected* area. Consider Fig. 4.5. The flux emitted by dA of the source into an element of solid angle $d\Omega$ about the direction $\mathbf{n}$ is written

$$\begin{aligned} d^2\Phi_e &= L_e d\Omega dA_\perp = L_e d\Omega dA\cos\theta \\ &= L_e\, d\Omega(\mathbf{n}\cdot d\mathbf{A}) \end{aligned} \tag{4.7}$$

Here $d\mathbf{A}$ is a vector pointing along the normal to the surface of the source with magnitude dA equal to the area of the surface element. In general, L_e will be a function of $\mathbf{n}$ and of position $\mathbf{r}$ on the surface of the source, that is, $L_e = L_e(\mathbf{r}, \mathbf{n})$.

In principle the radiance L_e may be measured by means of the apparatus illustrated in Fig. 4.6. The small opening or aperture has area dA_0. Project this area onto the source using straight lines from a point on the detector. The projected area at the source is then $dA_p = dA_0(l+s)^2/l^2$. The solid angle subtended by the detector as seen from the source is $d\Omega = dA_{det}/(l+s)^2$. Thus, $dA_p d\Omega = dA_0 dA_{det}/l^2$. Hence, if the detector is calibrated to read in watts, we take its output $d^2\Phi_e$ and obtain the radiance from the equation

$$L_e = \frac{d^2\Phi_e}{dA_p\cdot d\Omega} = \frac{d^2\Phi_e l^2}{dA_0 dA_{det}} \tag{4.8}$$

The concept of radiance may be extended to apply to objects that are not physical sources of radiation. For instance, we may speak of the radiance of an image formed in an optical system, or even of the radiance in a system away from object and image positions. In all cases Eq. (4.7) or Eq. (4.8) applies, and in principle the radiance may be measured by the apparatus of Fig. 4.6.

The radiant exitance M_e of a source may be obtained by integration of Eq. (4.7) over angle, for fixed dA:

$$d\Phi_e = \iint_{(\frac{1}{2})} d^2\Phi_e = dA \iint_{(\frac{1}{2})} L_e\cos\theta\, d\Omega$$

Hence

$$M_e = \iint_{(\frac{1}{2})} L_e\cos\theta\, d\Omega \tag{4.9}$$

Here $\iint_{(1/2)} (\cdots)\, d\Omega$ refers to an angular integration over one-half of the full solid angle, that is, over all outward directions from the surface.

Similarly the intensity I_e may be obtained by integration of Eq. (4.7) over area, for fixed **n** and $d\Omega$:

$$d\Phi_e = \iint_{\text{area}} d^2\Phi_e = d\Omega \iint_{\text{area}} L_e \cos\theta \, dA = \mathbf{n} d\Omega \cdot \iint_{\text{area}} L_e \, d\mathbf{A}$$

Hence

$$I_e = \iint_{\text{area}} L_e \cos\theta \, dA = \mathbf{n} \cdot \iint_{\text{area}} L_e \, d\mathbf{A} \tag{4.10}$$

Here the integral extends over that part of the surface which may radiate in direction **n**, for example; over the shaded area in Fig. 4.2.

We remind the reader that spectral versions of these quantities may be defined, for example, $L_e(\lambda)$, $I_e(\lambda)$.

2. *Lambert Source.* A *Lambert source* or *Lambert emitter* is one which has the same directional properties as a small hole in a cavity. Within a cavity, the light is completely randomized; it propagates everywhere in all directions with equal radiance. Thus, for a Lambert source we have

$$L_e(\mathbf{r}, \mathbf{n}) = L_e, \qquad \text{independent of } \mathbf{r} \text{ and } \mathbf{n} \tag{4.11}$$

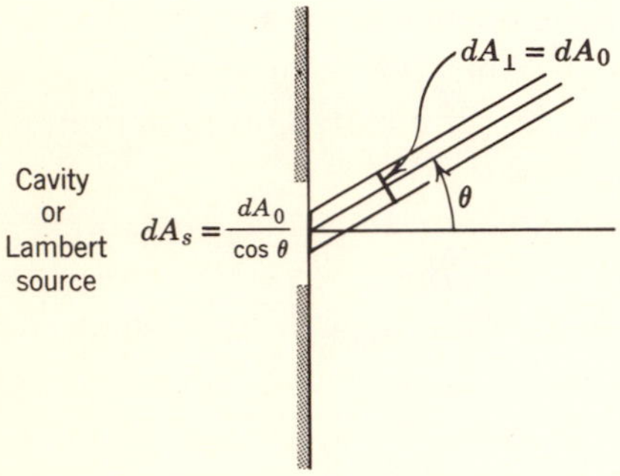

Fig. 4.7

For a Lambert source the apparatus in Fig. 4.6 would produce a detector output independent of position and angle θ, as long as the field of view of the detector is limited by the opening of area dA_0 and not by the edge of the source (Fig. 4.7).

For a Lambert source, we may take L_e outside of the integral in Eq. (4.9). The solid angle $d\Omega$ may be taken to be $2\pi \sin\theta \, d\theta$, the area of an annulus in the unit sphere, as shown in Fig. 4.8. We then obtain (*for a Lambert source only*)

$$M_{e,L} = L_e \int_0^{\pi/2} \cos\theta \, 2\pi \sin\theta \, d\theta = \pi L_e \int_0^{\pi/2} d(\sin^2\theta) = \pi L_e \tag{4.12}$$

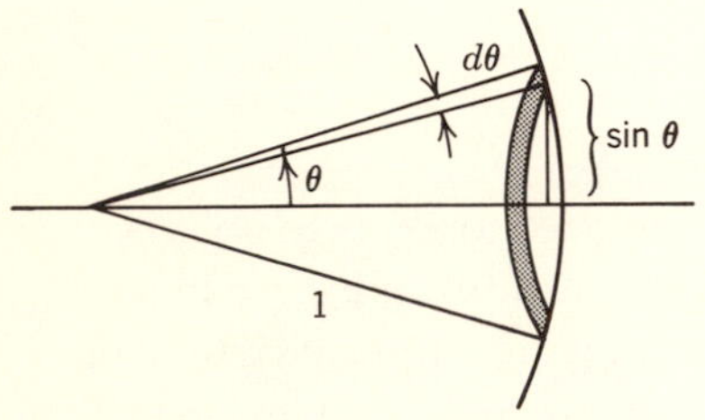

Fig. 4.8. The infinitesimal solid angle $d\Omega$ is the shaded area on the unit sphere and equals $2\pi \sin\theta \, d\theta$.

To obtain another relationship for a Lambert source, consider Fig. 4.9. Let a detector of area dA_{det} be a distance r from a small Lambert source of area dA_s. The orientations are as shown. The solid angle $d\Omega$ subtanded by dA_{det} at the source is given by

$$d\Omega = \frac{dA_\perp}{r^2} = \frac{dA_{\text{det}} \cos\phi}{r^2}$$

Thus the flux transferred is given by

$$d^2\Phi_e = L_e dA_s \cos\theta \, d\Omega = L_e \frac{dA_s \cos\theta \, dA_{\text{det}} \cos\phi}{r^2} \tag{4.14}$$

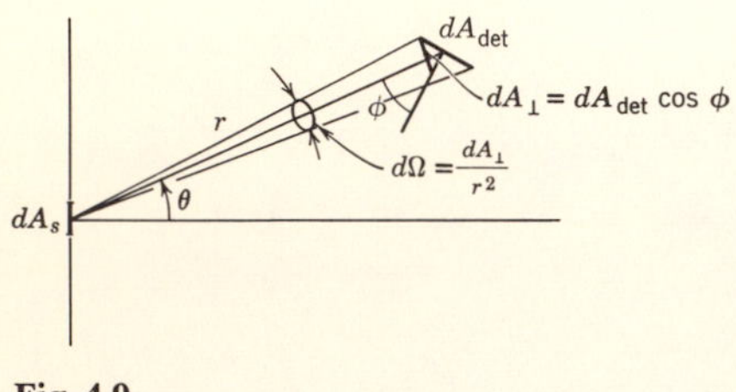

Fig. 4.9

For a Lambert source the $\cos\theta$ angular dependence of $d^2\Phi_e$ in Eq. (4.7) or (4.14) is known as *Lambert's law of cosines*. For a non-Lambert source we must use $L_e = L_e(\theta)$, but Eq. (4.14) is still valid.

Note the complete reciprocity between source and detector in Eq. (4.14). Because of it, the equation may be written

$$d^2\Phi_e = L_e dA_{\text{det}} \cos\phi \, d\Omega'$$

where

$$d\Omega' = dA_s \frac{\cos\theta}{r^2} \tag{4.14'}$$

Then L_e may be interpreted as the radiance incident onto the detector at the angle ϕ, as well as the radiance leaving the source at the angle θ. This interpretation of Eq. (4.14) is not restricted to Lambert sources.

When necessary, any of the above relationships may be applied to the corresponding spectral quantities.

3. Connection Between Radiance and Energy Density in a Cavity. Let U_e be the energy per unit volume in the cavity. Because of the complete randomness of the propagation direction, we can assume that in a given volume dV, light of energy

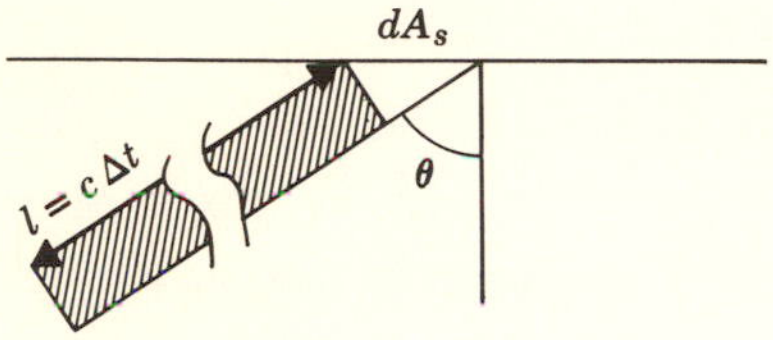

Fig. 4.10

$$U_e dV \frac{d\Omega}{4\pi}$$

is propagating into a cone of solid angle $d\Omega$ about any desired direction. Consider the volume to be the tube shown shaded in Fig. 4.10. In time Δt all the energy in the shaded volume propagating in the direction of the axis of the tube will have passed out of the cavity. The volume dV is clearly $dV = c\Delta t \cdot dA_s \cos\theta$. Thus the flux crossing dA_s in the direction of the tube is given by

$$d^2\Phi_e = \frac{1}{\Delta t} U_e \, dV \frac{d\Omega}{4\pi} = \frac{U_e}{\Delta t} c \, \Delta t \cdot dA_s \frac{\cos\theta}{4\pi} d\Omega$$

$$= \frac{U_e c}{4\pi} d\,A_s \cos\theta \, d\Omega \tag{4.15}$$

From a comparison with Eq. (4.7) we see that the radiance L_e is related to U_e by

$$L_e = \frac{U_e c}{4\pi} \tag{4.16}$$

We also have the relation

$$U_e = \frac{4M_e}{c} \tag{4.17}$$

These relationships also hold for the appropriate wavelength-dependent quantities $U_e(\lambda)$, $M_e(\lambda)$, $L_e(\lambda)$.

When the cavity is an isothermal enclosure at a temperature T, a small hole in it will emit *black body radiation*. The spectral exitance is then given by *Planck's formula*,

$$M_e(\lambda) = \frac{c_1\lambda^{-5}}{\exp(c_2/\lambda T) - 1} \tag{4.18}$$

where

$$c_1 = 2\pi hc^2 = 3.74 \times 10^{-5} \frac{\text{erg-cm}^2}{\text{sec}} = 3.74 \times 10^{-16} \text{ watt-m}^2$$

and

$$c_2 = \frac{hc}{k_B} = 1.438 \text{ cm-deg} = 1.438 \times 10^{-2} \text{ m-deg}$$

Here c is the velocity of light, h Planck's constant, and k_B Boltzmann's constant. The total emittance is given by the *Stefan-Boltzmann law*,

$$M_e = 5.672 \times 10^{-8}\, T^4 \text{ watt/m}^2 \qquad \text{when } T \text{ is in degrees Kelvin} \tag{4.19}$$

The spectral exitance $M_e(\lambda)$ peaks at a wavelength λ_m given by Wien's displacement law

$$\begin{aligned} \lambda_m T &= 2.8978 \times 10^6 \text{ (nanometers} \times \text{deg K)} \\ &= 2897.8 \text{ (microns} \times \text{deg K)} \end{aligned} \tag{4.20}$$

B. PSYCHOPHYSICAL OR PHOTOMETRIC NOMENCLATURE

1. Definitions. We now turn to luminance and other photometric quantities recorded by a human observer using visible light. This involves the science of photometry, in which comparisons are made of optical sources by an observer. These are then ultimately related to one or more standard sources. A system of photometric units has been developed which parallels exactly the physical units discussed above. This is shown in Table 4.1.

Other photometric units are often encountered in the literature depending on whether we use the meter, centimeter, foot, or inch as the unit of length. In all cases the unit of *luminous flux* ϕ_v is the *lumen* and that of *luminous intensity* I_v is the candle or lumen/steradian. One candle corresponds to a flux of 1 lumen through a solid angle of 1 steradian. Hence an isotropic source with an intensity of 1 candle will radiate a total flux of 4π lumens. We now discuss other quantities, list other units, and give their conversion factors.

1. Luminous incidence E_v, lumens/m^2 (lux) or luminous exitance M_v. (Both quantities represent luminous flux density or "illuminance.") Lux are also called *meter-candles*. Other units are the *foot-candle* (lumen/foot^2) and the *phot* (lumen/cm^2):

$$1 \text{ foot-candle} = 10.764 \text{ meter-candles (lux)} = 1.0764 \times 10^{-3} \text{ phot}$$
$$1 \text{ phot} = 10^4 \text{ lux}$$

Table 4.1 Correspondence of Radiometric and Photometric Nomenclature

Radiometry	Symbol	m.k.s. units
	Physical	
Radiant energy	Q_e	joule
Radiant density	U_e	joule/m^3
Radiant flux	Φ_e	watt
Radiant exitance	M_e	watt/m^2
Irradiance	E_e	watt/m^2
Radiant intensity	I_e	watt/ω
Radiance	L_e	watt/ω-m^2
Photometry	**Symbol**	**m.k.s. units**
	Psychophysical	
Luminous energy	Q_v	talbot
Luminous density	U_v	talbot/m^3
Luminous flux	Φ_v	lumen
Luminous exitance	M_v	lumen/m^2
Illuminance	E_v	lumen/m^2(lux)
Luminous intensity	I_v	lumen/ω(candle)
Luminance	L_v	lumen/ω-m^2

ω = unit solid angle (steradian)

2. Luminance, L_v, lumens/(m^2-ster). Like its physical counterpart, the radiance L_e, the luminance L_v is flux per unit solid angle per unit projected area of source. Since 1 lumen/ster = 1 candle, the m.k.s. units of L_v are also candles/m^2. One candle/m^2 is sometimes called a *nit*. Other units are the *stilb* or lumen/(cm^2-ster) or *candle/cm*^{2} = 10^4 candles/m^2; the *lambert* = $(1/\pi)$ candle/cm^2 = $(10^4/\pi)$ candle/m^2; the *millilambert* = $(1/1000\pi)$ candle/cm^2 = $(10/\pi)$ candle/m^2; and the *foot-lambert* = 0.0003426 candle/cm^2 = 3.426 candle/m^2.

The *primary standard* of the photometric system of units is the luminous exitance of the surface of a black body radiator at the freezing temperature of platinum (2043.5°K) which is *defined* to be 60 candles/cm^2 = 60 stilb = 60 lumen/(cm^2-ster).

2. Conversion between Luminous and Radiant Energy Units. The actual equivalence between physical and psychophysical units depends on several variables: the conditions of observation, the age and experience of the observer, where the light is falling on the retina of his eye, and the wavelengths present in the light. The last variable is most important—sources of equal physical spectral ra-

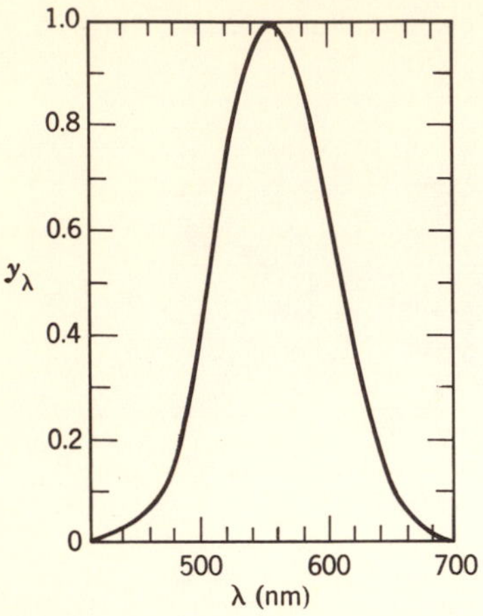

Fig. 4.11

diance will appear to an observer to have a different psychophysical spectral luminance as the wavelength varies. Results of experiments with many observers have been combined to yield the *standard luminosity curve* shown in Fig. 4.11. This curve is a function of λ, which we shall denote y_λ; it has a peak value of unity at $\lambda =$ 555 nm. This means that light of wavelength λ gives a factor of y_λ as much psychological sensation as light of wavelength 555 nm. The conversion at $\lambda = 555$ nm is standardized to be

$$K_m = 680 \text{ lumens/watt} \tag{4.21a}$$

This means that 1 watt of flux at 555 nm gives the same physical sensation as 680 lumens. For other wavelengths the conversion factor is

$$\begin{aligned} K &= K_m y_\lambda \\ &= 680\, y_\lambda \text{ lumens/watt} \end{aligned} \tag{4.21b}$$

that is, for $\lambda \neq 555$ nm, 1 watt produces less than 680 lumens.

We may use Eq. (4.21*b*) to convert any spectral physical quantity in Table 4.1 to the corresponding psychophysical quantity. For instance, if we have a spectral flux $\Phi_e(\lambda)$, then the luminous flux is given by

$$\Phi_v = 680 \int_0^\infty y_\lambda \Phi_e(\lambda)\, d\lambda \text{ lumens} \tag{4.22a}$$

The average conversion factor $\bar{K}$ is defined by

$$\Phi_v = \bar{K}\Phi_e \tag{4.22b}$$

where

$$\Phi_e = \int_0^\infty \Phi_e(\lambda)\, d\lambda$$

$\bar{K}$ is given by

$$\bar{K} = \frac{680 \int_0^\infty y_\lambda \Phi_e(\lambda)\, d\lambda}{\int_0^\infty \Phi_e(\lambda)\, d\lambda} \tag{4.22c}$$

If the intensity from the same source is desired, the conversion would be

$$I_v = 680 \int_0^\infty y_\lambda I_e(\lambda)\, d\lambda \text{ candles} \tag{4.23a}$$

$$= \bar{K} I_e \tag{4.23b}$$

where

$$I_e = \int_0^\infty I_e(\lambda)\, d\lambda$$

and

$$\bar{K} = \frac{680 \int_0^\infty y_\lambda I_e(\lambda)\, d\lambda}{\int_0^\infty I_e(\lambda)\, d\lambda} \frac{\text{candles}}{\text{watt/ster}} \tag{4.23c}$$

Equations (4.22*c*) and (4.23*c*) are equivalent, since for a given source and given geometry $I_e(\lambda)$ will be proportional to $\Phi_e(\lambda)$.

C. SOME EXAMPLES OF SOURCE BRIGHTNESS AND DETECTOR SENSITIVITY

The luminous incidence from the sun at noon at the equator is about 10^5 lux, which is 500,000 times greater than that from the full moon under similar circumstances.

One of the most intense incoherent sources, a super-high-pressure mercury lamp (PEK Labs, type 110) has a surface luminance of 140,000 candles/cm^2 or a radiance in the visible region of the spectrum of about 250 watts/(cm^2-ster).

The luminous flux density from a "60-watt" light is about 50 lux = 50 lumens/m^2 at a distance of 1 meter. The total flux would then be 50 lumens/m^2 $\times 4\pi$m^2 = 600 lumens or about 1 watt visible radiant flux.

Continuous gas lasers have a relatively low total power output, but it is highly directional. A typical example is furnished by a 4-milliwatt helium-neon laser emitting 6328 A radiation. Its output can be focused into a spot $r = 0.1$ mm with the beam making a cone of half-angle $\theta = 2 \times 10^{-3}$ radians. Hence the area is πr^2 and the solid angle is $2\pi(1-\cos\theta) \approx \pi\theta^2$, so that the product $dA \cdot d\Omega$ is

$$\pi r^2 \cdot \pi\theta^2 = \pi^2 (10^{-2})^2 \text{ cm}^2 \cdot (2 \times 10^{-3})^2 \text{ ster}$$

$$\approx 4 \cdot 10 \cdot 10^{-10} = 4 \times 10^{-9} \text{ cm}^2\text{-ster}$$

The radiance L_e is the power divided by $dA \cdot d\Omega$ or

$$\frac{4 \times 10^{-3} \text{ watts}}{4 \times 10^{-9} \text{ cm}^2\text{-ster}} = 10^6 \text{ watts/(cm}^2\text{-ster)}$$

4000 times brighter than the PEK mercury lamp.

A fully adapted human eye can see about 10^{-9} lux. If we assume that the pupil of the eye is 2 mm in radius, then the area of the eye receiving this light is $\pi \cdot 2^2 \cdot 10^{-6}$ m^2 $\sim 13 \times 10^{-6}$ m^2, which means that about 10^{-14} lumens can be detected by a human eye. This amounts to about 1.5×10^{-17} watts of visible power or about 30 photons per second of energy $3eV = 4.8 \times 10^{-19}$ joules.

A photomultiplier tube has a sensitivity as limited by "dark current" at room temperature of about 5×10^{-10} lumens. This can be reduced to as low as 10^{-15} lumens by cooling the tube.

A reasonably fast photographic film (Kodak Tri-X) will develop a very noticeable gray area if exposed to 10^{-2} lux for 1 second. If the resolution amounted to 30 lines/mm, then we could distinguish the darkening in an area of $1[(1/30) \cdot 10^{-3}]^2 \approx 10^{-9}$ m^2 produced in 1 second by $10^{-2} \cdot 10^{-9} = 10^{-11}$ lumens of light.

D. SOME EXAMPLES

The calculations that have to be performed in many cases involve complicated angular and surface integrals. But sometimes the geometry is particularly simple, as in the two examples to be discussed now. In both cases we assume the source to be a Lambert emitter with radiance L_e or luminance L_v (flux per unit area per unit solid angle). We now drop the subscripts e and v, so that our discussion may refer equally well to physical or psychophysical quantities.

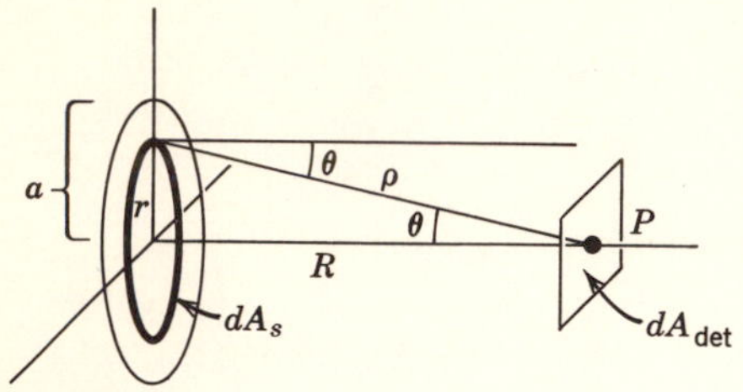

Fig. 4.12

1. Disk. We wish to calculate the incidence E at a point P a distance R from the center of the source disk shown in Fig. 4.12.

The element of source area dA_s may be set equal to

$$dA_s = 2\pi\, r\, dr = \pi\, dr^2 = \pi\, d\rho^2$$

where $\rho^2 = R^2 + r^2$ is taken to be our independent variable. The radiant or luminous flux $d^2\Phi$ reaching area dA_{det} at P from dA_s, obtained from Eq. (4.7) or (4.14), is

$$d^2\Phi = L d\Omega\, dA_s \cos\theta$$

$$d\Omega = \frac{dA_{\text{det}} \cos\theta}{\rho^2}$$

Thus

$$d^2\Phi = \pi L\, dA_{\text{det}} \frac{\cos^2\theta}{\rho^2}\, d\rho^2$$

but

$$\cos^2\theta = \frac{R^2}{\rho^2}$$

Thus the total flux reaching dA_d is

$$d\Phi = \pi L\, dA_{\text{det}} R^2 \int_{R^2}^{R^2+a^2} \frac{d\rho^2}{\rho^4}$$

$$= \pi L\, dA_{\text{det}} R^2 \left(\frac{1}{R^2} - \frac{1}{R^2+a^2}\right)$$

The flux density (radiant or luminous incidence) is

$$E = \frac{d\Phi}{dA_{\text{det}}} = \pi L\left(1 - \frac{1}{1 + a^2/R^2}\right) = \pi L \sin^2\theta_{\text{max}} \tag{4.24}$$

$(\sin\theta_{\text{max}} = a/\sqrt{a^2 + R^2})$

Note that as $a \to \infty$ for fixed R, E tends to πL, the analogue of Eq. (4.12). Note also that for fixed a and large R, E becomes

$$E \approx \frac{\pi L a^2}{R^2} = \frac{I}{R^2} \tag{4.24$'$}$$

where $I = L \cdot \pi a^2$ is the radiant or luminous intensity (flux/ster) of the source when emitting in a direction normal to its plane. The last result holds quite generally a large distance from a small source because then we can write

$$d\Phi = I\,d\Omega, \qquad \text{with } d\Omega = \frac{dA_{\text{det}}}{R^2}$$

Hence

$$d\Phi = \frac{I}{R^2}\,dA_{\text{det}}$$

or

$$E = \frac{d\Phi}{dA_{\text{det}}} = \frac{I}{R^2}$$

2. ***Sphere.*** The situation is shown in Fig. 4.13. In this case we have

$$dA_s = 2\pi\,a^2 \sin\beta\,d\beta, \qquad dA_\perp = dA_s \cos\theta$$

$$d\Omega = \frac{dA_{\text{det}}}{\rho^2}\cos\alpha$$

and

$$d^2\Phi = L\,dA_\perp\,d\Omega = L 2\pi a^2 \frac{dA_{\text{det}}}{\rho^2}\sin\beta\,d\beta\cos\theta\cos\alpha$$

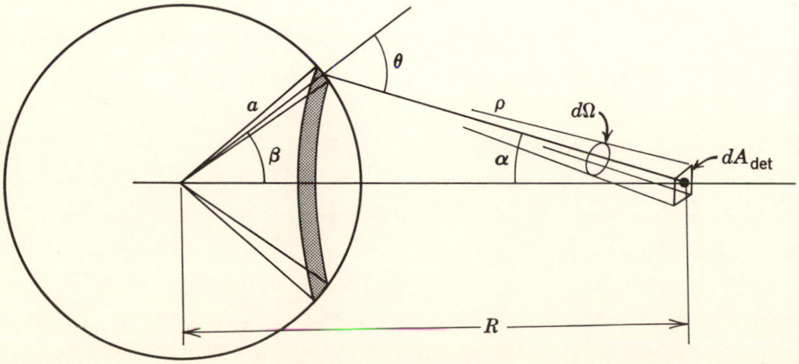

Fig. 4.13

This simplifies for $R \gg a$. Then $\alpha \approx 0$, $\rho \approx R$, and $\theta \approx \beta$. This gives

$$d^2\Phi \approx L\frac{2\pi a^2}{R^2}dA_{\text{det}}\frac{d\sin^2\beta}{2} = \frac{L}{R^2}dA_{\text{det}}\,dA_{\perp}$$

where $dA_{\perp} = d[\pi(a\sin\beta)^2]$ is the element of area of the source projected perpendicular to the line of observation (Fig. 4.14).

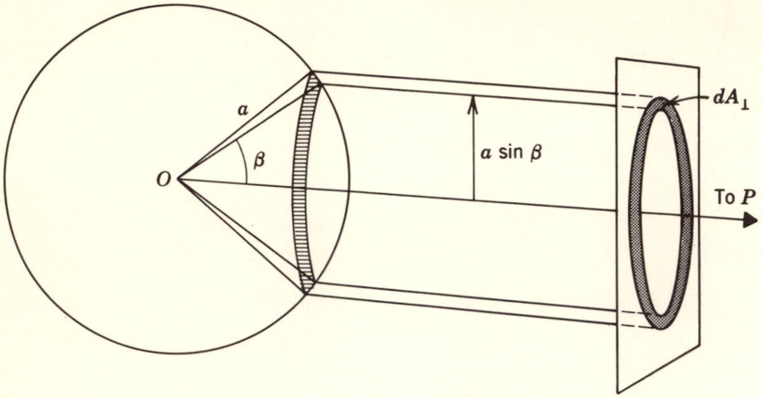

Fig. 4.14

Thus each element of the sphere contributes an amount proportional to $dA_{\perp}$.
Then

$$d\Phi = \frac{L}{R^2}dA_{\text{det}}A_{\perp} = \frac{L}{R^2}\pi a^2\,dA_{\text{det}}$$

and

$$E = \frac{d\Phi}{dA_{\text{det}}} = \frac{L}{R^2}\pi a^2$$

as in Eq. (4.24′). We also have, as for the disk

$$E = \frac{I}{R^2}$$

The derivation above has shown that a spherical Lambert emitter will appear to be a uniform disk to a distant observer.

E. IMAGE RADIANCE AND LUMINANCE

Using the basic nomenclature and equations of the previous section, we may now discuss the methods of computing the radiance or luminance at various places inside an optical system, particularly at locations of intermediate or final images. *We shall find that at such points the radiance or luminance obeys a conservation law.*

1. Contribution of an Elementary Beam—The Conservation Law. Recall our earlier definition of a pencil of light coming from a point. An *elementary pencil* is one subtending an infinitesimal solid angle $d\Omega = \sin\theta\, d\theta\, d\varphi$ and making an angle θ with the optical axis of a system. The azimuthal angle φ (Fig. 4.15*a*) will remain constant for the pencil as long as it passes through optical elements that are symmetrical about the main axis. Similarly the ray with azimuthal

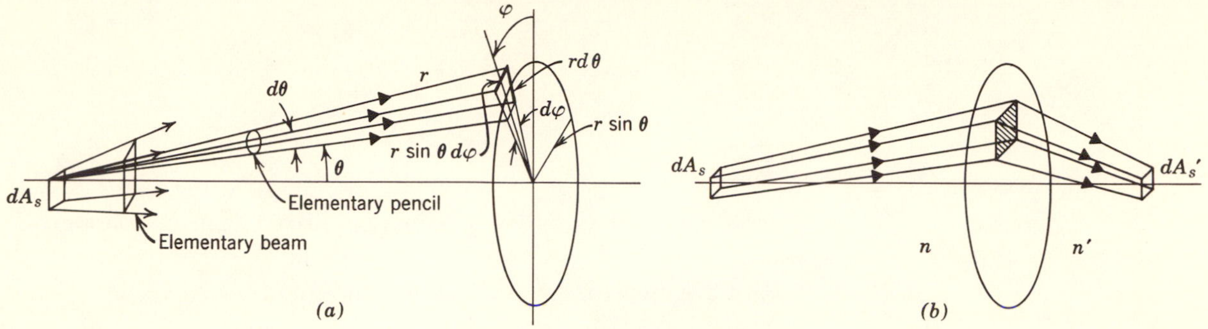

Fig. 4.15. (*a*) Elementary beam and pencil. (*b*) Focusing of an elementary beam.

angle $\varphi + d\varphi$ will retain that angle. Because of the action of the optical components, in general neither θ nor $d\theta$ will remain constant as the pencil passes through the system.

An *elementary beam* consists of all such pencils passing through the infinitesimal area dA_s, which is located perpendicular to the axis of the system. Let the radiance or luminance at dA_s be $L(\theta, \varphi)$. Then the flux in the elementary beam is given by Eq. (4.7) as

$$d^2\Phi = L(\theta, \varphi)\cos\theta\, dA_s\, d\Omega = L\, dA_s \sin\theta\cos\theta\, d\theta\, d\varphi$$

$$= \frac{L\, dA_s\, d\varphi}{2}\, d\sin^2\theta \tag{4.25}$$

We follow the elementary beam until it forms an image (Fig. 4.15*b*). Let the appropriate image parameters be denoted by primes. If no reflection or other losses are assumed, the beam should still have the same flux $d^2\Phi$ given by Eq. (4.25) written for primed quantities. Since by hypothesis $d^2\Phi' = d^2\Phi$, we have

$$L(\theta, \varphi)\, dA_s\, d\varphi\, d\sin^2\theta = L'(\theta', \varphi')\, dA_s'\, d\varphi'\, d\sin^2\theta' \tag{4.26}$$

where L' is the radiance or luminance of the image. We know that $d\varphi = d\varphi'$. We shall see later (Section 4.2C6) that for perfect imaging of slightly off-axis points the *sine condition* $ny\sin\theta = n'y'\sin\theta'$ must hold,* which when squared gives

$$n^2y^2\sin^2\theta = (n')^2(y')^2\sin^2\theta' \tag{4.27}$$

But the area magnification dA_s'/dA_s is simply $(y')^2/y^2$; thus Eq. (4.27) becomes

$$n^2\, dA_3 \sin^2\theta = (n')^2\, dA_s' \sin^2\theta' \tag{4.28}$$

*Such a system is said to be *aplanatic*.

Taking the differential yields

$$n^2\, dA_s\, d\sin^2\theta = (n')^2\, dA_s'\, d\sin^2\theta' \tag{4.29}$$

Taking the ratio of Eq. (4.26) to Eq. (4.29) will then give

$$\frac{L(\theta,\varphi)}{n^2} = \frac{L'(\theta',\varphi)}{(n')^2} \tag{4.30}$$

If the two indices of refraction are the same, then we obtain the result

$$\boxed{L = L'} \tag{4.31}$$

This is the law of conservation of luminance or radiance in an elementary beam.

We have therefore shown that no amount of focusing can increase the radiance or luminance (flux per unit area per unit solid angle) of each elementary beam from object to image. Imaging may separately change dA_s or $d\Omega = (\frac{1}{2}) d\sin^2\theta \cdot d\varphi$, but not the product $n^2\, dA_s\, d\Omega = (\frac{1}{2})\, d\varphi \cdot n^2\, dA_s\, d\sin^2\theta$. A reduced image will have more radiance or illuminance (power per unit area), but the light rays will subtend a greater solid angle at the image than at the object (Fig. 4.15*b*). There are numerous practical consequences of this conservation law.

2. *Total Flux Density.* The radiant or luminous incidence at the image will be the integral over the contributions of each elementary beam. A conical beam of half-angle θ' will contribute

$$\begin{aligned} L' \cos\theta'\, d\Omega' &= 2\pi L' \cos\theta' \sin\theta'\, d\theta' \\ &= \pi L'\, d\sin^2\theta' \end{aligned}$$

to the flux per unit area at the image. We integrate this expression from $\theta' = 0$ to $\theta' = \alpha'$, where α' is the half-angle subtended by the exit pupil at the image (Fig. 4.16). Using $L' = (n'/n)^2 L$ from (4.30)

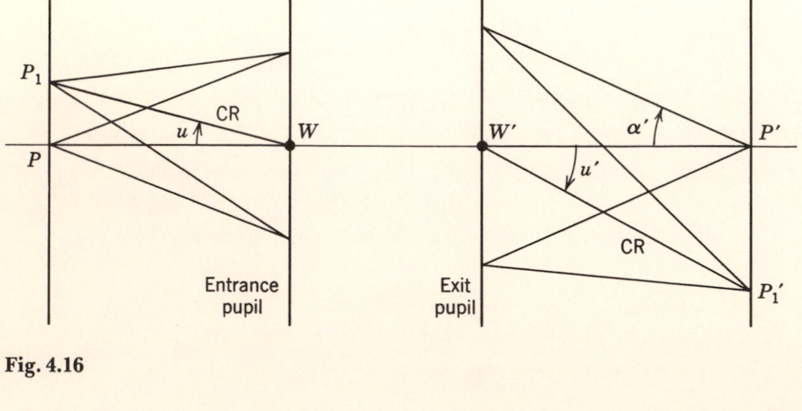

Fig. 4.16

and assuming L to be independent of θ (Lambert source), we obtain

$$E' = \left(\frac{n'}{n}\right)^2 L\pi \sin^2 \alpha' \tag{4.32}$$

The total solid angle subtended by the exit pupil at the image can be written

$$\Omega' = 2\pi \int_0^{\alpha'} \sin\theta \, d\theta = 2\pi(1 - \cos\alpha') = 4\pi \sin^2 \frac{\alpha'}{2} \tag{4.33}$$

For small angles α' we can approximate $\sin^2 \alpha' \approx 4 \sin^2(\alpha'/2)$. Then (4.32) becomes

$$E' \approx \left(\frac{n'}{n}\right)^2 L\Omega' \tag{4.34}$$

3. ***Behavior Off Axis.*** An off-axis point such as P_1 in Fig. 4.16 will be imaged at P_1'. If there is no vignetting, the limiting solid angle of rays reaching P_1' will be determined by the exit pupil and will be approximately

$$\frac{\sigma_\perp'}{|W'P_1'|^2} = \frac{\sigma' \cos u'}{|W'P_1'|^2} = \frac{\sigma' \cos^3 u'}{|W'P'|^2} \approx \Omega' \cos^3 u'$$

where σ' is the area of the exit pupil, $\sigma_\perp'$ is its projection perpendicular to $W'P_1'$, and $\Omega' = 4\pi \sin^2(\alpha'/2)$ is the solid angle as seen from P'. In this case of no vignetting, as long as u is less than a maximum value determined by the field stop (not shown in Fig. 4.16), all elementary beams leaving P_1 will reach P_1'. Furthermore, it can be shown that for these beams, Eq. (4.30) still holds if the system is aplanatic, that is, obeys the sine condition. Thus we can still use $(n'/n)^2 L$ in the off-axis version of Eq. (4.34).

If we want the flux density at P_1' falling on a screen perpendicular to the axis, we must insert another factor of $\cos u'$ to account for the orientation of the screen with respect to the chief ray. Thus for the flux density at the off-axis point P_1' we obtain

$$E'(\text{off-axis}) \cong \left(\frac{n'}{n}\right)^2 L\Omega' \cos^4 u' \tag{4.35}$$

This represents a fairly rapid decrease with u'. The flux density is even less if vignetting is present. The $\cos^4 u'$ decrease may be partially overcome by departing from an aplanatic system and deliberately introducing aberrations in the image of the aperture stop by the front part of the lens to form the entrance pupil.

F. TWO DEVICES FOR PROVIDING ILLUMINATION

1. ***The Searchlight.*** A searchlight is not the simple optical device that it appears at first sight to be. An explanation of its operation requires some tight reasoning. As shown in Fig. 4.17, a searchlight consists essentially of a small intense source S placed at the focal point of a reflecting mirror or a lens. If the source were truly a point, then a parabolic mirror or a corrected lens would give a perfectly collimated parallel beam having a flux density that would

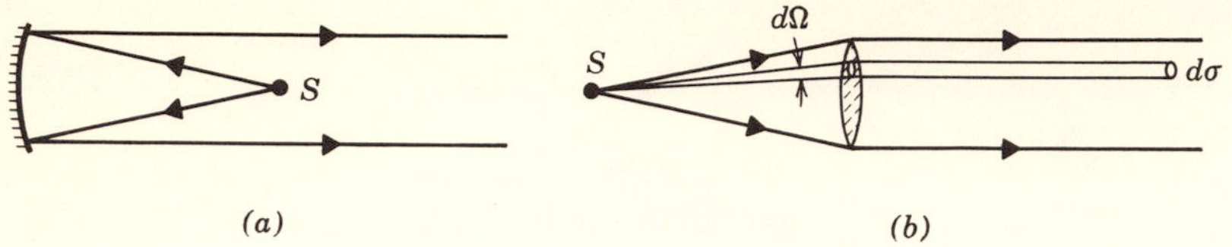

Fig. 4.17

be uniform, at least for small angular apertures. To see this, it is only necessary to realize that a pencil of rays from the source making a cone of solid angle $d\Omega = d\sigma/f^2$ can be found to illuminate area $d\sigma$, giving a constant flux density

$$E = E_0 = \frac{I}{f^2} \tag{4.36}$$

This result no longer holds with a source of finite size. For simplicity we let the source be a disk-shaped Lambert emitter of radius r_s. This disk is imaged by the lens at infinity to the left of the lens, where it subtends a half-angle $\delta = r_s/f$ (Fig. 4.18). Pick an observation point P to the right of the lens. The line from the center of the

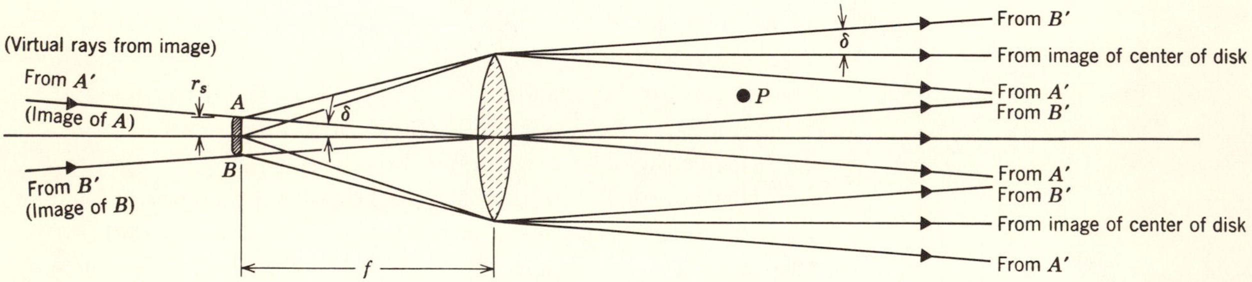

Fig. 4.18. The disk AB is imaged by the lens at infinity to the left.

image at infinity to P will remain parallel to the axis as P moves, but the rays from the image to P must pass through the lens. If P is off axis, they may not all be able to do so (Fig. 4.19).

It is useful to construct cones from P to the lens and from P to the image and project them onto a plane at a distance f to the left of P. The intersection of the cone to the image gives a disk of radius r_s, and the intersection of the cone to the lens gives a slightly elliptical disk of radius $R_L f/D$. The intersection of these two disks represents the image as seen from P. The area in the intersection equals A', the area of that part of the source that can radiate through the lens and reach P. The intersections of these cones are shown in the little figures at the bottom of Fig. 4.21.

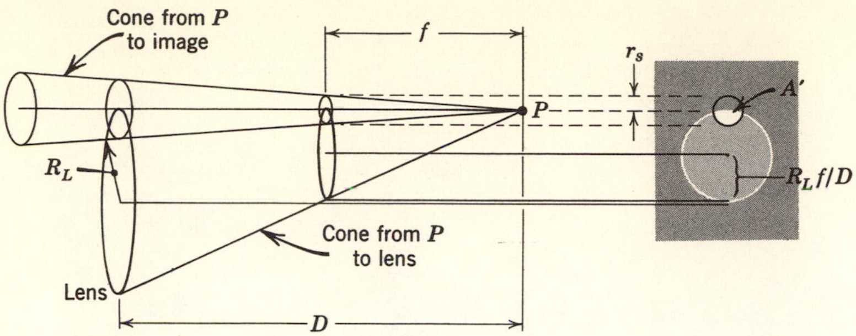

Fig. 4.19

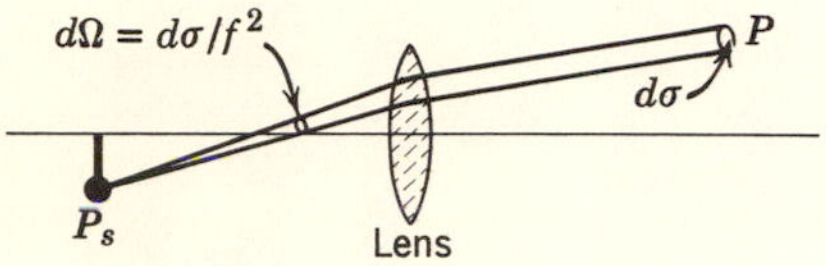

Fig. 4.20

The flux reaching area $d\sigma$ around P from area dA_s around a point P_s at the source is seen from Fig. 4.20 to be

$$d^2\Phi = L\,dA_s\,d\Omega = \frac{L\,dA_s\,d\sigma}{f^2}$$

where L is the radiance or luminance of the source. Here we assume the normal to $d\sigma$ to be along the axis, and neglect the distinction between $d\sigma$ and its perpendicular projection $d\sigma_\perp$. By integrating over the area A' of the source that can be "seen" from P, we may calculate the total flux through $d\sigma$:

$$d\Phi = \frac{LA'\,d\sigma}{f^2}$$

Then the flux density $d\Phi/d\sigma$ is

$$E = \frac{LA'}{f^2} \tag{4.37}$$

Within the shaded cone to the left of d in Fig. 4.21 the image subtends a smaller angle than does the lens, so that $r_s < R_L f/D$. The projected disk from the image is then completely inside the projected disk from the lens, so that $A' = A_s$. This will be the case at point a. Then

$$E = E_0 = \frac{LA_s}{f^2} = \frac{\pi L r_s^2}{f^2} \tag{4.38}$$

which is a constant. Such a constant flux density is expected in a searchlight beam with an infinitely small source. The length of the shaded cone from the lens to point d is $R_L f/r_s$, that is, inversely proportional to the size of the source.

When P moves laterally out of the shaded cone from a to b or c, A' becomes less than A_s and E drops from the constant value of E_0 of Eq. (4.38).

At the point d the projections of lens and image coincide, giving $R_L f/D = r_s$. When P is to the right of d, the projection of the lens is smaller than that of the image. When P remains in the shaded

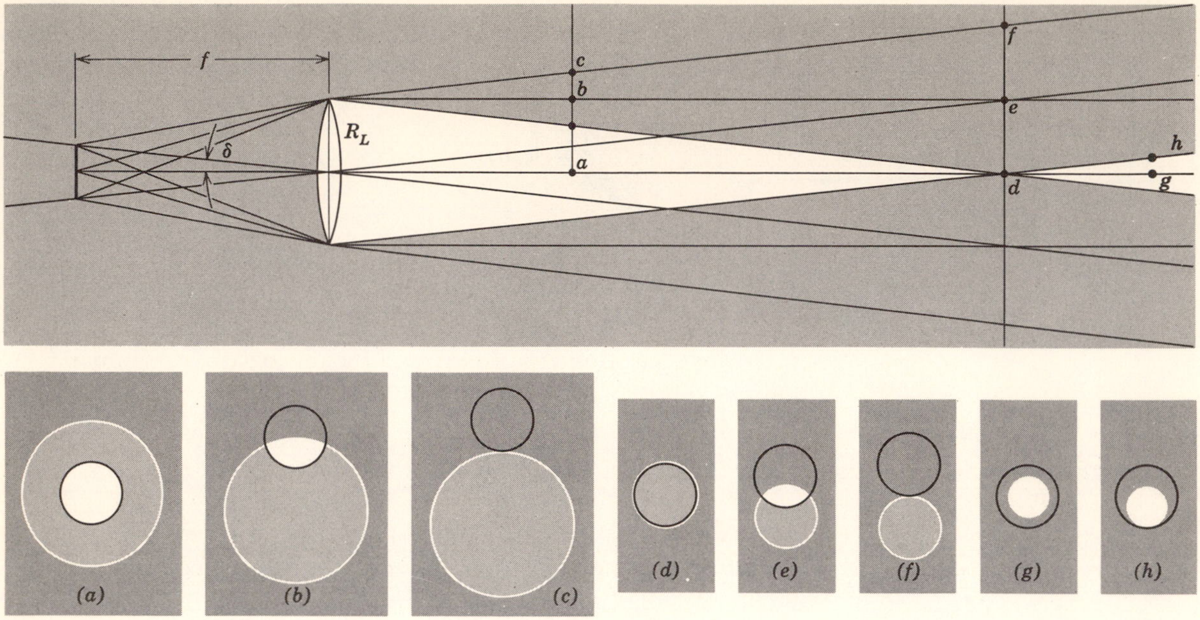

Fig. 4.21

cone to the right of d, the lens projection is entirely inside the image projection, and the common area is

$$A' = A(\text{lens projection}) = \pi\left(\frac{R_L f}{D}\right)^2$$

We may summarize these results by saying that when P is anywhere in the shaded cone of Fig. 4.21, the flux density is given by

$$E = E_0 = \frac{L\pi r_s^2}{f^2}, \qquad D < \frac{R_L f}{r_s}$$
$$E = \frac{LA'}{f^2} = \frac{L\pi R_L^2}{D^2}, \qquad D \geqslant \frac{R_L f}{r_s} \tag{4.39}$$

The second result in Eq. (4.39) gives an expression for E equivalent to that obtained by assuming that the lens were replaced by a source disk with luminance or radiance E and area πR_L^2. This disk would then have an intensity

$$I = L\pi R_L^2 \tag{4.40}$$

This is what is meant by the "intensity of a searchlight." It is usually expressed in candles (or "candle power," which means the same thing).

2. *The Projector.* A lantern slide projector or similar instrument has two lenses and performs two simultaneous image-formation processes. The projection lens PL images the semitransparent object O onto the screen S_C (Fig. 4.22). A properly used condenser

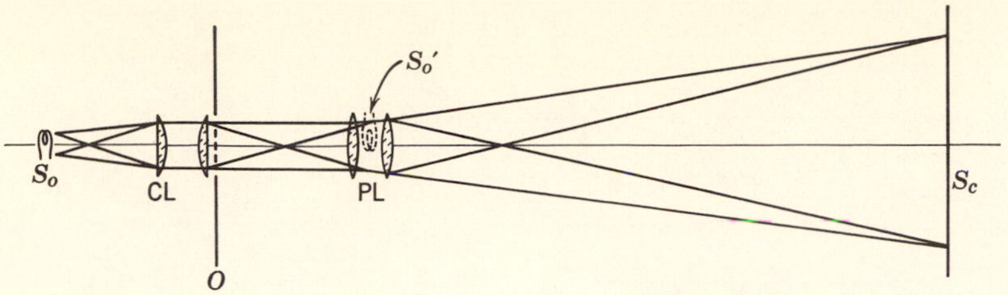

Fig. 4.22

lens CL will assure that the object is uniformly illuminated with an intense beam. The condenser forms an image S_0' of the source S_0 at the projection lens PL. Except for transmission losses this image will have the same luminance as the projection lamp and will uniformly illuminate those parts of the screen (except for the $\cos^4 u'$ factor) that correspond to the transparent parts of the object O.

REFERENCES FOR SECTION 4.1

The nomenclature used here is that recommended by international committees of IUPAP and CIE as reported in the *Journal of the Optical Society of America*, **57**, 854 (June, 1967).

W. J. Smith, *Modern Optical Engineering*. Chapter 8.

J. W. T. Walsh, *Photometry*.

PROBLEMS

4.1.1 Assume that the sun obeys Lambert's law. (It does not, because of a phenomenon known as "darkening of the limb.")

a. Find its luminance (MKS units).
b. Find the intensity of light from the sun (in lumens/ster). An image of the sun is formed on the earth by a lens of focal length 10 m and radius 1 m.
c. Find the luminous incidence in that image (neglect losses in the lens).
d. Find the total flux in lumens hitting the screen at the image.

Assume that the sun is overhead and reaches the earth with an illuminance of 10^5 lux. Radius of sun = $7 \cdot 10^8$ m; distance from earth to sun = $1.5 \cdot 10^{11}$ m.

4.1.2 A Lambert Source S with radiance $L_e = 10^6$ watt/m², 1 cm in diameter, is located 45 cm from a thin lens of +30 cm focal length and 10 cm diameter. A detector is located 120 cm to the right of the lens. It consists of 2 masks M_1 and M_2 each with 0.1-mm-diameter pin holes on axis 1 cm apart, followed closely by a large photocell PC with a sensitivity of 10^3 amp/watt. Neglect losses in the lens and find the value of the photocurrent coming from the cell.

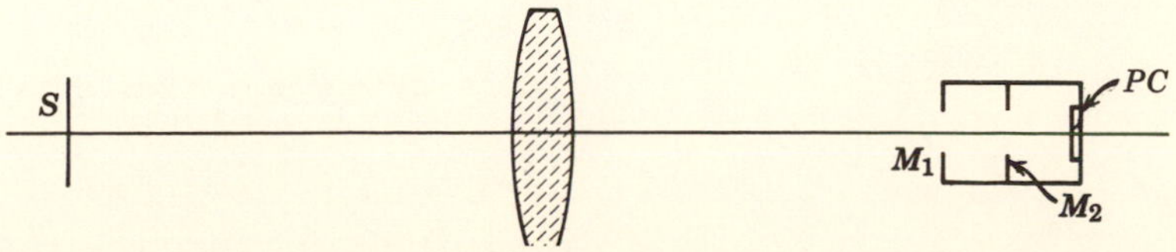

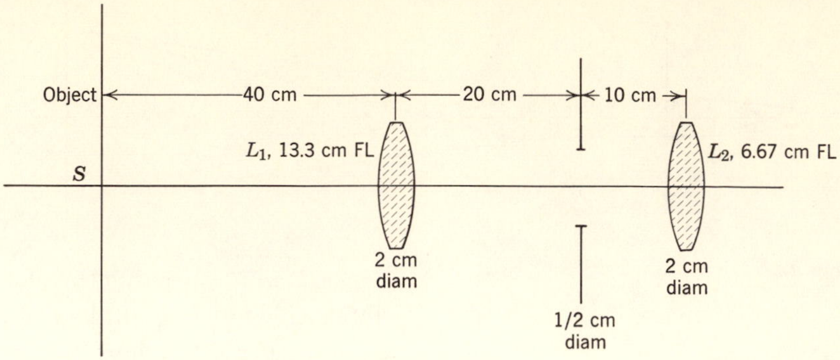

4.1.3 Let the object be a *large* uniform Lambert source of radiance 10 watt/(cm^2-ster). Assume 10 per cent loss in each lens and calculate the total power transmitted by the above system.

4.1.4 A projector uses a flat filament lamp with a luminance of 2000 candles/cm^2. The condenser lens completely fills the 1″ diameter aperture of the 2″ focal length projection lens with the image of the lamp filament. What is the luminous incidence on a screen 30 ft from the projection lens if each lens transmits 90 per cent of the light incident onto it?

4.1.5 What is the maximum luminous flux density (luminous incidence) at a distance of 1 km from a 100 million candlepower searchlight if, at this distance, the inequality

$$D < \frac{R_L f}{r_s}$$

is assumed to hold? If the reflector has a diameter of 1 meter, what is the required luminance of the arc lamp used as a source? Does the size of the arc affect the candle power of the searchlight? If not, what effect does the arc size have?

4.2 LENS ABERRATIONS

The ray-tracing equations used in the theory of Gaussian optics were correct to first order in the inclination angles of the rays and normals to refracting or reflecting surfaces. When higher-order approximations are used for the trigonometric functions of the angles, departures from the predictions of Gaussian optics will be found. No longer will it be generally true that all the rays leaving a point object will exactly meet to form a point image or that the magnification in a given transverse plane is a constant. In addition, the properties of the system may be wavelength-dependent. Such deviations from ideal Gaussian behavior are known as lens aberrations.

A. INTRODUCTION

Monochromatic aberrations occur with light of a fixed wavelength. They may be treated mathematically in lowest order by carrying out the ray-tracing calculations to third order in the angles or, equivalently, by carrying out the calculation of the optical path difference to fourth order, as we did for reflection and refraction by single spherical surfaces in Sections 2.2D, E. The resulting "third-order theory" is itself valid only for small angles, and for many real systems calculations must be carried out to still higher order, say fifth or seventh.* Because of their great complexity these higher-order aberrations are usually treated numerically.

Most compound lens systems contain enough degrees of freedom in their design so that if the third-order theory were exact, all aberrations could be eliminated. For real systems the residual higher-order aberrations would still be present, and there are not enough design parameters to eliminate all of them as well. An experienced optical designer strives for a balance between third-order and higher aberrations to give optimum performance with a given system. The performance must be judged according to the intended use of the instrument. The criteria for a telescope objective and for a camera lens for close-ups are quite different.

To go into the subject of lens aberrations in any detail would take us too far into the technicalities of optical design. Instead we shall merely describe the characteristic features of each type of aberration using the form of the third-order theory, and insert occasional comments about the methods of correcting these aberrations.

*For a centered system with rotational symmetry, only an odd combination of the powers of the angles will appear in ray-tracing formulas, and only an even combination of powers will appear in the expression for the optical path difference.

B. THE PRIMARY MONOCHROMATIC ABERRATIONS OF A SINGLE REFRACTING SURFACE

1. ***Review of Earlier Results.*** Primary aberrations are those that can be described by third-order theory. It is useful to recall some earlier results developed for a refracting surface in Section 2.2D. For an on-axis object point there is only spherical aberration. It may be described in terms of the aberration function $I(\rho) = -n'|\overrightarrow{AQ}|$ illustrated in Fig. 4.23, where ρ is essentially the distance

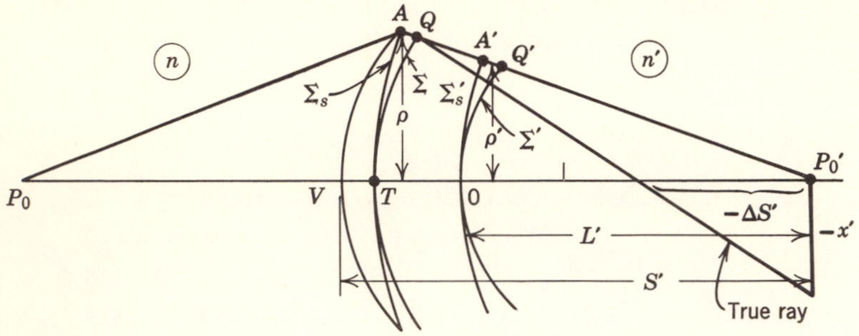

Fig. 4.23

from the axis to the point on the refracting surface where I is desired. The point Q is on the true wave surface Σ and A is on a spherical reference surface Σ_s centered at the Gaussian image point P_0'.

One of the important properties of the aberration function is its ability to give us expressions for the lateral aberrations. Consider, for example, the lateral spherical aberration, x' in Fig. 4.23. This is given by Eq. (2.52), which in the notation of Fig. 4.23 takes the form

$$x' = \frac{S'}{n'}\frac{dI}{d\rho} \tag{4.41}$$

The aberration function itself is given by Eq. (2.38):

$$I(\rho) = -\frac{C\rho^4}{4} \tag{4.42}$$

where C can be written

$$C = \frac{1}{2}\left(\frac{n'}{n}\right)^2\left(\frac{n'-n}{n'}\right)\left(\frac{1}{S'} - \frac{1}{R}\right)^2\left(\frac{n'}{R} - \frac{n+n'}{S'}\right) \tag{4.43}$$

2. ***Changing the Plane of Reference.*** Often it is not desired to know the aberration function at the refracting surface itself. The optical system may contain stops or other apertures such that it becomes more natural to refer to the aberration function in the plane of such apertures. With this in mind, let us compare the true wave surface Σ' that is a distance L' from P_0' with Σ_s', a sphere of radius L' centered on P_0' for a general value of $L' \neq S'$. The only requirement is that L' be large compared with the longitudinal aberration distance $\Delta S'$. The original definition of $I(\rho)$ was that of Eqs. (2.32), which in our present notation is

$$I(\rho) = \text{O.P.L.}(P_0AQA'Q'P_0') - \text{O.P.L.}(P_0TOP_0')$$
$$= \text{O.P.L.}(P_0AQ') + \text{O.P.L.}(Q'P_0') - \text{O.P.L.}(P_0O) - \text{O.P.L.}(OP_0')$$

Since Q' and O are on Σ', we have

$$\text{O.P.L.}(P_0AQ') = \text{O.P.L.}(P_0O)$$

We also have

$$\text{O.P.L.}(A'P_0') = \text{O.P.L.}(OP_0')$$

since O and A' are both on the reference sphere Σ_s'. Then we have

$$I(\rho) = -\text{O.P.L.}(A'Q') = -\text{O.P.L.}(AQ) = -n'|\overrightarrow{A'Q'}| = -n'|\overrightarrow{AQ}|$$

We can also write

$$\rho = \frac{\rho' S'}{L'}$$

Then

$$I = -n'|\overrightarrow{A'Q'}| = -\frac{C}{4}\left(\frac{S'}{L'}\right)^4 \rho'^4 \tag{4.44}$$

In Eq. (4.44) all quantities refer to the new reference surface at O. In making this change, we have not lost the ability to determine aberrations; Eq. (4.41) for the lateral spherical aberration becomes

$$x' = \frac{L'\,dI}{n'\,d\rho'} \tag{4.41$'$}$$

3. ***The Role of the Aperture Stop.*** If there is a stop in the system, the line from the center of curvature through the center of the stop defines a symmetry axis. If the object point lies off this axis, new types of aberration appear in addition to spherical aberration. To describe them it is convenient to work in image space and to make the comparison between the wave surface Σ' and the spherical reference surface Σ_s' at or near the plane of the exit pupil, as shown in Fig. 4.24. The emerging wavefronts lack rotational symmetry about the line PO.

If there were no stop, the ray PC through the center of curvature at C would be the center of a pencil of rays forming a Gaussian image at P'. For a ray reaching the plane of the exit pupil at Q' we would have approximately

$$I(Q') = \text{O.P.L.}(PQ'P') - \text{O.P.L.}(PCP') = -\frac{1}{4}C\left(\frac{S'}{L'}\right)^4 \rho'^4$$

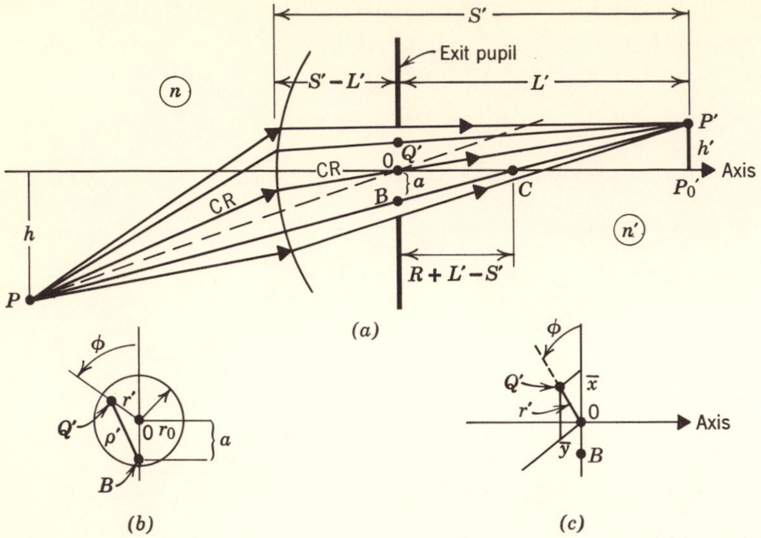

Fig. 4.24

where $\rho' = |BQ'|$, as shown in Fig. 4.24*b*. It is now natural to refer all path differences to the chief ray POP' through the center of the pupil since this ray forms the center of the bundle passed by the system. In this case the quantity of interest is the *off-axis aberration function*

$$W(Q') \equiv I(Q') - I(O) = \text{O.P.L.}(PQ'P') - \text{O.P.L.}(POP') \qquad (4.45)$$

In Eq. (4.45) for $W(Q')$, Q' was assumed to be in the plane of the exit pupil, but it can be moved a bit out of the plane along the ray without changing the result. Suppose that it is moved slightly onto the emerging wave surface that passes through the center of the exit pupil at O. Then we may write O.P.L.(PQ') = O.P.L. (PO). When we subtract the optical path length POP' from the optical path length $PQ'P'$, we are left with the simple result

$$W(Q') = \text{O.P.L.}(Q'P') - \text{O.P.L.}(OP')$$

Now construct a spherical reference surface centered at P' having radius $|OP'|$. Let A be the intersection point of this surface by the line $Q'P'$. Since $|OP'| = |AP'|$, we have

$$W(Q') = \text{O.P.L.}(Q'P') - \text{O.P.L.}(AP') = \text{O.P.L.}(Q'A)$$

If we introduce the *directed* distance $l(Q') = |\overrightarrow{Q'A}|$ from Q' to A, positive if Q' is to the left of A, then we have an expression for the off-axis aberration function,

$$W(Q') = n'|\overrightarrow{Q'A}| = n'l(Q') \qquad (4.46)$$

as the final refractive index times the separation of the wave surface from the reference sphere.

To obtain a useful expression we apply the law of cosines to triangle $OQ'B$ in Fig. 4.24*b*,

$$\rho'^2 = a^2 + r'^2 + 2ar' \cos\phi$$

where a is given by (Fig. 4.24*a*),

$$a = \frac{h'}{(S'-R)}(R+L'-S') \equiv bh'$$

Then $W = I(Q') - I(O) = -\frac{C}{4}\left(\frac{S'}{L'}\right)^4 (\rho'^4 - a^4)$

$$= -\frac{C}{4}\left(\frac{S'}{L'}\right)^4 [r'^4 + 4bh'r'^3 \cos\phi + 4b^3h'^3r' \cos\phi + 2b^2h'^2r'^2(2\cos^2\phi + 1)] \qquad (4.47)$$

We see from Eq. (4.47) that the form of the aberration function W depends on the position of the exit pupil through the variable L' and through $b = (R+L'-S')/(S'-R)$.

For a general image-forming system we may still refer to the aberration function $W(Q')$ at the exit pupil, and it will still have the interpretation associated with Eq. (4.46) as the optical path difference between the true wave surface at the exit pupil and a reference sphere centered at the Gaussian image point P'. The contributions of the individual refracting surfaces to the aberration function $W(Q')$ are additive, as long as W is small compared with ρ. The contribution to W from a single refracting surface is then the contribution from that surface acting alone, as though the other surfaces produced no aberrations. This result was proven in Eq. (2.60) for a special case, but it is generally true.

The general aberration function will have the form of Eq. (4.47), but with different coefficients, and can be written

$$W = I(Q') - I(O) = {}_0C_{40}r'^4 + {}_1C_{31}r'^3\cos\phi + {}_2C_{22}h'^2r'^2\cos^2\phi + {}_2C_{20}h'^2r'^2 + {}_3C_{11}h'^3r'\cos\phi \qquad (4.47')$$

The subscripts on the C coefficients refer to the powers of h', r', and $\cos\phi$, respectively, that appear in each term. These coefficients give the strength of the primary aberrations revealed by the system.

C. THE CHARACTERISTIC BEHAVIOR OF THE MONCHROMATIC ABERRATIONS

We now discuss the nature of the imperfect image produced when a single of the above monochromatic aberrations is present.

1. Spherical Aberration: $W = {}_0C_{40}r'^4$. We have discussed some of the features of primary spherical aberration in Chapter 2. Since W is independent of h', the aberration is the same for all image points in the field of view. In Fig. 4.25, which is similar to Fig. 2.23, we may use the earlier result for the angular aberration $-\delta \approx (1/n')(dW/dr')$ and obtain for the lateral spherical aberration

$$x' = -L'\delta = \frac{L'}{n'}\frac{dW}{dr'} = \frac{4L'}{n'}\,{}_0C_{40}r'^3 \tag{4.48a}$$

and for the longitudinal spherical aberration, often called simply "the spherical aberration,"

$$\Delta S' = \widehat{P'P'_r} = \frac{x'}{\tan u} \approx \frac{x'}{r'}L' = \frac{4L'^2}{n'}\,{}_0C_{40}r'^2 \tag{4.48b}$$

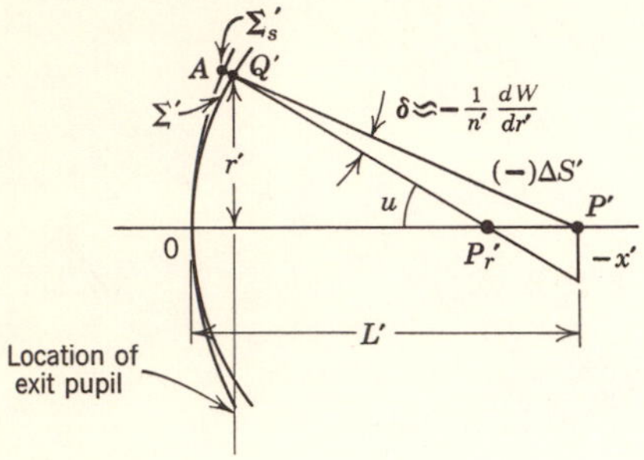

Fig. 4.25

2. *The Spherical Aberration of a Thin Lens.* For a simple positive lens the spherical aberration will be negative; that is, P'_r in Fig. 4.25 will be to the left of P', as was the case for a single refracting surface. Such a lens is said to have *undercorrected* primary spherical aberration.

We can conveniently plot the spherical aberration horizontally versus the zonal radius r' at the exit pupil. Such a plot can be used to draw the rays and the various zonal focal points as shown in Fig. 4.26.

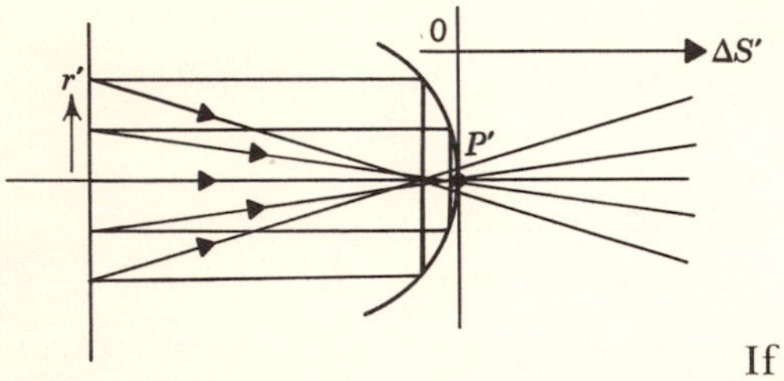

Fig. 4.26

If we perform a ray-tracing calculation to fifth order in the angles, or equivalently calculate W to sixth order in r', the result will be to add a term proportion to r'^4 to the expression for the longitudinal spherical in Eq. (4.48*b*). Thus it will take the form

$$\Delta S' = ar'^2 + br'^4$$

The plot of r' versus $\Delta S'$ might look like the curves shown in Fig. 4.27. We have assumed undercorrected primary spherical aberration ($a < 0$) and overcorrected fifth-order spherical aberration ($b > 0$).

We see from 4.27*c* that the resultant total spherical aberration is undercorrected if $r' < r_b \equiv \sqrt{-a/b}$, and overcorrected when $r' > r_b$. A system with an aberration curve like that in Fig. 4.27*c* is preferable to one with the curve of 4.27*b*, which is perfectly corrected for primary spherical aberration, since the curve in 4.27*c* represents a balance of the third-order aberration effects against those of higher order.

For a thin lens with focal length f and refractive index n, the following expression for the longitudinal spherical aberration may

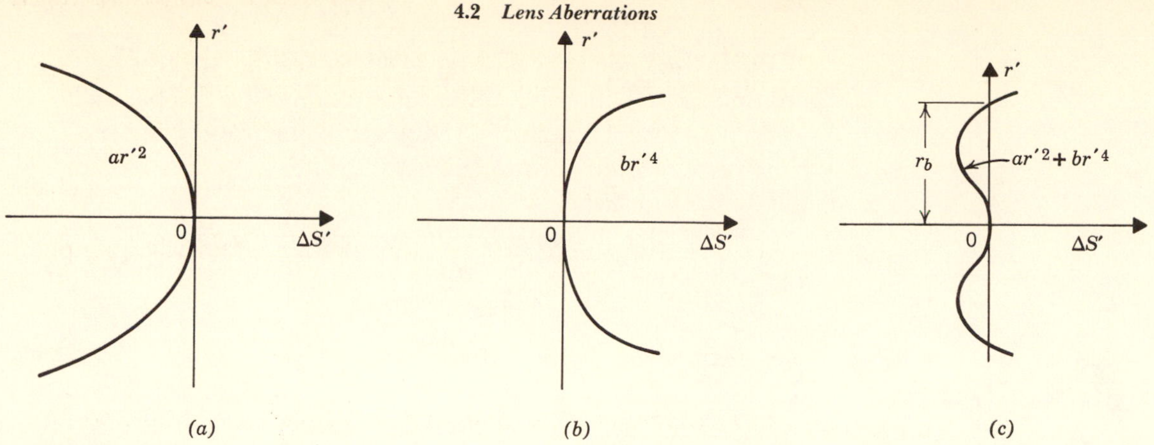

Fig. 4.27

be derived from the third-order theory:

$$\Delta\left(\frac{1}{S'}\right) = -\frac{\Delta S'}{S'^2} = \frac{r'^2}{8f^2 n(n-1)}\left[\frac{n+2}{n-1}q^2 + 4(n+1)pq + (3n+2)(n-1)p^2 + \frac{n^3}{n-1}\right] \tag{4.49}$$

Here we have assumed that the aperture stop is the lens itself ($L' = S'$). This gives

$${}_0C_{40} = -\frac{1}{32f^3 n(n-1)}\left[\frac{n+2}{n-1}q^2 + 4(n+1)pq + (3n+2)(n-1)p^2 + \frac{n^3}{n-1}\right] \tag{4.50}$$

In Eqs. (4.49) and (4.50), q is the so-called shape factor determined from the radii of curvatures of the two surfaces

$$q = \frac{R_2 + R_1}{R_2 - R_1} \tag{4.51}$$

and p is the so-called position factor

$$p = 1 - \frac{2f}{S'} \tag{4.52}$$

If we normalize all lengths to the focal length f, we obtain the convenient dimensionless formula

$$\frac{\Delta S'}{f} = 4f^3{}_0C_{40}\left[\left(\frac{r'}{f}\right)\left(\frac{S'}{f}\right)\right]^2 \tag{4.53}$$

with

$$4{}_0C_{40}f^3 = \frac{-1}{8n(n-1)}\left[\frac{n+2}{n-1}q^2 + 4(n+1)pq + (3n+2)(n-1)p^2 + \frac{n^3}{n-1}\right] \tag{4.53'}$$

From the functional form of Eq. (4.49) we see that for fixed-position factor p the longitudinal aberration will give a parabola when plotted as a function of the shape factor q, and vice-versa. Note further the proportionality to f^{-3}; when f changes sign, so will the spherical aberration.

In Fig. 4.28 we have plotted the coefficient $4_0C_{40}f^3$ as a function of q for three values of p, for $n = 1.5$. Sketches of the lens shapes are

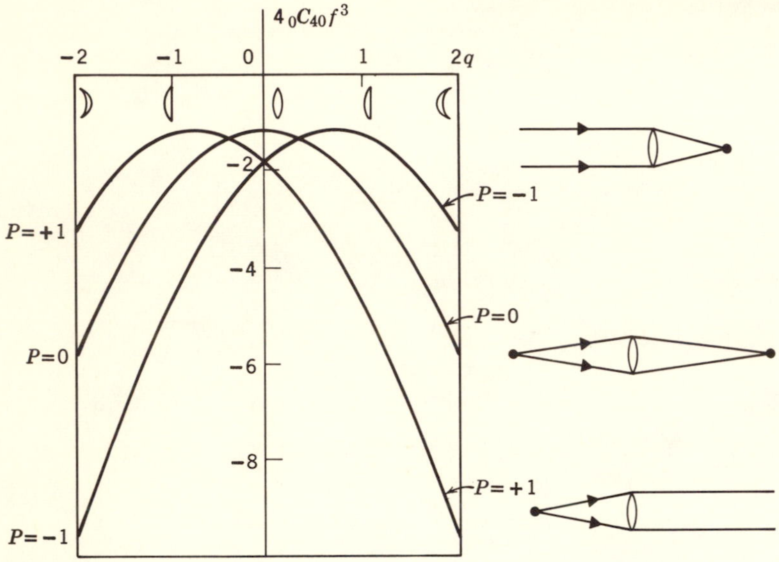

Fig. 4.28. Spherical aberration of a thin lens.

also included. The $p = 0$ curve applies when $S' = 2f = S$; the lens is used symmetrically and gives a magnification of (-1). The symmetrical equiconvex lens ($q = 0$) then has the smallest spherical aberration.

The $p = -1$ curve corresponds to $S' = f$. The minimum value of spherical aberration occurs when q is near the value -1. The minimum spherical aberration occurs when the two surfaces of the lens deviate the off-axis rays in the light beam by equal amounts. The minimum will not be zero for real objects and images (p between -1 and $+1$).

Primary spherical aberration can be eliminated by combining two thin lenses of opposite sign; the resulting expression for $\Delta(1/S')$ is the sum of the expressions (4.49) for the component lenses. We then exploit the fact that $\Delta S'$ is proportional to f^3, so that it is negative for a positive lens and positive for a negative lens. For a given position factor we choose a shape factor q so that the negative primary spherical aberration of the positive component has nearly its minimum magnitude. The shape factor of the negative component can then be chosen to give a relatively large amount of positive spherical aberration for its small negative power. The net power of the combination will be positive, the primary spherical aberration zero.

Such a lens is called a *doublet*. The two components are in contact in a *cemented doublet*. One of the surfaces will be plane in most inexpensive doublets.

3. ***Coma:*** $W = {}_1C_{31}h'r'^3 \cos\phi$. The aberration function in this case can also be written

$$W = {}_1C_{31}h'(\bar{x}^2+\bar{y}^2)\bar{x}$$

with $\bar{x} = r' \cos\phi$, $\bar{y} = r' \sin\phi$ (Fig. 4.24*c*). It is plotted as a surface in the exit pupil with its center perpendicular to the chief ray to the off-axis Gaussian image point P' in Fig. 4.29. The distance of this

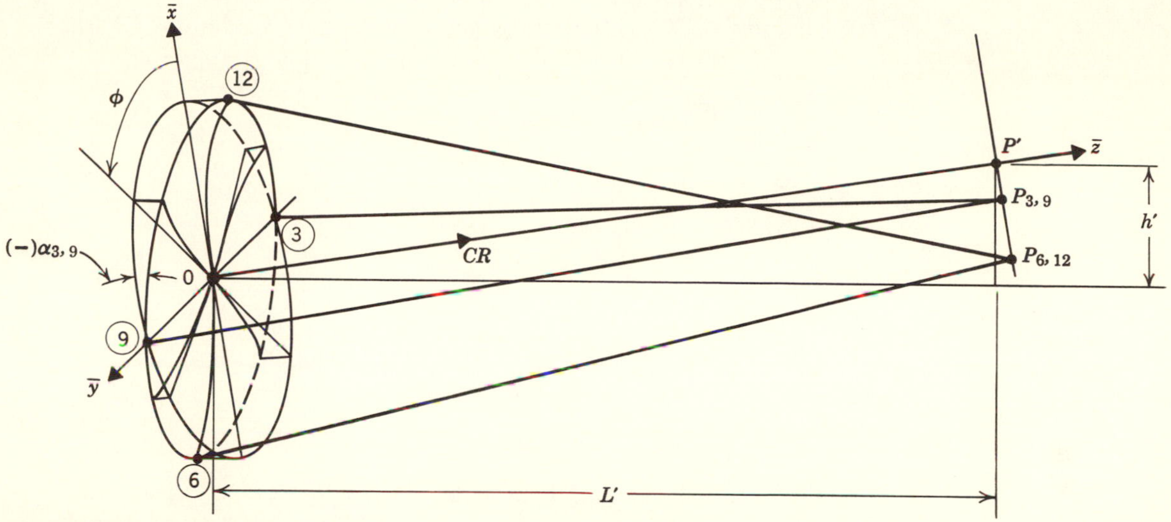

Fig. 4.29. The aberration function at the exit pupil for primary coma.

surface from the $\overline{xy}$ plane equals W. The surface contains the $\bar{y}$ axis, where it makes an angle $-\alpha_{3,9}$ with respect to the $\overline{xy}$ plane given by

$$\alpha_{3,9} = \frac{-1}{n'}\left(\frac{\partial W}{\partial \bar{x}}\right)_{\bar{x}=0} = \frac{{}_1C_{31}h'\bar{y}^2}{n'} = \frac{{}_1C_{31}h'r'^2}{n'}$$

Rays normal to the wave surface emerging from the exit pupil at points 3 and 9 will be inclined to the $\overline{yz}$ plane by $-\alpha_{3,9}$ and will intersect at the point $P_{3,9}$ at a distance $x'_{3,9} = \alpha_{3,9}L'$ below the Gaussian image point P'. Rays normal to the wave surface at points 6 and 12 will be inclined to the $\overline{yz}$ plane by an angle that differs from that of the normal to the spherical reference surface (not shown) by an amount

$$\alpha_{6,12} = \frac{-1}{n'}\left(\frac{\partial W}{\partial \bar{x}}\right)_{\bar{y}=0} = \frac{3{}_1C_{31}h'\bar{x}^2}{h'} = \frac{3{}_1C_{31}h'r'^2}{n'}$$

and will meet a distance $x'_{6,12} = \alpha_{6,12}L' = 3x'_{3,9}$ below P'.

*4. **Digression on Locating Ray Intercepts.*** So far we have used α and x' in a way similar to that of Fig. 2.23. We now present a more detailed justification for this procedure. Consider a surface given by the equation $\bar{z} = f(\bar{x}, \bar{y})$ or

$$\Phi \equiv \bar{z} - f(\bar{x}, \bar{y}) = 0$$

The unit vector normal to the surface is parallel to the gradient, $\nabla\Phi$, and for $|\partial f/\partial\bar{x}|, |\partial f/\partial\bar{y}| \ll 1$ it is approximately the gradient itself:

$$\mathbf{n} = \frac{[-(\partial f/\partial\bar{x}), -(\partial f/\partial\bar{y}), 1]}{[1 + (\partial f/\partial\bar{x})^2 + (\partial f/\partial\bar{y})^2]^{1/2}} \approx \left(-\frac{\partial f}{\partial\bar{x}}, -\frac{\partial f}{\partial\bar{y}}, 1\right)$$

Now let $\bar{z} = f_w(\bar{x}, \bar{y})$ be the equation of the emerging wave surface Σ and $\bar{z} = f_s(\bar{x}, \bar{y})$ that of the reference sphere Σ_s about P'. At the point $\mathbf{Q} = (\bar{x}, \bar{y}, 0)$ the normals to the two surfaces are given approximately by

$$\mathbf{n}_w = \left(-\frac{\partial f_w}{\partial\bar{x}}, -\frac{\partial f_w}{\partial\bar{y}}, 1\right)$$

and

$$\mathbf{n}_s = \left(-\frac{\partial f_s}{\partial\bar{x}}, -\frac{\partial f_s}{\partial\bar{y}}, 1\right)$$

If we translate a distance L' along the normal to the reference sphere Σ_s, we are essentially at the Gaussian image point P':

$$\mathbf{P}' = \mathbf{Q} + L'\mathbf{n}_s$$

If we translate along the normal to the wave surface Σ the same distance, we find where the true ray hits the image plane, namely at $\mathbf{P}'(Q) = \mathbf{Q} + L'\mathbf{n}_w$. These points will be separated by the vector

$$\mathbf{P}'(Q) - \mathbf{P}' = L'(\mathbf{n}_w - \mathbf{n}_r) = -L'\left[\frac{\partial(f_w - f_s)}{\partial\bar{x}}, \frac{\partial(f_w - f_s)}{\partial\bar{y}}, 0\right]$$

The aberration function is given by Eq. (4.46) as

$$W = n'l(Q)$$

Then $l(Q)$ is the distance along ray from the wave surface Σ to the reference sphere Σ_s. When all angles are small, this will be approximately the difference in $\bar{z}$ components or $(f_s - f_w)$. Thus we have approximately

$$W = n'(f_s - f_w)$$

The vector $(x', y', 0)$ from P' to $P'(Q)$ will then have components

$$x' = \frac{L'}{n'}\frac{\partial W}{\partial\bar{x}}, \qquad y' = \frac{L'}{n'}\frac{\partial W}{\partial\bar{y}} \tag{4.54}$$

5. The Comatic Circle and Flare Pattern. The pattern made by the "image" of a point source in the presence of primary coma will consist of the locus of all intersection points of pairs of rays, such as $P_{3,9}$ and $P_{6,12}$ in Fig. 4.29. For a general point Q at the exit pupil with lateral coordinates $\bar{x}$ and $\bar{y}$, application of Eqs. (4.54) gives, for the coordinates of the ray intercept in the image plane,

$$x' = {}_1C_{31}\frac{h'L'}{n'}(3\bar{x}^2+\bar{y}^2) = {}_1C_{31}\frac{h'r'^2}{n'}(2+\cos 2\phi)$$

$$y' = {}_1C_{31}\frac{h'L'}{n'}2\bar{y}\bar{x} = {}_1C_{31}\frac{h'r'^2}{n'}\sin 2\phi \tag{4.55}$$

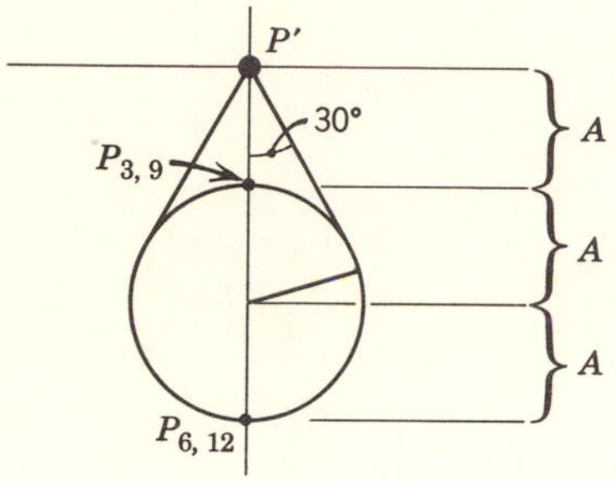

Fig. 4.30

Note that rays from points 180 deg apart, that is, with the same r' but with azimuth ϕ and $\pi-\phi$, have the same values of x' and y'.

Equations (4.55) describe a circle of radius

$$A = \frac{{}_1C_{31}h'L'r'^2}{n'} \tag{4.56}$$

centered at $x' = 2A$, $y' = 0$. Such a circle is shown in Fig. 4.30 for a negative value of A (as would be obtained with a single refracting surface – Eq. (4.46). This circle represents the locus of image "points" coming from an annular ring of radius r' at the exit pupil. The superposition of such rings for all values of r' from zero to a maximum value yields a comet-shaped flare pattern with its apex at the paraxial image point P'. Since A is proportional to h', we see that the size of this flare pattern is directly proportional to the amount of off-axis displacement of the image.

For a thin lens that is its own aperture stop, third-order theory gives the result

$${}_1C_{31} = \frac{1}{4nf^2}\left[(2n+1)p+\frac{(n+1)}{(n-1)}q\right] \tag{4.57}$$

where p and q are defined in Eqs. (4.51) and (4.52). The radius of the comatic circle is then given by

$$\frac{A}{f} = \frac{(r'/f)^2(h'/f)}{4n}\left[(2n+1)p+\frac{(n+1)}{n-1}q\right] \tag{4.58}$$

This is linear in both p and q. It is relatively easy to find reasonable values of p and q that will make this third-order coma zero for a thin lens.

6. Digression on the Abbe Sine Condition. We have just seen that primary coma produces a comet-shaped "image" with a size that varies linearly with the off-axis image distance h'. Unless this effect is small, the image of an extended object will be considerably blurred and distorted, even though the image of the on-axis point might be of higher quality. A similar effect can be produced by aberrations of higher order, namely, those with aberration functions W that depend on high powers of r', but are linear in h'. There is a general requirement, known as the *sine condition*, that must be met by an image-forming system if there is to be no such linear scaling of the size of an aberrant off-axis image with h'.

If the system in Fig. 4.31 has no spherical aberration for point P_1 on the symmetry axis, the imaging is perfect, and the system is Cartesian for the conjugate points P_1 and P_1'. Now consider the

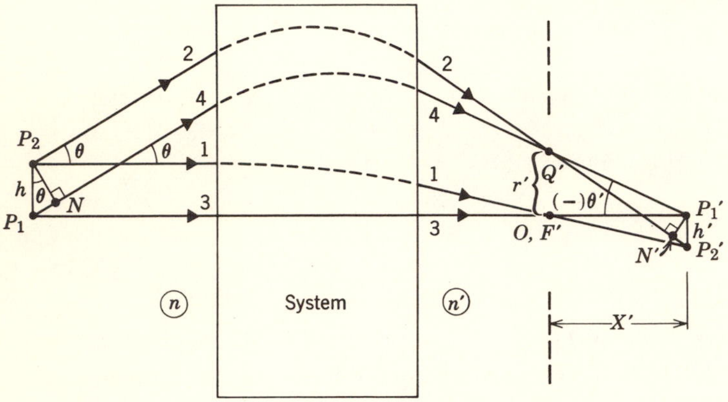

Fig. 4.31

point P_2 located off axis from P_1 by the small distance h. Let P_2' be its Gaussian image. Except for a plane mirror, no optical system that produces perfect imaging of P_1 onto P_1' can also produce perfect imaging of P_2 onto P_2'; that is, all optical path lengths from P_2 to P_2' cannot be made exactly equal. It is possible, however, to make the optical path lengths equal to first order in the off-axis image distance h'. In other words, the system can be designed so that the aberration function W for light coming from P_2 will contain no terms proportional to h'. This will be true if the sine condition holds.

It is convenient, but not necessary, to put the exit pupil of the system in Fig. 4.31 in the second focal plane through F'. Then F' may be identified with O in Fig. 4.24. Then the independent variable r' for the aberration function W represents the distance of a ray from the axis when it passes through the focal plane. Rays leaving the object point P_2 parallel to the axis pass through F' and have $r' = 0$. Ray 1 is such a ray. Ray 2 leaves P_2 at a finite angle θ and intersects the exit pupil at Q' a distance r' from the axis while making an angle $-\theta'$ with the axis. The optical path difference for these rays is given by the aberration function

$$\text{O.P.L.}(2) - \text{O.P.L.}(1) = W(Q') \tag{4.59}$$

and our requirement is that this be zero to first order in h'.

We first show that the parallel rays 1 and 3 have equal optical path lengths to first order in h'. Note that the optical path lengths from P_2 to F' and P_1 to F' are exactly equal. Thus

$$\begin{aligned} \text{O.P.L.}(1)-\text{O.P.L.}(3) &= n'\,|F'P_2'|-|F'P_1'| \\ &= n'(\sqrt{X'^2+h'^2}-X') = n'\frac{h'^2}{2X'}+\cdots \\ &= O(h'^2) \end{aligned} \tag{4.60}$$

The last equality results from an expansion of the square root in a power series in $(h'/X')^2$.

Keep in mind that according to Eq. (4.59) we ultimately want to compare the path lengths of rays 1 and 2. Rays 1 and 3 are equal to first order in h'. Rays 3 and 4 have exactly equal path lengths, because P_1' is a perfect image of P_1. Thus

$$\text{O.P.L.}(3) = \text{O.P.L.}(4) \tag{4.61}$$

Together with (4.60) this gives

$$\text{O.P.L.}(4)-\text{O.P.L.}(1) = O(h'^2) \tag{4.62}$$

To obtain (4.59) we need to compare path lengths of rays 2 and 4. In object space they are parallel, both making an angle θ with the axis. The optical path lengths from P_2 to Q' and from N to Q' are exactly equal, since P_2N is perpendicular to the ray direction. The path lengths from Q' to P_1' and Q' to N' are equal to first order in h'. Specifically,

$$\begin{aligned} |Q'P_1'| = \sqrt{|Q'N'|^2+|N'P_1'|^2} &= |Q'N'|\left(1+\frac{|N'P_1'|^2}{|Q'N'|^2}\right)^{1/2} \\ &= |Q'N'|\left(1+\frac{1}{2}\frac{|N'P_1'|^2}{|Q'N'|^2}+\cdots\right) \\ &= |Q'N'|+O(h'^2) \end{aligned}$$

since $|N'P_1'|$ is proportional to h' for small h'.

Thus, to first order in h' the part from P_2 to N' of ray 2 has the same optical path length as the part from N to P_1' of ray 1. This leaves the contribution from P_1N and $N'P_2'$:

$$\text{O.P.L.}(2)-\text{O.P.L.}(4) = n'|N'P_2'|-n|P_1N|+O(h'^2) \tag{4.63}$$

When we add Eqs. (4.62) and (4.63), we obtain the desired comparison of the path lengths of rays 1 and 2:

$$\text{O.P.L.}(2)-\text{O.P.L.}(1) = n'|N'P_2'|-n|P_1N|+O(h'^2) \tag{4.64}$$

This can be further simplified by use of the triangles P_1NP_2 and $P_1'N'P_2'$:

$$|P_1N| = h \sin\theta, \qquad |N'P_1'| = h' \sin\theta'$$

Thus

$$W(Q') = \text{O.P.L.}(2) - \text{O.P.L.}(1) = [n'h' \sin\theta' - nh\sin\theta] + O(h'^2) \tag{4.65}$$

We shall have our desired result that W be independent of h' to first order if and only if the *sine condition*

$$\boxed{n'h' \sin\theta' = nh \sin\theta} \tag{4.66a}$$

is satisfied. This represents a generalization of the Lagrange equation

$$n'h'\theta' = nh\theta$$

of Gaussian optics to large angles.

A system that (a) is Cartesean for a pair of points P_1 and P_1' and (b) satisfies the sine condition is called *aplanatic*.

Some implications of the sine condition will be explored in the problems. One way of stating it is that the ray angle magnification condition takes the form

$$\frac{\sin\theta'}{\sin\theta} = \frac{nh}{n'h'} = \frac{n}{n'm_x} = \text{const.} \tag{4.66b}$$

Here $m_x = h'/h$ is the transverse magnification. Thus, if a series of rays leaves P_1 at the angles $\theta_1, \theta_2, \ldots$ and arrives at P_1' at the angles $\theta_1', \theta_2', \ldots$, we must have

$$\frac{\sin\theta_1'}{\sin\theta_1} = \frac{\sin\theta_2'}{\sin\theta_2} = \cdots \tag{4.66c}$$

An interesting feature of Eq. (4.66*c*) is that as a test for an aplanatic system, that is, perfect imaging of slightly off-axis points, it involves only the behavior of rays from an on-axis point.

The sine condition has already been used to derive Eq. (4.31), the law of conservation of luminance or radiance in an elementary beam. It is possible to derive this conservation law from the laws of thermodynamics and then derive the sine condition from the conservation law. This was the approach taken by Clausius (1864) and Helmholtz (1874). Abbe gave a more general discussion of the sine condition (1879) as a by-product of his diffraction theory of image formation, to be discussed in the Gaussian approximation in Chapter 9. In other words, the Abbe theory of the image formation of coherent objects at large ray angles gives the sine condition as necessary for good imaging of an extended object.

Real optical systems can be made aplanatic or nearly aplanatic by a variety of devices such as the use of meniscus lenses (Problem 4.2.5). They will be useful only for small off-axis displacements h' and only for a single object plane (Problem 4.2.6). They can be used in microscopes and certain telescopes where the field of view is small and the object plane is fixed. For wider fields of view the aberrations that are quadratic or higher in h' become important, and one cannot reduce them in general without giving up the sine condition and accepting some coma and related effects.

7. *Astigmatism and Curvature of Field:* $W = ({}_2C_{22}\cos^2\phi + {}_2C_{20})\,h'^2 r'^2$. These two aberrations are closely related and are usually discussed together. We shall show that the emerging wave surface has two radii of curvature. The corresponding centers of curvature represent locations for two mutually perpendicular "line images" characteristic of astigmatism.

The aberration function W represents the index n' times the separation (in essentially the $\bar{z}$ direction) between the emerging wave surface Σ' at the exit pupil and a tangent reference sphere Σ_S' of radius L' centered about the paraxial or Gaussian image point P'. Let us take a new reference sphere Σ_R still tangent to the wave surface at the center of the exit pupil at O but having a radius $R \neq L'$. Let W_R be the new aberration function with respect to this new sphere Σ_R. It will be obtained from W by subtracting n' times the distance from the sphere Σ_S' to the sphere Σ_R. This distance is approximately the difference in $\bar{z}$ coordinates of points on the two spheres the same distance r' from the $\bar{z}$ axis (Fig. 4.32). For Σ_R we can write

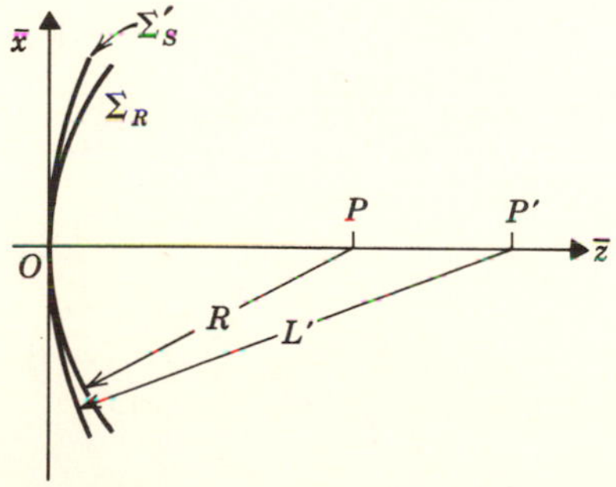

Fig. 4.32

$$\bar{z} = R - \sqrt{R^2 - r'^2} \approx \frac{r'^2}{2R}$$

and hence can express this separation approximately as

$$\frac{r'^2}{2}\left(\frac{1}{L'} - \frac{1}{R}\right)$$

This gives

$$W_R(r', \phi) = W(r', \phi) - \frac{n'r'^2}{2}\left(\frac{1}{L'} - \frac{1}{R}\right) = W(r', \phi) - \frac{n'r'^2(R - L')}{2L'^2} \tag{4.67}$$

The last equality follows when $|L' - R| \ll L'$, which we assume here.

For a fixed value of the aximuthal angle ϕ we can make $W_R = 0$ by proper choice of R. The extreme values of W occur when $\phi = 0$ and $\phi = \pi/2$, and the corresponding values of R that make $W_R = 0$

are obtained from Eq. (4.67):

$$\phi = 0: \qquad R = R_t \text{ with } R_t - L' = \frac{2({}_2C_{22} + {}_2C_{20})h'^2L'^2}{n'} \tag{4.68a}$$

$$\phi = \pi/2: \quad R = R_s \text{ with } R_s - L' = \frac{2\,{}_2C_{20}h'^2L'^2}{n'} \tag{4.68b}$$

These results can also be expressed by saying that the emerging wave surface intersects the planes $\phi = 0$ and $\phi = \pi/2$ to form arcs of circles having respective radii R_t and R_s.

We now show that the centers of these circles represent the centers of line images of the original point source. The new aberration function W_R expressed as a function of $\bar{x} = r' \cos \phi$ and $\bar{y} = r' \sin \phi$ can be used to tell us where the ray normal to the wave surface at x, y intersects a transverse plane at a distance $\bar{z} = R$ from the exit pupil. The reasoning is identical to that used to derive Eqs. (4.54). The intersection point has coordinates in the $\bar{x}, \bar{y}, \bar{z}$ system given by (x', y', R) with

$$x' = \frac{R}{n'}\frac{\partial W_R}{\partial \bar{x}} \approx \frac{L'}{n'}\frac{\partial W_R}{\partial \bar{x}} = 2L'\frac{a\bar{x}}{n'} = 2L'r'\,a\frac{\cos\phi}{n'} \tag{4.69a}$$

$$y' = \frac{R}{n'}\frac{\partial W_R}{\partial \bar{y}} \approx \frac{L'}{n'}\frac{\partial W_R}{\partial \bar{y}} = 2L'\frac{b\bar{y}}{n'} = 2L'r'\,b\frac{\sin\phi}{n'} \tag{4.69b}$$

where

$$a = {}_2C_{22}h'^2 + {}_2C_{20}h'^2 - \frac{n'(R-L')}{2L'^2} \tag{4.70a}$$

$$b = {}_2C_{20}h'^2 - \frac{n'(R-L')}{2L'^2} \tag{4.70b}$$

Note that for fixed radius r' at the exit pupil the coordinates x', y' sweep out an ellipse as ϕ varies. The ellipse has semiaxes of length

$$2L'r'\frac{a}{n'} \qquad \text{and} \qquad 2L'r'\frac{b}{n'}$$

When either b or a is zero, the ellipse degenerates into a straight line. The case $b = 0$ occurs when $R = R_s$ (Eq. 4.68*b*). Then we have

$$y' = 0, \quad x' = 2L'r'\,a\frac{\cos\phi}{n'}, \qquad \text{with } a = {}_2C_{22}h'^2$$

This line extends in the $\bar{x}$ direction and lies in the plane containing O, the lens axis, and P' (see Fig. 4.33). This is called the *sagittal* or *radial* image.

When $a = 0$ we have $R = R_t$ (Eq. 4.68*a*), and then

$$x' = 0, \quad y' = 2L'r'\,b\frac{\sin}{n'}\phi, \quad \text{where now } b = -{}_2C_{22}h'^2$$

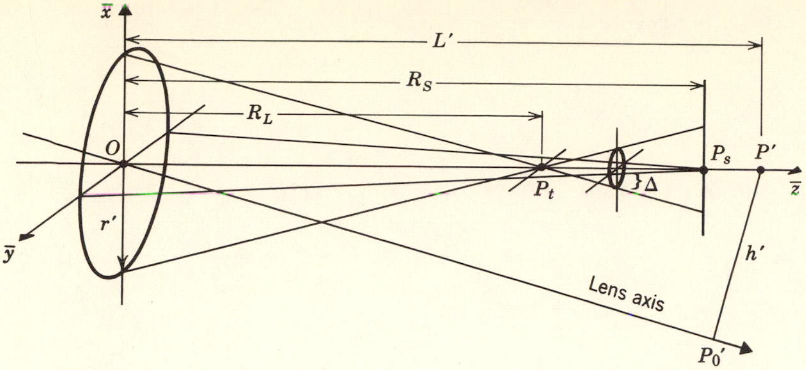

Fig. 4.33

This line image extends in the $\bar{y}$ direction and is called the *tangential* image.

When R is halfway between R_t and R_s, it is easy to show that $a = -b$. The ellipse represented by Eqs. (4.69) then is a circle of radius

$$\Delta = L'r'\frac{{}_2C_{22}h'^2}{n'} \tag{4.71}$$

which represents the best compromise between the two line images at P_t and P_s in Fig. 4.33.

The distances along the $\bar{z}$ axis from the Gaussian image point P' to the tangential and sagittal images P_t and P_s are given by

$$\Delta S_t' = R_t - L' = 2({}_2C_{22} + {}_2C_{20})h'^2\frac{L'^2}{n'} \tag{4.72a}$$

and

$$\Delta S_s' = R_s - L' = 2\,{}_2C_{20}h'^2\frac{L'^2}{n'} \tag{4.72b}$$

Note that these distances are proportional to h'^2. As the off-axis distance h' varies, the loci of these 2 image points sweep out two paraboloids of revolution σ_t, σ_s as shown in Fig. 4.34. The coefficient of h'^2 in Eqs. (4.72*a*,*b*) is the inverse of twice the radius of curvature

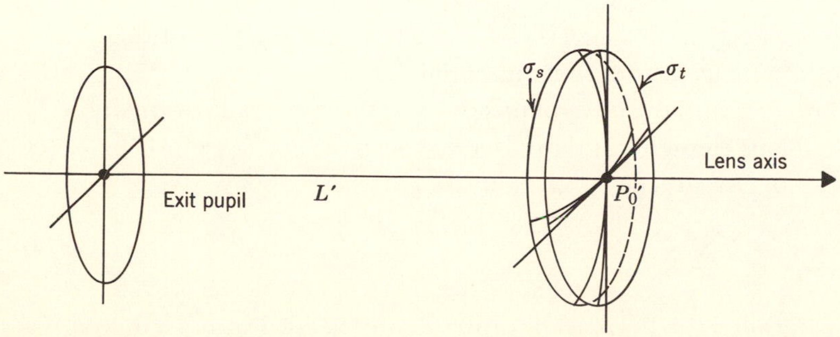

Fig. 4.34

(at the on-axis Gaussian image P'_0) of these paraboloids. Thus, if we write

$$S'_t = \frac{h'^2}{2r_t} \qquad \text{and} \quad S'_r = \frac{h'^2}{2r_s}$$

we find that the radii of curvature are given by

$$r_t = \frac{n'}{4({}_2C_{22} + {}_2C_{20})L'^2} \tag{4.73a}$$

$$r_s = \frac{n'}{4{}_2C_{20}L'^2} \tag{4.73b}$$

When ${}_2C_{22} = 0$ there is no astigmatism, and σ_t and σ_s coincide to form a single curved surface called the *Petzval surface*. This yields a curved image field from a plane object field. This *curvature of field*, when present in a photographic lens, requires a curved photographic film or plate to obtain the sharpest image.

When there is astigmatism, the compromise surface σ_c halfway between σ_t and σ_s generally gives the best overall image of a general object. Then σ_c will be a plane if σ_t and σ_s have equal and opposite curvature.

Even with astigmatism present, the Petzval surface can be defined. It has the property, within the framework of the third-order theory, of being three times as far from the tangential surface as from the sagittal surface. Examples are shown in Fig. 4.35. Notice that the Petzval surface p is always on the opposite side of the sagittal surface s from the tangential surface t.

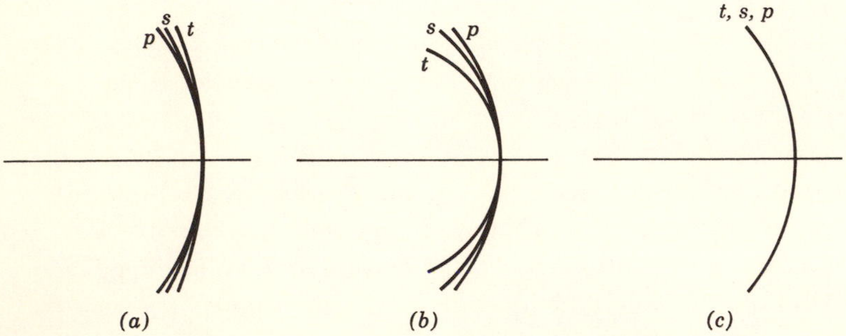

Fig. 4.35. (*a*) Overcorrected astigmatism. (*b*) Undercorrected astigmatism. (*c*) No astigmatism. Pure field curvature.

The Petzval surface depends only on the powers P_i of the individual lens elements and on their respective refractive indices n_i. Its radius of curvature r_p is given by

$$\frac{1}{r_p} = \sum_i \frac{P_i}{n_i} = \sum_i \frac{1}{n_i f_i} \tag{4.74}$$

This is called the Petzval curvature or the Petzval sum. In many systems the spacings of the components can be varied to eliminate primary astigmatism. Equation (4.74) says, however, that the Petzval surface will remain unchanged.

When a higher-order aberration calculation is performed, we find that σ_t and σ_s are no longer paraboloids of revolution, and they often bend back toward the plane through P'_0.

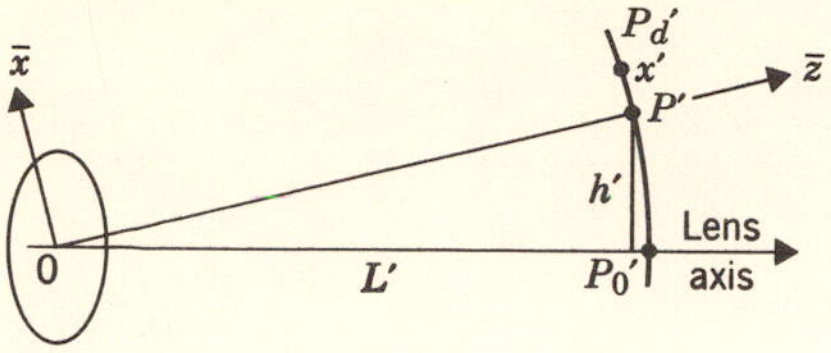

Fig. 4.36

8. Distortion: $W = {}_3C_{11}h'^3r'\cos\phi = {}_3C_{11}h'^3\bar{x}$. With a system that shows only primary distortion, we obtain a perfect point image from an off-axis point object, but the image is not at the Gaussian image point P'. Instead it is at P'_d a lateral distance x' away in the plane containing O, P', and the lens axis as shown in Fig. 4.36. From Eq. (4.54) we obtain the result

$$x' = \frac{L'}{n'}\frac{\partial W}{\partial \bar{x}} = \frac{L'_3C_{11}}{n'}h'^3 \tag{4.75}$$

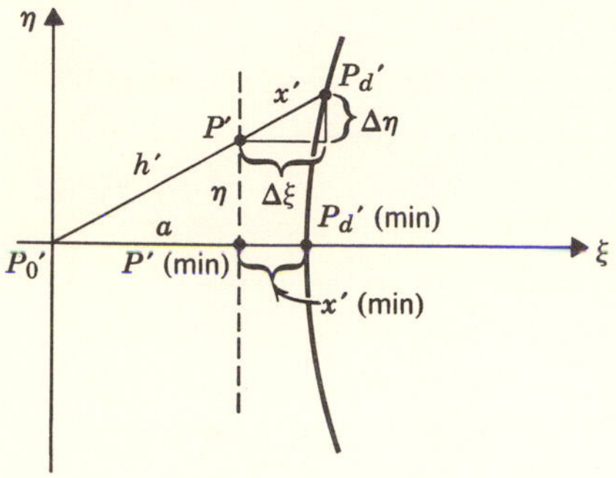

Fig. 4.37. The plane of the paper is perpendicular to the axis of the system.

Since x' is independent of r' and ϕ, all rays normal to the emerging wave surface will form the image at P'_d.

Now consider a line object that gives a Gaussian line image not passing through the lens axis. This is shown dotted in Fig. 4.37. Let a be the distance from the Gaussian line image to the axial image point P'_0. This is the minimum value of h', and it gives a minimum image displacement from $P'(\text{min})$ to $P'_d(\text{min})$ by an amount

$$x'_{\text{min}} = L'\,{}_3C_{11}\frac{a^3}{n'} \tag{4.76}$$

For each point P' on the Gaussian line image, there corresponds an image point P'_d that lies along the line from P'_0 through P'. We pick a Cartesian coordinate system (ξ, η) with origin at P'_0 and ξ axis along the line from P'_0 to $P'(\text{min})$ as shown in Fig. 4.37. Then P' will have coordinates (a, η) and P'_d will be obtained from P' by a displacement of amount x' along the line from P'_0 to P'. Thus we can write

$$\overrightarrow{P'P'_d} = (\Delta\xi, \Delta\eta) = \frac{x'}{h'}(a, \eta) = \frac{L'_3C_{11}h'^2}{n'}(a, \eta)$$

$$= \frac{L'_3C_{11}(a^2 + \eta^2)}{n'}(a, \eta)$$

or

$$\Delta\xi = x'_{\text{min}} + L'\frac{{}_3C_{11}a\eta^2}{n'} \tag{4.77a}$$

$$\Delta\eta = L'\frac{{}_3C_{11}}{n'}(a^2 + \eta^2)\eta \tag{4.77b}$$

For small ${}_3C_{11}$ the distorted image is a parabola with vertex at $P'_d(\text{min})$. It will curve as shown in Fig. 4.37 if ${}_3C_{11}$ is positive. This causes *pincushion distortion*. *Barrel distortion* occurs when ${}_3C_{11}$ is negative. The resultant images of a square grid are shown in Fig. 4.38.

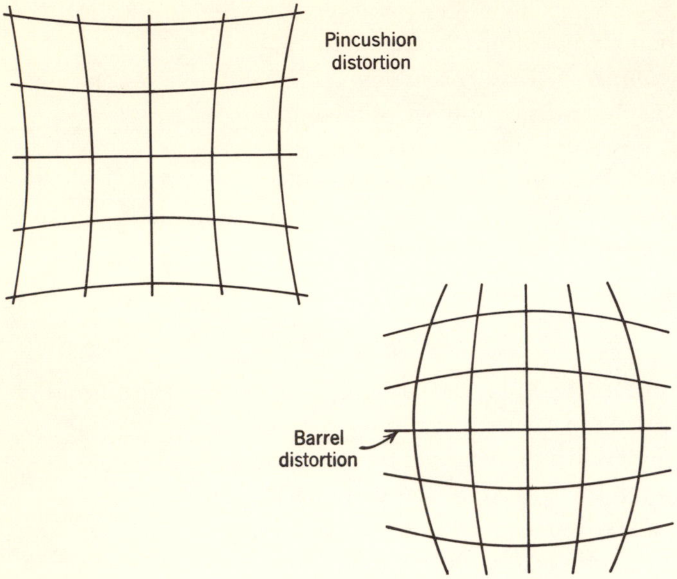

Fig. 4.38

*9. **Some Primary Aberrations of a Spherical Mirror.*** Consider the spherical mirror of Fig. 4.39. The sign convention for mirrors was discussed between Eqs. (3.22) and (3.23). Let the object be at infinity. An on-axis object will have a paraxial image at P_0', the focal point. An off-axis paraxial image is at P'. The exit pupil is a distance L' from the image plane and a distance $l = S' - L'| = |R/2| - L'$ from the mirror. The sign convention is such that distances in

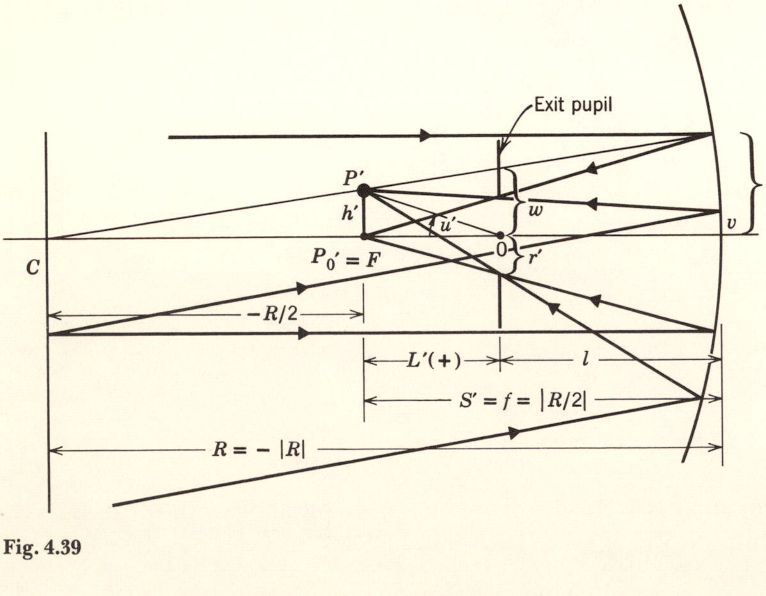

Fig. 4.39

image space are positive from right to left, consistent with the direction of the emerging rays. The radius of curvature R is negative if C is to the left of V, but the focal length $f = -R/2$ is positive.

From Eq. (2.54), for the aberration function referred to the mirror, we have

$$I(\rho) = -\frac{C\rho^4}{4}$$

$$C = \frac{1}{2}\left(\frac{1}{D_{01}} - \frac{1}{R}\right)^2 \left(\frac{1}{D_{01}} + \frac{1}{D_{12}}\right)$$

Letting $D_{01} \to \infty$, $D_{12} \to f = -R/2$, we obtain

$$C = -\frac{1}{R^3} = \frac{1}{8f^3} \tag{4.78}$$

The off-axis aberration function for the mirror is still given by Eq. (4.47):

$$W = -\frac{C}{4}\left(\frac{S'}{L'}\right)^4 [r'^4 + 4b\,h'\,r'^3 \cos\phi + 4b^3\,h'^3\,r' \cos\phi$$

$$+ 2b^2 h'^2 r'^2 \,(2\cos^2\phi + 1)] \tag{4.79a}$$

where

$$b \equiv \frac{w}{h'} = \frac{-R - l}{-R/2} = \frac{2f - l}{f} \tag{4.79b}$$

and where C is given by (4.78).

The coefficient of r'^4 is the spherical aberration coefficient ${}_0C_{40}$:

$${}_0C_{40} = -\frac{f}{32L'^4}$$

For the lateral spherical aberration from a circular zone at the exit pupil of radius r', Eq. (4.48*a*) gives

$$x' = 4L'\,{}_0C_{40} r'^3 = -\frac{1}{8} f \left(\frac{r'}{L'}\right)^3$$

The *F number* of the mirror is defined by

$$(F \text{ number}) = \frac{f}{2\rho'} = \frac{L'}{2r'}$$

This gives

$$\frac{x'(\text{max})}{f} = -\frac{1}{64(F \text{ number})^3} \tag{4.80}$$

The *comatic circle* from a zone of radius r' at the exit pupil will have a radius A given by Eq. (4.56):

$$A = {}_1C_{31} h'\,r'^2 L'$$

From Eqs. (4.79*a*) and (4.47′) we have

$$ {}_1C_{31} = 4b_0C_{40} = \frac{2f-l}{-8L'^4} $$

Hence

$$ A = -\frac{(2f-l)}{8}\left(\frac{h'}{L'}\right)\left(\frac{r'}{L'}\right)^2 $$

or

$$ \frac{A(\max)}{f} = -\frac{1}{16}\left(\frac{2f-l}{2f}\right)\frac{u'}{(F\text{ number})^2} \tag{4.81} $$

where $u' = h'/L'$ is the angle subtended by the off-axis Gaussian image point at the center of the exit pupil.

As for *astigmatism*, the compromise "best focus" gives a circle of radius

$$ \Delta = |L'r'\,{}_2C_{22}h'^2| $$

A comparison of Eqs. (4.47′) and (4.79*a*) gives

$$ {}_2C_{22} = 4b^2\,{}_0C_{40} = \frac{(2f-l)^2}{-8fL'^4} $$

Thus

$$ \Delta = \frac{(2f-l)^2}{8f}\left(\frac{h'}{L'}\right)^2\left(\frac{r'}{L'}\right) $$

and

$$ \frac{\Delta_{\max}}{f} = \frac{1}{4}\left(\frac{2f-l}{2f}\right)^2\frac{u'^2}{(F\text{ number})} \tag{4.82} $$

Note from Eqs. (4.81) and (4.82) that primary coma and astigmatism vanish when $l = 2f = |R|$. In this case the stop is at the center of curvature C.

D. CHROMATIC ABERRATIONS

Since most optical systems are used with light of varying wavelengths, the dispersion of the refraction media must be taken into account. The optical designer must consider the variation of refractive index with wavelength so that the resulting performance of the system is more or less wavelength-independent.

We can distinguish two general types of chromatic aberrations: (1) chromatic variations in the paraxial image-forming properties of a system, and (2) wavelength dependence of the monochromatic aberrations. We shall concentrate here primarily on the first type.

In Gaussian optics the properties of an image-forming system depend solely on the locations of the principal planes and the focal planes. Chromatic aberrations result when the position of any of these planes is wavelength-dependent. We can discuss the results in terms of a variation in image distance along the axis and a variation in transverse magnification. Since a single refracting surface shows behavior that can be observed in more complex systems, we shall start with it.

1. Paraxial Chromatic Aberrations of a Single Refracting Surface. Consider first an object point P on the axis of the system. With refractive indices as shown in Fig. 4.40, the image formation condition is

$$\frac{1}{S}+\frac{n}{S'}=\frac{n-1}{R} \tag{4.83}$$

Let us denote two different wavelengths by r and b (for red and blue, typically) and assume that the indices have the property $n(b) > n(r)$. The change in index $\delta n = n(b) - n(r)$ will be small compared with unity for real optical media, and we may use approximations to first order in δn. Then we let $\delta S' = S'(b) - S'(r)$ represent the longitudinal chromatic aberration, and by differentiating Eq. (4.83) we obtain

$$\delta\left(\frac{1}{S'}\right) = -\frac{\delta S'}{S'^2} = \frac{\delta n}{n}\left(\frac{1}{R}-\frac{1}{S'}\right) \tag{4.84}$$

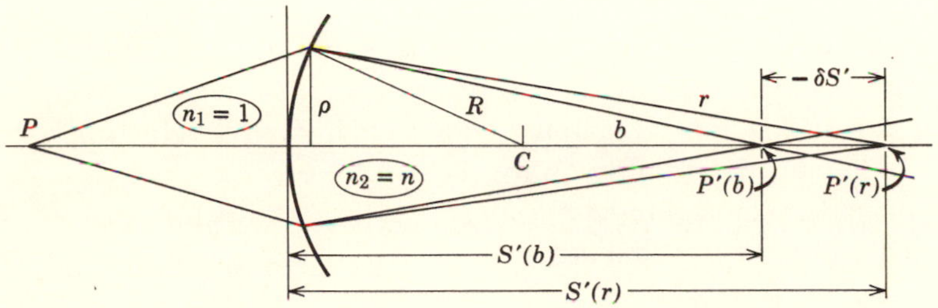

Fig. 4.40

Here n and S' can be evaluated at any convenient wavelength, but best accuracy is obtained with a wavelength between r and b.

At $P'(b)$, the focus for b light, the rays of r light will form a cone converging toward $P'(r)$. The cone will have a half-angle of approximately ρ/S', and it will intersect the b image plane in a disk of radius

$$x' = -\delta S' \frac{\rho}{S'}$$

$$= S' \rho \frac{\delta n}{n}\left(\frac{1}{R}-\frac{1}{S'}\right) \tag{4.85}$$

A disk of approximately the same radius is formed by b rays in the r image plane.

A stop in the system will define an axis through its center and through the center of curvature. An off-axis object point will produce images in r and b light at different heights from the axis. This gives a wavelength-dependent lateral magnification. This is shown in Fig. 4.41. The object point P has chromatic Gaussian

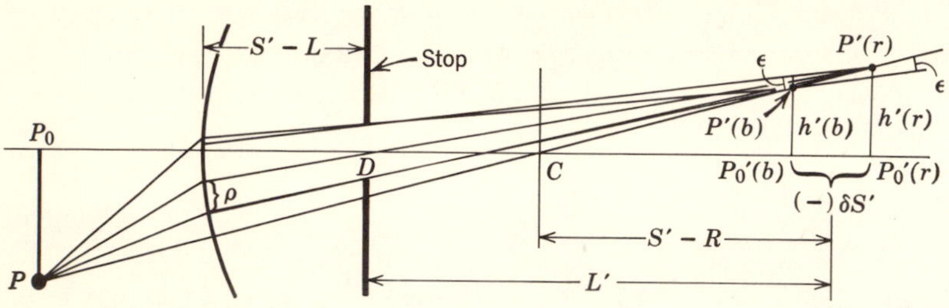

Fig. 4.41

point images at $P'(b)$ and $P'(r)$ that lie along the straight line from P through the center of curvature at C. The on-axis point P_0 gives images at $P_0'(b)$ and $P_0'(r)$ that lie on the axis of the system.

A cone of r light that fills the exit pupil will converge to $P'(r)$. If r' is the pupil radius, the cone will have a radius $\rho = r'S'/L'$ at the refracting surface. Equation (4.85) then gives the disk radius as

$$x' = r'S'^2 \frac{\delta n}{nL'}\left(\frac{1}{R} - \frac{1}{S'}\right) \tag{4.86}$$

The positions of these point and disk images are shown in the r and b image planes in Fig. 4.42. If the r disk is a height ϵ above $P'(b)$, then the b disk is approximately a height ϵ below $P'(r)$. The transverse magnification is different for the two colors.

We can determine ϵ and $\delta h' = h'(b) - h'(r)$ with the help of Fig. 4.43. Note first of all that $h'/(S' - r)$ is a constant independent

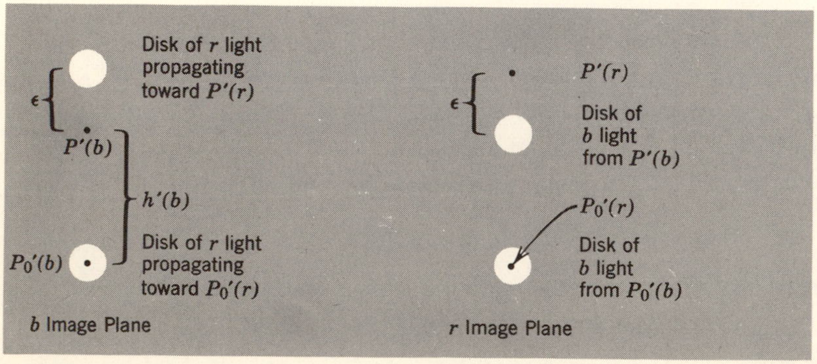

Fig. 4.42

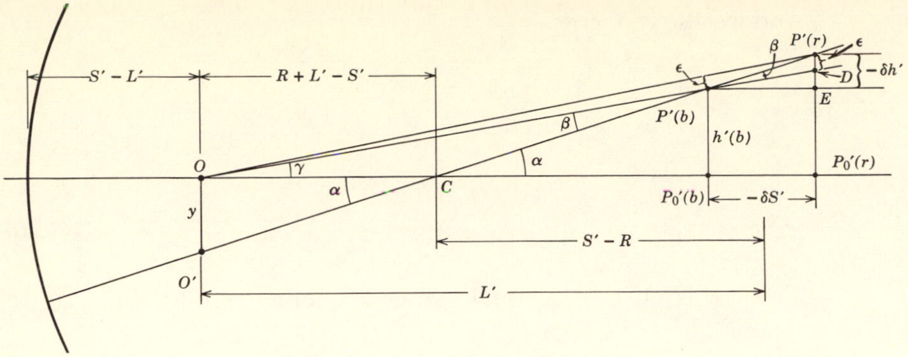

Fig. 4.43. An enlargement of part of Fig. 4.41.

of wavelength. Thus we have

$$\frac{\delta h'}{h'} = \frac{\delta S'}{S' - R}$$

or

$$\delta h' = \frac{h'}{S' - R}\delta S' = \alpha \delta S' \tag{4.87}$$

The last equality followed from triangle $(P'(b)EP'(r))$. From triangle $(OO'C)$ we have

$$\alpha = \frac{y}{R + L' - S'}$$

From triangle $(OO'P'(b))$ we have

$$\beta = \frac{y}{L'}$$

Thus

$$\beta = \frac{R + L' - S'}{L'}\alpha$$

Since from triangle $(P'(b)P'(r)D)$ we have $\beta = -\epsilon/\delta S'$, we can write

$$\epsilon = -\frac{(R + L' - S')}{L'}\delta S' \, \alpha = -\frac{(R + L' - S')}{L'} \frac{\delta S' h'}{(S' - R)} \tag{4.88}$$

The last equality follows from (4.87).

Equation (4.85) can be written

$$-\delta S' = \frac{\delta n}{n}\left(\frac{S' - R}{R}\right)S'$$

Then (4.88) becomes

$$\epsilon = \frac{h' \delta n}{n} \frac{S'(R - L - S')}{RL'} \tag{4.89}$$

And we also can write

$$\delta h' = \alpha\,\delta S' = \left(-\frac{\alpha}{\beta}\right)(-\beta\,\delta S') = -\frac{\alpha}{\beta}\epsilon$$

giving

$$-\delta h' = \frac{S'\,h'\,\delta n}{R\,n} \tag{4.90}$$

This yields a fixed fractional change in lateral magnification that is independent of h':

$$\frac{\delta h'}{h'} = -\frac{S'\,\delta n}{Rn}$$

In complex systems with several refracting surfaces we can often remove the longitudinal chromatic aberration and make $P'_0(b)$ and $P'_0(r)$ coincide for a particular object distance. Unless the different-colored off-axis rays themselves coincide, in general we still have $P'(b)$ and $P'(r)$ at different heights from the axis, giving a wavelength-dependent lateral magnification.

2. ***Two Thin Lenses in Contact.*** Given two thin lenses with essentially zero separation d, we can write for the power of the combination

$$\frac{1}{f} = P = P_1 + P_2$$

with

$$P_1 = \frac{1}{f_1} = (n_1 - 1)\left(\frac{1}{R_1} - \frac{1}{R'_1}\right) \equiv (n_1 - 1)\,K_1 \tag{4.91}$$

and

$$P_2 = \frac{1}{f_2} = (n_2 - 1)\left(\frac{1}{R_2} - \frac{1}{R'_2}\right) \equiv (n_2 - 1)K_2$$

The quantities K_1, K_2 are called the total curvature of each lens. The change in total power caused by a change in wavelength is

$$\delta P = \delta n_1\,K_1 + \delta n_2\,K_2$$

$$= \frac{\delta n_1}{n_1 - 1}\cdot(n_1 - 1)K_1 + \frac{\delta n_2}{n_2 - 1}\cdot(n_2 - 1)K_2 = \frac{P_1}{V_1} + \frac{P_2}{V_2} \tag{4.92}$$

We used (4.91) in the last equality and have introduced the *dispersion constants* of the glasses in the two lenses, which are defined as follows:

$$V_1 = \frac{n_1 - 1}{\delta n_1}, \qquad V_2 = \frac{n_2 - 1}{\delta n_2} \tag{4.93}$$

The wavelengths at which the n's and δn's in (4.93) are determined are taken conventionally to be the following:

$$\delta n = n(F) - n(C)$$
$$n = n(D),$$

where F represents the blue line of hydrogen at 486.1 nm, C the red line of hydrogen at 656.3 nm, and D the yellow lines of sodium at 589.3 nm. The D line is close to the wavelength of maximum response of the human eye. The F and C lines give a convenient spread across the visible spectrum. Data on indices of refraction and dispersion constants are supplied by the glass manufactures often as a 6-digit number; for example, 517645 means $n(D) = 1.517$, $V = 64.5$.

Some black-and-white film has peak sensitivity in the blue region of the spectrum; we would then want to select a camera lens achromatized for shorter wavelengths than the F and C lines.

We return now to Eq. (4.92). For achromatization at the two wavelengths used to define δn, we set $\delta P = 0$ and through use of Eq. (4.93) obtain

$$\frac{P_2}{P_1} = -\frac{V_2}{V_1} \tag{4.94a}$$

or

$$f_1 V_1 + f_2 V_2 = 0 \tag{4.94b}$$

By writing

$$P = P_1 + P_2 = P_2 - \left(\frac{V_1}{V_2}\right) P_2$$

we obtain

$$P_2 = \frac{V_2}{V_2 - V_1} P \qquad P_1 = \frac{-V_1}{V_2 - V_1} P \tag{4.95}$$

Equations (4.95) give the necessary condition for our thin lens to be achromatized for the two wavelengths used to define δn. The chromatic aberration that remains at other wavelengths is called the secondary spectrum and is often small enough to be neglected.

Note that P_1 and P_2 must have opposite signs. If P is to be positive, we need $P_1 > -P_2$, assuming that P_1 is positive. Then by (4.94*a*) we want

$$\frac{V_1}{V_2} = \frac{P_1}{-P_2} > 1$$

Thus the positive component must have the larger dispersion constant (and hence the smaller dispersion δn).

To control the secondary spectrum and to minimize the monochromatic aberrations, we want to keep the magnitudes of P_1 and P_2 small. This can be done if $|V_2 - V_1|$ is large.

The shape factors of the individual lenses do not influence Eqs. (4.95). These can be varied to reduce spherical aberration and coma. For inexpensive achromatic doublets the shape factors are more likely to be chosen for ease of manufacture. For instance, the positive element is often equiconvex; the negative element then becomes almost planoconcave for typical glasses used in a cemented doublet.

As an example, consider a cemented doublet made of borosilicate crown glass #517645, $n(D) = 1.517$, $V = 64.5$, as the positive element and dense flint glass #617366 as the negative element. Then if the focal length is to be 10 cm = 0.1 m, the power P will be 10 m^{-1} (diopter). Equations (4.88) then give $P_2 = -13.118$ diopters and $P_1 = 23.118$ diopters. If the positive element is equiconvex, the radii of curvature then are $R_1 = -R_1' = 4.47$ cm. Since the elements are in contact, we must have $R_2 = -4.47$ cm. Then we can calculate R_2' to be 91.2 cm. The lens is illustrated in Fig. 4.44.

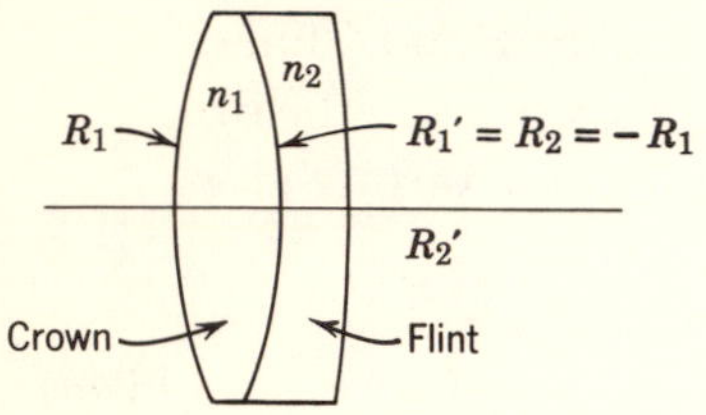

Fig. 4.44. Cemented doublet.

3. ***Separated Doublet.*** A separated doublet lens in air consisting of components of powers P_1 and P_2 a distance d apart will have a total power given by Eq. (3.91):

$$\frac{1}{f} = P = P_1 + P_2 - P_1 P_2 d$$

Then the change in P at two different wavelengths is given by

$$\delta P = \delta P_1 + \delta P_2 - d[P_2\,\delta P_1 + P_1\,\delta P_2]$$

Suppose, that the two lenses are both simple thin lenses. As above, we write

$$P_1 = (n_1 - 1)K_1 = (n_1 - 1)\left(\frac{1}{R_1} - \frac{1}{R_1'}\right)$$

$$P_2 = (n_2 - 1)K_1 = (n_2 - 1)\left(\frac{1}{R_2} - \frac{1}{R_2'}\right)$$

and obtain the result $\delta P_1 = P_1/V_1$, $\delta P_2 = P_2/V_2$ used in Eq. (4.92). Then we set $\delta P = 0$ and obtain

$$0 = \frac{P_1}{V_1} + \frac{P_2}{V_2} - dP_1P_2\left(\frac{1}{V_1} + \frac{1}{V_2}\right)$$

If the lenses are made from glass of the same index, we have $V_1 = V_2$, and we obtain

$$P_1 + P_2 - 2dP_1P_2 = 0$$

or

$$2d = \frac{1}{P_1} + \frac{1}{P_2} = f_1 + f_2 \tag{4.96}$$

Equation (4.96) is the condition that the doublet have a focal length independent of wavelength (to first order in δn). Doublets are often used as low-cost eyepieces in telescopes and microscopes, and we shall discuss some examples shortly. The function of the eyepiece is to magnify the intermediate image formed by the objective lens. The final virtual-image distance S' is much greater than the focal length of the eyepiece, and the object distance is very close to the focal length f. Under these conditions it is the angular size $\theta' = x/f$ of the image that is important as in Eq. (3.98), or alternatively its linear size at the conventional viewing distance of 250 mm. Thus the magnification of the eyepiece will be achromatized if f is. We say that the eyepiece is corrected for "lateral color."

4. Examples of Eyepieces. In general, when Eq. (4.96) is satisfied we have

$$f = \frac{f_1 f_2}{d} \tag{4.97}$$

The Petzval curvature of Eq. (4.74) is

$$\frac{1}{r_p} = \frac{1}{n}(P_1 + P_2) = \frac{2d}{n} P_1 P_2 = \frac{2d}{n f_1 f_2} = \frac{2}{nf} \tag{4.98}$$

$$\frac{1}{r_p} = \frac{1.33}{f} \qquad \text{for} \quad n = 1.5.$$

Now as an example of a doublet corrected for lateral color, consider the *Huygens eyepiece* of Fig. 4.45. For simplicity it is convenient to assume that the final virtual image is at infinity. Then the intermediate image formed by the objective lens of the

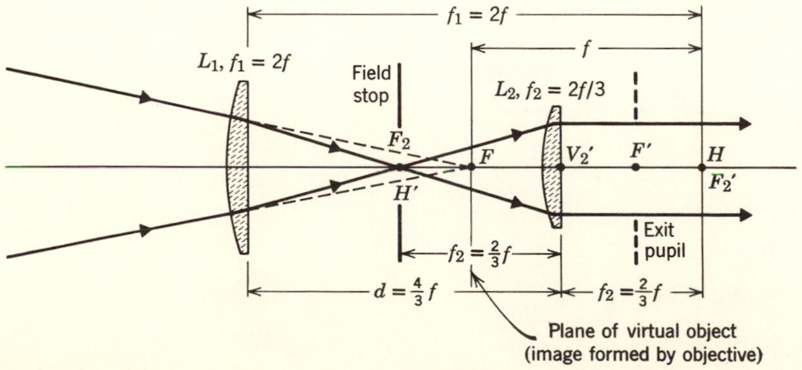

Fig. 4.45. Huygens eyepiece.

telescope or microscope must be in the first focal plane through F of the eyepiece. It acts as a virtual object for the eyepiece. The rays from the object are actually brought to a focus at F_2, the first focal plane of L_2. This is also the second principal plane for the eyepiece. This is where the stop that limits the field of view (field stop) should be placed. Here also should be put the cross hairs or reticles, so that they can be imaged by L_2 in the same plane as the final image. Unfortunately, this imaging is uncorrected, since L_2 alone is used; thus the Huygens eyepiece is not suitable for use with a reticle.

The eyepiece can be corrected for coma by proper selection of n and of f_2/f_1, but it has a relatively large amount of spherical aberration.

The objective lens of a telescope or microscope is the aperture stop. Its distance from the eyepiece is so much larger than the focal length of the eyepiece that the image of it formed by the eyepiece is almost in its second focal plane, through F' in Fig. 4.45. This image, of course, is the exit pupil, and it should coincide approximately with the pupil of the observer's eye. The distance $|V_2'F'|$ from the second vertex of the second lens is called the *eye relief*. For the lens shown in Fig. 4.45 this distance is $f/3$, if we may ignore the thickness of L_2.

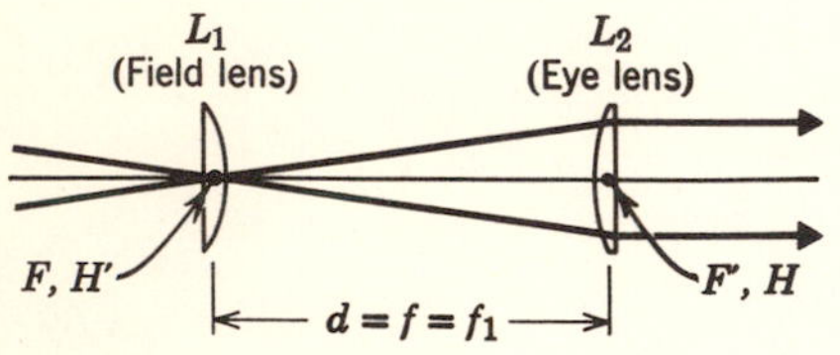

Fig. 4.46

Now let us pick $f_1 = f_2$. Then to correct for lateral color, Eq. (4.96) yields $d = f_1$ and (4.97) gives $f = f_1$. The resulting eyepiece is shown in Fig. 4.46. Here L_1 acts as a field lens, imaging the objective onto L_2 and preventing vignetting. The intermediate image is now coincident with L_1 and is inaccessible for placement of cross hairs or reticles. Another disadvantage comes with the scratches or dust on L_1 that is in focus along with the image. A third disadvantage lies in the zero eye relief, since the exit pupil is at L_2.

The disadvantages of the lens of Fig. 4.46 can be overcome if we sacrifice some correction for lateral color and move the lenses closer together. The case where $d = 2f_1/3$ is illustrated in Fig. 4.47. This is called a *Ramsden eyepiece* and will have $f = 3f_1/4$. The lenses

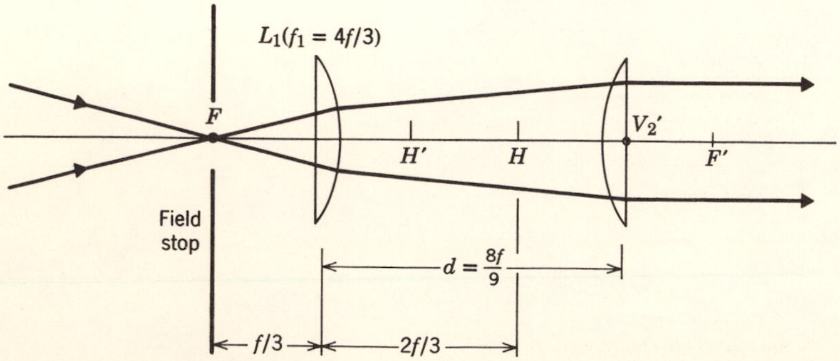

Fig. 4.47. Ramsden eyepiece.

are oriented as shown to reduce spherical aberration and eliminate coma. The eye relief is $f/3$. The Petzval curvature is

$$\frac{1}{r_p} = \frac{2}{nf_1} = \frac{3}{2nf}$$

$$\frac{1}{r_p} = \frac{1}{f} \qquad \text{for} \quad n = 1.5$$

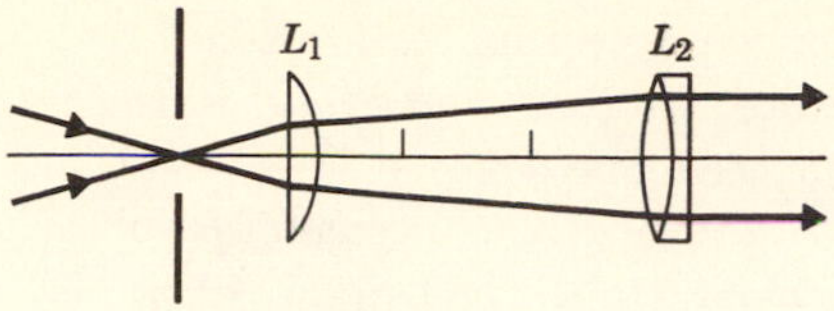

Fig. 4.48. Kellner eyepiece.

There is now some lateral chromatic aberration. The Kellner eyepiece (Fig. 4.48) is a Ramsden eyepiece with the eye lens L_2 achromatized to reduce this lateral color.

REFERENCES FOR SECTION 4.2

A. Cox, *Photographic Optics*. Pages 98–134.
F. A. Jenkins and H. E. White, *Fundamentals of Optics*. Chapter 9.
W. J. Smith, *Modern Optical Engineering*. Chapters 3, 11, 12, and 13.
A. C. Hardy and F. H. Perrin, *The Principles of Optics*.
L. C. Martin, *The Theory of the Microscope*. Sections 3.15 through 3.26.
M. Born and E. Wolf, *Principles of Optics*. Chapter 5 (section 4.5 for the sine condition).
L. Levi, *Applied Optics*. Chapter 9.

PROBLEMS

4.2.1 A uniform parallel beam 3 mm in diameter from a helium-neon laser of wavelength 633 nm is brought to a focus by a 1-m focal length planoconvex lens made with glass of index $n = 1.500$. The lens is oriented to give the best performance. Calculate the longitudinal and lateral spherical aberration as multiples of the wavelength.

4.2.2 A commercial spectrophotometer uses a $\frac{1}{2}$-*m* focal length mirror in an $F/10$ system (F number = 10) with an exit pupil $\frac{1}{4}$ *m* from the mirror. Parallel light incident on the mirror gives a paraxial point image 1 cm off axis. Discuss the size and shape of the true "image" resulting from primary spherical aberration, coma, and astigmatism.

4.2.3 The C coefficient for a combination of two or more thin lenses essentially in contact will be the sum of the individual C coefficients. Why? Consider the lens of Fig. 4.44, which was discussed as an example of an achromatic contact doublet. If we give up the contact requirement ($R_1' = R_2$) and the requirement that $R_1 = -R_1'$, then we can design a lens free from primary coma and primary spherical aberration that still is achromatic. Do this.

4.2.4 This is the spherical system already considered in Problem 2.2.2, where you were asked to show that P_2 is a perfect image of P_1 provided

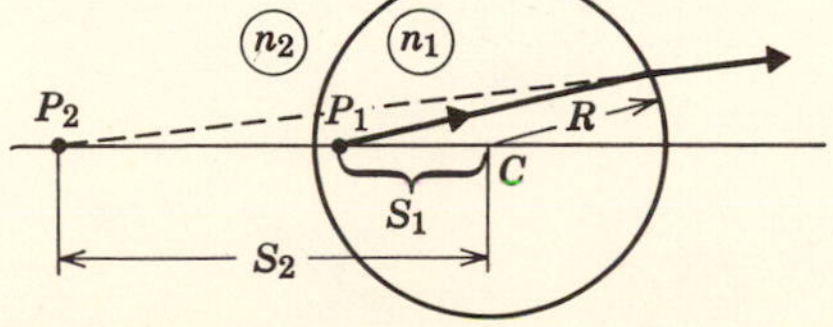

that $n_1^2 S_1 = n_2^2 S_2 = n_1 n_2 R$. Now show that P_1 and P_2 form an aplanatic pair; that is, show that the sine condition is obeyed.

4.2.5 One practical method of obtaining an object at P_1 in Problem 4.2.4 is to use a meniscus lens and to put the object at the center of curvature C_1 of the first surface. Then the first surface forms a perfect image of C_1 at C_1. The meniscus lens has a natural axis through the two centers of curvature C_1 and C_2. Show that the imaging by the first surface of an object O a short distance x above this axis from C_1 will satisfy the sine condition and that the image I lies along the line from C_1 to O. (The imaging of the second surface was shown to be aplanatic in Problem 4.2.4. The final image will be along the line $C_2 I$.) Hence the overall imaging is aplanatic. This property of meniscus lenses is often used in microscope objectives.

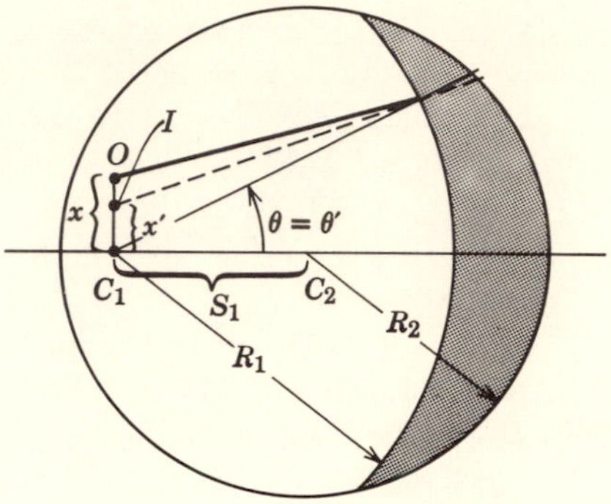

4.2.6 A given optical system has at most one pair of aplanatic points. In fact, the system gives perfect image formation only for the aplanatic pair of points and for points displaced laterally a small distance x. It will not give perfect image formation for points displaced axially a small distance z. Show that the condition for perfect image formation of P_1 onto P_1' and of P_2 onto P_2' gives this expression to first order in z:

$$n' z' \sin^2 \tfrac{1}{2}\theta' = n z \sin^2 \tfrac{1}{2}\theta$$

(*Hint.* Use the rays shown.)

Except for a plane mirror, it is incompatible with the sine condition. Why?

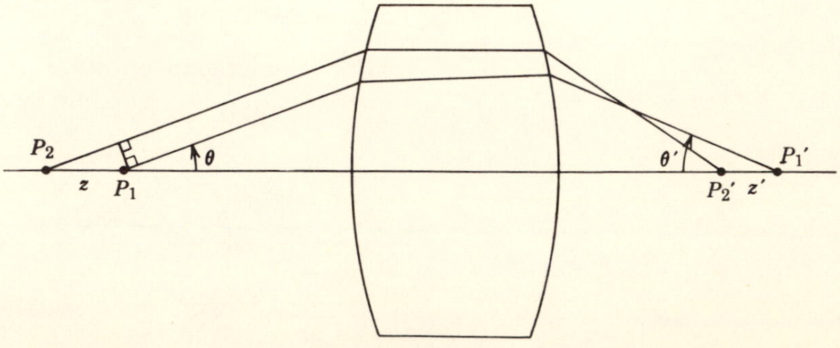

4.2.7 a. Consider an aplanatic pair of points P_1, P_1'. Draw a sphere of radius S about P_1 and one of radius S' about P_1', where S and S' are the Gaussian object and image distances measured from the principal points H and H'. The ray P_1Q comes out as the ray $Q'P_1'$. Show that $h' = h$.

b. It is tempting to say that the principal "planes are now the spheres HQ and $H'Q'$. But the spheres are not conjugate to each other for large θ; that is, Q' is not an approximate image of Q. Consider the expression

$$\Delta = [\text{O.P.L.}(P_2QQ'P_2') - \text{O.P.L.}(P_2Q) - \text{O.P.L.}(Q'P_2')]$$
$$- [\text{O.P.L.}(P_1QQ'P_1') - \text{O.P.L.}(P_1Q) - \text{O.P.L.}(Q'P_1')]$$

which is the optical path length from Q to Q' of the virtual path $P_2QQ'P_1'$ minus the path length of the true ray $P_1QQ'P_1'$. A measure of the angular deviation of the two paths is provided by the distance x. As mentioned in the discussion surrounding Eq. (2.31), Δ would have to be at least third order in x for Q' to be an approximate image of Q. Show that the leading term in Δ is

$$\frac{nx^2}{2S}[\sin^2\theta - \sin^2\theta'].$$

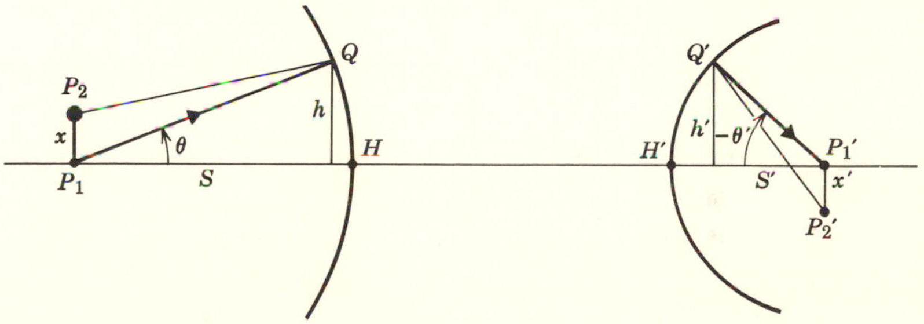

4.2.8 Show that to have an aplanatic pair of points with the object at infinity we must have $r = h'/\sin\theta' = \text{const} = f'$.

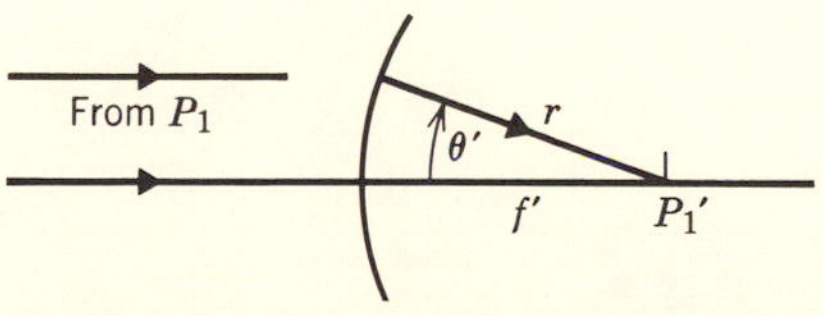

4.2.9 The Huygens eyepiece of Fig. 4.45 made with glass having an index $n(D) = 1.500$ with a dispersion constant $V = 65.0$ is adjusted to give an image of an achromatic object (virtual) that is at infinity for blue light. Where is the image for red light?

4.2.10 Calculate the lateral chromatic aberration, expressed as fractional change in magnification, for the Ramsden eyepiece of Fig. 4.47 if the glass used is the same as that of Problem 4.2.9.

We now begin the study of physical optics, *where the wave nature of light becomes apparent. One of the important properties of waves is the superposition principle—the algebraic sum of the amplitudes representing several disturbances also represents a disturbance. The other way of looking at this phenomenon gives us what might be called a* decomposition principle, *which says that a general solution of the wave equation can be decomposed into a sum of elementary solutions. The most elementary disturbance is the* sinusoidal plane wave, *since in media with little dissipation, but with appreciable dispersion (that is, where the velocity depends on wavelength), only a sine wave will propagate with no change in shape, as will be shown in Chapter 11. It is often convenient to decompose a given disturbance into elementary sine waves, by use of the Fourier integral, to be discussed later. One then describes the propagation of the sine waves, and recombines them to give the final propagated disturbance. Thus the three essential steps in this process are (1) decomposition, (2) propagation of sine waves, and (3) recombination. We begin by discussing the second step.*

CHAPTER 5

INTERFERENCE

CHAPTER 5 CONTENTS

5.1 SINUSOIDAL WAVE MOTION

This subject was introduced in Section 1.2. We now use the symbol $E(x, t)$ or $E(\mathbf{r}, t)$ for the disturbance. (It might be one component of the electric field in a light wave, for example.)

A. ELEMENTARY SINUSOIDAL AND EXPONENTIAL WAVES

*1. **Review of Sine Waves.*** The wave equation

$$\nabla^2 E - \frac{1}{v^2}\frac{\partial^2 E}{\partial t^2} = 0 \tag{5.1}$$

has a solution of the form of a sinusoidal wave running in the direction of interest, the x direction.

$$\begin{aligned} E(x, t) &= A \cos\left[2\pi\nu\left(t - \frac{x}{v}\right) + \phi\right] \\ &= A \cos(\omega t - kx + \phi) \end{aligned} \tag{5.2}$$

where ω is the angular frequency, $\omega = 2\pi\nu$; the wave number k obeys

$$vk = \omega \tag{5.3}$$

and is related to the wavelength λ by

$$k = \frac{2\pi}{\lambda} \tag{5.4}$$

Dispersion occurs when the wave or phase velocity v given by

$$v \equiv \frac{\omega}{k}$$

is dependent on k (or ω).

*2. **Standing Wave.*** An important type of interference between two sinusoidal waves of the same frequency moving in opposite directions will produce a *standing wave*. For example, we consider a superposition of the form

$$E(x, t) = A \cos\left[2\pi\nu\left(\frac{t - x}{v}\right) + \phi\right] + A \cos\left[2\pi\nu\left(\frac{t + x}{v}\right) + \psi\right]. \tag{5.5a}$$

We then write

$$\phi = \frac{\phi + \psi}{2} - \frac{\psi - \phi}{2}$$

$$\psi = \frac{\phi + \psi}{2} + \frac{\psi - \phi}{2}$$

and use the trigonometric identity

$$\cos(A \pm B) = \cos A \cos B \mp \sin A \sin B$$

to obtain

$$E(x, t) = 2A \cos\left(2\pi\nu t + \frac{\phi+\psi}{2}\right) \cos\left(\frac{2\pi\nu x}{v} + \frac{\psi-\phi}{2}\right) \quad (5.5b)$$

We thus have a sinusoidal factor in t times one in x. Then ν is the temporal frequency of the first factor: ν = number of oscillations per unit time; $\nu/v = 1/\lambda$ = number of oscillations per unit length. Standing waves can be easily produced on a string or in an organ pipe. Optical standing waves can also be produced upon reflection of well-collimated light from a high-quality mirror.

3. ***Exponential Solutions.*** Equation (5.1) also has complex solutions of the form

$$\tilde{E} = a\, e^{i(\omega t - kx)} \quad (5.6)$$

where $a = a_1 + ia_2$ is a complex number having real part a_1 and imaginary part a_2. The physical disturbances such as s, the transverse displacement of a string; p, the pressure in a fluid; and E, the electric field, are all describable as *real functions of real variables*. Equation (5.6) cannot be taken literally as a physical electrical field, although it is a proper mathematical solution of Eq. (5.1). We note that the *complex conjugate* of Eq. (5.6), obtained by replacing i by $(-i)$,

$$\tilde{E}^* = a^* e^{-i(\omega t - kx)} \quad (5.7)$$

where $a^* = a_1 - ia_2$, is also a solution. By the superposition principle, so is the *real part*

$$E = \text{Re}\, \tilde{E} = \tfrac{1}{2}(\tilde{E} + \tilde{E}^*) \quad (5.8)$$

Equation (5.8) is equivalent to Eq. (5.2), as we now show. The complex number $a = a_1 + ia_2$ can also be written

$$a = Ae^{i\phi} \quad (5.9)$$

where A is real and positive and ϕ is a real angle. Since we have

$$e^{i\phi} = \cos\phi + i\sin\phi \quad (5.10)$$

we can write Eq. (5.9) as

$$a_1 + ia_2 = A\cos\phi + iA\sin\phi \quad (5.11)$$

Equating real and imaginary parts gives us

$$\begin{aligned} a_1 &= A\cos\phi \\ a_2 &= A\sin\phi \\ A &= \sqrt{a_1^2 + a_2^2} \\ \tan\phi &= \frac{a_2}{a_1}. \end{aligned} \quad (5.12)$$

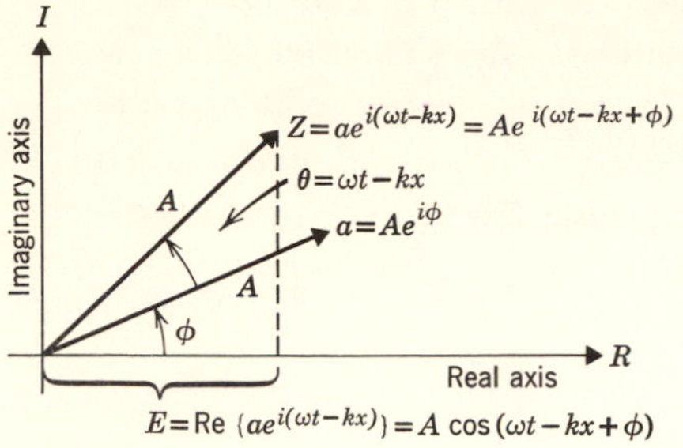

Fig. 5.1

Equation (5.8) then becomes

$$E = \text{Re}\,\tilde{E} = \text{Re}\,[Ae^{i\phi}e^{i(\omega t - kx)}]$$
$$= \text{Re}[Ae^{i(\omega t - kx + \phi)}] \tag{5.13}$$
$$= A\cos(\omega t - kx + \phi)$$

which is Eq. (5.2).

The above relationships are illustrated in Fig. 5.1, which represents a complex number as a two-dimensional vector in the "complex plane." As time progresses, the angle $\theta = \omega t - kx$ increases linearly, and the vector Z describes a circle in the complex plane of radius A with angular frequency ω. The projection of this *uniform circular motion* in the complex plane onto the real axis gives the physical electric field which is undergoing *simple harmonic motion* for a fixed value of x. These complex exponential solutions are useful to describe the superposition of several sine waves, as we now show.

B. COMBINATION OF TWO OR MORE SINE WAVES

1. Two Sine Waves. Consider two sine wave displacements at the same point in space having the same frequency, but with different amplitudes and phases. Let them be given by

$$E_1(x, t) = A_1\cos(\omega t - kx + \phi_1)$$
$$= \text{Re}[Z_1e^{i(\omega t - kx)}]$$

$$E_2(x, t) = A_2\cos(\omega t - kx + \phi_2) \tag{5.14}$$
$$= \text{Re}[Z_2e^{i(\omega t - kx)}]$$

where

$$Z_1 = A_1e^{i\phi_1} \tag{5.15}$$

and

$$Z_2 = A_2e^{i\phi_2}$$

The resultant electric field is given by

$$E_0(x, t) = E_1(x, t) + E_2(x, t) = \text{Re}[Z_1e^{i\theta} + Z_2e^{i\theta}]$$
$$= \text{Re}[(Z_1 + Z_2)e^{i\theta}] \tag{5.16}$$
$$(\theta = \omega t - kx)$$

The complex number $Z_0 \equiv Z_1 + Z_2$ can be obtained in the complex plane as the vector sum of $Z_1 + Z_2$. The resulting real electric field is given by

$$E_0(x, t) = \text{Re}\{Z_0e^{i\theta}\}$$
$$= A_0\cos(\omega t - kx + \phi_0) \tag{5.17}$$

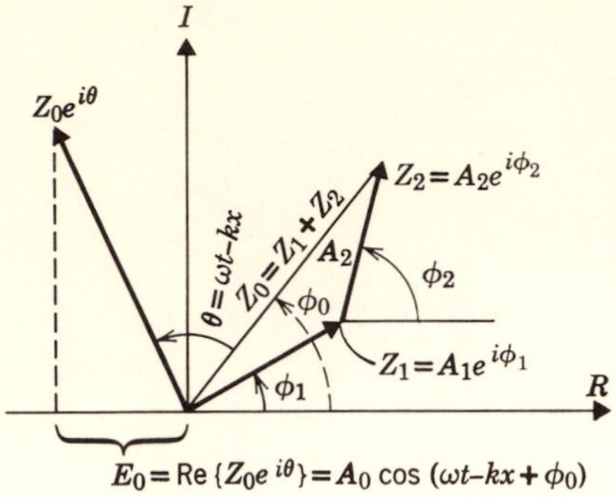

Fig. 5.2. Addition of two sine waves.

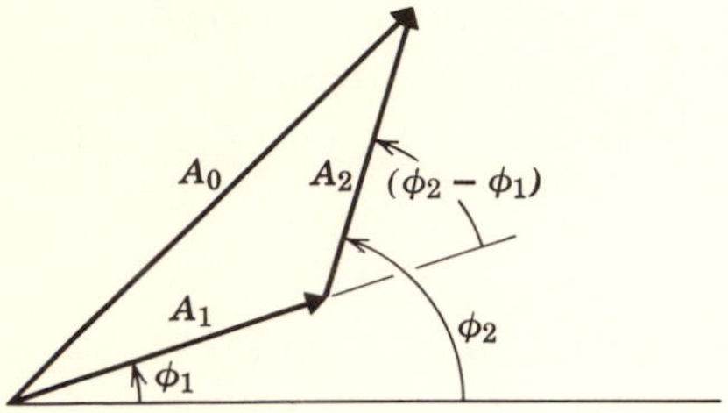

Fig. 5.3. Calculation of A_0 from the cosine law.

where

$$Z_0 = A_0e^{i\phi} \tag{5.18}$$

as shown in Fig. 5.2. The new resultant amplitude and phase can be obtained as follows. The law of cosines (Fig. 5.3) gives us

$$|Z_0|^2 = A_0^2 = A_1^2 + A_2^2 + 2A_1A_2\cos(\phi_2-\phi_1) \tag{5.19}$$

This can also be obtained algebraically:

$$|Z_0|^2 = Z_0Z_0^* = (Z_1+Z_2)(Z_1^*+Z_2^*) = Z_1Z_1^* + Z_2Z_2^* + Z_1Z_2^* + Z_1^*Z_2$$
$$= A_1^2 + A_2^2 + A_1A_2e^{i(\phi_1-\phi_2)} + A_1A_2e^{-i(\phi_1-\phi_2)}$$

To get ϕ_0, take the real and imaginary parts of the equation:

$$Z_0 = Z_1 + Z_2$$
$$A_0e^{i\phi_0} = A_1e^{i\phi_1} + A_2e^{i\phi_2}$$

$$\left.\begin{aligned} A_0\cos\phi_0 &= A_1\cos\phi_1 + A_2\cos\phi_2 \\ A_0\sin\phi_0 &= A_1\sin\phi_1 + A_2\sin\phi_1 \end{aligned}\right\} \tag{5.20}$$

Equations (5.19) and (5.20) will give A_0 and ϕ_0.

An important special case occurs when $A_1 = A_2$. Then A_0 is given by

$$|Z_0|^2 = A_0^2 = 2A_1^2[1+\cos(\phi_2-\phi_1)] \tag{5.19$'$}$$

2. ***Combination of Several Sine Waves.*** The procedure followed above can be generalized to find the sum of several sine waves. Suppose that we wish to find this sum:

$$\begin{aligned} E(x,t) &= \sum_{j=1}^{N} A_j\cos(\omega t - kx + \phi_j) \\ &= \sum_{j=1}^{N} \text{Re}\,[A_j\,e^{i(\theta+\phi_j)}] \\ &= \sum_{j=1}^{N} \text{Re}\,[A_j\,e^{i\phi_j}\,e^{i\theta}] \\ &= \text{Re}\left(\sum_{j=1}^{N} Z_j\right)e^{i\theta} \\ &= \text{Re}\,Z_0\,e^{i\theta} \\ &= A_0\cos(\omega t - kx + \phi_0) \end{aligned} \tag{5.21}$$

where

$$\begin{aligned} Z_0 &= \sum_{j=1}^{N} Z_j \\ &= A_0\,e^{i\phi_0} \end{aligned} \tag{5.22}$$

We can calculate A_0, the length of the complex vector Z_0, as follows. We write

$$A_0^2 = |Z_0|^2 = Z_0 Z_0^* \tag{5.23}$$

But we have

$$Z_0 = \sum_j Z_j$$

and

$$Z_0^* = \sum_{j'} Z_{j'}^* \tag{5.24}$$

so that we can write

$$A_0^2 = \sum_j \sum_{j'} Z_j Z_{j'}^* \tag{5.25}$$

We can separate the terms where $j = j'$ as follows:

$$A_0^2 = \sum_j |Z_j|^2 + \sum_{j<j'} (Z_j Z_{j'}^* + Z_j^* Z_{j'})$$

$$= \sum_j |A_j|^2 + \sum_{j<j'} A_j A_{j'} [e^{i(\phi_j - \phi_{j'})} + e^{-i(\phi_j - \phi_{j'})}]$$

$$A_0^2 = \sum_j |A_j|^2 + \sum_{j<j'} 2A_j A_{j'} \cos(\phi_j - \phi_{j'}) \tag{5.26}$$

Equations (5.20) generalize to

$$A_0 \cos \phi_0 = \sum_j A_j \cos \phi_j$$

$$A_0 \sin \phi_0 = \sum_j A_j \sin \phi_j \tag{5.27}$$

C. ENERGY IN A SINE WAVE

Quite generally, the instantaneous incidence or flux density $E_e(x, t)$ is proportional to the square of the amplitude. Explicitly, for the case of light we have

$$E_e(x, t) = CE(x, t)^2 \tag{5.28}$$

where C is a constant [see Eq. (7.38′) below].

The detectors of optical radiation (photographic plates, phototubes, the human eye, and so on) are sensitive to E_e, not to the field.* The instantaneous E_e is oscillating at a frequency of about 10^{15} cycles per second. No detector can follow such fast oscillations; the fastest of them have a response time of about 10^{-10} sec. In that time period there will be about 10^5 oscillations of E_e. The detector is thus able to respond only to an average of E_e

*Photochemical reactions, the photoelectric effect, and so on are sensitive to the number of photons per second, that is, to the radiant flux.

over a time period T that represents the response time of the detector. Hence we need to calculate

$$\bar{E}_e(x, t_0) = \frac{1}{T}\int_{t_0}^{t_0+T} E_e(x, t)\, dt \tag{5.29}$$

We combine this equation with Eqs. (5.21) and (5.28) to obtain

$$\begin{aligned}
\bar{E}_e(x, t_0) &= \frac{1}{T}\int_{t_0}^{t_0+T} CA_0^2 \cos^2(\omega t - kx + \phi_0)\, dt \\
&= \frac{CA_0^2}{\omega T}\int_{t_0}^{t_0+T} \cos^2(\omega t - kx + \phi_0)\, d\omega t \\
&= \frac{CA_0^2}{\omega T}\int_{t_0}^{t_0+T} \left[\frac{1}{2} + \frac{\cos 2(\omega t - kx + \phi_0)}{2}\right] d\omega t \\
&= \frac{CA_0^2}{2} + \frac{CA_0^2}{4\omega T}[\sin 2(\omega t_0 + \omega T - kx + \phi_0) \\
&\quad - \sin 2(\omega t_0 - kx + \phi_0)]
\end{aligned}$$

Because our detectors have a slow response time compared to the period $2\pi/\omega$ of the light oscillations, we have

$$\omega T \gg 1$$

and the term in square brackets is at most 2. This gives the result

$$\bar{E}_e(x, t_0) = \frac{CA_0^2}{2} \tag{5.30}$$

If two sine waves combine to produce $E(x, t)$, we can use Eq. (5.19) to obtain

$$\begin{aligned}
\bar{E}_e &= \frac{C}{2}(A_1^2 + A_2^2 + 2A_1A_2 \cos(\phi_2 - \phi_1)) \\
&= \bar{E}_{e1} + \bar{E}_{e2} + 2\sqrt{\bar{E}_{e1}\bar{E}_{e2}} \cos(\phi_2 - \phi_1)
\end{aligned} \tag{5.31}$$

where

$$\bar{E}_{e1} = \frac{C}{2}A_1^2 \qquad \text{and} \qquad \bar{E}_{e2} = \frac{C}{2}A_2^2$$

represent the flux densities of each separate sine wave. The presence of the *interference term*

$$2\sqrt{\bar{E}_{e1}\bar{E}_{e2}} \cos(\phi_2 - \phi_1)$$

prevents us from adding the separate flux densities. In the important case where $\bar{E}_{e1} = \bar{E}_{e2}$ we obtain, from Eq. (5.31),

$$\bar{E}_e = 2\bar{E}_{e1}[1 + \cos(\phi_2 - \phi_1)] \tag{5.31$'$}$$

When many sine waves add, we can use Eq. (5.26) to obtain

$$\bar{E}_e = \frac{C}{2}\sum_j A_j^2 + 2\sum_{j<j'} A_j A_{j'} \cos(\phi_j - \phi_{j'})$$

$$= \sum_j \bar{E}_{ej} + 2\sum_{j<j'} \sqrt{\bar{E}_{ej}\bar{E}_{ej'}} \cos(\phi_j - \phi_{j'}) \tag{5.32}$$

PROBLEMS

5.1.1 Suppose that a reflecting surface has a complex amplitude reflection coefficient $\rho = \rho_0 e^{i\delta}$. This means that if the incident electric field is given by

$$E_i(x, t) = A\cos\left[2\pi\nu\left(\frac{t-x}{v}\right) + \phi\right] \tag{1}$$

the reflected field will be given by

$$E_r(x, t) = \rho_0 A\cos\left[2\pi\nu\left(t - \frac{x}{v}\right) + \phi + \delta\right] \tag{2}$$

Suppose that $\rho_0 \neq 1$. Show that the combined field has a part consisting of a standing wave and a part consisting of a running wave.

5.1.2 Let a plane wave be incident at angle θ to a plane reflecting surface, where the reflection coefficient is $\rho = \rho_0 e^{i\delta}$ as in Problem 5.1.1.

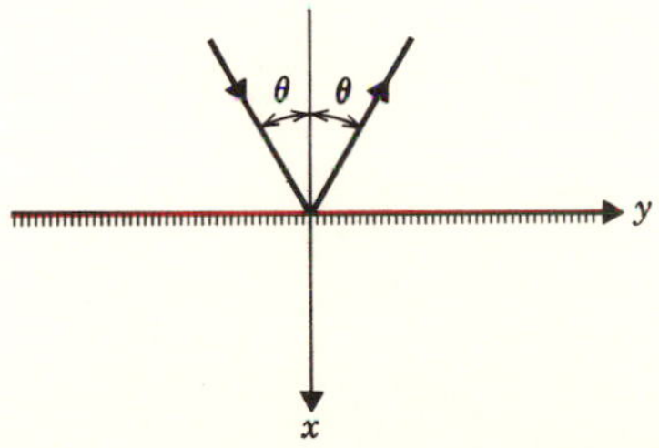

Take axes as shown on the left and write down the appropriate generalizations of Eqs. (1) and (2) of Problem 5.1.1 to describe $E_i(x, y, t)$ and $E_r(x, y, t)$. Show that the total field can be written as

$$E_{tot} = (1-\rho_0)A_0\cos\left[2\pi\nu\left(t - \frac{x\cos\theta - y\sin\theta}{v}\right) + \phi\right]$$

$$+ 2\rho_0 A_0 \cos\left[2\pi\nu\left(t - y\frac{\sin\theta}{v}\right) + \phi + \frac{\delta}{2}\right]\cos\left[2\pi\nu x\frac{\cos\theta}{v} + \frac{\delta}{2}\right]$$

Interpret this result. What is the spatial frequency of the standing wave component?

5.1.3 Find the amplitude A_0 of the resultant of these three complex vectors

$$Z_1 = 1 - i, \qquad Z_2 = 2 + \frac{i}{3}, \qquad Z_3 = 1.5 + i$$

5.2 TWO-BEAM INTERFERENCE

A wide variety of quite different-appearing interference experiments involve the combination of the fields in two different light beams (which in practically all cases are derived from the same source). A simple example occurs when the beams appear to be coming from two separate point sources.

A. INTERFERENCE OF LIGHT FROM TWO POINT SOURCES

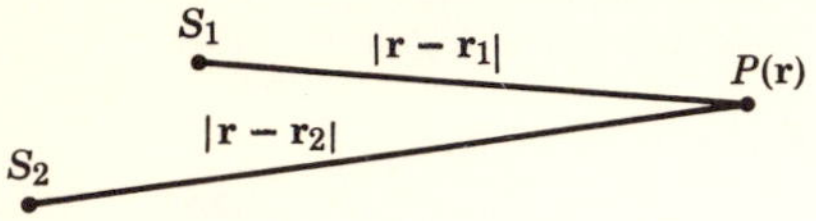

Fig. 5.4. Interference from two point sources.

*1. **General Discussion.*** Consider the situation shown in Fig. 5.4. Here S_1 and S_2 are two point sources of light each emitting a sinusoidal wave of the same angular frequency ω. They have position vectors $\mathbf{r}_1$ and $\mathbf{r}_2$. The field point P where we evaluate the intensity* has position vector $\mathbf{r}$. The electric field at P resulting from the two sources is assumed to be of the form

$$E_1 = \frac{B_1}{|\mathbf{r}-\mathbf{r}_1|}\cos(\omega t - k|\mathbf{r}-\mathbf{r}_1| + \phi_1) \approx A_1 \cos(\omega t - k|\mathbf{r}-\mathbf{r}_1| + \phi_1) \tag{5.33}$$

$$E_2 = \frac{B_2}{|\mathbf{r}-\mathbf{r}_2|}\cos(\omega t - k|\mathbf{r}-\mathbf{r}_2| + \phi_2) \approx A_2 \cos(\omega t - k|\mathbf{r}-\mathbf{r}_2| + \phi_2)$$

We absorb the slow r dependence of $|\mathbf{r}-\mathbf{r}_1|^{-1}$ and $|\mathbf{r}-\mathbf{r}_2|^{-1}$ into A_1 and A_2 and treat them essentially as constants.

We now use the results of Section 5.1, Eq. (5.31), to write the average intensity* at P as

$$I = \frac{C}{2}(A_1{}^2 + A_2{}^2 + 2A_1A_2 \cos \Psi_0)$$

$$= I_1 + I_2 + 2\sqrt{I_1 I_2}\cos \Psi_0 \tag{5.34}$$

where

$$\Psi_0 = \phi_2 - \phi_1 - k|\mathbf{r}-\mathbf{r}_2| + k|\mathbf{r}-\mathbf{r}_1|$$

or

$$\Psi_0 = \phi_2 - \phi_1 - \Delta \tag{5.35}$$

where

$$\Delta \equiv k|\mathbf{r}-\mathbf{r}_2| - k|\mathbf{r}-\mathbf{r}_1| \tag{5.35$'$}$$

The total relative phase Ψ_0 between the two waves at P thus consists of two parts: a part $\phi_2 - \phi_1$ coming from the relative phases at the two sources, and a part $-\Delta$ coming from the different retardation

*From now on, unless specific calculations are under consideration we shall use the generic term "intensity," I, instead of the more correct, and awkward, "flux density," "radiant incidence," or "irradiance," E_e.

in phase suffered by the two beams resulting from the propagation from S_2 to P and from S_1 to P.

An important special case occurs when $A_1 = A_2$. Then we can write

$$I = 2I_1(1+\cos(\phi_2-\phi_1-\Delta)) \tag{5.36}$$

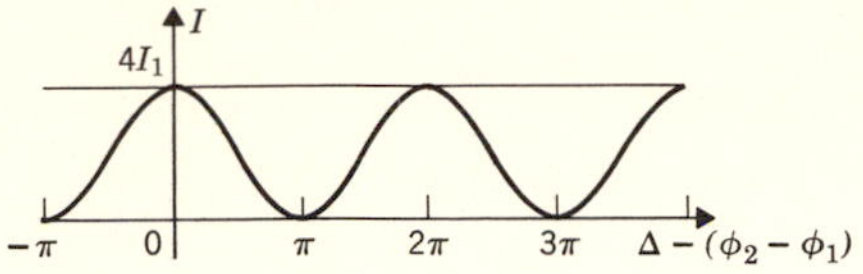

Fig. 5.5. Intensity distribution in two beam interference pattern.

This equation is schematically plotted in Fig. 5.5. Note that the average of Eq. (5.36) over an integral number of oscillations is

$$\langle I \rangle = 2I_1$$

which is the result we would obtain if there were no interference and if the energies were to add.

2. Introduction to the Concept of Coherence. We recall that I represents an average over a time period T that is essentially the time constant of the detection system. In order to observe the interference pattern, the total relative phase Ψ_0 should not change during the time of measurement. The uncontrolled variation of Δ can be eliminated by secure mechanical mounting of the components. This leaves the contribution $\phi_2-\phi_1$ coming from the sources. In order to observe an interference pattern as a function of Δ, the relative phase $\phi_2-\phi_1$ must not change much during the measurement time T. Let us define a mutual coherence time τ_c as the time necessary for the mean square value of $\phi_2-\phi_1$ to change by, say, 1 radian. (We can make this more precise later.) Then if $\tau_c \gg T$, we observe interference effects as P moves and hence Δ changes, whereas if $\tau_c \lessapprox T$, we do not. In the former case we say the two sources are *mutually coherent*; in the latter case they are *mutually incoherent*. Note that the concept of coherence is a quantitative one, since it depends on the relative sizes of the mutual coherence time and the response time of the measuring apparatus.

Consider the case of two radio transmitters as an example. Each can be made quite stable, but unless they are actually locked into phase by some kind of control circuit, they will drift out of phase with respect to one another. It is the same with the highly monochromatic modern sources of light known as *lasers*. Unless special synchronization schemes are used, two lasers will drift out of phase in a small fraction of a second ($\tau_c \sim 10^{-3}$ sec). Ordinary gas discharge sources of "monochromatic" radiation such as neon lamps or mercury lamps will have coherence times shorter than 10^{-8} sec. With fast photography (exposure times of 10^{-3} sec or less) we might be able to photograph the interference pattern from two separate unsynchronized lasers, but for ordinary sources this would be virtually impossible.

The traditional method of obtaining two sources S_1 and S_2 that are mutually coherent is to have the two sources related ultimately to the same single source. Two famous examples are provided by Young's experiment and by the Michelson interferometer.

B. YOUNG'S INTERFERENCE FRINGES

Figure 5.6 shows the basic elements of Young's experiment. Light from the sources S impinges on the two very small holes in

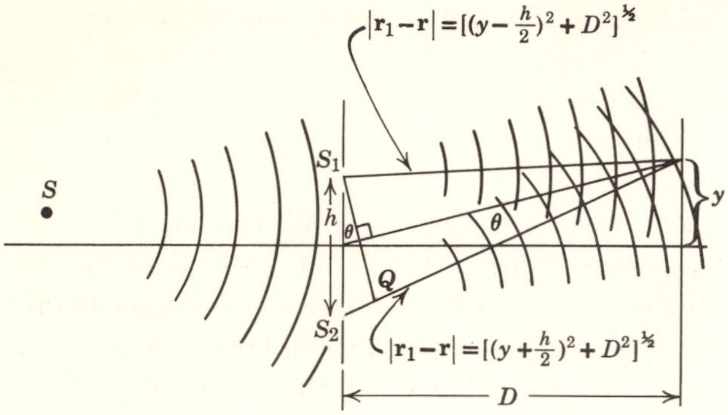

Fig. 5.6. Young's interference setup.

the screen. The secondary Huygens wavelets there act as if they originated at S_1 and S_2. The spreading of the light beyond the holes S_1 and S_2 will be pronounced if the holes are small. This is a manifestation of the phenomenon of *diffraction*, which we shall discuss in much detail later. For now we simplify by assuming S_1 and S_2 to be point sources with well-defined amplitudes and a definite phase difference $\phi_2 - \phi_1$. This phase difference will depend only on the difference in optical paths $\widehat{SS_2} - \widehat{SS_1}$, and hence remains constant if the mechanical mounts are sturdy and free from vibration.

1. ***Determination of the Phase Difference.*** We need to determine

$$\Delta = k[\mathbf{r}_2 - \mathbf{r}| - |\mathbf{r}_1 - \mathbf{r}|]$$

From Fig. 5.6 we can see that to lowest order in $|y|/D$ and h/D we have

$$|\mathbf{r}_2 - \mathbf{r}| = \left[\left(y + \frac{h}{2}\right)^2 + D^2\right]^{1/2} \approx D\left\{1 + \frac{1}{2}\frac{[y + (h/2)]^2}{D^2}\right\}$$

$$|\mathbf{r}_1 - \mathbf{r}| = \left[\left(y - \frac{h}{2}\right)^2 + D^2\right]^{1/2} \approx D\left\{1 + \frac{1}{2}\frac{[y - (h/2)]^2}{D^2}\right\}$$

Thus

$$|\mathbf{r}_2 - \mathbf{r}| - |\mathbf{r}_1 - \mathbf{r}| \approx \frac{1}{2D}\left[\left(y + \frac{h}{2}\right)^2 - \left(y - \frac{h}{2}\right)^2\right] = \frac{yh}{D} \tag{5.37}$$

Thus

$$\Delta = +\frac{kyh}{D} = +\frac{2\pi yh}{\lambda D}$$

An alternate derivation of this result can be made by using the angle θ shown in Fig. 5.6. When $|y|/D \ll 1$ and $h/D \ll 1$, we have $\theta \approx y/D \approx |S_2Q|/h$. Now $|S_2Q| \cong |\mathbf{r}_2 - \mathbf{r}| - |\mathbf{r}_1 - \mathbf{r}|$. Thus

$$\Delta = k|S_2Q| = \frac{khy}{D}$$

2. ***Result for Monochromatic Light of Wavelength*** $\boldsymbol{\lambda}$**.** We may assume S_1 and S_2 to be the same size and about the same distance from S. Then the amplitude A_1 of the electric field at P from S_1 approximates A_2, the amplitude from S_2. This gives $I_1 = I_2$, and hence from Eq. (5.36) we obtain

$$I = 2I_1 \left[1 + \cos\left(\frac{2\pi yh}{\lambda D} - \Delta\phi\right)\right] \tag{5.38}$$

Here $\Delta\phi$ represents $\phi_2 - \phi_1$, the inherent phase shift at the sources S_1 and S_2. This equation has already been plotted in Fig. 5.5. We have only to let $\Delta = 2\pi yh/(\delta D)$. The separation δy on the screen between two successive maxima (or minima) is obtained by setting $\delta\Delta = 2\pi$ and solving for δy:

$$\delta y = \frac{\lambda D}{h}$$

3. ***Result for Two Wavelengths.*** We next consider the case where the source S in Fig. 5.6 emits light of two wavelengths λ_1 and λ_2. Again we assume that the amplitudes obey

$$A_1(1) \approx A_2(1), \qquad A_1(2) \approx A_2(2)$$

and hence

$$I_1(1) \approx I_2(1), \qquad I_1(2) \approx I_2(2)$$

Here the subscript refers to the origin of the light (S_1 or S_2) and the term in parenthesis refers to λ_1 or λ_2. Thus, $I_1(1)$ represents the average flux density at the screen of wavelength λ_1 coming from S_1 (if S_2 is blocked) and $I_1(2)$ that of wavelength λ_2 from S_1, and so on. Now waves of different wavelengths usually do not interfere. (As is shown in Section 6.1, if $|\lambda_2 - \lambda_1|$ is very small and if the response time T of the detector is very short, it is possible to observe "beats.") Assuming this to be so, we may simply add the interference pattern for λ_1 to that for λ_2 to obtain

$$I = 2I_1(1)\left[1 + \cos\left(\frac{2\pi yh}{\lambda_1 D} - \Delta\phi_1\right)\right] + 2I_1(2)\left[1 + \cos\left(\frac{2\pi yh}{\lambda_2 D} - \Delta\phi_2\right)\right] \tag{5.39}$$

Here $\Delta\phi_1$ and $\Delta\phi_2$ represent the inherent phase shifts of S_2 with respect to S_1 for wavelengths λ_1 and λ_2. This equation readily generalizes to a finite number of wavelengths; $\lambda_1, \lambda_2, \ldots \lambda_N$:

$$I = \sum_{j=1}^{N} 2I_1(j)\left[1 + \cos\left(\frac{2\pi yh}{\lambda_j D} - \Delta\phi_j\right)\right] \tag{5.40}$$

We can go farther to infinitely many wavelengths and even change the sum to an integral, as we shall see later.

Let us now consider the case of two wavelengths in more detail. Let λ_2 be greater than λ_1, and assume $\Delta\lambda = \lambda_2 - \lambda_1 \ll \lambda$. For simplicity let $\Delta\phi_1 = \Delta\phi_2 = 0$, which would apply if S_1 and S_2 are equidistant from S. Then the two functions of y in Eq. (5.39) are almost identical as long as y is not too large. The two interference patterns reinforce. Eventually, however, y becomes large enough so that the λ_2 cosine term will get to be π radians out of phase with respect to the λ_1 cosine term. Then the two interference patterns add up to a constant. At still larger y values the two patterns will be back in phase and will again reinforce.

Take a simple fictitious numerical example that is not too different from the real case of the sodium yellow lines. Suppose that $\lambda_2 = 6000$ A, and $\lambda_1 = 5994$ A, and $I_1(1) = I_1(2)$. Then at y on the screen the λ_2 pattern will have made $yh/\lambda_2 D$ oscillations starting at $y = 0$, and the λ_1 pattern will have made $yh/\lambda_1 D$ oscillations. When $yh/\lambda_1 D = 500$, $yh/\lambda_2 D = 499.5$, and $\Delta_1 = 2\pi yh/\lambda_1 D = 500 \cdot 2\pi$, whereas $\Delta_2 = 2\pi yh/\lambda_2 D = 500 \cdot 2\pi - \pi$. Thus the λ_1 interference pattern will be shifted by exactly one half-fringe with respect to the λ_2 interference pattern. The intensity, which takes the general form

$$I = 2I_1(2 + \cos\Delta_1 + \cos\Delta_2) \tag{5.41}$$

is then simply $I = 4I_1$, since the cosine terms cancel. This situation will persist for y values in the vicinity of $500\ \lambda D/h$. When y increases to values near $1000\ \lambda D/h$, $\cos\Delta_1$ again practically equals $\cos\Delta_2$, and the enhanced interference fringes of the form $I \approx 4I_1(1 + \cos\Delta_1)$ are seen. For y near $1500\ \lambda D/d$, the fringes are gone, and so forth (see Fig. 5.7).

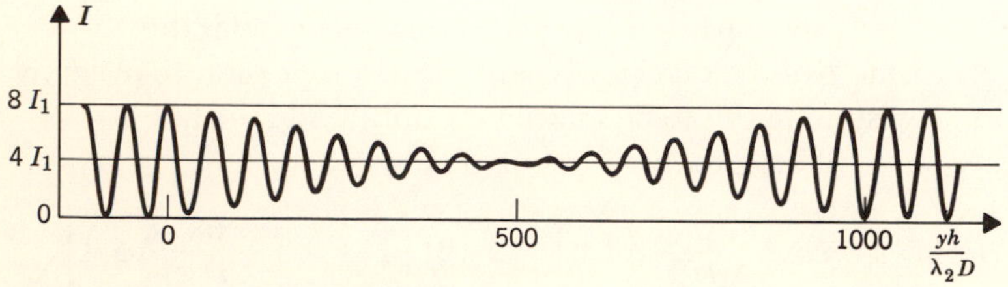

Fig. 5.7. Schematic illustration of two beam interference pattern when two closely spaced wavelengths are present.

The behavior of the *fringe contrast* is indicative of the different wavelengths that are present in the source. A similar description holds for the fringes in the Michelson interferometer, as we shall see, and this effect was exploited by Michelson himself to perform high-resolution optical spectroscopy.

There are more efficient methods of obtaining the split sources S_1 and S_2 than with the pinholes shown in Fig. 5.6. Slits will work instead of pinholes, and even more intensity can be obtained using the arrangement known as Lloyd's mirror or the Fresnel biprism (Fig. 5.8).

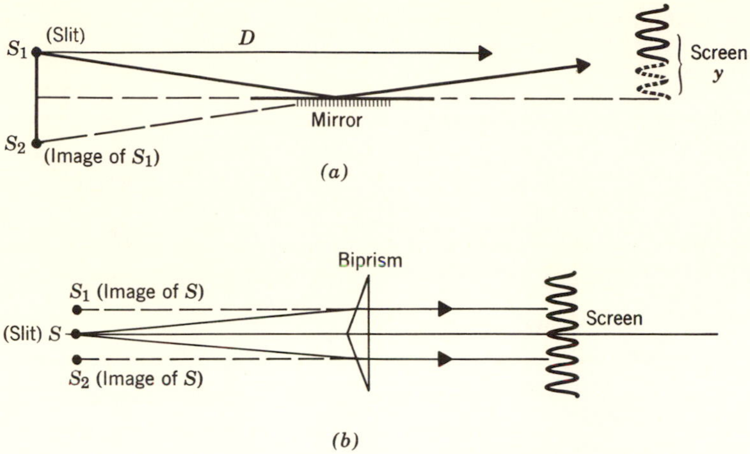

Fig. 5.8. (*a*) Lloyd's mirror. (*b*) Fresnel biprism.

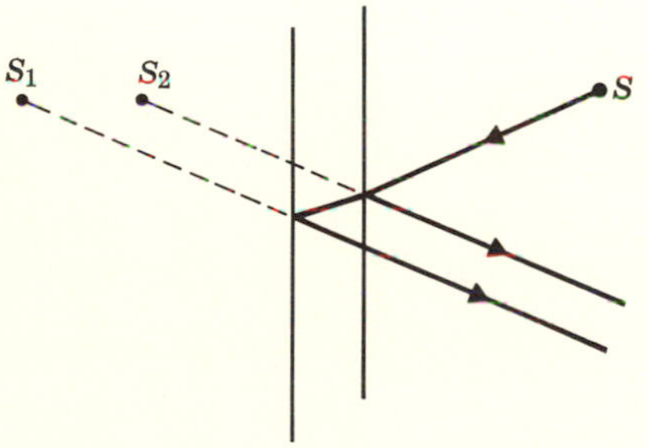

Fig. 5.9. Two-beam interference produced by reflection from a slab.

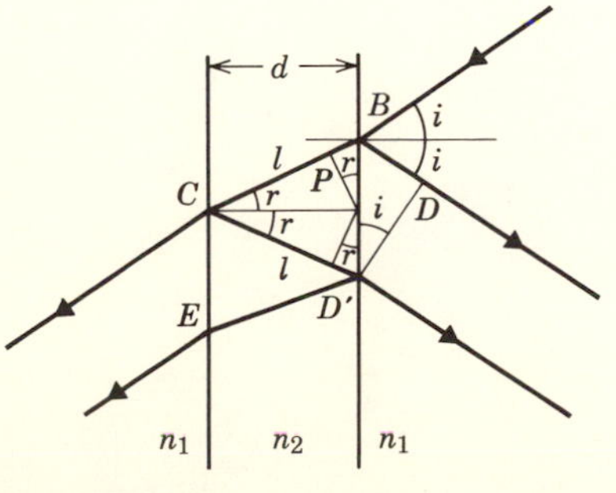

Fig. 5.10

C. INTERFERENCE VIA REFLECTIONS FROM A DIELECTRIC LAYER

Another method of obtaining effectively two coherent sources from the same physical source is to use the double reflection process illustrated in Fig. 5.9. Here S_1 and S_2 are virtual images of the original source S.*

The pertinent physical and geometrical relationships can be derived with the help of Fig. 5.10. The parallel sided slab of index n_2 is embedded in a medium with refractive index n_1. For simplicity, let the phase of the incident wave at B be zero. We shall employ the complex exponential notation. Then the situation at B is as follows:

$$\begin{aligned}&\text{incident wave: } E_0 e^{i\omega t}\\&\text{reflected wave: } \rho E_0 e^{i\omega t}\\&\text{transmitted wave: } \tau E_0 e^{i\omega t}\end{aligned} \tag{5.42}$$

The physical electric field is obtained from the real part of these equations, for example, for the incident wave, $E = E_0 \cos \omega t$. In the above equations ρ and τ denote the amplitude reflection and transmission coefficients when the wave is incident onto the interface from medium 1 to medium 2.

*This situation is not precisely equivalent to that with two sources at S_1 and S_2, the optical path lengths are not correctly obtained by reference to S_1 and S_2. The correct expression for the O.P.D. is given here. An analogous problem arises in the uncompensated Michelson interferometer. See Problem 5.3.3.

The wave transmitted at B will experience a decrease in phase* of amount $k_2\widehat{BC} = k_2 d/\cos r$ as it propagates from B to C. Here $k_2 = \omega/v_2 = \omega n_2/c$. Thus, at C we have the following situation:

$$\begin{aligned}&\text{incident wave: } \tau E_0 e^{i(\omega t - k_2 d/\cos r)}\\&\text{reflected wave: } \rho'\tau E_0 e^{i(\omega t - k_2 d/\cos r)}\end{aligned} \tag{5.43}$$

The passage of the wave from C to D' involves the same loss of phase as from B to C. Hence, at D we have

$$\begin{aligned}&\text{incident wave: } \rho'\tau E_0 e^{i(\omega t - 2k_2 d/\cos r)}\\&\text{transmitted wave: } \tau'\rho'\tau E_0 e^{i(\omega t - 2k_2 d/\cos r)}\end{aligned} \tag{5.44}$$

Here ρ' and τ' are the amplitude reflection and transmission coefficients when the incident wave is in medium 2. The original reflected wave at B will be retarded in phase by amount $k_1\widehat{BD} = k_1\widehat{BD'}\sin i = 2k_1 d\tan r\sin i$, where $k_1 = \omega/v_1 = \omega n_1/c$. Thus, at D the complex electric field $\tilde{E}_1$ is obtained from

$$\tilde{E}_1 = \rho E_0 e^{i\omega t} e^{-2ik_1 d\tan r\sin i} \equiv \rho E_0 e^{i\omega t} e^{-i\phi_1} \tag{5.45}$$

At D' we have, from Eq. (5.44),

$$\tilde{E}_2 = \tau'\rho'\tau E_0 e^{i\omega t} e^{-2ik_2 d/\cos r} \equiv \tau'\rho'\tau E_0 e^{i\omega t} e^{-i\phi_2} \tag{5.46}$$

These equations define ϕ_1 and ϕ_2.

If the two rays are brought together at the focal plane of a lens, they will interfere. Before we calculate the interference, we should mention that additional repeats of paths parallel to BC and CD' may occur by multiple reflection. We neglect these for now and shall return to them later.

We must know the relative phase Δ of E_1 and E_2 in Eqs. (5.45) and (5.46). We have

$$\Delta = \phi_2 - \phi_1 = 2d\left(\frac{k_2}{\cos r} - \frac{k_1\sin i\sin r}{\cos r}\right)$$

Use of $k_2 = \omega n_2/c$, $k_1 = \omega n_1/c$, and Snell's law gives

$$k_1\sin i\sin r = \frac{\omega n_1}{c}\sin i\sin r = \frac{\omega n_2}{c}\sin^2 r$$

Equation (5.47) then becomes

$$\boxed{\begin{aligned}\Delta &= \frac{2dn_2\omega}{c\cos r}\cdot(1-\sin^2 r)\\&= \frac{2n_2\omega d}{c}\cos r = \frac{4\pi n_2 d}{\lambda_0}\cos r\end{aligned}} \tag{5.48}$$

*This occurs because in the complex exponential $e^{i(\omega t - kx)}$ an *advance* by distance Δx introduces a factor $e^{-ik\Delta x}$, that is, *decreases* the phase by $k\Delta x$.

where λ_0 is vacuum wavelength obeying

$$c\frac{2\pi}{\lambda_0} = \omega \tag{5.49}$$

Equation (5.48) may also be obtained by geometrical arguments based on Fig. 5.10. The optical path length $n_2|BCD'|$ minus the optical path length $n_1|BD|$ is simply $2n_2l = 2n_2|PC|$, since by Snell's law $2n_2|BP| = n_1|BD|$. Then $\Delta = 2\pi(2n_2 l)/\lambda_0 = 4\pi n_2 d \cos r/\lambda_0$.

We shall derive later the following results for reflection from a dielectric interface:

$$\begin{aligned} \rho' &= -\rho \\ \tau\tau' &= 1-\rho^2 \end{aligned} \tag{5.50}$$

For an air-glass interface ρ is about 0.2 for nearly normal incidence; hence, $1-\rho^2 \approx 0.96$. From Eqs. (5.45), (5.46), and (5.50) we obtain

$$\begin{aligned} \tilde{E}_1 &= \rho E_0 e^{i\omega t} e^{-i\phi_1} \\ \tilde{E}_2 &= \tau\tau'\rho' E_0 e^{i\omega t} e^{-i\phi_2} = -(1-\rho^2)\rho E_0 e^{i\omega t} e^{-i\phi_1} e^{-i(\phi_2-\phi_1)} \\ \tilde{E}_2 &= -(1-\rho^2) e^{-i\Delta} \tilde{E}_1 \\ &\approx -e^{-i\Delta}\tilde{E}_1 = e^{-i(\Delta+\pi)}\tilde{E}_1 \end{aligned} \tag{5.51}$$

The last equality follows from $e^{-\pi i} = -1$. To the extent that the last approximation is true, we then have

$$\tilde{E} = \tilde{E}_1 + \tilde{E}_2 = \tilde{E}_1(1+e^{-i(\Delta+\pi)}) \tag{5.52}$$

We are thus adding two sine waves of equal energy and with relative phase $\phi_2 - \phi_1 = -(\Delta+\pi)$. Equation (5.31′) then applies and yields

$$I = 2I_1(1+\cos(\Delta+\pi)) = 2I_1(1-\cos\Delta) \tag{5.53}$$

$$\left(\Delta = \frac{4\pi n_2 d}{\lambda_0}\cos r\right)$$

Except for the minus sign, Eq. (5.53) is similar to Eq. (5.38). The minus sign comes from the extra phase shift of π radians upon reflection off a more dense medium. We met it in our discussion of transverse waves on a string for which Eqs. (5.50) hold. A consequence of Eq. (5.53) is that as $d \to 0$ (and hence $\Delta \to 0$), $I \to 0$, as it must.

PROBLEMS

5.2.1 In a Lloyd's mirror experiment with light of wavelength = 5000 A, it is found that the bright fringes on a screen 1 meter away from the source are 1 mm apart. Calculate the (perpendicular) height of the source slit from the mirror. (Answer: 1/4 mm.)

5.2.2 Discuss the "fictitious numerical example" worked out in part B of this section, but now assume that $I_1 = \frac{1}{2}I_2$. Discuss semiquantitatively the form the energy density takes as a function of Δ.

5.2.3 Derive an expression for the transmitted intensity resulting from interference between the beams at C and E in Fig. 5.10. (Answer: $I = I_i(1-\rho^2)^2(1+\rho^4+2|\rho|^2\cos\Delta)$, $I_i =$ incident intensity.)

5.2.4 In an experiment with Lloyd's mirror (see Fig. 5.8*a*) S_1 is 2 mm above the plane of the mirror. The mirror is halfway between S_1 and the screen and is 40 cm long. Also $D = 1.5$ m, $\lambda = 500$ nm. Calculate the fringe separation and the maximum and minimum values of y for which fringes are seen. How many fringes are observed?

5.2.5 Consider a Fresnel biprism with index of refraction n, where the prism angle A is small compared with unity. Show that a ray coming in at nearly normal incidence will be deviated a small angle $\delta = (n-1)A$. Now suppose that $h \ll D_1$ and suppose that the images S_1 and S_2 of S formed by the two halves of the biprism are a distance h from S. What then is the

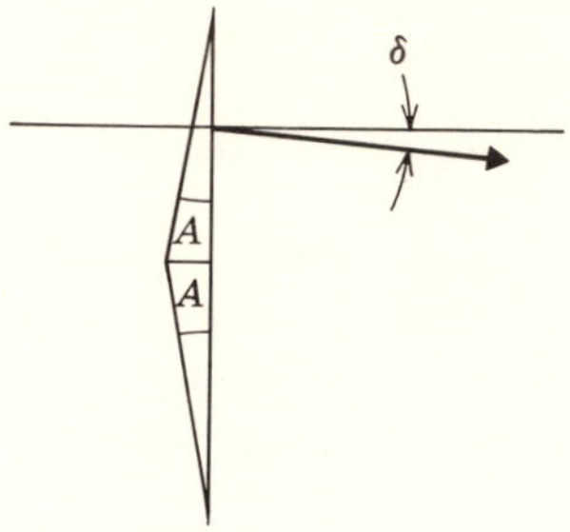

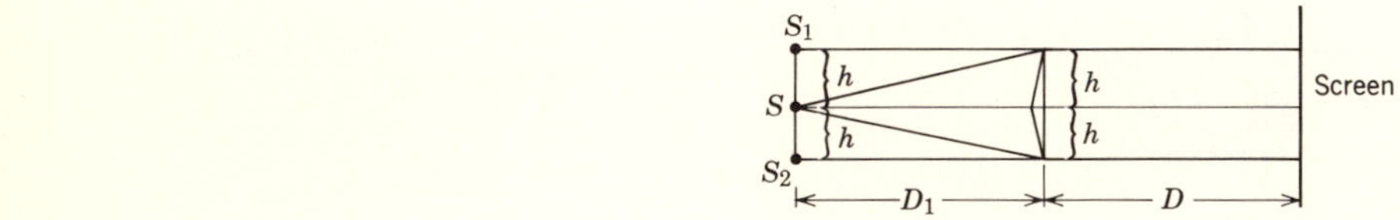

necessary value of D_1? Under these circumstances, and for a general value of $D > 0$, calculate the fringe separation on the screen. How many fringes are observed?

5.3 HAIDINGER INTERFERENCE FRINGES

Haidinger fringes are the type of interference pattern that results with an extended source where partial reflections occur from a plane parallel dielectric slab. The interference considered in Section 5.2C above then occurs for light rays coming from each separate part of the source. Two examples will now be considered.

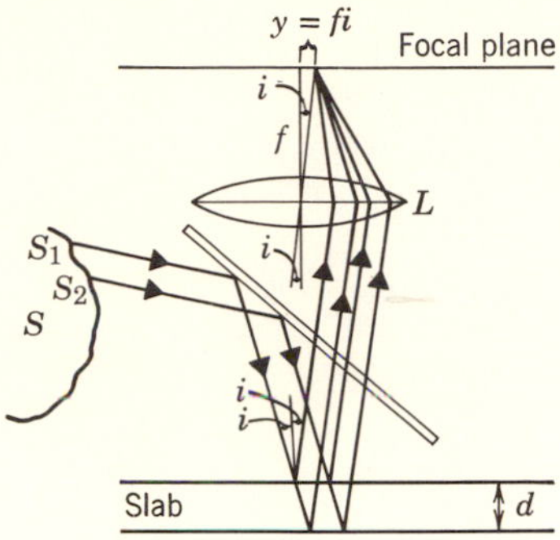

Fig. 5.11. Schematic illustration of the set-up for observation of Haidinger fringes.

A. PLANE PARALLEL SLAB

Figure 5.11 shows a typical experimental arrangement for the observation of Haidinger fringes with a plane parallel slab. The extended source S must be essentially monochromatic. The lens L could be in a telescope or in the eye of an observer. The fringes result from interference occurring in the focal plane of the lens.

The ray of light leaving the source at S_1 will be split into two rays by partial reflections off the slab. The two partial rays will leave at an angle i and will have a relative phase shift given by $-(\pi + \Delta)$ [see Eq. (5.53)], where Δ is given by Eq. (5.48):

$$\Delta = \frac{4\pi n_2 d}{\lambda_0} \cos r$$

The angle r is determined from the angle i by Snell's law.

The two partial rays will enter the lens L making an angle i with its axis and hence will be brought together in the focal plane at an off-axis distance $y = if$ (in the small angle approximation), where f is the focal length of the lens.

Another ray parallel to the first one leaving the source at a different point S_2 will also be split into two partial rays. Since they have the same value of i, and hence of r as the first ray (from S_1), there will be the same relative phase shift of $-(\pi + \Delta)$ as before. The two partial rays from S_2 will be parallel to those from S_1 and will be focused at the same place in the focal plane of the lens. There they will interfere in the same way as did the other partial rays from S_1. Rays from other parts of the source will behave in the same way; as long as they are parallel to one another, they will be focused at the same place by the lens. Furthermore, each original ray will be split into two rays with the same relative phase $-(\Delta + \pi)$ as before.

Because of rotational symmetry about the axis of the lens, the phase shift will be a constant on a circle in the focal plane having the lens axis at its center. The interference between the partial rays will be the same on this circle; if $\Delta + \pi = 2\pi \times$ integer, constructive interference produces a bright ring; if $\Delta + \pi = \pi \times$ odd integer, destructive interference produces a dark ring. The interference pattern then consists of a series of concentric rings.

The total number of multiples of 2π in the phase shift at a certain point in the pattern is called the *order* m of the interference pattern. We obtain the following result for m by setting $(\pi + \Delta) = 2\pi m$ in Eq. (5.48):

$$m = \frac{2n_2 d}{\lambda_0} \cos r + \frac{1}{2} \tag{5.54}$$

For the case under consideration an integral value of m corresponds to a bright band, a half-integral value to a dark band. Note that the highest order in the interference pattern occurs at the center where $r = 0$ and $\cos r = 1$.

This maximum-order number is thus given by

$$m_{\text{max}} = \frac{2n_2 d}{\lambda_0} + \frac{1}{2} \tag{5.55}$$

In terms of it, Eq. (5.54) becomes

$$m = (m_{\text{max}} - \tfrac{1}{2}) \cos r + \tfrac{1}{2} \approx m_{\text{max}} (1 - \tfrac{1}{2} r^2) \tag{5.56}$$

if the angle r is small and m_{max} is large. For small r, Snell's law becomes $n_2 r = n_1 i$, so that we can write

$$m = m_{\text{max}} \left[1 - \frac{1}{2} \left(\frac{n_1}{n_2} \right)^2 i^2 \right] \tag{5.57}$$

We can use the relation $i = y/f$ to obtain the following expression relating the order of the fringe to its radius in the focal plane in the lens:

$$m(y) \cong m_{\text{max}} \left[1 - \frac{1}{2} \left(\frac{n_1 y}{n_2 f} \right)^2 \right] \tag{5.58}$$

This can be solved for y as a function of m_{max}:

$$y = \frac{n_2 f}{n_1} \left[\frac{2(m_{\text{max}} - m)}{m_{\text{max}}} \right]^{1/2} \tag{5.59}$$

The number of visible bright fringes between $y = 0$ and y will be given by $p \equiv m_{\text{max}} - m$. If d is known, we can determine m_{max} from (5.55). This yields the following expression for the radius of the pth visible fringe:

$$y_p = \frac{f}{n_1} \left(\frac{p n_2 \lambda_0}{d} \right)^{1/2} \tag{5.59$'$}$$

From Eq. (5.58) we see that equal increments in m correspond to equal decrements in y^2; hence the area $\pi(\Delta y)^2$ enclosed between two successive rings is a constant for fixed d. In fact, we can write, from Eq. (5.58),

$$1 = \Delta m = -\frac{m_{\text{max}}}{2\pi} \left(\frac{n_1}{n_2 f} \right)^2 \Delta(\pi y^2)$$

Thus the area between the pth ring and the $(p+1)$th ring is

$$\Delta(\pi y^2) = 2\pi\left(\frac{n_2 f}{n_1}\right)^2 \frac{1}{m_{\max}} = \pi\left(\frac{f}{n_1}\right)^2 \frac{n_2\lambda_0}{d} \tag{5.60}$$

As d increases, so does $m_{\max}$; hence the area between fringes decreases, and the fringes get closer and closer together.

B. MICHELSON INTERFEROMETER

The Michelson interferometer is often used in such a way that it exhibits Haidinger fringes. This is shown in Fig. 5.12. Again S represents a broad monochromatic source. A given ray is split in two by the half-silvered mirror; the two halves are reflected at mirrors M_1 and M_2 and then recombined at the half-silvered mirror. The compensator plate is often inserted to make the optical paths symmetrical, so that each ray passes through the same thickness of glass.

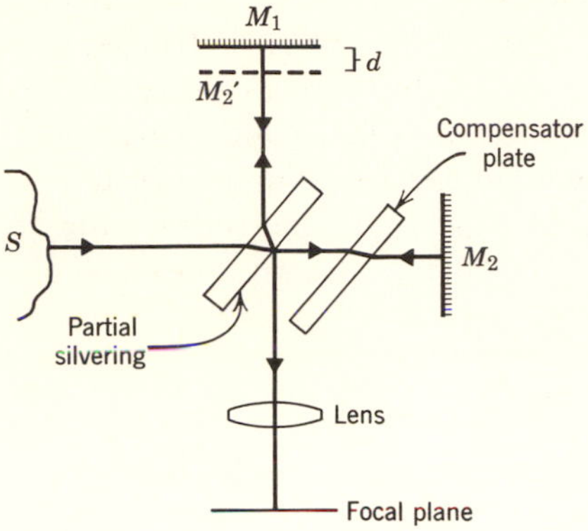

Fig. 5.12. Michelson interferometer.

You can think of an image of one of the mirrors as being formed in the space of the other (M_2', for instance). Then reflections off M_1 and M_2' lead to behavior similar to that off the sides of a plane parallel slab. If the internal and external reflections at the beam splitter produce the same phase shift, the phase difference between the two partial rays will be given by

$$\Delta = \frac{4\pi nd}{\lambda_0}\cos r \tag{5.61}$$

where n is the index of refraction of the air, λ_0 the vacuum wavelength of light, d the separation between M_1 and M_2', and where r is now the angle of incidence as well as the angle with the common normal to the mirrors M_1 and M_2' since no refraction actually takes place there. Note the absence of the extra phase shift by π; here we have the same kind of reflection off M_1 and M_2'. The phase Δ can be varied over a large range by simply moving one of the mirrors and thus changing d.

A slight change is necessary to carry over the expressions of the previous section (A) to the Michelson interferometer—we have only to set $n_1/n_2 = 1$ in Eqs. (5.57), (5.58), and (5.59) and drop the $\frac{1}{2}$ in Eqs. (5.54), (5.55), and (5.56).

PROBLEMS

5.3.1 Haidinger fringes are observed with a 2-mm-thick slab of index $n = 1.60000$ having accurately flat surfaces. The light used has $\lambda = 5000$ A. Calculate the maximum order at the center of the circular fringe pattern. How many bright fringes are observed within a cone of 1/30 radian of the normal to the surface? (Answer: $m_{\max} = 12{,}800.5$, 7.11 fringes.)

5.3.2 A Michelson interferometer is used with light of $\lambda = 6000$ A, and observations are made at two values of the effective plate separation d, namely, $d = 1/2$ mm and $d = 5$ mm.

a. Do the fringes at the center expand or contract as d is increased? Explain.
b. What is the angular separation between the fringes at the two values of d?
c. If the interferometer mirrors are round and 50 mm in diameter and observed with the naked eye at a distance of 400 mm, about how many fringes appear across the field of the mirror for the two values of d?

5.3.3 The insertion of an uncompensated dielectric slab in one arm of a Michelson interferometer changes the nature of the white light fringes, which occur when the optical path difference in the two arms is very close to zero. In a compensated interferometer the white light fringes have very large radii, when the plates are accurately parallel. In this problem you are asked to find expressions for the radii of the white light fringes in the uncompensated case. The setup is illustrated here. Assume the beam

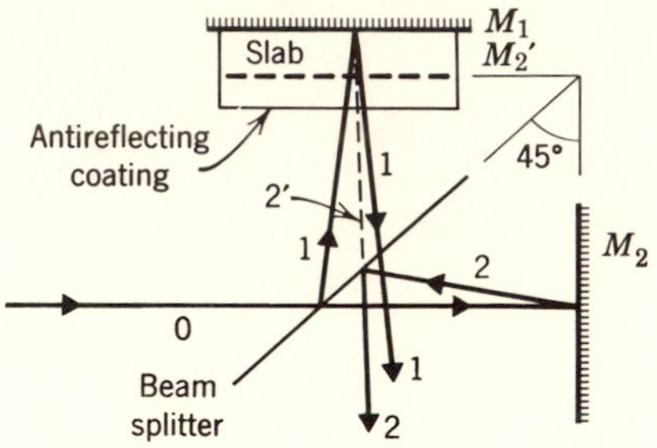

splitter to be a self-supporting semireflecting film. Mirror M_1 has a parallel dielectric slab in front of it. Mirror M_2' is the image of the mirror M_2 formed by the beam splitter. Ray 0 from the source splits into rays 1 and 2 at the beam splitter. Ray 2′, the image of ray 2, is reflected by M_2'. The situation may be analyzed as shown below. It is useful to refer the phases of rays 1 and 2′ to the reference ray that would be reflected off the surface of the slab if there were no antireflecting coating there. Discuss

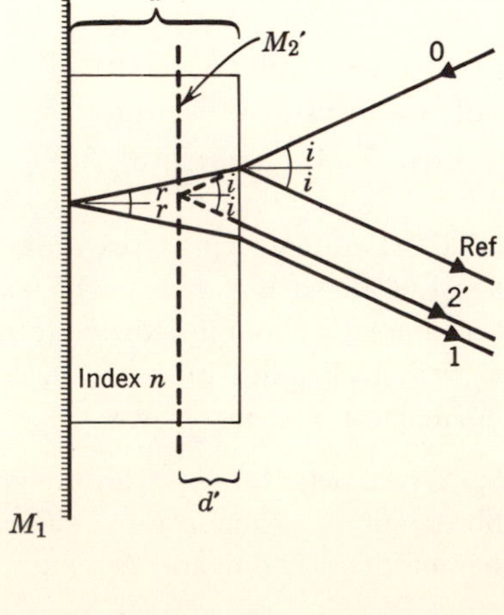

the order and radii of the fringes as a function of the equivalent separation d'. Where are the white light fringes located? What is their radius? You will need an auxiliary lens of focal length f to form images of the fringes.

5.3.4 Consider Haidinger fringes observed as in Fig. 5.11 with a slab of thickness d and index n. Let D be the diameter of the lens and f its focal length. Monochromatic light of wavelength λ is used. For d much larger than λ, the requirements on uniformity of n and on the flatness and parallelism of the surfaces of the slab are quite severe if the ideal Haidinger fringes described in the text are to be observed. Explain why this should be so, and estimate the tolerances on d and n if no observable departures from ideal behavior are to be noticed.

5.4 FIZEAU BANDS OR FRINGES

Fizeau fringes result from essentially two beam interference occurring near a thin layer. The mathematical description of these fringes is very similar to that for Haidinger fringes, but the methods of observation and the conditions for their appearance are quite different in the two cases. In particular, the requirements on uniformity, flatness, and parallelism necessary for Haidinger fringes may now be relaxed.

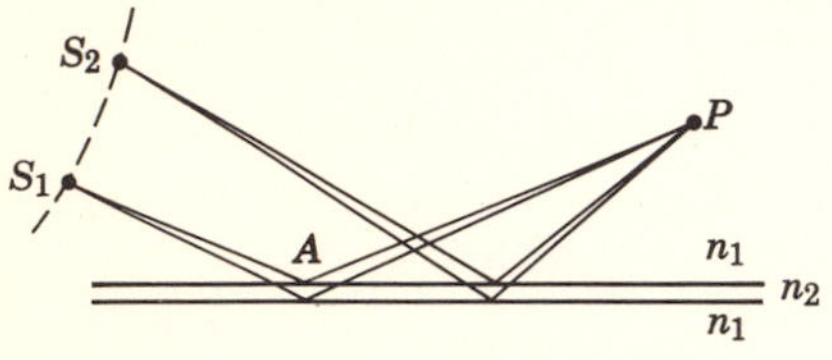

Fig. 5.13

A. GENERAL DISCUSSION

Consider monochromatic light falling on a thin dielectric film (Fig. 5.13). The existence of interference effects and their localization in space depends critically on the way the reflected (or transmitted) light is observed. Suppose that an observer has his eye focused at a point P, or is looking through a microscope focused at P, and let the film be illuminated by an extended source S. Consider a small element S_1 of S. The rays of interest then are those from S_1 to P reflected from the two surfaces of the film. If the film is very thin compared with the distances S_1A and AP and if the surfaces are almost parallel, our earlier result for the phase difference for the two rays will still hold, namely,

$$\pi + \frac{4\pi n_2 d}{\lambda_0} \cos r$$

From another part S_2 of the source there will be another pair of rays going to P. For these rays d and r will generally differ enough from those for the rays from S_1 so that the two different values of the phase shift will not be equal. The interference effects will not combine constructively, and as other parts of S are also considered, we generally obtain no net interference of any significance. We must find places on which to focus for which the phase difference is constant for all parts of the source that can be used.

However, if P is close to the surface of the film, the pairs of rays from both S_1 and S_2 will have practically the same values of d. They will have different values for the phase shift for a thin film only if their cos r values differ appreciably, provided that d is not very large. The variation in cos r can be reduced by either or both of the following devices. (1) If the eye or the optical instrument focused on the film has a small entrance pupil, all the rays that can pass from S to P off the film through the pupil will have just about the same value of r. (2) If the fringes are observed at nearly normal incidence ($r \approx 0$), cos r will be almost unity even if there is a moderate spread in r values about $r = 0$. Since one must be focused on (or very near) the film to see these Fizeau fringes, they are said to be "localized in the film."

As the eye, or observing microscope, moves over the surface of the film, cos r does not change rapidly, particularly near normal incidence, and the order m of the fringes will depend mainly on d. They can therefore be considered to be loci of constant optical film thickness. We observe a "contour map" of the cross section of the film.

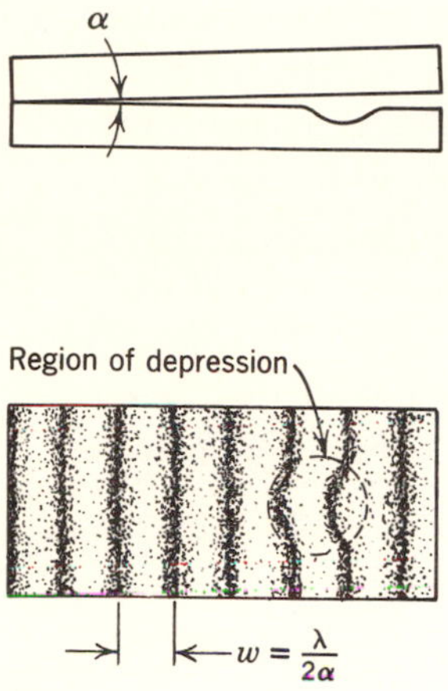

Fig. 5.14. Fizeau fringes in an air wedge.

Figure 5.14 illustrates this effect for an air wedge between two plates. The upper one is an optical flat, and the lower one may be a sample to be tested. If both pieces are very flat, a series of equidistant straight bands is observed. Between one fringe and the next the separation must change by $\lambda/2$ for a nearly perpendicular observation direction, since the light must go down and back. Hence, if α is the wedge angle, we have the following relation for the separation w between the fringes:

$$\alpha w = \frac{\lambda}{2} \tag{5.62}$$

Here $\lambda = \lambda_0/n_2$ represents the wavelength in the medium of the wedge. If air, $n_2 \approx 1$.

In the region of a slight depression in the lower plate, the gap thickness will increase sooner than usual, and the fringes will be convex toward the point of contact between the plates, as shown in Fig. 5.14.

B. NEWTON'S RINGS

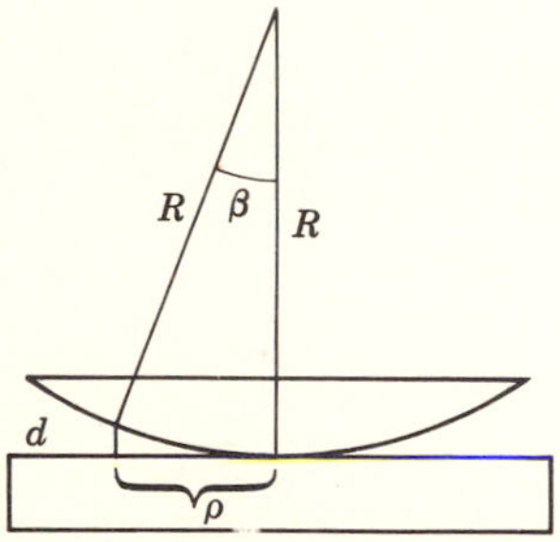

Fig. 5.15. Setup for the observation of Newton's rings.

A famous example of Fizeau fringes occurs when a concave spherical surface is placed on an optical flat. When we view from above with monochromatic illumination, we see a series of concentric rings around a central dark spot (Fig. 5.15). For the mth dark ring from the center we have

$$d_m = m\frac{\lambda}{2} \tag{5.63}$$

But d is approximately given by

$$d = R - R\cos\beta \approx \frac{R\beta^2}{2} = \frac{\rho^2}{2R}$$

Thus, $d_m = \rho_m{}^2/2R$, and the radius of the mth dark ring will be given by

$$\frac{\rho_m{}^2}{2R} = \frac{m\lambda}{2} \qquad \text{or} \qquad \rho_m = \sqrt{m\lambda R} \tag{5.64}$$

PROBLEMS

5.4.1 Prove the assertion made in the first paragraph of Section 5.4A.

5.4.2 Describe quantitatively the fringes observed when a cylindrical glass surface is placed in contact at one end with an optical flat as shown and observed from above with light of $\lambda = 4500$ A.

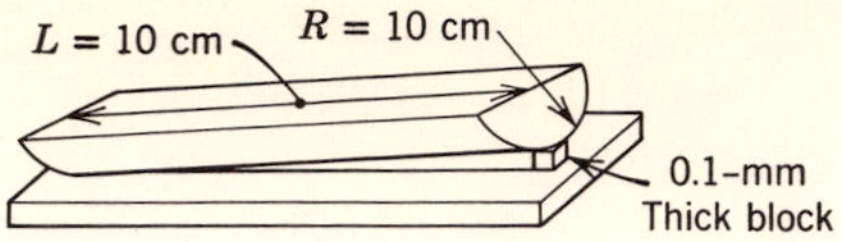

5.5 INTERFERENCE BY REFLECTIONS FROM NONIDENTICAL INTERFACES

Reflection from nonidentical interfaces occurs when the second surface is an interface with a third medium that is different from the first. Consider Fig. 5.16. Recall the derivation above where $n_3 = n_1$. We now use double subscripts on the amplitude reflection and transmission coefficients, so that, for example, ρ_{ij} is the amplitude reflection coefficient when a ray is incident from medium i onto an interface with medium j. We take the phase of the incident wave at B to be zero. The reflected complex electric field at D' will then be given by

$$\tilde{E}_1 = \rho_{12} E_0 e^{-ik_1 \widehat{BD'}} e^{i\omega t} \tag{5.65}$$

The second wave will then have amplitude

$$\tilde{E}_2 = \rho_{23} \tau_{12} \tau_{21} E_0 e^{-i\Delta} e^{-ik_1 \widehat{BD'}} e^{i\omega t} \tag{5.66}$$

where the phase difference Δ is given by the earlier result

$$\Delta = \frac{4\pi n_2 d}{\lambda_0} \cos r \tag{5.67}$$

If all three media are dielectrics, we have these inequalities holding

$$\rho_{12} > 0 \text{ if } n_1 > n_2; \qquad \rho_{12} < 0 \text{ if } n_1 < n_2;$$
$$\rho_{23} > 0 \text{ if } n_2 > n_3; \qquad \rho_{23} < 0 \text{ if } n_2 < n_3$$

Thus, $\rho_{12}\rho_{23} > 0$ if $n_1 > n_2 > n_3$ or $n_1 < n_2 < n_3$, and $\rho_{12}\rho_{23} < 0$ otherwise. We consider two applications of these ideas.

Fig. 5.16

A. REFLECTION REDUCING COATING

The idea here is to produce destructive interference between the two reflected beams. This is achieved by (1) having both reflections at less dense to more dense interfaces, and (2) having a π phase shift resulting from propagation in the second medium. Thus we choose the layer thickness d in Fig. 5.16 so that

$$\frac{4\pi n_2 d}{\lambda_0} = \pi \qquad \text{or} \qquad d = \frac{1}{4}\frac{\lambda_0}{n_2} = \frac{\lambda_2}{4}$$

(Hence the name "quarter-wave layer.") Often the wavelength λ_0 is chosen to be near 5500 A, at the maximum of the visual response curve. Then we choose the materials so that $n_1 < n_2 < n_3$. Then E_1 and E_2 will have opposite signs. Near normal incidence, using (5.65) and (5.66), we can write

$$e^{-i\omega t}\tilde{E}_1 = \rho_{12} E_0 e^{-ik_1 \widehat{BD'}}, \qquad \Delta = \pi$$

and

$$e^{-i\omega t}\tilde{E}_2 = (1-\rho_{12}^2)\rho_{23}E_0 e^{-\pi i} e^{-ik_1\widehat{BD}'}$$

or

$$e^{-i\omega t}\tilde{E}_2 \approx -\rho_{23}E_0 e^{ik_1\widehat{BD}'} \tag{5.68b}$$

where we have used the relations

$$\tau_{12}\tau_{21} = 1-\rho_{12}^2 \approx 1$$

Thus there will be no net reflection ($\tilde{E}_1+\tilde{E}_2 \approx 0$) if

$$\rho_{12} \approx \rho_{23} \tag{5.69}$$

We show later that near normal incidence

$$\rho_{12} = \frac{n_1-n_2}{n_1+n_2}$$

and

$$\rho_{23} = \frac{n_2-n_3}{n_2+n_3}$$

Thus the equation

$$\rho_{12} = \rho_{23}$$

implies that

$$\frac{n_1/n_2-1}{n_1/n_2+1} = \frac{n_1/n_3-1}{n_2/n_3+1}$$

or

$$\frac{n_1}{n_3}-\frac{n_2}{n_3}+\frac{n_1}{n_1}-1 = \frac{n_1}{n_3}+\frac{n_2}{n_3}-\frac{n_1}{n_2}-1$$

or

$$\frac{n_2}{n_3} = \frac{n_1}{n_2}, \qquad n_2 = \sqrt{n_1 n_3} \tag{5.70}$$

where n_2 is seen to be the geometric mean of n_1 and n_3. If $n_1 = 1$, the above requirement becomes

$$n_2 = \sqrt{n_3} \tag{5.71}$$

Often the second medium is a layer of MgF_2 ($n_2 = 1.38$). If we use heavy flint glass ($n_3 = 1.7$), then $\sqrt{n_3} = 1.3$. This meets the requirement closely enough to reduce the energy reflection coefficient from about 5 per cent for an uncoated simple lens to 1 per cent for a simple lens with the proper MgF_2 coating.

When multiple reflections are taken into account by methods to be described in the next section, this result must be modified somewhat. The result in the general case is as follows (see Fig. 5.16):

$$e^{-i\omega t}\tilde{E}_1 = \rho_{12}E_0 e^{-ik_1\widehat{BD}'}$$

$$e^{-i\omega t}\tilde{E}_2 = \tau_{12}\rho_{23}\tau_{21}e^{-i\Delta}E_0 e^{-ik_1\widehat{BD}'}$$

$$e^{-i\omega t}\tilde{E}_3 = \tau_{12}\rho_{23}\rho_{21}\rho_{23}\tau_{21}e^{-2i\Delta}E_0 e^{-ik_1\widehat{BD}'}$$

The total reflected complex electric field is then the sum of a geometrical series:

$$\tilde{E}_{\text{tot}} = \tilde{E}_1 + \tilde{E}_2 + \tilde{E}_3 + \cdots = \left[\rho_{12} + \frac{(1-\rho_{12}^2)\rho_{23}e^{-i\Delta}}{1-\rho_{21}\rho_{23}e^{-i\Delta}}\right]E_0 e^{-ik_1\widehat{BD}'}e^{i\omega t} \tag{5.72}$$

When $\Delta = \pi$, this becomes

$$\tilde{E}_{\text{tot}} = \left[\rho_{12} - \frac{(1-\rho_{12}^2)\rho_{23}}{1+\rho_{21}\rho_{23}}\right]E_0 e^{-ik_1\widehat{BD}'}e^{i\omega t}$$

When the reflection coefficients satisfy $\rho_{12} = \rho_{23}$, which implies the validity of Eq. (5.70), we obtain $\tilde{E}_{\text{tot}} = 0$. This occurs only when $\Delta = \pi, 3\pi \ldots$ or when $n_2 d = \lambda/4, 3\lambda/4, 5\lambda/4, \ldots$. For the quarter-wave thickness at a wavelength λ_0, the energy reflection coefficient given by

$$R = \left|\frac{\tilde{E}_{\text{tot}}}{E_0}\right|^2 = \left|\rho_{12} + \frac{(1-\rho_{12}^2)\rho_{12}e^{-i\Delta}}{1+\rho_{12}^2 e^{-i\Delta}}\right|^2$$

$$\approx |\rho_{12} + \rho_{12}e^{-i\Delta}|^2 = 2\rho_{12}^2\,(1+\cos\Delta)$$

remains small over a large wavelength region about λ_0, as shown in Fig. 5.17.

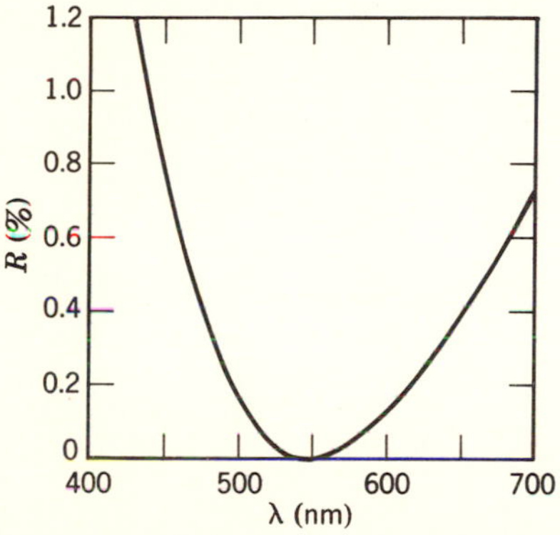

Fig. 5.17. Reflectivity with quarter wave antireflectivity coating ($\lambda_0 = 550$ nm, $n_2 = 1.3$, $n_3 = 1.7$).

B. REFLECTION ENHANCING COATING—THE QUARTER-WAVE STACK

We can construct an all-dielectric multilayer reflection enhancing coating from a so-called quarter-wave stack. This consists of alternating layers of high (H) and low (L) index material evaporated onto a substrate, as shown in Fig. 5.18. Let λ_0 be the wavelength at which maximum reflectivity is desired. Then control the thickness of the layers so that

$$n_H t_H = n_L t_L = \frac{\lambda_0}{4}$$

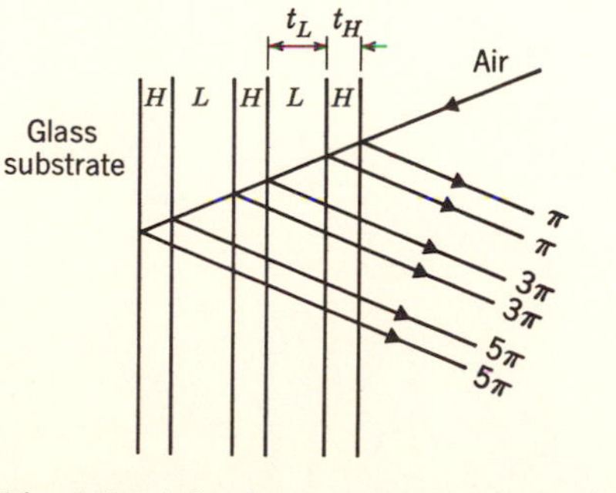

Fig. 5.18. A five-layer quarter-wave stack. The phase shifts due to partial reflection and propagation are indicated.

The phase differences for light of wavelength near λ_0 and close to normal incidence will be as shown. The partially reflected waves all interfere constructively, giving a large net reflection coefficient. The detailed theory is best worked out by use of matrices.

Quarter-wave stacks may be constructed with a large number of layers. One type that is used to construct mirrors for use in helium-neon gas laser cavities has 13 layers of zinc sulfide ($n_H = 2.32$) and magnesium fluoride ($n_L = 1.38$) with a peak reflectance of 98.9 per cent at 633 nm.

REFERENCES FOR SECTION 5.5

M. Born and E. Wolf, *Principles of Optics*, Section 7.5.
O. S. Heavens, *Optical Properties of Thin Solid Films*.
L. Holland, *Vacuum Deposition of Thin Films*.
P. Baumeister, "Interference, and Optical Interference Coatings", Chapter 8 of *Applied Optics and Optical Engineering*, edited by R. Kingslake, Vol. I.

An advanced discussion of multiple reflection starting with the boundary values for Maxwell's equations and using matrices is found with Born and Wolf, *Principles of Optics*, Section 1.6.

PROBLEM

5.5.1 Consider a reflection-reducing coating made with MgF_2 ($n_2 = 1.38$) on a glass of index $n = 1.70$ to work at nearly normal incidence. How thick should the coating be if the wavelength of the light is 5500 A? Calculate the net reflection at this wavelength as well as at 4000 A and 7500 A,
(a) considering only interference between first two reflected beams,
(b) considering interference among all reflected beams.

5.6 INTERFERENCE BY MULTIPLE REFLECTION

We now wish to treat in some detail the multiple reflections neglected in Section 5.2. They will be important in practice when the reflectivity of the interfaces is large. The effect can be studied by partially silvering the surfaces of a plane parallel dielectric slab or by use of a Fabry–Perot etalon or interferometer (where the "slab" is really an air gap). The "silvering" can be a metallic layer, or it can be a multilayer dielectric coating of the type mentioned in Section 5.5B.

A. GENERAL DERIVATION

1. Setting up the Equations Consider the situation shown in Fig. 5.19. Because of one extra pass through medium 2 the reflected ray 2 is retarded with respect to ray 1 by a phase Δ. Similarly ray 3 is retarded by the same amount with respect to ray 2. The transmitted ray 2′ will lag ray 1′ by Δ, and ray 3′ will lag ray 2′ also by Δ. And so on. If $\tilde{E}_0 e^{i\omega t}$ represents the complex field at B, the transmitted and reflected rays will be given by multiplying $E_0 e^{-ik_1\widehat{BD}} e^{i\omega t}$ by the coefficients in the following table.

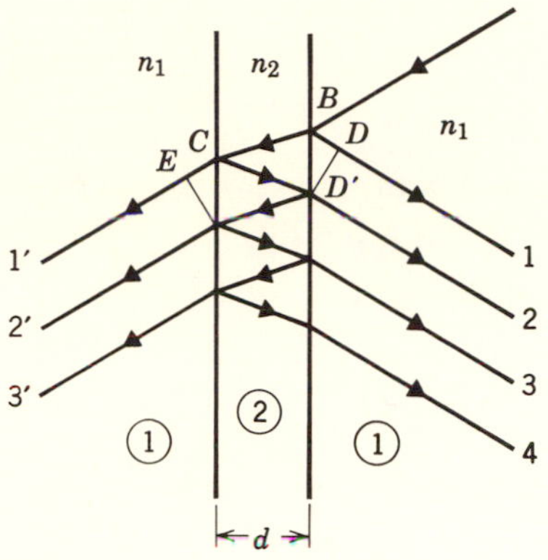

Fig. 5.19. Multiple reflection interference.

Reflected	*Transmitted*
1: ρ	1′: $\tau\tau' e^{-ik_2\widehat{BC}}$
2: $\tau\tau'\rho' e^{-i\Delta}$	2′: $\tau\tau'\rho'^2 e^{-ik_2\widehat{BC}} e^{-i\Delta}$
3: $\tau\tau'\rho'^3 e^{-2i\Delta}$	3′: $\tau\tau'\rho'^4 e^{-ik_2\widehat{BC}} e^{-2i\Delta}$
4: $\tau\tau'\rho'^5 e^{-3i\Delta}$	4′: $\tau\tau'\rho'^6 e^{-ik_2\widehat{BC}} e^{-3i\Delta}$
.	.
.	.
.	.

When the interfaces are layers of finite thickness, ρ and τ will be effective amplitude reflection and transmission coefficients through the entire layer with the ray incident in medium 1. Their values will depend on angle. Similar remarks apply to ρ' and τ', the coefficients in medium 2. We now combine each of the above components to obtain the total electric field reflected

$$\begin{aligned}\tilde{E}^R_{\text{tot}} &= E_0 e^{-ik_1\widehat{BD}} \{\rho + \tau\tau'\rho' e^{-i\Delta} + \tau\tau'\rho' \cdot \rho'^2 e^{-2i\Delta} + \cdots\} \, e^{i\omega t} \\ &= \{E_0\rho + E_0\tau\tau'\rho' e^{-i\Delta}(1 + \rho'^2 e^{-i\Delta} + \rho'^4 e^{-2i\Delta} + \cdots)\} \, e^{-ik_1\widehat{BD}} e^{i\omega t} \\ &= \left(E_0\rho + \frac{E_0\tau\tau'\rho' e^{-i\Delta}}{1 - \rho'^2 e^{-i\Delta}}\right) e^{-ik_1\widehat{BD}} e^{i\omega t} \end{aligned} \tag{5.73}$$

and the total electric field transmitted

$$\begin{aligned}\tilde{E}^T_{\text{tot}} &= \{E_0\tau\tau' + E_0\tau\tau'\rho'^2 e^{-i\Delta} + E_0\tau\tau'\rho'^4 e^{-2i\Delta} \cdots\} \, e^{-ik_1\widehat{BD}} e^{-ik_2\widehat{BC}} e^{i\omega t} \\ &= E_0\tau\tau' \, (1 + \rho'^2 e^{-i\Delta} + \rho'^4 e^{-2i\Delta} + \cdots) \, e^{-ik_1\widehat{BD}} e^{-ik_2\widehat{BC}} e^{i\omega t} \\ &= \left(\frac{E_0\tau\tau'}{1 - \rho'^2 e^{-i\Delta}}\right) e^{-ik_1\widehat{BD}} e^{-ik_2\widehat{BC}} e^{i\omega t} \end{aligned} \tag{5.74}$$

In the above formulas we have summed the infinite series $1+x+x^2+\cdots=(1-x)^{-1}$ and thus have implicitly assumed that the slab is large enough for a great many multiple reflections to occur.

2. The Energy Transmission Coefficient. We now want to discuss the transmitted flux density or radiant incidence, since these devices are usually used in transmission. For any given detection scheme, the flux density will be proportional to the square of the absolute value of the total complex electric field. Since the media on the two sides of the slab are the same, the proportionality constant is the same, and the energy transmission coefficient T will be given by

$$T=\frac{|\tilde{E}^T_{\text{tot}}|^2}{|E_0|^2}=\frac{|\tau\tau'|^2}{|1-\rho'^2e^{-i\Delta}|^2} \tag{5.75}$$

In general, the amplitude reflection coefficient from medium 2 to medium 1 will be a complex quantity. This is especially true if the interface is covered with a partially transmitting metallic layer. We can therefore write

$$\rho'=\rho_0'e^{i\theta}$$

where ρ_0' is real and positive and where θ is a real angle that represents the change in phase of a wave on reflection at the 2–1 interface. The denominator in Eq. (5.75) then becomes

$$\begin{aligned}|1-\rho'^2e^{-i\Delta}|^2&=[1-(\rho_0')^2e^{-i(\Delta-2\theta)}][1-(\rho_0')^2e^{i(\Delta-2\theta)}]\\&=1+(\rho_0')^4-2(\rho_0')^2\cos\delta\end{aligned} \tag{5.76}$$

Here $\delta=\Delta-2\theta$ represents the total decrease in phase of ray 2′ with respect to 1′ (or of 3′ with respect to 2′, or of 3 with respect to 2, and so on). Part of this phase lag results from propagation (Δ), part from reflection (-2θ).

It is convenient to introduce the intensity reflection coefficient for a single reflection at the 2–1 interface. It is given by

$$R_1=|\rho'|^2=(\rho_0')^2 \tag{5.77}$$

and represents the ratio of the intensity reflected in a single reflection to the incident intensity at the 2–1 interface. Using R_1, we can write

$$T=\frac{|\tau\tau'|^2}{1+R_1{}^2-2R_1\cos\delta} \tag{5.78}$$

3. Maximum and Minimum Transmission. Maximum transmission occurs when $\cos\delta=1$. The transmission coefficient is then

$$T_{\max}=\frac{|\tau\tau'|^2}{(1-R_1)^2} \tag{5.79}$$

For dielectric interfaces we have the relations (5.50) which imply

$$\tau\tau' = 1-\rho'^2 = 1-R_1 \tag{5.80}$$

Then T_{max} is given by

$$T_{max} = \frac{|\tau\tau'|^2}{(1-R_1)^2} = 1$$

However, such dielectric interfaces have small values of R_1, and the multiple reflection effects are not very important. Large values of R_1 are obtained either with metallic layers or with a multiple series of dielectric layers as discussed in Section 5.5B. For a metal, Eq. (5.80) is not valid, and for the multilayers the situation is much more complicated. Thus, in real situations T_{max} is unlikely to be equal to 1.

In the general case we can combine Eqs. (5.78) and (5.79) to obtain

$$T = T_{max}\frac{(1-R_1)^2}{(1+R_1^2-2R_1\cos\delta)} \tag{5.81}$$

The minimum transmission occurs when $\delta = \pi$. The minimum transmission coefficient is

$$T_{min} = T_{max}\frac{(1-R_1)^2}{(1+R_1)^2} \tag{5.82}$$

*4. **The Contrast.*** The value T_{min} can be very small when $R_1 \approx 1$, that is, if each surface is a good reflector. In this event, T stays small unless δ is close to $\pm p2\pi$. We can show this easiest after some rewriting:

$$T = T_{max}\frac{(1-R_1)^2}{1-2R_1\cos\delta+R_1^2} = \frac{T_{max}}{\dfrac{1+R_1^2-2R_1+2R_1-2R_1\cos\delta}{1+R_1^2-2R_1}}$$

$$= \frac{T_{max}}{1+\dfrac{2R_1(1-\cos\delta)}{(1-R_1)^2}}$$

Now use $1-\cos\delta = 2\sin^2\delta/2$. Then we can write

$$T = T_{max}\bigg/\left[1+\frac{4R_1\sin^2(\delta/2)}{(1-R_1)^2}\right] = \frac{T_{max}}{1+F\sin^2(\delta/2)} \tag{5.83}$$

where

$$F = \frac{4R_1}{(1-R_1)^2} \tag{5.84}$$

is called the contrast. Equation (5.83) is known as the Airy formula.

When $\delta = \pi + 2\pi p$, $\sin \delta/2$ equals ± 1, and T equals $T_{\min}$. The minima are halfway between the maxima. Thus we have

$$T_{\min} = \frac{T_{\max}}{1+F}$$

or

$$\frac{T_{\max}}{T_{\min}} = 1 + F \tag{5.85}$$

This ratio can be quite large when F is large, which in turn will be the case if R_1 is close to 1. For example, take $R_1 = 0.9$. Then $F = 3.6/(1-0.9)^2 = 3.6/0.01 = 360$.

5. Approximate Behavior of T near $T_{\max}$ for Large F. If $F \gg 1$, we can derive approximate expressions for T when δ is close to $2\pi p$, where p is an integer, that is, near a maximum in T. We can write

$$\sin^2 \frac{\delta}{2} = \sin^2 \left(\frac{\delta}{2} - p\pi\right) \approx \left(\frac{\delta}{2} - p\pi\right)^2 \tag{5.86}$$

Then we have

$$T = \frac{T_{\max}}{1 + F \sin^2 (\delta/2)} \approx \frac{T_{\max}}{1 + F\ (\delta/2 - p\pi)^2} = T_{\max} \frac{1/F}{1/F + (\delta/2 - p\pi)^2} \tag{5.87}$$

This approximation holds as long as

$$\left|\frac{\delta}{2} - p\pi\right| \ll 1$$

The functional dependence on δ in the last equality of Eq. (5.87) is called *Lorentzian*. Figure 5.20*a* shows the transmission as a function of δ.

The full-width at half-maximum W is twice the value of $\delta - 2\pi p$ at which $T = T_{\max}/2$. This will happen when

$$\left(\frac{\delta}{2} - p\pi\right)^2 = \frac{1}{F}$$

Thus

$$W = \frac{4}{\sqrt{F}} = \frac{2(1-R_1)}{\sqrt{R_1}} \quad (\approx 2(1-R_1) \text{ for } R_1 \text{ near } 1) \tag{5.88}$$

Figure 5.20 shows what the transmission curve looks like for high and low F.

The reason there is much more contrast when R_1 is close to unity than for small R_1 can be seen from the vector diagrams in Fig. 5.21. They show graphically the complex addition treated analytically in Eq. (5.74). When $R_1 \approx 1$, many small components add to form

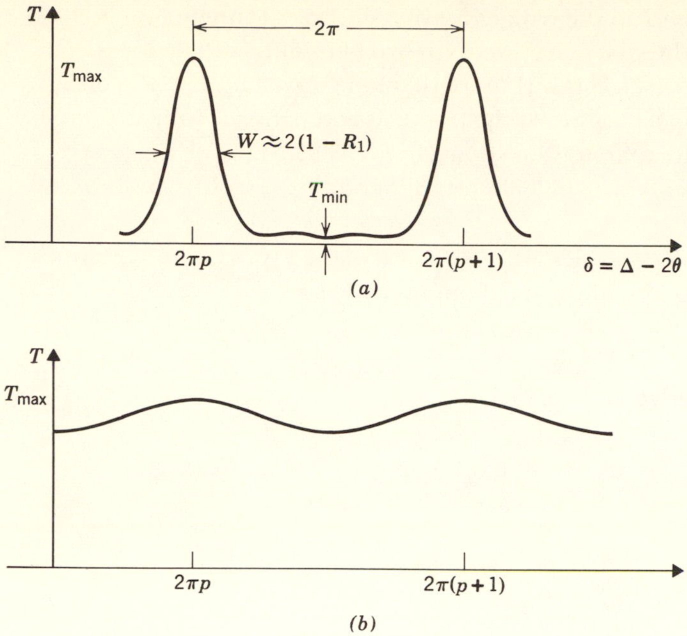

Fig. 5.20. (*a*) Large F(R_1 near 1). (b) Small $F(R_1 \ll 1)$.

E^T_{tot}. When they are in phase, the complex vectors all line up, giving a large resultant. When δ, the relative phase difference, is not equal to an integral multiple of 2π, the individual contributions lie on a polygon which almost closes, giving a small resultant. (We shall deal with similar "vibration curves" when we discuss diffraction.) When R_1 is small, the component vectors decrease rapidly in size, and E^T_{tot} varies only slightly with δ.

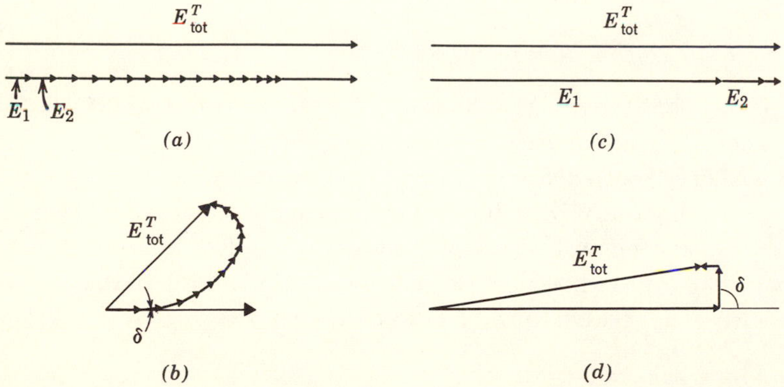

Fig. 5.21. Vector diagrams illustrating the addition of reflected components in transmission through (*a*) large F, $\delta = 2\pi p$; (*b*) large F, $\delta \neq 2\pi p$; (*c*) small F, $\delta = 2\pi p$; (*d*) small F, $\delta \neq 2\pi p$.

B. FABRY-PEROT INTERFEROMETER

The basic elements of a Fabry-Perot interferometer are indicated in Fig. 5.22. We often use it in visual operation by looking through it at a diffuse source. This produces fringes quite similar to the circular Haidinger fringes discussed earlier, but now the

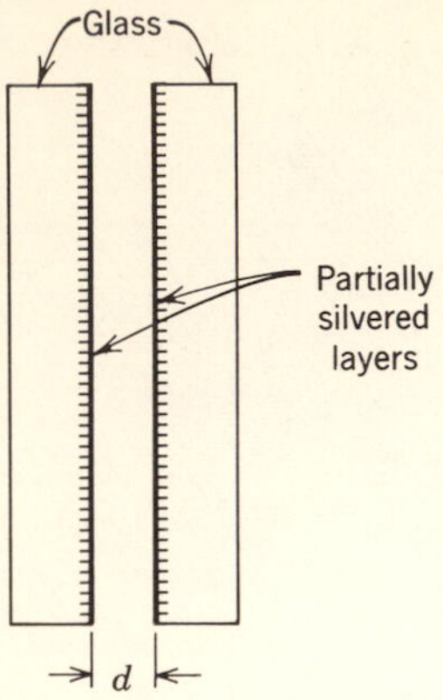

Fig. 5.22. Fabry-Perot interferometer.

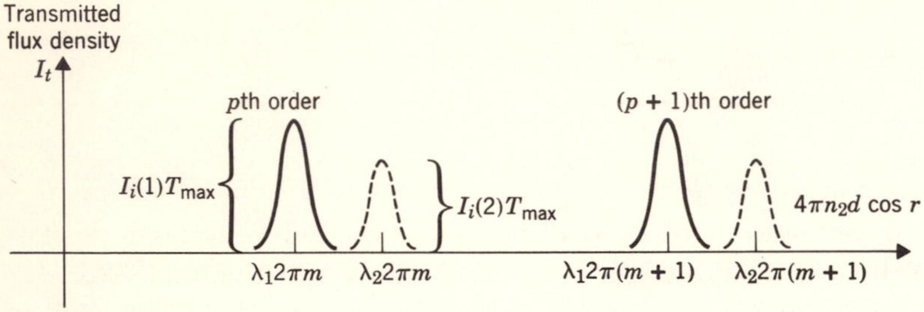

Fig. 5.23. Fabry-Perot transmission pattern when two nearby wavelengths are present.

transmission T in the fringes is not a sinusoidal function of δ, but shows the sharp maxima with broad minima of Fig. 5.20*a*.

*1. **Two Wavelengths.*** If more than one wavelength is present, we see a superposition of the transmission patterns for each wavelength separately, as shown in Fig. 5.23. Here we have set $\theta = 0$ or π and have taken our independent variable to be $4\pi n_2 d \cos r$.

For a single wavelength λ we would have a transmitted intensity I_t given in terms of the incident intensity I_i by

$$I_t = I_i T$$

where T is given by Eq. (5.83):

$$T = \frac{T_{max}}{1 + F\sin^2\delta/2}$$

with $\delta = \Delta - 2\theta = (4\pi n_2 d\cos r)/\lambda - 2\theta$. For two wavelengths λ_1 and λ_2 with incident intensities $I_i(1)$ and $I_i(2)$, we would add to obtain

$$I = \frac{I_i(1)T_{max}}{1 + F\sin^2(\delta_1/2)} + \frac{I_i(2)T_{max}}{1 + F\sin^2(\delta_2/2)}$$

with $\delta_1 = (4\pi n_2 d\cos r)/\lambda_1 - 2\theta$ and $\delta_2 = (4\pi n_2 d\cos r)/\lambda_2 - 2\theta$ and where T_{max} and F are essentially independent of λ.

*2. **The Free Spectral Range.*** Complications now arise. How can we be sure that the λ_2 peak has the same order as the nearby λ_1 peak? If λ_2 exceeds λ_1 by a certain amount $\Delta\lambda$, called the *free spectral range*, the mth-order fringe for λ_2 will fall on top of the $(m+1)$th-order fringe for λ_1. In this case we have

$$(m+1)\lambda_1 = m\lambda_2$$

Now let $\lambda_2 = \lambda_1 + \Delta\lambda$ and solve for the free spectral range $\Delta\lambda$:

$$(m+1)\lambda_1 = m(\lambda_1 + \Delta\lambda); \qquad \lambda_1 = m\Delta\lambda; \qquad \Delta\lambda = \frac{\lambda_1}{m} \tag{5.89}$$

Thus, if we are working in high order (large m), which usually means that d the plate separation is large, we must use wavelengths that are quite close together to avoid confusion of orders. Actually, to avoid all confusion, we should have $(\lambda_2 - \lambda_1) < \Delta\lambda/2$. Preliminary dispersion with a prism or grating monochromator can be used to separate out unwanted wavelengths.

3. ***Resolving Power.*** The *resolving power* of the interferometer is also of interest. How close together in wavelength can two lines be in order to be just resolvable? This depends on the criterion we use for resolution. We choose the following simple test. With two lines of equal intensity of wavelengths λ_1 and $\lambda_3 = \lambda_1 + \delta\lambda$, we say that they are just resolvable if the half-maximum point of one peak coincides with that of the other peak in the same order The peak separation corresponds to a phase change (in δ) of $W_{1/2}$. This is illustrated in Fig. 5.24. As a function of $4\pi n_2 d \cos r$, the peak

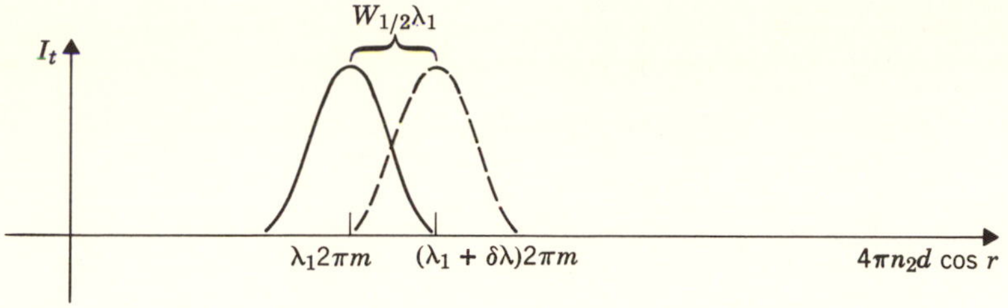

Fig. 5.24. Illustration of our criterion for resolving power determination.

separation is $W_{1/2}\lambda_1 = \delta\lambda\, 2\pi m$. The resolving power $\mathscr{R}$ is defined to be equal to $\lambda/\delta\lambda$ ($\lambda \approx \lambda_1 \approx \lambda_2$) and is thus given by

$$\mathscr{R} = \frac{\lambda}{\delta\lambda} = \frac{2\pi m}{W_{1/2}}$$

or

$$\mathscr{R} = \frac{\lambda}{\delta\lambda} = \frac{\pi m\sqrt{F}}{2} = \frac{\pi m\sqrt{R_1}}{1-R_1}\left(\approx \frac{\pi}{1-R_1} m \quad \text{if } R_1 \text{ is close to } 1\right) \tag{5.90}$$

This can be extremely large for large-order m, especially when the contrast F is also large.

The ratio $\Delta\lambda/\delta\lambda$ of the free spectral range (or fringe separation) to the minimum resolvable wavelength separation is called the *finesse* $\mathscr{F}$. Since $\Delta\lambda = \lambda/m$, we obtain

$$\mathscr{F} = \frac{\pi\sqrt{F}}{2}\left[\approx \frac{\pi}{1-R_1} \quad \text{if} \quad (1-R_1) \ll 1\right] \tag{5.91}$$

The Fabry-Perot interferometer is used in a similar way to the Michelson interferometer. A broad diffuse source is necessary to give the required spread in r and hence of Δ. The observing system (a telescope, or an unaided eye focused at infinity) brings rays with the same r values together to form a ring-shaped interference pattern, just like that seen with the Michelson interferometer; only now the fringes are not sinusoidal, but show much more contrast. The following equations, (5.92)–(5.94), describe the location of these fringes. (See Problem 5.6.2.)

C. SPECTROMETRY WITH THE FABRY-PEROT ETALON

A Fabry-Perot *etalon* is essentially a Fabry-Perot interferometer having a fixed separation d between the plates. This is achieved by means of a rigid mechanical separator. It is conventionally employed, with incident light covering a continuous range of values of the angle r. The spectrum is observed photographically in the focal plane of a camera lens. The equations of Section 5.3A apply to this case with changes similar to those needed for the Michelson interferometer. The order of a fringe is given by

$$m = \frac{2nd}{\lambda_0}\cos r, \qquad d = \text{plate separation}$$

$$n = \text{index of medium between plates}$$

$$\lambda_0 = \text{vacuum wavelength} \tag{5.92}$$

Here r represents the inclination angle both inside and outside the etalon. The maximum order is

$$m_{\text{max}} = \frac{2nd}{\lambda_0} \tag{5.93}$$

which does not have to be exactly an integer. The radius of the pth bright fringe from the center, where

$$p = m_{\text{max}} - m = \frac{2nd(1-\cos r)}{\lambda_0} \approx \frac{ndr^2}{\lambda_0} \approx \frac{nd\,y_p{}^2}{\lambda_0 f^2}$$

is given by

$$y_p = fr = f\left(\frac{\lambda_0 p}{nd}\right)^{1/2} \tag{5.94}$$

Here p need not be an integer either.

For both m_{max} and p the concept of a nonintegral number of fringes is useful, since in practice we can interpolate between fringes and can move out, say, 5.3 fringes from the center of the pattern.

Suppose only monochromatic green light is present. Then a green image of the slit will be formed at the place on the plate labeled "green" in Fig. 5.25. With the etalon inserted in the indicated position, the vertical position of these arcs is given by y_p of Eq. (5.94). If the green line actually consists of several closely spaced lines, they will appear separated in the vertical direction (out of the paper) because of the dispersion produced by the etalon,

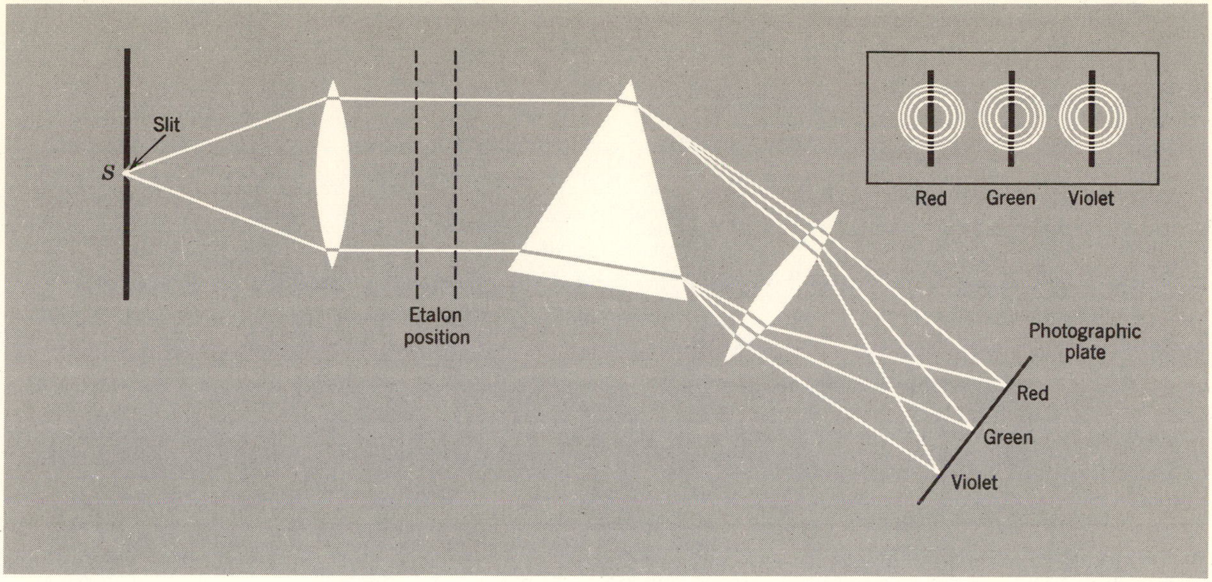

Fig. 5.25. Use of Fabry-Perot etalon with a prism spectrograph.

but the prism dispersion is too weak to disperse these closely spaced lines in the horizontal direction. What the prism does is perform a gross separation, so that there will be no confusion of orders. Thus, if there is a closely spaced group of lines in the violet (or in the red), they will be separated from the green group by the prism. The etalon then separates the individual lines in the violet group. In practice there sometimes is still confusion of orders, because the prism dispersion is often too weak and the etalon dispersion too strong for a group of lines of intermediate spacing.

Another way of using the etalon in spectroscopy is the *center spot scanning technique* shown in Fig. 5.26. It is applicable when only a

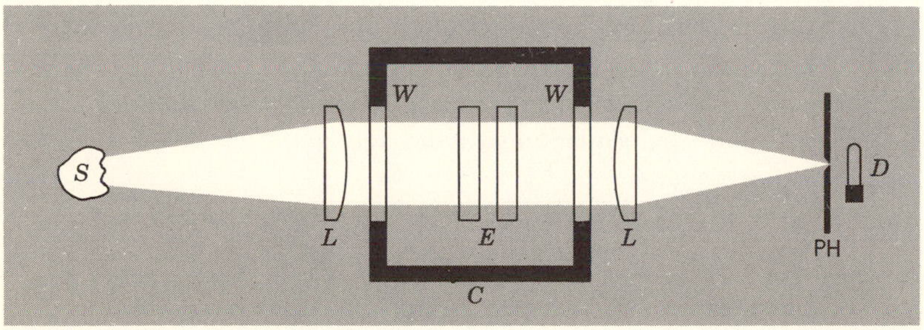

Fig. 5.26. Center spot scanning with a Fabry-Perot etalon: *S*, source; *L*, lenses; *C*, vacuum chamber for pressure scanning; *W*, windows; *E*, etalon; *PH*, pinhole; *D*, detector.

few wavelength groups are present, so that preliminary wavelength separation can be done with a filter or premonochromator. We either shine light from a diffuse source (after necessary filtering) directly into the etalon, or use collimated light from a broad source. Then, at the focal plane of a lens, we observe the circular fringe pattern.* A pinhole is placed at the center of the pattern through which light passes to a photomultiplier tube. We are therefore measuring the intensity of the "center spot" at $r = 0$. To see an intensity change, we must change either d or n. We can change the plate separation with an interferometer, but with the rigid etalon we must change n. This is done by placing the etalon in an evacuable enclosure and then changing the air pressure. The value of $n - 1$ for air at 15°C and 1 atmosphere is 2.78×10^{-4} at $\lambda = 5000$ A. This quantity is directly proportional to the air pressure. For example, for $\lambda = 5000$ A, $d = 3$ mm, from Eq. (5.93) we find that m_{max} changes by $\Delta m_{max} = 2d(n-1)/\lambda_0 = 3.34$ fringes as the pressure changes from 1 atmosphere to zero.

In the experiment we scan the fringe pattern by plotting the photomultiplier current versus time as the chamber is evacuated or as air is let back in. A better method, of course, is the more complicated one of plotting photocurrent versus pressure.

REFERENCE FOR SECTION 5.6

M. Born and E. Wolf, *Principles of Optics*. Section 7.6.

PROBLEMS

5.6.1 Calculate the ratio of full width at half-maximum to the separation between maxima (as a function of the phase difference Δ) for a Fabry-Perot etalon with $R_1 = 0.5, 0.8, 0.9, 0.98$. (Answers: 0.225, 0.071, 0.0336, 0.0064.)

5.6.2 A Fabry-Perot etalon is used with monochromatic light as described in Section C. Let the first bright fringe from the center occur at $p = \epsilon$ (not necessarily an integer), the first beyond that at $p = \epsilon + 1$, the second beyond that at $p = \epsilon + 2, \ldots$, the Nth beyond that at $p = \epsilon + N$. Let y_n be the radius of the Nth bright fringe beyond the first.

a. Show that ϵ is given by

$$\epsilon = \frac{y_N{}^2}{\Delta y^2} - N$$

where Δy^2 is the difference in y^2 between fringes, which does not depend on N. All the quantities on the right in the above equation can be directly measured.

b. Assuming n and λ to be known, find the plate separation in terms of Δy^2 and f, the focal length of the lens.

5.6.3 A Fabry-Perot interferometer is used with yellow light of two wavelengths of equal intensity $\lambda_1 = 6000$ A, $\lambda_2 = 5994$ A. At what approximate value of the total order number m will they be just resolvable according to our criterion if $R_1 = 0.85$? (Answer: 52.)

*The flux density in this pattern will be proportional to the square of the quantity (lens diam/focal length). Hence a short focal length is desirable. On the other hand, spherical aberration can reduce the contrast of the center spot so that a good-quality lens is needed.

5.6.4 In the center spot scanning technique, the question of the size of the pinhole in front of the detector often arises. If it is too small, we lose illuminance unnecessarily; if too large, resolution falls off. Let us assume that a good compromise is reached when the pinhole radius ρ corresponds to a change in phase Δ of amount $\frac{1}{2}W$, half the full width at half-maximum. Find an expression for ρ in terms of n, d, λ, R_1 (or F), and f, the focal length of the final lens.

5.6.5 Derive Eq. (5.72).

In Chapter 5 we discussed the interference of light beams from two sources having the same frequency. Interference could be observed if the phase difference remained constant during the time of measurement. The light was therefore said to be coherent. This approach to the two-beam interference experiment was not pursued. Later we discussed this experiment when two wavelengths were present, and found a periodic oscillation in the contrast or visibility of the fringe pattern. Now we shall develop these ideas further and discuss the spectral content of light, how it affects the two-beam interference experiment, and how the results of this experiment can be processed to yield a spectrum. We shall take up a description of coherence in terms of the correlation of two optical signals and relate this to the two-beam interference experiment.

We shall then discuss spatial coherence and the Michelson stellar interferometer. Finally there will be mention of interferometry using correlation of fluctuations.

CHAPTER 6

COHERENCE

CHAPTER 6 CONTENTS

6.1 INTERFERENCE SPECTROSCOPY

In Young's double-slit experiment two light beams obtained from the original source are brought together after they have divided and have traveled different optical path lengths. A similar result is achieved in the Michelson interferometer. In either case, if light of a well-defined frequency is present, we observe regular interference fringes as a function of the optical path difference. If more than one frequency is present, we must add the flux density that would be produced by each acting separately. Then if, for example, the total flux density is divided equally between two frequencies, we see a beat pattern ranging in a regular periodic way from fringes of maximum contrast to fringes of zero contrast. We shall show that, in general, the observed fringe pattern is uniquely determined by the distribution of flux density among the various component frequencies, and shall describe the method of recovering the spectrum from the interference pattern.

A. PRELIMINARY DISCUSSION

*1. **Monochromatic and Quasimonochromatic Light.*** The electric field from a truly monochromatic source would be sinusoidal and would have a constant amplitude A and phase ϕ:

$$E(t) = A \cos (2\pi \nu t + \phi) \tag{6.1}$$

Real sources are not truly monochromatic. Their output, which we shall sometimes refer to as "signals," can be represented as a superposition of terms like Eq. (6.1). A continuous distribution of frequencies is needed to describe this superposition mathematically. The technique involves the Fourier integral (Appendix A) and is discussed in Appendix B.

If the frequencies ν involved in the above superposition are strongly peaked about a certain frequency ν_0, the light is called *quasimonochromatic*. Then it becomes possible to write the signal in the form

$$E(t) = A(t) \cos [2\pi \nu_0 t + \phi(t)] \tag{6.2}$$

where $A(t)$ and $\phi(t)$ are slowly varying functions compared with $\cos 2\pi\nu_0 t$; that is, they change only slightly during one period $1/\nu_0$.

All light signals have a statistical or random nature. Certain properties cannot be predicted exactly, but mean values can be defined. Real quasimonochromatic light signals will have randomly varying functions $A(t)$ and $\phi(t)$. This is also true in other regions of the electromagnetic spectrum. Of course $A(t)$ is not always fluctuating—radio frequency oscillators and some lasers give signals with constant amplitudes—but the phase ϕ fluctuates because of inevitable drifts in the oscillator itself. These drifts, which are usually of thermal origin, can be reduced by lowering the temperature of certain components of the source, but ultimately for very stable systems these fluctuations will be a symptom of the fundamental quantum nature of light and matter.

The fluctuations inherent in light signals complicate the mathematical description of light. It helps to deal, wherever possible, with quantities that can be determined by experiment. Often in an experiment only average values are determined, and such averages can be handled rather easily in the theory. An important experimental quantity is the frequency distribution function or spectrum which can be determined by spectroscopy. We must therefore examine an idealized spectroscopy experiment.

2. ***Noninterference between Light Signals of Different Frequency.*** The basic fact that simplifies our considerations is that *under ordinary circumstances light signals of different frequency do not interfere.* The issue of interference versus noninterference can be clarified by examining the simplest possible case. Consider a fictitious optical signal that is the sum of two pure monochromatic components:

$$E = E_1 + E_2$$
$$E_1 = A_1 \cos (2\pi\nu_1 t + \phi_1) = A_1 \cos \theta_1$$
$$E_2 = A_2 \cos (2\pi\nu_2 t + \phi_2) = A_2 \cos \theta_2$$

Here A_1, A_2, ϕ_1, and ϕ_2 are constants.

We assume that the experimentally determined flux density at time t_0 is the average of the instantaneous flux density over the interval from $t_0 - T/2$ to $t_0 + T/2$, where T is the "measurement time." The instantaneous flux density is a constant times E^2. We denote the average of E^2 in the following way:

$$\langle E^2 \rangle \equiv \frac{1}{T} \int_{t\ -T/2}^{t_0 + T/2} E^2(t)\, dt$$

We thus obtain

$$\begin{aligned} \langle E^2 \rangle &= \langle E_1{}^2 \rangle + \langle E_2{}^2 \rangle + 2\langle E_1 E_2 \rangle \\ &= \tfrac{1}{2} A_1{}^2 + \tfrac{1}{2} A_2{}^2 + 2A_1A_2 \langle \cos \theta_1 \cos \theta_2 \rangle \end{aligned} \tag{6.3}$$

The factors of 1/2 come from taking the averages of $\cos^2 \theta_1$ and $\cos^2 \theta_2$, as discussed in connection with Eq. (5.30).

If there is to be any interference, it must come from the third term. Using the identity $\cos A \cos B = \frac{1}{2} \cos (A + B) + \frac{1}{2} \cos (A - B)$, it yields

$$2A_1A_2 \langle \cos \theta_1 \cos \theta_2 \rangle = A_1A_2 \langle \cos (\theta_1 - \theta_2) \rangle + A_1A_2 \langle \cos (\theta_1 + \theta_2) \rangle$$

For optical frequencies,

$$\cos(\theta_1+\theta_2)=\cos[2\pi(\nu_1+\nu_2)t+\phi_1+\phi_2]$$

oscillates about 10^{15} times per second and averages to zero no matter how short the measuring time T is. The difference term $\cos(\theta_1-\theta_2)=\cos[2\pi(\nu_1-\nu_2)t+\phi_1-\phi_2]$ ordinarily oscillates many times during a measurement also, because usually we have $(\nu_1-\nu_2)T \gg 1$. The interference term then is zero.

3. ***Digression on How to Produce Such Interference.*** Some insight into the above result may be gained by discussing cases where interference between two light beams having different frequencies *is* observed.

To see interference between E_1 and E_2 for real quasimonochromatic light signals requires special experimental techniques. As for the theory of the effect, we must let A_1, A_2, ϕ_1 and ϕ_2, have slow time dependence and obtain a slight generalization of Eq. (6.3).

$$\langle E^2\rangle=\tfrac{1}{2}\langle A_1(t)^2\rangle+\tfrac{1}{2}\langle A_2(t)^2\rangle+2\langle E_1E_2\rangle$$

The interference term then consists of

$$2\langle E_1E_2\rangle=\langle A_1(t)A_2(t)\cos[2\pi(\nu_1-\nu_2)t+\phi_1(t)-\phi_2(t)]\rangle \tag{6.4}$$

To see the beats at the difference frequency $\Delta\nu=\nu_1-\nu_2$ we need to satisfy two conditions: (1) a measuring time T that is short compared with the time between beats, $\Delta\nu^{-1}$, and (2) $\phi_1(t)-\phi_2(t)$ must be approximately constant during a period of one beat. Fluctuations in A_1 and A_2 will modulate the amplitude of the beats, but will not destroy them as will fluctuations in $\phi_1-\phi_2$. The phase condition requires that the *mutual* coherence time τ_{12}, the time for a fluctuation of $\phi_1-\phi_2$ of about one radian, be much longer than the beat period.

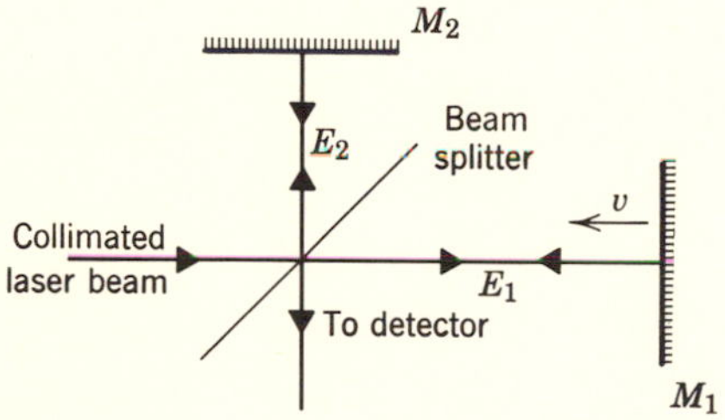

Fig. 6.1. Mirror M_1 moves in with velocity v, shifting the frequency of E_1.

One method of satisfying the two above conditions is to take two mutually coherent quasimonochromatic signals having the same frequency, such as would be derived by splitting the output from a single source and shifting the frequency of one beam by means of the Doppler effect. This can be achieved by moving one mirror in a Michelson interferometer, as shown in Fig. 6.1. There are two ways of analyzing this experiment.

(a) Moving fringes. If M_1 moves a distance Δd, this corresponds to a change in the optical path difference by $2\Delta d$, or by

$$\Delta n=\frac{2\Delta d}{\lambda}\quad \text{multiples of the wavelength}$$

and hence will shift the interference pattern by the same number Δn of fringes. This corresponds to Δn complete oscillations in the signal at the detector. If this occurs in Δt seconds, the frequency of these oscillations is

$$\Delta\nu = \frac{\Delta n}{\Delta t} = \frac{2}{\lambda}\frac{\Delta d}{\Delta t} = \frac{2v}{\lambda} = \frac{2\nu}{(\lambda\nu)}v = 2\nu\frac{v}{c} \tag{6.5}$$

where v is the velocity of M_1.

(b) Doppler effect. When a beam of light of frequency ν is reflected at normal incidence from a mirror moving towards it at velocity v, there is a change in frequency because of the Doppler effect from ν to $\nu + \Delta\nu$, where $\Delta\nu$ is given by (6.5). The phase is unchanged, apart from a constant phase shift, which also occurs at the stationary M_2. If we set $\Delta\nu = \nu_1 - \nu_2$, we then obtain the same result as Eqs. (6.3) and (6.4):

$$\langle E^2\rangle = \tfrac{1}{2}\langle A_1{}^2\rangle + \tfrac{1}{2}\langle A_2{}^2\rangle + \langle A_1(t)A_2(t)\cos\left[2\pi\Delta\nu t + \phi_1(t) - \phi_2(t)\right]\rangle \tag{6.6}$$

Another method of producing interference between light beams of different frequencies is much more difficult, since the beams in this case are independent. The coherence times τ_1 and τ_2, which are a measure of how long it takes the respective phases ϕ_1 and ϕ_2 to change by about one radian, must both be much longer than $1/\Delta\nu$. Then the relative phase will stay approximately constant during several beats, although it will fluctuate on a longer time scale. This can be seen if we approximate (6.4) by

$$2\langle E_1E_2\rangle = A_1(t_0)A_2(t_0)\cos\left[2\pi(\nu_1 - \nu_2)t_0 + \phi_1(t_0) - \phi_2(t_0)\right] \tag{6.7}$$

Then A_1 and ϕ_1 will fluctuate on a time scale given by τ_1 (A_1 may not fluctuate if the source is a laser). Similarly for A_2 and ϕ_2. We have to add this to

$$\tfrac{1}{2}\langle A_1{}^2\rangle + \tfrac{1}{2}\langle A_2{}^2\rangle \approx \tfrac{1}{2}A_1(t_0)^2 + \tfrac{1}{2}A_2(t_0)^2 \tag{6.8}$$

to obtain $\langle E^2\rangle$. The result is shown schematically in Fig. 6.2.

If A_1 and A_2 are constant, (6.8) gives a constant background on top of which (6.7) can be readily observed. This may be the case if the two beams come from intense lasers. But for ordinary quasi-monochromatic light sources of the so-called thermal type, the fluctuations of A take place on the same time scale as those of ϕ. Indeed this fluctuation is exploited in fluctuation-correlation interferometry, to be discussed later in this chapter. In the present context, however, the fluctuation in $A_1{}^2$ and $A_2{}^2$ acts as "noise" which tends to mask the beats in (6.7).

The first experiment of this type was performed with "thermal" light by Forrester, Gudmundsen, and Johnson using the two components of a Zeeman doublet in the 546.1 nm line of Hg^{202}.*

*A. T. Forrester, R. A. Gudmundsen, and P. O. Johnson, *Phys. Rev.*, **99**, 1691 (1955).

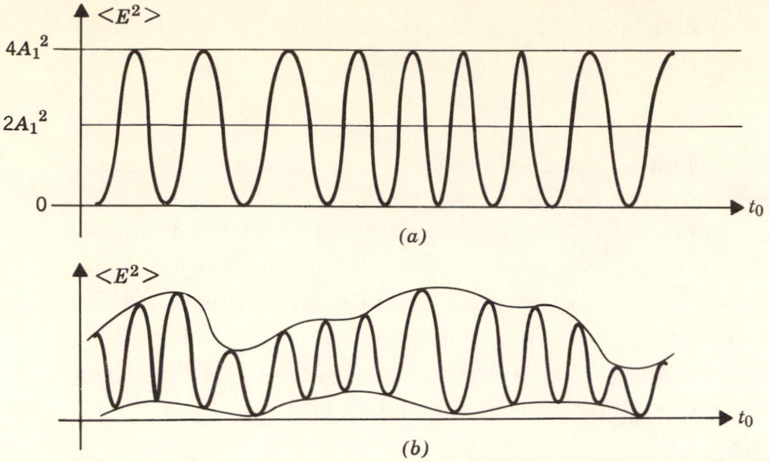

Fig. 6.2. Schematic representation of beats between two independent light sources. (*a*) Amplitudes $A_1 \approx A_2$ approximately constant. (*b*) A_1 and A_2 fluctuating. The number of beats per second is given by the difference frequency $\Delta\nu = \nu_1 - \nu_2$.

Filters removed all other lines but these. The frequency difference $\Delta\nu$ was proportional to the applied magnetic field and was adjusted to be about 10^{10} cycles per second. By using a special mercury lamp these workers were able to get $\tau_1 \approx \tau_2 \approx 10^{-9}$ sec, to satisfy the coherence condition. The 10^{10} cycle beat frequency was detected by coupling a microwave cavity tuned to it directly with a photomultiplier and using sophisticated techniques to improve an intrinsic signal to noise ratio of about 10^{-4}.

*4. **Spectrometers and Spectral Functions.*** The very difficulty of the Forrester experiment points up the extreme measures that must be taken to see beats between two independent optical sources. Ordinarily they are not observed, and we may assume that the squares of the individual electric fields simply add

$$\langle E^2 \rangle = \langle E_1^2 \rangle + \langle E_2^2 \rangle$$

To generalize, we say that when frequencies $\nu_1, \nu_2, \nu_3, \ldots$ are present in the optical signal, the measured optical intensity* will be proportional to

$$\langle E^2 \rangle = \langle E_1^2 \rangle + \langle E_2^2 \rangle + \langle E_3^2 \rangle + \cdots$$

This gives

$$I(\text{tot}) = I(1) + I(2) + \cdots \tag{6.9}$$

where $I(i)$ represents the intensity that would be measured if only frequency ν_i were present in the beam. Since real sources have a continuous distribution of frequency, we must generalize Eq. (6.9) further to

$$I(\text{tot}) = \int_0^\infty I(\nu)\, d\nu \tag{6.10}$$

*We use the term "intensity" in a general way to refer to any experimental quantity proportional to the flux density or intensity defined in the sense of Chapter 4.

Here $I(\nu)\,d\nu$ can be interpreted as the amount of radiant intensity present in the frequency range from ν to $\nu + d\nu$. Here $I(\nu)$ is called the *spectral intensity*. It can be measured by passing the beam through an ideal spectrometer. This can be defined as an instrument which passes all light in the frequency range ν to $\nu + \Delta\nu$ and none else. If $\Delta\nu$ is small enough, this will approximate $I(\nu)\,d\nu$.

Spectrometers of the ordinary type work because they contain an element such as a prism to disperse light spatially according to its frequency. A prism accomplishes this because the refractive index and hence the deviation angle depend on frequency—higher frequencies are deviated more than lower frequencies. This is illustrated in Figs. 6.3 and 6.4. If the slit widths are small, the output will be proportional to $I(\nu)$ (except for complications because of diffraction-limited resolving power, which we ignore for now—see Section 7.2D5). Gratings or other devices can also be used to achieve the desired dispersion. In real situations the efficiency of the instrument including that of the detector must be known if an absolute measure of $I(\nu)$ is to be obtained. Most of the time we are willing to settle for a relative measurement of $I(\nu)$. It will be described by the *spectral distribution function* $P(\nu)$,

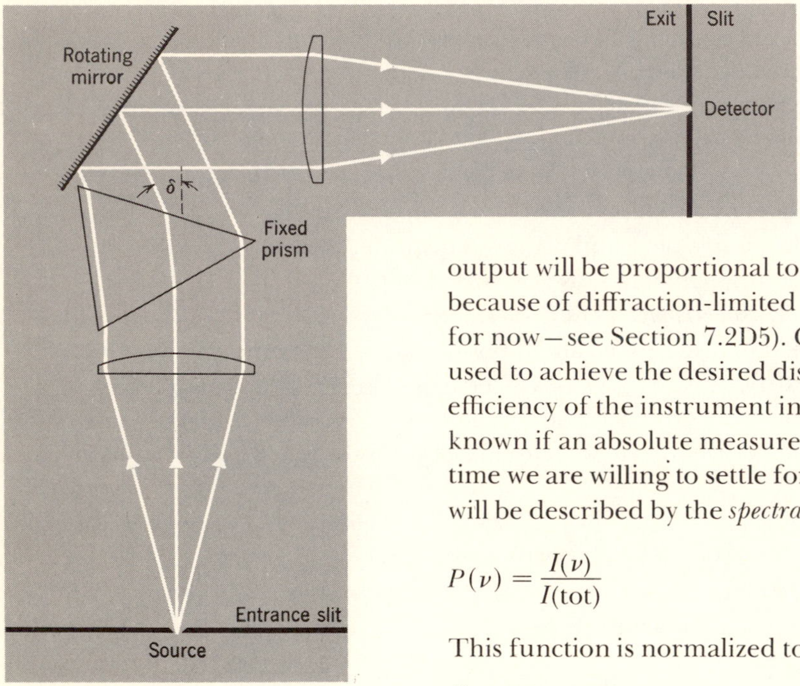

Fig. 6.3. Scanning prism spectrometer. The part of the instrument between source and detector is called a monochromator. In the instrument shown, the spectrum is moved across the exit slit by moving the mirror, which is called a Wadsworth mirror.

$$P(\nu) = \frac{I(\nu)}{I(\text{tot})} \qquad (6.11a)$$

This function is normalized to unity:

$$\int_0^\infty P(\nu)\,d\nu = 1 \qquad (6.11b)$$

The type of curves of $I(\nu)$ to be expected is indicated in Fig. 6.5.

The traditional method of measuring $I(\nu)$ is that discussed above; a prism or grating is employed to disperse the light as a function of ν; the signal produced when the dispersed light reaches a detector is then recorded photographically or electrically (Figs. 6.2, 6.3). Another method, that of *interference spectroscopy*, determines $I(\nu)$ from the fringes produced in a Michelson interferometer or similar instrument. In practice this method has proved particularly valuable in the far infrared region of the spectrum, but there have also been specialized applications in the

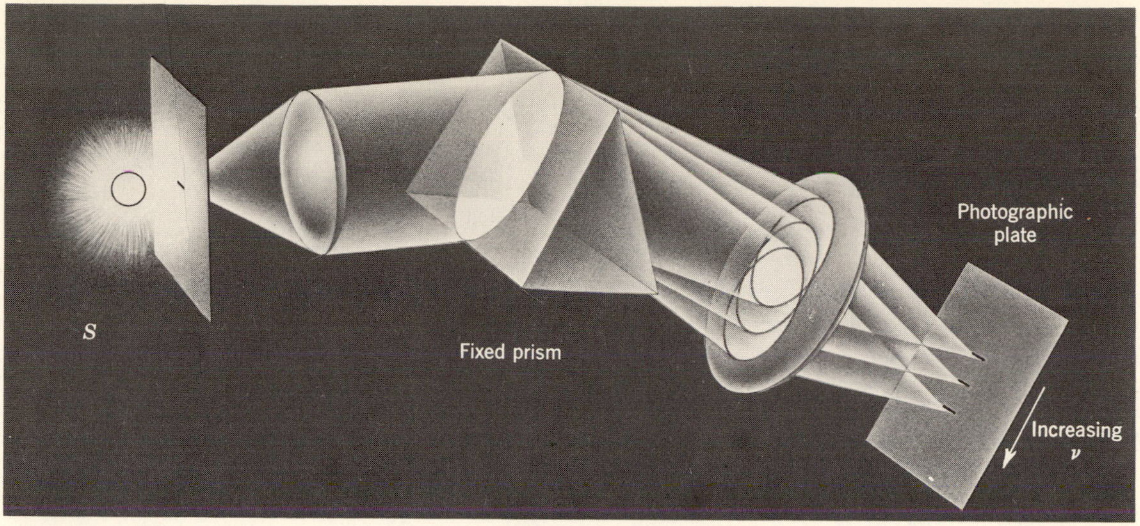

Fig. 6.4. The essential elements of a prism spectrograph.

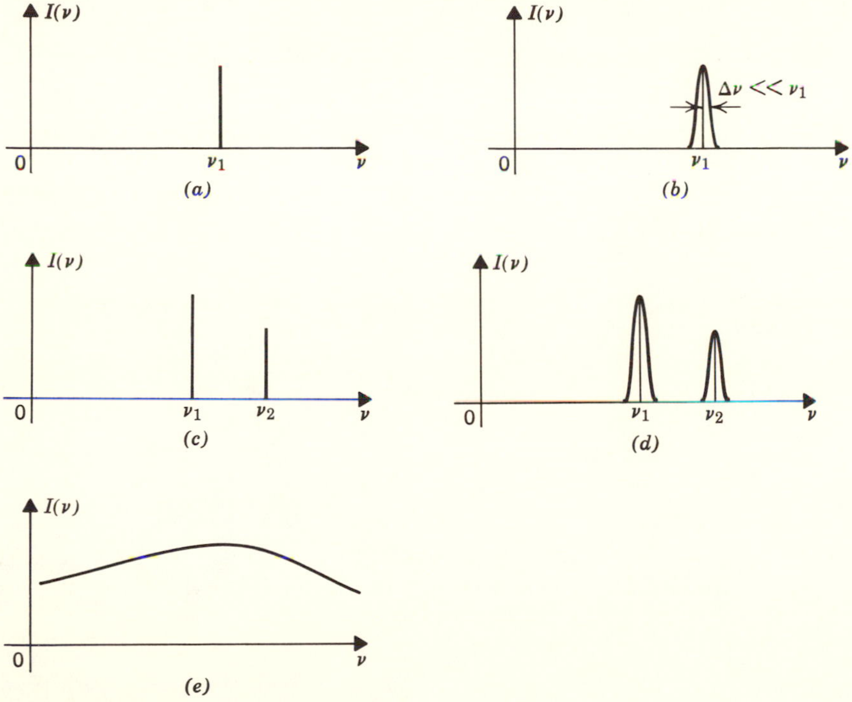

Fig. 6.5. The spectral intensity for several different types of source. (*a*) Ideal monochromatic source. (*b*) Quasi-monochromatic source. (*c*) Idealized two-frequency source. (*d*) Real two-frequency source. (*e*) Broad frequency distribution as from a thermal (blackbody) source.

visible and near infrared. Interference spectroscopy also forms an ideal basis for a discussion of coherence.

B. DERIVATION OF FORMULAS FOR INTERFERENCE SPECTROSCOPY

An interference spectrometer is a Michelson interferometer modified to use collimated light as shown in Fig. 6.6. An image of the aperture in front of the source S is formed in the focal plane of Lens L_2 at the detector D. The lenses allow us to set the angle r (discussed earlier) equal to zero and to write the following expression for the phase shift for light of wavelength λ or frequency ν reflected from the two mirrors:

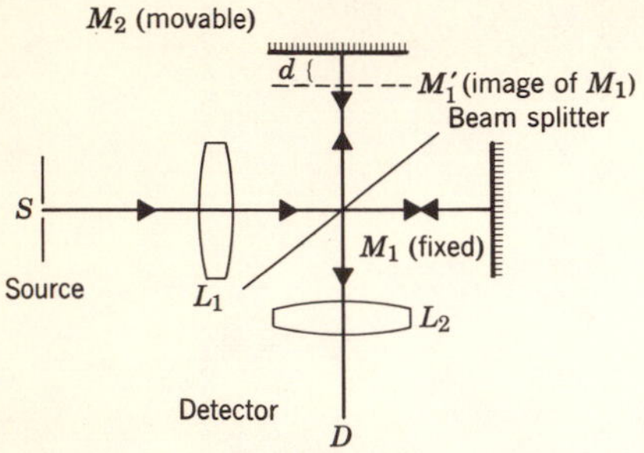

Fig. 6.6. Michelson interferometer for use in spectroscopy; S and D are in the focal planes of lenses L_1 and L_2, respectively.

$$\Delta = \frac{4\pi d}{\lambda} = \frac{4\pi \nu d}{c}$$

Here d is the effective separation of M_1 from M_2.

1. ***Single Frequency.*** If only one frequency ν_1 is present, then for the time-averaged intensity at the detector from Eq. (5.34) we can write

$$\begin{aligned} I_d &= I_1 + I_2 + 2\sqrt{I_1 I_2} \cos \frac{4\pi \nu_1 d}{c} \\ &= (I_1 + I_2)\left(1 + x \cos \frac{4\pi \nu_1 d}{c}\right) \\ &= I_0 \left(1 + x \cos \frac{4\pi \nu_1 d}{c}\right) \end{aligned} \tag{6.12}$$

where I_1 and I_2 are the intensities in each beam at the detector, $I_0 = I_1 + I_2$, and where

$$x = \frac{2\sqrt{I_1 I_2}}{(I_1 + I_2)}$$

An ideal beam splitter will give $I_2 = I_1$ and $x = 1$.

Light in the M_2 arm has traveled an extra distance $2d$ to reach the detector. This is the independent variable for this experiment. It is convenient to express it in the units of time. We therefore define

$$\tau = \frac{2d}{c} \tag{6.13}$$

which represents the retardation time suffered by the M_2 beam relatively to the M_1 beam.

Equation (6.12) may then be written

$$I_d = I_0 (1 + x \cos 2\pi \nu_1 \tau) \tag{6.14}$$

For simplicity we hereafter assume a perfect beam splitter—$x = 1$. When considered as a function of τ, I_d consists of an average term I_0 plus an oscillatory term

$$I_{osc} = I_0 \cos 2\pi\nu_1\tau$$

which averages to zero as τ is varied.
The oscillatory term may be normalized by dividing I_0 to obtain an important quantity

$$\gamma(\tau) = \frac{I_{osc}}{I_0} \tag{6.15}$$

For the present example,

$$\gamma(\tau) = \cos 2\pi\nu_1\tau \tag{6.16}$$

For this single frequency case the spectral distribution function $P(\nu)$ is an infinite spike having unit area at $\nu = \nu_1$; that is, $P(\nu)$ is a *delta function* (Appendix A):

$$P(\nu) = \delta(\nu - \nu_1) \tag{6.17}$$

The spectral intensity is simply

$$I(\nu) = I_0 P(\nu) \tag{6.18}$$

The functions $I(\nu)$, $P(\nu)$, $I_d(\tau)$, and $\gamma(\tau)$ are sketched in Fig. 6.7.

*2. **Two Frequencies.*** Consider two frequencies ν_1 and ν_2 to be present in the beam with total intensities I_1 and I_2. The intensity

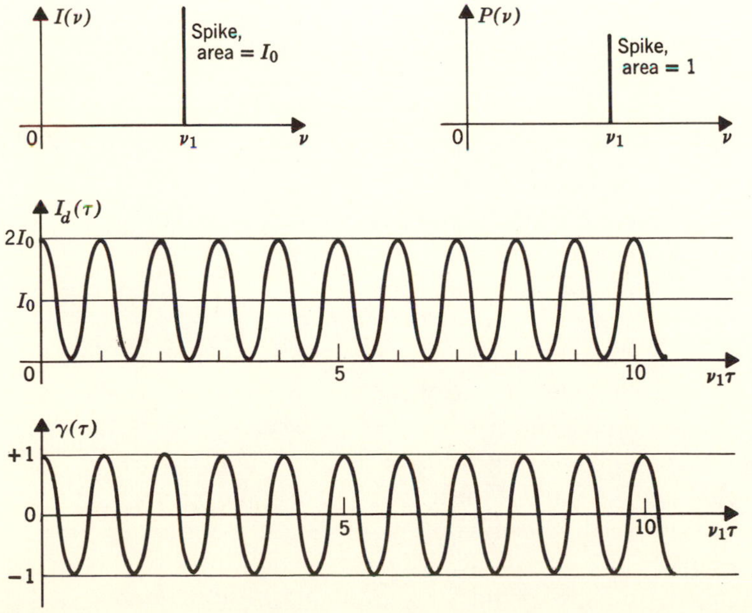

Fig. 6.7

at the detector is then the sum of two terms having the form of Eq. (6.14) with $x = 1$:

$$I_d = I_1(1 + \cos 2\pi\nu_1\tau) + I_2(1 + \cos 2\pi\nu_2\tau) \tag{6.19}$$

Again this consists of an average term

$$I_0 = I_1 + I_2 \tag{6.20}$$

plus an oscillatory term

$$I_{osc} = I_1 \cos 2\pi\nu_1\tau + I_2 \cos 2\pi\nu_2\tau$$

which may be written

$$I_{osc} = I_0\gamma(\tau) \tag{6.21a}$$

with

$$\gamma(\tau) = P_1 \cos 2\pi\nu_1\tau + P_2 \cos 2\pi\nu_2\tau \tag{6.21b}$$

and

$$P_1 = \frac{I_1}{I_0}, \qquad P_2 = \frac{I_2}{I_0}$$

Here P_1 and P_2 represent the fraction of the total intensity at frequencies ν_1 and ν_2.

The detector intensity has the same form found for the single frequency case, namely,

$$I_d(\tau) = I_0[1 + \gamma(\tau)] \tag{6.22}$$

but with $\gamma(\tau)$ now given by Eq. (6.21*b*).

For this case the spectral distribution function is a sum of two delta functions,

$$P(\nu) = P_1\delta(\nu - \nu_1) + P_2\delta(\nu - \nu_2) \tag{6.21c}$$

and the spectral intensity is still $I(\nu) = I_0P(\nu)$.

A comparison of Eq. (6.21*c*) with (6.21*b*) shows that the cosine terms and the delta functions have the same coefficients.

Figure 6.8 shows the behavior of the relevant functions for the case $P_1 = P_2 = \frac{1}{2}$; Fig. 6.9 shows what happens when $P_1 \neq P_2$.

If something should happen to destroy the interference—for example, a misalignment of one of the mirrors of the interferometer—there would be no fringes, γ would be zero, and the detector intensity I_d would be simply a constant, I_0. Then we learn nothing about the nature of the light in the original beam. Thus it is the function $\gamma(\tau)$ that really contains the useful information in this experiment.

The "beats" in Figs. 6.8 and 6.9 come about through the alternate reinforcement or cancellation of the two cosine terms at the

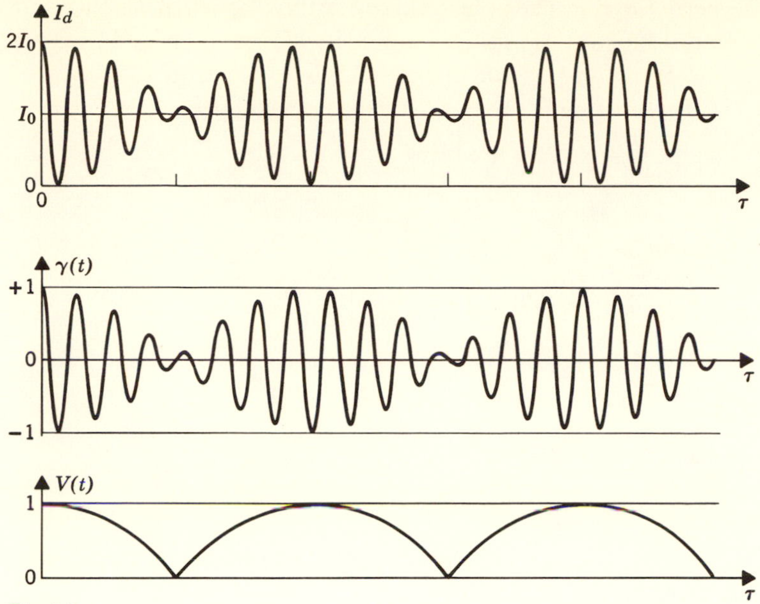

Fig. 6.8

different frequencies. Reinforcement occurs whenever $|\nu_2 - \nu_1|\tau$ is an even integer; then for a limited range of τ both cosines oscillate together between ± 1, giving a γ that also oscillates between ± 1. Cancellation occurs whenever $|\nu_2 - \nu_1|\tau$ is an odd integer and is complete if $P_1 = P_2$; otherwise the fringes are still present but with reduced strength.

The interference fringes in Figs. 6.7–6.9 continue periodically for all values of the retardation time τ. This behavior is unreal and is a property of the discrete nature of the distribution functions that we have used thus far.

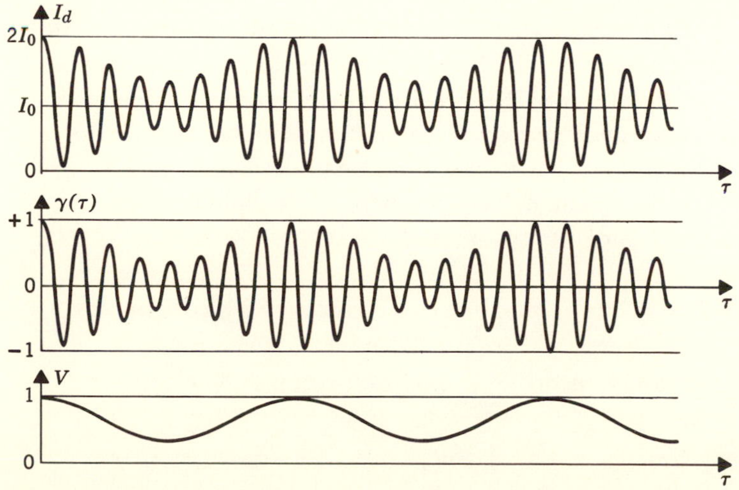

Fig. 6.9

3. General Case. In the general case for the Michelson interferometer of Fig. 6.6, the spectral intensity will be described by a continuous function $I(\nu)$ of the type shown in Fig. 6.5*b*, *d*, or *e*. Here $I(\nu)\,d\nu$ represents the intensity in the beam in the frequency range ν to $\nu + d\nu$. The sum in Eq. (6.19) then generalizes to an integral

$$I_d(\tau) = \int_0^\infty I(\nu)(1 + \cos 2\pi\nu\tau)\,d\nu \tag{6.23}$$

Again this result consists of an average term

$$I_0 = \int_0^\infty I(\nu)\,d\nu$$

plus an oscillatory term

$$I_{\text{osc}} = \int_0^\infty I(\nu)\cos 2\pi\nu\tau\,d\nu$$

and can be written in the form

$$I_d(\tau) = I_0[1 + \gamma(\tau)]$$

where

$$\gamma(\tau) = \int_0^\infty P(\nu)\cos 2\pi\nu\tau\,d\nu \tag{6.24}$$

is the *normalized oscillatory term* and where

$$P(\nu) = \frac{I(\nu)}{I_0} \tag{6.25}$$

is the *normalized spectral distribution function*.

Equation (6.24) expresses $\gamma(\tau)$ as a form of *Fourier transform* of $P(\nu)$; we can then recover P as a function of ν if we can determine γ as a function of τ for all τ, by Fourier inversion. Some details about Fourier integrals and transforms are given in Appendix A.

4. The Inversion Process to Recover P(ν). Suppose that we do know $\gamma(\tau)$ for all positive values of τ. (Negative values are unnecessary, since it is an even function of τ) Multiply Eq. (6.24) by $\cos 2\pi\mu\tau$ and integrate from $\tau = 0$, $\tau = \infty$:

$$\int_0^\infty \gamma(\tau)\cos 2\pi\mu\tau\,d\tau = \lim_{T\to\infty}\int_0^T \left[\int_0^\infty P(\nu)\cos 2\pi\nu\tau\,d\nu\right]\cos 2\pi\mu\tau\,d\tau$$

Now interchange the order of integration. This gives

$$\int_0^\infty \gamma(\tau)\cos 2\pi\mu\tau\,d\tau = \lim_{T\to\infty}\int_0^\infty P(\nu)\left[\int_0^T \cos 2\pi\nu\tau\cos 2\pi\mu\tau\,d\tau\right]d\nu \tag{6.26}$$

The factor in square brackets may be rewritten using

$\cos A \cos B = \frac{1}{2}\cos(A+B) + \frac{1}{2}\cos(A-B)$

and integrated, giving

$$\int_0^T \cos 2\pi\nu\tau \cos 2\pi\mu\tau\, d\tau = \frac{\sin 2\pi(\nu+\mu)T}{4\pi(\nu+\mu)} + \frac{\sin 2\pi(\nu-\mu)T}{4\pi(\nu-\mu)} \qquad (6.27)$$

At this point we make use of the result (Appendix A, Eq. A4)

$$\lim_{T\to\infty}\left[\frac{\sin \pi T(x-x_0)}{\pi(x-x_0)}\right] = \delta(x-x_0)$$

When we apply this result to (6.26), use (6.27), and assume $P(\nu)$ to be continuous, we obtain a contribution from only the second term in (6.27), since μ is nonnegative. This then gives $\frac{1}{4}P(\mu)$ for the right-hand side of Eq. (6.26). Thus

$$P(\mu) = 4\int_0^\infty \gamma(\tau)\cos 2\pi\mu\tau\, d\tau \qquad (6.28)$$

is our desired *inversion formula.*

The inversion is usually performed by a computer. There is a maximum value of τ given by $\tau_{max} = 2d_{max}/c$, and this introduces some complications that will be discussed later in one of the problems.

C. SOME PROPERTIES OF $\gamma(\tau)$

1. ***Limits.*** From its defining equation (6.24) we see that $\gamma(\tau)$ is always unity at $\tau = 0$. Since $P(\nu)$ is nonnegative, if we multiply it by $\cos 2\pi\nu\tau$ and integrate, we can never obtain a result greater in magnitude than the value obtained at $\tau = 0$. Thus

$$|\gamma(\tau)| \leq 1 \qquad \text{for all } \tau$$
$$\gamma(0) = 1 \qquad (6.29).$$

2. ***Superposition.*** Since $\gamma(\tau)$ is given as an integral transform of $P(\nu)$ and since the integral of a sum is the sum of the integrals, a spectral distribution function $P(\nu)$ that is a superposition of individual distribution functions will have a $\gamma(\tau)$ that is a similar superposition. Let

$$P(\nu) = \Sigma_j C_j P_j(\nu) \qquad (6.30a)$$

where each of the $P_j(\nu)$ is normalized to unity and where

$$\Sigma_j C_j = 1 \qquad (6.30b)$$

Then if $\gamma(\tau)$ is the normalized oscillatory term for $P(\nu)$ and $\gamma_j(\tau)$ is that for $P_j(\nu)$, we have

$$\gamma(\tau) = \Sigma_j C_j \gamma_j(\tau) \qquad (6.30c)$$

We saw an example of this in connection with Eqs. (6.21*b*, *c*)

3. $\gamma(\tau)$ for Quasimonochromatic Light. For quasimonochromatic light the spectral distribution function is peaked about $\nu = \nu_0$ and tends to zero rather rapidly as ν departs much from the vicinity of ν_0. If $\Delta\nu$ is a measure of the range of nonzero values of P, that is, from $\nu_0 - \Delta\nu/2$ to $\nu_0 + \Delta\nu/2$, then we must have $\Delta\nu \ll \nu_0$. It is helpful to write P in the form

$$P(\nu) = D(\nu - \nu_0) \tag{6.31a}$$

where $D(\mu)$ is small except for $-\Delta\nu/2 < \mu < \Delta\nu/2$. The normalization

$$\int_0^\infty P(\nu)\, d\nu = 1$$

of P then gives

$$1 = \int_0^\infty P(\nu)\, d\nu = \int_{-\nu_0}^\infty D(\mu)\, d\mu \approx \int_{-\infty}^\infty D(\mu)\, d\mu \tag{6.31b}$$

Then for γ we obtain

$$\gamma(\tau) = \int_0^\infty P(\nu) \cos 2\pi\nu\tau d\nu = \int_0^\infty D(\nu - \nu_0) \cos 2\pi\nu\tau d\nu$$

Let $\mu = \nu - \nu_0$. Then we have approximately

$$\begin{aligned}\gamma(\tau) &= \int_{-\nu_0}^\infty D(\mu) \cos 2\pi(\nu_0 + \mu)\tau d\mu \approx \int_{-\infty}^\infty D(\mu) \cos [2\pi(\nu_0 + \mu)\tau]\, d\mu \\ &= \cos 2\pi\nu_0\tau \int_{-\infty}^\infty D(\mu) \cos 2\pi\mu\tau d\mu \\ &\quad - \sin 2\pi\nu_0\tau \int_{-\infty}^\infty D(\mu) \sin 2\pi\mu\tau d\mu\end{aligned} \tag{6.32}$$

We define functions $C(\tau)$ and $S(\tau)$:

$$C(\tau) = \int_{-\infty}^\infty D(\mu) \cos 2\pi\mu\tau d\mu \tag{6.33a}$$

$$-S(\tau) = \int_{-\infty}^\infty D(\mu) \sin 2\pi\mu\tau d\mu \tag{6.33b}$$

(Note that $C(0) = 1$, $S(0) = 0$.) Then we may write

$$\gamma(\tau) = C(\tau) \cos 2\pi\nu_0\tau + S(\tau) \sin 2\pi\nu_0\tau \tag{6.34}$$

This result may also be written in the form

$$\gamma(\tau) = U(\tau) \cos [2\pi\nu_0 t + \Phi(\tau)] \tag{6.35a}$$

$$U(\tau) \cos \Phi(\tau) = C(\tau) \tag{6.35b}$$

$$-U(\tau) \sin \Phi(\tau) = S(\tau) \tag{6.35c}$$

$$U(\tau) = +\sqrt{C(\tau)^2 + S(\tau)^2} \tag{6.35d}$$

Since according to Eq. (6.29) $|\gamma(\tau)| \leq 1$, we always have

$$0 \leq U(\tau) \leq 1 \tag{6.36}$$

The functions C, S, U, and Φ are all slowly varying compared with $\cos 2\pi\nu_0\tau$ or $\sin 2\pi\nu_0\tau$. It is difficult to give a general justification for this remark that is both rigorous and simple. The basic idea is that rapid oscillations in C or S would require that they be superpositions of rapidly oscillating cosine or sine components. But since $D(\mu)$ is limited to a range of $\pm\Delta\nu/2$ that is much smaller than ν_0, the resulting C and S cannot oscillate nearly as rapidly as $\cos 2\pi\nu_0\tau$ or $\sin 2\pi\omega_0\tau$.

4. Symmetric Quasimonochromatic Spectrum. For quasimonochromatic light with a spectral distribution function $P(\nu)$ that is symmetric about $\nu = \nu_0$, we have $D(\mu) = D(-\mu)$. Now $\sin 2\pi\mu\tau$ is an odd function of μ, and the product of an odd function and an even function is an odd function. This means that the contribution to the integral

$$\int_{-\infty}^{\infty} D(\mu) \sin 2\pi\mu\tau \, d\mu$$

from positive values of μ will cancel that from negative values of μ, giving $S(\tau) = 0$. Then we have simply

$$\gamma(\tau) = C(\tau) \cos 2\pi\nu_0\tau \tag{6.37}$$

Examples of spectral distribution functions and D's are shown for symmetric and assymmetric cases in Fig. 6.10. Thus, $D(\mu)$ is simply $P(\nu)$ centered on zero frequency.

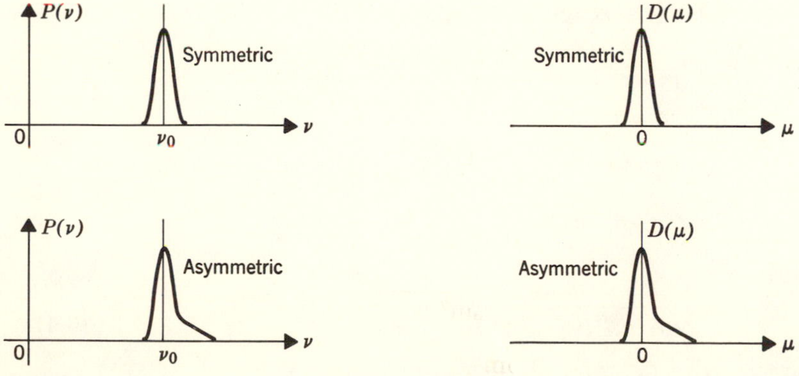

Fig. 6.10

5. Visibility of Fringes. The measured intensity is given by

$$I_d(\tau) = I_0[1 + \gamma(\tau)]$$

For quasimonochromatic light with γ given by (6.35*a*), there will be oscillations of γ between the limits $\pm U(\tau)$, and since U is slowly varying in comparison with these oscillations, we may consider them to be well-defined "fringes." In the fringe pattern the intensity varies between a maximum of

$$I_d(\text{max}) = I_0(1+U)$$

and a minimum of

$$I_d(\text{min}) = I_0[1-U] \tag{6.38}$$

The *visibility* V of the pattern is defined by

$$V = \frac{I_d(\text{max}) - I_d(\text{min})}{I_d(\text{max}) + I_d(\text{min})} \tag{6.39}$$

Equations (6.38) and (6.39) then yield simply

$$V = U(\tau) \tag{6.40}$$

6. ***Superposition of Two Monochromatic Lines.*** For two monochromatic lines at frequencies ν_1 and ν_2, Eq. (6.21) gives us for $\gamma(\tau)$:

$$\gamma(\tau) = P_1 \cos 2\pi\nu_1\tau + (1-P_1) \cos 2\pi\nu_2\tau \tag{6.41a}$$

Here P_1 represents the fraction of total flux density in the line of frequency ν_1, and $P_2 = 1 - P_1$ represents the fraction in the line of frequency ν_2.

Now suppose that $\nu_1 > \nu_2$, and write

$$\nu_1 = \nu_0 + \frac{\Delta\nu}{2}$$

$$\nu_2 = \nu_0 - \frac{\Delta\nu}{2}$$

where ν_0 is the average $\frac{1}{2}(\nu_1 + \nu_2)$ and $\Delta\nu$ is the difference,

$$\Delta\nu = \nu_1 - \nu_2$$

We then readily obtain

$$\gamma(\tau) = \cos \pi\Delta\nu\tau \cos 2\pi\nu_0\tau - (2P_1 - 1) \sin \pi\Delta\nu\tau \sin 2\pi\nu_0\tau$$

$$= U(\tau) \cos [2\pi\nu_0\tau + \Phi(\tau)] \tag{6.41b}$$

with

$$U(\tau) = +\sqrt{1 - 4P_1(1-P_1) \sin^2 \pi\Delta\nu\tau} \tag{6.41c}$$

and $\Phi(\tau)$ is obtainable from

$$U(t) \cos \Phi(\tau) = \cos \pi\Delta\nu\tau$$

$$U(t) \sin \Phi(\tau) = (2P_1 - 1) \sin \pi\Delta\nu\tau \tag{6.41d}$$

For $\Delta\nu \ll \nu_0$, Eq. (6.41*b*) gives a slowly varying beat pattern, plotted in Fig. 6.8 for $P_1 = \frac{1}{2}$ and in Fig. 6.9 for $P_1 \neq \frac{1}{2}$. The visibility $V = U(\tau)$ given by Eq. (6.41*c*) is also plotted in Figs. 6.8 and 6.9. The visibility periodically changes from $V_{max} = 1$ to

$$V_{min} = |1 - 2P_1| \tag{6.42}$$

7. Continuous Gaussian Spectral Distribution. When the Michelson interferometer is used with any real optical signal, there is a finite range of retardation time $\tau = 2d/c$ over which interference fringes can be observed. What happens is that eventually the maxima in the interference fringes from one frequency fill in between the minima from another frequency, and the normalized oscillatory function goes to zero.

This effect will be illustrated by a specific example. Assume quasimonochromatic light with a spectral distribution function in the form of a Gaussian

$$P(\nu) = D(\nu - \nu_0) = e^{-\pi^2(\nu-\nu_0)^2\tau_d{}^2}\sqrt{\pi}\tau_d \tag{6.43}$$

The parameter τ_d has the dimensions of time and characterizes the width, since $P(\nu)$ is down to $1/e$ of its value at $\nu = \nu_0$ when $\nu = \nu_0 \pm 1/\pi\tau_d$. For the light to be quasimonochromatic it is necessary that $\tau_d\nu \gg 1$.

To a good approximation, $P(\nu)$ in Eq. (6.43) is normalized, since

$$\sqrt{\pi}\tau_d \int_0^\infty e^{-\pi^2(\nu-\nu_0)^2\tau_d{}^2}\,d\nu$$

$$= \sqrt{\pi}\,\tau_d \int_{-\nu_0}^{\infty} e^{-\pi^2\mu^2\tau_d{}^2}\,d\mu \approx \sqrt{\pi}\,\tau_d \int_{-\infty}^{\infty} e^{-\pi^2\mu^2\tau_d{}^2}\,d\mu = 1$$

The last equality follows from a change of variables and use of

$$\int_{-\infty}^{\infty} e^{-x^2}\,dx = \sqrt{\pi} \tag{6.44}$$

Since the function $D(\mu)$ is symmetric about $\mu = 0$ in this case, Eq. (6.37) holds, and the normalized oscillatory term that describes the interference pattern is

$$\gamma(\tau) = C(\tau)\cos 2\pi\nu_0\tau$$

where $C(\tau)$ is given by Eq. (6.33):

$$C(\tau) = \int_{-\infty}^{\infty} D(\mu)\cos 2\pi\mu\tau\,d\mu = Re\int_{-\infty}^{\infty} D(\mu)e^{2\pi i\mu\tau}\,d\mu = Re\sqrt{\pi}\tau_d$$

$$\int_{-\infty}^{\infty} e^{-\pi^2\mu^2\tau_d{}^2}e^{2\pi i\mu\tau}\,d\mu$$

The integral may be evaluated by completing the square in the exponent:

$$\text{Exponent} = -\pi^2\tau_d^2\left[\mu^2 - \frac{2i\mu\tau}{\pi\tau_d^2}\right] = -\pi^2\tau_d^2\left[\mu^2 - \frac{2i\mu\tau}{\pi\tau_d^2} + \left(\frac{i\tau}{\pi\tau_d^2}\right)^2\right] + \pi^2\tau_d^2\left(\frac{i\tau}{\pi\tau_d^2}\right)^2$$

$$= -u^2 - \tau^2/\tau_d^2$$

where

$$u = \pi\tau_d\left(\mu - \frac{i\tau}{\pi\tau_d^2}\right)$$

Then

$$C(\tau) = \frac{1}{\sqrt{\pi}}e^{-\tau^2/\tau_d^2}Re\int_{-\infty-i\tau/\tau_d^2}^{+\infty-i\tau/\tau_d^2} e^{-u^2}\,du$$

The complex limits on the integral do not change its value from that given by Eq. (6.44). Thus

$$C(\tau) = e^{-\tau/\tau_d^2}$$

and

$$\gamma(\tau) = e^{-\tau^2/\tau_d^2}\cos 2\pi\nu_0\tau \tag{6.45}$$

The behavior of $P(\nu)$ and $\gamma(\tau)$ for this case is sketched in Fig. 6.11. Note that the envelope $e^{-\tau^2/\tau_d^2}$ is slowly varying in comparison with the cosine factor, since τ_d is much larger than the period ν_0^{-1}. For short retardation times compared with τ_d we have approximately

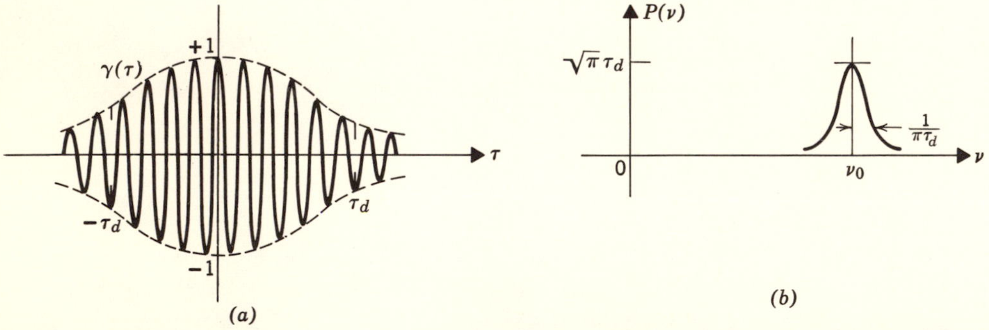

Fig. 6.11. Normalized oscillatory term (*a*) and spectral distribution function (*b*) for a quasimonochromatic line with a Gaussian shape.

$$\gamma(\tau) \approx \cos 2\pi\nu_0\tau \tag{6.46}$$

This, of course, is just the single-frequency result of Eq. (6.16). In order to see the effects of the finite spread $\Delta\nu$ in the spectral distribution, we must use retardation times τ of the order of τ_d, for then the envelope is noticeably reduced — in this case by the factor $1/e$.

Since $\Delta\nu$ is approximately $\Delta\nu \approx 1/\tau_d$, we have approximately

$$\Delta\nu\tau_d \sim 1 \tag{6.47}$$

The time τ_d may be interpreted as the *coherence time* for this example. The relation (6.47) is a general one connecting the coherence time and the frequency spread and will be discussed more fully at the end of Section 6.2.

REFERENCES FOR SECTION 6.1

G. A. Vanasse and J. Strong, "Applications of Fourier Transformations in Optics: Interferometric Spectroscopy." This is Appendix F of *Concepts of Classical Optics* by J. Strong.

M. Born and E. Wolf, *Principles of Optics*. Section 7.5.

Lawrence Mertz, *Transformations in Optics*.

G. A. Vanasse and H. Sakai, "Fourier Spectroscopy," in *Progress in Optics*, E. Wolf, editor, Vol. VI.

PROBLEMS

6.1.1 Show that the minimum visibility that will be observed with a source emitting two infinitely narrow spectral lines at frequencies ν_1, ν_2 with intensities I_1 and $I_2 < I_1$ is

$$\frac{I_1 - I_2}{I_1 + I_2}$$

Is it possible to determine which frequency is larger from a study of the visibility curve? Explain.

6.1.2 Discuss quantitatively the interference pattern that would be observed as a function of τ if the spectral distribution function is flat over an octave, but otherwise zero, that is, for

$$I(\nu) = A, \qquad \text{a constant for } \nu_0 < \nu \leqslant 2\nu_0$$
$$I(\nu) = 0, \qquad \text{otherwise}$$

Partial answer:

$$\gamma(\tau) = \frac{\sin 2\pi\nu_0\tau}{2\pi\nu_0\tau}(2\cos 2\pi\nu_0\tau - 1)$$

6.1.3 Consider a source of light consisting of two quasimonochromatic lines of the same shape centered at frequencies ν_1 and ν_2. The spectral distribution function will be

$$P(\nu) = \frac{1}{3}D(\nu - \nu_1) + \frac{2}{3}D(\nu - \nu_2)$$

Further, suppose that $D(\mu)$ is symmetric about $\mu = 0$ and that

$$\int_{-\infty}^{\infty} D(\mu)\cos 2\pi\mu\tau\, d\mu = e^{-\tau^2/\tau_0^2}$$

Let $\nu_1 > \nu_2$, $1/\tau_0 \ll (\nu_1 - \nu_2) \ll \nu_2$.

Calculate the normalized oscillatory function $\gamma(\tau)$ and sketch it. How does this function compare with that in Fig. 6.9?

6.2 THE AUTOCORRELATION FUNCTION

At the end of Section 6.1 we considered a specific example of a continuous spectral distribution function for quasimonochromatic light and showed by explicit calculation that the contribution of the various frequencies to the pattern of interference fringes eventually caused a decrease in the oscillatory part of the pattern. This occurred when the retardation time τ was comparable to the reciprocal of the width $\Delta\nu$ of the frequency distribution. This loss of fringe contrast is a general property of all true light sources, and its interpretation as a filling-in of the fringes from one frequency by those of another frequency is a perfectly valid one.

A. ALTERNATE INTERPRETATION OF $\gamma(\tau)$

There is another way of looking at the loss of interference in a Michelson interferometer. In this second method we consider the statistical aspects of the time dependence of the light signal. For quasimonochromatic light the light signal may be written in the form of Eq. (6.2):

$$E(t) = A(t) \cos [2\pi\nu_0 t + \phi(t)] \tag{6.2}$$

The phase $\phi(t)$ [and usually the amplitude $A(t)$] is not a well-defined function of time and must be described by its statistical properties. Consider, for instance, its mean square change (mean square fluctuation) after a time interval Δt, which we write in the form

$$\langle [\delta\phi(\Delta t)^2] \rangle = \langle [\phi(t+\Delta t) - \phi(t)]^2 \rangle = \langle [\phi(t) - \phi(t-\Delta t)]^2 \rangle$$

The average $\langle \cdot \cdot \cdot \rangle$ is over t for fixed Δt. To keep the present discussion simple, we assume that the average change in phase

$$\langle \delta\phi(\Delta t) \rangle = \langle \phi(t+\Delta t) - \phi(t) \rangle$$

is zero. Then positive fluctuations in ϕ are just as likely as negative fluctuations. Let τ_c be that value of Δt that gives a mean square phase fluctuation $\langle \delta\phi^2 \rangle$ of one radian.

Now imagine this light used in a Michelson interferometer (Fig. 6.6). Since the optical paths differ by amount d, the light in one arm is retarded by the time interval $\tau = 2d/c$ with respect to that in the other arm. If the retardation time is close to τ_c as just defined, there will be a one-radian mean square phase difference between the light in the two arms. This will substantially reduce the interference that can be observed when the two beams are brought together.

We now develop a quantitative description of the process just mentioned. The topic of mean square phase fluctuations is developed quantitatively in Problem 6.2.1.

1. ***Derivation.*** Suppose that the beam splitter in the interferometer performs ideally. Then if the light signal at the detector coming from the M_1 arm is $E(t)$, that from the M_2 arm is $E(t-\tau)$, namely E at the earlier time $t-\tau$. The detector intensity is then proportional to the time average of the total signal squared and, neglecting constants, is

$$I_d = \langle [E(t) + E(t-\tau)]^2 \rangle = \langle E(t)^2 \rangle + \langle E(t-\tau)^2 \rangle + 2\langle E(t)E(t-\tau) \rangle \tag{6.48}$$

The time averages are defined as follows:

$$\langle E(t)^2 \rangle \equiv \frac{1}{T} \int_{t_0+T/2}^{t_0-T/2} E(t)^2 \, dt \tag{6.49}$$

We assume that T is so large that the integral is independent of t_0. This is equivalent to taking the limit as $T \rightarrow \infty$.

The terms $\langle E(t)^2 \rangle$ and $\langle E(t-\tau)^2 \rangle$ represent the respective intensities received in the beams from M_1 and M_2, and for a steady source are each equal to $\frac{1}{2}I_0$, where I_0 is the total intensity at the detector if interference does not occur. The term $2\langle E(t)E(t-\tau) \rangle$ contains all the interference effects. If we normalize it by dividing by I_0, that is, if we define

$$\gamma(\tau) = \frac{\langle E(t)E(t-\tau) \rangle}{\langle E(t)^2 \rangle} \tag{6.50}$$

then we have

$$I_d = I_0[1 + \gamma(\tau)]$$

This last equation has the same form as Eq. (6.22) and is deliberately written using the same function $\gamma(\tau)$. In fact, the two definitions of $\gamma(\tau)$, Eq. (6.24) and (6.50), are equivalent, as is shown in Appendix B. When written in the form of (6.50), $\gamma(\tau)$ is called the normalized *autocorrelation function* of the signal $E(t)$. It describes the mean correlation of E with itself at an earlier time. The reader may be unfamiliar with such an expression, which is often used in various discussions of statistical physics. Some of its properties will be developed as we proceed.

The equivalence of the two expressions for $\gamma(\tau)$, namely the equation

$$\frac{\langle E(t)E(t-\tau) \rangle}{\langle E(t)^2 \rangle} = \int_0^\infty P(\nu) \cos 2\pi\nu\tau \, d\nu \tag{6.51}$$

is called the *Wiener-Khintchine theorem.* It says that the normalized autocorrelation function of a signal is the cosine Fourier transform of its spectral distribution function.

In many treatments of this subject, use is made of the *analytic signal* $\tilde{E}(t)$. This is a complex function of time that gives the ordinary light signal by

$$E(t) = \mathrm{Re}\,\tilde{E}(t) \tag{6.52}$$

Here $\tilde{E}(t)$ can be uniquely determined from $E(t)$. For example, for quasimonochromatic light with

$$E(t) = A(t)\cos[2\pi\nu_0 t + \phi(t)] \tag{6.2}$$

the analytic signal is

$$\tilde{E}(t) = A(t)\exp i[2\pi\nu_0 t + \phi(t)] \tag{6.53}$$

A complex autocorrelation function is then defined,

$$\tilde{\gamma}(\tau) = \frac{\langle \tilde{E}(t)\tilde{E}(t-\tau)^*\rangle}{\langle |\tilde{E}(t)|^2\rangle} \tag{6.54}$$

Here $\tilde{E}(t-\tau)^*$ is the complex conjugate of $\tilde{E}(t-\tau)$. The real function of Eq. (6.50) can be shown to be

$$\gamma(\tau) = \mathrm{Re}\,\tilde{\gamma}(\tau) \tag{6.55}$$

2. The Autocorrelation Function for Quasimonochromatic Light. Previously when we were concentrating on the spectral distribution function we defined quasimonochromatic light as light having a distribution function peaked at a certain frequency ν_0, and then showed that the normalized oscillatory interference term could be written in the form of Eq. (6.34) or (6.35*a*):

$$\gamma(\tau) = C(\tau)\cos 2\pi\nu_0\tau + S(\tau)\sin 2\pi\nu_0\tau \tag{6.34}$$

$$\gamma(\tau) = U(\tau)\cos[2\pi\nu_0\tau + \Phi(\tau)] \tag{6.35a}$$

We now want to show that these same results follow from our new point of view of concentrating on the light signal.

We start with

$$E(t) = A(t)\cos[2\pi\nu_0 t + \phi(t)]$$

The numerator of Eq. (6.50) for $\gamma(\tau)$ then becomes

$$\langle E(t)E(t-\tau)\rangle = \langle A(t)A(t-\tau)\cos A\cos B\rangle$$

where

$$A = 2\pi\nu_0 t + \phi(t), \qquad B = 2\pi\nu_0(t-\tau) + \phi(t-\tau)$$

Use of the identity $\cos A\cos B = \frac{1}{2}\cos(A+B) + \frac{1}{2}\cos(A-B)$ gives

$$\cos A\cos B = \tfrac{1}{2}\cos[2\pi\nu_0(2t-\tau) + \phi(t) + \phi(t-\tau)] + \tfrac{1}{2}\cos[2\pi\nu_0\tau + \phi(t) - \phi(t-\tau)]$$

The first term oscillates rapidly with t and gives zero when the time average is taken. This leaves

$$\text{Numerator of } \gamma(\tau) = \tfrac{1}{2}\langle A(t)A(t-\tau)\cos[2\pi\nu_0\tau + \phi(t) - \phi(t-\tau)]\rangle$$

The denominator of $\gamma(\tau)$ is simply the numerator at $\tau = 0$, which is $\frac{1}{2}\langle A(t)^2\rangle$. Thus

$$\gamma(\tau) = \frac{\langle A(t)A(t-\tau)\cos[2\pi\nu_0\tau + \phi(t) - \phi(t-\tau)]\rangle}{\langle A(t)^2\rangle} \tag{6.56}$$

Equation (6.56) expresses precisely what we previously tried to say imprecisely with words. It relates the interference pattern, represented by $\gamma(\tau)$, to the mean of a trigonometric function of the phase difference $[\phi(t) - \phi(t-\tau)]$. This is not surprising, since the phase is only defined to within an integral multiple of 2π, and only trigonometric functions of it are meaningful.

A qualitative interpretation of Eq. (6.56) may be given as follows. If the retardation time τ is short enough so that there is little fluctuation in phase between the times t and $t-\tau$, the cosine factor may be well approximated by $\cos 2\pi\nu_0\tau$ and taken in front of the averaging integral $\langle\cdots\rangle$. It is the nature of light signals that a small phase fluctuation implies a small amplitude fluctuation, so that in this case $A(t-\tau) \approx A(t)$, and (6.56) is approximately

$$\gamma(\tau) = \cos 2\pi\nu_0\tau$$

This represents simply the normalized oscillatory interference pattern for a monochromatic source.

When the mean square phase fluctuations are of the order of one radian or more, the cosine factor in (6.56) will fluctuate wildly and will average to something considerably smaller than $\cos 2\pi\nu_0\tau$. The interference has then been considerably reduced.

When the cosine in Eq. (6.56) is expanded, Eq. (6.34) is obtained, but with

$$C(\tau) = \frac{\langle A(t)A(t-\tau)\cos[\phi(t) - \phi(t-\tau)]\rangle}{\langle A(t)^2\rangle} \tag{6.57a}$$

and

$$-S(\tau) = \frac{\langle A(t)A(t-\tau)\sin[\phi(t) - \phi(t-\tau)]\rangle}{\langle A(t)^2\rangle} \tag{6.57b}$$

Equations (6.35) follow as before. Equations (6.33) then say that the cosine and sine Fourier transform of the function $D(\mu)$ are given by the right-hand sides of Eqs. (6.57).

The complex autocorrelation function for this case is

$$\tilde{\gamma}(\tau) = \frac{e^{2\pi i\nu_0\tau}\langle A(t)A(t-\tau)e^{i[\phi(t)-\phi(t-\tau)]}\rangle}{\langle A(t)^2\rangle} \tag{6.58a}$$

and its real part gives (6.56).

By writing

$$\frac{\langle A(t)A(t-\tau)e^{i[\phi(t)-\phi(t-\tau)]}\rangle}{\langle A(t)^2\rangle} = U(\tau)e^{i\Phi(\tau)} \tag{6.58b}$$

we find

$$\tilde{\gamma}(\tau) = U(\tau)e^{i[2\pi\nu_0\tau+\Phi(\tau)]} \tag{6.58c}$$

and the real part gives Eq. (6.35*a*). Furthermore, the real and imaginary parts of (6.58*b*) give Eqs. (6.35*b*, *c*).

This is as far as we can go unless we know more about the functions A and ϕ. For a radio signal or for the light signal from an ideal single mode laser, the amplitude A may be a constant. Then it will drop out of the averages in the above discussions, and the various functions γ, C, S, U, and Φ will depend only on the phase fluctuation through expressions such as

$$C(\tau) = \langle \cos [\phi(t)-\phi(t-\tau)]\rangle$$

and

$$S(\tau) = \langle \sin [\phi(t)-\phi(t-\tau)]\rangle$$

3. ***Thermal Light.*** At the other extreme from single mode laser light is another kind of quasimonochromatic light known as thermal light. The use of the word "thermal" does not imply that the light comes from a thermal black body source; it refers to the statistical properties of the fluctuating light amplitude, which are the same as those of black body light. Thermal light results from the superposition of the light signals from a large number of independent microscopic sources such as individual atoms or molecules. The resulting signal is what is known in the theory of probability as a *Gaussian random variable*. For this reason, thermal light is often called *Gaussian light*.

It is not appropriate for us to study in detail the statistical aspects of thermal light, but we can discuss a model that will bring out many of its important properties. Assume that the electric field is the sum of the fields from a large number of independent quasimonochromatic oscillators having the same dominant frequency ν_0. They will be excited by collisions or other mechanisms, emit radiation for a finite time, be reexcited, reemit, and so on. Each of these events occurs at a random time, and there is no loss of generality if we assume each oscillator to be excited and emit light only once.

To visualize what happens, consider the complex description where the light signal E is given as the real part of a complex analytic signal $\tilde{E}$. The complex signal from an individual source will be of the form

$$E_j = A_j e^{i\phi_j} e^{2\pi i\nu_0 t}$$

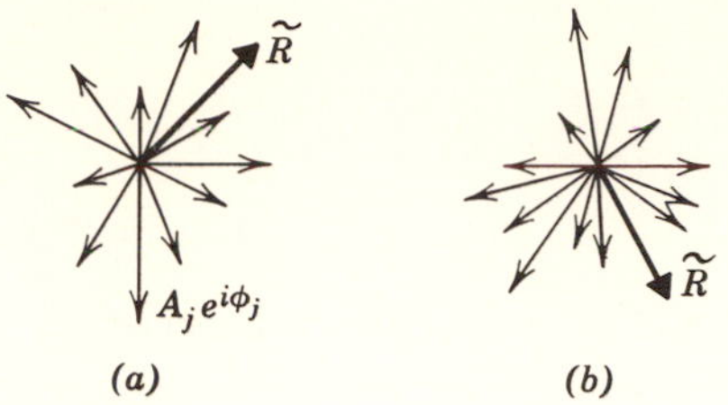

Fig. 6.12

All complex signals will have the common factor $e^{2\pi i\nu_0 t}$, which causes the vector diagram in the complex plane to rotate counterclockwise with angular velocity $2\pi\nu_0$. The total signal at any instant will be the real part of $\tilde{R}e^{2\pi i\nu_0 t}$, where $\tilde{R}$ has amplitude A and phase ϕ and is the sum

$$\tilde{R} = Ae^{i\phi} = \Sigma_j A_j e^{i\phi_j}$$

This is schematically shown in Fig. 6.12*a*. At a later time, different atoms are radiating with phases ϕ_j completely unrelated to the earlier phases. For steady sources the average number of oscillators is a constant, but the magnitude A of the resultant $\tilde{R}$ and its phase ϕ will have changed unpredictably, as in Fig. 6.12*b*. The time for an important fluctuation in the amplitude and phase of $\tilde{R}$, the coherence time, is essentially the duration of coherent radiation from an individual oscillator.

*4. **Calculation of $\gamma(\tau)$ for Thermal Light.*** Since the individual oscillators that contribute to a thermal light signal are independent, and since the Michelson interferometer experiment is done under conditions not capable of detecting interference between independent sources, the resulting interference pattern is essentially a superposition of patterns produced by each individual oscillator. In other words, *the light signal from each atomic or molecular oscillator interferes only with itself.*

A derivation of this result will now be given. The real light signal will be of the form

$$E(t) = \sum_{j=1}^{N} E_j(t) = \sum_{j=1}^{N} A_j(t) \cos\,[2\pi\nu_0 t + \phi_j(t)]$$

We calculate the autocorrelation function $\gamma(t)$ from Eq. (6.50). The numerator of that expression will be given by

$$\text{Numerator of (6.50)} = \int_{-\infty}^{\infty} E(t-\tau)E(t)\,dt$$

$$= \sum_{j=1}^{N}\sum_{k=1}^{N}\int_{-\infty}^{\infty} A_j(t-\tau)A_k(t)\cos\,[2\pi\nu_0(t-\tau) + \phi_j(t-\tau)]\cos\,[2\pi\nu_0 t + \phi_k(t)]\,dt$$

Again we use our trigonometric identity to obtain

$$\cos\,[2\pi\nu_0(t-\tau) + \phi_j(t-\tau)]\cos\,[2\pi\nu_0 t + \phi_k(t)]$$
$$= \tfrac{1}{2}\cos\,[2\pi\nu_0\tau + \phi_k(t) - \phi_j(t-\tau)] + \tfrac{1}{2}\cos\,[2\pi\nu_0(2t-\tau) + \phi_j(t-\tau) + \phi_k(t)]$$

and once again the integral over t gives a negligible contribution from the second term because of the rapid oscillations from the cosine factor. This gives

$$\text{Numerator} = \sum_{j=1}^{N}\sum_{k=1}^{N} \tfrac{1}{2} \int_{-\infty}^{\infty} A_j(t-\tau)A_k(t) \cos\left[2\pi\nu_0\tau + \phi_k(t) - \phi_j(t-\tau)\right] dt$$

The individual oscillators are assumed to be independent. There will then be no correlation between the phases ϕ_j and ϕ_k when $k \neq j$. Since there are a large number of oscillators, for every pair having a certain value of the cosine, there is likely to be another pair with the opposite value, and the net contribution will average to zero. This leaves only the terms with $k = j$. Hence

$$\text{Numerator} = \sum_{j=1}^{N} \tfrac{1}{2} \int_{-\infty}^{\infty} A_j(t-\tau)A_j(t) \cos\left[2\pi\nu_0\tau + \phi_j(t) - \phi_j(t-\tau)\right] dt$$

The contributions from the individual oscillators are seen to add. If we introduce the autocorrelation function for the jth oscillator,

$$\gamma(j;\tau) = \frac{\int_{-\infty}^{+\infty} A_j(t)A_j(t-\tau) \cos\left[2\pi\nu_0\tau + \phi_j(t) - \phi_j(t-\tau)\right] dt}{\int_{-\infty}^{+\infty} A_j(t)^2\, dt}$$

the numerator of Eq. (6.50) becomes

$$\text{Numerator} = \tfrac{1}{2} \sum_{j=1}^{N} \left[\int_{-\infty}^{\infty} A_j(t)^2\, dt\right] \gamma(j;\tau)$$

and the autocorrelation function is

$$\gamma(\tau) = \frac{(\text{Numerator})}{(\text{Numerator at } \tau = 0)}$$

$$= \frac{\sum_{j=1}^{N} \left[\int_{-\infty}^{\infty} A_j(t)^2\, dt\right] \gamma(j;\tau)}{\sum_{j=1}^{N} \int_{-\infty}^{+\infty} A_j(t)^2\, dt} \tag{6.59}$$

This expression is a weighted average of the individual autocorrelation functions. The weighting factor

$$\int_{-\infty}^{\infty} A_j(t)^2\, dt$$

is proportional to the total *energy* emitted by the jth oscillator.

If the individual oscillators are identical to the extent that each has the same $\gamma(j;\tau)$ we obtain simply

$$\gamma(\tau) = \gamma(j;\tau) \tag{6.60}$$

In this case the macroscopic autocorrelation function, which is the normalized oscillatory part of the interference pattern and can be measured directly, simply represents the autocorrelation function of the field from an individual microscopic oscillator. This result is sometimes known as Campbell's theorem.

It should be remarked that Eq. (6.59) can be shown to be valid for all sources of thermal light, not simply quasimonochromatic sources as assumed in the above derivation.

B. THE AUTOCORRELATION FUNCTION AND SPECTRAL DISTRIBUTION FUNCTION FOR SOME MODEL THERMAL SOURCES

The above results can be applied to simple models of the oscillators in radiating systems in order to calculate the oscillator autocorrelation functions $\gamma(j;\tau)$, the overall autocorrelation function, and the spectral distribution function. We make the simplifying assumption that the phase ϕ_j in the quasimonochromatic light signal is a constant. Then the amplitude A_j acts as a simple envelope modulating the fast cosine oscillations. Then Eq. (6.56) gives

$$\gamma(\tau) = \gamma(j;\tau) = \cos 2\pi\nu_0\tau \frac{\langle A_j(t)A_j(t-\tau)\rangle}{\langle A_j(t)^2\rangle} \tag{6.61a}$$

We drop the index j, as it is understood that we are now referring to a signal from an individual oscillator or to a collection of identical oscillators. Equation (6.61a) contains the factor $\cos 2\pi\nu_0\tau$ characteristic of ideal monochromatic light times the autocorrelation function for the envelope or amplitude $A(t)$:

$$\gamma_A(\tau) = \frac{\langle A(t)A(t-\tau)\rangle}{\langle A(t)^2\rangle} \tag{6.61b}$$

Thus, in the interference experiment $\gamma_A(\tau)$ acts as a slowly varying envelope for the $\cos 2\pi\nu_0\tau$ oscillations. Since the phase is constant, Eq. (6.57a) gives the results $C(\tau) = \gamma_A(\tau)$ and $S(\tau) = 0$, and by Eq. (6.33a) we find that the envelope autocorrelation function obeys

$$C(\tau) = \gamma_A(\tau) = \int_{-\infty}^{\infty} D(\mu)\cos 2\pi\mu\tau\, d\mu$$

For the spectral distribution function the inversion formulas from

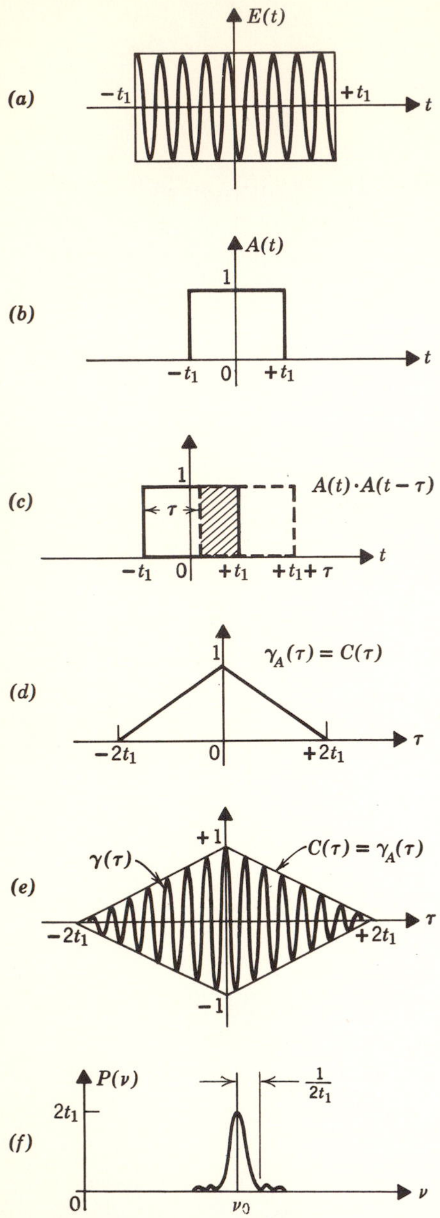

Fig. 6.13. (*a*) A square pulse, (*b*) its envelope, and (*d*) the autocorrelation function of the envelope. (*e*) Normalized autocorrelation function and (*f*) frequency distribution for a sine wave modulated by a square envelope.

Appendix A give

$$P(\nu) = D(\nu - \nu_0) = \int_{-\infty}^{\infty} C(\tau) \cos [2\pi(\nu - \nu_0)\tau]\, d\tau$$

$$= 2\int_0^{\infty} C(\tau) \cos\ [2\pi(\nu - \nu_0)\tau]\, d\tau \qquad (6.62)$$

1. Square Pulse. Suppose that the oscillator emits a sinusoidal signal for a limited time interval. This is sketched in Fig. 6.13*a*. The amplitude $A(t)$ is constant for $|t| \leq t_1$, and zero otherwise. Since Eq. (6.61*b*) is independent of the magnitude of the envelope $A(t)$, we take it to be either zero or unity, as shown in Fig. 6.13*b*. The function $A(t-\tau)$ is simply $A(t)$ shifted to the right a distance τ. This gives for $A(t) \cdot A(t-\tau)$ the behavior shown in Fig. 6.13*c*. The numerator of Eq. (6.61*b*) is proportional to the area of the shaded rectangle, which is $(2t_1 - |\tau|)$ for $|\tau| \leq 2t_1$, and zero otherwise. The denominator is the numerator at $\tau = 0$. This gives

$$C(\tau) = \gamma_A(\tau) = 1 - \frac{|\tau|}{2t_1}, \qquad |\tau| \leq 2t_1$$
$$= 0, \ |\tau| > 2t_1 \qquad (6.63)$$

The result is the triangle shown in Fig. 6.13*d*. The full autocorrelation function is plotted in Fig. 6.13*e*.

The spectral distribution function is obtained from Eq. (6.62),

$$P(\nu) = 2\int_0^{2t_1} (1 - \tau/2t_1) \cos [2\pi(\nu - \nu_0)\tau]\, d\tau = \frac{2t_1 \sin^2 [2\pi(\nu - \nu_0)t_1]}{[2\pi(\nu - \nu_0)t_1]^2} \qquad (6.64)$$

and is plotted in Fig. 6.13*f*.

2. Collision Broadening. In gas discharge lamps the quasimonochromatic spectral line width is often limited by what is known as *collision* or *pressure broadening*. The mechanism is as follows. Each radiating atom or molecule emits a cosine wave of frequency ν_0 with a square* envelope. The emission process is interrupted because of collisions with other molecules; it then continues with little if any correlation of the phase angle. The time between collisions t_c then equals the emission time $2t_1$ of the previous case, and the autocorrelation function for those atoms with collision time t_c is

$$\gamma(t_c; \tau) = C(t_c; \tau) \cos 2\pi\nu_0\tau$$

*This envelope should really be a slow exponential function because of the natural radiation damping of the oscillator, but we can often approximate it by a constant when the natural radiative lifetime is much longer than the collision time.

with

$$C(t_c;\tau) = \left(1 - \frac{|\tau|}{t_c}\right), \qquad \text{for } t_c \geqslant |\tau| \tag{6.63$'$}$$

$$= 0 \qquad \text{for } t_c < |\tau|$$

If all molecules are assumed to emit light with the same amplitude, the weight factor

$$\int_{-\infty}^{\infty} A_j(t')^2\, dt' = A_j^2 \cdot t_c$$

to be used in Eq. (6.59) is proportional to t_c. The sum over j must be replaced by the integral over t_c:

$$\sum_j (\cdot\cdot\cdot) \rightarrow \int_0^{\infty} t_c(\cdot\cdot\cdot)\, P(t_c)\, dt_c$$

Here $P(t_c)\, dt_c$ is the collision probability in the interval t_c to $t_c + dt_c$. And $P(t_c)$ is given by

$$-\frac{1}{N(0)}\frac{dN(t_c)}{dt_c}$$

where $N(t_c)$ is the number of molecules that have not yet made collisions at time t_c. It obeys the differential equation $dN/dt_c = -N/\bar{t}_c$ and has the solution $N = N(0)e^{-t_c/\bar{t}_c}$. Then

$$P(t_c)\, dt_c = e^{-t_c/\bar{t}_c}\frac{dt_c}{\bar{t}_c}$$

where

$$\bar{t}_c = \int_0^{\infty} t_c P(t_c)\, dt_c$$

is the mean time between collisions.

For the autocorrelation function we thus obtain

$$\gamma(\tau) = C(\tau) \cos 2\pi\nu_0\tau$$

where

$$C(\tau) = \frac{\displaystyle\int_0^{\infty} C(t_c;\tau) \cdot t_c \cdot P(t_c)\, dt_c}{\displaystyle\int_0^{\infty} t_c P(t_c)\, dt_c}$$

For positive t we have

$$C(\tau) = \int_{\tau}^{\infty} \left(1 - \frac{\tau}{t_c}\right)\left(\frac{t_c}{\bar{t}_c}\right) e^{-t_c/\bar{t}_c}\frac{dt_c}{\bar{t}_c} = e^{-\tau/\bar{t}_c}$$

Since $C(-\tau) = C(\tau)$, we have

$$C(\tau) = e^{-|\tau|/\bar{t}_c}$$

The final result is

$$\gamma(\tau) = e^{-|\tau|/\bar{t}_c} \cos 2\pi\nu_0\tau \tag{6.65}$$

The functions $C(\gamma)$ and $\gamma(\tau)$ are sketched in Fig. 6.14*b*.

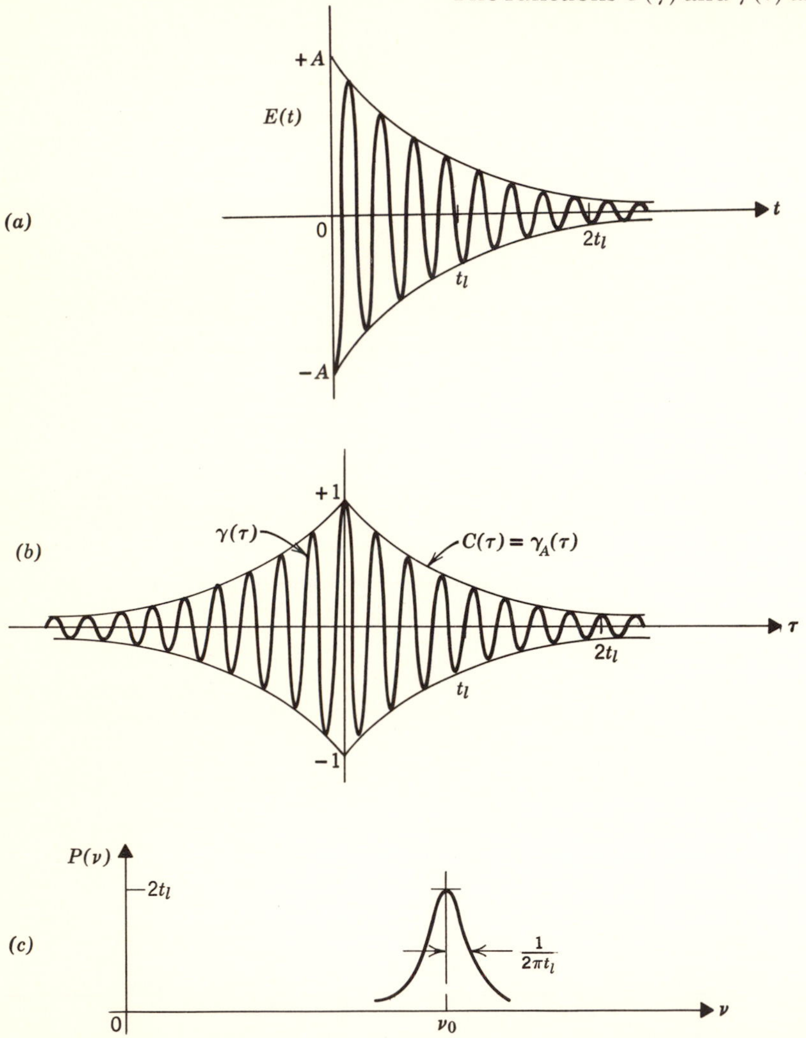

Fig. 6.14. Electric field, normalized autocorrelation function, and frequency distribution function for a sine wave modulated by an exponential envelope. The two lower curves also apply to a collision broadened line for which $t_c = t_l$.

For the frequency distribution function we require

$$C(\mu) = \int_{-\infty}^{\infty} C(\tau)\, e^{-2\pi i\mu\tau}\, d\tau = \int_{0}^{\infty} e^{-\tau/\bar{t}_c}\, e^{-2\pi i\mu\tau}\, d\tau + \int_{-\infty}^{0} e^{\tau/\bar{t}_c}\, e^{-2\pi i\mu\tau}\, d\tau$$

$$= \frac{1}{(1/\bar{t}_c) + 2\pi i\mu} + \frac{1}{(1/\bar{t}_c) - 2\pi i\mu} = \frac{2/\bar{t}_c}{(1/\bar{t}_c)^2 + (2\pi\mu)^2}$$

From Eq. (6.31*a*) we obtain

$$P(\nu) = \frac{2/\bar{t}_c}{(1/\bar{t}_c)^2 + 4\pi^2(\nu - \nu_0)^2} \tag{6.66}$$

We can write $\bar{t}_c$ as the mean free path l divided by the mean velocity $\bar{v}$. The kinetic theory of gases gives the following expression for $\bar{v}$ if the molecules are in thermal equilibrium at a temperature T:

$$\bar{v} = \left(\frac{8k_BT}{\pi m}\right)^{1/2}$$

Here k_B is Boltzmann's constant, and m is the molecular mass. The mean free path is given by

$$l = \frac{1}{\sigma N}$$

where σ is the collision cross section, and N is the number of gas molecules per unit volume. If an ideal gas law is assumed, N is given by

$$N = \frac{p}{k_BT}$$

where p is the pressure in the gas. The mean collision time is then

$$\bar{t}_c = \frac{l}{\bar{v}} = \frac{1}{\sigma p}\left(\frac{\pi m k_B T}{8}\right)^{1/2} \tag{6.67}$$

The function $P(\nu)$ in Eq. (6.66) is a Lorentzian function normalized to have unit area. It has a maximum value of $2\bar{t}_c$ at $\nu = \nu_0$; its half-width at half maximum is $(2\pi\bar{t}_c)^{-1}$. It is sketched in Fig. 6.14*c* (for $\bar{t}_c = t_l$).

3. ***Lifetime Broadened Line.*** If allowed to radiate undisturbed, an excited oscillator will radiate a cosine wave having an exponential envelope:

$$\begin{aligned} E(t) &= Ae^{-t/t_l}\cos(2\pi\nu_0 t + \phi), \qquad t \geq 0 \\ &= 0, \qquad t < 0 \end{aligned}$$

The parameter t_l then represents the natural radiative lifetime of the excited oscillator. The autocorrelation function of the envelope is readily calculated with a result that is the same as Eq. (6.65):

$$\gamma(\tau) = e^{-|\tau|/t_l}\cos 2\pi\nu_0\tau \tag{6.68}$$

The spectral frequency distribution function produced by a collection of such oscillators, each with the same t_l, is then given by Eq. (6.66) with $\bar{t}_c$ replaced by t_l. These results are illustrated in Fig. 6.14. Ordinarily t_l is much longer than $\bar{t}_c$, so that lifetime broadened lines are usually not seen in practice.

*4. **Doppler Broadening.*** Another important source of line width is the Doppler effect. When it dominates the broadening, the physical picture is as follows. Each atom or molecule emits a sharp, quasi-monochromatic line at frequency ν_0 *in its own rest frame*. But if the molecule is moving with respect to the observer with a relative velocity of v measured along the line connecting it to the observer, the observer will see a frequency

$$\nu = \nu_0\left(1+\frac{v}{c}\right)$$

If the molecules are in thermal equilibrium at temperature T, the velocities v will be distributed according to the Maxwell distribution law

$$P_m(v)\,dv = \left(\frac{\beta}{\pi}\right)^{1/2} e^{-\beta v^2}\,dv, \qquad \beta = \frac{m}{2k_BT} = \frac{3}{2\langle v^2\rangle}\;{}^*$$

This means that the spectral distribution function will be given by

$$P(\nu) = \left(\frac{\beta}{\pi}\right)^{1/2} e^{-\beta c^2(\nu-\nu_0)^2/\nu_0^2}\,\frac{c}{\nu_0} \tag{6.69a}$$

We shall show shortly that the autocorrelation function is given by

$$\gamma(\tau) = \cos\,[2\pi\nu_0\tau]\, e^{-\tau^2/\tau_d{}^2} \tag{6.69b}$$

$$\tau_d{}^2 \equiv \frac{\beta c^2}{\pi^2\nu_0{}^2} = \frac{3c^2}{2\pi^2\langle v^2\rangle\nu_0{}^2}$$

This has a Gaussian envelope that is down by $1/e$ when

$$\tau = \tau_d = \left(\frac{3}{2\pi^2}\right)^{1/2}\left(\frac{c}{v_{\rm rms}}\right)\frac{1}{\nu_0}$$

This is very much greater than the period $1/\nu_0$, since the velocity of light c is very much greater than the mean square velocity $v_{\rm rms} = \sqrt{\langle v^2\rangle}$. In terms of τ_d, we can write Eq. (6.69*a*) as

$$P(\nu) = \sqrt{\pi}\,\tau_d e^{-\pi^2(\nu-\nu_0)^2\tau_d{}^2} \tag{6.69c}$$

This is a Gaussian centered about $\nu = \nu_0$. It will be down by a fraction $1/e$ when $\nu - \nu_0 = \pm 1/\pi t_d$. These expressions for $P(\nu)$ and $\gamma(\tau)$ are sketched in Fig. 6.11.

To show that Eq. (6.69*c*) is the Fourier transform of (6.69*b*), it is sufficient to calculate the Fourier transform of the Gaussian envelope $C(\tau) = e^{-\tau^2/t_d{}^2}$. We obtain

$$\hat{C}(\mu) = \int_{-\infty}^{\infty} e^{-\tau^2/\tau_d{}^2} e^{-2\pi i\mu\tau}\,d\tau$$

*It is easy to show that the average of the square of this component of v is $(2\beta)^{-1}$. The average of the square of the velocity vector is then three times this or $\langle v^2\rangle = 3/(2\beta)$.

The evaluation of this integral in a somewhat different form has already been given above between Eq. (6.44) and (6.45). The result is

$$\hat{C}(\mu) = \sqrt{\pi}\tau_d\, e^{-\pi^2\mu^2\tau_d^2}$$

Then

$$P(\mu) = \hat{C}(\nu - \nu_0) = \sqrt{\pi}\tau_d\, e^{-\pi^2(\nu-\nu_0)^2\tau_d^2} \tag{6.69c}$$

5. ***Comments.*** These results have some general features in common. In the case of lifetime broadening, or of collision broadening, the respective mean time of emission t_l, or $\bar{t}_c$, furnishes a measure of the coherence time τ_c. The quasimonochromatic spectral frequency distribution function has a width that is of order τ_c^{-1}. The Doppler broadened case behaves in a similar way, although the individual oscillators emit for a much longer time than t_d. The frequency spread is still given by $\Delta\nu \sim \tau_c^{-1}$, and $\tau_c \sim t_d$ is a measure of how long $\gamma(\tau)$ stays nonzero.

C. DEFINITION OF THE COHERENCE TIME AND THE FREQUENCY SPREAD

We are now in a position to define quantitatively the notions of coherence time τ_c and frequency spread $\Delta\nu$. There is more than one way to do this.* We choose a method due to Mandel:†

$$\tau_c \equiv 4\int_0^\infty \gamma(\tau)^2\, d\tau \tag{6.70a}$$

$$\Delta\nu \equiv \frac{1}{\displaystyle\int_0^\infty P(\nu)^2\, d\nu} \tag{6.70b}$$

For quasimonochromatic light where the autocorrelation function takes the form

$$\gamma(\tau) = U(\tau)\cos\,[2\pi\nu_0\tau + \Phi(\tau)] = C(\tau)\cos 2\pi\nu_0\tau + S(\tau)\sin 2\pi\nu_0\tau$$

and where $C(\tau)$ and $S(\tau)$ are slowly varying with respect to the cosine, Eq. (6.70*a*) simplifies to

$$\tau_c = 2\int_0^\infty U(\tau)^2\, d\tau = 2\int_0^\infty [C(\tau)^2 + S(\tau)^2]\, d\tau \tag{6.70c}$$

This result is easily established once it is realized that the $\cos^2(2\pi\nu_0\tau)$ and $\sin^2(2\pi\nu_0\tau)$ oscillations will average to $\frac{1}{2}$ in a time interval over which U, Φ, C, and S are practically constant.

*For a discussion see L. Mandel and E. Wolf, *Proc. Phys. Soc.* (*London*), **80** 894 (1962).

†L. Mandel, *Proc. Phys. Soc.* (*London*), **74** 233 (1959).

Starting with Eq. (6.24),

$$\gamma(\tau) = \int_0^\infty P(\nu) \cos 2\pi\nu\tau \, d\nu \tag{6.24}$$

a direct calculation using Dirichlet's formula (Eq. A4) will then give

$$4\int_0^\infty \gamma(\tau)^2 \, d\tau = \int_0^\infty P(\nu)^2 \, d\nu \tag{6.71}$$

or by (6.70*a*, *b*):

$$\tau_c = \frac{1}{\Delta\nu} \tag{6.72}$$

Expressions for τ_c and $\Delta\nu$ are easily computed for the examples discussed above. For lifetime or collision broadening, we have a Lorentzian line shape,

$$P(\nu) = \frac{2/t_l}{(1/t_l)^2 + 4\pi^2(\nu - \nu_0)^2}$$

with an exponential envelope on the autocorrelation function,

$$\gamma(\tau) = e^{-|\tau|/t_l} \cos 2\pi\nu_0\tau$$

Then

$$\tau_c = t_l, \qquad \Delta\nu = \frac{1}{t_l} \tag{6.73}$$

For Doppler broadening where the line shape is Gaussian,

$$P(\nu) = \sqrt{\pi}\tau_d e^{-\pi^2(\nu-\nu_0)^2\tau_d^2}$$

the envelope of the autocorrelation function is also Gaussian:

$$\gamma(\tau) = e^{-\tau^2/\tau_d^2} \cos 2\pi\nu_0\tau$$

We can easily obtain

$$\tau_c = \tau_d\left(\frac{\pi}{2}\right)^{1/2}, \qquad \Delta\nu = \left(\frac{2}{\pi}\right)^{1/2}\frac{1}{\tau_d} \tag{6.74}$$

REFERENCES FOR SECTION 6.2

J. M. Stone, *Radiation and Optics*, Chapters 11 and 12. Stone's presentation is at about the same level as the above material.

M. Françon, *Optical Interferometry*, Chapters I, IV, and XIV.

M. Born and E. Wolf, *Principles of Optics*, Chapter 10. Their discussion is at quite a higher level of sophistication.

M. Garbuny, *Optical Physics*. Chapter 3 has a discussion of line shapes and related phenomena.

C. J. Troup, *Optical Coherence Theory—Recent Developments*.

PROBLEMS

6.2.1 An instructive application of Eqs. (6.57) is to a steady source ($A =$ constant) where the phase difference

$$\Delta\phi(\tau) \equiv \phi(t) - \phi(t-\tau)$$

is a Gaussian random variable. Using the theory of probability, we can show that

$$\langle[\Delta\phi(\tau)]^n\rangle = 0$$

for all odd powers n. For even powers the averages can be related to the mean square value of the phase difference as follows. If we define

$$\langle[\Delta\phi(\tau)]^2\rangle \equiv \sigma^2(\tau)$$

Then

$$\langle[\Delta\phi(\tau)]^4\rangle \equiv 3\sigma^4$$

$$\langle[\Delta\phi(\tau)]^6\rangle \equiv 3 \cdot 5\sigma^6$$

$$\langle[\Delta\phi(\tau)]^8\rangle \equiv 3 \cdot 5 \cdot 7\sigma^8, \text{ etc.}$$

Using the above results, show that

$$C(\tau) = \langle\cos \Delta\phi(\tau)\rangle = e^{-\sigma^2(\tau)/2}$$

and

$$S(\tau) = \langle\sin \Delta\Phi(\tau)\rangle = 0$$

(We can further show that $\sigma^2(\tau)$ can be written in the form

$$\sigma^2(\tau) = \frac{|\tau|}{\tau_1}$$

τ_1 then represents the retardation time for which the mean square phase change is one radian. The autocorrelation function is then

$$\gamma(\tau) = e^{-|\tau|/(2\tau_1)}\cos 2\pi\nu_0\tau)$$

6.2.2 Show that $C(\tau)$ as defined by Eq. (6.33*a*) is even, $C(-\tau) = C(\tau)$, and that $S(\tau)$ as defined by Eq. (6.33*b*) is odd, $S(-\tau) = -S(\tau)$. Show further that Eqs. (6.33*a,b*) can be combined into a single complex Fourier integral, which when inverted gives

$$D(u) = \int_{-\infty}^{\infty} U(\tau)e^{i\Phi(\tau)}e^{-2\pi i\mu\tau}\,d\tau = \int_{-\infty}^{\infty} [C(\tau) - iS(\tau)]e^{-2\pi i\mu\tau}\,d\tau$$

Now split $D(\mu)$ into even and odd parts:

$$D(\mu) = D_e(\mu) + D_0(\mu)$$

with

$$D_e(\mu) = \tfrac{1}{2}[D(\mu) + D(-\mu)] \quad \text{and} \quad D_0(\mu) = \tfrac{1}{2}[D(\mu) - D(-\mu)]$$

Then derive the results

$$D_e(\mu) = \int_{-\infty}^{\infty} C(\tau)\cos 2\pi\mu\tau\,d\tau = 2\int_{0}^{\infty} C(\tau)\cos 2\pi\mu\tau\,d\tau$$

$$D_0(\mu) = \int_{-\infty}^{\infty} S(\tau)\sin 2\pi\mu\tau\,d\tau = 2\int_{0}^{\infty} S(\tau)\sin 2\pi\mu\tau\,d\tau$$

6.2.3 Derive the results of the preceding problem by direct substitution of the equation

$$\gamma(\tau) = C(\tau)\cos 2\pi\nu_0\tau + S(\tau)\sin 2\pi\nu_0\tau$$

into the equation

$$P(\nu) = 4\int_0^\infty \gamma(\tau)\cos 2\pi\nu\tau\, d\nu$$

You will need to make some approximations that depend on $C(\tau)$ and $S(\tau)$ being slowly varying with respect to $\cos 2\pi\nu_0\tau$, which implies that $C(\mu)$ and $S(\mu)$ are nonzero only for a range of values $\Delta\nu \ll \nu_0$ about zero.

6.2.4 By using some of the results from Section 6.2, we can now give a quantitative discussion of the resolving power of an interference spectrometer. Start with the detector signal proportional to Eq. (6.22) and use Eq. (6.24) for $\gamma(\tau)$. Because τ must be less than $\tau_{\max} = 2d_{\max}/c$, what is actually measured is a truncated autocorrelation function

$$\gamma^T(\tau) = \gamma(\tau) \text{ if } |\tau| \leq \tau_{\max}, \qquad = 0 \text{ if } |\tau| > \tau_{\max}$$

a. Show that if γ^T is used in Eq. (6.28), a measured frequency distribution will be obtained that is given by

$$P_m(\nu) = \int_0^\infty P(\nu')\left[\frac{\sin 2\pi(\nu'-\nu)\tau_{\max}}{\pi(\nu'-\nu)} + \frac{\sin 2\pi(\nu'+\nu)\tau_{\max}}{\pi(\nu'+\nu)}\right]d\nu'$$

where $P(\nu')$ is the true frequency distribution function. (This type of expression is called a convolution.) Now consider $P(\nu)$ of the form shown

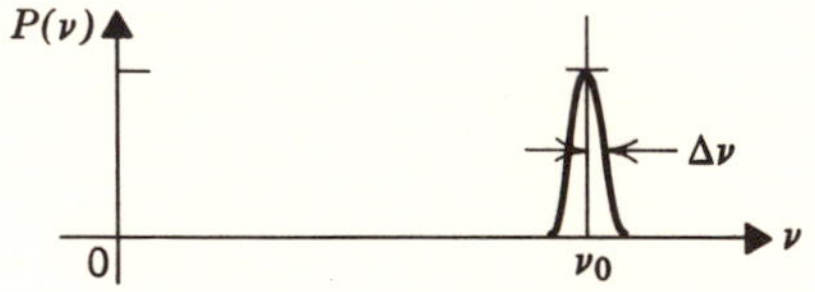

where $\nu_0\tau_{\max} \gg 1$. Then only the first term in the above integral is important for positive ν. You are asked (i) to discuss $P_m(\nu)$ with equations and (ii) to sketch its behavior in the following limiting cases.

b. When $\Delta\nu \gg 1/\tau_{\max}$.

c. When $\Delta\nu \ll 1/\tau_{\max}$. You will not require an explicit functional form for $P(\nu)$.

d. What does the result of part (c) have to say about the resolving power? By this we mean: How large does $\Delta\nu$ have to be before $P_m(\nu)$ mirrors the true shape of $P(\nu)$?

6.2.5 Suppose that a very narrow quasimonochromatic line of frequency ν_0 is being measured on an interference spectrometer. The line width is assumed to obey $\Delta\nu \ll 1/\tau_{\max}$. Instead of the abrupt truncation obtained by direct use of the experimental $\gamma^T(\tau)$ discussed in Problem 6.2.3, suppose that we smooth γ linearly to obtain a new function

$$\gamma^s(\tau) = \gamma_0(\tau)\left(1 - \frac{|\tau|}{\tau_{\max}}\right), \qquad |\tau| \leqslant \tau_{\max}$$

$$= 0, \qquad |\tau| \geqslant \tau_{\max}$$

where $\gamma_0(\tau) \approx \cos 2\pi\nu_0\tau$.

Discuss with equations and a sketch the frequency spectrum obtained if $\gamma^s(\tau)$ is used in Eq. (6.28), and compare with the result if $\gamma^T(\tau)$ of the preceding problem is used with $\gamma(\tau) = \gamma_0(\tau) \approx \cos 2\pi\nu_0\tau$. (Hint. See Fig. 6.13.) Which method do you think gives the greater resolving power? Which method gives the lesser spurious wiggles in the computed spectrum? (A smoothing operation such as the one discussed here is called apodization. This term was originally used to describe the analogous smoothing of diffraction patterns.)

6.2.6 Derive Eq. (6.71). This is one version of "Parseval's theorem."

6.2.7 Derive the results in (6.73) and (6.74). Find either $\Delta\nu$ or τ_c.

6.2.8 Consider a collection of exponentially damped oscillators, each emitting an electric field

$$e(t) = ae^{-t/t_l}\cos 2\pi\nu_0 t, \qquad t \geqslant 0$$

$$= 0, \qquad t < 0$$

but at random times. Find a rigorous expression for the Fourier transform $\hat{e}(\nu)$ of $e(t)$ and for $|\hat{e}(\nu)|^2$, which will be proportional to $I(\nu)$ (Appendix B). Show that for $\nu_0 t_l \gg 1$ your expression for $|\hat{e}(\nu)|^2$ agrees quite well with Eq. (6.68), except for normalization.

6.2.9 Consider the superposition of two cosine waves each having the same slowly varying envelope:

$$E(t) = A(t)\cos 2\pi\nu_1 t + A(t)\cos 2\pi\nu_2 t$$

$$= 2A(t)\cos 2\pi\nu_0 t\cos 2\pi f t$$

where

$$\nu_0 = \tfrac{1}{2}(\nu_1 + \nu_2), \qquad f = \tfrac{1}{2}(\nu_1 - \nu_2) > 0$$

Assume that $f \ll \nu_0$ and that $A(t)$ is slowly varying with respect to $\cos 2\pi f t$.

a. Find an approximate expression for the normalized autocorrelation function for $E(t)$ and sketch the result.
b. Using the result of (a), find an approximate expression for the spectral distribution function for this source by applying Eq. (6.62). The results should look like that shown in the figure.

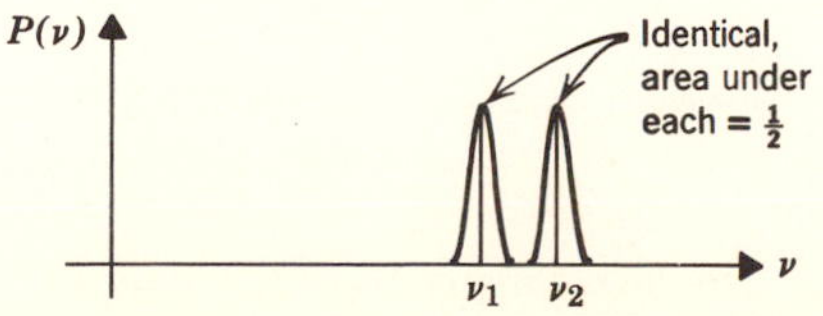

6.2.10 Consider a steady source emitting two steady quasimonochromatic lines so that the electric field is of the form

$$E(t) = A_1 \cos [2\pi\nu_1 t + \Phi_1(t)] + A_2 \cos [2\pi\nu_2 t + \Phi_2(t)]$$

where

$$\langle \cos [\Phi_1(t) - \Phi_1(t-\tau)] \rangle = e^{-\tau^2/t_1^2}$$

$$\langle \cos [\Phi_2(t) - \Phi_2(t-\tau)] \rangle = e^{-\tau^2/t_2^2}, \qquad A_1, A_2 \text{ constants}$$

and where

$$\nu_1 t_1 \gg 1, \quad \nu_2 t_2 \gg 1, \quad t_1 \sim t_2, \quad \nu_1 \gg \nu_1 - \nu_2 > 0$$

Find expressions for $\gamma(\tau)$ and for $P(\nu)$.

6.3 THE MICHELSON STELLAR INTERFEROMETER–TRANSVERSE COHERENCE

Often in optics we want to know if the electric fields at two different points in space are mutually coherent. Theoretical answers to this question are provided by the theory of partial coherence. The theory may be applied to the simple two-slit experiment of Young, which is the basic experimental method of measuring mutual coherence.

A. YOUNG'S EXPERIMENT AND ITS GENERALIZATIONS

1. Coherence Lengths. We may interpret the coherence time τ_c as a measure of how long the phase in a quasimonochromatic light beam retains a "memory" of the value it had at an earlier time. If the beam is perfectly collimated, we may move along it a distance

$$l_l = c\tau_c = \frac{c}{\Delta\nu}$$

and say that the phase remains correlated over this distance, which is called the *longitudinal coherence length*. For such a collimated beam the phase is the same on any plane oriented perpendicular to the direction of propagation. We say that there is perfect transverse coherence in the beam.

For an incompletely collimated beam the transverse coherence is no longer perfect. To describe it we introduce the *transverse coherence length* l_t. This represents a distance in a plane transverse to the main propagation direction over which the phases at two points remain correlated; the individual phases may fluctuate, but the two fluctuate together. Since for two points separated by a distance less than or about equal to l_t there is a definite phase relationship, interference effects should result if the light from these two points can be brought together. In principle we can do this with Young's experiment.

2. Young's Experiment. We thus return to Young's double-slit experiment. We had originally assumed that the light source was a point, and obtained a very simple interference pattern. How will this change when the source of light is more complicated?

What happens when a single point source of light is displaced? We assume for now that the light is quasimonochromatic with a frequency $\nu = c/\lambda$ and will later generalize the results to cover a distribution of frequencies. Assume at first that the source is at S in Fig. 6.15*a*. The zero order fringe is the one where the path lengths $|SS_1P_0|$ and $|SS_2P_0|$ are equal and is found on the axis at P_0. The angular fringe separation is $\delta\theta' = \lambda/h$. Now let the source point move down below the axis an angle θ to S'. The fringe pattern will

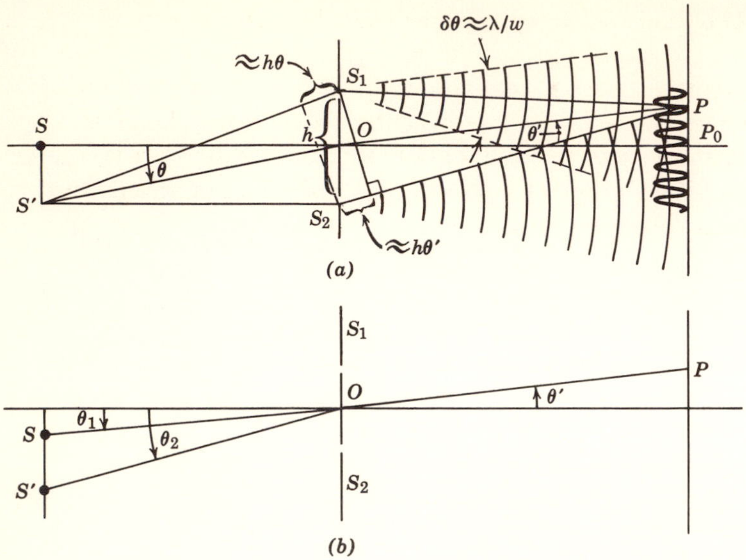

Fig. 6.15

then shift up an angle $\theta' = \theta$, and the zero order fringe is along the straight line from S' through 0. (These are approximate statements that hold when the angles θ and θ' are small and when h is much less than $|SO|$ and $|OP_0|$.)

In order to observe an interference pattern, it is necessary for the light beams emerging from S_1 and S_2 to spread because of *diffraction*. The relevant theory will be developed in Chapter 7. There it will be shown that the diffracted beam from a single slit of width w will be centered about $\theta' = \theta$. The flux density will decrease smoothly to zero at $\theta' = \theta \pm \lambda/w$ as shown very schematically in Fig. 6.15*a*. This will cover many interference fringes (of angular separation $\delta\theta' = \lambda/h$) provided that $h \gg w$. We assume this to be the case and will not discuss diffraction effects further in this chapter.

The fringes are perpendicular to the plane of the paper in Fig. 6.15, that is, perpendicular to the line joining S_1 and S_2. If the source point S is displaced *out* of the plane of the paper, the point P_0 at the center of the zero-order fringe will be displaced along that fringe *into* the plane of the paper, but the fringes will remain unchanged. A similar situation applies when S' is displaced out of the plane of the paper; there is no change in the fringes. Thus we may treat any three-dimensional distribution of source points as an equivalent two-dimensional distribution obtained by projecting the original distribution onto the plane of the paper.

If point sources having no mutual coherence are present at *both* S and S', the interference patterns will add, and there may be smoothing or loss of contrast of the fringes depending on the relative strength of the sources and on θ. With sources of equal intensity this occurs whenever the fringe pattern shifts by an *odd* number of half-wavelengths. The detailed argument goes as follows.

For a single source S' at angle θ (positive when S' is below the axis) and a point P on the screen at angle θ' (positive when P is above the axis), we can easily show that the path difference is given by $\widehat{S'S_2P} - \widehat{S'S_1P} = h(\theta' - \theta)$. The difference in travel times for the two beams is given by

$$\bar{\tau} = \frac{h(\theta' - \theta)}{c} \tag{6.75}$$

This case is quite familiar. We know that for quasimonochromatic light with a coherence time $\tau_c \gg \bar{\tau}$ the "intensity" (more properly, flux density) at the screen will be given by

$$I = I_1\,(1 + \cos 2\pi\nu\bar{\tau}) = I_1\left\{1 + \cos\left[\frac{2\pi\nu h(\theta' - \theta)}{c}\right]\right\} \tag{6.76}$$

Here $I_1/2$ represents the single beam intensity with only one slit open.

*3. **Two Point Sources.*** With two mutually incoherent* sources with the same frequency ν located at angles θ_1 and θ_2 as shown in Fig. 6.15*b*, we add the intensities to obtain

$$I = I_1\left\{1 + \cos\left[\frac{2\pi\nu h(\theta' - \theta_1)}{c}\right]\right\} + I_2\left\{1 + \cos\left[\frac{2\pi\nu h(\theta' - \theta_2)}{c}\right]\right\} \tag{6.77}$$

Note that the maxima of one pattern fall on the minima of the other when

$$\frac{2\pi\nu h}{c}\,(\theta_1 - \theta_2) = \pi \times \text{odd integer}$$

that is, when

$$\frac{h(\theta_1 - \theta_2)}{\lambda} = \text{half integer}$$

for then the cosine terms are equal and opposite. In this case the fringes have minimum visibility. On the other hand, if

$$\frac{2\pi\nu h}{c}\,(\theta_1 - \theta_2) = 2\pi \times \text{integer}$$

that is, if

$$\frac{h(\theta_1 - \theta_2)}{\lambda} = \text{integer}$$

*By "mutually incoherent" we mean that the mutual coherence time for the two sources is much shorter than the measurement time. By "quasimonochromatic" we mean that $2\pi\nu\tau_c \gg 1$ (many cycles within the envelope).

the maxima of both patterns coincide, and the fringes have maximum visibility.

4. Visibility of Fringes. At this point it is convenient to introduce a quantitative measure of visibility, denoted by V,* that will cover intermediate cases as well as the two above-mentioned extremes. We observe a few fringes on the screen as a function of θ' and note the maximum and minimum values of I. Then V is calculated from

$$V = \frac{I(\max) - I(\min)}{I(\max) + I(\min)} \tag{6.78}$$

Fringes of maximum visibility result when there are zeroes in the pattern, that is, when $I(\min) = 0$. Zero visibility occurs when there are no interference fringes. Then $I(\max) = I(\min)$.

For a general quasimonochromatic source, V will depend on both θ', the off-axis observation angle, and h, the slit separation. For the two-beam interference fringes observed with the Michelson interferometer we have already discussed how the visibility $V(\tau)$ of the fringes is equal to the envelope $U(\tau)$ of the autocorrelation function $\gamma(\tau)$. When h is small the same result applies here, with $\tau = h\theta'/c$. Diffraction effects limit the spread of the beams to small values of θ', corresponding to small values of τ. Thus, $\gamma(\tau)$ may safely be approximated by $\cos 2\pi\nu\tau$. In this situation the spectral composition of the light, which determines $\gamma(\tau)$, is not very important,† and *the slit separation h is the important independent variable.* This small τ limit is equivalent to the assumption of truly monochromatic light.

The general case is discussed below, and the resulting expression for $V(\tau, h)$ is given by Eq. (6.102′). The correction when τ is not small is simply obtained by multiplying the small τ result by $U(\tau)$.

Let us calculate V for the two-source case of Eq. (6.77).

Our purpose in the present discussion is to show how the spatial distribution of source brightness affects the visibility of the interference fringe pattern. We do this by expanding the cosines in Eq. (6.77) to separate the factors depending on the source angles θ_1 and θ_2 from those depending on the observation angle θ' at the screen. The result is

$$I = I_1 + I_2 + \left(I_1 \cos\frac{2\pi\nu h\theta_1}{c} + I_2 \cos\frac{2\pi\nu h\theta_2}{c}\right)\cos\frac{2\pi\nu h\theta'}{c} + \left(I_1 \sin\frac{2\pi\nu h\theta_1}{c} + I_2 \sin\frac{2\pi\nu h\theta_2}{c}\right)\sin\frac{2\pi\nu h\theta'}{c}$$

*See Eq. (6.39) for a discussion of the visibility of fringes in a Michelson interferometer.

†By this we mean that Young's experiment is a poor way to achieve large path differences necessary to probe the spectral composition.

The first and second parentheses can be written in the form $I_0 \cos \alpha$ and $I_0 \sin \alpha$ respectively, where

$$I_0 = \left[I_1^2 + I_2^2 + 2I_1 I_2 \cos \frac{2\pi\nu h(\theta_2 - \theta_1)}{c} \right]^{1/2} \tag{6.79}$$

(the value of α is not important here). Then

$$I = I_1 + I_2 + I_0 \cos\left(\alpha - \frac{2\pi\nu h\theta'}{c}\right)$$

Thus, as θ' is varied, the maximum intensity is seen to be

$$I_{\text{max}} = I_1 + I_2 + I_0$$

the minimum is

$$I_{\text{min}} = I_1 + I_2 - I_0$$

and the visibility is

$$V = \frac{I_0}{I_1 + I_2} \tag{6.80}$$

Note the periodic dependence on the slit separation h. As a function of h, V varies between

$$V_{\text{max}} = 1 \quad \text{when} \quad \frac{2\pi\nu h(\theta_1 - \theta_2)}{c} = 2\pi \cdot \text{integer}$$

that is, when

$$\frac{h(\theta_1 - \theta_2)}{\lambda} = \text{integer}$$

and

$$V_{\text{min}} = \frac{|I_1 - I_2|}{I_1 + I_2} \quad \text{when} \quad \frac{2\pi\nu(\theta_1 - \theta_2)}{c} = \pi \cdot \text{odd integer} \tag{6.81}$$

that is, when

$$\frac{h(\theta_1 - \theta_2)}{\lambda} = \text{half-integer}$$

These results are derived more simply in connection with Eq. (6.77).

For a fixed double point source we find that the visibility of the two-slit interference pattern reaches its first minimum when h is given by

$$h_1 = \frac{\lambda}{2(\theta_1 - \theta_2)} \tag{6.82}$$

Thus, if we can measure h_1, the angular separation is given by $(\theta_1 - \theta_2) = \lambda/(2h_1)$. This equation also follows directly from the reasoning we applied to Eq. (6.77).

According to the discussion following Fig. 6.15, all source points may be projected onto the plane of the paper in that figure without changing the interference pattern. If the two point sources lie out of the plane, the angles θ_1 and θ_2 refer to their projections.

Figure 6.18*b*, *c*, *d* shows the intensity on the measuring screen for a slightly different case, to be discussed below. The same general behavior holds here; the fringe separation is not quite the same, however.

This variation of the visibility V with slit separation h can be used to determine the nature of the source or distribution of sources. For instance, if V alternates periodically between a maximum of unity and a minimum V_{min}, then we may conclude that we have two point sources, whose projected angular separation $\theta_1 - \theta_2$ can be obtained from (6.82) and whose relative flux density can be obtained from (6.81) as the ratio

$$\frac{1 - V_{\text{min}}}{1 + V_{\text{min}}}$$

We now apply these arguments to a continuous distribution of source points.

5. ***Continuous Distribution of Source Points.*** Another simple case is a uniform line source lying between S and S' in Fig. 6.15*b*. This can result from the projection of a uniform rectangle onto the plane of the paper. For this type of source we can define a function $I(\theta)\,d\theta$ to represent twice the single slit intensity at the screen coming from the part of the source located between θ and $\theta + d\theta$. The intensity in the two-slit interference pattern coming from this part of the source is then given by a generalization of (6.76),

$$dI = I(\theta)\left\{1 + \cos\left[\frac{2\pi\nu h(\theta' - \theta)}{c}\right]\right\} d\theta$$

and the total intensity received at an angle θ' is then

$$I = \int_{-\infty}^{\infty} I(\theta)\left\{1 + \cos\left[\frac{2\pi\nu h(\theta' - \theta)}{c}\right]\right\} d\theta$$

$$= I(\text{tot})(1 + \gamma_{12}) \tag{6.83}$$

$$\gamma_{12} \equiv \frac{\int_{-\infty}^{\infty} I(\theta)\cos\left[\frac{2\pi\nu h(\theta' - \theta)}{c}\right] d\theta}{I(\text{tot})}$$

Here the total intensity with no interference is

$$I(\text{tot}) = \int_{-\infty}^{\infty} I(\theta)\, d\theta \tag{6.84}$$

6. ***Line Source.*** For a *uniform line source* of angular height $\theta_2 - \theta_1 = \Delta\theta$ and total intensity $I(\text{tot})$, we have

$$I(\theta) = \frac{I(\text{tot})}{\Delta\theta} \qquad \text{for} \quad \theta_1 < \theta < \theta_2$$

$$= 0 \text{ otherwise}$$

Let

$$\theta_0 = \frac{\theta_1 + \theta_2}{2}$$

represent the angular position of the center of the source. Then Eq. (6.83) becomes

$$I = \frac{I(\text{tot})}{\Delta\theta} \int_{\theta_0 - \Delta\theta/2}^{\theta_0 + \Delta\theta/2} \left\{1 + \cos\left[\frac{2\pi\nu h(\theta' - \theta)}{c}\right]\right\} d\theta$$

$$= I(\text{tot}) \left(1 - \frac{c}{2\pi\nu_0 h\Delta\theta} \left\{\sin\left[\frac{2\pi\nu h(\theta' - \theta_0 - \Delta\theta/2)}{c}\right]\right.\right.$$

$$\left.\left. - \sin\left[\frac{2\pi\nu h(\theta' - \theta_0 + \Delta\theta/2)}{c}\right]\right\}\right)$$

$$= I(\text{tot}) \left\{1 + \frac{2c}{2\pi\nu h\Delta\theta} \sin\left(\frac{2\pi\nu h\Delta\theta}{2c}\right) \cos\left[\frac{2\pi\nu h(\theta' - \theta_0)}{c}\right]\right\} \tag{6.85}$$

Form this we obtain

$$I(\max) = I(\text{tot}) \left\{1 + \left|\frac{\lambda}{\pi h\Delta\theta} \sin\left(\frac{\pi h\Delta\theta}{\lambda}\right)\right|\right\}$$

$$I(\min) = I(\text{tot}) \left\{1 - \left|\frac{\lambda}{\pi h\Delta\theta} \sin\left(\frac{\pi h\Delta\theta}{\lambda}\right)\right|\right\}$$

Here the visibility is given by

$$V = \left|\frac{\lambda}{\pi h\Delta\theta} \sin\left(\frac{\pi h\Delta\theta}{\lambda}\right)\right| \tag{6.86}$$

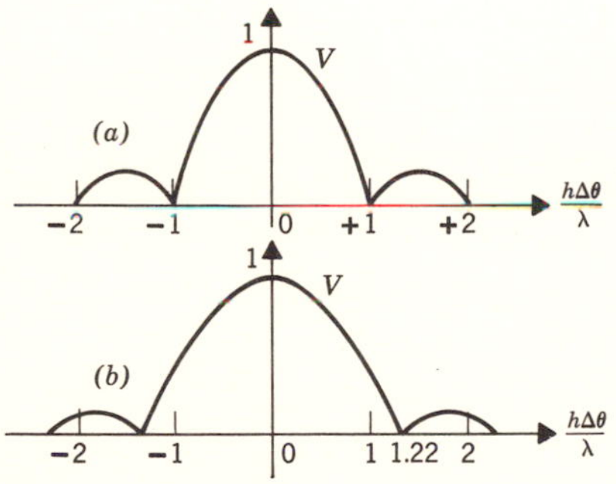

Fig. 6.16. Visibility of the fringe pattern from (*a*) a uniform line source, (*b*) a uniform disk.

This function is plotted in Fig. 6.16*a* (see also Fig. 6.18*a*). The value of h at which V is first equal to zero is given by

$$h_1 = \frac{c}{\nu\Delta\theta} = \frac{\lambda}{\Delta\theta} \tag{6.87}$$

just twice the value given in Eq. (6.82) for a double point source.

7. ***Extended Source as a Point Source.*** If we intended this line source to be used instead of a single point source and have the same optical effect, then values of h and $\Delta\theta$ must be chosen so that V in Eq. (6.86) is very close to unity; that is, we would want

$$\frac{\pi h\Delta\theta}{\lambda} \ll 1$$

which is equivalent to the condition

$$h\Delta\theta \ll \lambda$$

We can generalize this result somewhat and say that *when the condition $h\Delta\theta \ll \lambda$ is satisfied, an extended source of angular height $\Delta\theta$ will act as if it were a point source.*

8. ***Disk-Shaped Source.*** A similar but more complicated mathematical derivation can be made for a disk-shaped source (or a spherical Lambert source) of angular diameter $\Delta\theta$. The behavior of V as a function of h is quite similar to that for a line source. It is plotted in Fig. 6.16*b*. The value of h for which V is first equal to zero turns out to be

$$h_1 \text{ (disk)} = 1.22\frac{\lambda}{\Delta\theta} \tag{6.88a}$$

For the record, we state the formula for V without proof:

$$V = 2\left|\frac{J_1(\pi h\Delta\theta/\lambda)}{\pi h\Delta\theta/\lambda}\right| \tag{6.88b}$$

The function $J_1(\chi)$ is a first-order Bessel function. We discuss it further in Section 7.2C, where Fraunhofer diffraction by a circular aperture is treated. The mathematical similarity between these two phenomena is no accident. It is known formally as the *van Cittert-Zernicke* theorem and is covered in advanced treatments of partial coherence.

9. ***Michelson Stellar Interferometer.*** The phenomena just described were used by Michelson to measure the angular separation of double stars. He covered the objective lens of an astronomical telescope with a mask containing two slits, as shown in Fig. 6.17*a*. Because of the low light levels involved, he had to use white light. The effective frequency spread was that of his detector, and was large, perhaps one-half of the frequency itself. In spite of this, the autocorrelation function $\gamma(\tau)$ may be approximated by $\cos 2\pi\nu\tau$ for one or two cycles, and a few fringes could be seen near zero order.* Their visibility was studied as a function of h. For a double star the value of h at the first minimum of visibility gives a value for the angular separation $\Delta\theta = \theta_1 - \theta_2$ of the stars from Eq. (6.82)—$\Delta\theta = \lambda/2h_1$—and the intensity ratio may be obtained from the value of $V_{\min}$ in the form $(1 - V_{\min})/(1 + V_{\min})$. (The slits, of course, must be perpendicular to the line joining the two stars for this to hold.) Michelson was also able to measure the angular diameter of planetary satellites by measuring h_1 and using Eq. (6.88).

*An explicit discussion of the effects of a distribution of frequencies will be found in Part B, which follows.

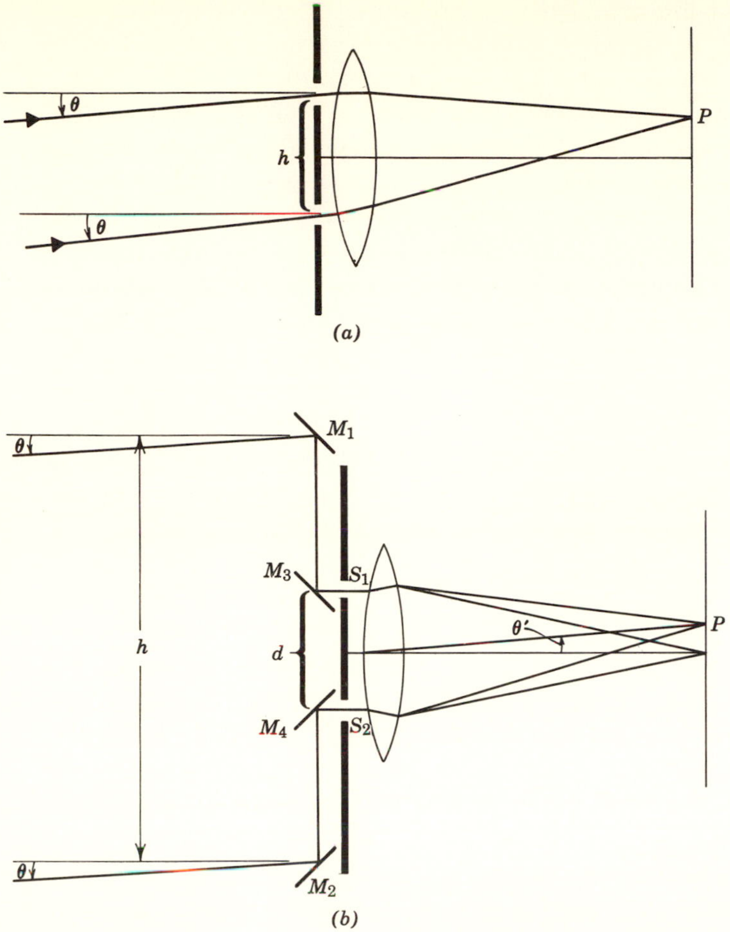

Fig. 6.17. Stellar interferometer.

No known star has a large enough angular diameter to be measured by the apparatus shown in Fig. 6.17*a* with its limited range of h. Michelson extended the effective width of the slits to more than six meters in his *stellar interferometer*, which is shown in Fig. 6.17*b*. Parallel light from a distant star incident at an angle θ will arrive at M_1 a time $h\theta/c$ later than at M_2. This retardation will be the same at the slits if $\widehat{M_1M_3S_1} = \widehat{M_2M_4S_2}$. At point P on the screen (where a photographic plate or an eyepiece is inserted) located at an angle θ', there will be an additional retardation of opposite sign of amount $d\theta'/c$. The net retardation is thus

$$\tau = \frac{d\theta'}{c} - \frac{h\theta}{c} \tag{6.89}$$

For this apparatus the fringe separation depends only on d, not h. But the fringe *contrast* depends on h, not d. Some of the above equations need to be changed. Equation (6.79) still holds if h is replaced by d in the argument of the cosine function. Equation

(6.80) is still valid. Again in (6.85) h must be replaced by d in the cosine function, but the subsequent equations are still valid.

Using the interferometer, we can clearly distinguish a double star with its sinusoidal $V(h)$ dependence [Eq. (6.80)] from a large single star that has an h dependence shown in Fig. 6.16*b*. Once the type of source is determined, its angular size can be determined from (6.82) or (6.88). Only the largest stars can be measured by this technique, namely, the so-called red giants. An example is Betelgeuse with an angular diameter of 0.047 seconds of arc and an actual diameter of 4.1×10^8 km, greater than the diameter of the earth's orbit around the sun!

The mirrors in the stellar interferometer must be mounted extremely carefully, so that path difference changes caused by mechanical vibration and thermal expansion are held to a small fraction of a wavelength. This becomes impossible for values of h larger than a few meters. The intensity fluctuation interferometers recently developed by Hanbury Brown and Twiss enable much larger values of h to be used. Some information is sacrificed, however, since these new techniques measure only V. This will be discussed in Section 6.4.

B. GENERALIZATIONS TO INCLUDE A DISTRIBUTION OF FREQUENCIES

Thus far the quasimonochromatic property of the light sources was used only to assure that the experiment was performed so that the various points in the extended source had complete mutual incoherence. In other words, we had to be sure that the measurement time was longer than the coherence time. Let us now see what other effects result from a finite spread in frequency.

1. Basic Derivation. We continue to use the notion of an angular distribution of intensity $I(\theta)$, where θ is defined in Fig. 6.15. Since the interference fringes are long and extend perpendicular to the line between S_1 and S_2, they are insensitive to (small) angular displacements of a source point out of the plane of the paper.

We define a normalized angular distribution function

$$i(\theta) = \frac{I(\theta)}{I(\text{tot})} \tag{6.90}$$

having the property

$$\int_{-\infty}^{\infty} i(\theta)\, d\theta = 1$$

We show the limits as $\pm\infty$, although in practice the range of θ will be quite small. The normalized oscillatory part of the interference pattern is then obtained from Eqs. (6.83) and (6.90) in the form

$$\gamma_{12} = \int_{-\infty}^{\infty} i(\theta) \cos\left[\frac{2\pi\nu h(\theta' - \theta)}{c}\right] d\theta \tag{6.91}$$

Now we also assume that we have a spread in frequency described by the normalized distribution function $P(\nu)$ of Section 6.1. We further assume that all the sources have the same distribution function $P(\nu)$. In the various equations and derivations given above in this section we must multiply by $P(\nu)$ and integrate. Any constant terms are unaffected; any frequency-dependent terms become averaged using the weighting function $P(\nu)$. In particular, I in Eq. (6.83) retains its form,

$$I = I(\text{tot})(1 + \gamma_{12})$$

with γ_{12} now given by

$$\gamma_{12} = \int_{0}^{\infty} P(\nu) \left\{ \int_{-\infty}^{\infty} i(\theta) \cos\left[\frac{2\pi\nu h(\theta' - \theta)}{c}\right] d\theta \right\} d\nu$$

$$= \int_{-\infty}^{\infty} i(\theta) \left\{ \int_{0}^{\infty} P(\nu) \cos\left[\frac{2\pi\nu h(\theta' - \theta)}{c}\right] d\nu \right\} d\theta \tag{6.92}$$

We recognize from Sections 6.1 and 6.2 that for $\tau = h(\theta' - \theta)/c$, the expression in curly brackets

$$\gamma(\tau) = \int_{0}^{\infty} P(\nu) \cos 2\pi\nu\tau \, d\nu$$

is the normalized autocorrelation function. Using it, we may then write

$$\gamma_{12} = \int_{-\infty}^{\infty} i(\theta)\, \gamma\left[\frac{h(\theta' - \theta)}{c}\right] d\theta \tag{6.93}$$

2. ***Assumption of Quasimonochromatic Light.*** To proceed, we assume that we have quasimonochromatic light by writing

$$\gamma(\bar{\tau}) = U(\bar{\tau}) \cos\left[2\pi\nu_0\bar{\tau} + \Phi(\bar{\tau})\right]$$

where U and Φ vary slowly with respect to the cosine. Then

$$\gamma_{12} = \int_{-\infty}^{\infty} i(\theta)\, U\left[\frac{h(\theta' - \theta)}{c}\right] \cos\left\{\frac{2\pi\nu_0 h(\theta' - \theta)}{c} + \Phi\left[\frac{h(\theta' - \theta)}{c}\right]\right\} d\theta \tag{6.94}$$

To continue the derivation we must further assume that U and Φ

remain essentially constant while θ varies over its range of nonzero values of Eq. (6.94). Let θ_0 be the angle about which the angular distribution $i(\theta)$ is centered, and let $\Delta\theta$ be the width of this distribution. Then our requirement is that U and Φ remain essentially constant when $\bar{\tau}$ changes by $h\Delta\theta/c$ about the value $h(\theta' - \theta_0)/c$. For this we need a coherence time τ_c much greater than $h\Delta\theta/c$. Now h will be of the order of the transverse coherence length l_t (see below). Thus

$$h \approx l_t \lessapprox \frac{\lambda}{\Delta\theta}$$

and our requirement becomes

$$\tau_c \gg \frac{h\Delta\theta}{c} \approx \left(\frac{\lambda}{\Delta\theta}\right)\left(\frac{\Delta\theta}{c}\right) = \frac{\lambda}{c} = \frac{1}{\nu_0}$$

This last inequality is the usual one,

$$\Delta\nu = \frac{1}{\tau_c} \ll \nu_0$$

used to define quasimonochromatic light. In practice it need not be strong, since the Michelson stellar interferometer works with "white" light where $\Delta\nu \sim \nu_0/2$.

Under these conditions we may replace θ by θ_0 in the arguments of U and Φ in Eq. (6.94):

$$\gamma_{12} = U\left[\frac{h(\theta' - \theta_0)}{c}\right] \int_{-\infty}^{\infty} i(\theta) \cos\left\{\frac{2\pi\nu_0 h(\theta' - \theta)}{c} + \Phi\left[\frac{h(\theta' - \theta_0)}{c}\right]\right\} d\theta$$

The natural angular variable to use at the screen becomes $(\theta' - \theta_0)$. Hence *we redefine* τ *as*

$$\tau = \frac{h(\theta' - \theta_0)}{c} \tag{6.95a}$$

Then

$$\gamma_{12} = U(\tau) \int_{-\infty}^{\infty} i(\theta) \cos\left\{2\pi\nu_0\tau + \Phi(\tau) - \frac{2\pi\nu_0 h(\theta - \theta_0)}{c}\right\} d\theta$$

The function $i(\theta)$ is centered about $\theta = \theta_0$. We can redefine it as follows

$$i(\theta) \rightarrow i(\theta - \theta_0) \tag{6.95b}$$

where the new function $i(\theta - \theta_0)$ is centered about $\theta - \theta_0 = 0$. Then finally

$$\gamma_{12} = U(\tau) \int_{-\infty}^{\infty} i(\theta) \cos\left\{2\pi\nu_0\tau + \Phi(\tau) - \frac{2\pi\nu_0 h\theta}{c}\right\} d\theta \tag{6.96}$$

It is useful to think of τ and h as the independent variables.

3. Use of Fourier Transform. When the cosine is expanded, Eq. (6.96) becomes

$$\gamma_{12}(\tau, h) = U(\tau)\cos\left[2\pi\nu_0\tau + \Phi(\tau)\right]\int_{-\infty}^{\infty} i(\theta)\cos\frac{2\pi\nu h\theta}{c}\,d\theta$$
$$+ U(\tau)\sin\left[2\pi\nu_0\tau + \Phi(\tau)\right]\int_{-\infty}^{\infty} i(\theta)\sin\frac{2\pi\nu h\theta}{c}\,d\theta \qquad (6.97)$$

The two integrals in Eq. (6.97) are related to the Fourier transform of the function $i(\theta)$. Appendix A discusses these transforms. The exponential version is given by

$$\hat{i}(u) = \int_{-\infty}^{\infty} i(\theta)e^{-2\pi i u\theta}\,d\theta \qquad (6.98a)$$

which can be written

$$\hat{i}(u) = \text{Re}\,\hat{i}(u) + i\,\text{Im}\,\hat{i}(u) = |\hat{i}(u)|e^{-i\psi(u)} \qquad (6.98b)$$

The real part is given by

$$\text{Re}\,\hat{i}(u) = |\hat{i}(u)|\cos\psi(u) = \int_{-\infty}^{\infty} i(\theta)\cos 2\pi u\theta\,d\theta \qquad (6.98c)$$

and the imaginary part is given by

$$\text{Im}\,\hat{i}(u) = -|\hat{i}(u)|\sin\psi(u) = -\int_{-\infty}^{\infty} i(\theta)\sin 2\pi u\theta\,d\theta \qquad (6.98d)$$

Equations (6.97) can then be written

$$\begin{aligned}
\gamma_{12}(\tau, h) &= U(\tau)|\hat{i}(u)|\{\cos\left[2\pi\nu_0\tau + \Phi(\tau)\right]\cos\psi(u) + \\
&\quad + \sin\left[2\pi\nu_0\tau + \Phi(\tau)\right]\sin\psi(u)\} \\
&= U(\tau)|\hat{i}(u)|\cos\left[2\pi\nu_0\tau + \Phi(\tau) - \psi(u)\right] \\
&= \text{Re}\,\{U(t)|\hat{i}(u)|e^{i[2\pi\nu_0\tau + \Phi(\tau) - \psi(u)]}\} \\
&= \text{Re}\,\{U(\tau)e^{i[2\pi\nu_0\tau + \Phi(\tau)]}\,\hat{i}(u)\}
\end{aligned} \qquad (6.99)$$

$$u = \frac{\nu_0 h}{c}$$

The complex valued function in curly brackets in Eq. (6.99) is called the complex degree of coherence:

$$\tilde{\gamma}_{12}(\tau, h) = U(\tau)e^{i[2\pi\nu_0\tau + \Phi(\tau)]}\hat{i}\left(\frac{\nu_0 h}{c}\right) \qquad (6.100a)$$

Since $\hat{i}(0) = 1$, we have

$$\tilde{\gamma}_{12}(\tau, 0) = \tilde{\gamma}(\tau) = U(\tau)e^{i[2\pi\nu_0\tau + \Phi(\tau)]} \qquad (6.100b)$$

and since $U(0) = 1$, $\phi(0) = 0$, we have

$$\tilde{\gamma}_{12}(0, h) = \hat{i}\left(\frac{\nu_0 h}{c}\right) = \left|\hat{i}\left(\frac{\nu_0 h}{c}\right)\right| e^{-i\Psi(\nu_0 h/c)}$$

$$= \int_{-\infty}^{\infty} i(\theta)\, e^{-2\pi i \nu_0 h\theta/c}\, d\theta \qquad (6.100c)$$

We can then write

$$\tilde{\gamma}_{12}(\tau, h) = \tilde{\gamma}_{12}(\tau, 0)\tilde{\gamma}_{12}(0, h) = \tilde{\gamma}(\tau)\tilde{\gamma}_{12}(0, h) \qquad (6.100d)$$

Equation (6.100*d*) expresses what is known as the reduction property of the complex degree of coherence.

The complex valued function $\tilde{\gamma}(\tau)$ in Eq. (6.100*b*) gives the normalized autocorrelation function via the equation $\gamma(\tau) = \text{Re}\,\tilde{\gamma}(\tau)$. If we are interested in transverse spatial coherence only, we do not need the full functional form of $\gamma(\tau)$, but only its value for small τ, $\gamma(\tau) \approx \cos 2\pi\nu_0\tau$.

4. Visibility or Degree of Coherence. Once again we consider short retardation times τ and calculate the visibility

$$V(h) = \frac{I(\text{max}) - I(\text{min})}{I(\text{max}) + I(\text{min})}$$

of the fringes near the center ($\theta' = \theta_0$) of the pattern. The maximization and minimization are taken with respect to small changes in τ. We may then set $U(\tau) \approx 1$ and $\Phi(\tau) \approx 0$ and obtain, from the second equality in Eq. (6.99),

$$\tilde{\gamma}_{12}(\tau, h) \approx \left|\hat{i}_s\left(\frac{\nu_0 h}{c}\right)\right| \cos\left[2\pi\nu_0\tau - \psi\left(\frac{\nu_0 h}{(c)}\right)\right] \qquad (6.101)$$

With I given by

$$I = I(\text{tot})(1 + \gamma_{12})$$

it is easy to derive the result

$$V(h) = \left|\hat{i}\left(\frac{\nu_0 h}{c}\right)\right|$$

$$= |\tilde{\gamma}_{12}(0, h)| \qquad (6.102)$$

This provides a quantitative measure of mutual coherence and is called the degree of coherence.

In the more general situation where $U(\tau) \neq 1$, it is easy to show from Eq. (6.99) that the visibility is then given by

$$V(\tau, h) = U(\tau)\left|\hat{i}\left(\frac{\nu_0 h}{c}\right)\right| = U(\tau)|\tilde{\gamma}_{12}(0, h)|$$

$$= U(\tau)V(h) \qquad (6.102')$$

*5. **Determination of the Phase of*** $\hat{\imath}$. A measurement of the visibility $V(h)$ will give us the absolute value of $\hat{\imath}(\nu_0 h/c)$. The phase angle $\psi(\nu_0 h/c)$ can be obtained if accurate measurements of the flux density in the interference pattern are available for retardation times t near zero. According to Eq. (6.101) the zero order fringe of maximum intensity will be shifted if $\psi \neq 0$ from the position $\tau = h(\theta' - \theta_0)/c = 0$* to one where

$$\tau = \frac{h(\theta' - \theta_0)}{c} = \frac{1}{2\pi\nu_0}\psi\left(\frac{\nu_0 h}{c}\right)$$

This allows us, in principle, to determine $\psi(\nu_0 h/c)$ and to construct the complex valued function $\tilde{\gamma}_{12}(0, h) = \hat{\imath}(\nu_0 h/c)$.
A typical $V(h)$ visibility curve is shown in Fig. 6.18*a*. This is Fig. 6.16*a* replotted and applies to a uniform line source of angular width θ. The corresponding interference pattern is shown in parts

*This position is given by the center of the interference pattern observed for small values of h. It is easier to determine with a relatively broad spectral distribution than with nearly monochromatic light.

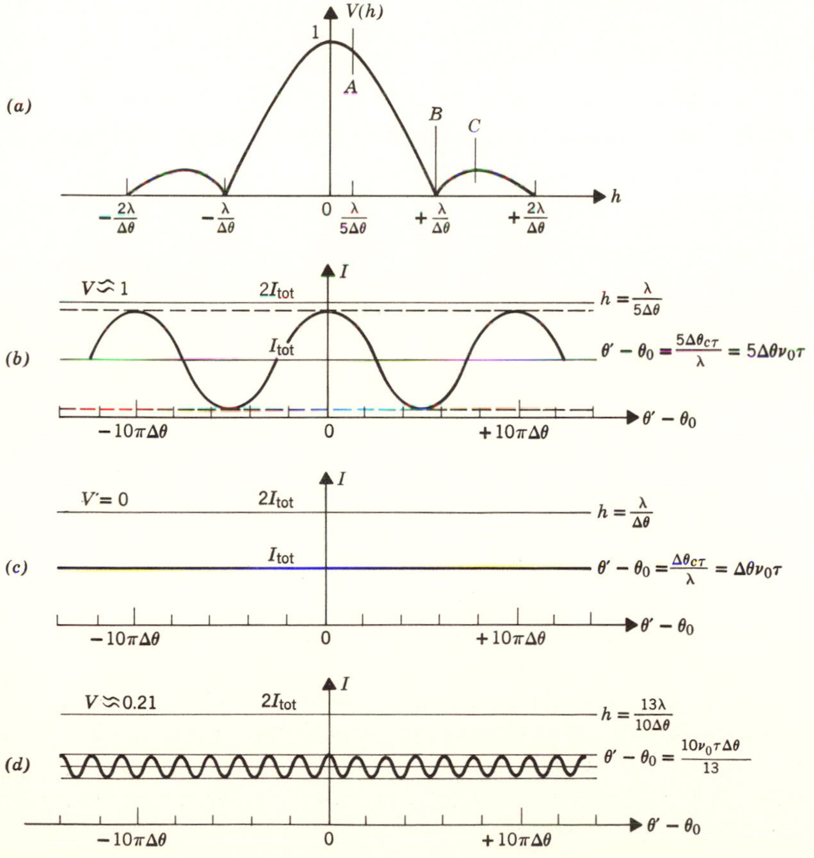

Fig. 6.18

b–d of Fig. 6.18 for three different values of h. The flux density is plotted as a function of $c\tau/h = (\theta' - \theta_0)$.

C. TRANSVERSE COHERENCE

1. The Transverse Coherence Length. We now consider the two-slit experiment from a different but equally correct point of view, namely, that of correlations between the fluctuations in the fields at S_1 and S_2. Fringes with full visibility ($V = 1$) will occur only if these fluctuations have a definite correlation, that is, only if the light at S_1 and S_2 in Fig. 6.15 is *mutually coherent*.

For extended sources we saw that $V(h)$ decreases toward zero, perhaps with a few oscillations. Small values of V correspond to a lack of mutual coherence. It is useful to define a *transverse coherence length* l_t such that for $h \gg l_t$, $V \approx 0$ and there is no coherence; for $h \ll l_t$, $V \approx 1$ and there is coherence; and for $h \sim l_t$, V is somewhere between 0 and 1 and there is partial coherence. There is no unique definition of l_t, but one possibility is discussed in Problem 6.3.10. This length l_t is quite distinct from the longitudinal coherence length $l_l = c\tau_c$, where τ_c is the coherence time of the light.

The transverse coherence length is finite because the light is emitted from a finite-sized distribution of sources. Even if the light beams from all the individual sources were to arrive in phase at S_1, they could not all arrive in phase at S_2, unless h were very small, because of the different propagation distances. When h is small, the phase at S_1 is completely correlated with that at S_2 (that is, the phase difference is a well-defined, constant quantity). When h equals the transverse correlation length, this phase difference has average fluctuations in time of about ± 1 radian. (These fluctuations occur in time intervals of order τ_c seconds. A detector whose response time T is much less than τ_c could still record an interference pattern on the screen for h greater than the transverse coherence length, but this pattern would fluctuate with time and shift, on the average, by one-half fringe in a time equal to τ_c. Hence there will be no long-term interference pattern.)

For a line source or a disk we can take the transverse coherence length l_t to be about equal to h_1 in Eq. (6.87) or Eq. (6.88). In general, we obtain the important result

$$l_t \sim \frac{\lambda}{\Delta\theta} \tag{6.103}$$

for this length, where $\Delta\theta$ is the angular width of the source. Exact definitions of l_t and $\Delta\theta$ usually yield exact equality: $l_t = \lambda/\Delta\theta$.

2. The Correlation Function. The notion of transverse coherence can be explored from a slightly different point of view, which lets us formulate the notion of correlation of fluctuations in a correct mathematical way. The electric field at a point P on the screen (Fig. 6.15*b*) at time t' will, in general, be proportional to the electric field $E_1(t' - |S_1P|/c)$ at S_1 at a time $|S_1P|/c$ earlier plus the field $E_2(t' - |S_2P|/c)$ at S_2 at a time $|S_2P|/c$ earlier; that is, we have

$$E_P(t) \propto E_1\left(t' - \frac{|S_1P|}{c}\right) + E_2\left(t' - \frac{|S_2P|}{c}\right)$$

The difference in retardation times will be denoted by τ

$$\tau = \frac{|S_2P| - |S_1P|}{c} \approx \frac{h\theta'}{c}$$

For convenience we introduce the time variable $t = t' - |S_1P|/c$, which represents the arrival time at S_1. In terms of it we obtain

$$E_P\left(t + \frac{S_1P}{c}\right) = E_1(t) + E_2(t - \tau) \tag{6.104}$$

The radiant intensity at P is proportional to the average of $E_P{}^2$. We denote this average by $\langle E_P{}^2\rangle$, and for a measurement time T we shall assume it is given by

$$\langle E_P{}^2\rangle = \frac{1}{T}\int_{t_0-T/2}^{t_0+T/2} E_P(t)^2\,dt = \langle [E_1(t) + E_2(t-\tau)]^2\rangle$$

$$= \langle E_1(t)^2\rangle + \langle E_2(t-\tau)^2\rangle + 2\langle E_1(t)E_2(t-\tau)\rangle \tag{6.105}$$

Note that the first term $\langle E_1(t)^2\rangle$ is proportional to I_1 the single-beam intensity that would be measured at P if only slit S_1 were open. A similar statement holds for the second term, which we can assume to be equal to the first term. The third term is the interference term. In the case treated earlier by these techniques, namely, that of spectroscopy with the Michelson interferometer, E_2 was the same function as E_1. Now these functions will, in general, be different.

By writing Eq. (6.105) in the form

$$\langle E_P{}^2\rangle = \langle E_1{}^2\rangle + \langle E_2{}^2\rangle + 2\sqrt{\langle E_1{}^2\rangle\langle E_2{}^2\rangle}\left[\frac{\langle E_1(t)E_2(t-\tau)\rangle}{\sqrt{\langle E_1{}^2\rangle\langle E_2{}^2\rangle}}\right]$$

we can express the intensity as follows:

$$I = I_1 + I_2 + 2\sqrt{I_1 \cdot I_2}\,\gamma_{12} \tag{6.106a}$$

here

$$\gamma_{12} = \frac{\langle E_1(t)E_2(t-\tau)\rangle}{\sqrt{\langle E_1(t)^2 E_2(t)^2\rangle}} \tag{6.106b}$$

The function γ_{12} is known as the *normalized correlation function* of E_1 and E_2. Note that it is independent of the relative magnitude of

E_1 and E_2. Its dependent variables are both the retardation time τ and the slit separation h. Since $E_2(t-\tau)$ becomes a constant times $E_1(t-\tau)$ as $h \to 0$, we must have

$$\lim_{h\to 0} \gamma_{12}(\tau, h) = \gamma(\tau) \tag{6.107}$$

where $\gamma(\tau)$ is the autocorrelation function.

For the case to be discussed shortly, we shall find $I_1 = I_2$ and hence $\langle E_1^2\rangle = \langle E_2^2\rangle$. This gives

$$I = 2I(1)\,(1+\gamma_{12})$$

$$\gamma_{12} = \frac{\langle E_1(t)E_2(t-\tau)\rangle}{\langle E_1(t)^2\rangle} \tag{6.108}$$

3. ***Some Comments.*** The definition of γ_{12} in Eq. (6.108) is equivalent to that in Eq. (6.93), as we shall show below.

In advanced treatments of the theory of partial coherence, the fields $E_1(t)$ and $E_2(t-\tau)$ are obtained from complex valued fields $\tilde{E}_1$ and $\tilde{E}_2$ known as analytic signals. Complex valued correlation functions are then defined. For example,

$$\tilde{\Gamma}_{12} = \langle \tilde{E}_1(t)\tilde{E}_2(t-\tau)^*\rangle$$

is called the mutual coherence function. Its normalized counterpart

$$\tilde{\gamma}_{12} = \frac{\langle E_1(t)E_2(t-\tau)^*\rangle}{\sqrt{\langle |E_1|^2\rangle\langle |E_2|^2\rangle}}$$

is called the complex degree of coherence. The real valued function γ_{12} is simply the real part of $\tilde{\gamma}_{12}$. [See Eq. (6.100*a*).]

The absolute value $|\tilde{\gamma}_{12}|$ is called the degree of coherence. It is identical with the visibility V discussed in Sections 6.2A and 6.3B4. The reason for this nomenclature resides in the fact that $|\tilde{\gamma}_{12}|$ gives a quantitative measure of the mutual coherence between the fields E_1 and E_2. When $|\tilde{\gamma}_{12}| = 1$, coherence is complete, and the interference term in Eq. (6.106*a*) has its maximum amplitude. On the other hand, there is complete incoherence with no interference when $|\tilde{\gamma}_{12}| = 0$. In the intermediate case when $0 < |\tilde{\gamma}_{12}| < 1$, there is partial coherence with some interference, but not the full amount a priori possible.

4. ***Evaluation of γ_{12} for an Extended Source.*** Let us consider an extended source composed of a collection of distant point sources making angle θ_j with the normal to the plane of our two slits S_1 and S_2. The fields at S_1 and S_2 will be given by sums of the form

$$E_1(t) = \sum_{j=1}^{N} E_{1,j}(t)$$

$$E_2(t-\tau) = \sum_{j=1}^{N} E_{2,j}(t-\tau)$$

where, for instance, $E_{2,j}$ is the instantaneous field at S_2 resulting from the jth source. The correlation function γ_{12} is to be calculated from Eq. (6.108):

$$\gamma_{12} = \frac{\langle E_1(t)E_2(t-\tau)\rangle}{\langle E_1(t)^2\rangle} \tag{6.108}$$

The averaging time is assumed long compared with the coherence time of the light. This means that averages of the form

$$\langle E_{1,j}(t)E_{2,k}(t-\tau)\rangle$$

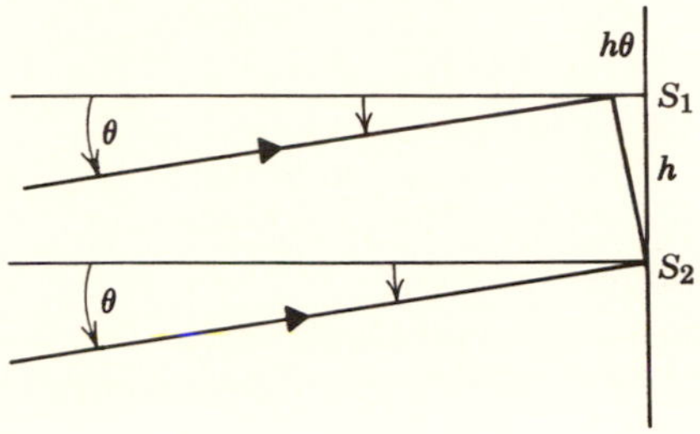

Fig. 6.19

will give zero unless $j=k$, because the phase difference will change by many radians during the measurement. The light from the jth source can then interfere only with itself. Equation (6.108) gives

$$\gamma_{12} = \frac{\sum_j \langle E_{1,j}(t)E_{2,j}(t-\tau)\rangle}{\sum_j \langle [E_{1,j}(t)]^2\rangle}$$

From Fig. 6.19 we see that for light from a distant point source at angle θ there is an extra propagation time $h\theta/c$ for the light to reach S_1. Thus

$$E_{1,j}(t) = E_{2,j}\left(t - \frac{h\theta_j}{c}\right)$$

This gives

$$\gamma_{12} = \frac{\sum_j \langle E_{2,j}(t-h\theta_j/c)E_{2,j}(t-\tau)\rangle}{\sum_j \langle [E_{2,j}(t-h\theta_j/c)]^2\rangle}$$

The precise meaning of the average used here is defined in Eq. (6.105). For long averaging times T, a steady state is achieved in which each average is independent of t_0. This means that we can replace t by $t+\text{const}$ in each term without changing the result. Thus we may replace t by $t+h\theta_j/c$ and obtain

$$\gamma_{12} = \frac{\sum_j \langle E_{2,j}(t)E_{2,j}(t-\tau+h\theta_j/c)\rangle}{\sum_j \langle E_{2,j}(t)^2\rangle} \tag{6.109}$$

The autocorrelation function $\gamma_j(\bar{\tau})$ for the jth source is defined by

$$\gamma_j(\bar{\tau}) = \frac{\langle E_{2,j}(t)E_{2,j}(t-\bar{\tau})\rangle}{\langle [E_{2,j}(t)]^2\rangle}$$

If we assume that the spectral distribution functions for all the sources are the same, they will have a common $\gamma_j(\bar{\tau}) = \gamma(\bar{\tau})$.

Equation (6.109) then becomes

$$\gamma_{12} = \frac{\sum_j \langle [E_{2,j}(t)]^2 \rangle \gamma(\tau - h\theta_j/c)}{\sum_j \langle [E_{2,j}(t)]^2 \rangle} \tag{6.110}$$

This is a weighted average of the various $\gamma(\tau - h\theta_j/c)$. The weight factor

$$\frac{\langle [E_{2,j}(t)]^2 \rangle}{\sum_j \langle [E_{2,j}(t)]^2 \rangle}$$

gives the relative flux density at S_2 and hence at the observing screen if one slit is closed.

For a continuous distribution the sum goes over to the integral

$$\gamma_{12} = \int_{-\infty}^{\infty} i(\theta)\gamma(\tau - h\theta/c)\, d\theta$$

which is identical with Eq. (6.93) since $\tau = h\theta'/c$.

REFERENCES FOR SECTION 6.3

A. T. Forrester, "On Coherence Properties of Light Waves," *Am. J. Phys.*, **24**, 192 (1956).

M. Françon, *Optical Interferometry*. Chapter 8.

F. A. Jenkins and H. E. White, *Fundamentals of Optics*. Sections 16.7 and 16.8.

J. M. Stone, *Radiation and Optics*. Sections 7-13, 13-9, and 13-10.

C. J. Troup, *Optical Coherence Theory—Recent Developments*.

L. Levi, *Applied Optics*. Volume I, Chapter 2.

M. Born and E. Wolf, *Principles of Optics*, Section 7.3 and Chapter 10.

PROBLEMS

6.3.1 Show explicitly that the zero-order fringe moves up an angle $\theta' = \theta$ when the source moves down an angle θ, as described in the second paragraph of 6.3A2. Include a detailed derivation of Eq. (6.75).

6.3.2 Derive the equivalent of Eq. (6.79) for three point sources.

6.3.3 A screen S_1 is placed in front of a large diffuse incoherent source of quasimonochromatic light (wavelength λ). It has two openings in the form of long narrow slits a distance h_1 apart. At a distance $L_1 \gg h_1$ from S_1 is another screen S_2 with a pair of similar long, narrow slits with separation h_2 oriented parallel to the first pair of slits. The interference fringes are observed on a screen S_3 a distance $L_2 \gg h_2$ from S_2. For fixed values of L_2, h_1, and h_2, at what values of L_1 will the fringes disappear?

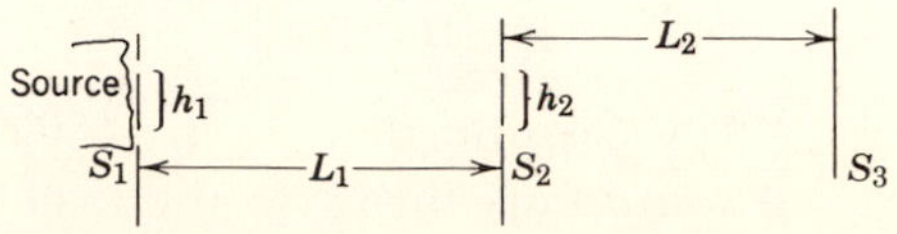

6.3.4 The need for having long distances L_1 and L_2 in Problem 6.3.3 can be avoided by using lenses with S_1 and S_3 at the focal planes as shown. For fixed h_2, f_1, and f_2, λ, at what values of h_1 will the fringes on S_3 disappear?

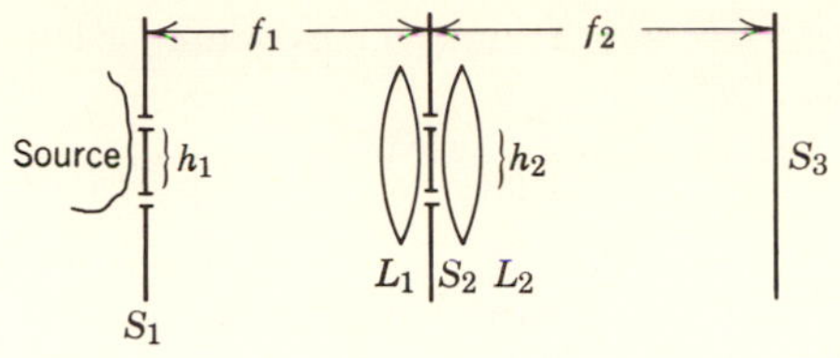

6.3.5 A zirconium lamp has an arc 10^{-3} inches in diameter. It is used with a filter that limits the wavelength to the vicinity of 550 nm. You ask: Will it be suitable as a source in an optical image-processing system?

Some of the features of such a system are indicated in the figure. One requires almost complete transverse coherence of the incident light across the object. Suppose that the focal length f of L_1 is 20 cm and its diameter is 3 cm. Will the source have the necessary coherence?

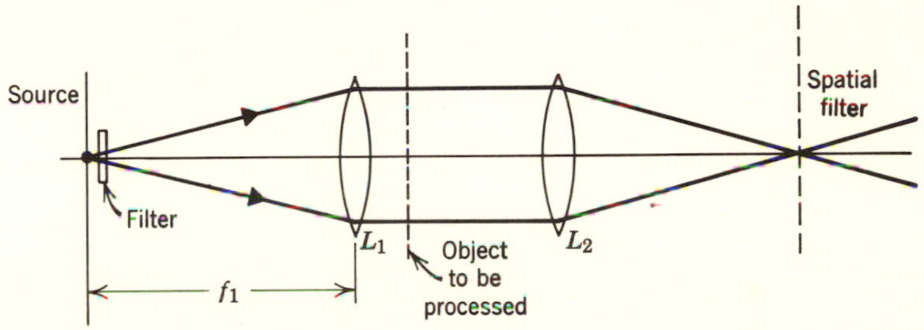

6.3.6 Consider a double star. The components have equal intensities. Their separation is so small that the visibility of the fringes at the maximum-slit separation h_{max} is still close to unity. What would this visibility be if h_{max} is 1/10 of the value needed to give the first zero in visibility?

6.3.7 What is the smallest angular diameter of stars that can be measured by the Michelson stellar interferometer using light of $\lambda \approx$ 550 nm and $h_{max} = 6$ m?

6.3.8 Find the complex degree of coherence $\tilde{\gamma}_{12}(0, h)$ for (a) two equally intense point sources with angular separation $\Delta\theta$, (b) a uniform line source having width $\Delta\theta$. Assume monochromatic light of wavelength λ.

6.3.9 Show that the equal displacement of all sources in a Young's double-slit experiment by the angle θ_0 results in a change in phase ψ of the complex degree of coherence given by

$$\Delta\psi = \frac{2\pi\nu\phi_0 h}{c}$$

6.3.10 We can define the angular spread of the source by

$$\Delta\theta \equiv \frac{1}{\int_{-\infty}^{\infty} i^2(\theta)\, d\theta}$$

and the transverse coherence length by

$$l_t = 2\int_0^{\infty} \left|\hat{i}\left(\frac{\nu_0 h}{c}\right)\right|^2 dh = 2\int_0^{\infty} V(h)^2\, dh$$

a. Show that $l_t = \lambda/\Delta\theta$.
b. Show that for a uniform line source extending from θ_1 to θ_2, $\Delta\theta$ as defined above equals $(\theta_2 - \theta_1)$.

6.3.11 In the text we showed that in principle one may determine the amplitude and phase of the complex valued function

$$i(u), \qquad u = \frac{\nu_0 h}{c} = \frac{h}{\lambda}$$

where $i(\theta)$ is the angular distribution function for the source distribution as projected onto the plane of the paper in Fig. 6.16. From the known function $\hat{i}(u)$, we can compute $i(\theta)$ by calculating a Fourier integral

$$\hat{i}(\theta) = \int_{-\infty}^{\infty} \hat{i}(u) e^{2\pi i u\theta}\, du$$

Since h is positive, so is u. The above integral can be converted to a sum of integrals involving $\mathrm{Re}\,\hat{i}(u)$ and $\mathrm{Im}\,\hat{i}(u)$ for positive u only. Do this.

6.3.12 This problem uses the methods of calculation of optical path length used for Fraunhofer diffraction in Section 7.2 below. The notation will be the same as that in Section 7.2.

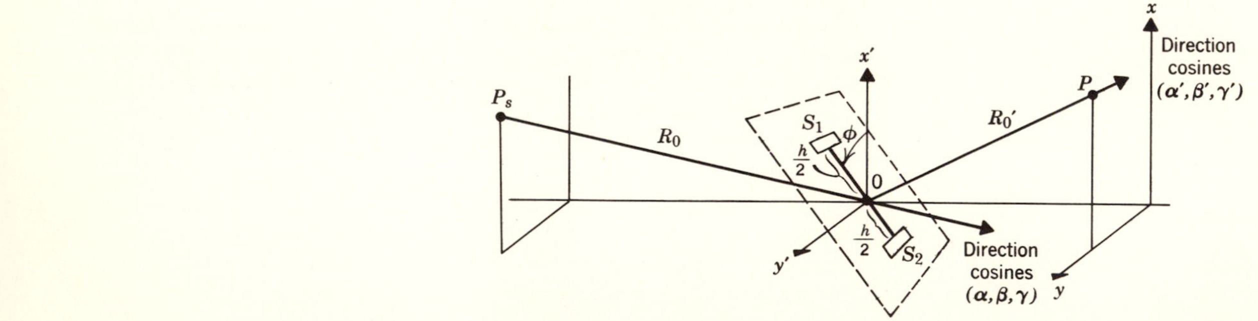

Let a representative source point be at P_s. The line P_sO has direction cosines (α, β, γ). The observation point is at P. The line OP has direction cosines $(\alpha', \beta', \gamma')$. The line between slits S_1S_2 is rotated out of the plane of the paper through the angle ϕ. The limit $R_0 \gg h$, $R_0' \gg h$ will apply.

a. Show that

$$\tau \equiv \frac{|S_2P| - |S_1P|}{c}$$

equals $h(\alpha' \cos\phi + \beta' \sin\phi)$

b. Show that

$$\tau_s \equiv \frac{|P_sS_1| - |P_sS_2|}{c}$$

equals $h(\alpha \cos \phi + \beta \sin \phi)$.

c. Find the loci of constant $\bar{\tau} = \tau - \tau_s$ in the xy plane.

d. Show that for quasimonochromatic light of frequency ν a normalized distribution of sources $i(\alpha, \beta)$ will yield a flux density in the xy plane proportional to $1 + \gamma_{12}(\tau, h, \phi)$, where $\gamma_{12} = \text{Re}\, \tilde{\gamma}_{12}(\tau, h\ \phi)$, $\tilde{\gamma}_{12}(\tau, h, \phi) = \tilde{\gamma}(\tau)\tilde{\gamma}_{12}(0, h, \phi)$, with our previous definition of $\tilde{\gamma}(\tau)$ and

$$\gamma_{12}(0, h, \phi) = \hat{i}\left(\frac{h}{\lambda}\cos\phi, \frac{h}{\lambda}\sin\phi\right)$$

Here $\hat{i}(u, v)$ is the two-dimensional Fourier transform of $i(\alpha, \beta)$:

$$\hat{i}(u, v) = \iint^{\infty} i(\alpha, \beta) e^{-2\pi i(\alpha u + \beta v)}\, d\alpha\, d\beta$$

You may assume that $i(\alpha, \beta)$ is centered about $\alpha = 0, \beta = 0$.

e. Discuss how you would determine the real valued function $i(\alpha, \beta)$ from a study of the interference pattern as a function of h and ϕ.

6.3.13 The "uncertainty principle" of quantum mechanics says that the combined uncertainty in a given component of momentum, say p_x, and of position, x, cannot be less than Planck's constant h. Thus, for minimum uncertainty we have $\Delta x \Delta p_x = h$, and similarly for the other components. The magnitude of the momentum of a photon is given by $p = h/\lambda = h\nu/c$. Let a nearly collimated beam of photons be propagating along the z axis. If the frequency spread is $\Delta\nu$, show that the minimum position spread Δz is $l_l = c\tau_c$.

Suppose that the lack of collimation is due to a distant source of width $\Delta\theta_x$. Show that the minimum transverse position spread Δx is l_t.

6.4 INTENSITY FLUCTUATIONS IN INTERFERENCE SPECTROSCOPY

Gaussian light results from the superposition of the signals from a great many independent optical sources. For quasimonochromatic Gaussian light the instantaneous amplitude and phase of the resultant signal will fluctuate because of the randomly varying phase (and perhaps amplitude) relationships among the individual component signals. A large fluctuation in the phase of the resultant signal is likely to be accompanied by a large fluctuation in the amplitude, as shown in Fig. 6.12. Thus we are led to expect that the behavior of the amplitude fluctuations will tell something about the phase fluctuations, or more precisely about the correlation functions for the optical signals.

This expectation is correct for Gaussian light, as we shall show explicitly below. If the light has mutual coherence at two points S_1 and S_2 in space, there is a positive correlation in intensity fluctuations. If there is a positive fluctuation about the mean at S_1, then there is more likelihood of a positive fluctuation about the mean at S_2 than of a negative fluctuation. Similarly a negative fluctuation at S_2 will accompany a negative fluctuation at S_1 more often than will a positive fluctuation at S_2.

We can use this effect to measure spatial (and temporal) coherence. In astronomy the spatial coherence of the light from a stellar source depends on its size and shape. The technique of correlation of intensity fluctuations can be used to great advantage to measure the angular diameter of stars.

A. THE INTENSITY FLUCTUATION CORRELATION INTERFEROMETER

1. ***Basic Experimental Arrangement***. A photomultiplier or other energy-sensitive detector of light will give an output signal proportional to the optical flux incident onto it averaged over a time interval T. We shall continue to abandon the precise terminology of Chapter 4 and refer to the flux as "intensity." We assume that we have full spatial coherence across the light-sensitive area of each detector. It is then possible to speak of a well-defined "instantaneous" intensity $I(t)$ at a detector given by an extremely fast average of E^2. We assume at first that the averaging time T is much less than the coherence time τ_c of the light. We block the dc or average component of the detector output, which will be proportional to the long-term average intensity $\langle I \rangle$, and detect only the ac or fluctuating component, which is proportional to the intensity fluctuation $\delta I(t) = I(t) - \langle I \rangle$ (Fig. 6.20).

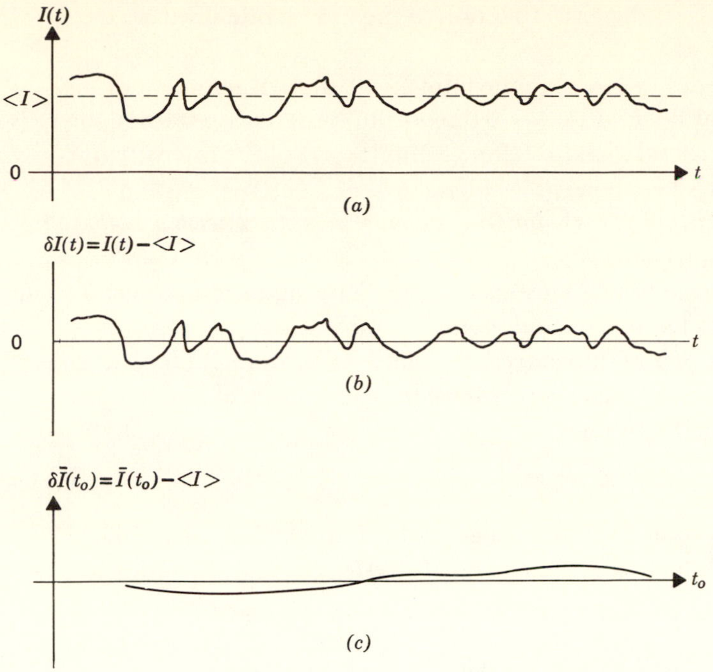

Fig. 6.20. Schematic representation of (*a*) the instantaneous intensity $I(t)$, (*b*) the instantaneous intensity fluctuation $\delta I(t)$, and (*c*) the smoothed intensity fluctuation.

Formerly when we wanted to study the mutual coherence between two points S_1 and S_2 we set up pinholes or slits and tried a Young interference experiment (Fig. 6.15). Now we put detectors PM1 and PM2 at S_1 and S_2, as shown in Fig. 6.21*a*. The output from

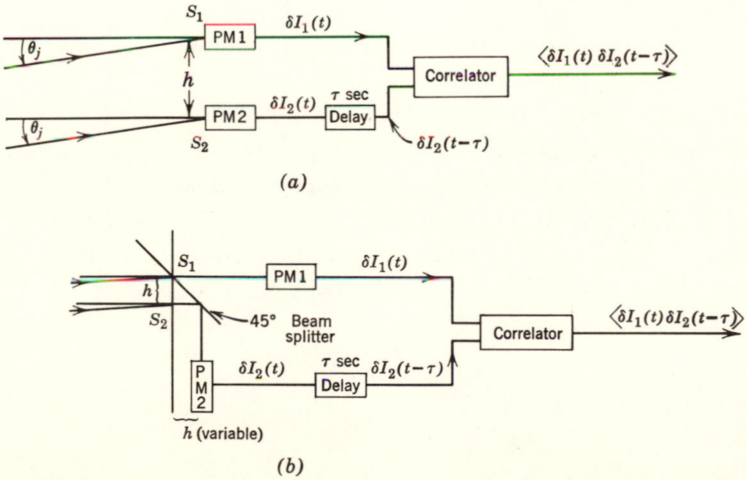

Fig. 6.21. Devices to correlate intensity fluctuations at S_1 and S_2. (*a*) S_1 and S_2 separated; $h > 0$. (*b*) S_1 and S_2 identical as $h \to 0$.

PM2 is delayed τ seconds by passing through a variable delay line. The resulting signals, which will be proportional to $\delta I_1(t)$ and $\delta I_2(t-\tau)$, are then fed into a correlator, which is a device that takes the long-term average of their product. Hence its output is proportional to

$$\langle \delta I_1(t)\delta I_2(t-\tau)\rangle$$

and this is studied as a function of the independent variables τ and h.

If it is desired to go to the limit $h = 0$, the arrangement of Fig. 6.21*b* may be used. At $h = 0$ the output is then proportional to the autocorrelation function $\langle \delta I(t)\delta I(t-\tau)\rangle$ of the intensity fluctuations, since then $I_1(t) = I_2(t) = I(t)$.

2. $\langle \delta I_1(t)\, \delta I_2(t-\tau)\rangle$ *for Quasimonochromatic Thermal Light.* The calculation of the correlation function is quite involved and will not be given here. One expresses the light signal as a superposition of signals from many independent sources and proceeds to calculate δI and the desired correlation function. Under the assumption that all sources have the same spectral distribution function, the result is

$$\langle \delta I_1(t)\delta I_2(t-\tau)\rangle = \langle I\rangle^2 |\tilde{\gamma}_{12}(\tau, h)|^2 \tag{6.111}$$

Here $\tilde{\gamma}_{12}$ is the "complex degree of coherence" of Eq. (6.100*a*). By the reduction property, Eq. (6.100*d*),

$$\tilde{\gamma}_{12}(\tau, h) = \tilde{\gamma}(\tau)\tilde{\gamma}_{12}(0, h)$$

From Eqs. (6.102′) and (6.100*b*) we have the result

$$|\tilde{\gamma}_{12}(\tau, h)| = U(\tau)V(h)$$

Thus, Eq. (6.111) may also be written

$$\langle \delta I_1(t)\delta I_2(t-\tau)\rangle = \langle I\rangle^2 U(\tau)^2 V(h)^2 \tag{6.112}$$

This is the basic result relating intensity fluctuation correlations to coherence properties.

3. Extension to Long Averaging Times. There is a practical difficulty neglected in the derivation of Eq. (6.112). It stems from our inability to perform the required "instantaneous" average for the intensity assumed in the above derivation. The electronic circuits simply cannot respond rapidly enough. Instead we are more likely to measure a smoothed intensity given by an average similar to

$$\bar{I}(t_0) \equiv \frac{1}{T}\int_{-T/2}^{+T/2} I(t+t_0)\, dt \tag{6.113}$$

where $I(t+t_0)$ is the instantaneous average used above. The deviation $\delta\bar{I}$ of $\bar{I}$ from its long-term average $\langle I\rangle$ is much smoother, as shown in Fig. 6.20*c*. It represents the signal correlated by the apparatus of Fig. 6.21,

$$\langle \delta\bar{I}_1(t_0)\delta\bar{I}_2(t_0-\tau)\rangle \equiv \langle [\bar{I}_1(t_0) - \langle I\rangle][\bar{I}_2(t_0-\tau) - \langle I\rangle]\rangle$$

where $\langle\cdot\cdot\cdot\rangle$ now represents a very long-term average over t_0.

When these changes are made, it can be shown that Eq. (6.112) must be replaced by the expression

$$\frac{\langle \delta\bar{I}_1(t_0)\delta\bar{I}_2(t_0-\tau)\rangle}{\langle I\rangle^2} = V(h)^2\,\frac{\xi(T,\tau)}{T} \tag{6.114a}$$

where

$$\xi(T,\tau) = T^{-1}\int_{-T}^{+T} (T-|r|)\,[U(\tau+r)]^2 dr \tag{6.114b}$$

In most real situations the measurement time T is much longer than the coherence time τ_c. Then Eq. (6.114*b*) gives an approximate result

$$\begin{aligned}\langle \delta\bar{I}_1(t_0)\delta\bar{I}_2(t_0-\tau)\rangle &= \langle I\rangle^2 V(h)^2\frac{\tau_c}{T}, \quad \text{for} \quad \tau < T \\ &= 0, \qquad\qquad\qquad \text{for} \quad \tau > T\end{aligned} \tag{6.114c}$$

These last relations hold except when $|(|\tau| - T)| \sim \tau_c$.

The reduction in Eq. (6.114*c*) by the factor $\tau_c/T \ll 1$ makes the smoothed fluctuations much more difficult to observe than the instantaneous fluctuations given by Eq. (6.112). This conclusion is also suggested by a comparison of Fig. 6.20*c* to 6.20*b*. But note that Eq. (6.114*c*) says that we can still measure the visibility function

$$V(h) = \left|\hat{i}\left(\frac{\nu_0 h}{c}\right)\right|$$

4. Some Experiments. The correlation of intensity fluctuations in a light beam was first demonstrated experimentally by R. Hanbury Brown and R. Q. Twiss in 1956. Their apparatus was similar to that shown in Fig. 6.21*b*. The source was a mercury lamp-filter combination with a coherence time τ of about 1 nanosecond (10^{-9} sec), and the electronics had an effective integration time T of about 40 nsec. They verified the behavior predicted by Eq. (6.114*c*).

Once they had established the existence of correlated intensity fluctuations using laboratory light sources, Hanbury Brown and Twiss applied the technique to stellar interferometry. The arrangement is essentially that of Fig. 6.21*a*, with the phototubes placed at the focal points of two large mirrors as shown in Fig. 6.22. (Mirror diameters up to 7 m have subsequently been used.) The mirrors form images of the desired star at the two photocathodes. These images need not be of high optical quality. The telescope separation h may be varied by moving them on rails. Baselines as long as 25 m have been used so far, but much larger values are practical.

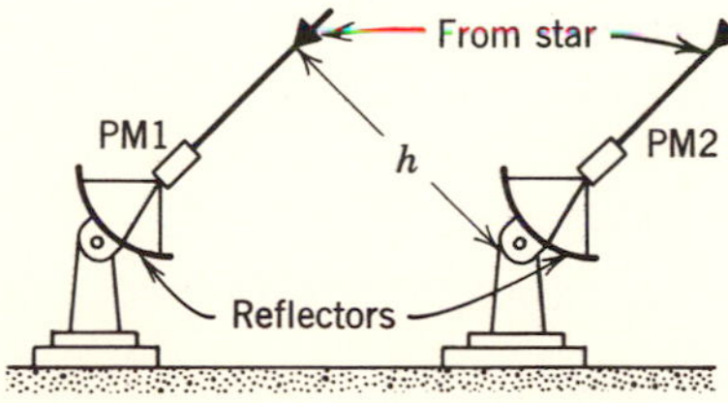

Fig. 6.22

In the Michelson stellar interferometer we must hold dimensional tolerances to a fraction of a wavelength; otherwise the interference pattern "washes out." In an intensity fluctuation interferometer the requirements are enormously relaxed. Mechanical dimensions should be maintained to within a small fraction of l_t or of cT, whichever is less. Timing errors from transit time fluctuations in the electronic circuitry should be much less than T sec in duration.

B. QUANTUM CONSIDERATIONS

A second effect neglected in the treatment of Section 6.4A involves the inherent quantum nature of light and of the process by which light intensities are measured. Electromagnetic radiation arrives at a detector in quanta of energy of amount $h\nu$, where h is Planck's constant. This fact causes extra fluctuations in intensity beyond those already discussed and introduces an important source of "noise."

1. Quantum Counter. For definiteness, assume that we use a photoelectric detector, such as a photomultiplier, as a quantum counter where it counts individual photoelectrons. We assume that all noise sources such as dark current, Johnson noise, and amplifier shot noise have been eliminated, so that the only source of "noise" or fluctuations in the photoelectric current is the quantum nature of the photoelectric process or, what is the same thing, the fact that the current results from the flow of individual charged particles.

We let $I(t)$ be the short-time average flux reaching the detector. Then in a small time interval dt energy $I(t)\,dt$ has arrived at the detector. Because of the quantum nature of light, this energy arrives in quanta or "photons" of size $h\nu$, where ν is the frequency of the light. Each quantum may emit a photoelectron, which we shall assume will be detected with unit probability by the associated electronic circuitry. In a photomultiplier each photoelectron is multiplied to give a charge pulse of about 10^6 electrons that can be collected in a few nanoseconds (1 nsec $= 10^{-9}$ sec). The probability that 1 photon will yield 1 photoelectron is called the "quantum yield" α.

When energy $I(t)\,dt$ arrives at the detector, the number $dn(t)$ of photoelectrons emitted will fluctuate because of the inherent statistical nature of the photoelectric process. The "expectation value" or mean value over many experiments will be denoted by $[dn(t)]$ and is given by

$$[dn(t)] = \frac{\alpha I(t)\,dt}{(h\nu)} \tag{6.115a}$$

The mean number of photoelectrons detected over a time interval equal to the averaging time T of the detector is

$$[\bar{n}(t_0)] = \int_{t_0-T/2}^{t_0+T/2} [dn(t')] = \frac{\alpha}{h\nu} \int_{t_0-T/2}^{t_0+T/2} I(t')\, dt' = \frac{\alpha}{h\nu} T\bar{I}(t_0) \qquad (6.115b)$$

For thermal light the intensity $I \sim A^2$ is a Gaussian random variable, as discussed in connection with Fig. 6.12. Detailed investigations of photon statistics have shown that in this case it is not necessary to distinguish the instantaneous value $\bar{n}(t_0)$ from the mean value $[\bar{n}(t_0)]$. The fluctuations in $\bar{n}(t_0)$ may then be treated as due entirely to fluctuations in $\bar{I}(t_0)$. We therefore proceed by using

$$\bar{n}(t_0) = \frac{\alpha T\bar{I}(t_0)}{(h\nu)} \qquad (6.115c)$$

The long-term average of Eq. (6.115c) gives

$$\langle n \rangle = \langle n \rangle = \frac{\alpha}{h\nu} T\langle \bar{I} \rangle$$

We now consider the number of counts n_1 and n_2 measured by the counters PM1 and PM2 in Fig. 6.21. The fluctuations will be given by

$$\delta\bar{n}_1(t_0) = \bar{n}_1(t_0) - \langle n \rangle = \frac{\alpha}{h\nu} T[I_1(t_0) - \langle I \rangle] = \frac{\alpha T}{h\nu} \delta\bar{I}_1(t_0)$$

and

$$\delta\bar{n}_2(t_0 - \tau) = \frac{\alpha T}{h\nu} \delta\bar{I}_2(t_0 - \tau)$$

We may then calculate the correlation function for deviation of electron counts

$$\langle \delta\bar{n}_1(t_0)\delta\bar{n}_2(t_0 - \tau)\rangle = \left(\frac{\alpha T}{h\nu}\right)^2 \langle \delta\bar{I}_1(t_0)\delta\bar{I}_2(t_0 - \tau)\rangle$$

If we use Eq. (6.114*a*), this becomes

$$\begin{aligned} \langle \delta n_1(t_0)\delta n_2(t_0 - \tau)\rangle &= \left(\frac{\alpha T\langle I \rangle}{h\nu}\right)^2 V^2(h)\, \frac{\xi(T, \tau)}{T} \\ &= \langle n \rangle^2\, V^2(h)\, \frac{\xi(T, \tau)}{T} \end{aligned} \qquad (6.116)$$

Equation (6.116) says that for values of h and τ giving $V^2\xi > 0$, there is a positive correlation between the excess counts at PM1 and those at PM2. Thus a count at PM1 is more likely to be accompanied by a count at PM2 than would be expected on the basis of pure chance. Such correlated counts are called *coincidences*.

2. ***A Coincidence Experiment.*** Suppose that T is short enough so that the average number of counts $\langle n\rangle$ in time T is much less than unity. The function $\bar{n}_1(t_0)$ is then shown schematically in the top plot of Fig. 6.23. If a photoelectron is emitted at time t_1, then $\bar{n}_1(t_0)$ is unity over a time interval of length T centered about $t_0 = t_1$.

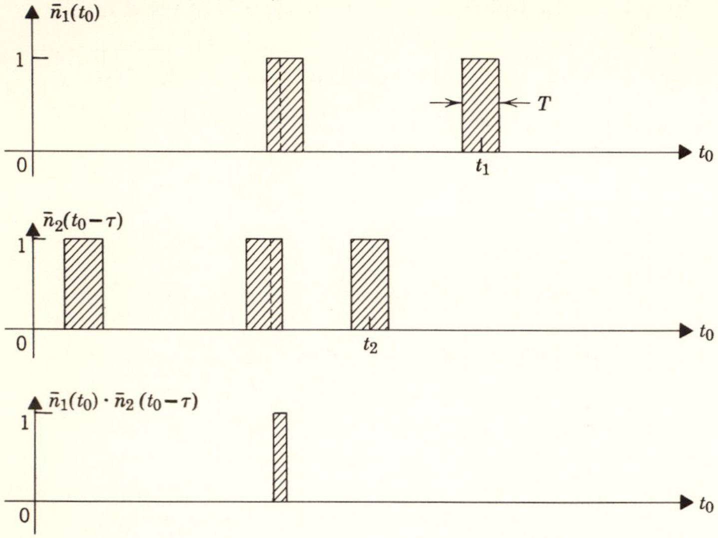

Fig. 6.23

Under these conditions the product $\bar{n}_1(t_0)\bar{n}_2(t_0-\tau)$ then gives unity when both PM1 and PM2 have a count, and zero otherwise. Such a double count is called a coincidence. It will occur whenever $|t_1-t_2| \leqslant T$ (Fig. 6.23), and the width of the overlap pulse on a plot of $\bar{n}_1(t_0)\bar{n}_2(t_0-\tau)$ versus t_0 will be $T-|t_1-t_2|$, which ranges from 0 to T and has an average of $T/2$. The average number of coincidences per second will therefore be

$$R_c = \frac{2\langle \bar{n}_1(t_0)\bar{n}_2(t_0-\tau)\rangle}{T}$$

$$= \frac{2\langle n\rangle^2}{T}\left[1+V(h)^2\frac{\xi(T,\tau)}{T}\right] \tag{6.117a}$$

If $\bar{n}_1$ and $\bar{n}_2$ were purely random rather than correlated, there would be an accidental coincidence rate given by

$$R_a = \frac{2\langle n\rangle^2}{T} = 2TR_1^{\,2} \tag{6.117b}$$

where

$$R_1 = \frac{\langle n\rangle}{T} \tag{6.117c}$$

is the average rate of detection of single photoelectrons. Then we can write

$$R_c = R_a[1+\eta], \qquad \text{where } \eta = V(h)^2 \frac{\xi(T,\tau)}{T} \tag{6.117d}$$

Here $[1+\eta]$ is the enhancement factor resulting from correlations.

If we count coincidences for S seconds, there would be an average of

$$\overline{N}_a = SR_a = \frac{2S\langle n\rangle^2}{T} = 2STR_1^2 \tag{6.118a}$$

accidental coincidences and

$$\overline{N}_c = SR_c = \overline{N}_a[1+\eta] \tag{6.118b}$$

coincidences overall.

How accurate will a given determination of N_a or N_c be? We expect the coincidences themselves to be uncorrelated random events. Then the standard deviation for a single determination of N_a or N_c is conventionally estimated to be $\sqrt{N_a}$ or $\sqrt{N_c}$, respectively. When N is not too small, approximately 67 per cent of the individual determinations of N will be within the range $N \pm \sqrt{N}$ and over 95 per cent within the range $N \pm 2\sqrt{N}$.

3. ***Noise Considerations.*** In order to measure η within any certainty, we would like to have the difference $\overline{N}_c - \overline{N}_a$ be an order of magnitude larger than $\sqrt{\overline{N}_c}$. (If N_a has been previously determined to high accuracy, then the error in $(N_c - N_a)$ will be that of N_c, namely, $\pm\sqrt{\overline{N}_c}$.) This result can be expressed in the form

$$N_c - N_a = \eta\overline{N}_a \pm \sqrt{1+\eta}\,\sqrt{\overline{N}_a} \tag{6.119}$$

The *signal-to-noise* ratio SNR is defined as the mean value of the signal divided by its standard deviation:

$$\text{SNR} = \frac{\eta}{\sqrt{1+\eta}}\sqrt{\overline{N}_a} \tag{6.120}$$

In the slow detector limit where $\tau_c \ll T$, $\eta = V(h)^2\tau_c/T \ll 1$, and we have

$$\text{SNR} = V(h)^2\left(\frac{2S}{T}\right)^{1/2}(R_1\tau_c) \tag{6.121}$$

For nonlaser thermal light the product $R_1\tau_c$ will be 10^{-2} or less.

For a numerical estimate, take $R_1\tau_c = 10^{-3}$, $V(h) = 0.1$, $T = 10$ nsec. Then for SNR $= 10$, $S = 5000$ sec $\cong 1\frac{1}{2}$ hours. This is a typical value for the counting time S. But note that for SNR $= 1$, $S = 50$ sec.

In the fast detector limit when $T \ll \tau_c$ we have $\eta = V(h)^2U(\tau)^2$, and (6.120) can be written

$$\text{SNR} = \frac{[V(h)U(\tau)]^2}{\sqrt{1+V^2U^2}}\left(\frac{2ST}{\tau_c^2}\right)^{1/2}(R_1\tau_c) \tag{6.122}$$

This uncertainty in the result of a single measurement of $(N_c - N_a)$ is due to the quantum particle nature of light and is often called photon shot noise. Since the same intrinsic quantum processes are involved in measurements of the correlation of intensity fluctuations, where individual photoelectrons are not explicitly counted, this type of noise must nevertheless be present. Since a measurement of $(N_c - N_a)$ is essentially equivalent to a determination of $\langle \delta \bar{I}_1(t') \delta \bar{I}_2(t'-t) \rangle$, the latter will also have a signal-to-noise ratio given by Eq. (6.120), (6.121), or (6.122), where S is the time interval for the determination of the average $\langle \cdots \rangle$ and T is that for determination of $\bar{I}$.*

4. The Degeneracy Parameter. The average "intensity" will be given by

$$\langle \bar{I} \rangle = E_e\,(\text{inc})\,A$$

where E_e(inc) is the incident radiant flux density (radiant incidence) and A is the sensitive area of the detector projected onto a plane perpendicular to the light beam. If a light-gathering mirror or lens is used as in Fig. 6.22, then A must be its projected area. By analogy with Eq. (6.117*c*) we may write

$$R_1 = \frac{\alpha}{h\nu} \langle \bar{I} \rangle = \frac{\alpha}{h\nu} E_e(\text{inc})\,A$$

If the source subtends the solid angle $\Delta\Omega$ as seen from the detector and has a total radiance L_e for light radiated in the direction of the detector, then E_e is simply given by

$$E_e = L_e \Delta\Omega$$

We may then relate $\Delta\Omega$ to the so-called area of coherence A_c at the detector by the equation

$$A_c = \frac{\lambda^2}{\Delta\Omega} \tag{6.123}$$

This is a generalization to two dimensions of the notion of coherence length $l_t = \lambda/\Delta\theta$ for a source of angular width $\Delta\theta$. There will be a definite correlation between the electric fields at two points within the area of coherence.

Thus we may write.

$$R_1 \tau_c = \alpha \left(\frac{L_e}{h\nu} \tau_c \right) \lambda^2 \frac{A}{A_c}$$

$$= \alpha \cdot \delta \cdot \frac{A}{A_c} \tag{6.124a}$$

*This result, of course, still assumes that all other sources of noise, except photon shot noise, are negligible.

This last equality introduces the *degeneracy parameter* δ. Since L_e is the rate of radiant energy emitted by the source per unit time, per unit area, and per unit solid angle, we may interpret

$$\delta = \left(\frac{L_e}{h\nu}\right) \tau_c \lambda^2 \tag{6.124b}$$

as the average number of light quanta or photons (of a definite polarization) emitted into unit solid angle in the direction of the detector from an area of the source equal to λ^2 in a time interval equal to the coherence time τ_c. Since $\lambda^2 = \Delta\Omega A_c$, we can also interpret δ as the average number of photons that arrive on a surface of area A_c at the detector from the entire source in the time interval τ_c, or as the number of photons at any given time that fill the "volume of coherence"

$$V_c = A_c \tau_c c \tag{6.125}$$

For quasimonochromatic light that is in thermal equilibrium at temperature T (black body radiation) the degeneracy parameter is given by a formula due to Planck,

$$\delta = \frac{1}{\exp\,(h\nu/k_B T) - 1} \tag{6.126}$$

where h = Planck's constant and k_B = Boltzmann's constant. This can also be written as in Eq. (4.18),

$$\delta = \frac{1}{\exp\,(c_2/\lambda T) - 1} \tag{6.126$'$}$$

$c_2 = hc/k_B = 1.438$ cm-deg. Here δ is ordinarily very much less than unity.

For light from a laser δ can be as high as 10^{12}. Laser light, however, is ordinarily not thermal light in the technical sense required for the above discussion of fluctuations to apply. There is a method of converting light from a laser into "quasithermal light" that has both a large value of δ and the right statistical properties by passing it through a moving random medium such as a moving ground glass plate. This technique is described in the paper by Martienssen and Spiller mentioned in the references.

REFERENCES FOR SECTION 6.4

Treatments at about the same level as this text

A. T. Forrester, "On Coherence Properties of Light Waves," *Am. J. Phys.*, **24**, 192 (1956).

W. Martienssen and E. Spiller, "Coherence and Fluctuations in Light Beams," *Am. J. Phys.*, **32**, 919 (1964).

Max Garbuny, *Optical Physics*, Section 6.2.5.

M. Françon, *Optical Interferometry*, Chapter IX.

Advanced reviews of the subject of coherence

L. Mandel and E. Wolf, "Coherence Properties of Optical Fields," *Rev. Mod. Phys.*, **37**, 231 (1965). Parts of this article are at the level of the text. Many references to original papers.

J. R. Klauder and E. C. G. Sudarshan, *Fundamentals of Quantum Optics*.

C. J. Troup, *Optical Coherence Theory – Recent Developments*.

Some original papers

R. Hanbury Brown and R. Q. Twiss, "A New Type of Interferometer for Use in Radio Astronomy," *Phil. Mag.*, **45**, 663 (1954); "Correlation between Photons in Two Coherent Beams of Light," *Nature*, **177**, 27 (1956); "A Test of a New Type of Stellar Interferometer on Sirius," *Nature*, **178**, 1046 (1956); "The Question of Correlation Between Photons in Coherent Light Rays," *Nature*, **178**, 1447 (1956). See also the Appendix to this last article, written by E. M. Purcell.

PROBLEMS

6.4.1 Calculate degeneracy parameters for light from the following sources, assumed to be black bodies:

a. A star at 10,000°K for λ near 500 nm.
b. A gas discharge at 3000°K for λ at 546 nm.

6.4.2 An intensity fluctuation interferometer is used to measure stellar diameters, in particular the diameter of the star in Problem 6.4.1A. The detectors each have area $A = 10^{-2}A_c$ (A_c = area of coherence), quantum yield $\alpha = 0.10$, and resolving time $T = 15$ nsec. What is the minimum value of V that can be measured in a time $S = 3000$ sec with a signal-to-noise ratio of unity? Assume the slow detector limit.

6.4.3 The derivations in this section assumed that the light was in a definite state of polarization. Unpolarized light is an equal mixture of two mutually incoherent components in two independent states of polarization. For instance, these two states may be taken to be plane-polarized along two mutually perpendicular directions. Let R_1 (unpol) be the average rate of emission of photoelectrons from each phototube when unpolarized light is used and R_1(pol) the rate when a polarizer is inserted in the beam. Then

$$R_1(\text{unpol}) = 2R_1(\text{pol}) \tag{1}$$

and

$$\eta(\text{unpol}) = \tfrac{1}{2}\eta(\text{pol}) \tag{2}$$

a. Explain why Eq. (2) should hold on the basis of the statement at the beginning of this problem.
b. Find the expression corresponding to Eq. (6.121) for the SNR for unpolarized light.

Diffraction of light refers to the propagation of light waves in a manner not predicted by the laws of geometrical optics. The wave nature of light is involved in an essential way, but completely rigorous solutions of the wave equation are not necessary for an understanding of diffraction. Many of its phenomena can be described quantitatively by a mathematical application of Huygens' principle that is an approximate consequence of the wave equation. In this chapter we develop the theory and then apply it to Fraunhofer diffraction by a wide variety of apertures. More material on diffraction will be found in Chapters 8 and 9.

CHAPTER 7

DIFFRACTION I—MAINLY FRAUNHOFER DIFFRACTION

CHAPTER 7 CONTENTS

7.1 THE FRESNEL-KIRCHOFF FORMULATION OF HUYGENS' PRINCIPLE

Huygens' principle says that each part of the optical disturbance on a reference surface in space acts as a source of spherical wavelets which combine to yield the disturbance at a distant point. Earlier we assumed that the resultant disturbance consisted of only the envelope of the secondary wavelets, and we discussed how this prescription gave results equivalent to geometrical optics. We now abandon the envelope prescription and describe the superposition of the wavelets in more detail. The intuitive approach used here is that of Fresnel, with one feature due to Kirchoff. We continue to treat light waves in the scalar approximation.

A. INTRODUCTION

1. The Superposition of Monochromatic Waves. Diffraction phenomena may be discussed by considering the propagation of monochromatic light having a well-defined frequency ν. The irradiance at a given point associated with this frequency is proportional to the square of the component of the electric field having this frequency, and the total irradiance is obtained as a superposition of the contributions from the various frequencies. In more complicated experiments, such as interference experiments using diffracted light, we may need to know the relative phase of the electric field at a given point $\mathbf{r}$. This requires more explicit knowledge of the time dependence, which can always be described as a superposition of exponentials, that is, as a Fourier integral:*

$$E(\mathbf{r}, t) = \int_{-\infty}^{\infty} \hat{E}(\mathbf{r}, \nu) e^{2\pi i\nu t}\, d\nu \tag{7.1}$$

This topic is covered in Appendices A and B. Some important results will be summarized here. The Fourier coefficient $\hat{E}(\mathbf{r}, \nu)$ still depends on position. Because the original electric field $E(\mathbf{r}, t)$ is real, $\hat{E}$ must obey the condition

$$\hat{E}(\mathbf{r}, \nu)^* = \hat{E}(\mathbf{r}, -\nu)$$

If we write

$$\hat{E}(\mathbf{r}, \nu) = |\hat{E}(\mathbf{r}, \nu)| e^{i\phi(\mathbf{r}, \nu)}$$

we may derive

$$E(\mathbf{r}, t) = 2\int_{0}^{\infty} |\hat{E}(\mathbf{r}, \nu)| \cos\,[2\pi\nu t + \phi(\mathbf{r}, \nu)]\, d\nu$$

$$= \mathrm{Re}\int_{0}^{\infty} 2\hat{E}(\mathbf{r}, \nu) e^{2\pi i\nu t}\, d\nu \tag{7.2}$$

*As described in Appendix B, the steady nature of the light signal requires that it be truncated before Fourier analysis can be performed. We assume that this has been done.

Equation (7.2) expresses the field at **r** as the real part of a superposition of complex exponentials with positive frequency ν and with a spatially dependent complex coefficient $2\hat{E}(\mathbf{r}, \nu)$, which still must be determined. Once we know $\hat{E}(\mathbf{r}, \nu)$, (7.2) will give us $E(\mathbf{r}, t)$.

2. *The Spatially Dependent Electric Field for a Monochromatic Wave.* It follows then that we need to concentrate on finding the spatial dependence of a single monochromatic omponent, namely, the integrand appearing in Eq. (7.2). We write this in the following way in terms of a complex spatially dependent, time-independent "wave function" $\tilde{E}(\mathbf{r})$:

$$E(\mathbf{r}, t) = \mathrm{Re}\, 2\hat{E}(\mathbf{r}, \nu)e^{2\pi i\nu t} \equiv \mathrm{Re}\, \tilde{E}(\mathbf{r})e^{2\pi i\nu t} \tag{7.3}$$

For convenience we drop ν as an explicit variable. The term $E(\mathbf{r}, t)$ in Eq. (7.3) is not the same as that in (7.2). If we substitute (7.3) into the time-dependent three-dimensional wave equation

$$\nabla^2 E - \frac{1}{v^2}\frac{\partial^2 E}{\partial t^2} = 0$$

we easily derive the so-called time-independent wave equation or *Helmholtz equation*:

$$\nabla^2\tilde{E} + k^2\tilde{E} = 0 \tag{7.4}$$

with $k = 2\pi\nu/v = 2\pi/\lambda = 2\pi n/\lambda_0$.

The elementary sinusoidal solutions of the time-dependent wave equation discussed in Section 1.2 all have their exponential counterparts, which are solutions of Eq. (7.4). Consider, for instance, a plane wave propagating in the $+x$ direction:

$$E(\mathbf{r}, t) = A\cos\left[2\pi\nu\left(\frac{t-x}{v}\right) + \phi\right]$$

$$= \mathrm{Re}\,[(Ae^{i\phi}e^{-ikx})e^{2\pi i\nu t}]$$

If we identify the factor in parentheses with $E(\mathbf{r})$, we find it obeys (7.4). Thus

$$\tilde{E} = Ae^{i\phi}e^{-ikx}$$

where $Ae^{i\phi}$ is a complex constant.

Similarly an exponential of the form

$$\tilde{E} = Ae^{i\phi}e^{-i\mathbf{k}\cdot\mathbf{r}}$$

is a solution of Eq. (7.4) that represents the spatial part of a plane wave propagating in the direction of the vector **k**, where

$$|\mathbf{k}| = k = \frac{2\pi}{\lambda}$$

The time-dependent counterpart is of course

$E(\mathbf{r}, t) = A \cos(2\pi\nu t - \mathbf{k}\cdot\mathbf{r} + \phi).$

Another important example is a spherical wave propagating outward from a point source. In time-dependent form it can be written

$$E(\mathbf{r}, t) = \frac{A}{\rho}\cos[2\pi\nu t - k\rho + \phi]$$

where $\mathbf{r}_s$ is the center of the wave and $\rho = |\mathbf{r} - \mathbf{r}_s|$ is its radius. This can be written as the real part of

$$\left(Ae^{i\phi}\frac{e^{-ik\rho}}{\rho}\right)e^{2\pi i\nu t}$$

The term in parenthesis obeys the Helmholtz equation (except at $\rho = 0$) and hence represents the spatial dependence $E(\mathbf{r})$ of the spherical wave.

3. Interference by Superposition of Spherical Waves. The interference effects studied in Chapter 5 required the superposition of a limited number of these elementary plane and spherical waves. For instance, we began our discussion of Young's interference experiment by adding two spherical waves to obtain an expression which was equivalent to the following time-independent one:

$$\tilde{E}(\mathbf{r}) = \frac{A_1}{|\mathbf{r}-\mathbf{r}_1|}e^{-ik|\mathbf{r}-\mathbf{r}_1|} + \frac{A_2}{|\mathbf{r}-\mathbf{r}_2|}e^{-ik|\mathbf{r}-\mathbf{r}_2|} \tag{7.5}$$

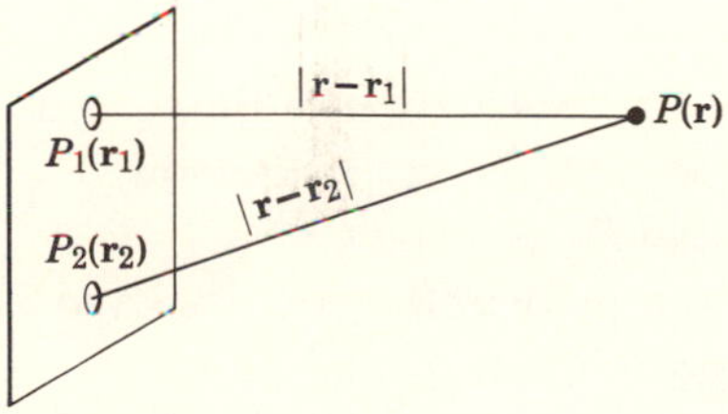

Fig. 7.1

An expression like (7.5) will also describe diffraction by two pinholes in a screen (Fig. 7.1). The pinholes are so small that each radiates a single Huygens' wavelet; for example, from P_1 the wave at P is of the form

$$\frac{A_1}{|\mathbf{r}-\mathbf{r}_1|}e^{-ik|\mathbf{r}-\mathbf{r}_1|}$$

The complex coefficient A_1 determines the strength and phase of the wavelet. Huygens' principle says that this wavelet is reradiated by the incident wave at P_1 impinging on the screen from the left. Thus we would expect A_1 to be proportional to the incident field there, $\tilde{E}_{\text{inc}}(\mathbf{r}_1)$. We also expect A_1 to be proportional to $\Delta\sigma_1$, the element of area of the pinhole, provided that it is so small that $\tilde{E}_{\text{inc}}(\mathbf{r}')$ and $k|\mathbf{r}-\mathbf{r}'|$ can be treated as constant* for $\mathbf{r}'$ anywhere in the pinhole. Thus we expect

$$A_1 = C\tilde{E}_{\text{inc}}(\mathbf{r}_1)\Delta\sigma_1 \tag{7.6}$$

*Since $k|\mathbf{r}-\mathbf{r}'|$ appears as the phase angle of an exponential, it will be effectively constant if it varies by much less than 2π radians.

The factor C in Eq. (7.6) turns out to be a slowly varying function of the angle θ' between the normal to the incident wavefront and the normal to the screen, and of the angle θ between the vector $\mathbf{r}-\mathbf{r}_1$ and the normal to the screen. If these angles are small, which is usually the case, it turns out that we must set

$$C = \frac{ik}{2\pi} = \frac{i}{\lambda} \tag{7.7}$$

We shall show in Section 8.2B4 that this value of C predicts the correct result in the trivial case of no diffraction. Thus

$$A_1 \approx \frac{i}{\lambda} \tilde{E}_{\text{inc}}(\mathbf{r}_1)\Delta\sigma_1$$

Similarly from the other pinhole we have

$$A_2 \approx \frac{i}{\lambda} \tilde{E}_{\text{inc}}(\mathbf{r}_2)\Delta\sigma_2$$

Hence the spatial part of the electric field at the point P is given by

$$\tilde{E}(\mathbf{r}) = \frac{i}{\lambda}\left[\frac{\tilde{E}_{\text{inc}}(\mathbf{r}_1)}{|\mathbf{r}-\mathbf{r}_1|} e^{-ik|\mathbf{r}-\mathbf{r}_1|}\ \Delta\sigma_1 + \frac{\tilde{E}_{\text{inc}}(\mathbf{r}_2)}{|\mathbf{r}-\mathbf{r}_2|} e^{-ik|\mathbf{r}-\mathbf{r}_2|}\Delta\sigma_2\right] \tag{7.8}$$

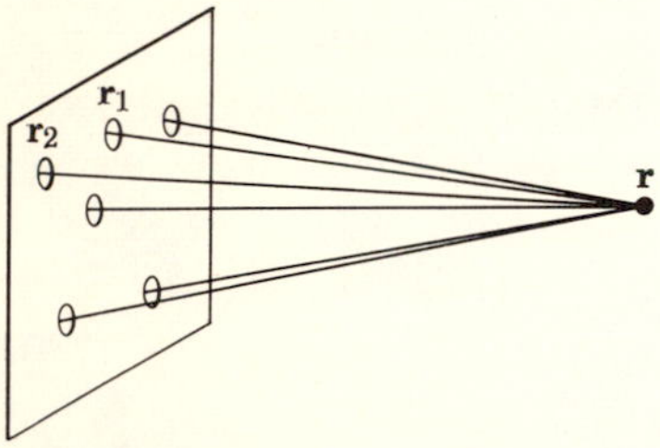

Fig. 7.2

For N pinholes at $\mathbf{r}_j$, $j = 1,2, \ldots, N$ with areas $\Delta\sigma_j$ (Fig. 7.2) we may generalize Eq. (7.8) to obtain

$$\tilde{E}(\mathbf{r}) = \frac{i}{\lambda}\sum_{j=1}^{N} \frac{\tilde{E}_{\text{inc}}(\mathbf{r}_j)e^{-ik|\mathbf{r}-\mathbf{r}_j|}\Delta\sigma_j}{|\mathbf{r}-\mathbf{r}_j|} \tag{7.9}$$

4. Transition to a Large Opening in a Screen. Consider the screen shown in Fig. 7.3. Let $\mathbf{r}'$ be a general point in the opening. How shall we treat an opening Σ_0 so large that $\tilde{E}_{\text{inc}}(\mathbf{r}')$ and $k|\mathbf{r}-\mathbf{r}'|$ may not be considered constant as $\mathbf{r}'$ ranges over the opening? We break up the total area into smaller components $\Delta\sigma_1$ located at $\mathbf{r}_1$, $\Delta\sigma_2$ located at $\mathbf{r}_2$, and so on, so that $k|\mathbf{r}-\mathbf{r}'|$ is essentially constant over each component. (See Fig. 7.3.) The contribution from the jth component will be

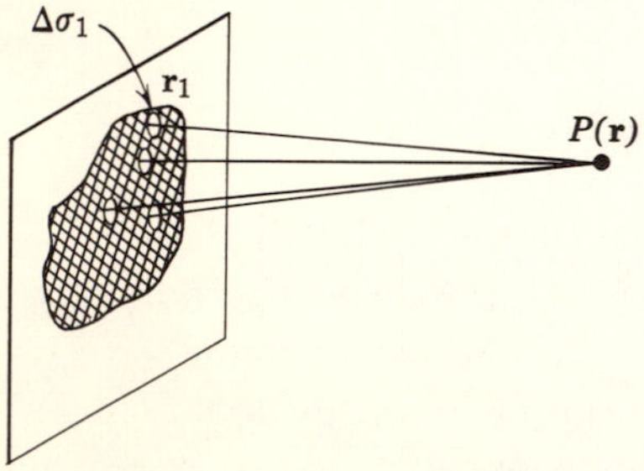

Fig. 7.3

$$\frac{i}{\lambda} \tilde{E}_{\text{inc}}(\mathbf{r}_j) \frac{e^{-ik|\mathbf{r}-\mathbf{r}_j|}\Delta\sigma_j}{|\mathbf{r}-\mathbf{r}_j|}$$

and Eq. (7.9) will be a good approximation to the total electric field at P. In the limit as all the $\Delta\sigma$'s tend to zero we obtain the surface integral

$$\boxed{\tilde{E}(\mathbf{r}) = \frac{i}{\lambda}\iint_{\Sigma_0} \tilde{E}_{\text{inc}}(\mathbf{r}') \frac{e^{-ik|\mathbf{r}-\mathbf{r}'|}}{|\mathbf{r}-\mathbf{r}'|} d\sigma} \tag{7.10}$$

where $d\sigma$ represents an infinitesimal element of surface area at the point $\mathbf{r}'$ in the opening Σ_0.

*5. **The Inclination Factor.*** Equation (7.10) forms the basis for the Fresnel theory of diffraction. Occasionally the weak dependence on angles mentioned above must be taken into account; this is done by inserting a slowly varying function of angles $Q(\theta, \theta')$ called the inclination factor into the integrand of Eq. (7.10). Here Q is unity in the forward direction and then drops off smoothly. According to the Kirchoff version of the theory to be discussed in Section 8.1, this function takes the form

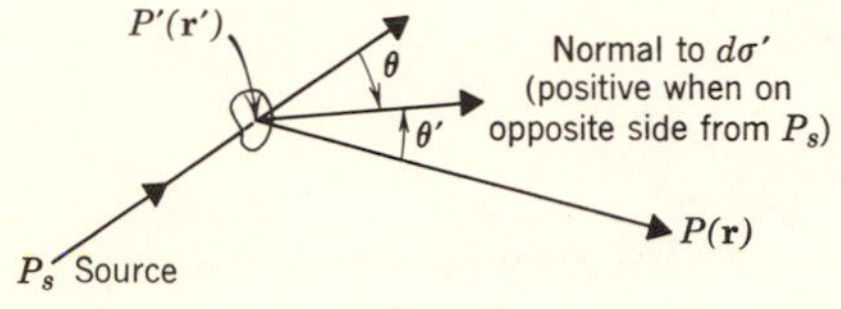

Fig. 7.4

$$Q = \tfrac{1}{2}(\cos\theta + \cos\theta') \tag{7.11}$$

where θ and θ' are defined in Fig. 7.4. Then θ is the angle between the direction of the incident light and the normal, that is, the angle of incidence. The term θ' is the angle between the normal and the line from $P'(\mathbf{r}')$ to the observation point $P(\mathbf{r})$.

We shall discuss in Section 8.1 how Eqs. (7.10) and (7.11) may be derived from the wave equation if the proper assumptions are made.

B. DISCUSSION

Equation (7.10) with or without Eq. (7.11) has been applied to a wide variety of diffraction phenomena with results in a great many cases that are surprisingly good, considering that it is not a rigorous deduction from the wave equation, but only an approximation. It works best when both source and observation points are a large number of wavelengths away from the diffracting aperture and when the openings in the aperture are somewhat larger than a wavelength. It does not take any polarization effects into account. These can be important in some experiments.

*1. **Choosing the Surface*** Σ_0. The question often arises of how to choose the surface over which to integrate in Eq. (7.10). After all, the opening represents the absence of a physical surface. How do we decide what Σ_0 shall be? It is usually a matter of convenience. If the aperture is a plane, we often take Σ_0 to be the continuation of that plane. But in principle it can be any reasonably smooth surface. Sometimes it is chosen to make $\tilde{E}_{\mathrm{inc}}(\mathbf{r}')$ or $|\mathbf{r}-\mathbf{r}'|$ constant. This is illustrated in Fig. 7.5, where the incident light is from the point source P_s. The surface Σ_s would make $\tilde{E}_{\mathrm{inc}}(\mathbf{r}')$ constant, whereas Σ_p would make $|\mathbf{r}-\mathbf{r}'|$ constant. In the "parabolic approximation," in which calculations are carried out to second order in h, the results with Σ_s or Σ_p or any intermediate surface will all be the same.

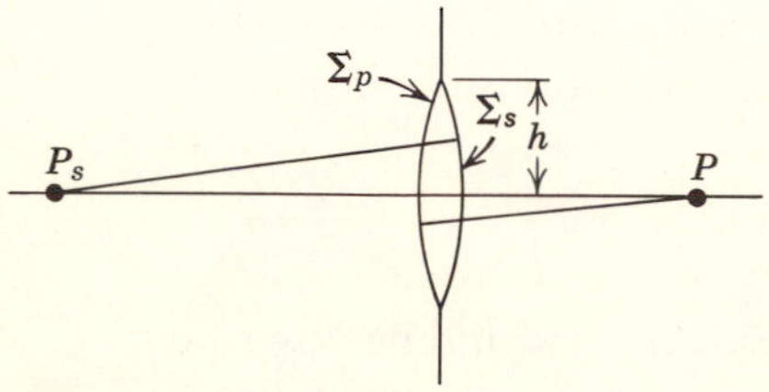

Fig. 7.5

*2. **Use of a Transmission Function.*** If the opening is covered by a partly transmitting film with amplitude transmission factor $t(r')$

so that

$$\tilde{E}_{\text{trans}}(\mathbf{r}') = t(\mathbf{r}')\tilde{E}_{\text{inc}}(\mathbf{r}') \tag{7.12}$$

represents the transmitted light, we should replace $\tilde{E}_{\text{inc}}(\mathbf{r}')$ by $\tilde{E}_{\text{trans}}(\mathbf{r}')$ in Eq. (7.10), or equivalently insert the function $t(\mathbf{r}')$ into the integrand.

This use of a function $t(\mathbf{r}')$ can be generalized in the following way to handle the original case described by Eq. (7.10). Let Σ represent the entire surface of the aperture, which is divided into an open part Σ_0 and a closed part Σ_c. Thus $\Sigma = \Sigma_0 + \Sigma_c$. Then Eq. (7.10) becomes

$$\tilde{E}(\mathbf{r}) = \frac{i}{\lambda}\iint\limits_{\Sigma_0} \tilde{E}_{\text{inc}}(\mathbf{r}')\frac{e^{-ik|\mathbf{r}-\mathbf{r}'|}}{|\mathbf{r}-\mathbf{r}'|}\,d\sigma + \frac{i}{\lambda}\iint\limits_{\Sigma_0} 0 \cdot d\sigma \tag{7.10$'$}$$

Now define a transmission function $t(\mathbf{r}')$ for $\mathbf{r}'$ on Σ which is a step function:

$$t(\mathbf{r}') = 1 \quad \text{for } \mathbf{r}' \text{ on } \Sigma_0$$
$$t(\mathbf{r}') = 0 \quad \text{for } \mathbf{r}' \text{ on } \Sigma_c$$

Then Eq. (7.10′) can be written as

$$\tilde{E}(\mathbf{r}) = \frac{i}{\lambda}\iint\limits_{\Sigma} t(\mathbf{r}')\tilde{E}_{\text{inc}}(\mathbf{r}')\frac{e^{-ik|\mathbf{r}-\mathbf{r}'|}}{|\mathbf{r}-\mathbf{r}'|}\,d\sigma \tag{7.13}$$

More generally, any function $t(\mathbf{r}')$ for which Eq. (7.12) holds may be inserted into Eq. (7.13). This applies also to functions which change the phase, that is, functions of the form

$$t(\mathbf{r}') = |t(\mathbf{r}')|e^{i\Delta(\mathbf{r}')} \tag{7.14}$$

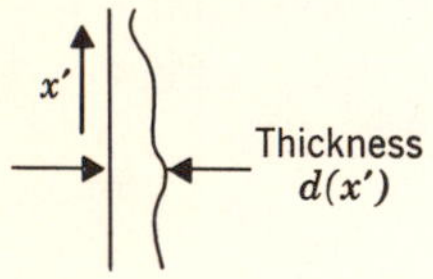

Fig. 7.6. A phase plate produced by a dielectric plate of varying thickness.

with nonzero phase angle Δ. Such can be provided by a transparent plate with varying thickness as shown in Fig. 7.6. For nearly normal incidence and small changes in d, the phase change produced by the plate referred to the case with no plate is

$$\Delta(x') = \frac{-2\pi}{\lambda}(n-1)\,d(x')$$

and the effect of this plate in a diffracting aperture would be approximately described by the transmission function

$$t(x') = |t(x')|\exp\left[-\frac{2\pi i}{\lambda}(n-1)\,d(x')\right]$$

Losses caused by reflection and absorption would be described by a value of $|t(x')|$ less than unity.

3. ***Babinet's Principle.*** An important and useful conclusion can be drawn from the fact that the Fresnel-Kirchoff formula (7.10) expresses the field $\tilde{E}(\mathbf{r})$ as an integral over the open aperture

Σ_0. If Σ_0 is split into two parts Σ_1 and Σ_2 so that $\Sigma_0 = \Sigma_1 + \Sigma_2$ (Fig. 7.7), then the field at the point $P(\mathbf{r})$ will be the sum of integrals over Σ_1 and over Σ_2. But the integral over Σ_1 represents the field at P if Σ_1 is the only open aperture, and the integral over Σ_2

Fig. 7.7. The open surfaces Σ_1 and Σ_2 together comprise the open surface Σ_0

represents the field at P if Σ_2 is the only open aperture. Thus, if we denote these fields by

$$\tilde{E}_1(\mathbf{r}) = \frac{i}{\lambda} \iint_{\Sigma_1} (\cdots)\, d\sigma$$

and

$$\tilde{E}_2(\mathbf{r}) = \frac{i}{\lambda} \iint_{\Sigma_2} (\cdots)\, d\sigma$$

and also denote the field if Σ_0 is the open aperture by

$$\tilde{E}_0(\mathbf{r}) = \frac{i}{\lambda} \iint_{\Sigma_0} (\cdots)\, d\sigma$$

where $(\cdots)$ denotes the integrand in Eq. (7.10), we obtain

$$\tilde{E}_0(r) = \tilde{E}_1(r) + \tilde{E}_2(r) \tag{7.15}$$

This equation expresses a principle due to Babinet. The apertures Σ_1 and Σ_2 are said to be *complementary*. The fields observed with complementary apertures add to give the field observed with the combined open surface of both.

Equation (7.15) can be very helpful when, say, Σ_1 is a relatively complicated aperture composed of the difference $\Sigma_0 - \Sigma_2$ of two simple apertures. Then we calculate the (generally complex) field at $P(\mathbf{r})$ resulting from Σ_0 and Σ_2 separately, and subtract to obtain the field resulting from Σ_1.

Babinet's principle is particularly useful when the field at $P(\mathbf{r})$ is zero (or very small) with the aperture Σ_0. Then we obtain

$$\tilde{E}_1(\mathbf{r}) = -\tilde{E}_2(\mathbf{r})$$

By taking the absolute value squared, we obtain for the intensity

$$I_1(\mathbf{r}) = I_2(\mathbf{r}) \tag{7.16}$$

Thus the intensity at a point that would be in the dark region with aperture Σ_0 has the same value with Σ_1 as with its complement Σ_2.

C. EXAMPLES

A systematic discussion of diffraction will begin in Section 7.2. Before we get too deeply involved in details, some particular examples are worth discussing.

1. Field at the Focus of a Converging Spherical Wave. Our first example is shown in Fig. 7.8. Here a spherical wave brought to a focus (by a lens) at P_i is partly interrupted by a screen. The radius of the wave at the screen is R. What is the electric field at P_i? We take our surface of integration Σ_0 to be that part of the spherical surface of radius R that is in the aperture. On this surface the incident field is constant, having the value E_{inc}, and the other important factor is

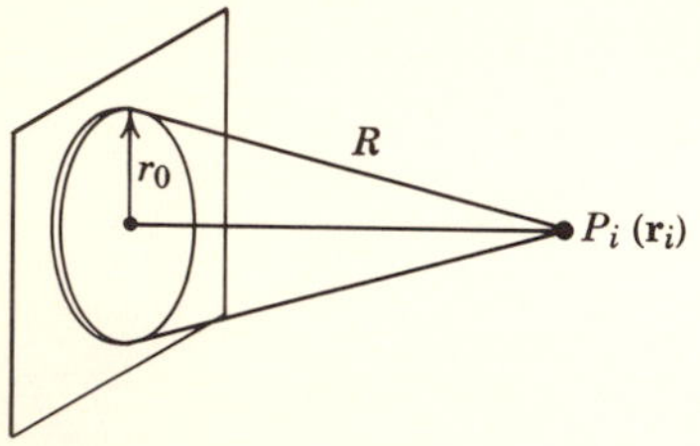

Fig. 7.8. The spherical wave converging to a focus at P_i is limited by the aperture.

$$\frac{e^{-ik|\mathbf{r}_i-\mathbf{r}'|}}{|\mathbf{r}_i-\mathbf{r}'|} = \frac{e^{-ikR}}{R}, \qquad \text{also a constant}$$

Equation (7.10) then gives

$$\tilde{E}(\mathbf{r}_i) = \frac{i}{\lambda}\iint_{\Sigma_0} \tilde{E}_{\text{inc}} \frac{e^{-ikR}}{R}\, d\sigma = \frac{i}{\lambda}\tilde{E}_{\text{inc}}\frac{e^{-ikR}}{R}\sigma \tag{7.17}$$

where $\sigma = \int\int_{\Sigma_0} d\sigma$ is the area of Σ_0, approximately given by πr_0^2 if r_0 is the radius of the opening. This result applies only at the geometrical focal spot P_i; at other points the calculation is more difficult.

The spatial dependence of the field in the incoming spherical wave can be written

$$\tilde{E}_{\text{inc}}(\mathbf{r}') = \frac{A}{|\mathbf{r}'-\mathbf{r}_i|}e^{+ik|\mathbf{r}'-\mathbf{r}_i|} \tag{7.18}$$

(The plus sign occurs because the wave is propagating toward P_i.) Hence, on Σ_0 we have

$$E_{\text{inc}} = \frac{A}{R}e^{ikR} \tag{7.19}$$

Equation (7.17) can then be written

$$\tilde{E}(\mathbf{r}_i) = \frac{i}{\lambda}A\left(\frac{\sigma}{R^2}\right) = \frac{iA}{\lambda}\Omega \tag{7.20}$$

where $\Omega = \sigma/R^2$ is the solid angle subtended by the hole in the aperture as seen from P_i.

Equation (7.20) shows that the singularity in (7.19) at $R = 0$ is removed by taking proper account of diffraction. As far as the amplitude is concerned, the effective value of R that occurs at the focal spot is $R_{\text{eff}} = \lambda/\Omega$, which can be quite small, but which is certainly not zero.

2. *Diffraction by a Long Narrow Slit.* As a second example we consider diffraction by a long narrow slit (Fig. 7.9). We assume

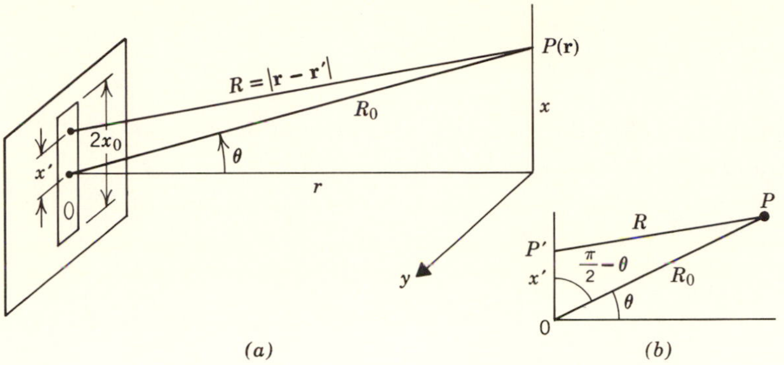

Fig. 7.9

that the incident light is collimated and propagates parallel to the normal to the plane of slit so that

$$\tilde{E}_{\text{inc}}(\mathbf{r}') = E_0, \quad \text{a constant}$$

Let the width y' of the slit be so small that R in Fig. 7.9 is independent of y'. We evaluate the field in the plane containing the slit and the normal to the aperture. Equation (7.10) then becomes

$$\tilde{E}(\mathbf{r}) = \frac{i}{\lambda}\,\Delta y'\,E_0 \int_{-x_0}^{x_0} \frac{e^{-ikR}}{R}\,dx' \tag{7.21}$$

Here $R = |\mathbf{r} - \mathbf{r}'|$. We must find a suitable approximate expression for R. There are two general ways of proceeding.

Case 1. Fraunhofer case. The slit height $2x_0$ is assumed much smaller than the distance R_0. We apply the law of cosines to triangle $OP'P$ in Fig. 7.9*b* to obtain

$$R^2 = R_0^2 + x'^2 - 2x'R_0 \sin\theta$$

$$= R_0^2\left[1 + \left(\frac{-2x'}{R_0}\sin\theta + \frac{x'^2}{R_0^2}\right)\right]$$

Then writing

$$R = R_0\sqrt{1+\epsilon}$$

where

$$\epsilon = \left(-\frac{2x'}{R_0}\sin\theta + \frac{x'^2}{R_0^2}\right)$$

we may expand, using the binomial theorem:

$$\sqrt{1+\epsilon} = 1+\tfrac{1}{2}\epsilon-\tfrac{1}{8}\epsilon^2+\cdots$$

We take only the first three terms and calculate them to second order in x':

$$R = R_0\left[1-\frac{x'}{R_0}\sin\theta+\frac{x'^2}{2R_0^{\,2}}-\frac{x'^2\sin^2\theta}{2R_0^{\,2}}+0(x'^3)\right]$$

$$= R_0 - x'\sin\theta+\frac{x'^2}{2R_0}\cos^2\theta+0(x'^3)$$

The product $kR = 2\pi R/\lambda$ appears in the exponential in the integrand of Eq. (7.21). We may approximate it by

$$kR_0 - kx'\sin\theta$$

provided that the next term

$$\frac{x'^2}{2R_0}\frac{2\pi}{\lambda}\cos^2\theta$$

is much less than unity. Thus, if we require

$$x_0 \ll \sqrt{\lambda R_0}$$

only the linear or x' term will be important. *This is the Fraunhofer limiting case.* For $1/R$ in Eq. (7.21) we may use simply $1/R_0$.

We then obtain

$$\tilde{E}(\mathbf{r}) = \frac{i}{\lambda}\Delta y' E_0 \frac{e^{-ikR_0}}{R_0}\int_{-x_0}^{+x_0} e^{ikx'\sin\theta}\,dx'$$

The integral gives

$$\frac{e^{ikx_0\sin\theta}-e^{-ikx_0\sin\theta}}{ik\sin\theta} = 2x_0\frac{\sin(kx_0\sin\theta)}{kx_0\sin\theta}$$

Thus

$$\tilde{E}(\mathbf{r}) = \frac{i}{\lambda}\sigma\tilde{E}_0\frac{e^{-ikR_0}}{R_0}\frac{\sin(kx_0\sin\theta)}{(kx_0\sin\theta)} \qquad (7.22)$$

with $\sin\theta = x/R_0$ and $\sigma = \Delta y'\cdot 2x_0 =$ area of slit.

The maximum value of $|\tilde{E}|$ occurs in the straight-ahead direction when $x = 0$ and $\theta = 0$. Then, since

$$\lim_{t\to 0}\frac{\sin t}{t} = 1$$

we obtain

$$\tilde{E}_{\max} = \frac{i}{\lambda}\sigma\tilde{E}_0\frac{e^{-ikr}\,{}^{*}}{r}\ . \tag{7.23}$$

Case 2. Fresnel case. Often we cannot afford to neglect the terms proportional to x'^2 occurring in the expansion of R. A consistent treatment that avoids $x'y'$ cross terms† demands that we treat x' and x on an equal footing and expand R to second order in both. We must then expand about $R = r$ in Fig. 7.9:

$$R = \sqrt{r^2 + (x-x')^2} \cong r + \frac{1}{2r}(x-x')^2$$

Then Eq. (7.10) gives

$$E(\mathbf{r}) = \frac{i}{\lambda}\Delta y' E_0 \frac{e^{-ikr}}{r}\int_{-x_0}^{+x_0} e^{-(i\pi/\lambda r)(x-x')^2}\,dx' \tag{7.24}$$

The integral appearing here is called a *Fresnel integral.* It will be discussed in much more detail later, since it always appears in one-dimensional Fresnel diffraction problems.

The Fresnel integral generally behaves quite differently from the simple integral that led to Eq. (7.22), but when the condition for Fraunhofer diffraction is satisfied, it gives the same result as Eq. (7.22) if we are willing to approximate

$$R_0 = \sqrt{r^2 + x^2} \qquad \text{by} \quad r + \frac{x^2}{2r}$$

REFERENCE FOR SECTION 7.1

M. Françon, *Diffraction: Coherence in Optics.* Chapter I.

PROBLEMS

7.1.1 A helium-neon laser emits a uniform circular beam 2 mm in diameter and well collimated with a flux of 1 mW at $\lambda = 633$ nm. The beam is interrupted by a mask containing two pinholes 1 mm apart, each 10

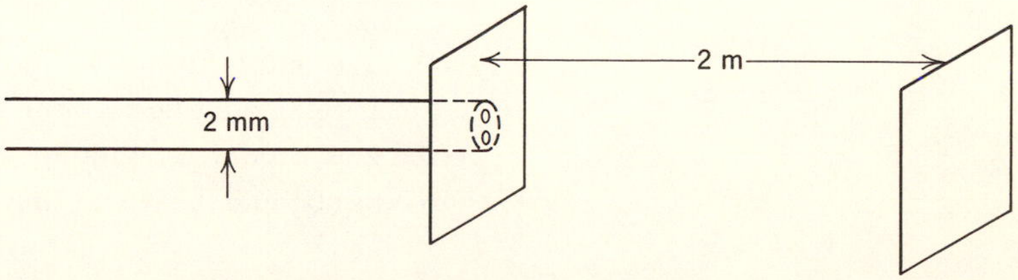

*This has the same form as Eq. (7.17). This is no accident, for we shall show later that Fraunhofer diffraction occurs in the plane of a converging point image. The use of converging spherical wavefronts in Fig. 7.8 allows us to relax the condition $x_0 \ll \sqrt{\lambda R_0}$ necessary for Fraunhofer diffraction without a lens.

†More details will be given later, in Chapter 8.

microns = 0.01 mm in diameter. The resulting Young's fringes are studied on a screen 2 m away.

a. What is the separation of the fringes?
b. What is the maximum irradiance (in watts/mm^2) in a fringe near the center of the pattern?

7.1.2 For the spherical surface of integration Σ_0 used in Fig. 7.8 the inclination factor Q obeys the equation

$$Q = \tfrac{1}{2}(\cos\theta + \cos\theta') = 1$$

since the normal to the surface is pointing along the radius vector toward P_i as are the incoming and outgoing ray directions.

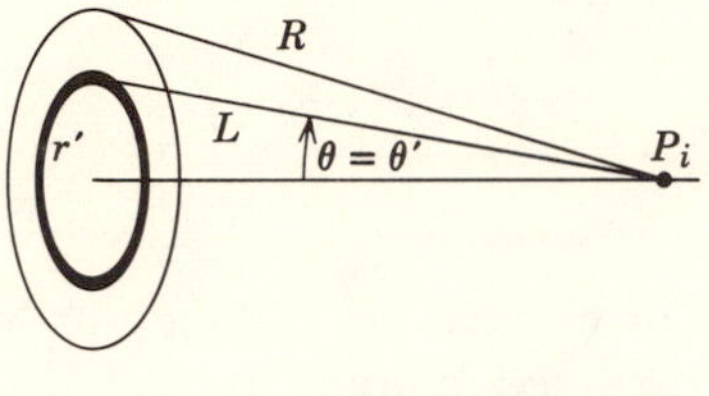

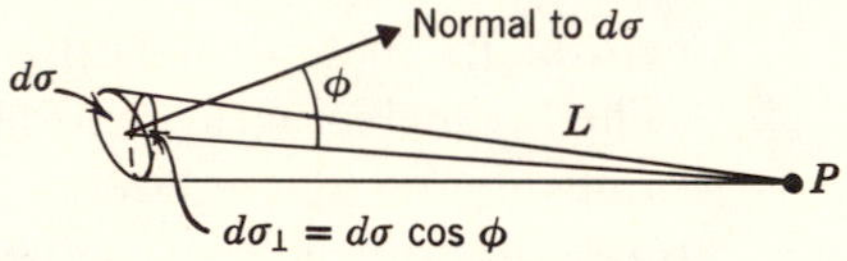

The problem is to show that Eq. (7.20) is also obtained if the surface of integration Σ_0 is the plane of the aperture. Use an incident field given by

$$\tilde{E}_{\text{inc}} = \frac{A}{L}e^{+ikL}, \qquad L = |\mathbf{r}' - \mathbf{r}_i|$$

For the element of area, try the annular ring shown above. Use the correct Q. The correct definition of an infinitesimal solid angle subtended at P by $d\sigma$ is

$$d\Omega = \frac{d\sigma_\perp}{L^2} = \frac{d\sigma\cos\theta}{L^2}$$

7.1.3 The output from a 1-watt Argon ion laser at $\lambda = 488$ nm forms a uniform collimated beam 6 cm in diameter. It is focused by a well-corrected lens to a diffraction-limited "point" at P_i. A 4-cm-diameter circular stop is just behind the lens as shown.

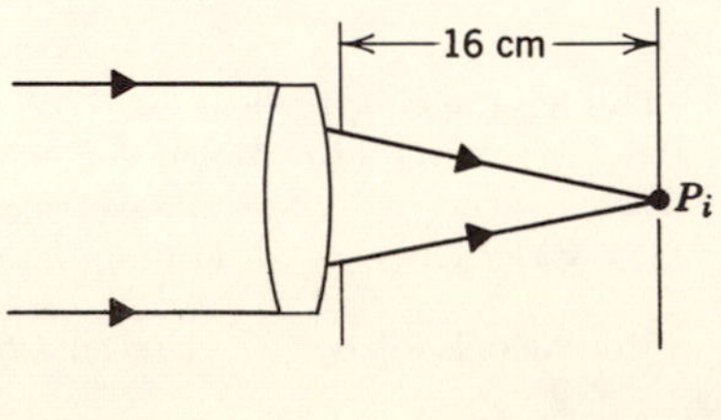

a. Calculate the magnitude of the ratio of the electric field at P_i to that in the converging beam at the stop.

b. Calculate the irradiance at P_i.

[Answers: (a) $1.61 \cdot 10^4$, (b) $5.7 \cdot 10^6$ w/cm^2.]

7.1.4 An experimenter is studying Fraunhofer diffraction by a slit using the setup shown in Fig. 7.9. The slit height $2x_0$ is 1 mm; the wavelength is 500 nm. How large does r have to be if the maximum value of the neglected x'^2 term in the phase kR is to be 0.1 radians?

7.1.5 Consider Fraunhofer diffraction by a small hole of irregular shape under the same conditions as for Fig. 7.9. Show that for $\theta = 0$, Eq. (7.17) still holds. What is the approximate condition on h, the maximum dimension of the aperture for this result to be valid?

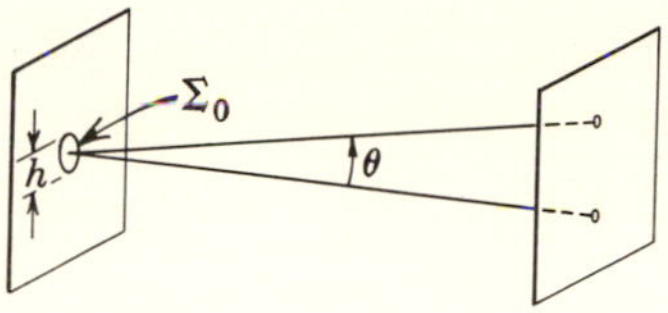

7.1.6 Prove the statement made in the last paragraph of the above text about the Fresnel integral giving the Fraunhofer result in the appropriate limit.

7.2 FRAUNHOFER DIFFRACTION BY A SINGLE APERTURE

Fraunhofer diffraction is a special phenomenon that occurs when the optical path from points in the diffraction aperture to the observation point depends at most linearly on the coordinates of the aperture point. This can be achieved by having very distant source and observation points or by proper use of lenses. Fraunhofer diffraction produced by lenses will be discussed in detail in Section 7.2D1 and Chapter 9.

A. DERIVATION OF FORMULAS

Figure 7.10*a* defines the geometry for our calculation. The aperture is in the plane through the origin O normal to the z axis. Here

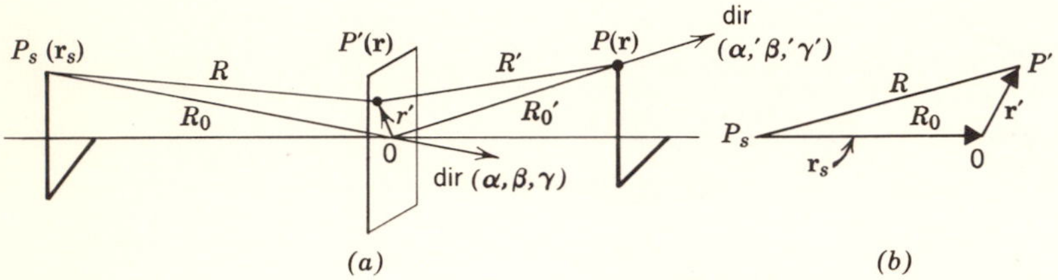

Fig. 7.10

P' with position vector $\mathbf{r}' = (x', y', 0)$ is a typical point in the aperture. The point source of monochromatic light is at P_s with position vector $\mathbf{r}_s$. The observation point is at $P(\mathbf{r})$. Our starting equation is the basic diffraction integral of Eq. (7.10). The incident field will be a spherical wave centered at P_s. Letting $|\mathbf{r}' - \mathbf{r}_s| = R$ and $|\mathbf{r} - \mathbf{r}'| = R'$, we obtain

$$\tilde{E}(\mathbf{r}) = \frac{i}{\lambda} A \iint_{\Sigma_0} \frac{e^{-ik(R'+R)}}{RR'} \, d\sigma \tag{7.25}$$

where A is the amplitude of the incident wave at unit distance from P_s.

1. The Linear Approximation for the Path Differences. We now must find approximate expressions for R and R' good when x' and y' are very small compared with R_0' and R_0. From Fig. 7.10*b* we obtain

$$\overrightarrow{P_sP'} = \overrightarrow{P_sO} + \mathbf{r}'$$

Hence (law of cosines)

$$|\overrightarrow{P_sP'}|^2 = |\overrightarrow{P_sO}|^2 + |\mathbf{r}'|^2 + 2\overrightarrow{P_sO} \cdot \mathbf{r}'$$

or

$$R^2 = R_0{}^2 + r'^2 + 2R_0(\alpha x' + \beta y') \tag{7.26}$$

In Eq. (7.26) we have set $\mathbf{r}' = (x', y', 0)$ and have introduced direction cosines for the direction $\overrightarrow{P_sO}$: $\overrightarrow{P_sO} = R_0(\alpha, \beta, \gamma)$. We also may write

$$\alpha = -\frac{x_s}{R_0} \qquad \beta = -\frac{y_s}{R_0} \tag{7.27}$$

where x_s, y_s are the transverse coordinates of the source point P_s.

Now we expand R about R_0,

$$R = R_0\sqrt{1+\epsilon} \approx R_0(1+\tfrac{1}{2}\epsilon - \tfrac{1}{8}\epsilon^2)$$

with

$$\epsilon = \frac{2(\alpha x' + \beta y')}{R_0} + \frac{r'^2}{R_0{}^2}$$

and calculate to second order in x' and y',

$$R = R_0\left[1 + \frac{\alpha x' + \beta y'}{R_0} + \frac{1}{2}\frac{r'^2}{R_0'^2} - \frac{1}{2}\frac{(\alpha x' + \beta y')^2}{R_0{}^2}\right]$$

or

$$R = R_0 + \alpha x' + \beta y' + \frac{1}{2}\left[\frac{x'^2 + y'^2 - (\alpha x' + \beta y')^2}{R_0}\right]$$

In the Fraunhofer approximation the path difference $(R - R_0)$ is linear in x' and y', and hence the last term is small compared with λ. Thus we must have

$$|x'| \ll \sqrt{R_0\lambda} \qquad |y'| \ll \sqrt{R_0\lambda} \tag{7.28}$$

Then we obtain

$$R \approx R_0 + \alpha x' + \beta y' \tag{7.29}$$

A similar derivation yields a similar expression for R'. The starting equation is

$$\overrightarrow{\mathrm{P'P}} = \overrightarrow{\mathrm{OP}} - \mathrm{r}'$$

which gives

$$R'^2 = R_0'^2 + r'^2 - 2R_0'(x'\alpha' + y'\beta')$$

where α', β' are direction cosines of $\overrightarrow{OP}$: $\overrightarrow{OP} = R_0'(\alpha', \beta', \gamma')$ with

$$\alpha' = \frac{x}{R_0'} \qquad \beta' = \frac{y}{R_0'} \tag{7.30}$$

where x, y are the transverse coordinates of the observation point P. Then to second order in x' and y' we have

$$R' = R_0' - \alpha' x' - \beta' y' + \frac{1}{2}\left[\frac{x'^2 + y'^2 + (\alpha' x' + \beta' y')^2}{R_0'}\right] \tag{7.31}$$

Thus, if in addition to (7.28) we require

$$\begin{aligned} |x'| &\ll \sqrt{R_0'\lambda} \\ |y'| &\ll \sqrt{R_0'\lambda} \end{aligned} \tag{7.32}$$

we may write

$$R' \approx R_0' - \alpha' x' - \beta' y'$$

and the quadratic terms in Eq. (7.31) will be small compared with λ.

Because of the above calculations, we may approximate $(RR')^{-1}$ in (7.25) by $(R_0 R_0')^{-1}$ and $k(R + R')$ by

$$k(R_0 + R_0') + \frac{2\pi}{\lambda}[(\alpha - \alpha')x' + (\beta - \beta')y']$$

The result is

$$\tilde{E}(\mathbf{r}) = \frac{i}{\lambda} A \frac{e^{-ik(R_0 + R_0')}}{R_0 R_0'} \iint_{\Sigma_0} \exp\left\{-\frac{2\pi i}{\lambda}[(\alpha - \alpha')x' + (\beta - \beta')y']\right\} dx'\, dy'. \tag{7.33}$$

If we need to insert the inclination factor Q into this integral, we note by comparing Figs. 7.4 and 7.10*a* that, to a good approximation when x' and y' are small, we can write $\cos\theta = \gamma$ and $\cos\theta' = \gamma'$. Hence, by Eq. (7.11) we have

$$Q \cong \tfrac{1}{2}(\gamma + \gamma') \tag{7.33$'$}$$

We often find it convenient to write

$$A \frac{e^{-ikR_0}}{R_0} = \tilde{E}(\mathrm{O})$$

where $\tilde{E}(\mathrm{O})$ represents the complex amplitude of the incident field at the origin O.

2. *Transmission Functions and Fourier Transforms.* As mentioned in Section 7.1, we can insert a transmission function

$$t(\mathbf{r}') = t(x', y')$$

into the diffraction integral and then let the integration cover the entire $x'\, y'$ plane. Use of a step function recovers the result of Eq. 7.33. We can also generalize to a variety of apertures. Use of $t(x'\, y')$ gives

$$\tilde{E}(\mathbf{r}) = \frac{i}{\lambda}\tilde{E}(\mathrm{O}) \frac{e^{-ikR_0'}}{R_0'} \iint_{-\infty}^{\infty} t(x', y') e^{-2\pi i(ux' + vy')}\, dx'\, dy' \tag{7.34}$$

with $u = (\alpha - \alpha')/\lambda$, $v = (\beta - \beta')/\lambda$.

There is an important special case where the transmission function $t(x', y')$ factors into a product of x-dependent and y-dependent functions:

$$t(x', y') = t_1(x')t_2(y') \tag{7.35}$$

Examples of apertures which have this property are shown in Fig. 7.11*a–d*. The light areas represent regions where both t_1 and t_2 are unity. The aperture in Fig. 7.11*e* does *not* have this

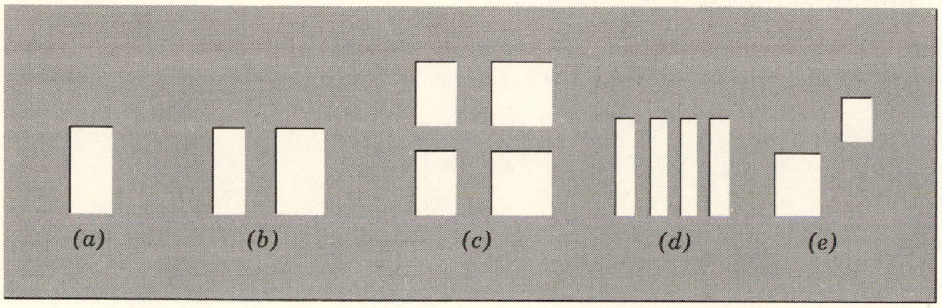

Fig. 7.11

property. Under the assumption of Eq. (7.35) the integral in Eq. (7.34) also factors,

$$\iint_{-\infty}^{\infty} t(x', y')e^{-2\pi i(ux' + vy')}\, dx'\, dy' = \hat{t}_1(u)\hat{t}_2(v) \tag{7.36a}$$

where

$$\hat{t}_1(u) = \int_{-\infty}^{\infty} t_1(x')e^{-2\pi iux'}\, dx' \tag{7.36b}$$

is the Fourier transform of the function $t_1(x')$. A similar expression holds for $\hat{t}_2(v)$ ($\hat{t}_1$ and $\hat{t}_2$ have the units of length; t in [7.37] has units of area).

In the case of a general function $t(x', y')$ the integral may be *defined* to be a two-dimensional Fourier transform:

$$\hat{t}(u, v) = \int_{-\infty}^{\infty} t(x', y')e^{-2\pi i(ux' + vy')}\, dx'\, dy' \tag{7.37}$$

This two-dimensional transform can be introduced by applying the one-dimensional transform in two steps. First hold y' fixed. Then $t(x', y')$ considered as a function of x' may be transformed to a function we shall call $\hat{T}(u; y')$:

$$\hat{T}(u; y') = \int_{-\infty}^{\infty} t(x', y')e^{-2\pi iux'}\, dx'$$

The expression for $t(x', y')$ in terms of $T(u; y')$ is

$$t(x', y') = \int_{-\infty}^{\infty} \hat{T}(u; y')e^{+2\pi iux'}\, du$$

On the other hand, for fixed u, $\hat{T}(u; y')$ may be considered to be a function of y' and may be Fourier-transformed to a function $\hat{t}(u, v)$:

$$\hat{t}(u, v) = \int_{-\infty}^{\infty} \hat{T}(u; y) e^{-2\pi i v y'} \, dy'$$

The inverse is

$$\hat{T}(u; y') = \int_{-\infty}^{\infty} \hat{t}(u, v) e^{2\pi i v y'} \, dv$$

Combining these equations yields

$$\hat{t}(u, v) = \iint_{-\infty}^{\infty} t(x', y') e^{-2\pi i (u x' + v y')} \, dx' \, dy'$$

$$t(x', y') = \iint_{-\infty}^{\infty} \hat{t}(u, v) e^{+2\pi i (u x' + v y')} \, du \, dv$$

We now apply our general Fraunhofer formulas to some important special cases.

B. FRAUNHOFER DIFFRACTION BY RECTANGULAR APERTURES

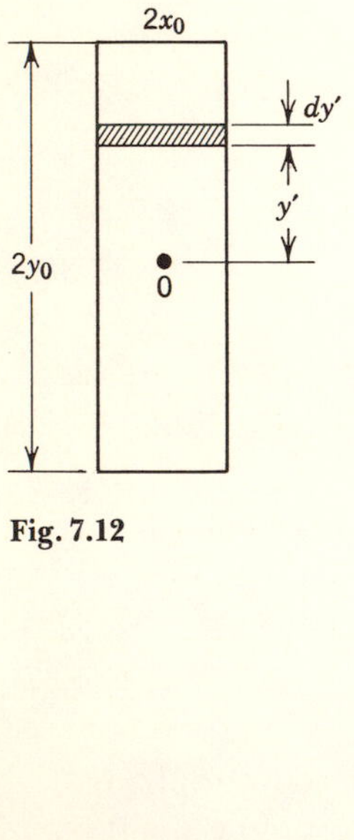

Fig. 7.12

1. Mathematical Derivation. Let the aperture be $2x_0$ by $2y_0$ in area as shown in Fig. 7.12. Then the transmission function is unity inside the aperture and zero outside. Thus we have

$$t(x', y') = 1 \qquad \text{if } |x'| < x_0 \text{ and } |y'| < y_0$$
$$t(x', y') = 0 \qquad \text{otherwise}$$

It is convenient to write the integral in Eq. (7.36) in the form

$$\hat{t}_1(u)\,\hat{t}_2(v) = 4x_0 y_0 \left(\frac{1}{2x_0} \int_{-x_0}^{+x_0} e^{-2\pi i u x'} \, dx' \right) \left(\frac{1}{2y_0} \int_{-y_0}^{+y_0} e^{-2\pi i v y'} \, dy' \right)$$
$$\equiv 4x_0 y_0 I_x I_y$$

This factorization corresponds to integrating over strips. The contribution from the shaded strip in Fig. 7.12 is

$$e^{-2\pi i v y'} \, dy' \int_{-x_0}^{x_0} e^{-2\pi i u x'} \, dx'$$

The mean radiant incidence or flux density at $P(\mathbf{r})$ is given by

$$E_e(\mathbf{r}) = nC|\tilde{E}(\mathbf{r})|^2 \tag{7.38}$$

Here $|\tilde{E}|$ represents the peak electric field strength E_{peak}, and n is the index of refraction and C is a constant. If E_e is expressed in watt/m^2 and $|\tilde{E}|$ in volt/m, then

$$C = \frac{1}{240\pi} \left(\frac{3 \cdot 10^8 \text{ m/sec}}{c \text{ m/sec}} \right) \approx \frac{1}{240\pi} = 1.327 \cdot 10^{-3} \text{ watt/volt}^2 \tag{7.38$'$}$$

Thus, $E_e = 1$ watt/m^2 $\Longleftrightarrow$ $E_{\text{peak}} = \sqrt{754}$ v/m = 27.45 v/m. For an example, take the focused laser beam of Problem 7.1.3*b* where $E_e = 5.7 \cdot 10^6$ w/cm^2. This corresponds to $E_{\text{peak}} = 1.57 \cdot 10^8$ v/cm. Equations (7.34), (7.36), and (7.38) then combine to give

$$E_e(\mathbf{r}) = E_e(\mathrm{O})\,\frac{16x_0^{\,2}y_0^{\,2}}{\lambda^2(R_0')^2}\,I_x^{\,2}I_y^{\,2} \tag{7.39}$$

where $E_e(\mathrm{O}) = nC\,|\tilde{E}(\mathrm{O})|^2$ is the flux density in the incident beam at the center of the aperture O.

The function I_x is readily evaluated to give

$$I_x = \frac{1}{2x_0}\int_{-x_0}^{+x} e^{-2\pi i u x'}\,dx' = \frac{\sin 2\pi u x_0}{2\pi u x_0} \tag{7.40}$$

The corresponding result for the Fourier transform is

$$\hat{t}_1(u) = 2x_0 I_x = \frac{\sin 2\pi u x_0}{\pi u} \tag{7.40$'$}$$

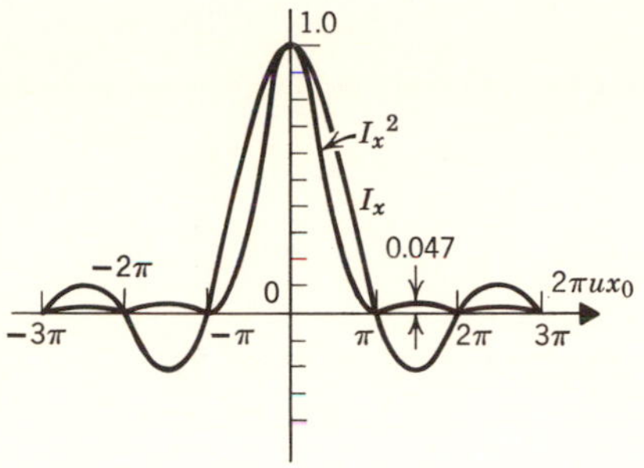

Fig. 7.13

The function I_x and its square are plotted in Fig. 7.13. Its maximum value is unity when $u = 0$, that is, at $\alpha' = \alpha$. Zeros occur when ux_0 is a half-integer, that is, for

$$ux_0 = \tfrac{1}{2}m, \qquad m = \pm 1, \pm 2, \ldots$$

or when

$$\frac{x_0(\alpha - \alpha')}{\lambda} = \frac{m}{2}$$

or when

$$2x_0(\alpha - \alpha') = m\lambda \tag{7.41}$$

Because of the Fraunhofer diffraction condition, $x_0^{\,2} \ll \lambda R_0'$ and $y_0^{\,2} \ll \lambda R_0'$, and because $I_x^{\,2}$ and $I_y^{\,2}$ are never greater than unity, Eq. (7.39) says that $E_e(\mathbf{r})$ is always smaller than $E_e(0)$.

2. ***Elementary Argument.*** An elementary argument may be used to obtain a simplified version of the result in Eq. (7.41). Assume $\beta = \beta' = \alpha = 0$ and m $= -1$. Then (7.41) gives

$$2\alpha' x_0 = \lambda$$

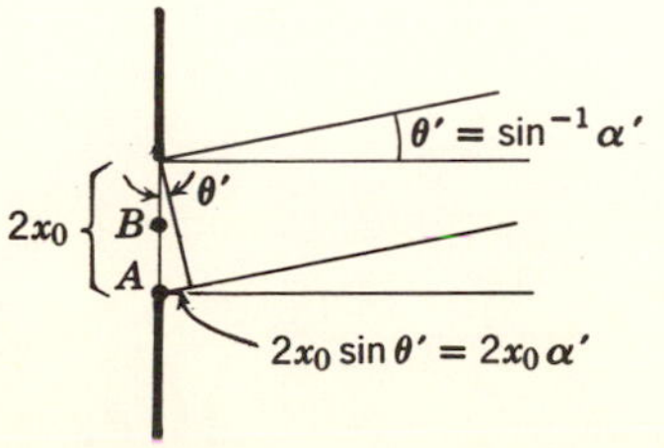

Fig. 7.14

Now refer to Fig. 7.14. When $\beta = 0$, the direction cosine α' is the same as the sine of θ'. When

$$2x_0 \sin\theta' = 2x_0\alpha' = \lambda$$

the electric fields in Huygens wavelets from points A and B headed in the direction θ will be π radians out of phase. When they are added, they will mutually cancel. We thus get pairwise cancellation of the contribution from a point at x' and from the point at $x' + x_0$. In this case we have divided the aperture into two equal parts, A and B, and paired off the contributions from each part. When

$$2\alpha' x_0 = 2\lambda$$

we can imagine the aperture divided into four equal parts, A, B, C, and D, with the contributions from A and B canceling each other and the same for C and D. The same can be done for six equal parts, eight equal parts, and so on.

3. ***Vibration Curves.*** The integral in Eq. (7.40) can be represented by means of a vibration curve in the complex plane. Such a representation will also be quite useful for other diffraction problems. We can write

$$I_x = \int_{-x_0}^{+x_0} dI_x$$

where

$$dI_x = e^{-2\pi i u x'} \frac{dx'}{2x_0} = e^{i\theta(x')}\, ds$$

can be represented as a vector in the complex plane of length $ds = (dx'/2x_0)$ and phase angle $\theta(x') = -2\pi u x' = -kx'(\alpha - \alpha')$.

The general form of the integral for I_x is then

$$I_x = \int ds e^{i\theta(x)}$$

which represents the vector sum of all the infinitesimal vectors $dse^{i\theta}$. We can approximate the infinitesimal vectors $dse^{i\theta}$ by finite, but small, vectors $\Delta se^{i\theta}$. Then I_x is approximated by the sum $\Sigma \Delta se^{i\theta}$. By the polygon method of vector addition, this sum is given by a vector along the chord of the (usually open) polygon whose sides are formed by the vectors $\Delta se^{i\theta}$. As $\Delta s \rightarrow 0$ the polygon becomes a smooth curve, the *vibration curve*, and the integral for I_x is given by the complex vector along the chord of this curve from the start of the curve to the end. See Fig. 7.15.

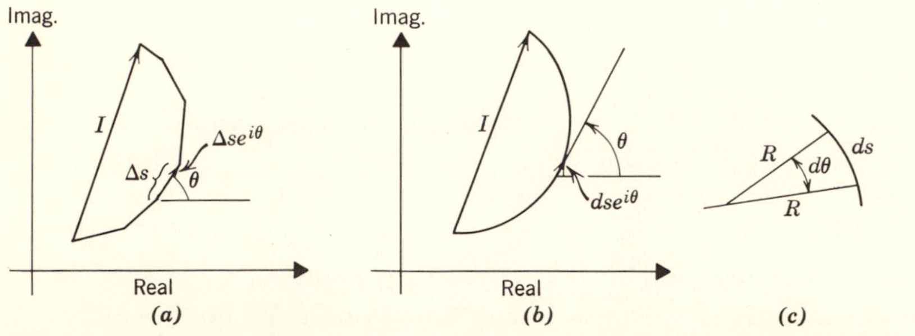

Fig. 7.15

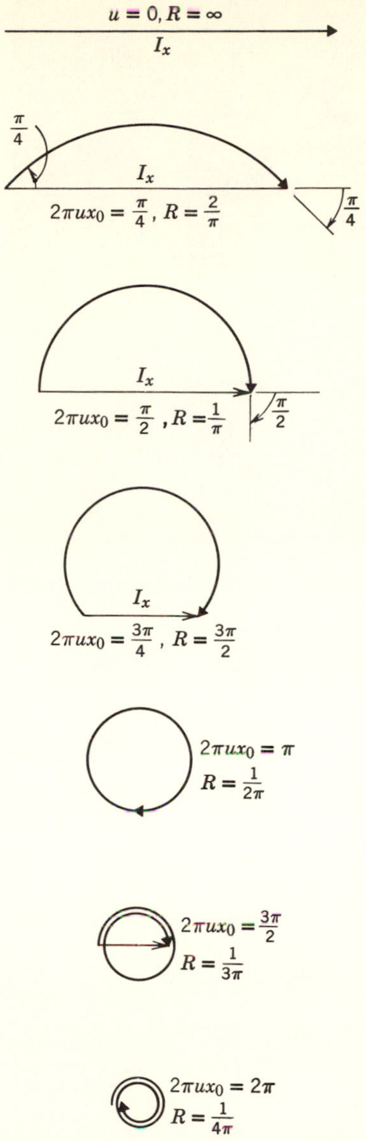

Fig. 7.16. Vibration curves representing the diffraction integrals for Fraunhofer diffraction by a rectangular opening.

The total *arc length* of the vibration curve is given in general by

$$S = \int ds$$

and, in the case of the vibration curve for I_x, it is

$$S = \int_{-x_0}^{x_0} \frac{dx'}{2x_0} = 1$$

The radius of curvature of a vibration curve is given in general by

$$R = \frac{ds}{d\theta} \tag{7.42}$$

(see Fig. 7.15*c*). For our special case this becomes

$$R = -\frac{dx'/(2x_0)}{2\pi u dx'} = -\frac{1}{4\pi u x_0}$$

which is a constant along the vibration curve. Hence the vibration curve in this case is circular, and the radius of the circle is R. The minus sign means that the curve is concave downward. The initial and final phase angles θ_i and θ_f are given by

$$\theta_i = 2\pi u x_0 \qquad \text{and} \qquad \theta_f = -2\pi u x_0$$

Some of these circular vibration curves are shown in Fig. 7.16.

4. The Size and Shape of the Diffraction Pattern. The behavior exhibited by the function I_y is exactly analogous to that of I_x. When Eq. (7.40) is used in (7.39) the flux density at P becomes

$$E_e(\mathbf{r}) = E_e(\mathrm{O}) \frac{16x_0^2y_0^2}{\lambda^2(R_0')^2} \frac{\sin^2 2\pi u x_0}{(2\pi u x_0)^2} \frac{\sin^2 2\pi v y_0}{(2\pi v y_0)^2}$$
$$u = \frac{(\alpha - \alpha')}{\lambda}, \qquad v = \frac{(\beta - \beta')}{\lambda} \tag{7.43}$$

The maximum value of E_e occurs when $\alpha' = \alpha$ and $\beta' = \beta$. This means that the source point P_s, the center of the aperture O, and the observation point P are all collinear (Fig. 7.17). This causes all the Huygens wavelets from the open aperture to arrive in phase at P so that the integral in Eq. (7.33) becomes

$$\iint_{\Sigma_0} d\sigma = \int_{-x_0}^{x_0} \int_{-y_0}^{y_0} dx'\, dy' = 4x_0 y_0$$

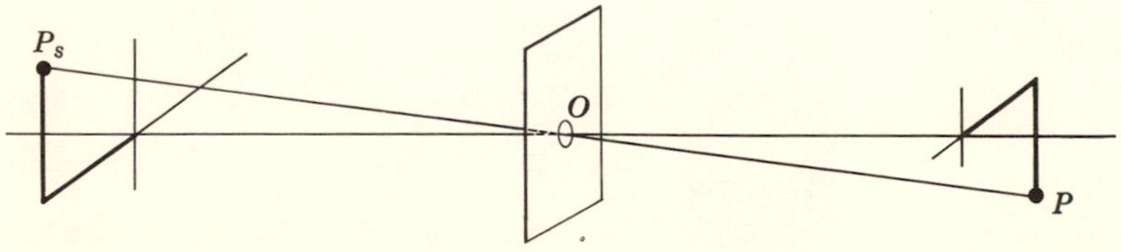

Fig. 7.17

The nature of the two-dimensional distribution of flux density is indicated in Fig. 7.18. The width of the central maximum from

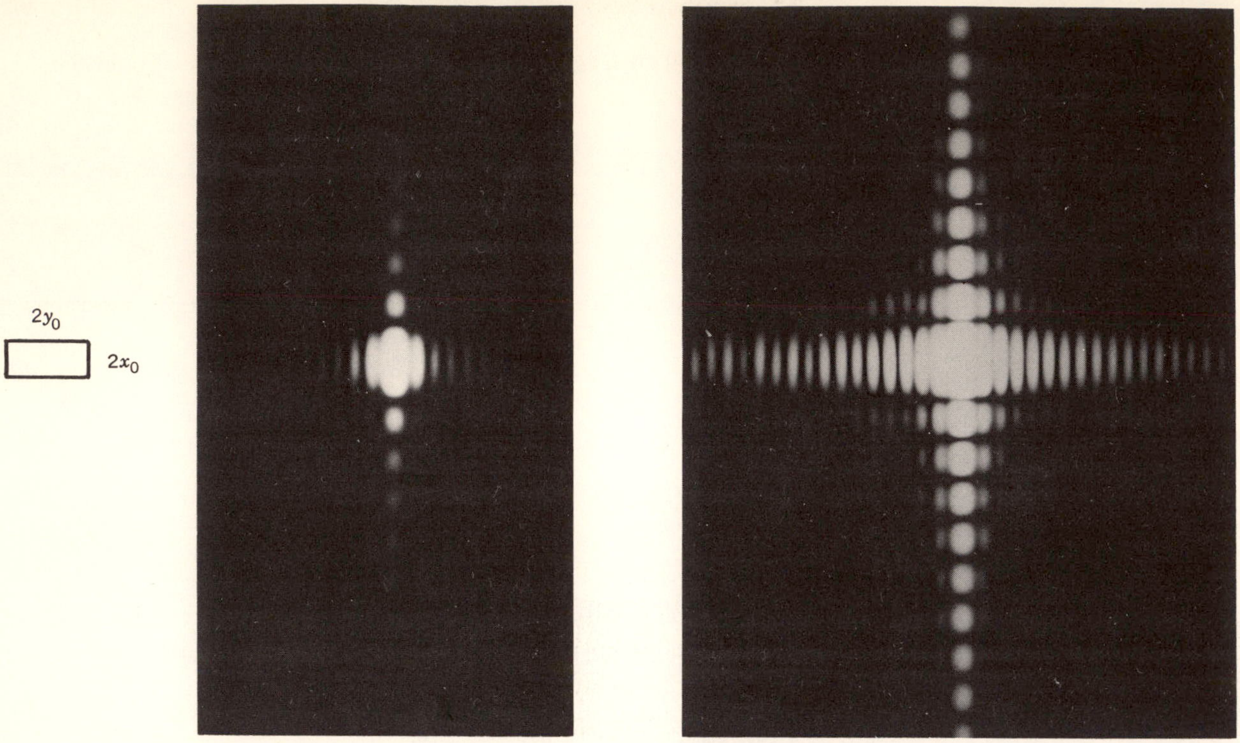

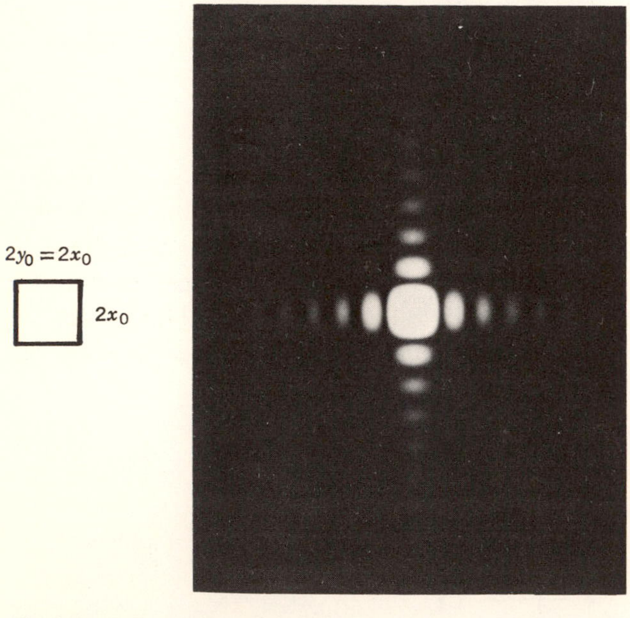

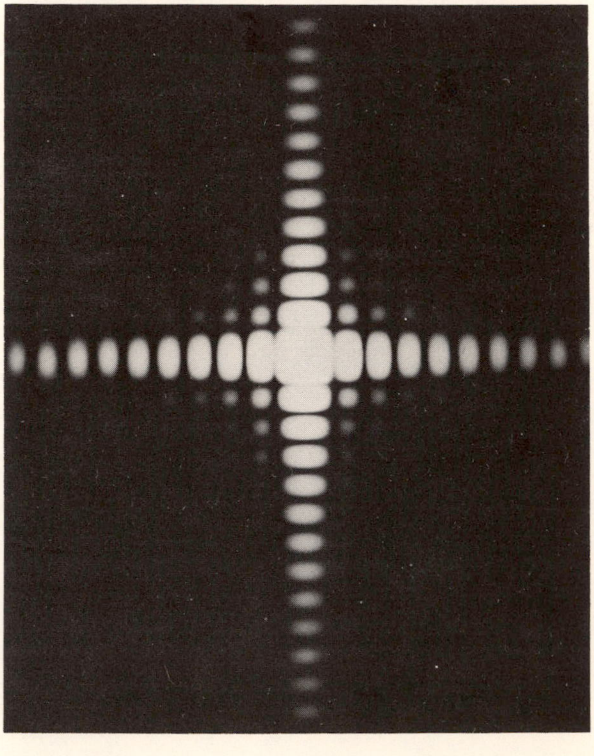

Fig. 7.18. Rectangular and square apertures and their Fraunhofer diffraction patterns at two different exposures.

zero to zero is

$$\Delta\alpha' = \frac{\lambda}{x_0}$$
$$\Delta\beta' = \frac{\lambda}{y_0} \tag{7.44}$$

Note the inverse relationship to the aperture dimensions!

There are several other versions of Eq. (7.44). Noting that $\Delta\alpha' = \Delta x/R_0'$ and $\Delta\beta' = \Delta y'/R_0'$, we can write (7.44) as

$$\Delta x = \frac{\lambda R_0'}{x_0}$$
$$\Delta y = \frac{\lambda R_0'}{y_0} \tag{7.44'}$$

This gives the linear width of the central maximum on the observation screen. By multiplying by $1/x_0$ and $1/y_0$, we obtain

$$\frac{\Delta x}{x_0} = \frac{\lambda R_0'}{x_0^2} \gg 1$$

and (7.45)

$$\frac{\Delta y}{y_0} = \frac{\lambda R_0'}{y_0^2} \gg 1$$

The inequalities follow from the Fraunhofer diffraction condition, Eq. (7.32). We see that the dimensions of the Fraunhofer diffraction pattern are much greater than those of the diffracting aperture.*

For small α', β' we can interpret $\Delta\alpha'$ and $\Delta\beta'$ as the angular size of the central maximum. For larger values of α' and β' a short calculation is necessary to give a precise interpretation of $\Delta\alpha'$ and $\Delta\beta'$. Suppose that x_0 and y_0 are sufficiently larger than λ so that $\Delta\alpha'$ and $\Delta\beta'$ in Eq. (7.44) are small. Then the diffraction pattern is concentrated close to its center at P_0 where $\overrightarrow{OP_0}$ has coordinates $R_0'(\alpha, \beta, \gamma)$. The three points P_s, O, and P_0 are colinear as shown in Fig. 7.19.

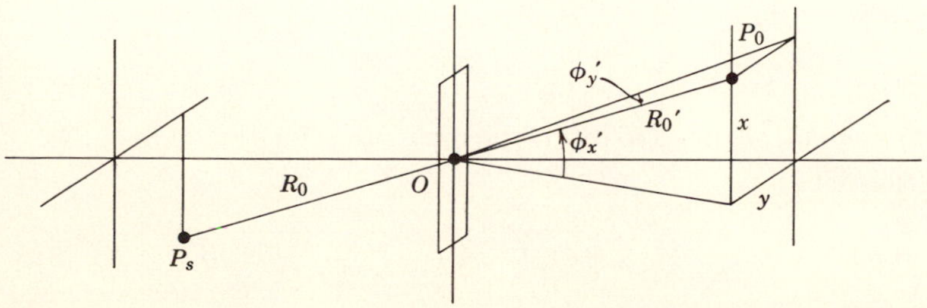

Fig. 7.19

*This result holds only for Fraunhofer diffraction occurring at great distances without the use of a lens.

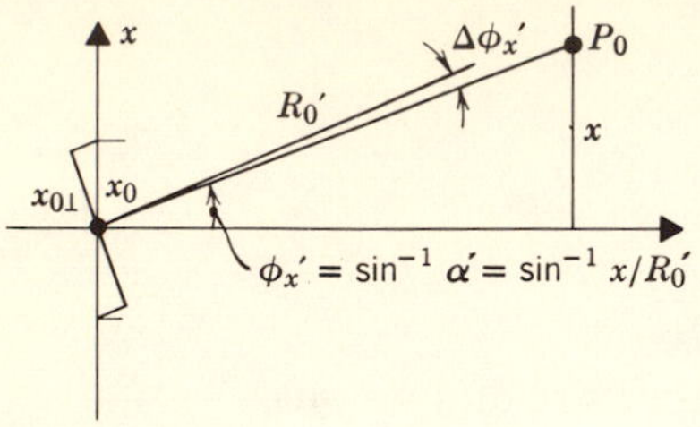

Fig.7.20

We show the plane containing OP_0 and the x axis in Fig. 7.20. The projection of the aperture dimension in the plane normal to OP_0 is

$$x_{0\perp} = x_0 \cos \phi_x' = x_0\sqrt{1-\alpha'^2} \tag{7.46a}$$

A small angular spread in the x direction is given by

$$\Delta\phi_x' = \Delta(\sin^{-1}\alpha') = \frac{\Delta\alpha'}{\cos\phi_x'} = \frac{\Delta\alpha'}{\sqrt{1-\alpha'^2}}$$

Thus the angular spread associated with the first of Eqs. (7.44) is

$$\Delta\phi_x' = \frac{\Delta\alpha'}{\cos\phi_x'} = \frac{\lambda}{x_0\cos\phi_x'} = \frac{\lambda}{x_{0\perp}} \tag{7.47a}$$

In a similar way we may define

$$\phi_y' = \sin^{-1}\beta' = \sin^{-1} y/R_0'$$

and the projected dimension becomes

$$y_{0\perp} = y_0 \cos\phi_y' = y_0\sqrt{1-\beta'^2} \tag{7.46b}$$

so that the angular spread in the y direction becomes

$$\Delta\phi_y' = \frac{\lambda}{y_{0\perp}} \tag{7.47b}$$

*5. **Use of Projected Apertures.*** From the foregoing Eqs. (7.47*a*,*b*) we expect that various apertures having the same dimensions when projected onto the plane perpendicular to the line OP_0 will have the same diffraction pattern. This can be seen for a general aperture from Eq. (7.33). If we write $\alpha' - \alpha = \Delta\alpha'$ and define the projected coordinate

$$x_\perp' = x'\cos\phi_x' = x'\sqrt{1-\alpha'^2}$$

then we obtain

$$\begin{aligned}(\alpha'-\alpha)x' = \Delta\alpha' x' &= \frac{\Delta\alpha'}{\sqrt{1-\alpha'^2}}x'\sqrt{1-\alpha'^2} \\ &= \Delta\phi_x' x_\perp'\end{aligned} \tag{7.48a}$$

Similarly

$$(\beta'-\beta)y' = \Delta\phi_y' y_\perp' \tag{7.48b}$$

At this point we need to insert the inclination factor Q into (7.33). By Eq. (7.33′) it is

$$Q = \tfrac{1}{2}(\gamma+\gamma') \simeq \gamma = \sqrt{1-\alpha'^2-\beta'^2} \approx \sqrt{(1-\alpha'^2)(1-\beta'^2)}$$

Then $Q\,dx'\,dy'$ becomes approximately

$$\sqrt{(1-\alpha'^2)}\,dx'\sqrt{1-\beta'^2}\,dy' = dx_\perp'\,dy_\perp'$$

With these changes Eq. (7.33) is transformed into an integral over the projection of the aperture onto the plane perpendicular to OP_0:

$$\tilde{E}(r) = \frac{iAe^{-ik(R_0+R_0')}}{R_0 R_0'} \iint_{\Sigma_0} \exp\left[\frac{2\pi i}{\lambda}(\Delta\phi_x' x_\perp' + \Delta\phi_y' y_\perp')\right] dx_\perp'\, dy_\perp' \qquad (7.49)$$

Any small rotation of the aperture about an axis lying in its plane will not change its projection very much; hence the diffraction pattern will remain essentially the same. This behavior is often observed in practice.

C. FRAUNHOFER DIFFRACTION BY CIRCULAR APERTURES

Circular apertures occur in many optical instruments, and the Fraunhofer diffraction they produce can play an important role in the performance of these instruments. We discuss first the Fraunhofer diffraction produced at great distances without lenses.

The effective shape of the aperture is determined by its projection onto the plane perpendicular to the line from the source P_s to the center of the aperture at O. The projection of a circle is an ellipse, but results obtained from elliptical projected apertures of small eccentricity will be quite similar to those from circular projected apertures. Because of its higher symmetry, the circular case is easier to compute. Without loss of generality we assume that the aperture itself is circular with its normal parallel to the line P_sO. This is equivalent to setting $\alpha = \beta = 0$. Figure 7.21 shows the arrangement and defines the coordinates. Here r' and δ are polar coordinates in the aperture. When $\delta = 0$, the lines OP', OP and the symmetry axis all lie in the plane of the paper.

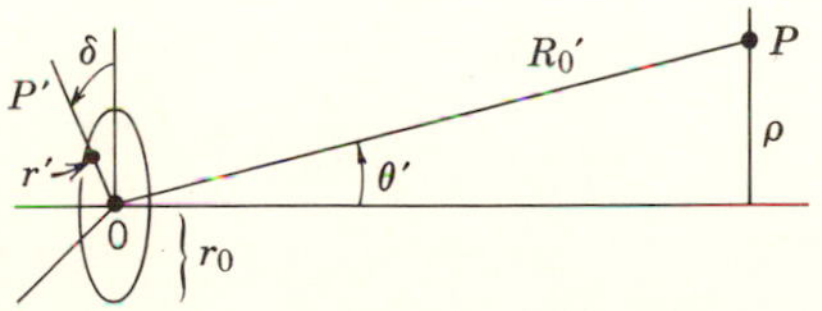

Fig. 7.21

The exponent in the integrand of Eq. (7.34) can be written

$$\text{exponent} = \frac{2\pi i}{\lambda}(\alpha' x' + \beta' y') = \frac{2\pi i}{\lambda R_0'}(xx' + yy')$$

Now we can write

$$xx' + yy' = (OP \cdot OP') = r'\rho\cos\delta = r'R_0'\sin\theta'\cos\delta$$

Then, using $d\sigma' = r'dr'd\delta$, we can obtain from Eq. (7.34) the result

$$\tilde{E}(\mathbf{r}) = \frac{i}{\lambda}\tilde{E}(\mathrm{O})\frac{e^{-ikR_0'}}{R_0'}\int_0^{r_0} r'\,dr' \int_0^{2\pi} e^{-ikr'\sin\theta'\cos\delta}\,d\delta \qquad (7.50)$$

The evaluation of Eq. (7.50) will be shown below to yield the following result:

$$\tilde{E}(\mathbf{r}) = \tilde{E}(\mathrm{O})\frac{e^{-ikR_0'}}{R_0'}\,ikr_0^2\,\frac{J_1(kr_0\sin\theta')}{kr_0\sin\theta'} \qquad (7.51)$$

Let

$$t = kr_0 \sin\theta' = \frac{2\pi r_0 \rho}{(\lambda R_0')} = 2\pi r_0 w$$

The connection with our former coordinates u, v is

$$w = \sqrt{u^2 + v^2} = \sqrt{\frac{\alpha'^2 + \beta'^2}{\lambda}}$$

Then

$$\tilde{E}(\mathbf{r}) = \tilde{E}(\mathrm{O})\frac{e^{-ikR_0'}}{R_0'} ikr_0{}^2 \frac{J_1(t)}{t} \tag{7.51$'$}$$

The function $J_1(t)$ is known as a *first-order Bessel function*. The flux density is given by

$$E_e(\mathbf{r}) = E_e(\mathrm{O})\frac{k^2}{R_0'^2} r_0{}^4 \left[\frac{J_1(t)}{t}\right]^2 \tag{7.52}$$

The last factor in this equation is plotted in Fig. 7.22. The general behavior is similar to that of $(\sin t/t)^2$, but the "feet" beyond the

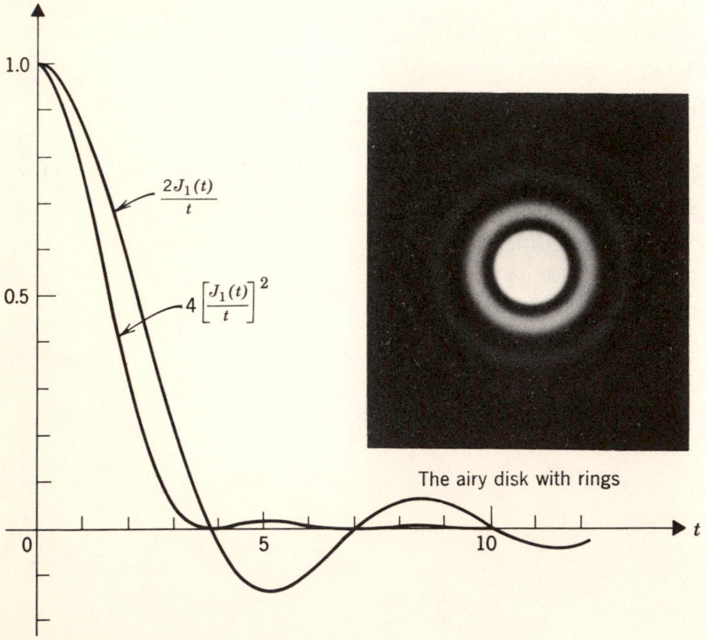

Fig. 7.22. Fraunhofer diffraction pattern for a circular aperture.

first zero are weaker. Eighty-four per cent of the total area in the pattern is in the first maximum. The secondary maxima are considerably weaker. The main maximum gives a circular pattern called the *Airy disk*. Its radius corresponds to the value of t at the first zero, which is $t = 1.22\pi = 3.83$. The radius of the disk is

$$r_{\text{disk}} = \frac{0.61\lambda R_0'}{r_0} \tag{7.53}$$

The angular radius of the disk is

$$\Delta\theta' \simeq \Delta \sin\theta' = \frac{0.61\lambda}{r_0} \tag{7.53$'$}$$

These results should be compared with the value of $\Delta x/2$ for a $2x_0$-by-$2x_0$ square at the first zero that we obtain from Eq. (7.44′),

$$\frac{\Delta x}{2} = \frac{0.50\lambda R_0'}{x_0} \tag{7.54}$$

or corresponding angular width of

$$\frac{\Delta\theta_x'}{2} = \frac{0.50\lambda}{x_{0_\perp}} \tag{7.54$'$}$$

We now derive Eq. (7.51). The θ' integral in Eq. (7.50) may be evaluated with the help of the formula

$$\frac{1}{2\pi}\int_0^{2\pi} e^{it\cos\delta}\,d\delta = J_0(t)$$

where $J_0(t)$ is called a *Bessel function of zero order*. Thus, in our case we have

$$\int_0^{2\pi} e^{ikr'\sin\theta'\cos\delta}\,d\delta = J_0(t')\cdot 2\pi$$

where $t' = kr'\sin\theta'$. Now there are higher-order Bessel functions $J_n(t)$ given by

$$J_n(t) = \frac{(i^{-n})}{2\pi}\int_0^{2\pi} e^{it\cos\theta}e^{in\theta}\,d\theta$$

These Bessel functions can be shown to be related by the recursion formula:

$$t\frac{dJ_n}{dt} + nJ_n = tJ_{n-1}$$

If we set $n = 1$, we get

$$t\frac{dJ_1}{dt} + J_1 = tJ_0, \qquad \text{or} \qquad \frac{d}{dt}(tJ_1) = tJ_0$$

This can be integrated to give

$$tJ_1(t) = \int_0^t t'J_0(t')\,dt' \tag{7.55}$$

We have thus shown that the θ' integration of our diffraction problem gives J_0:

$$\tilde{E}(\mathbf{r}) = \tilde{E}(\mathrm{O})\,\frac{ike^{-ikR_0}}{R_0'}\int_0^{r_0} r'\,dr'\,J_0(kr'\sin\theta')$$

The r' integration may now be performed via (7.55) with $t = kr_0 \sin\theta'$, $t' = kr'\sin\theta'$ to give

$$\begin{aligned}\tilde{E}(\mathbf{r}) &= \tilde{E}(\mathrm{O})\,\frac{ike^{-ikR_0'}}{R_0'}\,\frac{1}{(k\sin\theta')^2}\,J_1(kr_0\sin\theta')\,kr_0\sin\theta' \\ &= \tilde{E}(\mathrm{O})\,\frac{ike^{-ikR_0'}}{R_0'}\,r_0^2\,\frac{J_1(kr_0\sin\theta')}{kr_0\sin\theta'} \qquad (7.51)\end{aligned}$$

Numerical values of $J_1(t)$ may be found tabulated in reference books.

D. LIMITS ON THE PERFORMANCE OF OPTICAL INSTRUMENTS BECAUSE OF DIFFRACTION

1. Fraunhofer Diffraction with Lenses. The finite size of the Airy disk or similar diffraction patterns will place an ultimate limit on the performance of optical instruments. Before we discuss some specific examples, we shall show how *Fraunhofer diffraction can occur in the image plane of an optical instrument with a point source.* In the cases examined so far the distance R_0' was so large that all light rays from the diffracting aperture to the observation point P were essentially parallel, as shown in Fig. 7.23*a*. A linear variation of the phase proportional to $\alpha'x' + \beta'y'$ (that is, linear in the direction cosines of these rays) was found.

These parallel rays propagating in direction $(\alpha', \beta', \gamma')$ can also be brought together in a finite distance with the help of a lens L_2, as shown in Fig. 7.23*b*. The point P having coordinates (x, y, f_2) with respect to the second principal point of the lens, where f_2 is the focal length, will be a geometrical focus for all rays with direction cosines $\alpha' = x/R_0'$, $\beta' = y/R_0'$, with $R_0' = \sqrt{f_2^2 + x^2 + y^2}$. If the aperture is close to the lens, the resulting pattern is mathematically the same as that produced with no lens, but it now involves this very much smaller value of R_0' that need not satisfy the lensless Fraunhofer condition (7.32).

A more detailed discussion of this sort of problem will be given in Chapter 9, where the above statements will be proved more rigorously. It turns out that the above remarks about the effect of the lens apply equally well if the aperture is placed just to the right of the lens as if it is placed just to the left. Diffraction occurs because the wavefront emerging from the system is essentially part of a spherical surface.

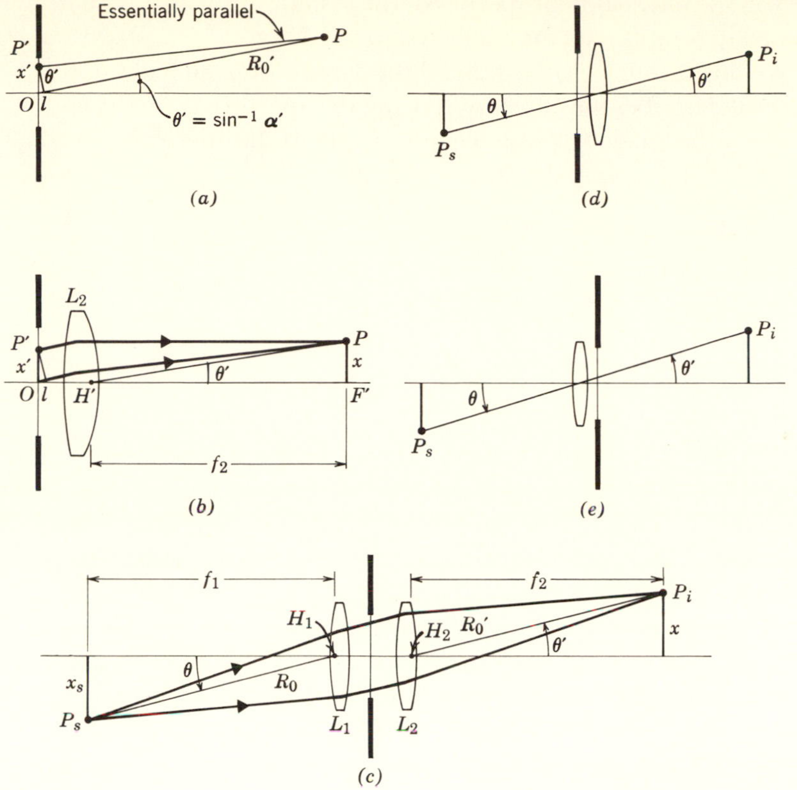

Fig. 7.23

Another lens L_1 may be similarly used to bring the point source P_s in from what is essentially infinity while preserving the above mathematical results. The emerging wavefront is the same partial sphere. The appropriate variables here are the direction cosines $\alpha = -x_s/R_0$, $\beta = -y_s/R_0$ and the distance $R_0 = \sqrt{f_1^2 + y_s^2 + x_s^2}$. The source at P_s is placed in the focal plane of L_1, as shown in Fig. 7.23*c*. Here L_1 and L_2 are both very close to the diffracting aperture. After passing through L_1, the light from P_s is collimated parallel to the direction of the vector $P_sH_1 = (-x_s, -y_s, f_1) = R_0(\alpha, \beta, \gamma)$.

The combined effect of the two lenses is to produce an "image" of P_s at P_i. (Note that $\overrightarrow{P_sH_1}$ and $\overrightarrow{H'P_i}$ are parallel if the index of refraction is the same on both sides of the system.) But instead of a true point image, what is observed near P_i is the Fraunhofer diffraction pattern of the aperture, for example, the Airy disk with rings if the aperture is a circle. The effect is the same if the two lenses are combined into a single lens on either side of the aperture, as shown in Fig. 7.23*d*,*e*.

2. ***Effects of Moving the Aperture.*** The aperture we are discussing is the aperture stop of the system. In addition to its geometrical role in limiting the angular spread of rays from the source point that may pass through the system, it is now seen to play a dominant role in the diffraction properties of the system, a role that is maintained in complicated multilens systems.

We now quote part of a general result to be derived in Section 9.3 that describes the effects of moving the aperture from the lens towards the image P_i, as shown in Fig. 7.24. If the aperture size is changed so that it has the same angular shape as seen from P_i (that is, so that it has a constant projection from P_i on any convenient plane, say the second principal plane at H' of the lens), then the intensity distribution in the diffraction pattern of the aperture will be unchanged. (The phases in various parts of the pattern will change as the aperture is moved, however.) This result is also true if the aperture is to the left of the lens in position 4 or 5. Then its image in image space, that is, the exit pupil, must have the proper projection on the H' plane. In all cases, then, we may say that *if the projection of the exit pupil through the point P_i onto the second principal plane remains the same, the intensity distribution near P_i will be unchanged.*

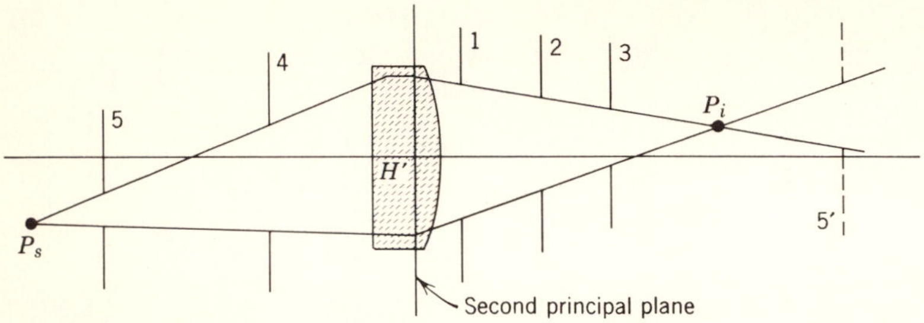

Fig. 7.24. Positions 1, 2, 3, 4, 5 label possible locations for the diffracting aperture; 5′ is the image of 5.

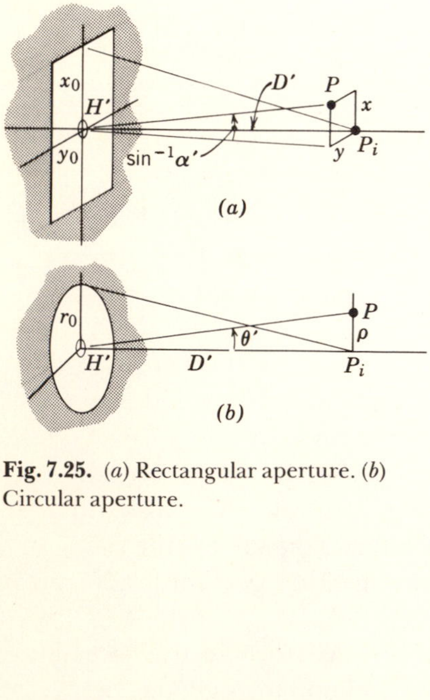

Fig. 7.25. (*a*) Rectangular aperture. (*b*) Circular aperture.

Thus it is the size and shape of this projected exit pupil that determine the intensity distribution in the "image" of P_s at P_i. Projected pupils are shown explicitly in Fig. 7.25 for two aperture shapes. The geometrical point image is a distance D' away from the second principal plane. It serves as an origin for the Cartesian coordinates x, y of the observation point P in the image plane. For small x and y, D' plays the role of the distance R_0 used earlier. For the rectangular aperture of Fig. 7.25*a* the flux density at P will, in analogy with Eqs. (7.39) and (7.40), be proportional to

$$E_e(P) \propto \frac{\sin^2 \xi}{\xi^2}\frac{\sin^2 \eta}{\eta^2}$$

with

$$\xi = 2\pi u x_0 = 2\pi\left(\frac{x}{D'}\right)\left(\frac{x_0}{\lambda}\right) = 2\pi\left(\frac{x_0}{D'}\right)\left(\frac{x}{\lambda}\right)$$

and (7.56)

$$\eta = 2\pi v y_0 = 2\pi\left(\frac{y}{D'}\right)\left(\frac{y_0}{\lambda}\right) = 2\pi\left(\frac{y_0}{D'}\right)\left(\frac{y}{\lambda}\right)$$

Here (x/D') and (y/D') represent the angular coordinates of the observation point P as seen from the principal point H'. On the other hand, we note that for small x_0 and y_0, (x_0/D') and (y_0/D') represent the angular half-width of the aperture as seen from P_i. For the circular opening of Fig. 7.25*b* we have

$$E_e(P) \propto \left[\frac{J_1(t)}{t}\right]^2$$

with (7.57)

$$t = \frac{2\pi r_0 \sin\theta'}{\lambda} \cong 2\pi\left(\frac{r_0}{D'}\right)\left(\frac{\rho}{\lambda}\right) = 2\pi\theta'\left(\frac{r_0}{\lambda}\right)$$

$$\rho = \sqrt{x^2+y^2}$$

Here θ' is the angular coordinate of P as seen from the principal point, and from the other point of view for small r_0 we may interpret (r_0/D') as the angular radius of the aperture as seen from P_i.

With this general background we now turn to a discussion of specific systems.

*3. **Telescopic Resolving Power.*** The finite size of the Airy disk will place an ultimate limit on the angular resolving power of a telescope. Suppose, for example, that we are looking at a double star through a telescope having an objective with focal length f and aperture diameter D. The two components of the double star will be assumed to be equally intense and to have an angular separation $\Delta\phi$ as shown in Fig. 7.26*a*. The diffraction pattern

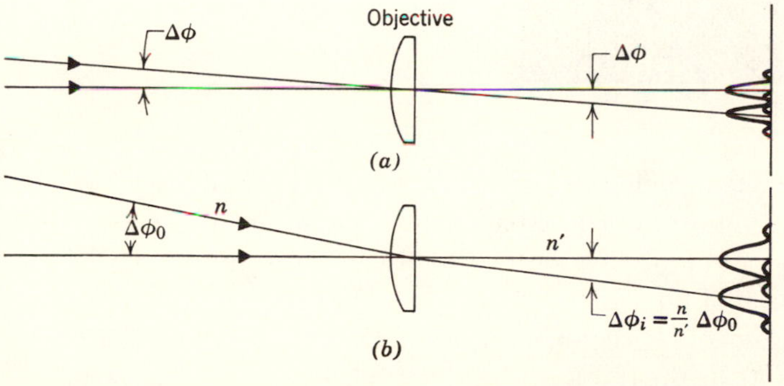

Fig. 7.26

appearing in the focal plane of the objective will consist of the superposition of two Airy disks having an angular separation $\Delta\phi$ together with their surrounding rings.* They will be resolved as two patterns if $\Delta\phi$ is much larger than the angular width of the

*For normal visual observation these diffraction patterns will be magnified without appreciable additional diffraction by the eyepiece, since the aperture stop of the overall system is the telescope objective.

disk and not resolved if $\Delta\phi$ is much smaller. The critical value of $\Delta\phi$ for which resolution is just possible is usually taken to be that given by Rayleigh. It occurs when $\Delta\phi$ equals the angular radius of the Airy disk, which from Eq. (7.53′) gives

$$\Delta\phi_{\min} = \frac{0.61\lambda}{r_0} = \frac{1.22\lambda}{D} \tag{7.58}$$

A photograph of this double diffraction pattern at the Rayleigh limit is shown in Fig. 7.27. Many telescopes are not of high enough

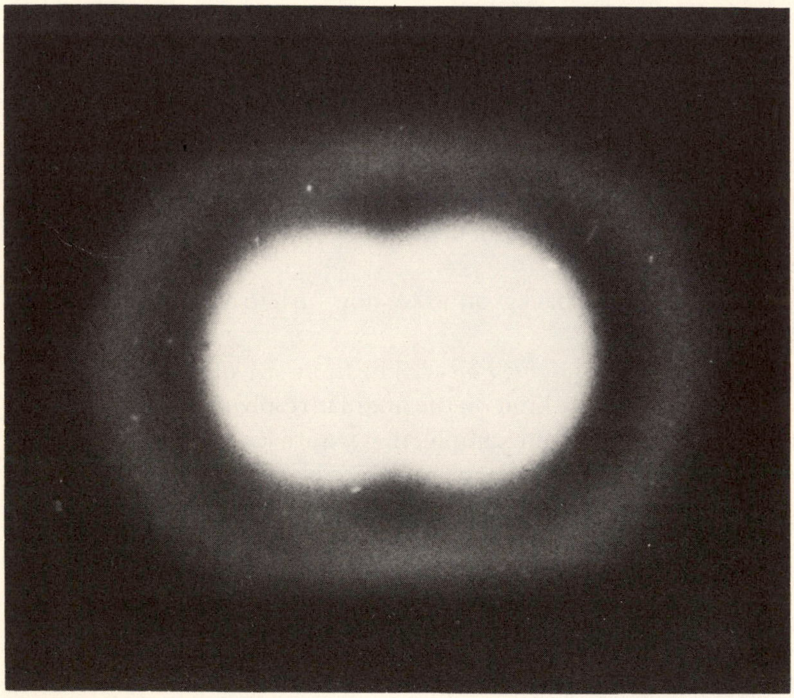

Fig. 7.27. Image of a double star at the Rayleigh limit.

optical quality to yield this "diffraction-limited" resolution. Even if they are, atmospheric fluctuations may cause the image to move slightly in a random way, so that the practical resolution may be less than that of Eq. (7.58).

If the index of refraction n differs from unity, λ in Eq. (7.58) is assumed to be the reduced wavelength $\lambda = \lambda_0/n$. Thus, for the situation in Fig. 7.26*b* in object space the angle would be

$$\Delta\phi_0 = \frac{1.22\lambda_0}{nD}$$

whereas in image space it would be

$$\Delta\phi_i = \frac{1.22\lambda_0}{n'D}$$

4. Light-Gathering Power of Telescopes. For visual observation under so-called *normal magnification*, which makes most efficient use of both the telescope and the eye, the exit pupil of the telescope is the same size as the pupil of the observer's eye (Fig. 7.28).

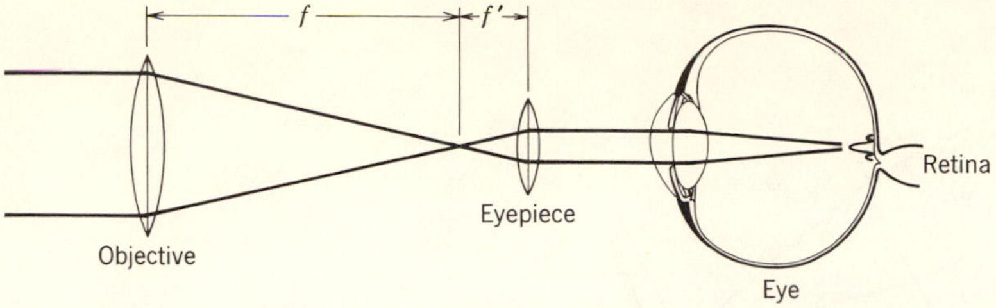

Fig. 7.28. Telescope plus eye as a combined optical system.

For the combined system of telescope plus eye, the aperture stop may be considered to be either the objective lens or the eye pupil—it makes no difference—and the diffraction pattern on the retina of the eye will be the same Airy disk that would be obtained without the telescope.

Thus the telescope cannot change the size of the retinal diffraction pattern from a distant point object. What it can do is increase the radiant flux density or irradiance (in watts/mm^2) (or the luminous flux density or illuminance [in lumens/mm^2]) falling on the retina in the Airy disk. With the unaided eye the flux passing through the pupil is focused onto the disk, whereas with the telescope the flux passing through its objective is focused onto the disk, and this is greater by the ratio of the area of the objective to the area of the pupil, that is, by the ratio of the squares of the diameters,

$$\left(\frac{d_{\text{obj}}}{d_{\text{pupil}}}\right)^2 = m^2$$

where $m = f/f'$ is the magnification of the telescope. The telescope has m^2 times more *light-gathering power* than the unaided eye.

We can say that the use of a telescope for visual astronomical observations has two important advantages: (1) it increases the separation of diffraction patterns of individual stars on the retina by the ratio m, and (2) it increases the flux in each pattern by the ratio m^2. If the stars are seen on a continuous background such as the daytime sky, then the irradiance in the retinal image of this background will be independent of the magnification m (m^2 times more flux is incident on m^2 times more retinal area). Thus the use of a telescope will increase the contrast of the irradiance in the star's Airy disk over that of the background by the ratio m^2. Bright stars then become visible in daytime.

5. Spectrograph. Diffraction ultimately limits the resolving power of a prism spectrograph, monochromator, or similar instrument of the type shown in Fig. 7.29. Because of its high cost the prism itself usually forms the aperture stop of such instruments. Light from the entrance slit is collimated by the collimating lens L_1, dispersed by the prism, and then focused on the photographic plate

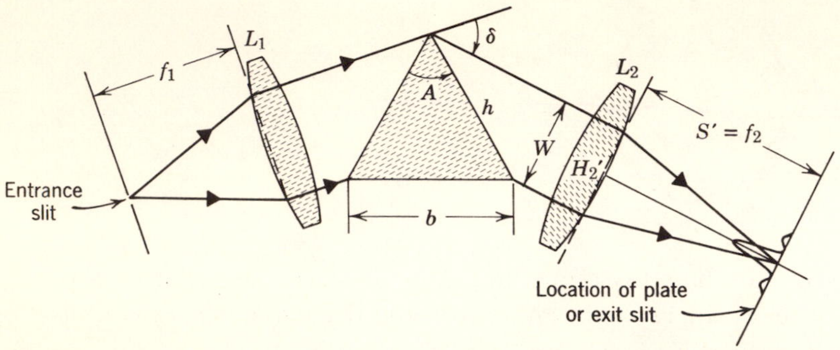

Fig. 7.29. Essential elements of a prism spectrograph.

by a camera lens L_2 or, in a monochromator, light is focused on an exit slit by a lens as in Fig. 6.3.

If the entrance slit were replaced by a pinhole, and if the light were truly monochromatic, the light would be exactly collimated and deviated through a well-defined angle δ that depends on the wavelength λ. In the focal plane of lens L_2 we would observe the Fraunhofer diffraction pattern of the rectangular prism as projected onto a plane through H_2' normal to the final direction of propagation. This pattern would be something like that in Fig. 7.18.

With a *narrow slit* instead of a *pinhole* at the entrance, we obtain a superposition of infinitely many such patterns displaced along the direction of the slit which is out of the paper in Fig. 7.29. The resulting intensity pattern will vary in the direction normal to the slit exactly as does the "single slit" pattern of Fig. 7.13. The effective "slit width" of this pattern is the projected height of the prism denoted by W in Fig. 7.29. Thus the angular distance $\Delta\phi = \Delta x/(2S')$ from the center of the diffraction pattern to the first zero, which is also the minimum resolvable angle by the Rayleigh criterion, is given by Eq. (7.44′) with $S' = R_0$, $W = 2x_0$ to be

$$\Delta\phi = \frac{\lambda}{W} \tag{7.59}$$

Suppose that the prism is operated near minimum deviation so that the deviation angle δ is close to the minimum value δ_m which is given in terms of the prism angle A and the refractive index n by

$$n = \frac{\sin\frac{1}{2}(A+\delta_m)}{\sin\frac{1}{2}A} \tag{7.60}$$

(See Problem 1.3.6 for an explanation.)

The angular spread $\Delta\phi$ implies an uncertainty in the wavelength determination for a given spectral line of an amount $\Delta\lambda = (d\lambda/d\delta)\Delta\phi$, where $d\lambda/d\delta$ is the reciprocal of the angular dispersion of the instrument. The spectroscopic resolving power $\lambda/\Delta\lambda$ is then given by

$$\frac{\lambda}{\Delta\lambda} = W\frac{d\delta}{d\lambda}$$

The wavelength dependence of the deviation angle δ is given implicitly for δ near δ_m by the $n(\lambda)$ dependence through Eq. (7.60). In fact, we can write $d\delta/d\lambda \approx d\delta_m/d\lambda$ and obtain

$$\left(\sin\frac{1}{2}A\right)\frac{dn}{d\lambda} \cong \frac{1}{2}\cos\frac{1}{2}(A+\delta_m)\frac{d\delta_m}{d\lambda}$$

$$\frac{\lambda}{\Delta\lambda} = W\frac{d\delta_m}{d\lambda} = \frac{2W\sin\frac{1}{2}A}{\cos\frac{1}{2}(A+\delta_m)}\frac{dn}{d\lambda}$$

From Fig. 7.30 we see that the slant height h of the prism is given by

$$h = \frac{W}{\cos\frac{1}{2}(A+\delta_m)}$$

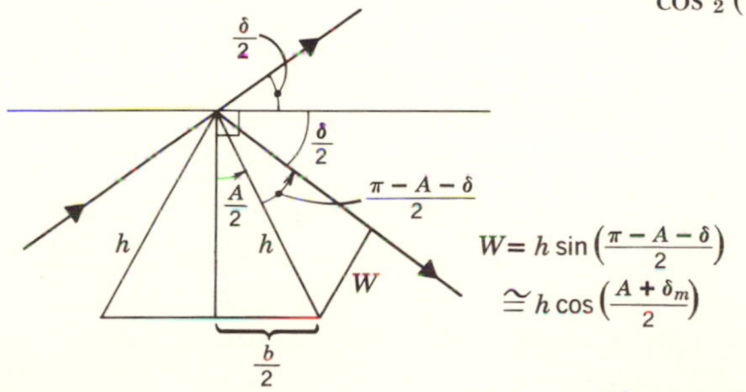

Fig. 7.30

Thus we have $\lambda/\Delta\lambda = 2h(\sin\frac{1}{2}A)\,dn/d\lambda$. But $h\sin\frac{1}{2}A$ can be seen from Fig. 7.30 to equal $b/2$, where b is the length of the prism base. Thus the resolving power near minimum deviation is simply

$$\frac{\lambda}{\Delta\lambda} = b\frac{dn}{d\lambda} \tag{7.61}$$

6. ***Microscope with Incoherent Illumination.*** So far we have assumed that the light from the two point sources to be resolved is incoherent. This will be the case for self-luminous sources. In a microscope the object is illuminated by a condensing system and is usually not self-luminous. The transverse coherence length of the light illuminating the object is usually larger than the separation between the two points we wish to resolve. We must then treat the light from such an object as coherent. The discussion of resolving power changes somewhat in this case and will be postponed until we discuss the Abbe theory of image formation.

But for a microscope with incoherent illumination we can find the minimum resolvable separation as follows. Consider Fig. 7.31, which shows the objective lens. The objects P_s and P_s' are separated a distance x_s and are imbedded in a medium with index of refraction n (in an oil-immersion microscope n will be larger than 1). The

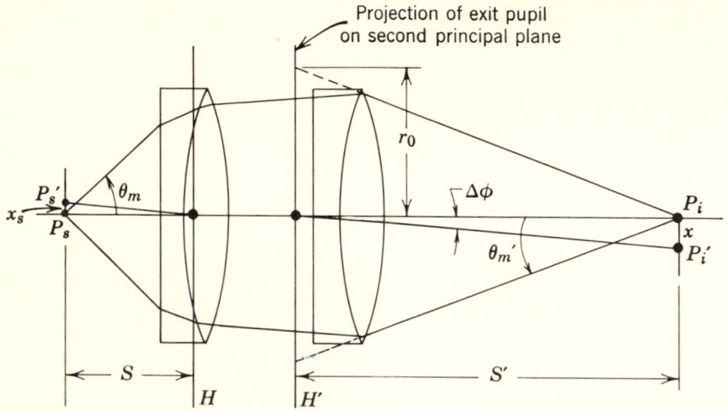

Fig. 7.31. Geometry pertinent to the calculation of the resolving power of a microscope.

images P_i and P_i' are found in air. The limiting angular aperture is θ_m', as seen from the image. The Abbe sine condition (Section 4.2B6) gives

$$nx_s \sin \theta_m = x \sin \theta_m' \tag{7.62}$$

Now the limiting angular resolution is given by Eq. (7.58) as

$$\Delta\phi_{\text{min}} = \frac{x_{\text{min}}}{S'} = \frac{0.61\,\lambda}{r_0} \tag{7.63}$$

Because of the large magnification produced by the objective, we have $S' \gg S$; hence, θ_m' will be small even if θ_m is not. Thus we can write

$$\theta_m' \approx \frac{r_0}{S'} \tag{7.64}$$

Thus, from (7.63) and (7.64) we obtain

$$x_{\text{min}} = \frac{0.61\,\lambda}{\theta_m'} \tag{7.65}$$

Then Eq. (7.62) becomes

$$nx_{s,\,\text{min}} \sin \theta_m = x' \sin \theta_m' \approx x'\theta_m' = 0.61\,\lambda$$

or

$$x_{s,\,\text{min}} = \frac{0.61\,\lambda}{n \sin \theta_m} \tag{7.66}$$

The quantity $n \sin \theta_m$ appearing here is called the *numerical aperture*. It can be made as large as 1.6 for an immersion objective. With $\lambda = 560$ nm this will give $x_{s,\min} = 210$ nm.

REFERENCES FOR SECTION 7.2

M. Françon, *Diffraction: Coherence in Optics*, Chapters II and VI.
J. M. Stone, *Radiation and Optics*, Chapter 9.
M. Born and E. Wolf, *Principles of Optics*, Sections 8.3, 8.5, and 8.6.

PROBLEMS

7.2.1 A rectangular aperture is illuminated at normal incidence by parallel light of $\lambda = 500$ nm. The dimensions of the aperture are 1×3 mm. What are the dimensions of the main maximum of the diffraction pattern formed on a screen 50 m away oriented parallel to the aperture?

7.2.2 Show that the maxima in the intensity of the Fraunhofer diffraction pattern of a slit of width $2x_0$ occur at values of α' given by the solution of the equation

$$\tan t = t$$

$$t = 2\pi u x_0 = 2\pi(\alpha' - \alpha)\frac{x_0}{\lambda}$$

Show that for large t these solutions approximately coincide with the maxima of $\sin^2 t$.

7.2.3 Consider Fraunhofer diffraction by a square aperture $2x_0$ by $2x_0$. Let the center of the pattern be at $\alpha' = \beta' = x = y = 0$. What is the expression for irradiance as a function of distance along the line $x = y$?

7.2.4 Another way to do the integral for Fraunhofer diffraction by a circular aperture is to stay in Cartesian coordinates. Let $P(\mathbf{r})$ in Fig. 7.21

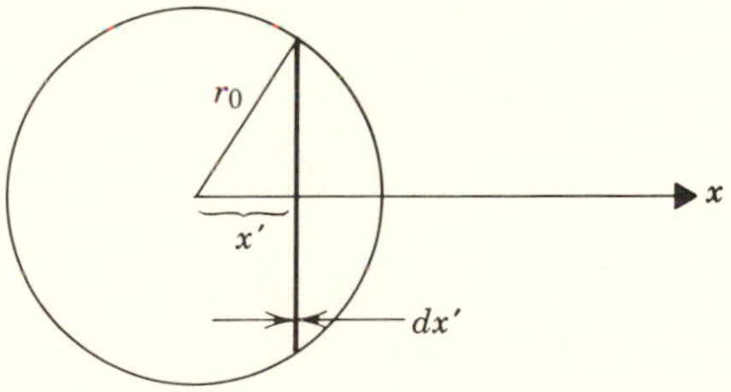

be along the x axis so that $x = \rho$, $y = 0$. By integrating along strips as indicated above, show that the field at P can be written as

$$\tilde{E}(\mathbf{r}) = \frac{i}{\lambda}\tilde{E}(0)\frac{e^{-ikR_0'}}{R_0'}4r_0^2\int_0^1 \sqrt{1-u^2}\cos tu\, du$$

where $t = 2\pi\rho r_0/(\lambda R_0)$. (By comparison with Eq. (7.51′) we obtain another representation of the Bessel function $J_1(t)$, namely,

$$J_1(t) = \frac{2t}{\pi}\int_0^1 \sqrt{1-u^2}\cos tu\, du$$

7.2.5 Show by using Parseval's theorem for Fourier integrals,

$$\int_{-\infty}^{\infty} |f(x, y)|^2\, dx\, dy = \int_{-\infty}^{\infty} |\hat{f}(u, v)|^2\, du\, dv$$

that the total power at the observing screen in the Fraunhofer diffraction pattern of an arbitrary aperture equals the total power passing through the aperture.

7.2.6 Estimate the resolution in lines/mm required of a photographic plate to be used at the focus of a diffraction-limited telescope objective 6″ in diameter if it is to reveal any structure in the diffraction pattern at the "image" of a star.

7.2.7 In the main text we argued that for visual observation with a telescope at normal magnification the contrast of the irradiance in the Airy disk of a star image over that of a continuous background is proportional to $m^2 \propto d_{\text{obj}}^2$. Show that this proportionality to d_{obj}^2 also holds for a photographic telescope.

7.2.8 What is the theoretical angular resolution of the 200″ diameter Palomar telescope? Take an effective value of $\lambda = 550$ nm. (Atmospheric fluctuations keep large telescopes from attaining the theoretical resolution.)

7.2.9 In a spectrograph such as that shown in Fig. 7.29 the effective width W of the prism is 40 mm, and the focal lengths of L_1 and L_2 are 300 and 150 mm, respectively. Estimate an upper limit on the width of the entrance slit that will not appreciably reduce the ultimate resolution of the instrument.

7.2.10 Show that for the spectrograph in Fig. 7.29 used at minimum deviation with $A = 60°$ the answer to Problem 7.2.9 is

$$\text{width} \ll \frac{2\lambda f_1}{b\sqrt{4-n^2}}$$

7.2.11 The dispersion of transparent media can often be approximately described by an equation for the refractive index of the Cauchy type

$$n = A + \frac{B}{\lambda^2}$$

For Barium flint glass, $n = 1.59825$ at 486.1 nm and $n = 1.60870$ at 298.8 nm. From these data you can determine A and B. Calculate the resolving power at 400 nm of a spectrograph that uses a prism of this glass with a base 50 mm long.

7.2.12 Describe semiquantitatively the Fraunhofer diffraction pattern observed with an aperture consisting of two narrow slits of equal width oriented at right angles.

The following problems can be answered with the help of Babinet's principle.

7.2.13 Describe analytically the intensity in the Fraunhofer diffraction pattern of an open rectangle with a centered rectangular obstruction. Sketch qualitatively the results.

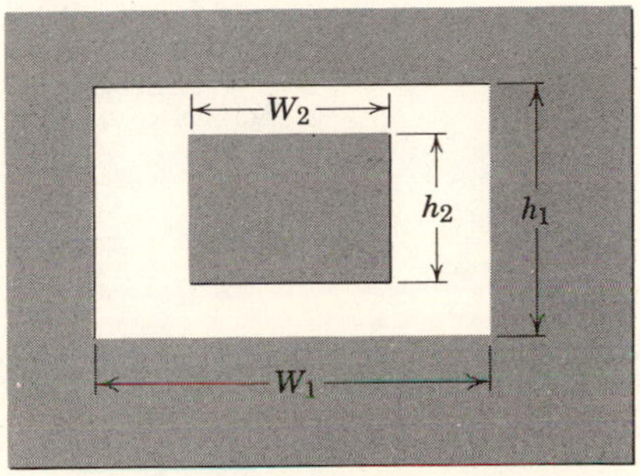

7.2.14 Describe analytically the intensity in the Fraunhofer diffraction pattern of this annular aperture. Discuss the qualitative effect on the height and width of the main maximum when $r_1 \ll r_2$.

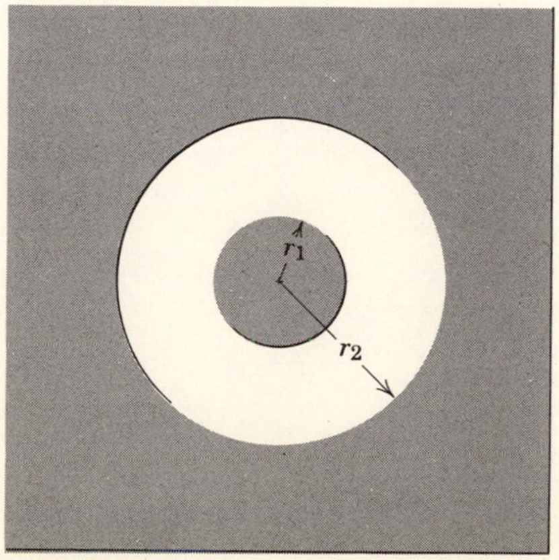

7.2.15 Describe quantitatively the intensity in the Fraunhofer diffraction pattern of this aperture in regions away from the center of the pattern where without the bar the irradiance would be essentially zero. Sketch the appearance of the two-dimensional pattern.

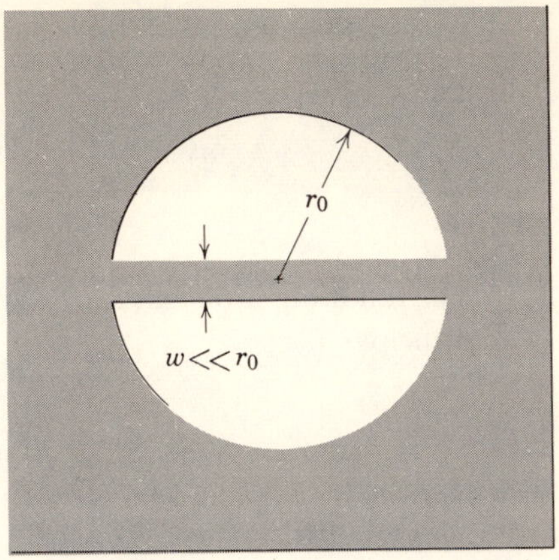

7.2.16 Same as 7.2.14, but with this aperture.

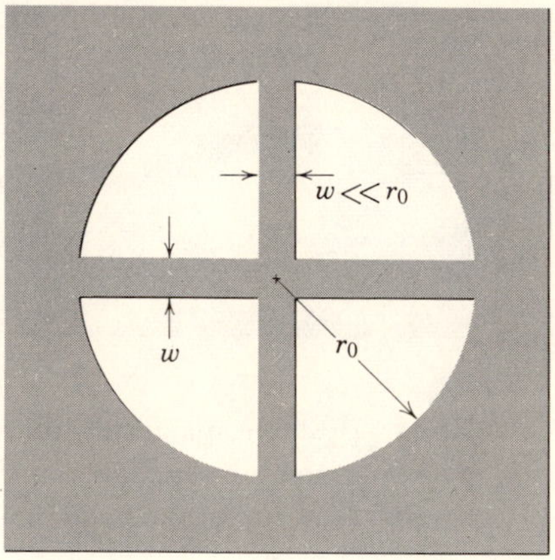

7.3. FRAUNHOFER DIFFRACTION BY AN ARRAY OF IDENTICAL APERTURES

By the superposition principle the diffracted electric field resulting from an array of apertures is the sum of the fields from the individual apertures. In the Fraunhofer case with identical apertures a great simplification occurs: The fields resulting from each aperture are the same to within one phase factor. The total field is then given by the sum of the phase factors times the field from a single aperture.

A. GENERAL FORMULATION

1. The Effect of Moving a Single Aperture in the Diffracting Plane. Consider the situation shown in Fig. 7.32 with the usual arrangement of point source at P_s, an observation point at P, and a diffracting aperture centered at O. Recall from our earlier discussion in Section 7.2 that OP and O$'P$ are essentially parallel, either because R_0' satisfies the Fraunhofer condition $R_0' \gg \sqrt{\lambda \cdot (\text{aperture dimension})}$ or because of the intervention of a lens. The same remark applies to P_sO and P_sO$'$.

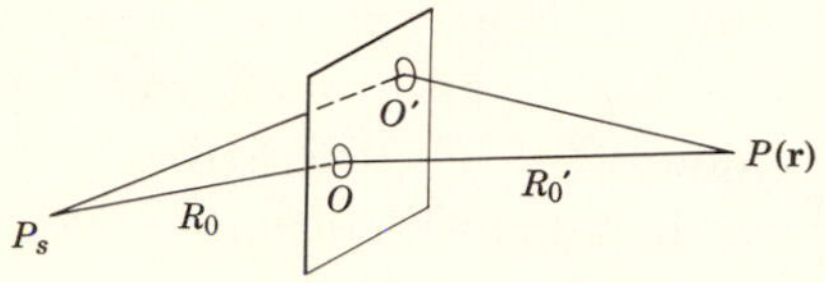

Fig. 7.32

What is the effect at P of moving the aperture to O$'$? The original field can be described by Eq. (7.33) with a transmission function inserted:

$$\tilde{E}(\mathbf{r}) = \frac{iA}{\lambda} \frac{e^{-ik(R_0+R_0')}}{R_0 R_0'} \iint_{-\infty}^{\infty} t(x', y') \exp \left\{-\frac{2\pi i}{\lambda}[(\alpha-\alpha')x' + (\beta-\beta')y']\right\} dx'\, dy'$$

Here x' and y' have O as origin. Now move the aperture and the origin to O$'$. The x', y' integral is unchanged since the relative position of the aperture and the origin is unchanged. The only important effect is in the phase factor

$$e^{-ik(R_0+R_0')} = e^{-ik(\widehat{P_sOP})}$$

This changes to

$$e^{-ik(\widehat{P_sO'P})} = e^{-ik(\widehat{P_sOP})}\, e^{ik(\widehat{P_sOP}-\widehat{P_sO'P})}$$

its old value times $e^{i\Delta}$, where

$$\Delta = \frac{2\pi}{\lambda}(\widehat{P_sOP} - \widehat{P_sO'P})$$

Let (l_1, l_2) be Cartesian coordinates of O$'$ with respect to O. Then the same arguments that led to the original phase factor in Eq. (7.33),

$$\exp\left\{-\frac{2\pi i}{\lambda}[(\alpha-\alpha')x' + (\beta-\beta')y']\right\}$$

yield

$$\Delta = \frac{2\pi}{\lambda}[l_1(\alpha' - \alpha) + l_2(\beta' - \beta)]$$

Hence $\tilde{E}(\mathbf{r})$ changes to its old value multiplied by

$$e^{i\Delta} = \exp\left\{\frac{2\pi i}{\lambda}[l_1(\alpha' - \alpha) + l_2(\beta' - \beta)]\right\} \tag{7.67}$$

For a single aperture this phase factor will not show up in the observed irradiance. It is important for interference effects when more than one aperture is present.

An elementary derivation of the above result may be given with the help of Fig. 7.33 when $\alpha = \beta = \beta' = 0$. The path from O′ to P is less than that from O to P, by the amount $l_1\alpha'$. Thus, with the aperture at O′ the phase at P is greater than that with the aperture at O by the amount $\Delta = 2\pi l_1\alpha'/\lambda$.

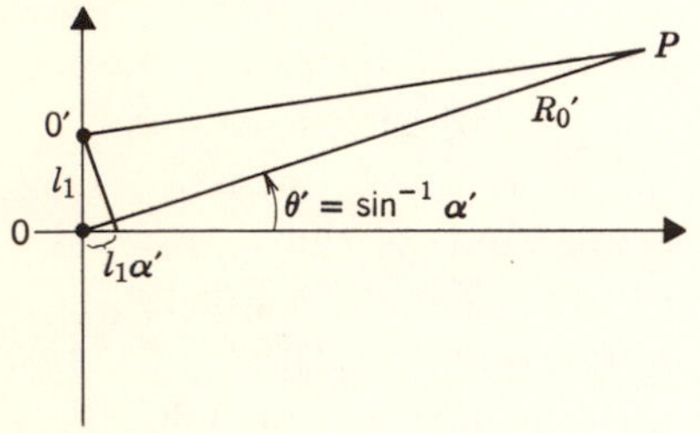

Fig. 7.33

2. *The Superposition of the Fields from Several Apertures.* We start with two identical apertures in Fig. 7.32, one at O for which the field is $\tilde{E}(\mathbf{r})$ and one at O′ for which the field is $\tilde{E}(\mathbf{r})e^{i\Delta}$, with Δ given by Eq. (7.67). The resultant field is then

$$\tilde{E}(\mathbf{r}) + \tilde{E}(\mathbf{r})e^{i\Delta} = \tilde{E}(\mathbf{r})\,(1 + e^{i\Delta})$$

Now $\tilde{E}(\mathbf{r})$ will be written $\tilde{E}(\mathbf{r}) = D\tilde{E}(\mathrm{O})$ with

$$E(\mathrm{O}) = \frac{Ae^{-ikR}}{R_0}$$

the incident field at O, and by (7.66),

$$D = \frac{i}{\lambda}\frac{e^{-ikR_0'}}{R_0'}\iint_{-\infty}^{\infty} t(x', y')\exp\left\{-\frac{2\pi i}{\lambda}[(\alpha - \alpha')x' + (\beta - \beta')y']\right\}dx'\,dy' \tag{7.68a}$$

Here $t(x', y')$ is the transmission function for a *single aperture*, the origin for x' and y' being at O. Thus, for the total field at $P(\mathbf{r})$ we have

$$\tilde{E}_{\mathrm{tot}}(\mathbf{r}) = \tilde{E}(\mathrm{O}) \cdot D \cdot I \tag{7.68b}$$

where

$$I = 1 + e^{i\Delta} \tag{7.68c}$$

The factor D describes the diffraction from a single aperture; the factor I describes the interference between apertures.

The flux density at $P(\mathbf{r})$ will be given by

$$E_e(\mathbf{r}) = E_{e,\mathrm{inc}}(\mathrm{O}) \cdot |D|^2 \cdot |I|^2 \tag{7.68d}$$

These results can be immediately generalized to cover more than two apertures—we need only add more terms to the expression for I.

B. DIFFRACTION BY A REGULAR ARRAY OF SLITS

1. Derivation of Formulas. Consider an array of N identical slits as shown in Fig. 7.34. The centers of the slits are at $O_1, O_2, \ldots, O_N$, separated in the x' direction by a distance l so that $|O_2 - O_1| = l$, $|O_3 - O_1| = 2l$, and so on. The total electric field at the observation point is given by

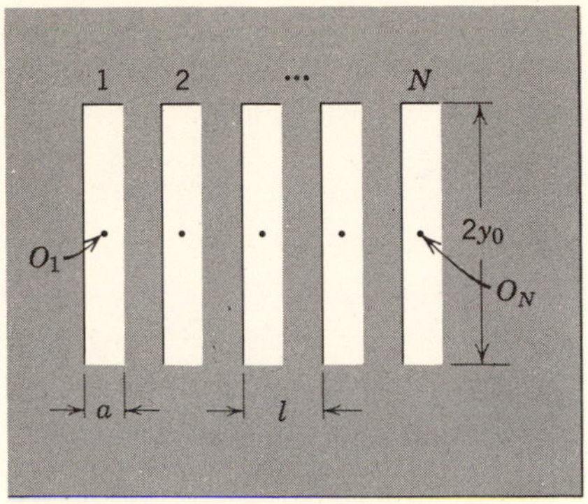

Fig. 7.34. An array of N identical slits.

$$\tilde{E}_{\text{tot}}(\mathbf{r}) = E(\mathrm{O}) \cdot D \cdot I$$

where D is the diffraction factor for a single slit at O_1 and

$$I = [1 + e^{(2\pi i/\lambda)\,l(\alpha' - \alpha)} + e^{(4\pi i/\lambda)\,l(\alpha' - \alpha)} + \cdots + e^{(2\pi i/\lambda)\,l(\alpha' - \alpha)(N-1)}]$$

or

$$I = [1 + e^{-2\pi iul} + e^{-4\pi iul} + \cdots + e^{-2\pi iul(N-1)}]$$

Here we have set $u = (\alpha - \alpha')/\lambda$, as usual. This expression for I can be written

$$1 + \rho + \rho^2 + \cdots + \rho^{N-1}, \qquad \rho = e^{-2\pi iul}$$

which can be summed to give

$$\frac{1 - \rho^N}{1 - \rho}$$

Thus the interference factor is

$$I(u) = \frac{1 - e^{-2\pi iNul}}{1 - e^{-2\pi iul}} \tag{7.69}$$

For the irradiance or intensity we need $|I(u)|^2$, which can be written several ways. One of the most useful is

$$|I(u)|^2 = \left| \frac{e^{-i\pi Nul}(e^{\pi iNul} - e^{-\pi iNul})}{e^{-\pi iul}(e^{\pi iul} - e^{-\pi iul})} \right|^2 = \frac{\sin^2 \pi uNl}{\sin^2 \pi ul} \tag{7.70}$$

Many of the properties of this function can be understood if it is realized that $I(u)$ is simply the vector sum of N unit vectors in the complex plane, each making an angle $-\Delta = -2\pi ul$ with the preceding vector (Fig. 7.35).

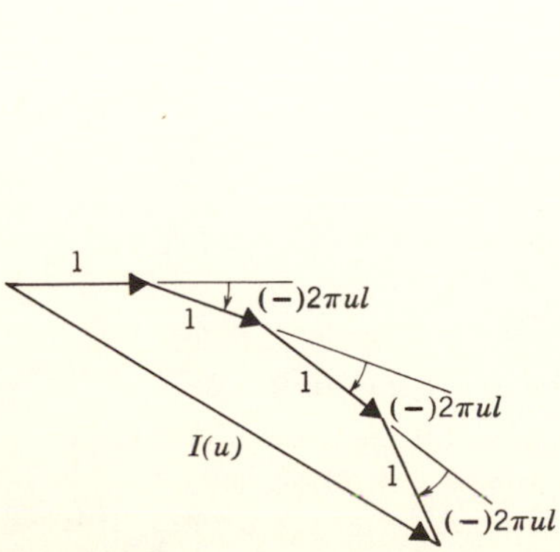

Fig. 7.35

We now need the single-slit diffraction factor D. The slit is a units wide by $2y_0$ units tall (Fig. 7.34). From Eqs. (7.34)–(7.38) and the accompanying discussion it is easily seen that D is given by

$$D = \frac{i}{\lambda} \frac{e^{-ikR_0'}}{R_0'} 2y_0 a d_x d_y \tag{7.71a}$$

with

$$d_y = \frac{\sin 2\pi v y_0}{2\pi v y_0}, \qquad v = \frac{\beta - \beta'}{\lambda} \tag{7.71b}$$

$$d_x = \frac{\sin \pi u a}{\pi u a}, \qquad u = \frac{\alpha - \alpha'}{\lambda} \tag{7.71c}$$

Then, with a little rearranging we obtain the irradiance

$$\begin{aligned} E_e(\mathbf{r}) &= E_e(\mathrm{O})|D|^2 \cdot |I|^2 \\ &= \frac{E_e(\mathrm{O})}{\lambda^2 R_0'^2} (2y_0 aN)^2 d_y^2 \left(\frac{\sin \pi u a}{\pi u a}\right)^2 \left(\frac{\sin \pi u l N}{N \sin \pi u l}\right)^2 \end{aligned} \tag{7.72}$$

The factor $\sigma = 2y_0 aN$ represents the open area of the entire aperture. The maximum of $E_e(\mathbf{r})$ occurs in the straight-ahead direction when $\alpha' = \alpha$, $\beta' = \beta$ and has the value $E_e(\mathrm{O})\sigma^2/(\lambda R_0')^2$.

2. *Example:* $N = 2$. For two slits it is more convenient to write the last factor in (7.72) as

$$\left|\frac{I(u)}{2}\right|^2 = \frac{\sin^2 2\pi u l}{4 \sin^2 \pi u l} = \cos^2 (\pi u l) = \frac{1 + \cos 2\pi u l}{2}, \qquad u = \frac{\alpha - \alpha'}{\lambda}$$

This is plotted in Fig. 7.36 along with the diffraction factor $|d_x(u)|^2$ and the product $|d_x \cdot I|^2/4$.

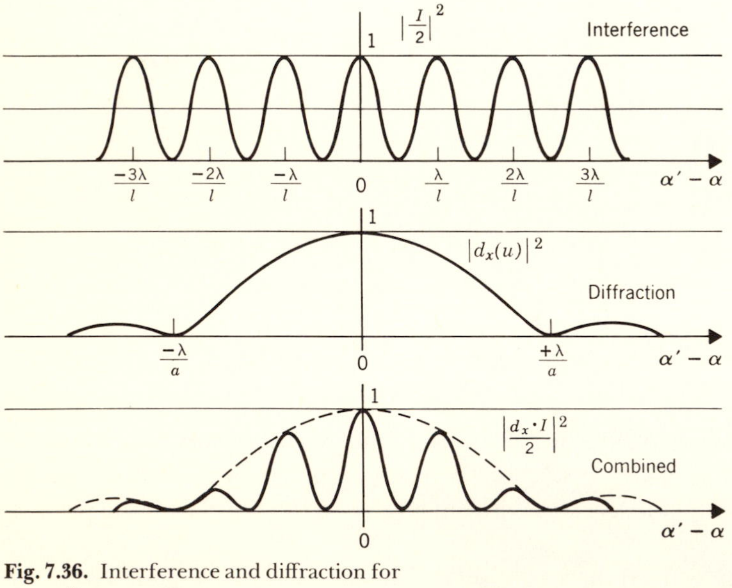

Fig. 7.36. Interference and diffraction for a two-slit aperture.

Interference maxima will occur whenever u is an integer times $1/l$, or when $(\alpha' - \alpha)l =$ integer $\cdot\ \lambda$. This is the condition that the path difference via adjacent slits be a multiple of λ (Fig. 7.37). The integer multiplying λ here is called the *order* of the interference maximum.

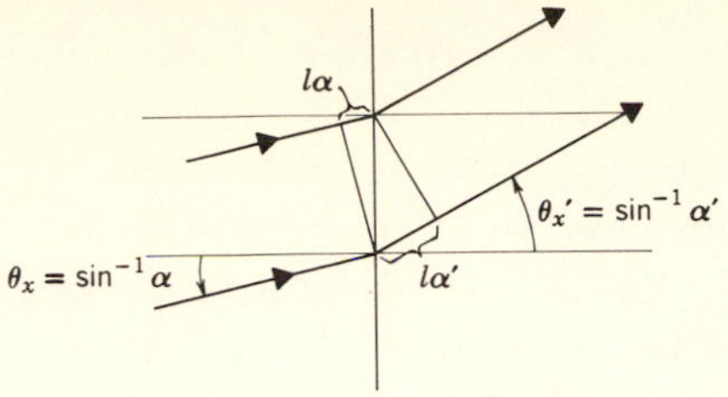

Fig. 7.37

The width a of the slits must be less than the separation l between slits. Hence $1/a$ must be greater than $1/l$, so that by Fig. 7.36 the first zero in the diffraction factor $|d_x|^2$ must occur at a larger value of u than the first subsidiary maximum in the interference factor. The diffraction pattern forms an envelope for the sinusoidal interference fringes. The combined pattern then will have all the zeros of the single-slit diffraction pattern. Zeros in the diffraction pattern can coincide with interference maxima causing the phenomenon of *missing orders*, but since $1/a > 1/l$, the first-order interference maximum will always be present.

3. Example: N = 4. For more than two slits the condition for a principal maximum in the interference pattern remains the same. The N unit vectors are all parallel. This can also be seen in the interference factor

$$\left|\frac{I(u)}{N}\right|^2 = \frac{\sin^2 \pi u l N}{N^2 \sin^2 \pi u l}$$

because its denominator is zero whenever $u =$ integer. But then the numerator is also zero, and application of l'Hospital's rule gives $|I(u)/N|^2 = 1$. The numerator can be zero without the denominator being zero. This happens $(N-1)$ times before both are zero again. The first zeros after a main maximum always occur at $ul =$ integer $\pm 1/N$, and the main maxima occur with a separation $\Delta(ul) = 1$. There are $(N-2)$ minor maxima between adjacent principal maxima.

The case for $N = 4$ is shown in Fig. 7.38. Vector diagrams are also given for various values of the phase shift $-\Delta = -2\pi ul$.

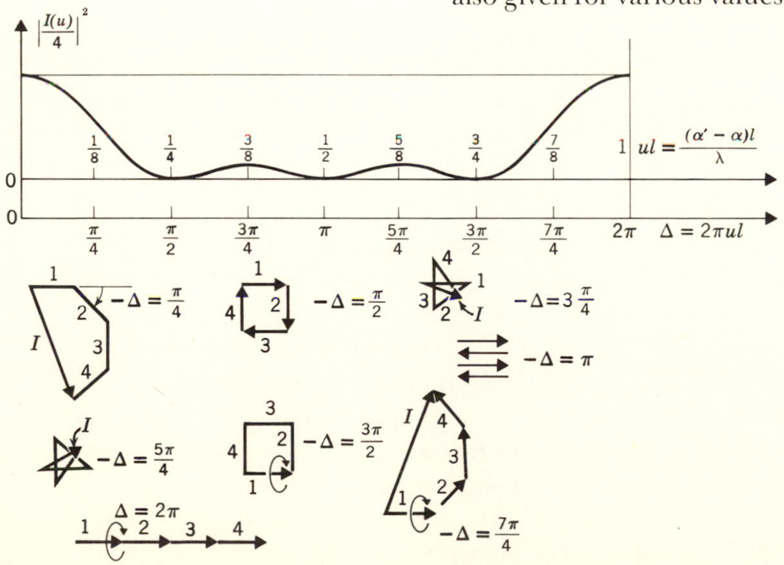

Fig. 7.38. Interference pattern for four slits.

4. Case of Large N. For large N the first zero after a main maximum occurs for a very small change in ul of $1/N$, corresponding to a phase difference in the contribution from neighboring slits of $-\Delta = 2\pi/N$. The vector diagram describing $I(u)$ then approximates a circle of circumference N (Fig. 7.39). We can also see this analytically. Let ul be close to the integer m so that $|ul - m|$ is much less than unity (but not necessarily less than $1/N$). Then we have

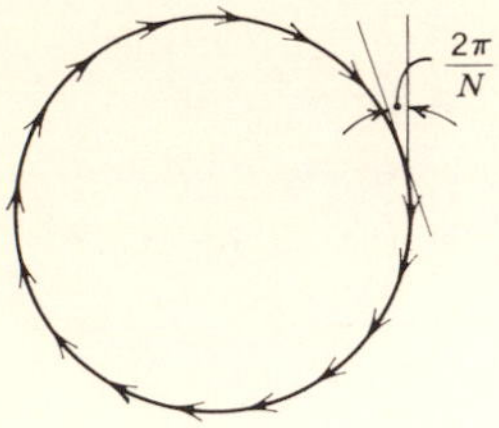

Fig. 7.39

$$\sin^2 \pi ulN = \sin^2 \pi (ul - m)N$$

and

$$\sin^2 \pi ul \approx [\pi(ul - m)]^2$$

so that

$$\left|\frac{I(u)}{N}\right|^2 \cong \left[\frac{\sin \pi(ul-m)N}{\pi(ul-m)N}\right]^2$$

When we compare this with our earlier result for the diffraction pattern from a slit of width $2x_0$,

$$\left(\frac{\sin 2\pi u x_0}{2\pi u x_0}\right)^2$$

we see that $I(u)/N^2$ has the same form as a single-slit diffraction pattern centered about $ul = m$ for a slit of width Nl, the total width of our array of small slits!

The form of the interference pattern for large N is sketched in Fig. 7.40. Near $ul =$ integer we see the narrow single-slit patterns

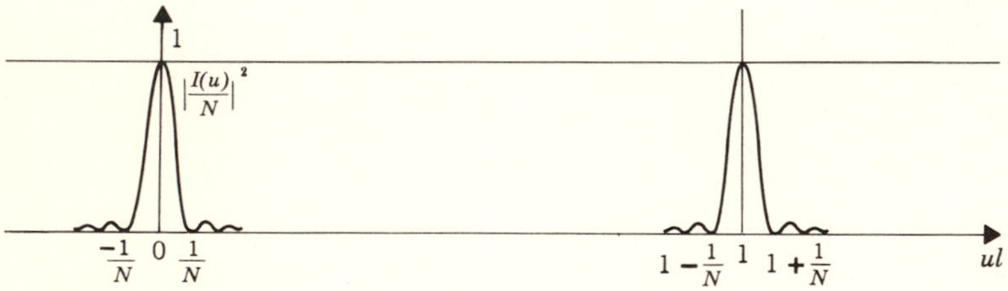

Fig. 7.40. Interference pattern for a large number N of slits.

corresponding to the full width of the array. There are small wiggles between these main maxima; for a small distance in ul from the main maxima these look like those in the $(\sin \xi/\xi)^2$ single-slit pattern. This holds as long as the phase difference is small in magnitude compared with 2π, for then the vector diagram looks like a smooth circle, and $\sin^2 \pi ul$ may be approximated by $[\pi(ul - m)]^2$. The approximation breaks down toward the center of the region between maxima, and the full formula must be used.

C. THE DIFFRACTION GRATING

1. Grating Equation and Dispersion. An equally spaced array of a large number of slits forms a diffraction grating. Such gratings have been extremely useful in spectroscopy because of their ability to produce angular dispersion of light of different wavelengths. The condition for a diffraction maximum is

$$ul = \text{integer}$$

or

$$(\alpha' - \alpha) = \frac{m\lambda}{l}, \qquad \text{where} \quad m = 0, \pm 1, \pm 2, \ldots \tag{7.73}$$

We define the angles

$$\phi_x = \sin^{-1} \alpha, \qquad \phi'_x = \sin^{-1} \alpha'$$
$$\phi_y = \sin^{-1} \beta, \qquad \phi'_y = \sin^{-1} \beta'$$

which have the interpretation shown in Fig. 7.41. If ϕ'_y is small, ϕ'_x approximately equals the angle of diffraction θ'. But in any case

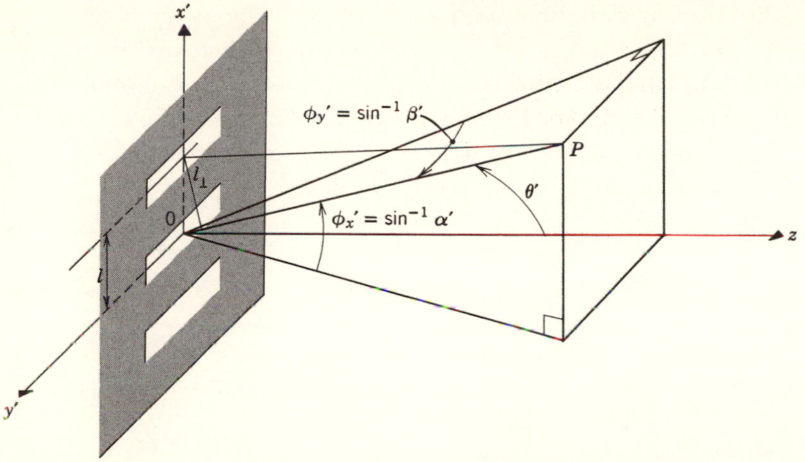

Fig. 7.41

Eq. (7.73) can be written in the form of the familiar *grating equation*

$$l(\sin \phi'_x - \sin \phi_x) = m\lambda \tag{7.73'}$$

The dispersion implied here can be seen from the wavelength dependence of the diffracted angle ϕ'_x for a fixed incident angle ϕ_x and nonzero order m. Differentiating, we obtain

$$l \cos \phi'_x d\phi'_x = m\, d\lambda$$

$$\frac{d\phi'_x}{d\lambda} = \frac{m}{l \cos \phi'_x} = \frac{m}{l_\perp} \tag{7.74}$$

where $l_\perp = l \cos \phi'_x$ is the projected slit separation.

2. ***Some Geometrical Features.*** Now a word about the dependence on the other angle ϕ'_y. The y part of the diffraction pattern is represented by $d_y{}^2$ in Eq. (7.72):

$$d_y{}^2 = \frac{\sin^2 2\pi v y_0}{(2\pi v y_0)^2}$$

This function is peaked about the value $v = 0$ or $\beta' = \beta$ with first zeros at $v = \pm 1/(2y_0)$, or at $\beta' = \beta \pm \lambda/(2y_0)$. The slits are long compared with λ (of order 10 mm = 10,000 μ). Then the angular width of the pattern is very small:

$$\frac{\lambda}{2y_0} \sim \frac{\frac{1}{2}\mu}{10{,}000\,\mu} = \frac{1}{20{,}000}\,\text{rad}$$

For most practical purposes we can take it as zero and set $\beta' = \beta$.

From Fig. 7.41 we see that the diffracted ray OP must lie along a cone about the y axis making a half-angle $B = \cos^{-1}\beta' = \cos^{-1}\beta$. A similar cone is obtained as the locus of solutions for ϕ_x' of the grating equation (7.73) or (7.73′). Some of the implications of this behavior will be explored in the problems.

*3. **More on Dispersion — Grating Resolving Power.*** Suppose that light of two wavelengths λ_1 and λ_2 is present in the incident beam propagating in direction (α, β, γ). The zeroth-order diffracted beam occurs in the same direction, $\alpha' = \alpha$, $\beta' = \beta$, and all other orders keep $\beta = \beta'$. For λ_1 the first-order diffracted beam occurs at $\alpha' - \alpha = \lambda_1/l$, for λ_2 at λ_2/l. The second-order beam for λ_1 is $2\lambda_1/l$, that for λ_2 at $2\lambda_2/l$, and so on as indicated in Fig. 7.42.

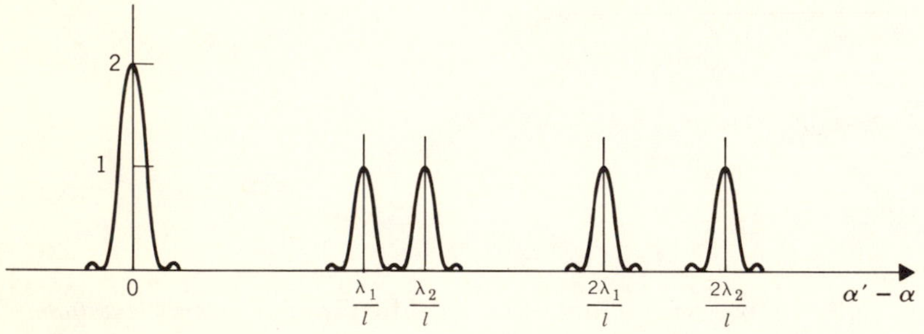

Fig. 7.42. Interference pattern when two wavelengths are present.

Note that the separation of the two peaks in first order is

$$\delta \sin \phi_x' = \delta\alpha' = \frac{\lambda_2 - \lambda_1}{l}$$

In second order it is

$$\delta \sin \phi_x' = \delta\alpha' = \frac{2(\lambda_2 - \lambda_1)}{l}$$

In general, in mth order the separation is

$$\delta \sin \phi_x' = \delta\alpha' = \frac{m(\lambda_2 - \lambda_1)}{l} = \frac{m}{l}\,\delta\lambda \tag{7.75}$$

Now we ask: How close together may λ_1 and λ_2 be to be just resolvable? If the two lines are equally intense, the criterion usually applied is to say that they are just resolved in mth order if the mth-

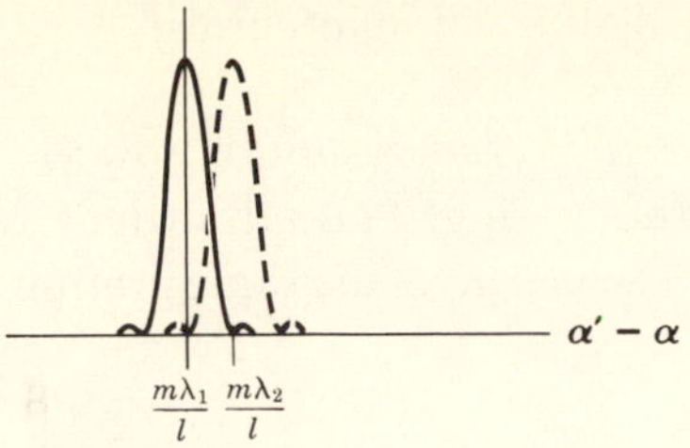

Fig. 7.43. Interference in *m*th order at the limit of resolution.

order maximum of λ_2 occurs at the first zero just beyond the *m*th maximum of λ_1 (Fig. 7.43). From maximum to first minimum involves a change in u of $1/(Nl)$, and hence the change in α' is $\delta\alpha' = \lambda/(Nl)$ where λ could be either λ_1 or λ_2, since they are almost equal. Equating this expression for $\delta\alpha'$ with that in Eq. (7.75), we obtain

$$\frac{\lambda}{Nl} = \frac{m\,\delta\lambda}{l}$$

Thus the resolving power $\mathscr{R} = \lambda/\delta\lambda$ is given by

$$\mathscr{R} = \frac{\lambda}{\delta\lambda} = Nm \tag{7.76}$$

This number can be quite large. Today moderate-priced grating monochromators are available that use gratings having typically 1200 lines/mm with a ruled area 100 mm long. Thus

$$N = 100\text{ mm} \times 1200\text{ lines/mm} = 120{,}000\text{ lines}$$

This would give a theoretical resolving power in first order of 120,000. If $\lambda = 600$ nm, $\delta\lambda = 600/120{,}000 = 1/200$ nm.

4. ***Reflection Gratings.*** In actual use we find more diffraction gratings of the reflection type than of the transmission type. The simplest type of reflection grating is an exact analogue of the simplest type of transmission grating—where the latter has an opening, the former has a reflecting surface, and where the latter is opaque, the former is either transmitting or absorbing, as indicated in Fig. 7.44.

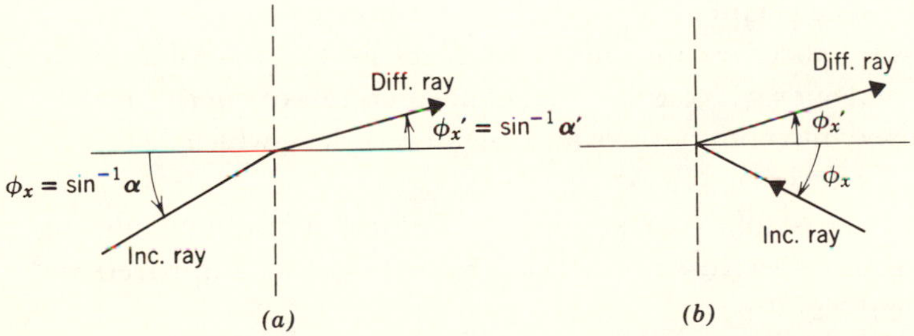

Fig. 7.44. Equivalent transmission grating (*a*) and reflection grating (*b*).

Few changes are necessary to convert our mathematical description of a transmission grating to a reflection grating. The incident light now propagates from right to left, but the first two direction cosines remain the same; that is, the equivalent direction cosines are

transmission:	(α, β, γ),	γ pos
reflection:	$(\alpha, \beta, -\gamma)$,	γ pos

We retain our former notation for the diffracted beam direction $(\alpha', \beta', \gamma')$. The transmission function $t(x', y')$ is replaced by a reflection function $r(x', y')$; otherwise all diffraction integrals remain the same.

Zero-order diffraction where $\alpha' = \alpha$, $\beta' = \beta$ corresponds now to specular reflection from the plane of the grating. The mth-order diffraction maxima still occur at $ul = m$, which gives the same grating equation

$$m\lambda = l(\alpha' - \alpha) = l(\sin\phi'_x - \sin\phi_x)$$

*5. **The Diffraction Factor.*** The full expression for the irradiance includes the diffraction factor $|D|^2$ for a single unit in the array of apertures. For a simple rectangular opening of width a, Eq. (7.72) says that $|D|^2$ contains the factor

$$|d_x(u)|^2 = \left(\frac{\sin \pi ua}{\pi ua}\right)^2$$

Its effect on the overall pattern is indicated in Fig. 7.45, which should be compared with Fig. 7.36. Since a must be less than l, the

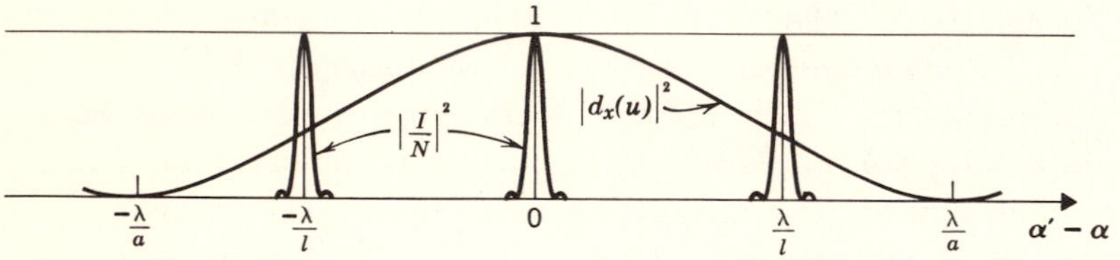

Fig. 7.45

main maximum of the diffraction pattern will cover at least three interference maxima, including the one at zero order, which is useless for spectroscopy and which receives the greatest amount of diffracted energy. Note also that equal amounts of energy are diffracted into both positive and negative orders, which again is a waste of energy.

This state of affairs makes this simple kind of grating inefficient for practical spectroscopy. Efficiency can be greatly improved by blazing the grating.

*6. **Blazed Diffraction Gratings.*** Often it is a disadvantage for a simple slit-type transmission grating or the equivalent reflection grating to disperse large amounts of light of a given wavelength simultaneously into several different orders. Another drawback is the necessary absorption of light in the opaque, nonreflecting parts of the grating. Both disadvantages can be overcome by the technique known as blazing which makes the grating into an array of facets each inclined the same amount to the plane of the grating as a whole. We shall discuss the blazed reflectance grating since it is the most commonly employed, and at first we shall assume for simplicity that $\beta = \beta' = 0$.

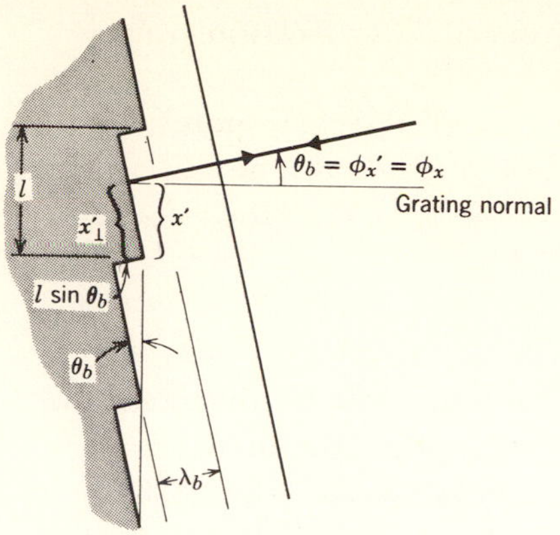

Fig. 7.46. Detail of blazed reflection diffraction grating with blaze angle θ_b.

The inclination angle of the facets $\theta_b = \sin^{-1} \alpha_b$ shown in Fig. 7.46 is called the *blaze angle*. The effectiveness of blazing and the role of the blaze angle can be seen by the following argument. Let the angle of incidence ϕ_x equal $-\theta_b$. If the step height $l \sin \theta_b$ is an integral number of half-wavelengths, that is, if

$$l\alpha_b = \frac{p\lambda}{2}, \qquad p = 1, 2, \ldots \tag{7.77}$$

then the light reflected off one facet will be in phase with that reflected off the next. In fact, for these values of α and λ the light will behave exactly as if it were reflected off a plane mirror inclined at the angle θ_b and would be reflected right back on itself. This conclusion neglects the small amount of diffraction that may occur off the short edge of the facet, but to the extent that this neglect is justified, we see that the entire beam is so reflected; no energy can be diffracted into other orders.

Since $-\alpha = \alpha' = \alpha_b$, the grating equation in this special case becomes

$$m\lambda = 2l\alpha_b = 2l \sin \theta_b$$

which when compared with Eq. (7.77) yields $m = p$. The value of λ which satisfies (7.77) with $p = 1$ is called the *blaze wavelength* λ_b. It is given by

$$\lambda_b = 2l\alpha_b = 2l \sin \theta_b \tag{7.78}$$

This means that the blaze wavelength is twice the step height. Then Eq. (7.78) can be written

$$\lambda = \frac{\lambda_b}{p} \tag{7.77$'$}$$

On the basis of the argument below Eq. (7.77), we expect that for λ close to λ_b practically all of the light is diffracted into first order ($m = 1$). Similarly, for λ near λ_b/m, practically all of the light will be diffracted into mth order. The following mathematical treatment will justify this expectation.

The important result achieved by blazing is the great enhancement in diffracted energy when both the grating equation and the condition for specular reflection from a facet are satisfied. Specular reflection can also occur when $\phi'_x \neq -\phi_x$. The condition is

$$\phi'_x - \phi_x = 2\theta_b \tag{7.79}$$

as illustrated in Fig. 7.47.

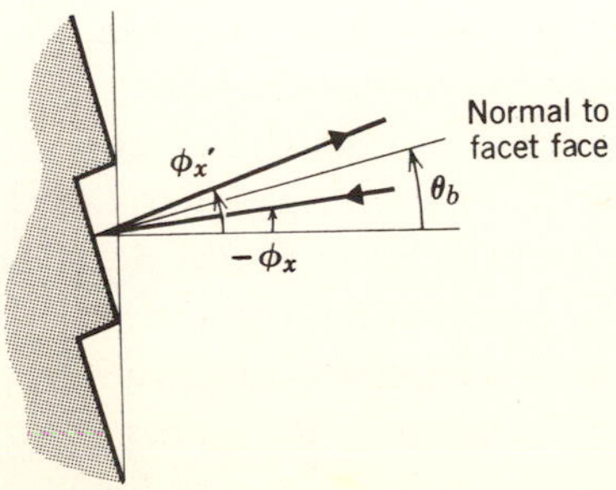

Fig. 7.47

7. ***An Approximate Theory of the Blazed Reflectance Grating.*** We now work out an approximate theory which will hold for small values of β and β' and hence of ϕ_y and ϕ'_y. The discussion of Fig. 7.41 showed that we then have

$$\phi'_x \approx \theta' = \text{angle of diffraction}$$

and

$$\phi_x \approx \theta = \text{angle of incidence}$$

We next assume that ϕ'_x and $-\phi_x$ are not far from the blaze angle θ_b.* We do not assume that the blaze angle is small, because with a bit of care we can treat the case for moderate to large values of θ_b on the same footing as the case of small θ_b. This is necessary since most diffraction gratings have θ_b in the 15° to 30° range, but some important special gratings have $\theta_b \approx 60°$.

For the diffraction factor we have to integrate over one facet of the grating. We start with an equation like (7.34) but replace $t(x', y')$ with a reflection coefficient and insert the inclination factor $Q = \frac{1}{2}(\cos\theta + \cos\theta') \approx \cos\theta_b$:

$$D = \frac{i}{\lambda}\frac{e^{-ikR'_0}}{R'_0}\int\int_{-\infty}^{\infty} r(x', y')e^{-2\pi i(ux'+vy')}\,dy' \cos\theta_b\,dx' \tag{7.80}$$

The y' dependence of $r(x', y')$ sets the limits on the y' integral as $-y_0$ and $+y_0$. Hence the y' integral gives, as before,

$$2y_0 d_y = 2y_0\left(\frac{\sin 2\pi v y_0}{2\pi v y_0}\right), \qquad v = \frac{\beta - \beta'}{\lambda} \tag{7.81a}$$

and as before this has the effect of forcing $v = 0$ or $\beta' = \beta$. In the x' part of the integral the natural coordinate to use is

$$x'_\perp = x' \cos\theta_b$$

which represents the distance along the reflecting facet, as shown in Fig. 7.46. The use of projected coordinates like $x'_\perp$ was described earlier in Section 7.2.B5.

Consider optical path lengths for a single facet in Fig. 7.46. A ray incident from direction $\phi_x \sim -\theta_b$ and diffracted into a direction $\phi'_x \sim \theta_b$ at height x' from the apex will travel an extra distance $2x'_\perp \tan\theta_b$ compared with the distance with an unblazed grating. This will produce a phase lag of

$$\Delta = \frac{4\pi x'_\perp \tan\theta_b}{\lambda}$$

If the amplitude reflection coefficient at the facet surface is ρ_0,

*This theory is also not very accurate quantitatively unless l is at least several multiples of λ. It should also be pointed out that the scalar theory of diffraction used here cannot discuss the polarizing properties of real gratings.

then the effect of the facet can be described by an effective complex reflection coefficient

$\rho_0 e^{-i\Delta}$

The x' integral for D then becomes

$$\rho_0 l d_x = \rho_0 \int_{-l_\perp/2}^{+l_\perp/2} e^{-2\pi i \gamma x'_\perp}\, dx'_\perp = \rho_0 l_\perp \left(\frac{\sin \pi\gamma l_\perp}{\pi\gamma l_\perp}\right) \tag{7.81b}$$

where

$$-\gamma = \frac{(\alpha' - \alpha) - 2 \sin \theta_b}{\lambda \cos \theta_b}, \qquad l_\perp = l \cos \theta_b$$

Equations (7.80) and (7.81) then combine to give

$$D = \frac{e^{-ikR'_0}}{R'_0} \rho_0 \,(2y_0 l_\perp)\, d_y d_x \tag{7.81c}$$

We can also write d_x as

$$d_x = \frac{\sin \pi u' l}{\pi u' l}, \qquad u' = u + \frac{2 \sin \theta_b}{\lambda} = \frac{(\alpha - \alpha') + 2 \sin \theta_b}{\lambda} \tag{7.81d}$$

This set of equations should be carefully compared with the set (7.71*a, b, c*) for a slit-type diffraction grating. We see that now d_x describes a single-slit pattern for a slit of width l (instead of width $a < l$) centered about

$$\alpha' - \alpha = 2 \sin \theta_b = 2\alpha_b$$

(instead of $\alpha' - \alpha = 0$). In Fig. 7.48 we have sketched this new diffraction factor d_x^2 together with the interference factor $|I|^2$.* The spacing between the peaks of the latter and the zeros of the former are now both λ/l. (Compare with Fig. 7.45 for a nonblazed grating.) The combined pattern is now asymmetrical about zero order, with much more useful energy available in the dispersive region.

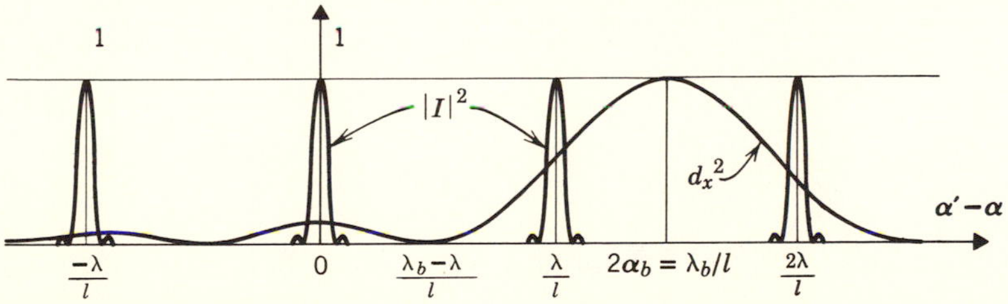

Fig. 7.48. Diffraction and interference factors for a blazed diffraction grating. (Here $1/\lambda \approx 3/2\lambda_b$.)

*The analytic form for d_x in Eq. (7.81*d*) used in Fig. 7.48 was derived on the assumption that $-\alpha'$ and α' are close to α_b. The theory is only qualitatively correct if this assumption is violated.

When $\lambda = \lambda_b, \lambda_b/2, \lambda_b/3$, and so on, the peak of the diffraction pattern coincides with an interference maximum; all other interference maxima are eliminated by zeros in the diffraction pattern. This behavior is shown in Fig. 7.49 for various values λ at or near λ_b and $\lambda_b/2$.

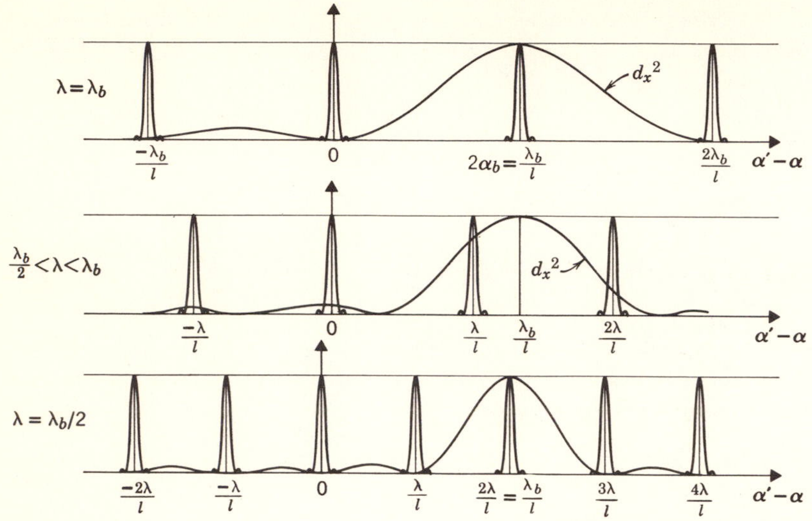

Fig. 7.49

At the peak of the mth-order interference maximum where $m\lambda = l(\alpha' - \alpha)$, we may write, for $u'l$ appearing in Eq. (7.81d),

$$u'l = m - \lambda_b/\lambda$$

using Eqs. (7.78). Then we obtain

$$d_x = \frac{\sin \pi(\lambda_b/\lambda - m)}{\pi(\lambda_b/\lambda - m)} \tag{7.82}$$

We may use Eq. (7.82) to predict the relative energy (proportional to d_x^2) in various orders m as a function of $1/\lambda$. Usually we work in mth order over the range

$$\frac{1}{\lambda} = \frac{1}{\lambda_b}\left(m - \frac{1}{2}\right) \quad \text{to} \quad \frac{1}{\lambda_b}\left(m + \frac{1}{2}\right)$$

At the extreme ends of this range d_x^2 is down from its maximum value of unity to

$$\left(\frac{\sin \pi/2}{\pi/2}\right)^2 = 0.41$$

We close this discussion by pointing out the necessity, when working with diffraction gratings, of eliminating light of unwanted wavelengths that would otherwise be diffracted in higher order and then detected. This can be done with auxiliary filters or with a prism monochromator.

*8. **The Infinite Grating—Fourier Series.*** It is instructive to consider the fictitious limiting case when the number of repeating elements in a diffraction grating goes to infinity. We can see what happens qualitatively from Fig. 7.40. Each of the interference maxima becomes an infinitely high, infinitely narrow "spike."

To discuss this case mathematically we have to change our origin of coordinates. So far we have referred all phases of the scattered waves to that scattered from the center O_1 of the first slit in Fig. 7.34. If there are N slits, we can shift our reference point to the *center* of the grating by multiplying $I(u)$ in Eq. (7.69) by $e^{\pi i(N-1)ul}$. The new interference factor that results is

$$\begin{aligned} \bar{I}(u) &= \frac{e^{\pi i(N-1)ul}(1-e^{-2\pi iNul})}{1-e^{-2\pi iul}} \\ &= \frac{\sin \pi uNl}{\sin \pi ul} \end{aligned} \tag{7.83}$$

In the neighborhood of the mth-order diffraction maximum where $ul = m$, we have approximately

$$\bar{I}(u) = \frac{\sin \pi(ul-m)N}{\pi(ul-m)} \tag{7.84}$$

In the limit as $N \to \infty$ this becomes a delta function (Appendix A)

$$\bar{I}(u) \to \delta(ul-m) = \frac{1}{l}\delta\left(u-\frac{m}{l}\right)$$

When $ul \neq m$, the limit as $N \to \infty$ of $\bar{I}(u)$ is zero. Thus, $\bar{I}(u)$ takes on the limiting value of a *sum* of delta functions:

$$\lim_{N\to\infty} \bar{I}(u) = \sum_{m=-\infty}^{\infty} \frac{1}{l}\delta\left(u-\frac{m}{l}\right) \tag{7.85}$$

Before we can use this expression, we have to reinterpret early results. Equations (7.34) and (7.36*a*) give the Fraunhofer diffraction pattern for a general aperture with a factorizable transmission function $[t(x', y') = t_1(x')t_2(y')]$:

$$\begin{aligned} \tilde{E}(r) &= \frac{i}{\lambda}\tilde{E}(O)\frac{e^{-ikR_0'}}{R_0'}\iint_{-\infty}^{\infty} t(x', y')e^{-2\pi i(ux'+vy')}dx'dy' \\ &= \frac{i}{\lambda}\tilde{E}(O)\frac{e^{-ikR_0'}}{R_0'}\hat{t}_1(u)\,\hat{t}_2(v) \\ u &= \frac{\alpha-\alpha'}{\lambda} \\ v &= \frac{\beta-\beta'}{\lambda} \end{aligned} \tag{7.86}$$

Here $\hat{t}_1(u)$ is the Fourier transform of the x-dependent part of the transmission function $t_1(x')$. The field $\tilde{E}(\mathrm{O})$ is the incident field at the center of the aperture.

For an infinite grating the function $t_1(x')$ extends from $-\infty$ to $+\infty$ and is *periodic* with spatial period l. By this we mean

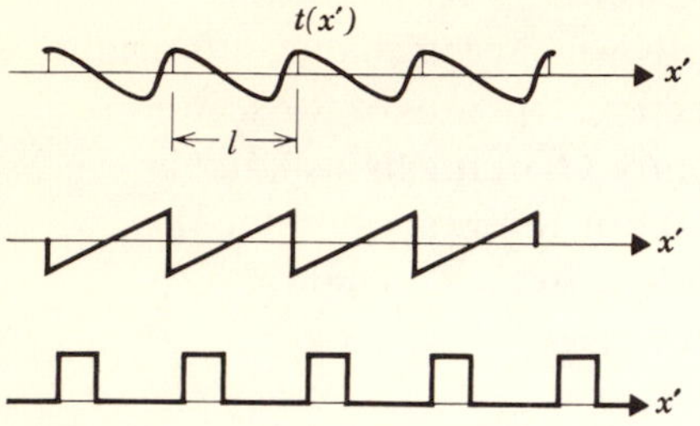

Fig. 7.50. Examples of periodic functions.

$$t_1(x' + ml) = t_1(x') \tag{7.87}$$

for all x' and all integers m. Examples of such a function are sketched in Fig. 7.50. It need be specified only over an interval of length l, say from $x' = -l/2$ to $x' = +l/2$. Equation (7.87) then defines it everywhere else.

A factorization may then be performed on the integral that gives $\hat{t}_1(u)$. This is equivalent to the factorization of Eqs. (7.68) into a single-slit diffraction factor $D(u)$ times an interference factor $I(u)$. In the present case we have

$$\hat{t}_1(u) = \int_{-\infty}^{\infty} t_1(x')e^{-2\pi iux'}dx'$$

From here on we shall drop the subscript 1. Now split the infinite domain of integration into strips of length l. There is no unique way to do this, but we assume that the strip at the origin will run from $x' = -l/2$ to $x' = +l/2$. Other strips will be centered at $x' = \pm ml$, $m = 1, 2, \ldots$. This gives

$$\hat{t}(u) = \sum_{m=-\infty}^{\infty} \int_{ml-l/2}^{ml+l/2} t(x')e^{-2\pi iux'}dx'$$

Now change variables. Let $z = x' - ml$. Because of the *periodicity* $t(x') = t(z)$. We can write $e^{-2\pi iux'} = e^{-2\pi iml}e^{-2\pi iuz}$. Then our integral becomes

$$\hat{t}(u) = \left(\lim_{N\to\infty} \sum_{m=-N/2}^{+N/2} e^{-2\pi iml}\right)\left[\int_{-l/2}^{+l/2} t(z)e^{-2\pi iuz}dz\right] \tag{7.88}$$

It is easy to show that the first factor is just

$$\lim_{N\to\infty} \bar{I}(u) = \lim_{N\to\infty} \frac{\sin \pi uNl}{\sin \pi ul} = \sum_{m=-\infty}^{+\infty} \frac{1}{l}\delta\left(u - \frac{m}{l}\right)$$

Thus we have

$$\hat{t}(u) = \sum_{m=-\infty}^{+\infty} \frac{1}{l}\delta\left(u - \frac{m}{l}\right)\left[\int_{-l/2}^{+l/2} t(z)e^{-2\pi iuz}dz\right] \tag{7.88$'$}$$

Equation (7.88′) says that the Fourier transform $\hat{t}(u)$ of a periodic function contains only discrete frequencies which are integral multiples of the *fundamental* (spatial) *frequency* $u = 1/l$.

The significance of this result can be seen when we write $t(x')$ in terms of its Fourier transform and use Eq. (7.88′):

$$t(x') = \int_{-\infty}^{\infty} \hat{t}(u)e^{2\pi iux'}\,du = \sum_{m=-\infty}^{\infty} \frac{1}{l}\int_{-\infty}^{\infty} \delta\left(u-\frac{m}{l}\right)e^{2\pi iux'}$$

$$\cdot\left[\int_{-l/2}^{+l/2} t(z)e^{-2\pi iuz}\,dz\right]du$$

$$= \sum_{m=-\infty}^{\infty}\left[\frac{1}{l}\int_{-l/2}^{+l/2} t(z)e^{-2\pi imz/l}\,dz\right]e^{2\pi imx'/l}$$

This can be written in the form of the *Fourier series*

$$t(x') = \sum_{m=-\infty}^{\infty} t_m e^{2\pi imx'/l} \tag{7.89a}$$

with the *Fourier coefficient* given by

$$t_m = \frac{1}{l}\int_{-l/2}^{+l/2} t(z)e^{-2\pi imz/l}\,dz \tag{7.89b}$$

For a large but finite grating the function $\hat{t}(u)$ is continuous, but it is strongly peaked near $u = m/l$ and takes a value there proportional to the function $d_x(u)$ which describes diffraction from a single element of the grating. Except for a constant factor, $d_x(u)$ is the integral

$$\frac{1}{l}\int_{-l/2}^{+l/2} t(z)e^{-2\pi iuz}\,dz$$

which for $u = m/l$ becomes t_m. Then in the limit as $N \to \infty$, $\hat{t}(u)$ becomes the discontinuous function of Eq. (7.88′).

REFERENCES FOR SECTION 7.3

M. Françon, *Diffraction: Coherence in Optics*. Section 6.2.

J. M. Stone, *Radiation and Optics*. Section 9.5.

M. Born and E. Wolf, *Principles of Optics*. Section 8.6.1.

Madden and Strong, "Diffraction Gratings," Appendix P of *Concepts of Classical Optics* by J. Strong. A detailed discussion of blazed gratings, including a comparison with experiment. Also a discussion of concave gratings.

G. W. Stroke, "Diffraction Gratings," in *Encyclopedia of Physics*, S. F. Flügge, ed., Springer-Verlag, Berlin, 1967, Vol. 29.

PROBLEMS

7.3.1 Describe quantitatively the electric field in the Fraunhofer diffraction pattern of this double circular aperture. Sketch roughly the two-dimensional distribution of irradiance when $r_0 = l/2,\ l/5,\ l/100$.

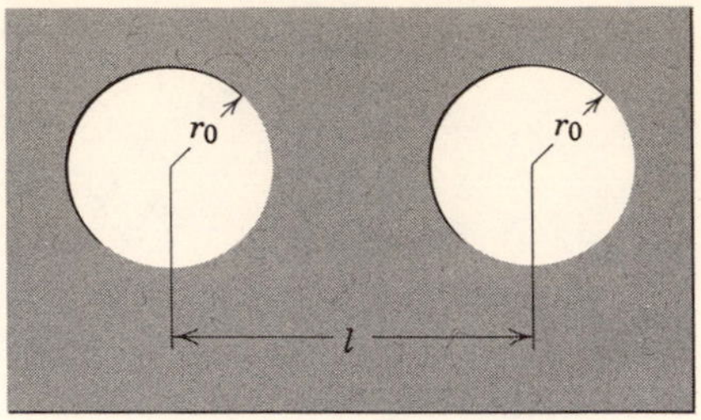

7.3.2 Describe quantitatively the field in the Fraunhofer diffraction pattern of this aperture. The slits are identical. Make a rough sketch of the two-dimensional distribution of irradiance. Assume that the aperture is just in front of a lens of focal length $f = 10^6\,\lambda$.

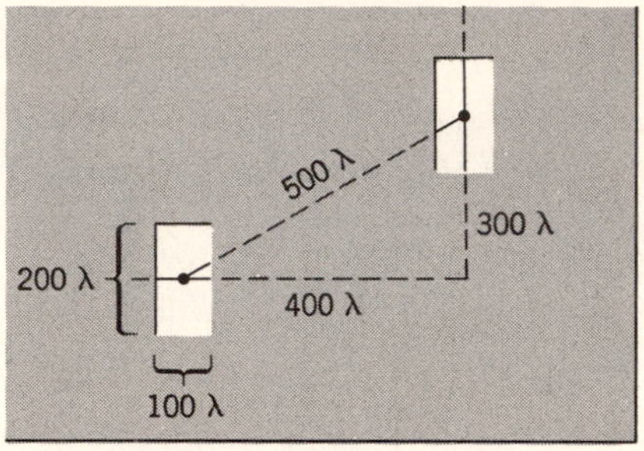

7.3.3 In the Michelson stellar interferometer discussed in Chapter 6 an aperture like that shown in (*a*) in the figure is put over the end of a telescope. The pattern observed in the focal plane with collimated monochromatic light is not the simple fringe pattern assumed in Chapter 6. Describe what it actually is analytically. The expanded interferometer of (*b*) makes the effective slit separation h, not l. What does this do to the pattern?

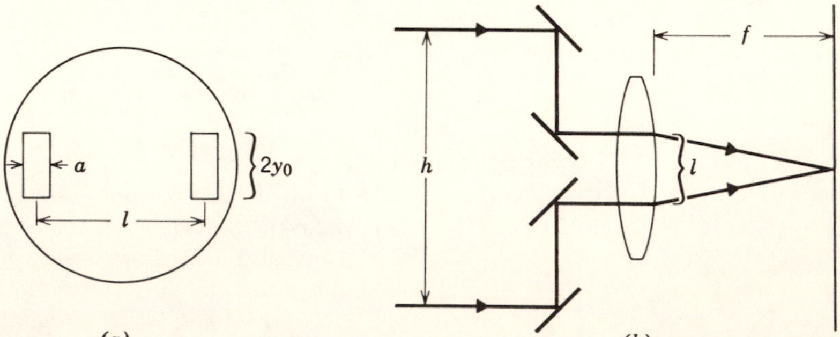

(*a*) (*b*)

7.3.4 A man focuses his eye at a point source of light emitting at wavelength $\lambda = 500$ nm, 100 m away. He holds up his handkerchief and stretches it in a plane in front of his face perpendicular to the light rays with the threads running horizontal and vertical. Instead of the point source he now sees a rectangular array of bright spots having the apparent separation (as seen against a background 100 m away) as shown:

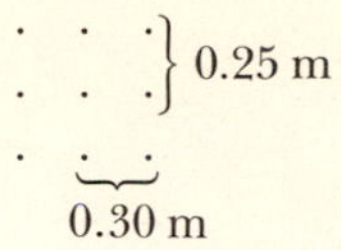

a. Find the horizontal and vertical separations of the threads of the handkerchief. To do this you will have to treat the eye as focused essentially at infinity.
b. What determines the number of spots that can be seen? (Try this experiment using a distant street light.)

7.3.5 Consider a diffraction grating made up of a series of slits of width $d/4$ separated a distance d.

a. Discuss the intensity pattern observed on the focal plane of a lens just to the right of the grating with light of wavelength λ for a total of 2 slits, 4 slits, and 100 slits. Include a discussion of the positions and widths of main maxima and subsidiary maxima. Are any orders missing?
b. What is the effect on the observed pattern of varying the angle of incidence of the light?
c. Discuss semiquantitatively the effect on the patterns in (a) if the incident light, instead of being accurately collimated, was collimated into a beam of total angular width $\Delta\theta = \lambda/10d$.

7.3.6 A "cosine grating" is one with a constant transmission as a function of y and a transmission depending on x as

$t(x) = \frac{1}{2}A(1 + B\cos\gamma x), \qquad |A| < 1$
and $|B| < 1$

Let the grating extend a distance $2x_0$ in the x direction and let $N = x_0\gamma/\pi$ be the total integral number of repetitions of the cosine pattern. For light of wavelength λ, calculate the relative energy as a function of the diffracted angle α. How many orders are seen?

7.3.7 Compare the resolving power of the prism in Problem 7.2.11 with the resolving power of a 50-mm-wide diffraction grating of 1200 lines/mm used in first order at 400 nm.

7.3.8 Calculate the blaze angle of a 600 line/mm grating that is to be used in first order at 750 nm. What would be the theoretical resolving power of a 52-mm-wide grating of this type?

7.3.9 A grating monochromator is shown schematically in the figure. The grating can be rotated, which will change both the angle of incidence $\theta \approx \phi_x$ and the diffracted angle $\theta' \approx \phi'_x$. The fixed slits and fixed lens guarantee that the incident and diffracted ray directions cannot change; hence $\theta + \theta' = \gamma$ a constant. Let d be the grating spacing.

a. Write down the appropriate equation connecting θ_0 with the wavelength(s) that pass through the exit slit.

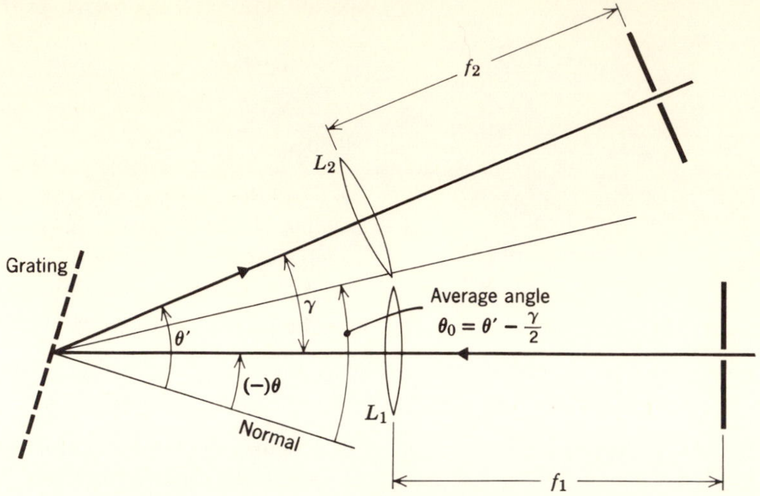

b. Find the dispersion factor $d\lambda/d\theta_0 = d\lambda/d\theta'$ which tells how fast the wavelength at the exit slit changes as the grating is rotated.

c. Fix θ and suppose that a spread $\Delta\lambda$ in λ is present at the entrance slit. Calculate the resulting spread $\Delta\theta'$ in θ'.

d. Find numerical results for (b) and (c) when $\lambda = 400$nm, $d = (1200)^{-1}$ mm, $\gamma = 10°$, and the grating is used in first order. In part (c), what will be the linear spread of the beam at the exit slit if $\Delta\lambda = 0.1$ nm and the focal length f of $L_2 = 1$ m?

7.3.10 When a slit is used with a monochromator as shown here, we obtain a series of collimated beams whose directions (parallel to OP) have a projection in the xz plane along the line O$\bar{z}$ through the center of the slit and the center of the lens. The "coordinates" of the direction OP in the past have been the direction cosines α, β or alternatively the angles $\phi_x = \sin^{-1}\alpha$, $\phi_y = \sin^{-1}\beta$. Instead of ϕ_x we may take the angle ϕ_{x0} that the projection of OP in the xz plane makes with the z axis. This angle ϕ_{x0}

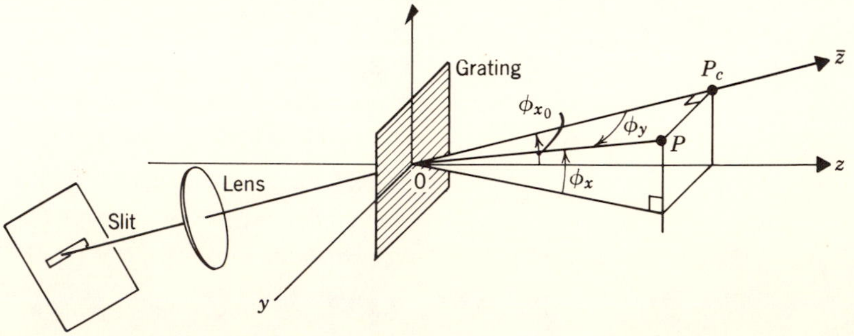

will be a constant if the slit and lens are used as shown above, but ϕ_x and α will not be constant. Show that they are given by

$$\alpha = \sin\phi_x = \sin\phi_{x0}\cos\phi_y \tag{1}$$

A transmission diffraction grating with spacing l and rulings parallel to the y axis lies in the xy plane. It will diffract monochromatic light moving parallel to OP into a beam in mth order moving parallel to OP', which will have a projection OP'_c onto the xz plane and "coordinates" $\phi'_y = \angle(P'OP'_c)$ and $\phi'_{x0} = \angle(P'_cOz)$. Find expressions for ϕ'_y and ϕ'_x in terms of ϕ_y and ϕ_{x0}, and show that for small ϕ_y, ϕ'_{x0} changes quadratically with ϕ_y.

Another lens is used to the right of the grating to bring this mth-order beam to a focus. Show that the resulting monochromatic image of the slit will be curved, and derive a formula for the sagitta, that is, the deviation of the curve from a straight line. Is this an important effect? Compared to what? Discuss, using realistic numbers for the focal lengths, grating constant, wavelength, and slit height.

7.3.11 Consider a diffraction grating with spacing l for which the ruled area is *circular* with radius $r_0 \gg l$. Discuss the quantitative behavior of the interference pattern for such a grating as a function of $(\alpha' - \alpha)$ when $\beta' = \beta$. (This problem is somewhat related to Problem 7.2.4.)

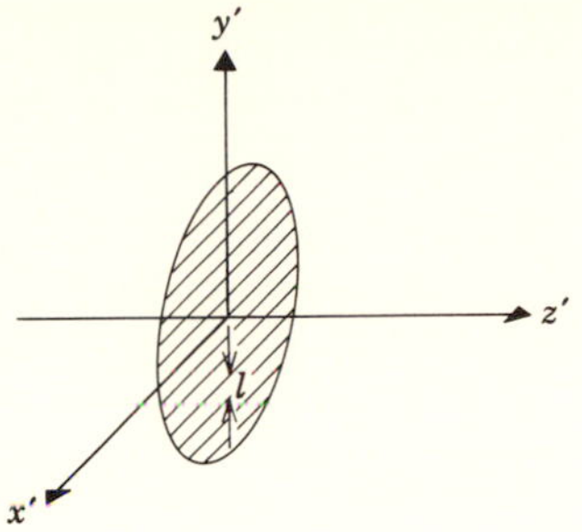

This chapter is a continuation of Chapter 7. We first give a more rigorous derivation of the basic Fresnel–Kirchoff diffraction integral than that of Section 7.1. Next follows a discussion of Fresnel diffraction by rectangular and circular apertures. We then take up holography, a method of photographically recording both the amplitude and the phase of an optical signal so that the entire wavefront may be reconstructed.

CHAPTER **8**

DIFFRACTION II–MAINLY FRESNEL DIFFRACTION

CHAPTER 8 CONTENTS

8.1 THE FOUNDATIONS OF THE THEORY OF DIFFRACTION

In this section we shall put the theory of diffraction on a mathematical foundation firmer than that provided by the intuitive Fresnel approach of Section 7.1. The line of reasoning, which is due to Kirchoff, is based ultimately on the time-independent wave equation (Helmholtz equation).

A. DERIVATION OF THE FRESNEL–KIRCHOFF INTEGRAL FROM THE HELMHOLTZ–KIRCHOFF THEOREM

The basic integral of the Fresnel–Kirchoff scalar theory of diffraction that is used in this book is obtained by combining Eqs. (7.10) and (7.11):

$$\tilde{E}(\mathbf{r}) = \frac{i}{\lambda} \iint_{\Sigma_0} \tilde{E}_{\text{inc}}(\mathbf{r}') \frac{e^{-ik|\mathbf{r}-\mathbf{r}'|}}{|\mathbf{r}-\mathbf{r}'|} \frac{(\cos\theta + \cos\theta')}{2}\, d\sigma' \tag{8.1}$$

The reader is reminded of the discussion at the beginning of Chapter 7—we are dealing with a monochromatic component $E(\mathbf{r}, t)$ of the physical electrical field that is obtained from the complex function $\tilde{E}(\mathbf{r})$ by Eq. (7.3):

$$E(\mathbf{r}, t) = \text{Re}\tilde{E}(\mathbf{r})e^{2\pi i\nu t}$$

We now discuss the extent to which this equation can be obtained from first principles. To what more basic equation is it an approximation? It is an approximation to the following equation, known as the Helmholtz–Kirchoff theorem:

$$\tilde{E}(\mathbf{r}) = \frac{1}{4\pi} \iint_{\Sigma} \left[\frac{e^{-ik|\mathbf{r}-\mathbf{r}'|}}{|\mathbf{r}-\mathbf{r}'|} \frac{\partial \tilde{E}(\mathbf{r}')}{\partial n} - \tilde{E}(\mathbf{r}') \frac{\partial}{\partial n} \frac{e^{-ik|\mathbf{r}-\mathbf{r}'|}}{|\mathbf{r}-\mathbf{r}'|} \right] d\sigma' \tag{8.2}$$

This is a rigorous mathematical expression following from the wave equation.

Equation (8.2) gives the electric field at a point $\mathbf{r}$ in terms of the field $E(\mathbf{r}')$ and its normal derivative $\partial E(\mathbf{r}')/\partial n$ evaluated on a closed surface Σ that either surrounds the point $P(\mathbf{r})$ (Fig. 8.1*b*) or excludes P (Fig. 8.1*a*). The normal $\mathbf{n}$ is assumed to point away from

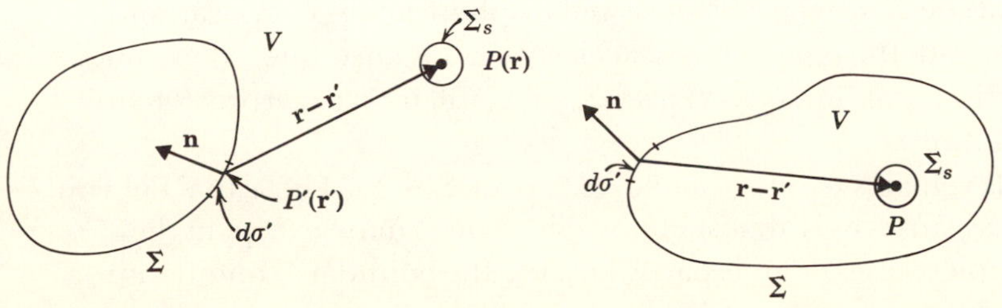

(*a*) (*b*)

Fig. 8.1

the region that contains P. Furthermore, the region containing P must be free of any sources of optical radiation;* that is, the sources must be on one side of Σ and P must be on the other side. We shall derive Eq. (8.2) in the section that follows.

It is important to realize that $E(\mathbf{r}')$ and $\partial E(\mathbf{r}')/\partial n$ represent the field and its derivative at the closed surface Σ *in the presence of the diffracting aperture*, if any. To obtain Eq. (8.1) from (8.2) we must let Σ include Σ_0, the open part of the diffracting aperture. We also let it include Σ_c, a surface just in back of the nontransmitting part of the diffracting aperture. Since Σ must be a closed surface, we must add a mathematical surface Σ_a or Σ_b as shown in Fig. 8.2.

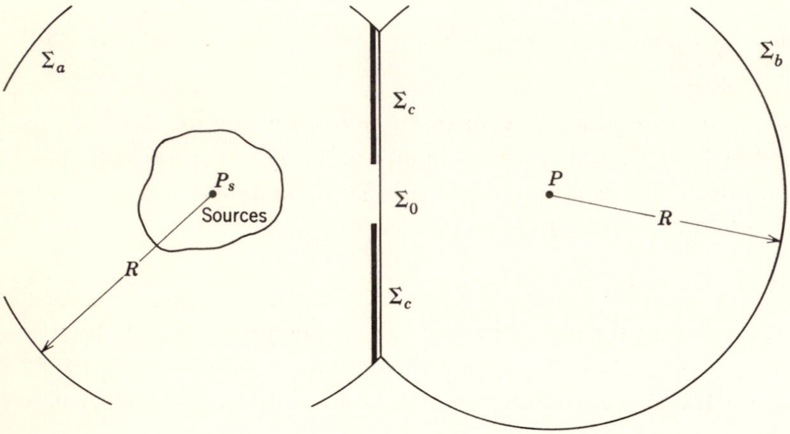

Fig. 8.2

Thus, corresponding to Fig. 8.1*a* we have

$$\Sigma = \Sigma_0 + \Sigma_c + \Sigma_a$$

and corresponding to Fig. 8.1*b* we have

$$\Sigma = \Sigma_0 + \Sigma_c + \Sigma_b$$

Here Σ_a is useful if all the primary and secondary optical sources except the aperture are located in a finite region of space. It can be taken to be part of a large sphere of radius R about these sources; Σ_b can be part of a similar sphere about P. We can show that with the type of physical electric fields encountered in optics, the integrals in Eq. (8.2) over Σ_a or Σ_b will tend to zero as R tends to infinity.

It remains to integrate Eq. (8.2) over $\Sigma_0 + \Sigma_c$. The actual field on these surfaces is almost impossible to determine except in the simplest cases. This is partly because the boundary conditions for

*This includes secondary sources of reradiated or scattered light. Thus, strictly speaking, the region must be free of optical material of any kind.

E and its derivatives are apt to be complicated for real physical apertures, which will partially absorb, partially reflect, and partially scatter the incident light. The first and key approximation that we make to go from Eq. (8.2) to (8.1) "cuts the Gordian knot" posed by the real boundary conditions by applying the so-called *Kirchoff boundary conditions.* These replace the actual field in the presence of the aperture by the incident field wherever the aperture is open and by zero wherever the aperture is nontransmitting. Thus we assume

$$\tilde{E}(\mathbf{r}') = 0 \tag{8.3a}$$

$$\frac{\partial \tilde{E}(\mathbf{r}')}{\partial n} = 0 \tag{8.3b}$$

for $\mathbf{r}'$ on Σ_c and

$$\tilde{E}(\mathbf{r}') = \tilde{E}_{\text{inc}}(\mathbf{r}') \tag{8.3c}$$

$$\frac{\partial \tilde{E}(\mathbf{r}')}{\partial n} = \frac{\partial \tilde{E}_{\text{inc}}(\mathbf{r}')}{\partial n} \tag{8.3d}$$

for $\mathbf{r}'$ on Σ_0 the open part of Σ. Then Eq. (8.2) becomes

$$\tilde{E}(\mathbf{r}) = \frac{1}{4\pi} \iint_{\Sigma_0} \left[\frac{e^{-ik|\mathbf{r}-\mathbf{r}'|}}{|\mathbf{r}-\mathbf{r}'|} \frac{\partial \tilde{E}_{\text{inc}}(\mathbf{r}')}{\partial n} - \tilde{E}_{\text{inc}}(\mathbf{r}') \frac{\partial}{\partial n} \frac{e^{-ik|\mathbf{r}-\mathbf{r}'|}}{|\mathbf{r}-\mathbf{r}'|} \right] d\sigma' \tag{8.4}$$

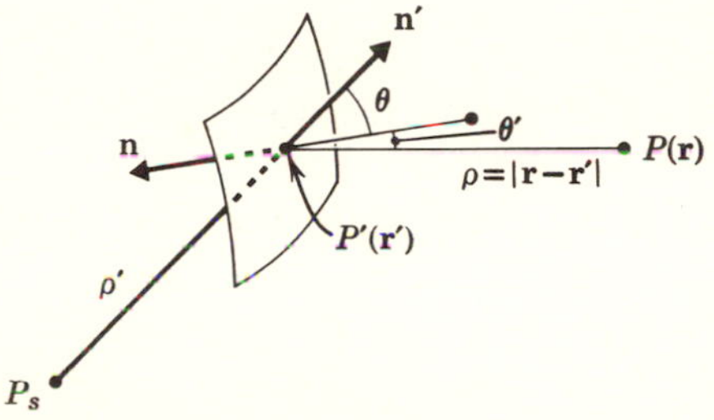

Fig. 8.3

Equation (8.4) is unnecessarily complicated for use with light. A simplification occurs if the incident light has a well-defined direction of propagation $\mathbf{n}'$ at the point P' on the surface Σ_0. This would be the case if the light were coming from a point source P_s (Fig. 8.3). Then the gradient of $\tilde{E}$ at P' is in the direction of $\mathbf{n}'$, and we can write

$$\begin{aligned} \frac{\partial \tilde{E}_{\text{inc}}(\mathbf{r}')}{\partial n} &= \mathbf{n} \cdot \nabla \tilde{E}_{\text{inc}}(\mathbf{r}') \\ &= \frac{(\mathbf{n} \cdot \mathbf{n}')\partial \tilde{E}_{\text{inc}}}{\partial n'} = -\cos\theta \frac{\partial \tilde{E}_{\text{inc}}}{\partial n'} \end{aligned} \tag{8.5}$$

where θ is the angle between $(-\mathbf{n})$ and $\mathbf{n}'$. The other normal derivative appearing in Eq. (8.2) can also be rewritten. Set $\rho = |\mathbf{r}-\mathbf{r}'|$. Then we have

$$\frac{\partial}{\partial n} \frac{e^{-ik\rho}}{\rho} = \cos\theta' \frac{\partial}{\partial \rho}\left(\frac{e^{-ik\rho}}{\rho}\right) = -\cos\theta' \left(ik + \frac{1}{\rho}\right) \frac{e^{-ik\rho}}{\rho} \tag{8.6}$$

Here θ' is the angle between $(\mathbf{r}-\mathbf{r}')$ and $(-\mathbf{n})$. For light waves the distance ρ from $P'(\mathbf{r}')$ to $P(\mathbf{r})$ is usually much larger than the wavelength $\lambda = 2\pi/k$. Thus

$$\frac{1}{\lambda} \quad \frac{1}{\rho} \qquad \text{and} \quad k = \frac{2\pi}{\lambda} \gg \frac{1}{\rho}$$

and the second term in Eq. (8.6) is negligible compared with the first. This gives

$$\frac{\partial}{\partial n}\frac{e^{-ik\rho}}{\rho} \approx -ik\cos\theta'\,\frac{e^{-ik\rho}}{\rho} \tag{8.7}$$

We can simplify Eq. (8.5) in a similar way. Light with a well-defined direction $\mathbf{n}'$ of propagation at P' will have the approximate local spatial dependence

$$\tilde{E}_{\text{inc}}(\mathbf{r}') \approx \text{const}\cdot e^{-ik\rho'}$$

where ρ' is a distance measured in the direction of $\mathbf{n}'$. Hence

$$\frac{\partial}{\partial n'}\tilde{E}_{\text{inc}}(r') = \frac{\partial \tilde{E}_{\text{inc}}}{\partial \rho'} \approx -ik\tilde{E}_{\text{inc}}(\mathbf{r}') \tag{8.8}$$

Equation (8.8) is valid provided that the source P_s is very many wavelengths from the surface Σ_0.

We may use Eqs. (8.7) and (8.8) to write Eq. (8.4) in the form

$$\tilde{E}(\mathbf{r}') = \frac{i}{\lambda}\iint_{\Sigma_0}\frac{e^{-ik\rho}}{\rho}\tilde{E}_{\text{inc}}(\mathbf{r}')\left(\frac{\cos\theta+\cos\theta'}{2}\right)d\sigma'$$

which is Eq. (8.1). The angle θ' is well defined in all cases, but θ, the angle between the direction of propagation from the source and $\mathbf{n}$, may not be well defined if a spatial distribution of sources is present instead of just one at P_s. If this distribution is present, the integral in Eq. (8.1) will have to be replaced by a sum of integrals, with a different source point and a different value of $\cos\theta'$ appearing in each one. Fortunately this extra complexity is unnecessary when the range of values of θ' is small.

B. DERIVATION OF THE HELMHOLTZ–KIRCHOFF THEOREM FROM THE WAVE EQUATION

The derivation of Eq. (8.2) starts with Green's theorem, which can be written

$$\iiint_V (\psi\nabla^2\phi - \phi\nabla^2\psi)\,d^3v = \iint_{\Sigma_t}\left(\psi\frac{\partial\phi}{\partial n} - \phi\frac{\partial\psi}{\partial n}\right)d\sigma$$

Here the volume integral extends throughout a volume V which is bounded by the surface Σ_t. The normal derivitives $\partial\phi/\partial n$ and $\partial\psi/\partial n$ are taken with respect to the outward normal of Σ_t. Then ψ and ϕ may be arbitrary functions of position that are sufficiently well behaved. Green's theorem may be proved by applying Gauss's theorem

$$\iiint_V \nabla\cdot\mathbf{A}\,d^3v = \iint_{\Sigma_t}\mathbf{A}\cdot d\boldsymbol{\sigma}$$

successively to $\mathbf{A} = \psi\nabla\phi$ and $\mathbf{A}' = \phi\nabla\psi$ and subtracting.

We choose $\phi(\mathbf{r}') = \tilde{E}(\mathbf{r}')$ and choose the volume V to contain no primary or secondary sources of the electric field, so that the free-space time-independent wave equation

$$\nabla^2\phi = -k^2\phi$$

holds there. We also set

$$\psi(\mathbf{r}') = \frac{e^{-ik|\mathbf{r}-\mathbf{r}'|}}{|\mathbf{r}-\mathbf{r}'|}$$

From Chapter 1 we know that this is a spherically symmetric solution of our wave equation, so that

$$\nabla^2\psi = -k^2\psi$$

except when $\mathbf{r}' = \mathbf{r}$, and we must exclude this point from our integration volume V. We do this by surrounding the point $P(\mathbf{r})$ by a small sphere of radius ϵ, and exclude this volume from V. The surface Σ_t is then given by the sum $\Sigma + \Sigma_s$ where Σ has already been defined and where Σ_s is the surface of the small sphere, as shown in Fig. 8.1. This choice of V and of Σ guarantees that both ϕ and ψ satisfy the Helmholtz equation everywhere in V. Thus, in V we have

$$\psi\nabla^2\phi - \phi\nabla^2\psi = k^2\psi\phi - k^2\phi\psi = 0$$

Then Green's theorem gives

$$0 = \iint_{\Sigma_t} \left(\psi\frac{\partial\phi}{\partial n} - \phi\frac{\partial\psi}{\partial n}\right) d\sigma' = \iint_{\Sigma} \left(\psi\frac{\partial\phi}{\partial n} - \phi\frac{\partial\psi}{\partial n}\right) d\sigma + \iint_{\Sigma_s} \left(\psi\frac{\partial\phi}{\partial n} - \frac{\partial}{\partial n}\right) d\sigma'$$

We now explicitly evaluate the integral over the little sphere Σ_s'. The sphere is outside the region V; thus the outward normal to Σ_s' is along its radius, directed inward toward $P(\mathbf{r})$. We can write

$$\iint_{\Sigma_s} \left(\psi\frac{\partial\phi}{\partial n} - \phi\frac{\partial\psi}{\partial n}\right) d\sigma' = \iint_{\Sigma_s} \left[\frac{e^{-ik\epsilon}}{\epsilon}\frac{\partial\tilde{E}}{\partial n} + \tilde{E}\left(-ik - \frac{1}{\epsilon}\right)\frac{e^{-ik\epsilon}}{\epsilon}\right] d\sigma'$$

$$= \iint_{\Sigma_s} \frac{e^{-ik\epsilon}}{\epsilon}\frac{\partial\tilde{E}}{\partial n} d\sigma' + \iint_{\Sigma_s} (-ik)\,\tilde{E}\,\frac{e^{-ik\epsilon}}{\epsilon} d\sigma' + \iint_{\Sigma_s} \left(-\frac{1}{\epsilon^2}\right)\tilde{E}e^{-ik\epsilon} d\sigma'$$

The value of $\partial E/\partial n$ will be bounded in the neighborhood of $\mathbf{r}$, so that the absolute value of the first term is less than

$$\left|\frac{\partial\tilde{E}}{\partial n}\right|_{\max} \iint_{\Sigma_s} \frac{d\sigma'}{\epsilon} = \left|\frac{\partial\tilde{E}}{\partial n}\right|_{\max} \cdot 4\pi\epsilon$$

This tends to zero as $\epsilon \to 0$. The second term is less in absolute value than

$$k|\tilde{E}|_{\max} 4\pi\epsilon$$

which also tends to zero. For the third term, since $\tilde{E}(\mathbf{r}')$ is regular at $\mathbf{r}' = \mathbf{r}$ (there is no source there), we may treat it as a constant for small ϵ and take it outside the integral. The last terms then gives

$$-\frac{\tilde{E}(\mathbf{r})}{\epsilon^2} \iint_{\Sigma_s} d\sigma' = -4\pi\tilde{E}(\mathbf{r})$$

as $\epsilon \to 0$. Recalling that

$$\phi(\mathbf{r}') = \tilde{E}(\mathbf{r}') \qquad \text{and} \qquad \psi(\mathbf{r}') = e^{-ik|\mathbf{r}-\mathbf{r}'|}/|\mathbf{r}-\mathbf{r}'|$$

We have proved $\iint_{\Sigma} [\psi(\partial\phi/\partial n) - \phi\,(\partial\psi/\partial n)]\, d\sigma' = 4\pi\tilde{E}(\mathbf{r})$, which is Eq. (8.2).

REFERENCES AND COMMENTS PERTAINING TO SECTION 8.1.

A somewhat more detailed discussion of the foundations of diffraction theory on about the same level as this discussion will be found in Chapter 8 of J. M. Stone, *Radiation and Optics.*

The reader will find still more detail in Sections 8.2 and 8.3 of M. Born and E. Wolf, *Principles of Optics.* These authors give references to original papers and to other treatments of the subject, and provide a critique of the foundations of the theory and its underlying assumptions.

A recent paper, "Consistent Formulation of Kirchoff's Diffraction Theory" by E. Wolf and E. W. Marchand, in the *Journal of the Optical Society of America*, **56**, 1712 (1966), presents experimental and theoretical evidence for the correctness of the predictions of Eq. (8.4) in a specific case even when $P(\mathbf{r})$ is very close to the diffracting aperture. See also their earlier paper "Comparison of the Kirchoff and the Rayleigh-Sommerfeld Theories of Diffraction at an Aperture," *loc. cit.*, **54**, 587 (1964).

8.2 FRESNEL DIFFRACTION

Fresnel diffraction occurs when the condition for Fraunhofer diffraction is not satisfied, that is, when there are terms in the optical path difference that are second order in the aperture coordinates and which are not smaller than a wavelength and hence cannot be ignored. The explicit calculations of Fresnel diffraction patterns are very complicated except for a few important special cases.

A. INTRODUCTION

The arrangement is still that of Fig. 7.10 with a point source at $P_s(\mathbf{r}_s)$, an observation point at $P(\mathbf{r})$, and a diffracting aperture Σ_0 at O. The choice of Σ_0 was discussed in Section 7.1B.1. There are important special cases where we can conveniently take Σ_0 to be part of the incident wave surface, but a more general treatment can be given if Σ_0 is a plane through O.

1. Derivation of the Fresnel Diffraction Integrals. Figure 8.4 defines the geometry. The starting equation is (7.25), and we ignore the inclination factor Q for the time being:

$$\tilde{E}(\mathbf{r}) = \frac{iA}{\lambda} \iint_{\Sigma_0} \frac{e^{-ik(R+R')}}{RR'}\, d\sigma \tag{8.9}$$

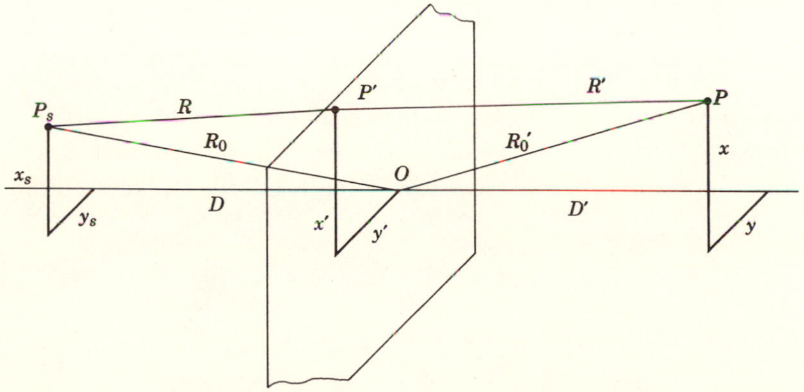

Fig. 8.4

For Fraunhofer diffraction we expanded R about R_0 and R' about R_0' and obtained for R

$$R = R_0 + \alpha x' + \beta y' + \frac{1}{2}\left[\frac{x'^2 + y'^2 + (\alpha x' + \beta y')^2}{R_0}\right] + \cdots$$

To keep the last term small we made the Fraunhofer approximation $|x'| \ll \sqrt{R_0\lambda}$, $|y'| \ll \sqrt{R_0\lambda}$, but we did not have to assume that α and β were small. This was an advantage because α and β often are not small for diffraction gratings.

The quadratic term in the above expansion about R_0 is unwieldy because of the presence of the $x'y'$ cross term obtained by multiplying out

$$\frac{(\alpha x' + \beta y')^2}{2R_0}$$

This can be eliminated only by assuming that α and β are small. When we do this, we should also expand R_0 about D. This is equivalent to treating x_s, y_s, x', and y' on an equal footing and expanding R about D. Thus we write

$$R = \sqrt{D^2 + (x_s - x')^2 + (y_s - y')^2} = D\left[1 + \frac{(x_s - x')^2}{D^2} + \frac{(y_s - y')^2}{D^2}\right]^{1/2}$$

and expand to obtain

$$R = D + \frac{(x_s - x')^2}{2D} + \frac{(y_s - y')^2}{2D} - \frac{1}{8}\frac{[(x_s - x')^2 + (y_s - y')^2]^2}{D^3} + \cdots \tag{8.10a}$$

We intend to drop the last term in Eq. (8.10*a*). It must therefore be much less than λ, and it almost always is in the cases of interest.

Similarly we obtain the result

$$R' = \sqrt{D'^2 + (x - x')^2 + (y - y')^2}$$

$$= D' + \frac{(x - x')^2}{2D'} + \frac{(y - y')^2}{2D'} - \frac{1}{8}\frac{[(x - x')^2 + (y - y')^2]^2}{D'^3} + \cdots \tag{8.10b}$$

We shall crop the last term here also.

In the denominator of Eq. (8.9) we shall replace R, R' by D, D' since the denominators are slowly varying. The exponential varies much more rapidly, and more care is required. In the exponent we shall put

$$|\widehat{P_s P' P}| = R + R' = D + D' + \left[\frac{(x - x')^2}{2D'} + \frac{(x_s - x')^2}{2D}\right] + \left[\frac{(y - y')^2}{2D'} + \frac{(y_s - y')^2}{2D}\right] \tag{8.11}$$

The integral in Eq. (8.9) can be reduced to standard form by completing the squares in (8.11). Consider

$$[\cdots]_x \equiv \left[\frac{(x - x')^2}{2D'} + \frac{(x_s - x')^2}{2D}\right] = \frac{x^2}{2D'} + \frac{x_s{}^2}{2D} + \frac{x'^2}{2}\left(\frac{1}{D} + \frac{1}{D'}\right) - x'\left(\frac{x}{D'} + \frac{x_s}{D}\right)$$

The coefficient of $(-x')$ is

$$\frac{x}{D'} + \frac{x_s}{D} = \frac{Dx + D'x_s}{DD'} = \left(\frac{Dx + D'x_s}{D + D'}\right)\left(\frac{D + D'}{DD'}\right)$$

We define two new variables

$$\rho = \frac{DD'}{D+D'}, \qquad \frac{1}{\rho} = \frac{D+D'}{DD'} = \frac{1}{D} + \frac{1}{D'} \tag{8.12}$$

and

$$x_m = \frac{Dx + D'x_s}{D+D'} \tag{8.13a}$$

Then the coefficient of $(-x')$ is x_m/ρ. The coefficient of x'^2 is

$$\frac{1}{2}\left(\frac{1}{D} + \frac{1}{D'}\right) = \frac{1}{2\rho}$$

Thus

$$[\cdots]_x = \frac{x^2}{2D'} + \frac{x_s^2}{2D} + \frac{x'^2}{2\rho} - \frac{x'x_m}{\rho} = \frac{x^2}{2D'} + \frac{x_s^2}{2D}$$

$$+ \frac{1}{2\rho}(x'^2 - 2x'x_m + x_m^2) - \frac{x_m^2}{2\rho}$$

$$= \frac{(x'-x_m)^2}{2\rho} + \frac{Dx^2 + D'x_s^2 - (Dx + D'x_s)^2/(D+D')}{2DD'}$$

$$= \frac{(x'-x_m)^2}{2\rho} + \frac{(x-x_s)^2}{2(D+D')} \tag{8.14a}$$

Similarly, if we define

$$y_m = \frac{Dy + D'y_s}{D+D'} \tag{8.13b}$$

algebraic manipulations lead to

$$[\cdots]_y = \left[\frac{(y-y')^2}{2D'} + \frac{(y_s-y')^2}{2D}\right] = \frac{(y'-y_m)^2}{2\rho} + \frac{(y-y_s)^2}{2(D+D')} \tag{8.14b}$$

Equations (8.14*a*, *b*), when inserted into Eq. (8.11), give

$$|\widehat{P_sP'P}| = R + R' = \left[D + D' + \frac{(x-x_s)^2 + (y-y_s)^2}{2(D+D')}\right]$$

$$+ \left[\frac{(x'-x_m)^2 + (y'-y_m)^2}{2\rho}\right] \tag{8.15}$$

The first bracketed term in (8.15) is a good approximation to the distance from P_s to P:

$$|P_sP| = \sqrt{(D+D')^2 + (x-x_s)^2 + (y-y_s)^2}$$

$$= (D+D')\left[1 + \frac{(x-x_s)^2 + (y-y_s)^2}{(D+D')^2}\right]^{1/2}$$

$$\approx (D+D') + \frac{(x-x_s)^2 + (y-y_s)^2}{2(D+D')} \tag{8.16}$$

Then (8.15) can be written

$$|\widehat{P_sP'P}| = |P_sP| + \left[\frac{(x'-x_m)^2 + (y'-y_m)^2}{2\rho}\right] \tag{8.15$'$}$$

and (8.9) becomes

$$\tilde{E}(\mathbf{r}) = \frac{iA}{\lambda DD'} e^{-ik|P_sP|} \iint_{\Sigma_0} \exp\left\{-\frac{ik}{2\rho}[(x'-x_m)^2 + (y'-y_m)^2]\right\} dx'\, dy'$$

One further change is made with the use of (8.12):

$$\frac{1}{DD'} = \frac{(D+D')}{DD'} \frac{1}{(D+D')} = \frac{1}{(D+D')\rho} \approx \frac{1}{|P_sP|\rho}$$

Then the factor in front of the integral becomes

$$\frac{iA}{\lambda DD'} e^{-ik|P_sP|} = \frac{i}{\lambda\rho} \frac{Ae^{-ik|P_sP|}}{|P_sP|} = \frac{i}{\lambda\rho} \tilde{E}_{\text{na}}(\mathbf{r})$$

where

$$\tilde{E}_{\text{na}}(\mathbf{r}) \equiv \frac{Ae^{-ik|P_sP|}}{|P_sP|} \tag{8.17}$$

is the field that would be observed at $P(\mathbf{r})$ if there were no aperture at all. Our Fresnel diffraction integral then becomes

$$\tilde{E}(\mathbf{r}) = \frac{i}{\lambda\rho} \tilde{E}_{\text{na}}(\mathbf{r}) \iint_{\Sigma_0} \exp\left\{-\frac{\pi i}{\lambda\rho}[(x'-x_m)^2 + (y'-y_m)^2]\right\} dx'\, dy' \tag{8.18}$$

Before we evaluate it we should interpret the coordinates x_m, y_m.

2. *Interpretation of the Coordinates* x_m, y_m. Define a point P_m in the plane of the aperture having coordinates $\mathbf{r}_m = (x_m, y_m, 0)$. Equations (8.13*a*, *b*) may then be written as the single vector equation

$$\mathbf{r}_m = \frac{D\mathbf{r} + D'\mathbf{r}_s}{D+D'} \tag{8.19}$$

We prove that P_m lies on the straight line from P_s to P as illustrated in Fig. 8.5. We have only to rewrite (8.19):

$$\mathbf{r} - \mathbf{r}_m = \frac{(D+D')\mathbf{r} - (D\mathbf{r} + D'\mathbf{r}_s)}{D+D'} = \frac{D}{D+D'}(\mathbf{r} - \mathbf{r}_s)$$

Since $\overrightarrow{P_mP} = (\mathbf{r} - \mathbf{r}_m)$ is a nonzero scalar times $\overrightarrow{P_sP} = (\mathbf{r} - \mathbf{r}_s)$, P_mP and P_sP must be parallel. Thus, P_s, P_m, and P are colinear. If we let $\mathbf{r}' = (x', y', 0)$ be the coordinates of P', we may write (8.15$'$) in the form

$$|\widehat{P_sP'P}| - |\widehat{P_sP_mP}| = \frac{|\mathbf{r}' - \mathbf{r}_m|^2}{(2\rho)} = \frac{|P'P_m|^2}{(2\rho)} \tag{8.15$''$}$$

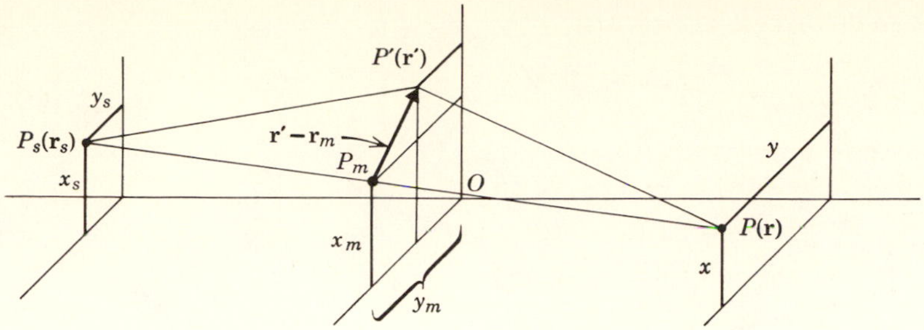

Fig. 8.5

We also see that the argument of the exponent in Eq. (8.18) is $-ik$ times (8.15″) or

$$\frac{-\pi i}{\lambda\rho}|\mathbf{r}'-\mathbf{r}_m|^2=\frac{-\pi i}{\lambda\rho}|P'P_m|^2$$

3. Use of Dimensionless Coordinates. It is convenient and conventional to define dimensionless coordinates $\boldsymbol{\eta}=(\eta_x,\eta_y,0)$ by the equation

$$\boldsymbol{\eta}=\sqrt{\frac{2}{\lambda\rho}}\,(\mathbf{r}'-\mathbf{r}_m) \tag{8.20}$$

Then Eq. (8.18) becomes

$$\tilde{E}(\mathbf{r})=\frac{i}{2}\tilde{E}_{\mathrm{na}}(\mathbf{r})\iint e^{-(\pi i/2)|\boldsymbol{\eta}|^2}\,d\eta_x d\eta_y \tag{8.21}$$

The limits on the η_x, η_y integrations have to be determined from those on x' and y' by Eq. (8.20). For instance, if $\mathbf{r}_l'=(x_l',y_l',0)$ represents the coordinates of a limit point P_l' at the edge of the aperture Σ_0, the corresponding limits on η_{lx} and η_{ly} are the components of $\boldsymbol{\eta}_l=(\eta_{lx},\eta_{ly},0)$ given by

$$\boldsymbol{\eta}_l=\left(\frac{2}{\lambda\rho}\right)^{1/2}(\mathbf{r}_l'-\mathbf{r}_m) \tag{8.22}$$

4. Projection of the Aperture onto the Observation Plane. Often the most convenient method of describing Fresnel diffraction refers everything to the plane of observation, rather than to the plane of the aperture. So, instead of $P_m(\mathbf{r}_m)$ we return to $P(\mathbf{r})$, where from (8.19) $\mathbf{r}$ is given by

$$\mathbf{r}=\frac{D+D'}{D}\mathbf{r}_m-\frac{D'}{D}\mathbf{r}_s$$

and instead of the limiting point P_l' on the edge of the aperture we have its projection P_l having coordinates $\mathbf{r}_l=(x_l,y_l,D')$ illustrated in Fig. 8.6 and given by

$$\mathbf{r}_l=\frac{D+D'}{D}\mathbf{r}_l'-\frac{D'}{D}\mathbf{r}_s \tag{8.23}$$

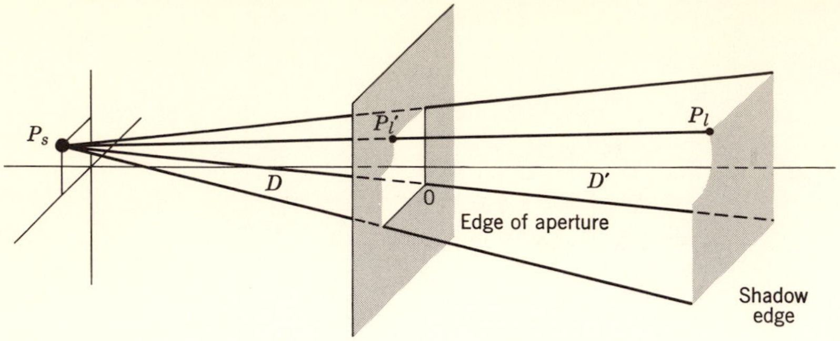

Fig. 8.6

Here P_l is at the edge of the geometrical shadow of the aperture produced by the light from the point source P_s. We may then write

$$\mathbf{r}_l' - \mathbf{r}_m = \frac{D}{D+D'}(\mathbf{r}_l - \mathbf{r})$$

and then rewrite (8.22) as

$$\boldsymbol{\eta}_l = \left(\frac{2}{\lambda}\right)^{1/2}\left[\frac{D+D'}{DD'}\frac{D^2}{(D+D')^2}\right]^{1/2}(\mathbf{r}_l - \mathbf{r}) = \left[\frac{2D}{\lambda D'(D+D')}\right]^{1/2}(\mathbf{r}_l - \mathbf{r})$$

$$= \frac{(\mathbf{r}_l - \mathbf{r})}{S_F} \tag{8.24}$$

where

$$S_F \equiv \left[\frac{\lambda D'(D+D')}{2D}\right]^{1/2} \tag{8.25}$$

is a scale factor having the dimension of length.

When the limits on the η_x and η_y integrations in Eq. (8.21) are mutually independent, the integral factors giving

$$\tilde{E}(\mathbf{r}) = \tfrac{1}{2}i\tilde{E}_{\text{na}}(\mathbf{r})I_x I_y$$

$$I_x = \int e^{-(\pi i/2)\eta_x^2}\,d\eta_x, \qquad I_y = \int e^{-(\pi i/2)\eta_y^2}\,d\eta_y \tag{8.26}$$

If $E_{e.\text{na}}(\mathbf{r})$ represents the flux density with no aperture, then the flux density is

$$E_e(\mathbf{r}) = E_{e.\text{na}}(\mathbf{r})\,\frac{|I_x|^2|I_y|^2}{4} \tag{8.26$'$}$$

B. APPLICATION TO A RECTANGULAR APERTURE AND ITS LIMITING CASES

In this section we discuss Fresnel diffraction by a rectangular aperture and the limiting cases obtained when one or more of the edges go to infinity. The aperture and its shadow in the observation plane are illustrated in Fig. 8.7 for two dimensions. Because

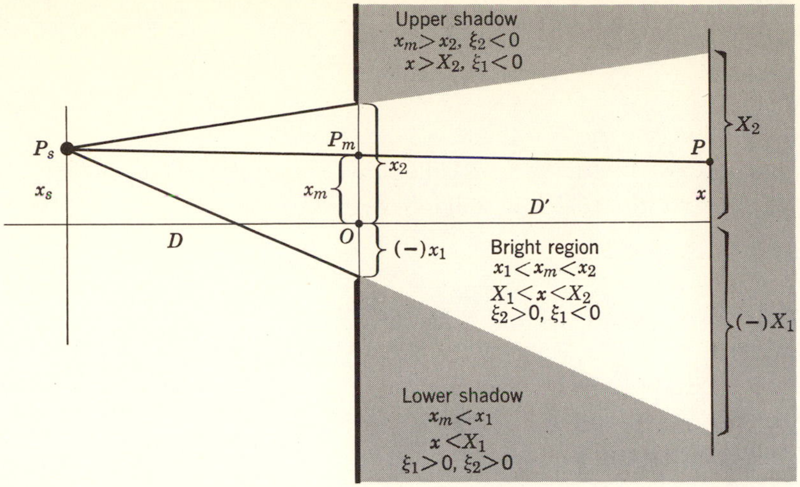

Fig. 8.7

of the factorization we may discuss the x and y dependence separately. The open area of the aperture is defined by the inequalities

$$x_1 \leqslant x' \leqslant x_2$$
$$y_1 \leqslant y' \leqslant y_2$$

and the bright region of the shadow by

$$X_1 \leqslant x \leqslant X_2$$
$$Y_1 \leqslant y \leqslant Y_2$$

Here X_1, Y_1, X_2, and Y_2 can all be obtained from (8.23). For example,

$$X_1 = \frac{D+D'}{D} x_1 - \frac{D'}{D} x_s$$

We may now put limits on the integrals in (8.26),

$$I_x = \int_{\eta_{x1}}^{\eta_{x2}} e^{-(\pi i/2)u^2} du, \qquad I_y = \int_{\eta_{y1}}^{\eta_{y2}} e^{-(\pi i/2)u^2} du \tag{8.27a}$$

where from (8.24) we obtain

$$\eta_{x1} = \frac{(X_1 - x)}{S_F}, \qquad \eta_{y1} = \frac{(Y_1 - y)}{S_F}$$

$$\eta_{x2} = \frac{(X_2 - x)}{S_F}, \qquad \eta_{y2} = \frac{(Y_2 - y)}{S_F}$$

$$S_F = \left[\frac{\lambda D'(D+D')}{2D}\right]^{1/2} \tag{8.27b}$$

1. Fresnel Integrals. The integrals I_x and I_y in Eqs. (8.27*a*) can be more easily evaluated if we write them as follows:

$$I_x = I(\eta_{x2}) - I(\eta_{x1}) \tag{8.28a}$$
$$I_y = I(\eta_{y2}) - I(\eta_{y1}) \tag{8.28b}$$

where

$$I(\eta) = \int_0^\eta e^{-(\pi/2)iu^2}\,du \tag{8.28c}$$

The integral in Eq. (8.28*c*) is called the *complex Frenel integral.* It can be split into real and imaginary parts as follows:

$$I(\eta) = C(\eta) + iS(\eta) \tag{8.29a}$$

where

$$C(\eta) = \int_0^\eta \cos\left(\frac{\pi}{2}u^2\right) du \tag{8.29b}$$

and

$$S(\eta) = -\int_0^\eta \sin\left(\frac{\pi}{2}u^2\right) du \tag{8.29c}$$

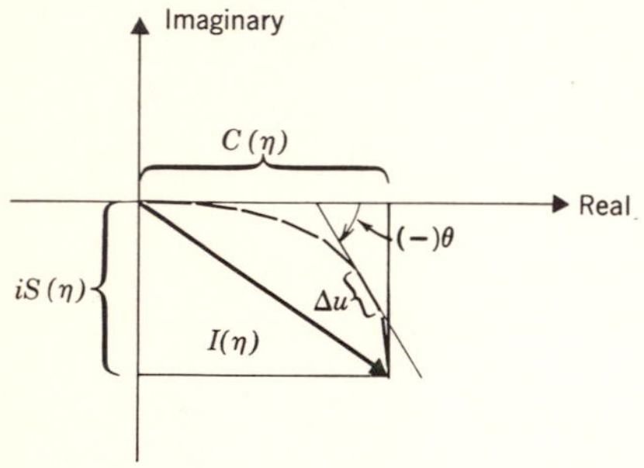

Fig. 8.8. Polygon approximation to vibration curve.

The integrals in Eqs. (8.28*a*, *b*, *c*) and (8.27*a*) have a graphical interpretation in terms of a *vibration curve* in the complex plane. We break the interval from 0 to η into N equally spaced segments of length $\Delta u = \eta/N$. Then we write

$$I(\eta) = \lim_{N\to\infty} \sum_{j=0}^{N-1} e^{-(\pi i/2)u_j^2}\Delta u$$

where $u_j = j\Delta u$. Before the limit is taken, the vector sum can be represented as the chord of a polygon in the complex plane. As $N \to \infty$ the polygon in Fig. 8.8 becomes a vibration curve, and $I(\eta)$ represents the complex vector lying along the chord of this curve with its tail at the origin and its head at the head of the last infinitesimal vector $\exp(-\pi i u^2/2)\,du$.

The total arc length S of the curve is given by

$$S = \int_0^\eta du = \eta$$

The inclination angle θ of Fig. 8.8 is the phase angle of the complex vector

$$du\, e^{-(1/2)\pi i u^2}$$

Hence

$$\theta(u) = -\frac{\pi u^2}{2}$$

This quadratic dependence of θ on arc length, which is characteristic of Fresnel diffraction, comes originally from the quadratic dependence of the optical path difference $|\widehat{P_sP'P}| - |\widehat{P_sP_mP}|$ on

$|\mathbf{r}' - \mathbf{r}_m|$ (Eq. 8.15″). In the Fraunhofer case the phase angle depends linearly on arc length, and the vibration curve was a circle. Here the curve coils up more rapidly than a circle, and the result is the *Cornu spiral* of Fig. 8.9.

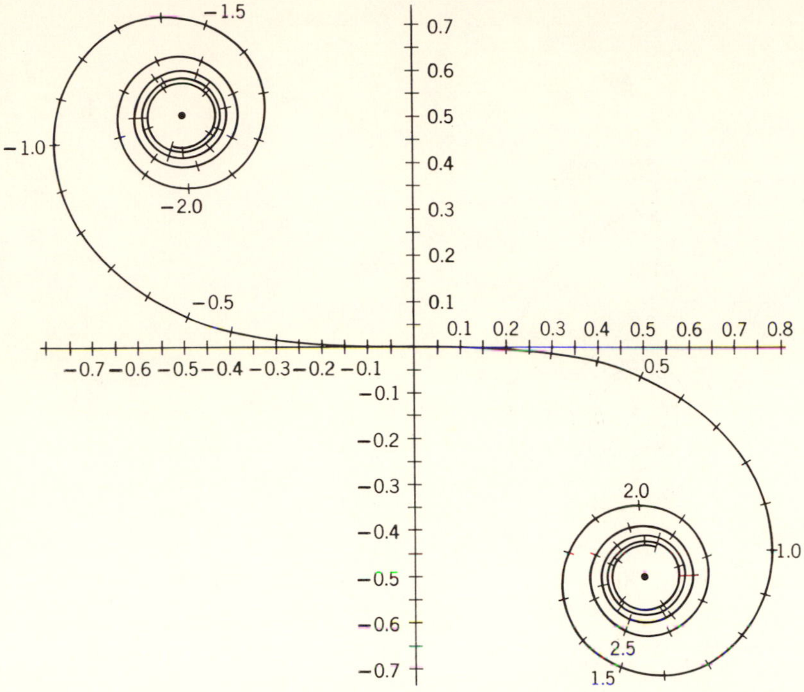

Fig. 8.9. The Cornu spiral.

2. ***Use of Cornu Spiral.*** For a given value of η, $I(\eta)$ is obtained from the spiral as shown in Fig. 8.10. The total arc length of the spiral starting from the origin is labeled by the numbers on the curve. The sign convention is such that η is positive in the lower right quadrant and negative in the upper left quadrant. Then $I(\eta)$ is the complex number represented by a vector with its tail at the origin and its head at the place on the spiral marked "η," whose total arc length from the origin amounts to η units.

Then $I(\eta_2) - I(\eta_1)$ is a complex vector whose head lies at η_2 with its tail at η_1 on the spiral. If $\eta_2 > 0$ and $\eta_1 < 0$, the head of the vector is in the fourth quadrant and the tail is in the second quadrant. If $\eta_2 > 0$ and $\eta_1 > 0$, both head and tail lie in the fourth quadrant. If $\eta_2 < 0$ and $\eta_1 < 0$, both lie in the second quadrant.

Another way to obtain numerical values for $I(\eta_2) - I(\eta_1)$ is to use Eq. (8.28*a*) and look up $[C(\eta_2) - C(\eta_1)] + i[S(\eta_2) - S(\eta_1)]$in tables of the real Fresnel integrals $C(\eta)$ and $S(\eta)$.*

*Jahnke–Emde–Lösch, *Tables of Higher Functions*, 6th ed. (McGraw–Hill, New York/Teubner Verlag, Stuttgart, 1960).

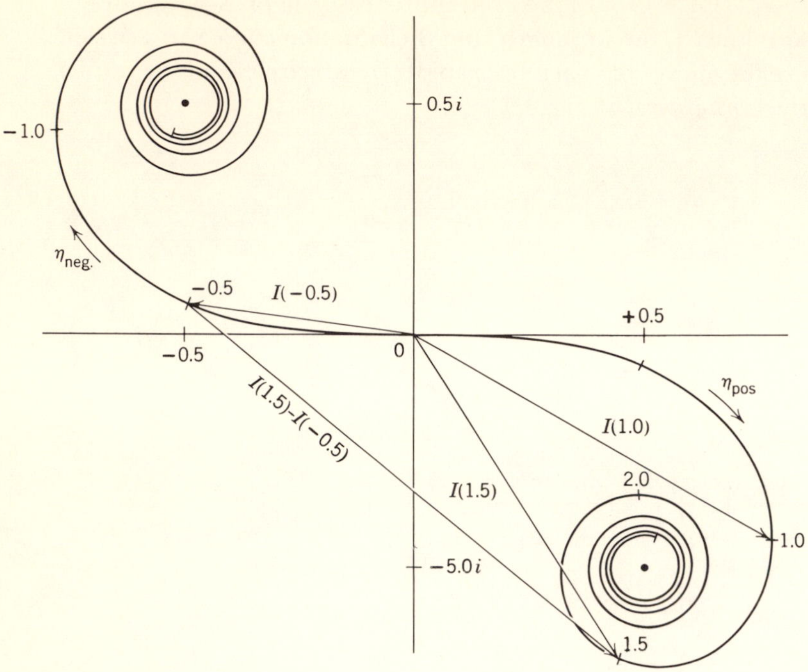

Fig. 8.10. Use of Cornu spiral.

The limiting values of the integrals as $\eta \to \pm\infty$ can be shown to be

$$C(+\infty) = -C(-\infty) = \tfrac{1}{2}$$
$$S(+\infty) = -S(-\infty) = -\tfrac{1}{2} \tag{8.30a}$$

Thus the "eye" of the spiral in the fourth quadrant is at

$$I(+\infty) = C(+\infty) + iS(+\infty) = \tfrac{1}{2}(1-i)$$

and the eye in the second quadrant is at

$$I(-\infty) = C(-\infty) + iS(-\infty) = -\tfrac{1}{2}(1-i) \tag{8.30b}$$

We now turn to some limiting cases of the rectangular aperture.

3. ***Open Aperture.*** Here there is no obstruction and hence $X_1 = Y_1 = X_2 = Y_2 = +\infty$. Then $\eta_{x2} = \eta_{y2} = +\infty$ and $\eta_{x1} = \eta_{y1} = -\infty$, and from Eq. (8.30) we obtain

$$I_x = I_y = I(+\infty) - I(-\infty) = (1-i)$$

Equation (8.26) then gives

$$\tilde{E}(\mathbf{r}) = \tfrac{1}{2}i(1-i)^2\,\tilde{E}_{\text{na}}(\mathbf{r}) = \tfrac{1}{2}i(1-1-2i)\,\tilde{E}_{\text{na}}(\mathbf{r}) = \tilde{E}_{\text{na}}(\mathbf{r})$$

This hardly unexpected result justifies the choice of $C = i/\lambda$ used in Eq. (7.6). The vector representing $I_x = I_y$ goes from eye to eye of the Cornu spiral, as shown in Fig. 8.11.

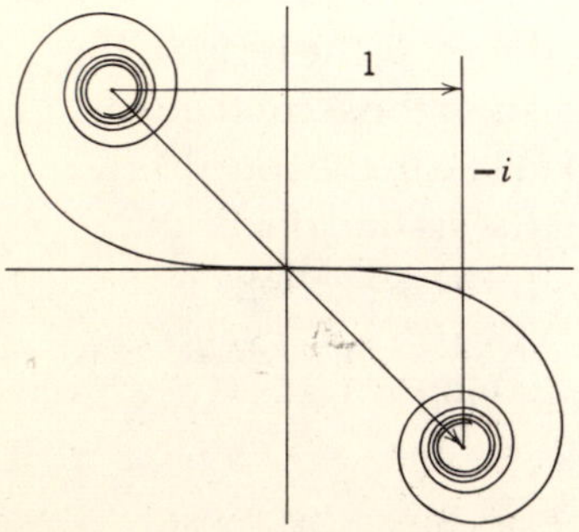

Fig. 8.11

The values in (8.30*a*, *b*) imply the following important result for the limiting value of the integral in Eq. (8.18):

$$\frac{i}{\lambda\rho}\iint_{-\infty}^{\infty} \exp\left\{-\frac{\pi i}{\lambda\rho}[(x'-x_m)^2+(y'-y_m)^2]\right\} dx'\,dy' = 1 \tag{8.30c}$$

independent of the values of x_m, y_m, and ρ.

4. ***Straight Edge.*** The straight edge runs from $x' = -\infty$ to $x' = x_1$ and is then open, as shown in Fig. 8.12. The geometrical bright region in the observation plane runs from $x = X_1$ to $x = +\infty$. Since

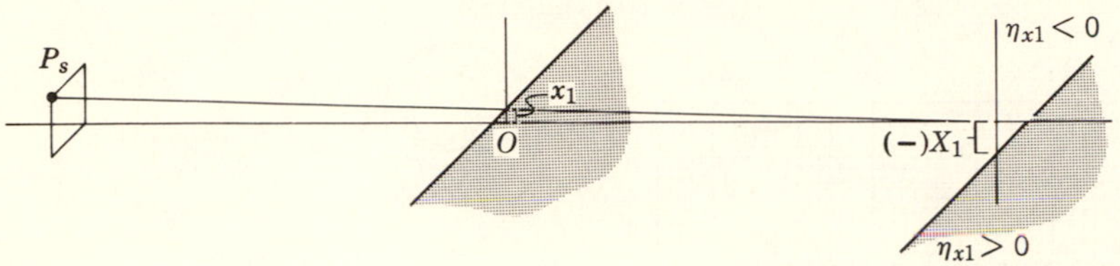

Fig. 8.12

we still have $-Y_1 = Y_2 = +\infty$, the integral I_y is still $(1-i)$. For I_x the upper limit remains $+\infty$, but the lower limit is

$$\eta_{x1} = \frac{X_1 - x}{S_F}$$

The head of the complex vector representing I_x remains in the lower right eye of the Cornu spiral. The position of the tail of the vector depends on the lower limit η_{x1}. If x is more positive than X_1 by an amount that is many multiples of the scale factor

$$S_F = \left[\frac{\lambda D'(D+D')}{2D}\right]^{1/2}$$

then η_{x1} is negative enough to be practically $-\infty$, and $I_x \approx (1-i)$, giving

$$\tilde{E}(\mathbf{r}) \approx \tilde{E}_{\text{na}}(\mathbf{r})$$

This is point A in Fig. 8.13.

As x decreases towards X_1, the lower limit, while still negative, is decreasing in magnitude. The tail of the vector for I_x comes out of the eye in the second quadrant of the spiral and traverses a larger and larger arc. This starts at values of η_{x1} near -5 and corresponds to the behavior near point B in Fig. 8.13.

When $\eta_{x1} = 0$ then $I_x = \frac{1}{2}(1-i)$, corresponding to the point C. This value is just half that for no aperture, giving for the flux density or irradiance

$$E_e(X_1) = \tfrac{1}{4}E_{e\,\text{na}}(x > X_1)$$

Point C corresponds to the exact edge of the geometric shadow.

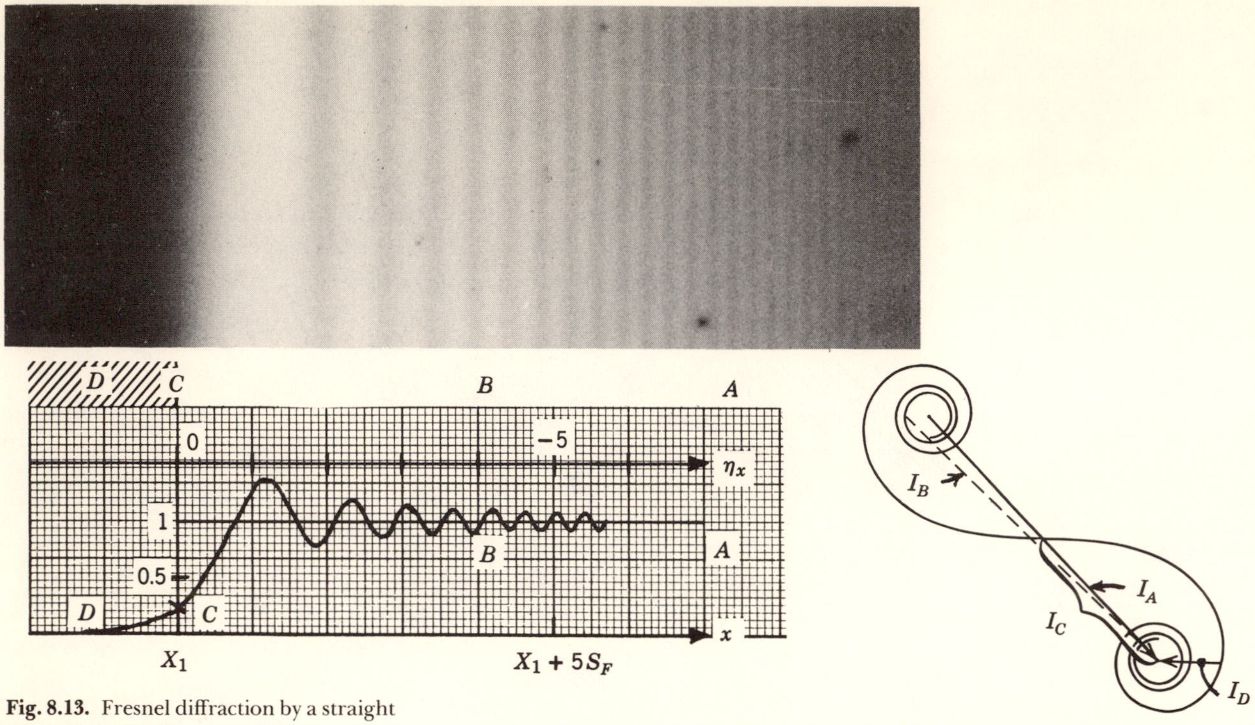

Fig. 8.13. Fresnel diffraction by a straight edge. At the top is an actual photograph. At the bottom the intensity is plotted. The vectors on the Cornu spiral at the right correspond to the labeled points.

For $x < X_1$, η_{x1} is positive, and the tail of the vector for I_x is on the lower right half of the spiral. As x continues to decrease η_{x1} grows, and the tail spirals in toward the eye. This is region D. The magnitude of I_x decreases monotonically and for $\eta_{x1} > +5$, say, I_x is quite small. This is the region of the geometrical shadow.

5. ***Wide Slit.*** Let the edges of the slit be at $x' = \pm x_0$. Then the shadow edges in the observation plane are at $x = X_1$ and $x = X_2$, where

$$X_1 = -\frac{(D+D')}{D}x_0 - \frac{D'}{D}x_s, \qquad X_2 = \frac{(D+D')}{D}x_0 - \frac{D'}{D}x_s$$

and the shadow region has width

$$\Delta X = X_2 - X_1 = \frac{(D+D')}{D}2x_0 \tag{8.31}$$

The integral for I_y still has infinite limits and gives $I_y = (1-i)$.

A "wide" slit is one that has a total arc length on the Cornu spiral

$$\Delta\eta = \eta_{x2} - \eta_{x1}$$

that is very large compared with unity. This condition can be rewritten using Eq. (8.27*b*):

$$1 \ll (\eta_{x2} - \eta_{x1}) = \frac{X_2 - X_1}{S_F} = 2x_0\left(\frac{2}{\lambda\rho}\right)^{1/2} \tag{8.32}$$

$$S_F = \left[\frac{\lambda D'(D+D')}{2D}\right]^{1/2}, \qquad \frac{1}{\rho} = \frac{1}{D} + \frac{1}{D'}$$

When x is in the center area of the geometrical bright region, we have

$$X_2 - x \gg S_F \qquad \text{or} \qquad \eta_{x2} \gg 1, \qquad \text{that is,} \qquad \eta_{x2} \sim +\infty$$

and

$$x - X_1 \gg S_F \qquad \text{or} \qquad \eta_{x1} \ll -1, \qquad \text{that is,} \qquad \eta_{x1} \sim -\infty$$

Then I_x is essentially equal to its value for no aperture $(1-i)$ with its vector running practically from eye to eye on the Cornu spiral.

As x moves toward the left edge of the shadow at X_1, still in the bright region, η_{x2} becomes even larger, and the head of the vector stays in the lower right eye. Then η_{x1} becomes less negative, and the tail of the vector begins to unwind from the upper left eye, just as though there were no right edge to the slit. Similarly as x approaches the right edge of the shadow at X_2, the tail of the vector for I_x remains very close to the upper left eye, and its head begins to spiral out the lower right branch, just as we would expect if there were no left edge of the slit. The resulting diffraction pattern is essentially a superposition of two straight-edge patterns as indicated in Fig. 8.14.

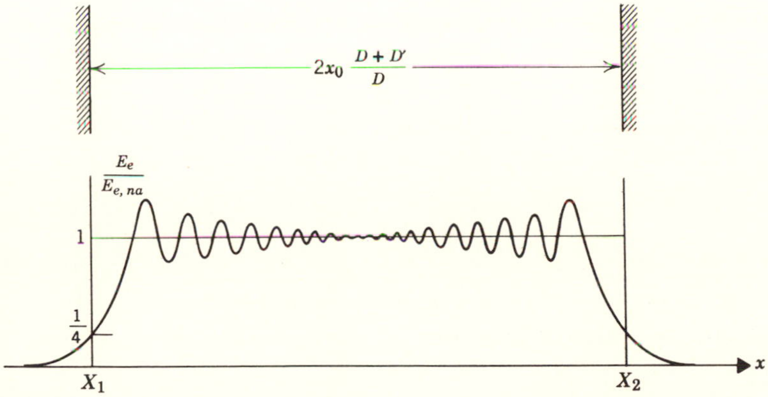

Fig. 8.14. Fresnel diffraction by a wide slit.

6. ***Narrow Slit.*** If the slit width $2x_0$ is of order $\sqrt{\lambda\rho/2}$ so that $\Delta\eta$ is of order unity or perhaps a bit larger, then one end of the vector for I_x will come out of one eye of the Cornu spiral before the other end has a chance to wind up into the other eye. The form of the diffraction pattern is quite sensitive to the size of $\Delta\eta$, as shown in Fig. 8.15.

In the limiting case where $\Delta\eta \ll 1$ we go over to Fraunhofer diffraction. The arc length is so small that it must be wound up way into an eye of the Cornu spiral before the chord representing I_x becomes smaller than the arc length. This makes the width of the

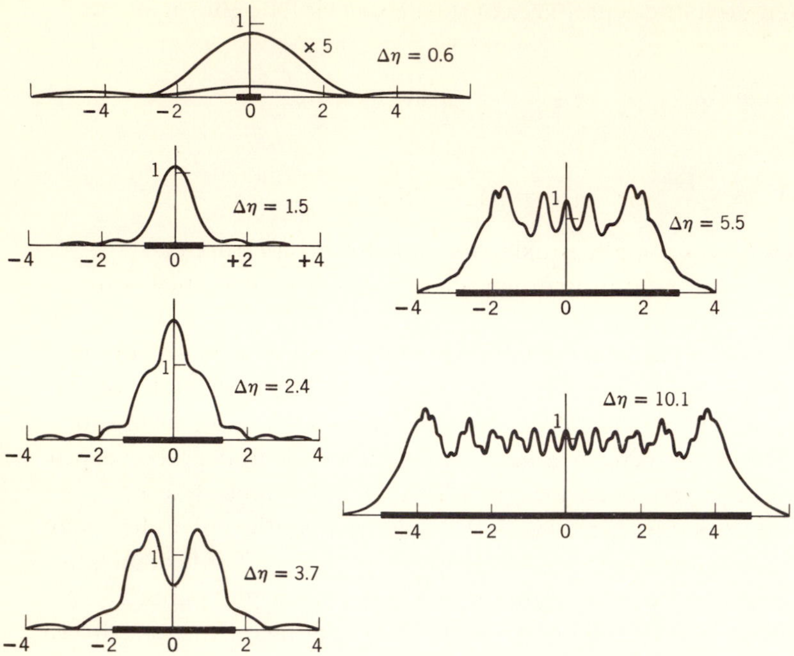

Fig. 8.15. Fresnel diffraction by slits of varying width. We plot the ratio of irradiance or flux density $E_e/E_{e,na}$ versus x/S_F and assume the pattern to be symmetrical about $x = 0$. Heavy bars denote the regions of geometrical brightness.

diffraction pattern much greater than that of the shadow. We can show that in the limit $\Delta\eta \rightarrow 0$ the Fresnel integrals do give the Fraunhofer diffraction expression.

7. The "Limit" of Geometrical Optics. Consider a slit that is very wide. The requirement is (Eq. [8.32])

$$1 <<< \Delta\eta = \eta_{x2} - \eta_{x1} = 2x_0 \left(\frac{2}{\lambda \rho}\right)^{1/2}$$

No matter how large $\Delta\eta$ is, there are always departures from the predictions of geometrical optics near the edge of the geometrical shadow. This occurs for $|\eta_{x1}|$ or $|\eta_{x2}|$ of the order of 10, for then the oscillations in region B and the nonzero intensity in region D of Fig. 8.14 occur. These diffraction effects take place when x is within a distance Δx of the shadow edge given by

$$\Delta x \sim 10 S_F = 10 \left[\frac{\lambda D'(D + D')}{2D}\right]^{1/2}$$

Hence, for fixed D and D', Δx can be made arbitrarily small by decreasing the wavelength sufficiently, although the size of the oscillations in E_e will not diminish. In this sense geometrical optics can be obtained from physical, or wave optics in the limit as $\lambda \rightarrow 0$.

8. The Two-Dimensional Pattern for Apertures of Rectangular Symmetry. An expression for the irradiance from (8.26) is

$$E_e(\mathbf{r}) = E_{e.\mathrm{na}}(\mathbf{r}) \; \frac{|I_x|^2 \, |I_y|^2}{4}$$

When the limits on the y integration are not infinite, the integral I_y will behave just like I_x discussed above, and the two-dimensional diffraction can be obtained from the product $|I_x I_y|^2$.

The case of a rectangular obstacle and its limiting configuration of a needle can be treated by an obvious generalization of the above methods. It is sometimes convenient to use Babinet's principle (Eq. 7.15), which can be written in the form

$$\tilde{E}_{\mathrm{na}}(\mathbf{r}) = \tilde{E}_{\mathrm{aperture}}(\mathbf{r}) + \tilde{E}_{\mathrm{obst}}(\mathbf{r})$$

With rectangular symmetry and the resulting factorization of the integral, we obtain a corresponding equation for the integrals I_x and I_y, for example,

$$I_{x.\mathrm{aperture}} + I_{x.\mathrm{obst}} = I_{x.\mathrm{na}} = (1-i) \tag{8.33}$$

C. FRESNEL DIFFRACTION BY A CIRCULAR APERTURE AND RELATED PHENOMENA

Fresnel diffraction from circular apertures and obstacles has some surprising properties. An essential ingredient is the rotational symmetry about an axis through the center of the circle involved. We shall therefore assume that the source point P_s lies on this axis, the z axis. Without loss of generality we may assume that the observation point P lies in the xz plane. The geometry is shown in Fig. 8.16.

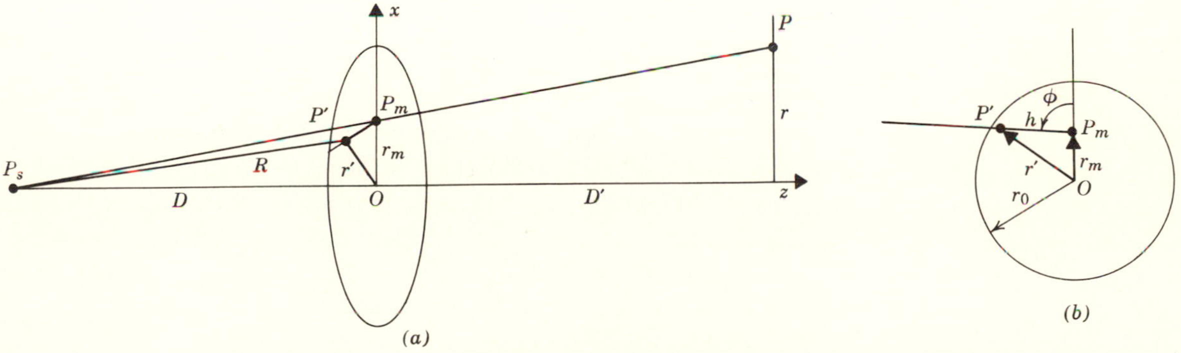

Fig. 8.16. Geometry for circular aperture. (*a*) In three dimensions. (*b*) In the plane of the aperture.

1. *Setting up the Integral.*

We start with Eq. (8.21), with the inclination factor Q inserted:

$$\tilde{E}(\mathbf{r}) = \frac{i}{2}\tilde{E}_{\mathrm{na}}(\mathbf{r}) \iint e^{-(\pi i/2)|\boldsymbol{\eta}|^2} Q\, d\eta_x\, d\eta_y \tag{8.34}$$

The exponential will be written $e^{-i\psi}$, where by Eq. (8.15″) and the equation below it we may put

$$\psi = \frac{2\pi}{\lambda}\left[|\widehat{P_s P' P}| - |\widehat{P_s P_m P}|\right] = \frac{\pi}{\lambda\rho}|\mathbf{r}' - \mathbf{r}_m|^2 = \frac{\pi}{2}|\boldsymbol{\eta}|^2 \tag{8.35}$$

$$\frac{1}{\rho} = \frac{1}{D} + \frac{1}{D'}$$

We let P_m be our origin in the aperture for polar coordinates h, ϕ where $h = |\mathbf{r}' - \mathbf{r}_m|, = |P_m P'|$, as shown in Fig. 8.16*b*. Then (8.35) becomes

$$\psi = \frac{\pi h^2}{\lambda\rho} \tag{8.35$'$}$$

We may then scale h by the factor $\sqrt{2/(\lambda\rho)}$ to obtain polar coordinates η, ϕ for the dimensionless vector variable $\boldsymbol{\eta}$:

$$\eta = |\boldsymbol{\eta}| = h\left[\frac{2}{\lambda\rho}\right]^{1/2} \tag{8.36}$$

If we set $d\eta_x\, d\eta_y = \eta\, d\eta\, d\phi$, Eq. (8.34) becomes

$$\tilde{E}(P) = \frac{i}{2}\tilde{E}_{\text{na}}(P) \iint e^{-(\pi i/2)\eta^2} Q\eta\, d\eta\, d\phi \tag{8.37}$$

At this point we let P_m and O coincide so that $r_m = 0 = r$. Then P is along the symmetry axis and will remain so until we explicitly state otherwise. There is no restriction on the variable ϕ which may range from 0 to 2π independently from h. Then since

$$\eta d\eta \int_0^{\pi} d\phi = 2\pi\eta\, d\eta = \pi\, d\eta^2 = 2d\psi,$$

we obtain

$$\tilde{E}(P) = i\tilde{E}_{\text{na}}(P) \int_0^{\psi_0} e^{-i\psi} Q(\psi)\, d\psi \tag{8.38}$$

where $\psi_0 = \pi r_0{}^2/(\lambda\rho)$ is the value of the phase difference $k[|\widehat{P_sP'P}| - |P_sOP|]$ at the edge of the aperture of radius r_0.

2. ***Vibration Curve.*** The vibration curve for the complex integral

$$I(\psi_0) = \int_0^{\psi_0} Q(\psi)\, e^{-i\psi}\, d\psi \tag{8.39}$$

is shown in Fig. 8.17. The inclination factor $Q(\psi)$ is a smooth monotonically decreasing function of ψ, as shown in Fig. 8.18. If it were to keep its initial value of unity, we could immediately perform the integration in Eq. (8.38); the vibration curve would be a circle as in Fraunhofer diffraction. Furthermore, the discussion surrounding Eq. (7.42) tells us that this circle would have unit radius. Since the vibration curve starts out horizontally at the origin with $\psi = 0$ and then curves into the fourth quadrant as the phase becomes negative, we can conclude that the center of the circle would be at the point $(0 - i)$. The slow decrease in Q will cause the curve to spiral in gradually toward the center. The departure from the original circle is slow because $Q(\psi)$ is very slowly varying, especially

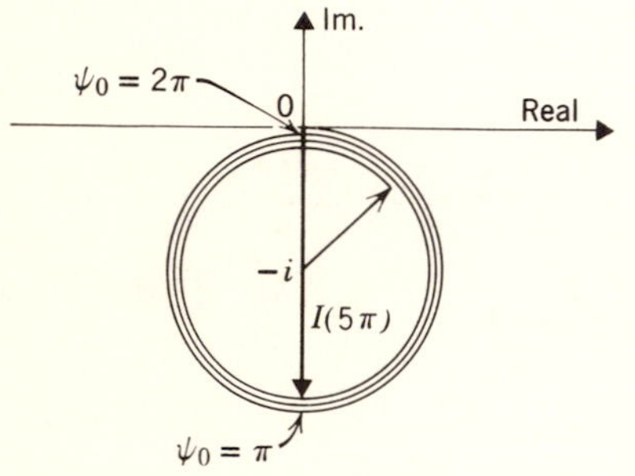

Fig. 8.17. Vibration curve appropriate for Fresnel diffraction by a circular aperture (P and P_s are on the axis).

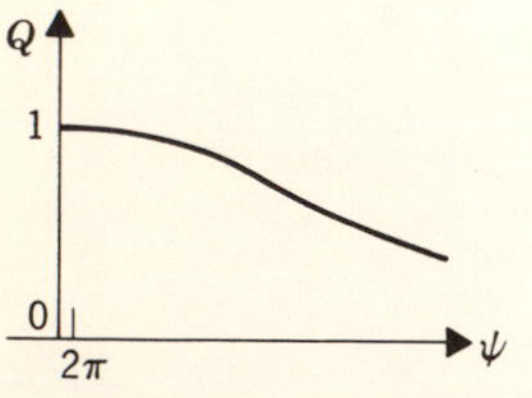

Fig. 8.18. Behavior of the function $Q(\psi)$. A large phase change is necessary before Q will depart noticeably from unity.

for the first few multiples of 2π. Then $I(\psi_0)$ will initially oscillate between the extremes of zero and $-2i$.

In the limit as ψ_0 gets very large, Q will tend to zero. (In this limit the angles θ and θ' approach $\pi/2$; hence $Q = \frac{1}{2}(\cos\theta + \cos\theta') \to 0$.) The vibration curve gradually spirals in to the point $-i$, and $\tilde{E}(P)$ approaches the value $\tilde{E}_{na}(P)$.

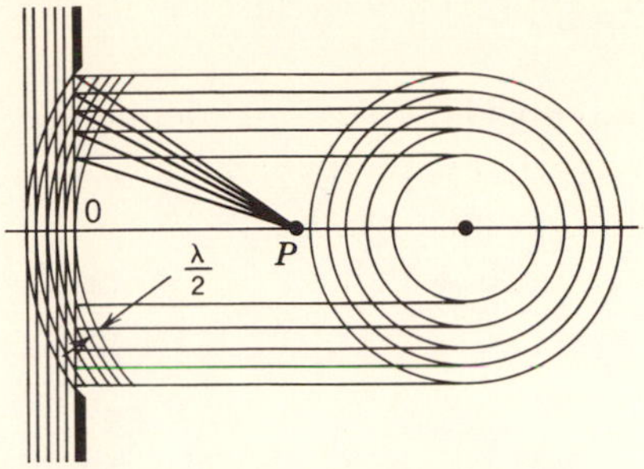

Fig. 8.19. Construction of Fresnel zones with the point source at infinity. The spheres about P have radii differing by $\lambda/2$. Their intersections with the plane of the aperture form the circular zones.

3. Fresnel Zones. It is useful to think of the open area of the diffracting aperture as divided into zones, the so-called *Fresnel zones*, which are centered at P_m.* These are defined so that the phase difference ψ changes by π across one zone and equals $n\pi$ at the edge of the nth zone. This is equivalent to changes in the path difference $|\widehat{P_sP'P}| - |\widehat{P_sP_mP}|$ by $\lambda/2$ and $n\lambda/2$, respectively. An illustration of the method of constructing these zones is found in Fig. 8.19 for the case where the source point P_s is at infinity. The successive spheres drawn from the observation point P have radii differing by $\lambda/2$.

Equation (8.35′) gives for the radius of the nth zone

$$r_n = \sqrt{n\lambda\rho} \tag{8.40}$$

Note that the area of a zone $\Delta\sigma$ is independent of n (at least within the quadratic approximation for ψ that we are using here):

$$\Delta\sigma = \pi(r_{n+1}^2 - r_n^2) = \pi\lambda\rho = \pi r_1^2 \tag{8.41}$$

As we integrate over a total of N zones in the diffracting aperture, the vibration curve in Fig. 8.17 makes $N/2$ revolutions. If N is not too large and there are an even number of zones inside the aperture, the integral $I(\psi_0)$ is very small; if there are an odd number of zones, $I(\psi_0)$ is close to $-2i$.

We may take the inclination factor Q into account in the following approximate way. The contribution to the integral from just the nth zone is essentially

$$\Delta I_n \approx -2iQ_n, \qquad n \text{ odd}$$

and

$$\Delta I_n \approx +2iQ_n, \qquad n \text{ even}$$

where Q_n is the average value of the inclination factor over the nth zone. The integral $I(\psi_0)$ is then given approximately by

$$I = \sum_{n=1}^{N} \Delta I_n = -2i[Q_1 - Q_2 + Q_3 - Q_4 + \cdots \pm Q_N]$$

$$= -2i\left[\frac{Q_1}{2} + \left(\frac{Q_1}{2} - Q_2 + \frac{Q_3}{2}\right) + \left(\frac{Q_3}{2} - Q_4 + \frac{Q_5}{2}\right) + \cdots\right]$$

*In some treatments of the subject the wavefront is divided into zones. The results are the same, but a discussion of some topics such as zone plates is more awkward.

The terms in parentheses are essentially zero, giving*

$$I \cong -i(Q_1 + Q_N), \qquad N \text{ odd}$$
$$I \cong -i(Q_1 - Q_N), \qquad N \text{ even}, \qquad (Q_1 \cong 1)$$

For large N, Q_N approaches zero, and I approaches $-i$, just *half the contribution from the first Fresnel zone.*

4. Behavior Off Axis. The Fresnel zones will help us to understand qualitatively or semiquantitatively what happens when P is off axis so that $r_m \neq 0$. The limits on the variable ϕ in Eq. (8.37) depend on h in a complicated way. We can show that

$$\int d\phi = 2\pi, \qquad \text{for } h < (r_0 - r_m)$$

and

$$\int d\phi = 2 \arccos \frac{r_m^2 + h^2 - r_0^2}{2hr_m} \qquad \text{for } (r_0 - r_m) < h < (r_0 + r_m) \tag{8.42}$$

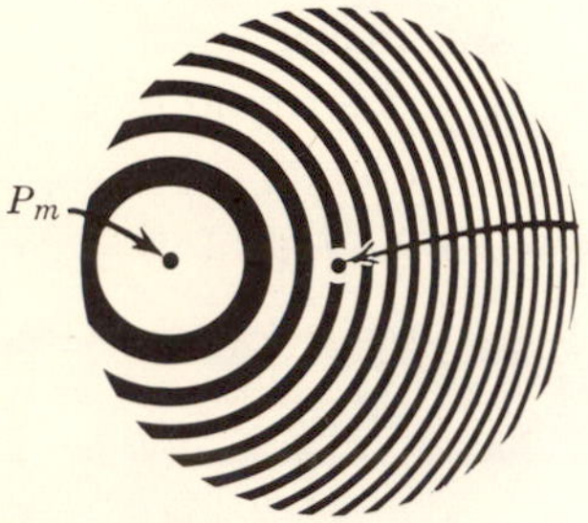

Fig. 8.20. The nature of the Fresnel zones inside a circular diffracting aperture when the center of the zones at P_m, which is along the line from the point source to the observation point, is not at the center of the aperture O. Every other zone is shown dark. The contributions from adjacent zones are of opposite sign and almost cancel.

We thus get the full contribution from all zones of radius less than $(r_0 - r_m)$ and decreasing contributions from zones of radius between $(r_0 - r_m)$ and $(r_0 + r_m)$, as can be seen from Figs. 8.16*b* and 8.20.

The contribution to the arc length of the vibration curve from each of these partially obstructed zones is proportional to the unobstructed area and hence is less than the value of $Q_n\pi$ that comes from the full nth zone. The phase difference still changes by π across the zone. The result is that the vibration curve coils up more rapidly. The kind of diffraction pattern that is observed is indicated in Plate 1*a*.

The number of rings in the pattern will increase as the plane of observation moves toward the diffracting aperture; the pattern will expand outward. As the plane of observation moves away from the aperture, the pattern will have fewer rings and contract. Finally, when only a small fraction of a zone covers the aperture for P on axis, we pass over to the Fraunhofer limit.

For a large aperture diameter r_0 and/or a short distance D', the total number of zones involved will be quite large. *The behavior of the diffraction pattern for P_m near O depends on the aperture being very accurately a circle.* If it has ragged edges with a roughness scale comparable to the width of the last or Nth zone, which from Eq. (8.41) is given by

$$r_N - r_{N-1} = \Delta r_N = \frac{r_1^2}{r_N + r_{N-1}} \approx \frac{r_1^2}{2r_N} = \frac{r_1}{2\sqrt{N}}, \qquad r_1 = \sqrt{\lambda\rho} \tag{8.43}$$

then the last turn or so of the vibration curve in Fig. 8.17 will be modified, producing important changes in the diffraction pattern

*For details see M. Born and E. Wolf, *Principles of Optics*, Section 8.2.

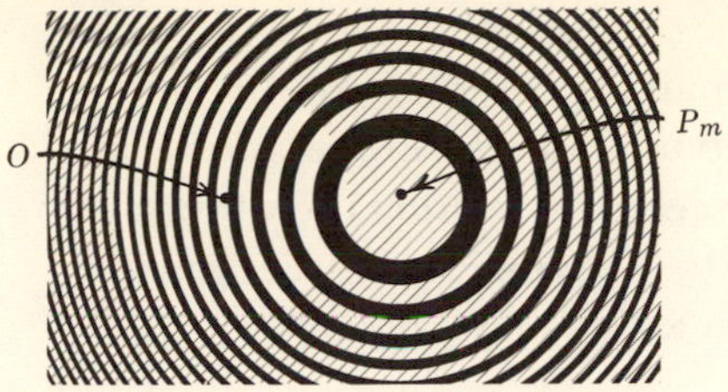

Fig. 8.21

and tending to smooth out the rings. Since any real aperture will be rough on a small enough scale, at a sufficiently small wavelength this effect will take over. In this way the limit of geometrical optics is reached as $\lambda \to 0$.

When the observation point P moves into the geometrical shadow, the point P_m moves into the opaque part of the aperture, as shown in Fig. 8.21. The contributions from the many partial zones that are exposed tend to cancel one another, giving a very small resultant electric field. There is one last broad bright diffraction ring formed just before the point P passes into the shadow, because of incomplete cancellation of the contribution from the central zone. This ring appears in Plate 1*a*. Its width is about equal to that of the central zone projected onto the plane of observation at P. This is of the order of the scale factor $S_F = \sqrt{\lambda D'(D + D')/(2D)}$ used in our discussion of straight edges and rectangular apertures, and the first bright bringe for a straight edge was about S_F wide. Near the shadow edge of a large diameter circular aperture the diffraction pattern should be quite similar to that of a straight edge.

5. *Fresnel Diffraction by a Circular Obstacle.* Consider an accurately round circular obstacle of radius r_i. Then for P on the axis Eq. (8.38) becomes

$$\tilde{E}(P) = i\tilde{E}_{\text{na}}(P)\text{I}, \qquad I = \int_{\psi_i}^{\infty} e^{-i\psi} Q(\psi)\, d\psi \tag{8.44a}$$

where the initial phase angle ψ_i is given by

$$\psi_i = \frac{\pi}{\lambda\rho} r_i^2 \tag{8.44b}$$

In this case the section of the vibration curve in Fig. 8.17 from $\psi = 0$ to $\psi = \psi_i$ is missing. The tail of the vector for I will be on the spiral where the phase equals ψ_i. Its head will be at the center at $-i$. Thus, if ψ_i is not too large, so that Q is still approximately unity, we find that I should have an absolute value of one, and the flux density should be given by

$$E_e = E_{e.\text{na}}$$

Thus the flux density at the center of the diffraction pattern of a circular obstacle equals the flux density when there is no obstacle! This surprising prediction is strictly a wave phenomenon.

For this effect to be observed, the departure of the outline of the obstacle from a true circle must be much less than the width Δr_N of the last or Nth Fresnel zone that is covered by the obstacle. This is given by Eq. (8.43) as

$$\Delta r_N = \frac{1}{2}\left(\frac{\lambda\rho}{N}\right)^{1/2}$$

Furthermore, the width Δr of the bright spot in the plane of observation will be approximately half the width Δr_N of this last zone projected from the source point onto the observation plane. That this is roughly true can be seen by sketching the vibration curve with about one-fourth of the Nth zone and three-fourths of the $(N+1)$th zone open. From Eq. (8.43) we obtain

$$\Delta r \sim \frac{1}{2}\Delta r_N \frac{(D+D')}{D} = \frac{\sqrt{2}S_F}{4\sqrt{N}}, \qquad S_F = \frac{\lambda D'(D+D')}{2D} \tag{8.45}$$

The resulting pattern is shown in Plate 1*b*. The spot is known as *Poisson's bright spot*, after the French mathematician who predicted it.

There is an interesting story behind Poisson's prediction of the bright spot. In an effort to discredit the Fresnel theory of diffraction, he used the theory to predict the existence of the bright spot. This, he said, violated both common sense and the results of the crude experiments that had been done up to that time. When more refined experiments were tried, the spot was indeed seen. Instead of discrediting Fresnel's theory, Poisson's prediction lead to a vindication of it.

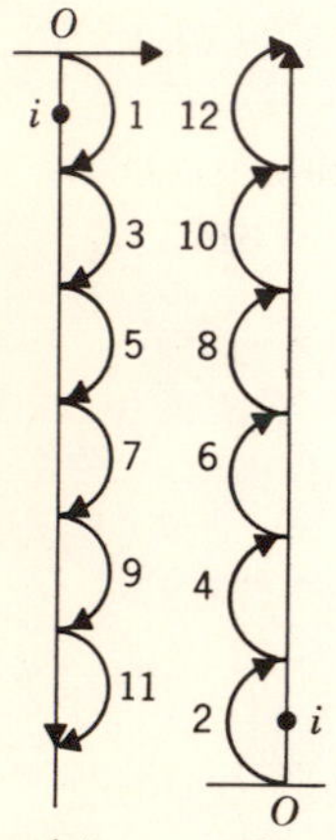

Fig. 8.22. The vibration curve for the diffraction pattern of a zone plate having 12 zones (6 open zones). In (*a*) the odd zones are transparent; in (*b*) the even zones are transparent.

D. ZONE PLATES

Suppose that we have an aperture with alternating opaque and transparent circular regions that just coincide with the Fresnel zones appropriate to the values of D, D', and λ being used. This is one type of zone plate. Suppose that such a plate is centered and used as the diffracting aperture in Fig. 8.16. The radius of its first zone will be

$$r_1 = \sqrt{\lambda\rho} = \left(\frac{\lambda DD'}{D+D'}\right)^{1/2}$$

as given by Eq. (8.40). If the plate has $N/2$ open zones and is then opaque after that, and if we neglect the inclination factor Q, we find that the vibration curve consists of a column of $N/2$ semicircles of unit radius, as shown in Fig. 8.22. The integral I is the sum of the chords of these circles and is $-Ni$ for Fig. 8.22*a* and $+Ni$ for Fig. 8.22*b*. The field is $\pm N\tilde{E}_{\text{na}}$ in the two cases, and the irradiance is

$$E_e = N^2 E_{e,\text{na}} \qquad (N = \text{twice the number of open zones})$$

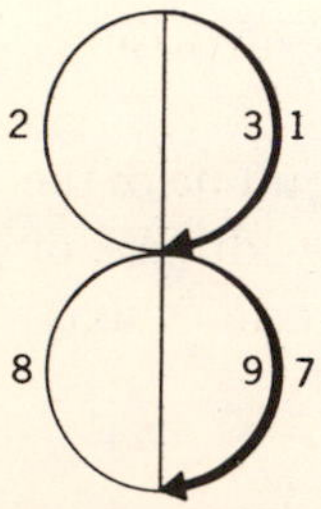

Fig. 8.23

Now suppose that for given D, D', and ρ the zone plate is so large that it blocks or transmits the zones three at a time. If the center is open, it will let through the light from the first three zones, then block the light from the next three and so on. The vibration curve of Fig. 8.22*a* would then become that of Fig. 8.23; the resultant electric field would be reduced by a factor of 3 and the irradiance reduced by a factor of 9. Similar results would be obtained with a large enough zone plate that blocked or passed 5, 7, and so on, zones at a time. With an even number of zones blocked and passed, however, the field is approximately zero.

The condition that the first circle in the zone plate cover n zones is that its radius be given by

$$r_1 = \sqrt{n\lambda\rho} \tag{8.46}$$

1. Lens Action of Zone Plate. We may rewrite Eq. (8.46) in the form

$$\frac{n\lambda}{r_1^2} = \frac{1}{\rho} = \frac{1}{D} + \frac{1}{D'} \tag{8.47}$$

where n is an odd integer. This has the appearance of the Gaussian lens equation with a focal length of

$$f_n = \frac{r_1^2}{n\lambda} \tag{8.48}$$

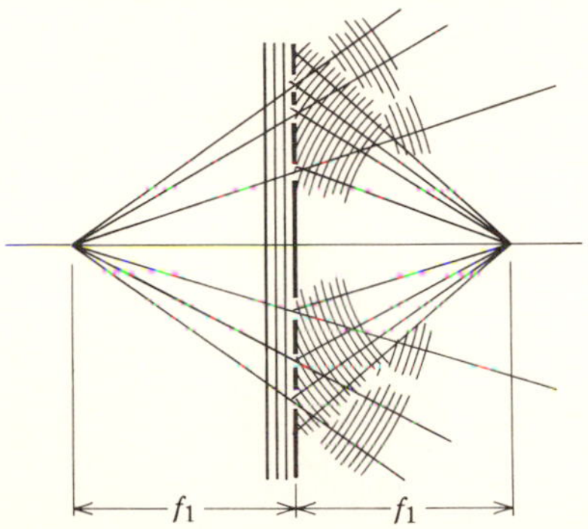

Fig. 8.24. Illustration of the simultaneous positive and negative lens action of a zone plate.

The properties of the zone plate are similar to those of a simple lens except for the multiple values of f. For instance, if plane waves of wavelength λ are incident normally onto the zone plate, there will be a bright spot a distance f_1 to the right of the plate and other spots of decreasing brightness at distances of $f_1/3, f_1/5$, and so on.

The zone plate acts simultaneously as a positive and negative lens with focal lengths $\pm f_1, \pm f_1/3$, and so on. In addition to the series of spots mentioned just above, formed to the right of the plate with parallel incident light, a similar series of "virtual spots" will be found at distances $f_1, f_1/3, f_1/5, \ldots$ to the left of the lens. We try to illustrate this effect in Fig. 8.24. A rigorous proof of this dual positive and negative property will be given for another type of zone plate in the next subsection.

2. Starting with a Partial Zone. Figure 8.25*a* shows the transmission function of the zone plate having odd-numbered zones open and having the vibration curve of Fig. 8.22*a*. We can also start with a partial zone open; say a fraction α of the area of the first zone is open. The radius of its edge will be $\sqrt{\alpha}\, r_1$, and the phase angle across it will vary from 0 to $\alpha\pi$. Then we alternate opaque and open rings, each having an area πr_1^2. The first dark ring will extend from $r' = \sqrt{\alpha}\, r_1$ to $r' = \sqrt{\alpha + 1}\, r_1$; the first full bright ring from $\sqrt{\alpha+1}\, r_1$ to $\sqrt{\alpha+2}\, r_1$; and so on. The transmission function for the case $\alpha = \frac{1}{2}$ is shown in Fig. 8.25*c*, and the vibration curve is drawn in Fig. 8.26 for a total of 12 zones. This should be compared with the vibration curves of Fig. 8.22*a, b* which correspond to $\alpha = 0$ and $\alpha = 1$, respectively.

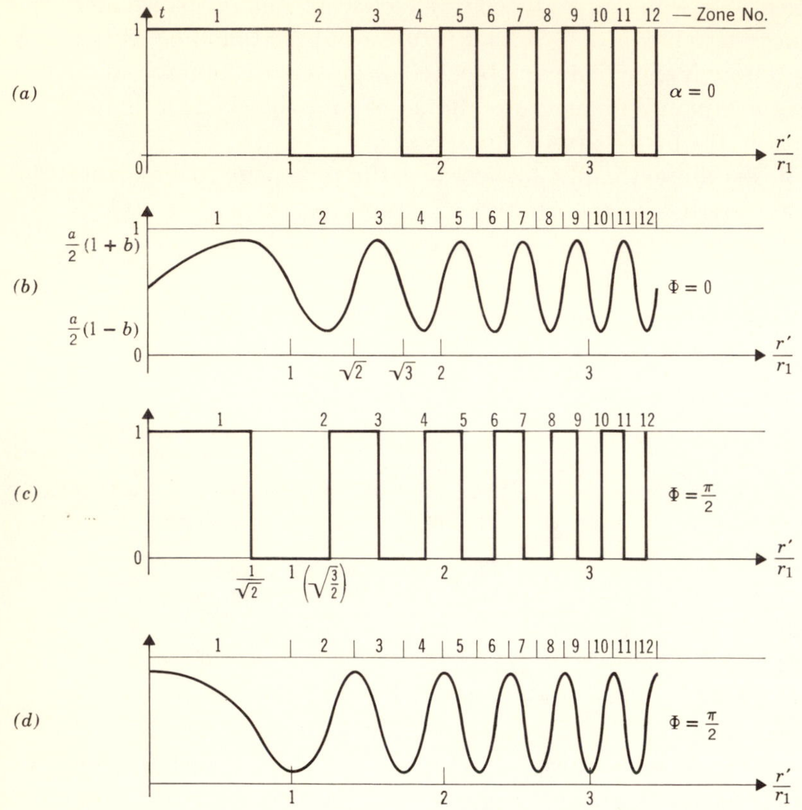

Fig. 8.25

It is easily seen that the effect of the partial zone is to change the resulting electric field from $N\tilde{E}_{\text{na}}$ to $Ne^{\pi i\alpha}\tilde{E}_{\text{na}}$ (if we set $Q = -1$), where N is the total number of zones.

3. Sinusoidal Zone Plate. There are other types of zone plate that do not have the rectangular transmission function shown in Fig. 8.25*a*, *c*. The most general property shared by all zone plates is that the transmission function is periodic as a function of r'^2 with a period equal to $R_0^2 \equiv 2r_1^2$. The simplest periodic function is a sine or cosine, and the sinusoidal zone plate is therefore the simplest kind of zone plate. It is also the most important kind because of its application to holography and related phenomena.

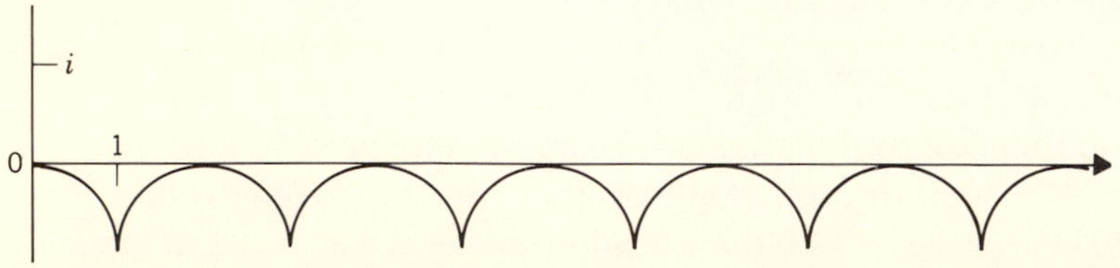

Fig. 8.26

Consider a plate with a transmission function of the form

$$t(r') = \frac{a}{2}\left[1 + b \sin\left(\frac{2\pi r'^2}{R_0^2} + \Phi\right)\right] \tag{8.49}$$

where $0 < a \leqslant 1$ and $0 < b \leqslant 1$. This is plotted in Fig. 8.25*b* for $\Phi = 0$ and in Fig. 8.25*d* for $\Phi = \pi/2$ for comparison with the transmission functions of the corresponding rectangular plates in Fig. 8.25*a* and 8.25*c*.

A sinusoidal zone plate has just two values for its focal length: $f = \pm r_1^2/\lambda$. To show this we study the effect of such a plate centered and used as the diffracting aperture in Fig. 8.16*a* on the field at an axial point P.

In our case, ψ in Eq. (8.35) can be written

$$\psi = \frac{\pi r'^2}{\lambda\rho}$$

Here, t in Eq. (8.49) becomes

$$t(\psi) = \frac{a}{2}\left[1 + b \sin\left(\frac{\lambda\rho\psi}{r_1^2} + \Phi\right)\right]$$

$$= \frac{a}{2} + \frac{ab}{4i} e^{i\Phi} e^{i\psi\lambda\rho/r_1^2} - \frac{ab}{4i} e^{-i\Phi} e^{-i\psi\lambda\rho/r_1^2}$$

We put this expression inside the integral in Eq. (8.38) and obtain three terms for $\tilde{E}$:

$$\tilde{E} = \tilde{E}_0 + \tilde{E}_+ + \tilde{E}_- \tag{8.50a}$$

where

$$\tilde{E}_0 = \frac{ia}{2}\tilde{E}_{\text{na}} \int_0^{\psi_0} e^{-i\psi} Q(\psi)\, d\psi = \frac{a}{2}\tilde{E}_{\text{na}} \qquad \text{for large } \psi_0 \tag{8.50b}$$

$$\tilde{E}_+ = \frac{ab}{4} e^{\pi i\alpha} \tilde{E}_{\text{na}} \int_0^{\psi_0} e^{i\psi(\lambda\rho/r_1^2 - 1)} Q(\psi)\, d\psi \tag{8.50c}$$

$$\tilde{E}_- = -\frac{ab}{4} e^{-\pi i\alpha} \tilde{E}_{\text{na}} \int_0^{\psi_0} e^{-i\psi(\lambda\rho/r_1^2 + 1)} Q(\psi)\, d\psi \tag{8.50d}$$

Here $\tilde{E}_0$ is the field we would obtain through a uniform gray sheet having a transmission function of $a/2$, which is the average over its area of $t(r')$. The other two terms yield approximate point images when the exponents in the integrands are zero. For $\tilde{E}_+$ this occurs when

$$\frac{1}{\rho} = \frac{1}{D} + \frac{1}{D'} = \frac{\lambda}{r_1^2} \equiv \frac{1}{f_+} \tag{8.51a}$$

and the integral becomes large, as all contributions to it add in phase. If we approximate Q by unity, we obtain

$$\tilde{E}_+ = \frac{ab}{4} e^{\pi i \alpha} \tilde{E}_{\mathrm{na}} \psi_0 = \frac{ab}{4} e^{\pi i \alpha} \tilde{E}_{\mathrm{na}} N\pi \qquad (8.52a)$$

(N = total number of zones).

A similar situation occurs with E_- for

$$\frac{1}{\rho} = \frac{1}{D} + \frac{1}{D'} = -\frac{\lambda}{r_1^2} \equiv \frac{1}{f_-} = -\frac{1}{f_+} \qquad (8.51b)$$

Then (with $Q \approx 1$) we obtain

$$\tilde{E}_- = -\frac{ab}{4} e^{-\pi i \alpha} \tilde{E}_{\mathrm{na}} N\pi \qquad (8.52b)$$

Thus, D' must be negative to satisfy (8.51*b*) with positive D; the point image is virtual.

These "point images" are really diffraction-limited Airy disks like those produced by a lens with a point source. The size of the Airy disk depends on the size of the limiting aperture in a way similar to that for a lens.

4. Production of Sinusoidal Zone Plates. A practical method of making a sinusoidal zone plate (with a small value of b) is to expose a photographic plate to the interference fringes produced by two mutually coherent monochromatic point sources P_1, P_2, as illustrated in Fig. 8.27. Assume that the wavelength is $\lambda' = 2\pi/k'$,

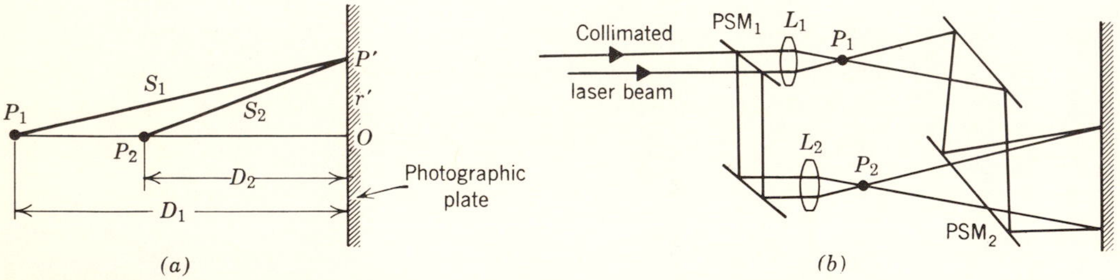

Fig. 8.27. Photographic production of sinusoidal zone plate. (*a*) Ideal geometry. (*b*) Practical setup using laser beam. PSM_1 and PSM_2: partially silvered mirrors. L_1 and L_2: microscope objective lenses. P_1 and P_2: diffraction limited focal spots that act like point sources.

and that the amplitudes A_1 and A_2 of the fields are constant across the plate. The total field at P' is then given by

$$\tilde{E}(P') = \tilde{E}_1(P') + \tilde{E}_2(P') = A_1 e^{i\phi_1} e^{-ik'S_1} + A_2 e^{i\phi_2} e^{-ik'S_2}$$

where ϕ_1 and ϕ_2 represent the phases at P_1 and P_2. The flux density at P' is given by

$$E_e(P') = C\{A_1^2 + A_2^2 + 2A_1A_2 \cos [k'(S_1 - S_2) - \phi_1 + \phi_2]\} \qquad (8.53)$$

where C is a constant. We may use the approximations

$$S_1 = D_1 + \frac{r'^2}{2D_1}, \qquad S_2 = D_2 + \frac{r'^2}{2D_2}$$

The argument of the cosine function in (8.53) can then be approximated by

$$\frac{k'r'^2}{2}\left(\frac{1}{D_1}-\frac{1}{D_2}\right)+[\phi_2-\phi_1+k'(D_1-D_2)]=\frac{\pi r'^2}{\lambda'\rho'}+\Delta\phi \tag{8.54}$$

where

$$\frac{1}{\rho'}\equiv\frac{1}{D_1}-\frac{1}{D_2}, \qquad \Delta\phi=\phi_2-\phi_1+k'(D_1-D_2)$$

We assume that one of the sources, say P_1, is brighter than the other, so that $A_1 \gg A_2$. Then we neglect A_2^2 in Eq. (8.53) and obtain

$$E_e(P')=E_{e,0}(P')\left[1+\frac{2A_2}{A_1}\cos\left(\frac{\pi r'^2}{\lambda'\rho'}+\Delta\phi\right)\right] \tag{8.55}$$

where $E_{e,0}=CA_1^2$ is the flux density from source P_1.

The plate is exposed for an appropriate time and developed. To describe what this does to the plate, we must digress a bit into photography. The photographic sensitivity of each type of film or plate under given conditions of development is described by the so-called Hurter-Driffield or H-D curve, which is a plot of the density of the developed film or plate versus the logarithm of the product of the radiant incidence or flux density and the exposure time.* This is shown in Fig. 8.28. The density is defined by $D=-2\log\tau$, where τ is the amplitude transmission coefficient of the developed plate. There is a linear region of the H-D curve with a slope γ that usually has values from 1 to 2. The exposure time is adjusted so that ($E_{e,0}\cdot$ time) corresponds to the middle of the linear region where the density is D_0 and transmission coefficient τ_0.

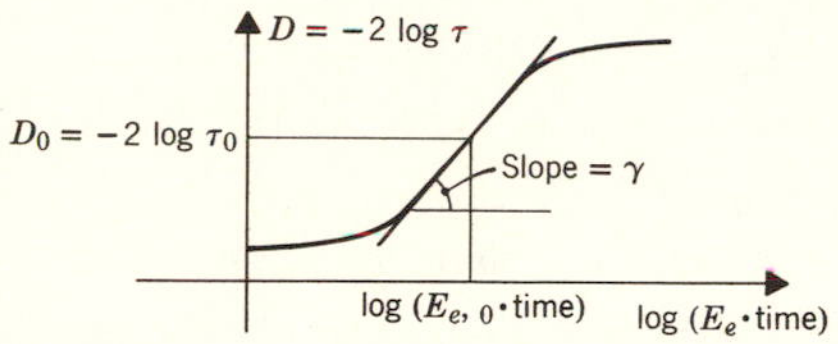

Fig. 8.28. The H–D curve.

It is easy to show that for small excursions ΔE_e of E_e about $E_{e,0}$ we may write approximately

$$\tau=\tau_0\left(1-\frac{1}{2}\frac{\gamma\Delta E_e}{E_{e,0}}\right)$$

From Eq. (8.55) we see that the fractional change in E_e is

$$\frac{\Delta E_e}{E_{e,0}}=\frac{2A_2}{A_1}\cos\left(\frac{\pi r'^2}{\lambda'\rho'}+\Delta\phi\right)$$

and this is small since $A_2 \ll A_1$. We then obtain

$$\tau(r')=\tau_0\left[1-\frac{\gamma A_2}{A_1}\cos\left(\frac{\pi r'^2}{\lambda'\rho'}+\Delta\phi\right)\right] \tag{8.56}$$

*In practice the equivalent psychophysical unit of luminous incidence in lux (meter-candles) is used, but this does not affect our argument.

This has the same form as Eq. (8.49), provided we make the following identifications:

$$\frac{a}{2} = \tau_0, \qquad b = \frac{\gamma A_2}{A_1}$$

$$r_1{}^2 = \lambda'\rho', \qquad \Phi = \Delta\phi - \frac{\pi}{2}$$

The focal lengths of the zone plate are, by Eqs. (8.51*a*, *b*),

$$\frac{1}{f_\pm} = \pm\frac{\lambda}{r_1{}^2} = \pm\frac{\lambda}{\lambda'\rho'} = \pm\frac{\lambda}{\lambda'}\left(\frac{1}{D_1} - \frac{1}{D_2}\right) \tag{8.57}$$

Here λ' is the wavelength used to make the plate, and λ is the wavelength used later with the developed plate. Equations (8.51*a*, *b*) and (8.57) may be combined to give

$$\frac{1}{D} + \frac{1}{D'} = \pm\frac{\lambda}{\lambda'}\left(\frac{1}{D_1} - \frac{1}{D_2}\right)$$

(Here D_1 and D_2 are the distances of the two point sources to the left of the plate during exposure; D, distance [to the left] of the source used subsequently; D', distance [to the right] of the resulting point image.)

All the points P_1, P_2, P_s, and the image points in the above arrangement lie on the normal to the plate at its center. More general setups are possible without this colinearity. The process just described forms the heart of a lensless method of image formation with coherent light known as *holography*, which will be discussed systematically in Section 8.3. A zone plate is the hologram for a point object.

REFERENCES FOR SECTION 8.2

M. Born and E. Wolf, *Principles of Optics*, Sections 8.2 and 8.7.
F. A. Jenkins and H. E. White, *Fundamentals of Optics*, Chapter 18.
J. M. Stone, *Radiation and Optics*, Chapter 10.
J. Strong, *Concepts of Classical Optics*, Chapter IX.

PROBLEMS

8.2.1 A plane monochromatic wave of wavelength $\lambda = 4000$ A is incident perpendicularly upon an opaque plane screen bounded by a straight edge. Using Cornu's spiral, determine the positions of the minima and maxima in the diffraction pattern observed on a plane parallel to the screen at a distance of 2 m.

8.2.2 Estimate the requirements on the displacements $(x_s - x') \sim (y_s - y') \sim (x - x') \sim (y - y') \sim \Delta$ for $D = D'$, if the last terms in Eqs. (8.10) may be safely neglected. Do this numerically for the case $D = D' = 2\text{m}$, $\lambda = 500$ nm.

8.2.3 The behavior of the Fresnel diffraction pattern of a straight edge way inside the bright region may be obtained by studying the Fresnel integral

$$J(\eta) = \int_{-\infty}^{\eta} e^{-(\pi i/2)u^2}\, du \qquad \text{for large values of } \eta$$

Let $\eta = \eta_0 + \Delta\eta$, where $\eta_0 \gg 1$ and $\Delta\eta < 1$. Show that one can then obtain an approximate expression for J which oscillates corresponding to a sinusoidal oscillation of $E_e(\eta)$ about $E_{e,\text{na}}$ with an amplitude $\Delta E_e = E_e(\eta) - E_{e,\text{na}} = E_{e,\text{na}}[\sqrt{2}/(\pi\eta_0)]$ and with a period $2/\eta_0$.

8.2.4 In the laboratory we can observe the Fresnel diffraction pattern of a straight edge using the setup in the figure, with a diffuse source behind a narrow slit parallel to the straight edge. Explain how this will work by considering the open region of the slit to be composed of many incoherent

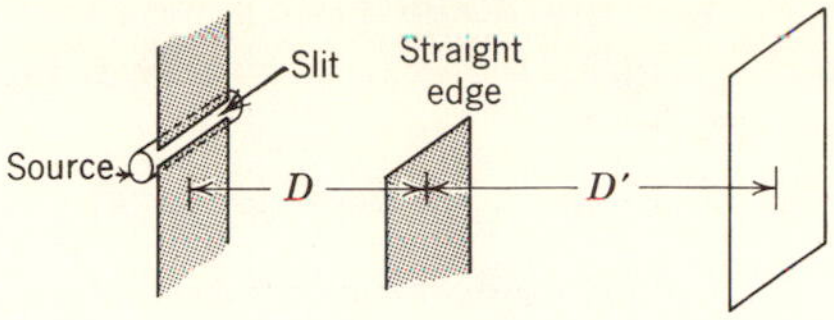

point sources. Describe semiquantitatively the effect on the diffraction pattern of each of the following acting separately.

a. A small spread $\Delta\lambda$ in the wavelength.
b. A finite width w in the slit.
c. A small angular misalignment ϕ between slit and straight edge.
d. A rough edge on the straight edge on a scale Δx.

Your answers can be made more precise if you use the results of Problem 8.2.3.

8.2.5 A straight wire 1 mm in diameter lies along the path of a plane light wave of wavelength 5000 A, perpendicular to the direction of propagation. Plot the distribution of intensity on a screen 2 m from the wire. Mark the edges of the geometrical shadow.

8.2.6 Show that for small values of ψ the inclination factor $Q(\psi)$ in Eq. (8.38) may be written

$$Q(\psi) \approx 1 - \frac{\lambda}{4\pi} \frac{D^2 + (D')^2}{DD'(D + D')} \psi$$

8.2.7 A monochromatic point source S emits light of wavelength 6000 A. It falls onto an aperture A 10 cm away and then onto a screen 20 cm beyond A.

a. What is the value of r_1, the radius of the first Fresnel half-period zone at A.
b. The aperture A is a circle of radius 1 cm. How many Fresnel half-period zones does it contain?

c. A zone plate with every other zone blocked out and with the radius of its first zone = r_1 (answer to [a]) is placed at A. This "focuses" the light from S onto the screen. The screen is moved toward A. Where will additional images be formed?

8.2.8 Monochromatic plane waves of wavelength λ are incident normally on a circular opening in a screen of radius r_0, and they then fall onto an observing screen a distance S away. Assume that $r_0 = \sqrt{2N\lambda S}$ where N is an integer. Let P be the point on the second screen that is in the center of the geometrical bright region.

a. How many half-period zones are there in the opening as seen from P? Call this answer M. In terms of M: What is the ratio of the electric field at P to the electric field with no first screen at all:
b. When a zone plate is put in the opening that blocks out every other half-period zone and lets all the light from the other zones through.
c. When a zone plate is put in the opening that shifts the phase of every other half-period zone by π radians and does not affect the other zones.
d. When a perfect lens is put in the opening that focuses the rays at P?

You may neglect inclination factors.

8.2.9 What is the electric field at an axial point P in Fig. 8.16a if the aperture is an open section of wedge angle γ?

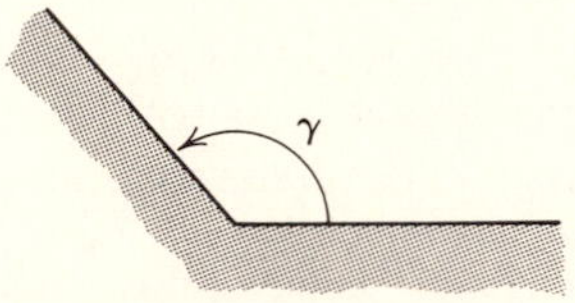

8.2.10 a. Estimate a value for the integral $I(\psi_0)$ in Eq. (8.39) by integrating by parts.
b. Use the approximation of Problem 8.2.6 to obtain an explicit expression for $I(\psi_0)$.

8.2.11 Consider a zone plate which has the transmission function shown in the figure (plotted versus r'^2/r_1^2). Each open region has a width of $C < 2$.

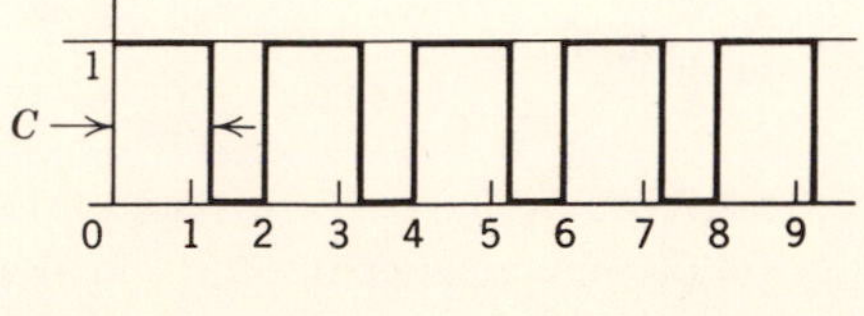

Sketch the vibration curve at an axial point P when such a zone plate is used as an aperture with $r_1 = \sqrt{\lambda\rho}$. What is the field at P if there are N zones in the plate (N is even)? When $C \neq 1$ the zone plate can give a focus at a point P for which $r_1 = \sqrt{2\lambda\rho}$, unlike the zone plate of Fig. 8.25c. Explain.

8.2.12 Derive the equation

$$\tau = \tau_0\left(1 - \frac{1}{2}\frac{\gamma \Delta E_e}{E_{e,0}}\right)$$

(Stated without proof between Eqs. [8.55] and [8.56] in the text.)

8.2.13 With reference to Fig. 8.28, show that when the point corresponding to $E_{e,0}$ and τ_0 is not centered in the middle of the linear region and/or when $\Delta E_e / E_{e,0}$ is not small, we still obtain a developed plate with a $\tau(r')$ that is periodic with a period in $r'^2 = 2r_1^2$.

8.2.14 An important practical consideration in the photographic construction of a sinusoidal zone plate, or other zone plate, is the required resolution of the photographic plate used. Estimate this resolution for a 20-cm primary focal length zone plate 4 cm in diameter to be used with light of $\lambda = 450$ nm.

8.3 HOLOGRAPHY

The essential elements of the theory of holography were worked out by Dennis Gabor in 1948. He described the recording and reproduction of the complete optical signal coming from an object and performed experiments to illustrate the theory, but the light sources then available were inadequate to realize the full potential of the technique.

The big experimental advance came in 1963 when Emmett Leith and Juris Upatnieks applied the method to three-dimensional objects using light from a gas laser. This new field has grown rapidly during the past few years.

A. INTRODUCTION

In ordinary photography one reproduces the intensity distribution associated with the light coming from an object. In holography one reconstructs the wavefronts themselves. They contain all the optical information contained in the original light waves. This is indicated schematically in Fig. 8.29, which shows an object

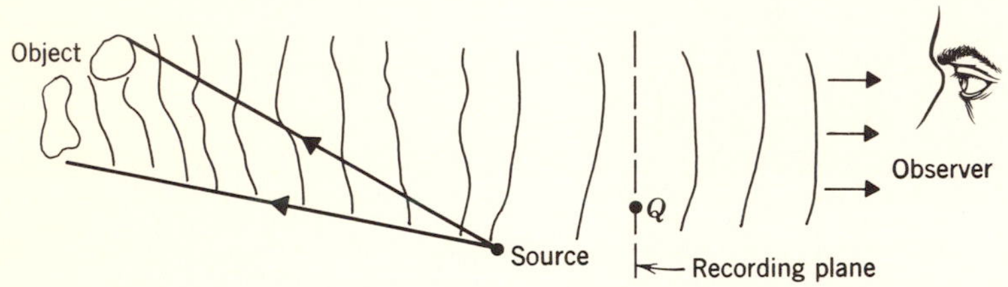

Fig. 8.29

being illuminated by light from a source. The scattered light waves then fall onto the eyes of an observer. If we have some way of recording these waves at, say, the plane shown dotted in the figure and then reproducing them at a later time so that the same waves proceed toward the observer, then what he will see is an image of the original object. It will appear in three dimensions with the proper relative position of its various parts. The essence of holography then is *wavefront reconstruction.*

1. Assumptions. In this discussion of wavefront reconstruction we have implicitly used the following important assumption. If we can reproduce the wavefront on a single plane in space, the subsequent wavefront propagated beyond this plane will be the same as the original one. From the mathematical point of view this represents the statement of a boundary-value problem associated with the time-independent wave equation (Helmholtz equation). An approximate proof may be provided using the theory of diffraction as applied below, in Chapter 9.

For simplicity we shall treat just one transverse dimension x in the recording plane; the results are readily generalized to two transverse dimensions. We assume quasimonochromatic light of frequency ν and angular frequency $\omega = 2\pi\nu$, and for the electric field at the point $Q(x)$ coming from the object we write

$$E(x) = A(x) \cos [\omega t + \phi(x)] \tag{8.58}$$

Here $\phi(x)$ represents the phase of the light at x. It is necessary for a full description of the light waves there. In writing Eq. (8.58) we assume that $\phi(x)$ is a well-defined quantity or, more correctly, we assume that differences in phase can be defined precisely. This brings us to the concept of *coherence*, the subject of Chapter 6.

We must distinguish between two kinds of coherence. Both are required in holography. We need *transverse coherence*, which means that at two points A and B in Fig. 8.30 having a separation perpendicular or transverse to the main direction of propagation there is a

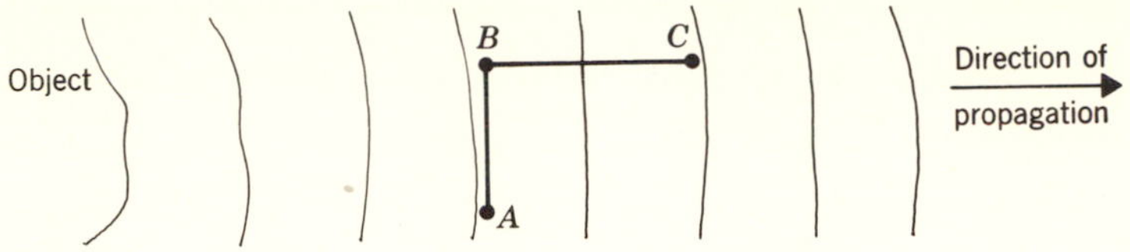

Fig. 8.30

definite, persistent phase relationship. This means that the phase difference $\phi_A - \phi_B$ is a well-defined quantity that does not fluctuate randomly as a function of time. Note that we do not require that *the phase* at A or B be well defined; indeed the concept of an absolute phase is meaningless—only phase differences can be defined. For holography we need transverse coherence across the recording plane.

The other kind of coherence necessary is *longitudinal coherence*. When we have longitudinal coherence, roughly speaking there is a definite value of the phase difference between points B and C, where C is a point along the direction of propagation from B. For holography we need longitudinal coherence over distances at least as great as the depth of our three-dimensional object.

Both kinds of coherence can be obtained with light from a gas laser operated in the so-called *uniphase mode*. Such a laser has perfect transverse coherence and longitudinal coherence over distances at least as great as the distance separating the mirrors in the laser cavity. If light from such a laser illuminates our object, then we can express the electric field as in Eq. (8.58) without worrying about statistical fluctuations in any relative phases.

2. Interference, the Key Notion. Suppose that we insert a photographic plate in the recording plane and expose it to the light from the object. The plate is sensitive to the integrated flux density (incidence) $(E_e \cdot T) \equiv \int E_e \, dt$, where E_e is proportional to the time

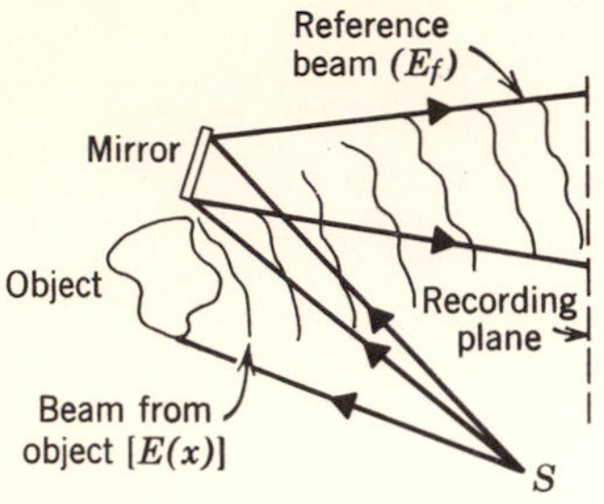

Fig. 8.31

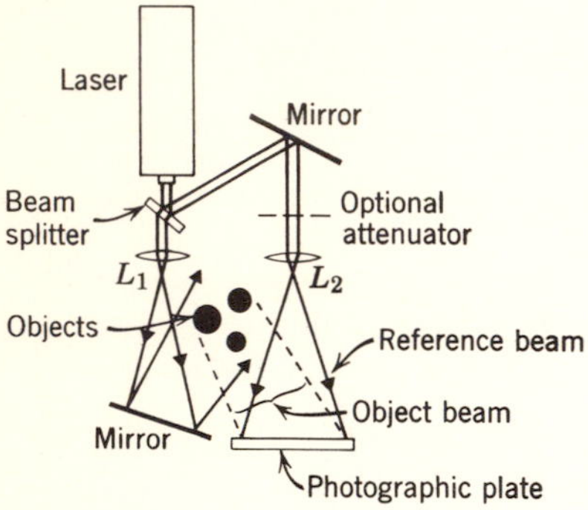

Fig. 8.32

average of the square of the electric field. Thus

$$(E_e \cdot T) \propto \overline{E(x)^2} = \tfrac{1}{2}[A(x)]^2 \tag{8.59}$$

since

$$\overline{\cos^2[\omega t + \phi(x)]} = \tfrac{1}{2}$$

Thus we have lost all information about the phase $\phi(x)$ in the recording process, and there is no hope of reconstructing the complete electric field $E(x)$.

The key notion of holography is to record the result of interference between E(x) and a coherent reference beam E_f. This can be done by reflecting some of the original light from the laser source S directly onto the recording plane, as shown in Fig. 8.31. Then Eq. (8.59) is changed to

$$(E_e \cdot T) \propto \overline{E^2_{\text{tot}}} \tag{8.60a}$$

where E_{tot} represents the superposition of the light from the object and the light in the reference beam,

$$E_{\text{tot}} = E(x) + E_f \tag{8.60b}$$

Before we show how this interference allows us to reconstruct $E(x)$, we shall describe an actual setup (used in our laboratory) to produce holograms. Figure 8.32 illustrates the arrangement. The beam splitter is an optical flat—the 4 per cent reflection off the front face gives a reference beam of about the right intensity. Lenses L_1 and L_2 are microscope objectives: L_1 is 10 power, and L_2 is 40 power. They provide effective point sources at their focal points. Pinholes placed at these focal points provide spatial filtering (Section 9.2) and give better results, but they are not absolutely necessary. To guarantee longitudinal coherence, the path lengths in the sample and reference beams from beam splitter to photographic plate should be equal to within the length of the laser, unless a special laser is used that has only a single longitudinal mode. To avoid having to use extremely high-resolution photographic plates (with their extremely low speed) one should keep the angle between sample and reference beams small and have the focal point of L_2 about the same distance from the plate as the objects are. The reason for this is discussed below.

3. Mathematical Proof of the Reconstruction Process. Let the field in the beam from the object at the recording plane be given by Eq. (8.58). Let the field in the reference beam be given by

$$E_f = A_f \cos[\omega t + \phi_f(x)] \tag{8.61}$$

For the time being it is not necessary to assume any special form for A_f, which might be a function of x, or of $\phi_f(x)$. We must add the two contributions, square, and integrate over time. The result is that the integrated flux density can be written

$$E_e T = k\overline{[(E_f + E(x)]^2}$$

where k is a constant. This can be written

$$\overline{[E_f + E(x)]^2} = \overline{E_f^2} + \overline{E(x)^2} + \overline{2E_f E(x)}$$

We now substitute from Eqs. (8.58) and (8.61). We use the trigonometric identity

$$\cos A \cos B = \tfrac{1}{2}\cos(A+B) + \tfrac{1}{2}\cos(A-B)$$

to obtain

$$\overline{E_f^2} = \tfrac{1}{2}A_f^2$$

$$\overline{E(x)^2} = \tfrac{1}{2}A(x)^2$$

and $\overline{E_f E(x)} = \tfrac{1}{2}A_f A(x)\cos(\phi - \phi_f)$.

For the integrated flux density we then obtain

$$E_e \cdot T = k\{\tfrac{1}{2}A_f^2 + \tfrac{1}{2}A(x)^2 + A_f A(x)\cos[\phi(x) - \phi_f(x)]\} \tag{8.62}$$

At this point it is convenient to adjust the relative intensities so that

$$A_f^2 \gg A(x)^2 \tag{8.63}$$

A three- to five-fold inequality is sufficient in practice. Use of (8.63) then allows us to approximate Eq. (8.62) by

$$E_e(x) \cdot T \cong E_{e0}T\left\{1 + \frac{2A(x)}{A_f}\cos[\phi(x) - \phi_f(x)]\right\} \tag{8.64}$$

where

$$E_{e0} \cdot T = \tfrac{1}{2}kA_f^2$$

is the integrated flux density in the reference beam alone.

The reader may have noticed the similarity between this discussion and that of the photographic production of a sinusoidal zone plate at the end of Section 8.2. One should compare Figs. 8.27 and 8.32 and Eqs. (8.55) and (8.64). The photographic plate is exposed and developed just as it was earlier. The resulting amplitude transmission function of the developed plate is then given by an analogue of Eq. (8.56):

$$\tau(x) = \tau_0\left\{1 - \frac{\gamma A(x)}{A_f}\cos[\phi(x) - \phi_f(x)]\right\} \tag{8.65}$$

The developed plate is called a *hologram* and was first given this name by Gabor. A photomicrograph of a hologram is shown in Plate 2*b*. The small scale fringes carry the information. Plate 2*a* shows a photograph of some three dimensional objects and Plate 2*c* a photograph of the holographically reconstructed image. The graininess in the bottom photograph of Plate 2*c* is characteristic of objects illuminated with coherent laser light and was present when the hologram was made.

To obtain exact reconstruction of the original beam from the object, we take the developed plate, put it back in its original place, and illuminate with the original reference beam. In other words, we keep the original geometry but eliminate the real object. The light incident on the plate will have an electric field given by

$$E_i = E_f = A \cos [\omega t + \phi_f(x)]$$

Using (8.65), we obtain for the electric field transmitted by the plate

$$E_t = \tau E_i = \tau_0 \left[1 - \frac{\gamma A(x)}{A_f} \cos (\phi - \phi_f) \right] A_f \cos (\omega t + \phi_f) \tag{8.66}$$

Equation (8.66) can be reworked to give three contributions to E_t. The first is simply

$$E_r = \tau_0 E_f$$

which is what the plate would transmit if it had been exposed only to the reference beam. The other terms are obtained by applying our trigonometric identity to the product of cosines in Eq. (8.66):

$$\cos (\phi - \phi_f) \cos (\omega t + \phi_f) = \tfrac{1}{2} \cos (\omega t + \phi) + \tfrac{1}{2} \cos (\omega t + 2\phi_f - \phi)$$

This gives

$$E_t = E_r - \frac{\gamma \tau_0}{2} A(x) \cos [\omega t + \phi(x)] - \frac{\gamma \tau_0}{2} A(x) \cos [\omega t - \phi(x) + 2\phi_f(x)] \tag{8.67}$$

The second term in Eq. (8.67) is proportional to the original electric field $E(x)$ and represents the reconstructed light waves. If only it were present, an observer would see a virtual image of the original object in its original position. The third term often gives a real image of the object, as we shall see later. It and the first term can be separated from the desired second term because of the different directions traveled by the beams they represent. It is best to illustrate these statements by discussing particular experimental arrangements.

B. POINT SOURCE INTERFERENCE PATTERNS FOR PARTICULAR ARRANGEMENTS OF OBJECTS AND OF THE REFERENCE BEAM

To describe the effect of the third term in Eq. (8.67) and to handle cases where the hologram is viewed in a different geometry from that of the reference beam to which it was exposed, it is convenient to change our point of view.

We note that the interference term in Eq. (8.62) is linear in the electric field coming from the object. Thus, if $E(x)$ is given by the superposition of several contributions,

$$E(x) = \Sigma_j E_j(x) = \Sigma_j A_j(x) \cos [\omega t + \phi_j(x)] \tag{8.68}$$

the interference term is given by the superposition

$$\Sigma_j A_f A_j(x) \cos [\phi_j(x) - \phi_f(x)]$$

The second and third terms of E_t in Eq. (8.67) would preserve the superposition. The reconstructed light waves from the object can be described as the sum of the waves from the component parts.

It is convenient to break up the object into a superposition of point objects. We then describe how the electric field from each point affects the hologram and the reconstructed wave. We then add the result to obtain the total effect from the actual object.

1. Behavior of the Phase in the Recording Plane for Light from Point Sources. Monochromatic, coherent light moving from left to right across the recording plane is uniquely characterized by its value on this plane. For light from point sources that are not too close, the amplitude is essentially constant across the useful part of the plane and the key variable is the phase $\phi(x)$. Over small regions of space the waves will be approximately plane, and plane traveling waves have a temporal dependence of the form

$$E(r, t) \sim \cos\left(\omega t - \frac{2\pi r}{\lambda} + \phi_0\right)$$

where r is a distance measured along the direction of propagation. If we refer to an origin O in the recording plane, the phase $\phi(x)$ will thus be of the form

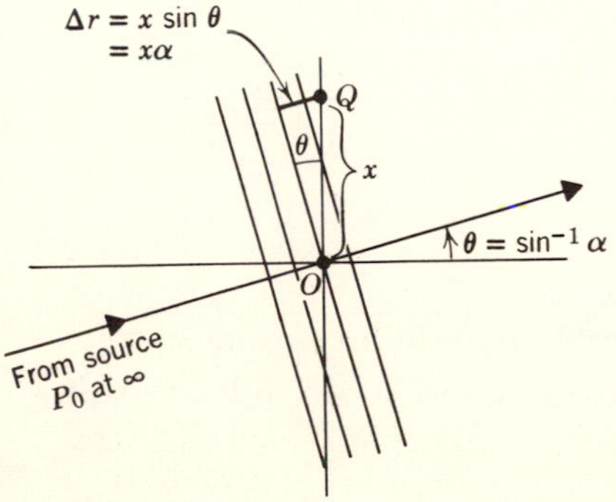

Fig. 8.33

$$\phi(x) = -\frac{2\pi\Delta r}{\lambda} + \phi(\mathrm{O})$$

where

$$\Delta r(x) = r(x) - r(\mathrm{O})$$

represents the change in propagation distance associated with a move from O to x. This will be brought out in the examples that now follow.

(a) *Plane waves from infinity.* Figure 8.33 gives the necessary information. We can think of the light as coming from a distant point source at an angle $-\theta$. From the figure we obtain

$$\Delta r = x \sin\theta$$

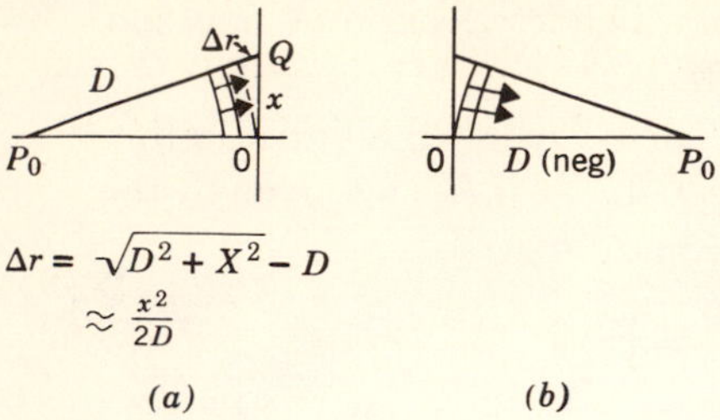

Fig. 8.34

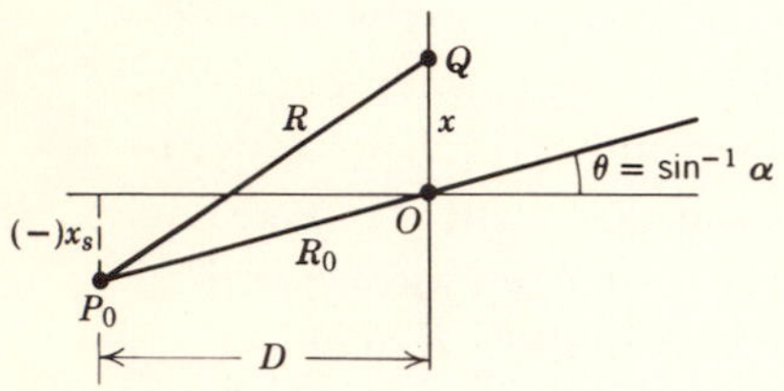

Fig. 8.35

and hence

$$[\phi(x)-\phi(O)] = -\frac{2\pi x \sin\theta}{\lambda} = -\frac{2\pi x\alpha}{\lambda} \tag{8.70}$$

Here $\alpha = \sin\theta$ is the x direction cosine of the wave normal.

(b) *Spherical waves from a point source.* Figure 8.34a shows the situation for a real source at P_0 a distance D (positive) to the left of the recording plane. We obtain

$$\Delta r \approx \frac{x^2}{2D}$$

and hence

$$[\phi(x)-\phi(O)] \approx -\frac{2\pi x^2}{2\lambda D} \tag{8.71}$$

If the light is converging toward a virtual source a distance D (negative) to the right of the recording plane as shown in Fig. 8.34b, Eq. (8.71) is still valid.

(c) *Spherical waves from an off-axis source point.* This is exactly the situation we encountered in our discussion of Fresnel diffraction—compare Fig. 8.35 with Fig. 8.4. If we assume that $|x_s|$ and $|x|$ are small with respect to D, we may use the result of Eq. (8.10a).

$$R = D + \frac{(x_s - x)^2}{2D}$$

$$R_0 = D + \frac{{x_s}^2}{2D}$$

$$[\phi(x)-\phi(0)] = -\frac{2\pi}{\lambda}(R-R_0) \cong -\frac{2\pi}{\lambda}\frac{(x^2-2x\,x_s)}{2D} \approx -\frac{2\pi}{\lambda}\left(\frac{x^2}{2D}+\alpha x\right) \tag{8.72}$$

with

$$\alpha = -\frac{x_s}{R_0} \approx -\frac{x_s}{D}$$

The three-dimensional version of Eq. (8.72) is

$$[\phi(x)-\phi(0)] \cong -\frac{2\pi}{\lambda}\frac{(x^2+y^2-2x\,x_s-2y\,y_s)}{2D} \approx -\frac{2\pi}{\lambda}\left[\frac{x^2+y^2}{2D} + (\alpha x+\beta y)\right] \tag{8.72$'$}$$

Equation (8.72′) is the most general situation for a point source within the so-called parabolic approximation.* It is valid for small values of $|\alpha|$ and $|x/D|$ for both positive and negative values of D and includes sections (a) and (b) as special cases.

*This approximation will also be used in Chapter 9, where we discuss the diffraction theory of image formation. It bears the same relationship to the true equations for the optical path lengths as do the equations of Gaussian ray tracing to the true ray tracing equations.

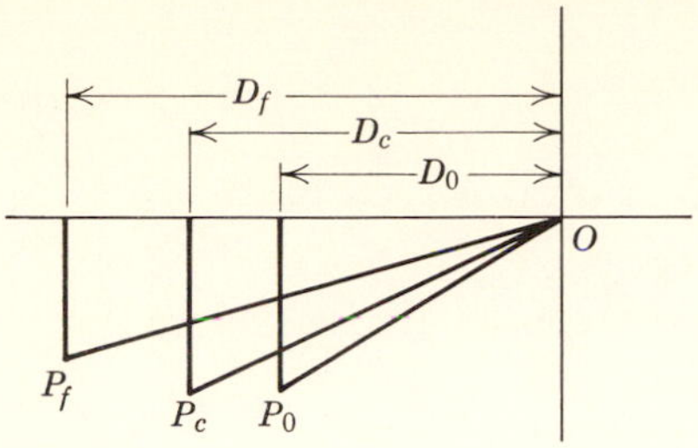

Fig. 8.36

2. ***The Interference Term.*** In the exposure of the hologram, suppose that there is a point object at P_0 (Fig. 8.36). Suppose further that the reference beam comes from a point source at P_f. Then its phase in the recording plane will be

$$\phi_f(x) - \phi_f(0) = -\frac{2\pi}{\lambda}\left[x\alpha_f + \frac{x^2}{2D_f}\right] \tag{8.73}$$

where α_f is the x direction cosine of the line P_fO. Interference between the two beams leads to an interference term of the form

$$A\,A_f \cos\,[\phi(x) - \phi_f(x)]$$

and to a transmission function for the developed photographic plate given by

$$\tau = \tau_0\left[1 - \gamma\frac{A(x)}{A_f}\cos\,(\phi - \phi_f)\right]$$

where $A(x)$ is the amplitude due to P_0.

3. ***The Reconstruction Process.*** The method of reconstruction described above used a beam of light that was identical to the reference beam, and the reconstruction was perfect. We now consider reconstruction where the point source P_c for the reconstruction beam is not oriented exactly the same way as was the source P_f for the reference beam. We further let the wavelength $\lambda' = 2\pi\,c/\omega'$ of the light used for reconstruction be different from that used to expose the hologram ($\lambda = 2\pi\,c/\omega$). With the source distance and direction cosine denoted by D_c and α_c, the incident electric field at the hologram will then be

$$E_i = A_c \cos\,(\omega' t + \phi_c)$$

with

$$\phi_c(x) \cong \phi_c(0) - \frac{2\pi}{\lambda'}\left(x\alpha_c + \frac{x^2}{2D_c}\right) \tag{8.74}$$

and the transmitted light will be given by

$$E_t = \tau E_i = \tau_0 E_i - \gamma\frac{\tau_0}{2}\frac{AA_c}{A_f}\cos\,(\omega t + \phi - \phi_f + \phi_c) - \gamma\frac{\tau_0}{2}\frac{AA_c}{A_f}\cos\,(\omega t + \phi_c - \phi + \phi_f) \tag{8.75}$$

The phase of the second term in Eq. (8.75) may be written approximately

$$\begin{aligned}[\phi - \phi_f + \phi_c] &- [\phi(0) - \phi_f(0) + \phi_c(0)] \\ &= -\frac{2\pi}{\lambda'}\left\{x\left[(\alpha - \alpha_f)\left(\frac{\lambda'}{\lambda}\right) + \alpha_c\right] + \frac{x^2}{2}\left[\left(\frac{1}{D} - \frac{1}{D_f}\right)\frac{\lambda'}{\lambda} + \frac{1}{D_c}\right]\right\} \\ &\equiv -\frac{2\pi}{\lambda'}\left[x\alpha' + \frac{x^2}{2D'}\right]\end{aligned} \tag{8.76}$$

This is equivalent to the phase of light from a point source P' having direction cosine α' and distance D' where

$$\alpha' = (\alpha - \alpha_f)\frac{\lambda'}{\lambda} + \alpha_c \tag{8.77a}$$

and where

$$\frac{1}{D'} = \left(\frac{1}{D} - \frac{1}{D_f}\right)\frac{\lambda'}{\lambda} + \frac{1}{D_c} \tag{8.77b}$$

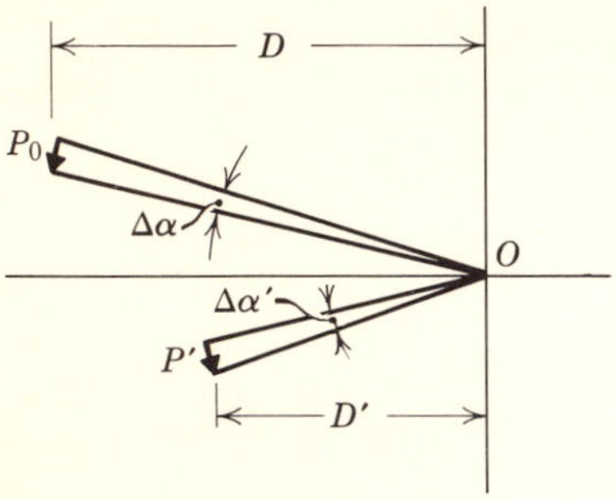

Fig. 8.37

If P_f coincides with P_c and λ with λ', then P' will coincide with P_0.

If the object point is moved laterally a distance $\Delta x_0 \approx \Delta\alpha D$, then P' moves laterally a distance $\Delta x' \approx \Delta\alpha' D'$. We may then define a transverse magnification that to first order in α and x/D is given by

$$m = \frac{\Delta x'}{\Delta x_0} = \frac{\lambda' D'}{\lambda D} \tag{8.78a}$$

as indicated in Fig. 8.37.

A small change in D will result in a corresponding change in D' obtained by differentiating Eq. (8.77*b*). For the longitudinal magnification we then obtain

$$\frac{\Delta D'}{\Delta D} = \frac{\lambda' D'^2}{\lambda D^2} \tag{8.78b}$$

A similar treatment can be given the third term in Eq. (8.75). Its phase angle is given by

$$[-\phi + \phi_f + \phi_c] - [-\phi(0) + \phi_f(0) + \phi_c(0)]$$

$$= -\frac{2\pi}{\lambda'}\left\{x\left[-(\alpha - \alpha_f)\left(\frac{\lambda'}{\lambda}\right) + \alpha_c\right] + \frac{x^2}{2}\left[\left(-\frac{1}{D} + \frac{1}{D_f}\right)\frac{\lambda'}{\lambda} + \frac{1}{D_c}\right]\right\}$$

$$\equiv -\frac{2\pi}{\lambda'}\left[x\alpha'' + \frac{x^2}{2D''}\right] \tag{8.79}$$

This corresponds to light from a point source at P'' with direction cosine

$$\alpha'' = (\alpha_f - \alpha)\left(\frac{\lambda'}{\lambda}\right) + \alpha_c \tag{8.80a}$$

and distance D'' given by

$$\frac{1}{D''} = \left(\frac{1}{D_f} - \frac{1}{D}\right)\frac{\lambda'}{\lambda} + \frac{1}{D_c} \tag{8.80b}$$

In this case the angular orientation of the light tends to be opposite to that from P'. For, if both α_f and α_c are zero, we have

$$\alpha'' = -\alpha\lambda'/\lambda = -\alpha'.$$

The transverse magnification is

$$m = \frac{\Delta x''}{\Delta x_0} = \frac{D''\Delta\alpha''}{D\Delta\alpha} = -\frac{D''\lambda'}{D\lambda} \tag{8.81a}$$

and the longitudinal magnification is

$$\frac{\Delta D''}{\Delta D} = -\frac{\lambda'}{\lambda}\frac{D''^2}{D^2} \tag{8.81b}$$

4. Discussion of Particular Geometries. We now apply the general results for a point object to some important special cases.

(a) *Fourier transform holography.* In this holographic method the object and reference sources are equidistant from the recording plane, so that $D = D_f$. The phase difference $[\phi(x) - \phi_f(x)] - [\phi(0) - \phi_f(0)]$ has only the linear term $-(2\pi/\lambda)(\alpha - \alpha_f)x$, and the hologram consists of equispaced sinusoidal fringes with a separation

$$\Delta x = \frac{\lambda}{\alpha - \alpha_f} \tag{8.82}$$

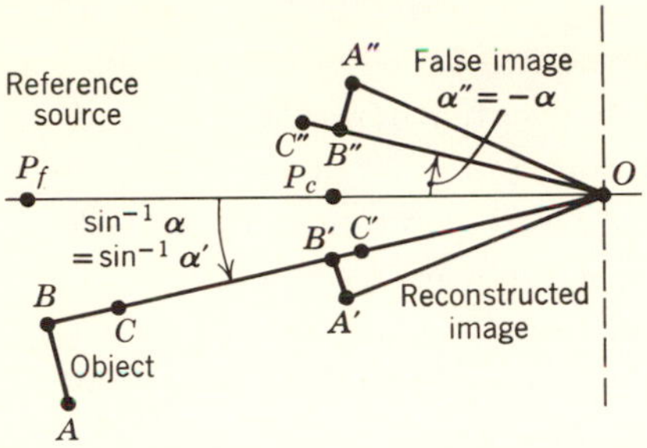

Fig. 8.38. Fourier transform holography. Here $D_c = \frac{1}{2}D_f$.

Equations (8.77*b*) and (8.80*b*) then give us

$$D' = D'' = D_c$$

The situation is particularly simple when $\lambda = \lambda'$ and $\alpha_f = \alpha_c = 0$. Then $\alpha' = \alpha$ and $\alpha'' = -\alpha$. Figure 8.38 shows schematically the two images of a three-point object A, B, C. Note that the three-dimensional reconstruction of the double-primed objects has the wrong perspective; that is, B'' would appear to be in front of C'', even if it is larger!* This is characteristic of the "false image" coming from the third term in Eq. (8.75). The reconstructed image A', B', C' is distorted—the lateral magnification is one-half, from Eq. (8.78*a*), but the longitudinal magnification is one-fourth, from Eq. (8.78*b*).

The name "Fourier transform" is applied to this type of holography because the transmission function of the hologram is related to the Fourier transform of the electric field at the object. This notation will be examined in Section 9.2.

(b) *Fresnal holography in parallel light.* When the condition $D = D_f$ is not satisfied, so that there are terms in x^2 in the expression for the phase difference $\phi - \phi_f$, we have what is known as *Fresnel transform holography.* A simple example occurs when both reference and reconstruction beams are parallel, that is, $D_f = D_c = \infty$. For simplicity we again assume that $\lambda' = \lambda$, $\alpha_f = \alpha_c = 0$. Then

*For point C the condition $D = D_f$ is not satisfied, and $\phi - \phi_f$ will contain a term quadratic in x.

$\alpha' = \alpha$ and $\alpha'' = -\alpha$. We also have $D' = D$ and $D'' = -D$. This means that the "false image" P'' is to the right of the recording plane, which implies that the light actually comes to a focus at P''. The situation is illustrated in Fig. 8.39. Again notice the wrong perspective in the "false image" A'', B'', C''.

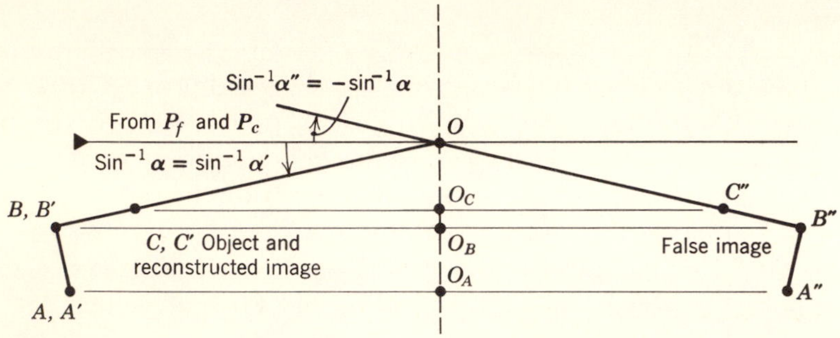

Fig. 8.39

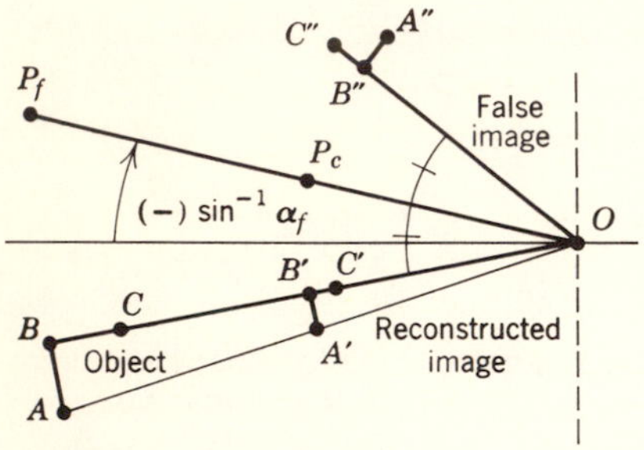

Fig. 8.40

Another interesting modification occurs when reference and reconstruction beams come from the same direction $\alpha_f = \alpha_c \neq 0$ and when we still have $\lambda' = \lambda$. Then $\alpha' = \alpha$, but now $\alpha'' + \alpha = 2\alpha_f$. This means that the false reconstructed beam is obtained by reflecting the original object beam about the direction of the reference and reconstruction beams. Figures 8.40 and 8.41 show how Figs. 8.38 and 8.39 become changed.

When $\alpha_f \neq \alpha_c$, but $\lambda' = \lambda$, the reconstructed image is shifted by an angle: $\alpha' = \alpha + (\alpha_c - \alpha_f)$. The false image is obtained by reflection about a line at an angle $\frac{1}{2}(\alpha_f + \alpha_c)$.

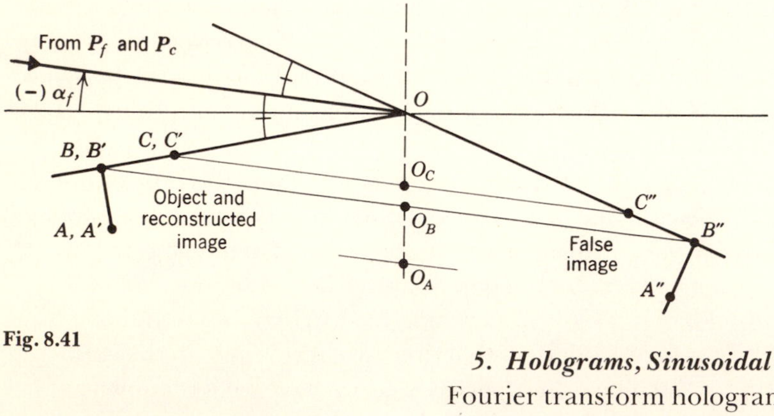

Fig. 8.41

5. Holograms, Sinusoidal Gratings, and Sinusoidal Zone Plates. A Fourier transform hologram of a point object is a sinusoidal diffraction grating with a transmission function

$$\tau(x) = \tau_0\left[1 - \frac{\gamma A}{A_f}\cos\frac{2\pi}{\lambda}(\alpha - \alpha_f)x\right] \tag{8.83}$$

Such a grating was the subject of Problem 7.3.6. The present grating has a spacing given by Eq. (8.82). It diffracts an incident beam into just three orders: −1, 0, and +1. These correspond to the three terms in Eq. (8.75).

A Fresnel transform hologram is closely related to a sinusoidal zone plate. For example, consider Fig. 8.27*a*, let P_1 go to infinity along the line OP_2, and compare with Fig. 8.39. The hologram of the object *A*, *B*, *C* will consist of a superposition of three sinusoidal zone plates having their circular fringe patterns centered about O_A, O_B, and O_C. The lines AO_A, BO_B, and CO_C are parallel to the line P_fO. When illuminated with collimated monochromatic light, each of these zone plates will give two images, one virtual (the "reconstructed image") and one real (the "false image"), plus an attenuated collimated beam.

The zone plates formed as part of the hologram of Fig. 8.41 will have elliptical fringes. Additional properties of the fringe patterns of holograms will be brought out in Problem 8.3.5.

C. SOME ADDITIONAL ASPECTS OF HOLOGRAPHY

1. Comments on the Perfection of Reconstructed and False Images. The perfection of the reconstructed image and the false image, that is, their freedom from aberrations, depends on the extent to which the higher-order terms beyond the parabolic approximation for the optical path lengths can be neglected, or on the extent to which they cancel. We shall not go into any detail here beyond saying that neglect of these terms is similar to the approximations made in Gaussian optics.

We have seen already that exact, aberration-free reconstruction can be obtained when the reconstruction beam is the same as the original reference beam.* Even when this is true for the reconstructed image, the false image will usually have some sort of aberration. An exception is the situation shown in Fig. 8.39, where $\lambda = \lambda'$ and $\alpha_f = \alpha_c = 0$, for there the real false image is perfect in the sense that the electric field that comes to a focus to form it is equal to the time-reversed electric field coming from the original object. What we mean by this can be seen from an examination of the third term in Eq. (8.67). In the present case we have $\phi_f \equiv 0$, and hence this term is proportional to

$$A(x)\cos[\omega t - \phi(x)] = A(x)\cos[-\omega t + \phi(x)]$$

whereas the original field was

$$A(x)\cos[\omega t + \phi(x)]$$

This means that if the reconstructed light were going from right to left across the recording plane, the third term would give the exact

*See Problem 8.3.4 for a generalization of this statement.

time reverse of the original electric field. Since the light is in fact passing from left to right, we obtain the *mirror image* at A'', B'', C'' in Fig. 8.39.

2. Photographic Recording of Images Reconstructed from Holograms. If the "false image" is real as in Fig. 8.39 or 8.41, it can be recorded directly on a photographic plate inserted at the image position, for example, in the plane containing A'' and B'' in Fig. 8.41. It is interesting to consider the effect of the size of the hologram on the resulting photograph. Decreasing the size of the hologram has the same effect as stopping down a camera lens, in terms of brightness of the final image, depth of field, and resolving power. This is so because in both cases the effect on the angular size of the converging wavefronts is the same (Fig. 8.42).

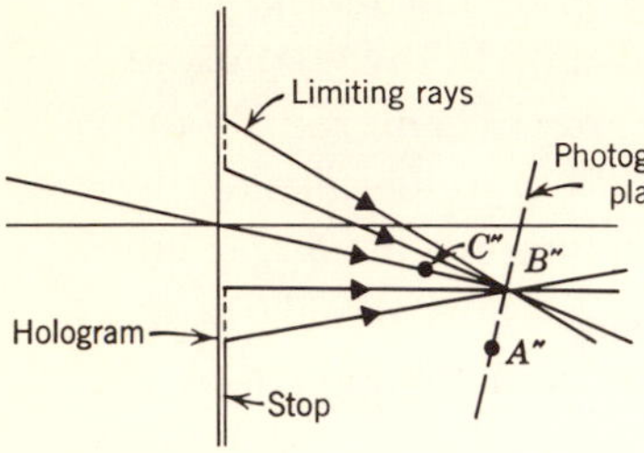

Fig. 8.42

One can also photograph the virtual reconstructed image by means of the usual camera with a lens, and the result should be essentially like photographing the original object.

3. Mechanical and Optical Requirements for the Manufacture and Reconstruction of Holograms. The requirements that must be met to make and reconstruct holograms can be worked out if one has a clear understanding of the essential mechanism of holography. We must record the pattern produced by interference between object and reference beams, and in order to do so we must meet the kind of requirements needed for other types of interferometry. The effects of mechanical and acoustical vibration, thermal fluctuations, and air currents must be reduced so that during an exposure, the interference pattern moves by, say, less than $\frac{1}{8}$ fringe. This means, for instance, that the various lenses, mirrors, objects, and plate holder in Fig. 8.32 must not move by more than about an eighth of a wavelength during the exposure. One way to achieve this is to have a very rigid and solid mechanical platform, such as a granite slab, that all components are mounted on. It helps to have a powerful laser, so that exposure times can be reduced. It also helps, for the same reason, to have a fast photographic film or plate.

But for ordinary holographic work we cannot use fast film or plates, because they do not have high enough resolution. The required resolution will depend on the angular and spatial separation of the object and the source of the reference beam, for these quantities determine the separation of the interference fringes in the recording plane. For object and reference beams coming from point sources, Eqs. (8.72) and (8.73) give for the phase difference as a function of x

$$\phi(x) - \phi_f(x) = -\frac{2\pi}{\lambda}\left[(\alpha - \alpha_f)x + \left(\frac{1}{D} - \frac{1}{D_f}\right)\frac{x^2}{2}\right]$$

The fringe spacing is the value of a change in x needed to change this phase difference by $\pm 2\pi$. Thus we obtain

$$\pm 2\pi = -\frac{2\pi}{\lambda}\left[(\alpha - \alpha_f)\,\delta x + \left(\frac{1}{D} - \frac{1}{D_f}\right)x\,\delta x\right]$$

or

$$\delta x = \frac{\lambda}{|\alpha - \alpha_f + (x/D) - (x/D_f)|} \tag{8.84}$$

Unless special precautions are taken, we would ordinarily have the angle in the denominator of Eq. (8.84) be a reasonable fraction of a radian. Then the fringe spacing might be, say, $2\lambda \approx 1$ micron. This would require a photographic plate with a resolution of 1000 lines/mm. Special plates have this resolution, but they are extremely slow. If we wish to use ordinary films or plates, special geometries must be used in the holographic recording process. We first of all set $D \approx D_f$, that is, we use the Fourier transform method, and then put object and reference sources as close as possible to minimize $|\alpha - \alpha_f|$ in Eq. (8.85) and hence maximize δx.

The requirements for reconstruction of a hologram are less severe than those for making it. If the reconstruction beam comes from a gas laser and closely approximates the original reference beam, we automatically satisfy the requirements for reconstruction. They can also be satisfied, at least approximately, by less coherent sources. Loss of transverse spatial coherence is obtained when the reconstruction beam comes from a source of finite size instead of a point source. We can treat this source as an incoherent superposition of many point sources located in the finite-sized area. Let the source subtend an angle $\Delta\alpha_c$ at the hologram. Then we must locate the image for each point in the reconstruction source and add up the intensities. Equation (8.77*a*) or (8.80*a*) then tells us that the point object will be smeared out over an angle

$$\Delta\alpha' = \Delta\alpha_c \tag{8.85}$$

If the effect is to be completely negligible, then the angular spread in Eq. (8.85) must be less than that produced by diffraction at the exit pupil of the optical system* that "looks at" the reconstructed image. This uncertainty because of diffraction is approximately

$$\Delta\alpha_d = \frac{\lambda}{D}$$

where D is the diameter of the exit pupil as projected onto the second principal plane. Our requirement then becomes

$$\frac{\lambda}{\Delta\alpha_c} > D$$

*For unaided visual observation, this is the pupil of the eye.

This can also be written

$$l_t = \frac{\lambda}{\Delta\alpha} > D \tag{8.86}$$

where l_t is the transverse coherence length of the light used for reconstruction.

A similar argument can be applied to study the effect of limited longitudinal or temporal coherence in the light used for reconstruction. Such limitations would come from a spread in frequency or wavelength. We can image the source made up of the incoherent superposition of oscillators of various frequencies or wavelengths having a wavelength spread $\Delta\lambda'$ about the mean wavelength λ'. Light of each wavelength would then produce an image at an angle given by Eq. (8.77*a*) or (8.80*a*). The spread in $\Delta\lambda'$ would then produce a spread in angle given by

$$\Delta\alpha' = |\alpha - \alpha_f| \frac{\Delta\lambda'}{\lambda} \tag{8.87}$$

To be completely negligible this should be less than the spread resulting from diffraction. This gives

$$|\alpha - \alpha_f| \frac{\Delta\lambda'}{\lambda} < \frac{\lambda'}{D}$$

or

$$\frac{\lambda'}{\Delta\lambda'} > \frac{D}{\lambda} |\alpha - \alpha_f| \tag{8.88}$$

A rough estimate for the case of wide-angle holography where $|\alpha - \alpha_f| \sim \frac{1}{2}$, $\lambda = 500$ nm would then be $\lambda'/\Delta\lambda' \gtrsim 1000^{-1} D$ (D in nm). For small-angle holography, where the resolution of the plate need not be too high (Eq. 8.84), we might have $|\alpha - \alpha_f| \sim 1/20$. Then $\lambda'/\Delta\lambda' \gtrsim 10^{-4} D$.

When the conditions of Eqs. (8.86) and (8.88) are not satisfied, some blurring of the reconstructed image will be observed, but this may be quite tolerable, especially when this is the price we must pay to obtain a reconstruction source of sufficient intensity.

REFERENCES ON HOLOGRAPHY

Gabor's original papers

D. Gabor, "Microscopy by Reconstructed Wave-Fronts," *Proc. Roy. Soc.*, **A197**, 454 (1949); "Microscopy by Reconstructed Wave Fronts: II," *Proc. Phys. Soc.*, **64**, 449 (1951).

Early papers of Leith and Upatnieks

E. N. Leith and J. Upatnieks, "Wavefront Reconstruction with Continuous-Tone Objects," *J. Opt. Soc. Am.*, **53**, 1377 (1963); "Wavefront Reconstruction with Diffused Illumination and Three-Dimensional Objects," *J. Opt. Soc. Am.*, **54**, (1964).

Books

J. B. DeVelis and G. O. Reynolds, *Theory and Applications of Holography*.
J. W. Goodman, *Introduction to Fourier Optics*.
H. M. Smith, *Principles of Holography*.
G. W. Stroke, *An Introduction to Coherent Optics and Holography*. Contains reprints of Gabor's papers.

PROBLEMS

8.3.1 We have some holograms obtained from a commercial firm in which the reference and reconstruction beams make an angle of about 45° with the hologram plate. The object and reconstructed image are in a direction roughly normal to the hologram plane. What is the required resolution of the hologram plate if laser light of 633 nm wavelength is used to make the hologram? Explain why no "false image" is observed in this case. See Fig. 8.41.

8.3.2 Describe the nature of the reconstructed images when a hologram is exposed as in Fig. 8.38 but reconstructed as in Fig. 8.39 and vice versa. Assume $\lambda' = \lambda$.

8.3.3 Show that a photographic negative of a hologram made with unit magnification will give the same observable images as the original hologram.

8.3.4 Suppose a hologram is made using light of wavelength λ with a reference source-to-hologram distance D_f. A photographic copy is now made at magnification m. The copy is then used with a reconstruction source of wavelength $\lambda' = m\lambda$, same angle, and distance $D_c = mD_f$. Show that a perfect reconstructed virtual image is formed that is magnified in all dimensions by a factor m.

8.3.5 Show that with a point object at P_0 and a point reference source at P_f the fringe pattern on the hologram is obtained as the intersection in its plane of a family of hyperboloids of revolution having P_0 and P_f as foci.

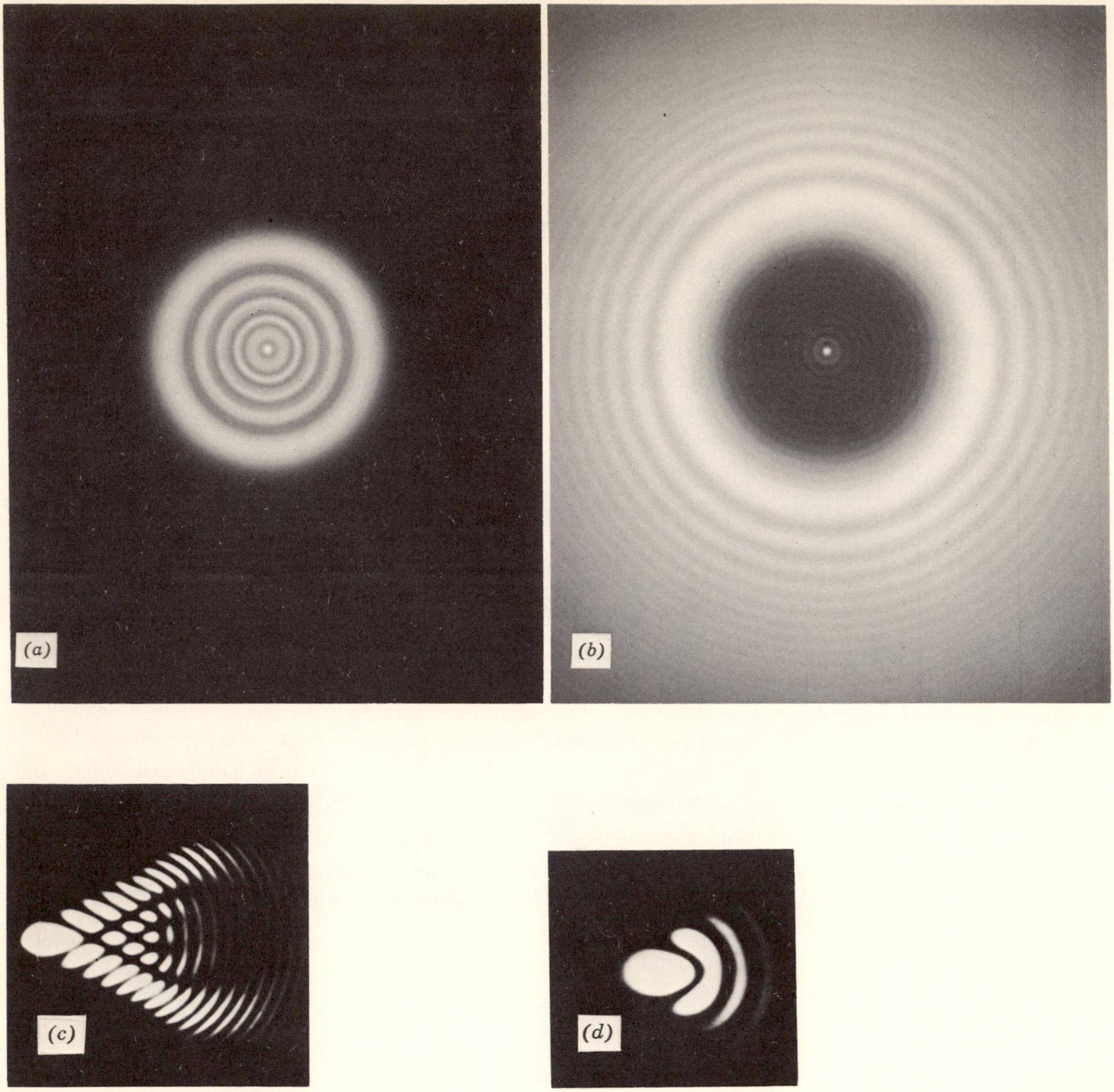

Plate 1. (*a*) Fresnel diffraction pattern for a circular aperture. (*b*) Fresnel diffraction pattern for a circular obstacle. (*c*) Monochromatic image of a point source with about five wavelengths of primary coma. (Photograph courtesy of R. S. Longhurst.) (*d*) Monochromatic image of a point source with about $\frac{1}{2}$ wavelength of primary coma. (Photograph courtesy of R. S. Longhurst.)

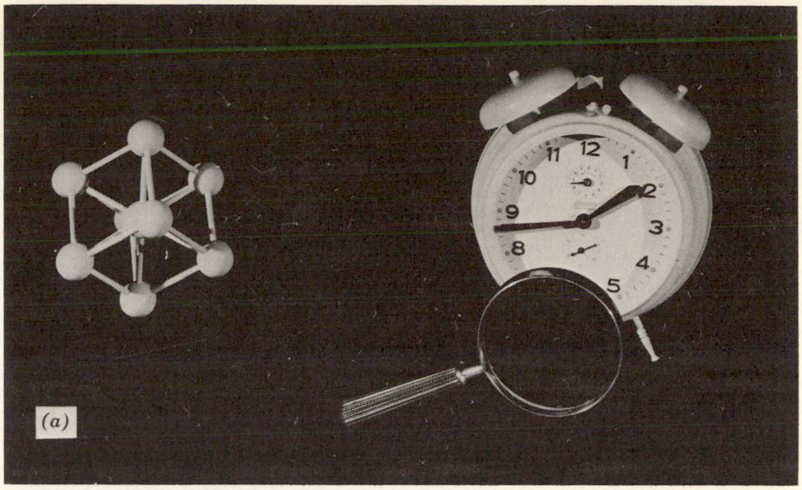

Plate 2*a*. Photograph of a three-dimensional object.

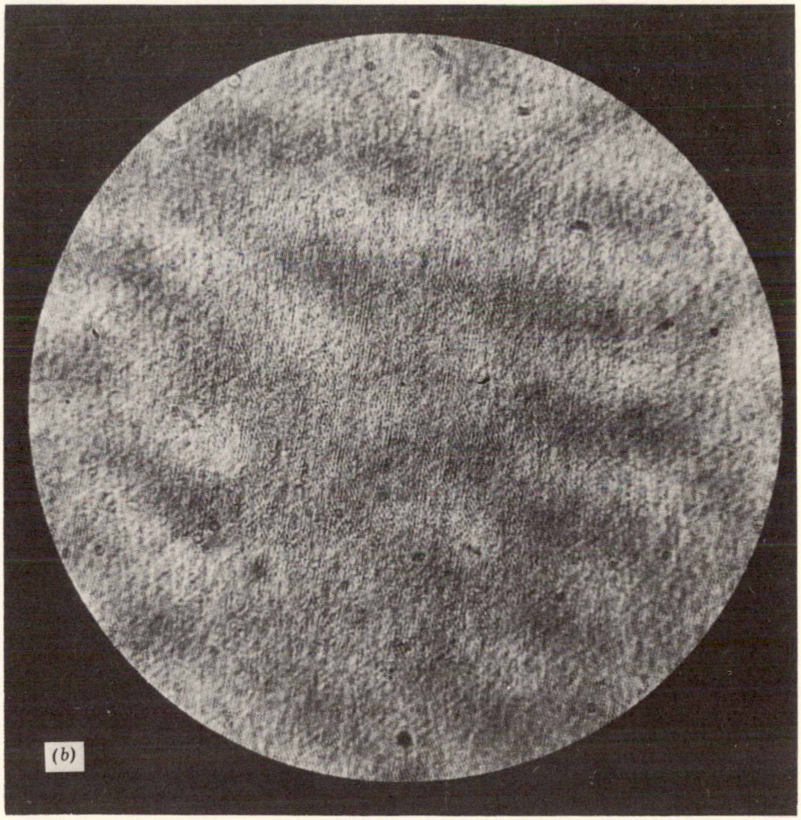

Plate 2*b*. Photomicrograph of hologram.

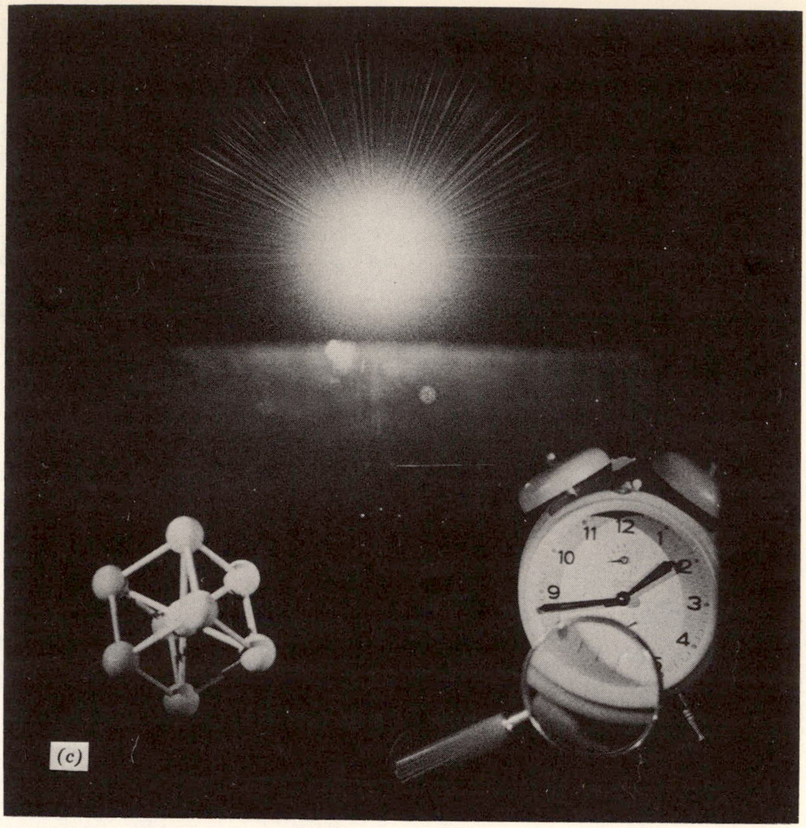

Plate 2*c*. Holographic reconstruction. (Photographs courtesy of Condustron Corp.)

Plate 3. Photograph of the surface of the moon radioed from Lunar Orbiter.
Top: Before spatial filtering.
Bottom: After spatial filtering. (Photographs courtesy of Conductron Corp.)

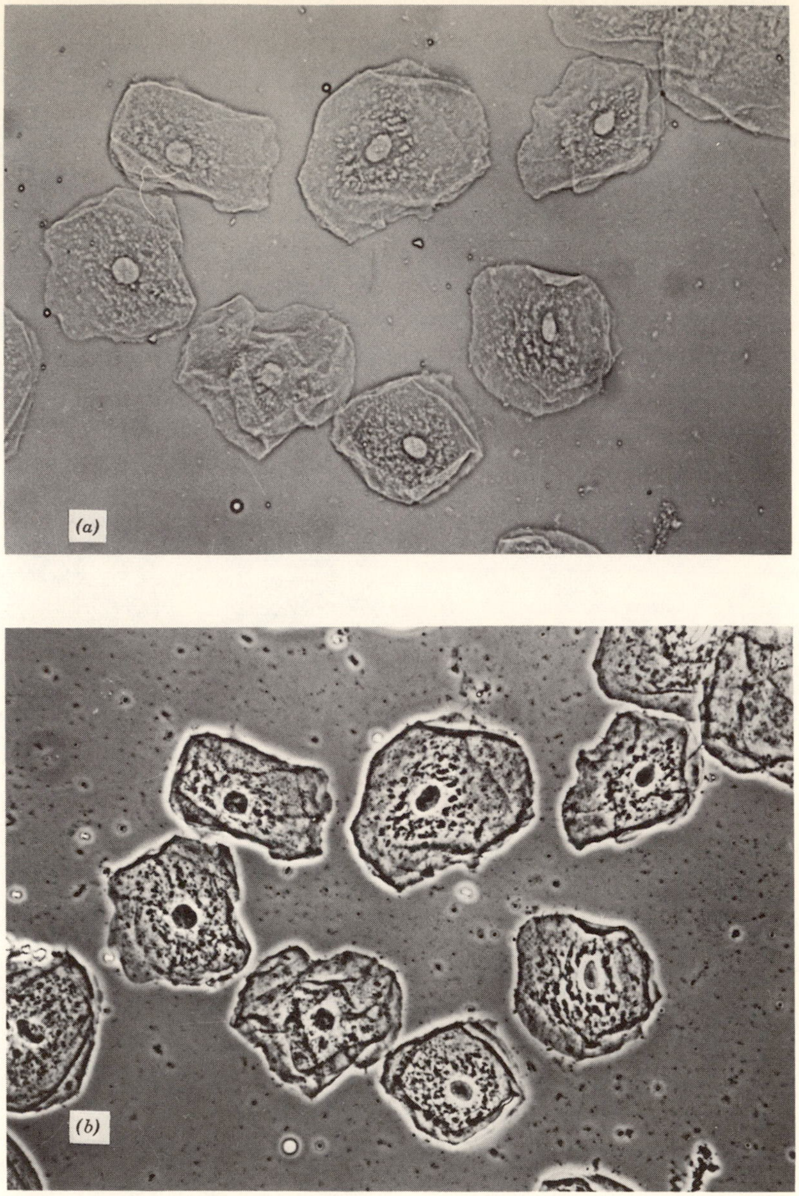

Plate 4. Epithelial cells of the mouth as revealed by three kinds of microscopy. (*a*) Ordinary bright field. (*b*) Phase contrast.

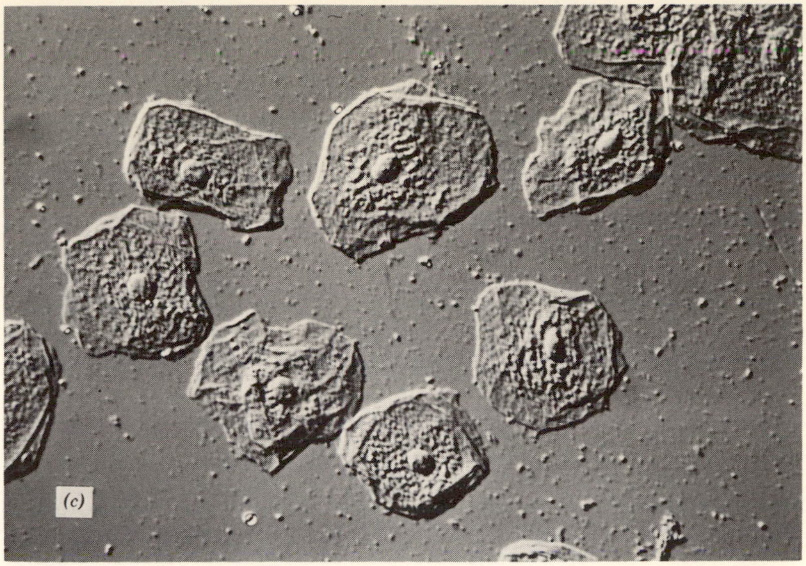

(*c*) Interference contrast. (Photographs courtesy of Carl Zeiss, Inc.)

In this chapter we apply diffraction theory to image formation and related phenomena, and we thereby discuss several recent developments in modern optics. As in Chapter 6 the material here is more sophisticated than that elsewhere in this book.

We first take up Fresnel and Fraunhofer diffraction when lenses are involved. This includes the Abbe theory of image formation. Some applications of the theory to spatial filtering of optical signals will be discussed. Finally diffraction theory will be applied to the evaluation of the performance of optical systems.

CHAPTER 9

DIFFRACTION III

CHAPTER 9 CONTENTS

INFORMAL PREVIEW

The detailed derivations to be given below are sufficiently complex that it is easy to lose sight of the essential physical features of the diffraction and image-forming phenomena that are taking place. We therefore first give an informal qualitative discussion of the topics to be treated in more depth below.

If perfect transverse coherence is assumed, it is meaningful to talk about the phase (strictly, phase difference) at various lateral parts of a surface perpendicular to the direction of propagation. At first we neglect lens aberrations and the diffraction effects from finite lens apertures. This means that the edges of light beams must be left unspecified and will be shown as dotted lines. Consider a semitransparent object that is illuminated by a plane parallel beam as shown in Fig. 9.1. With no object in place the light beam is

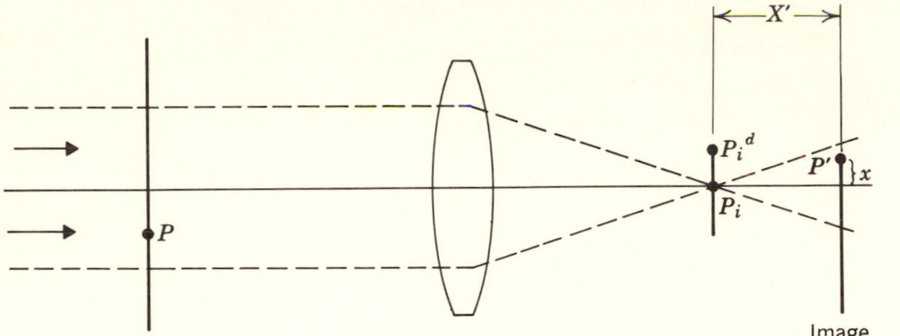

Fig. 9.1

focused onto a spot at P_i, the second focal point of the lens. The light continues to propagate beyond P_i, producing in the image plane a uniform flux density, to a first approximation. The phase across the image plane is not constant, however, since the distance

$$|P_iP'| \approx X' + \frac{1}{2}\frac{x^2}{X'}$$

depends on the distance x from the axis. Thus, for a constant electric field across the object plane, the field at the image plane is

$$\text{const} \cdot e^{-2ikx^2/X'}$$

Now put in a semitransparent object with a transmission function $\tau(P)$ for light passing through it at the point P. Let P' be the Gaussian image of P. Then the complex electric field at P' will be

$$\tilde{E}(P') = \text{const} \cdot \tau(P) e^{-2ikx^2/X'}$$

The light incident on the object is a plane wave that the lens focuses to a spot at P_i. The light transmitted by the object can be considered to consist of the attenuated incident plane wave plus a superposition of other plane waves traveling in various directions and making not too large an angle with the axis of the system. These other plane waves can be regarded as arising from the incident wave from scattering by the object. Each of them will be focused to a point such as P_i^d in the focal plane. The spherical waves diverging from these spots interfere to produce the field at the image $\tilde{E}(P')$.

This argument forms the essence of the *Abbe theory of image formation*. We can block or otherwise alter some of the spots in the focal plane of the lens and thereby alter the image. This is called *spatial filtering*.

The electric field distribution on a plane just past the object can be regarded as a coherent superposition of the fields from infinitely many point sources; the strength of the source at P is proportional to the field there $\tilde{E}(P)$. It is therefore possible to discuss imaging of the light from a single one of the "point sources," including the effects of aberrations and diffraction by the aperture stop, and then obtain the electric field distribution at the image by a coherent superposition of the image fields resulting from all the point sources.

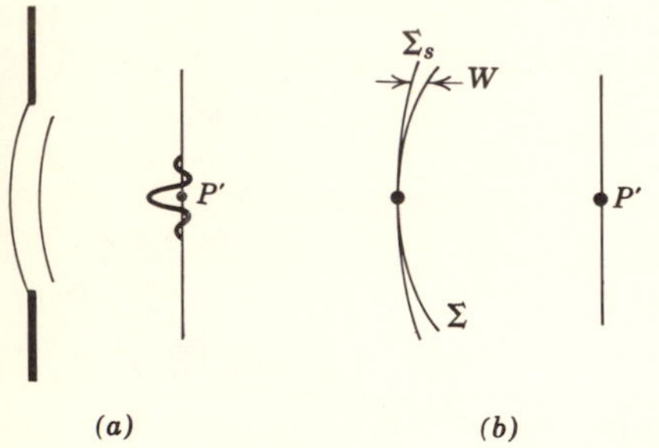

Fig. 9.2

Diffraction by the aperture stop of the light from a single point source, regardless of where it actually occurs in the optical system, can be treated as occurring in image space. There the spherical waves converging to a focus are partially interrupted by the exit pupil. The remaining partial spherical waves form a diffraction pattern at P', as shown schematically in Fig. 9.2*a*. The mathematical form of the field distribution can be obtained via a Kirchoff–Fresnel diffraction integral applied to the spherical wave surface in the pupil. The total field in the image plane is the superposition of all these individual diffraction patterns with proper accounting of relative phases.

Aberrations produce an emerging surface of constant phase Σ at the exit pupil that is not spherical (Fig. 9.2*b*). The distance between Σ and Σ_s, a sphere centered about the Gaussian image point P', defines the aberration function W. Then if we apply the diffraction integral to the surface Σ_s, we must insert an extra factor e^{-ikW} into the integrand. This alters the field distribution in the image plane, which can be calculated analytically in simple cases.

Imaging of light from an *incoherent* object can be treated by calculating the flux density at the image plane from each point in the object and then forming a superposition for all such points. In this way all phase information at the object is lost. The distribution of flux density at the image from a point source is called the *point spread function*. It will include effects of diffraction and aberrations and can be calculated analytically in simple cases, or by computer in more complex cases. Most importantly, it, or functions derived from it, can be measured experimentally.

It is often convenient to express the distribution of exitance (flux density) leaving the incoherent object as a superposition of cosine terms by means of a Fourier transform. Each cosine term is imaged independently as another cosine term with, in general, a reduced amplitude, corresponding to a loss in contrast and a shifted phase, corresponding to distortion. The relative amplitude and phase combine to form the *optical transfer function* $\hat{O}(u)$ of the system. This turns out to be the Fourier transform of the point spread function and is very useful in characterizing the performance of the optical system. It can be determined experimentally by measuring the change in contrast and the phase shift in the image of a sinusoidal test pattern, that is, one whose transmittance varies as

$$(A + B \cos 2\pi u x_0)$$

If the optical transfer function is written in terms of its amplitude $|\hat{O}(u)|$ and phase $\psi(u)$, then the flux density in the image plane will be proportional to

$$\left\{A + |\hat{O}(u)| B \cos\left[\frac{2\pi u x}{m} + \psi(u)\right]\right\}$$

where m is the magnification.

9.1 THE RELATION OF DIFFRACTION TO IMAGE FORMATION

The scalar theory of diffraction may be used to describe the formation of images of coherent objects. These are defined to be objects smaller than the transverse coherence length of the light coming from them, as often occurs in microscopes. The foundations were laid independently by Abbe and Helmholtz in the early 1870's. The basis for a modern treatment was given by Rayleigh in 1896. In recent years the ideas of the theory have been applied to optical information processing.

A. BASIC INGREDIENTS OF THE THEORY

*1. **Assumptions.*** The basic method in this discussion will be repeated application of a form of the Fresnel–Kirchoff diffraction integral. This tells us how the electric field distribution in a given transverse plane is related to that in another plane. The necessary formulas are complicated, but in many cases the final results are rather easy to interpret. We assume the light to have perfect *transverse and longitudinal coherence.* This can be achieved by using light from a gas laser operating in a single transverse mode.

The *parabolic approximation,* where spherical waves are replaced by paraboloids, will be used to describe the spatial dependence of the phase shifts resulting from propagation and lens action. This approximation yields the wave equivalent of the Gaussian theory of geometrical optics. It must be extended to include the diffraction theory of aberrations, but it can be generalized rather easily to large aperture systems which satisfy the Abbe sine condition.*

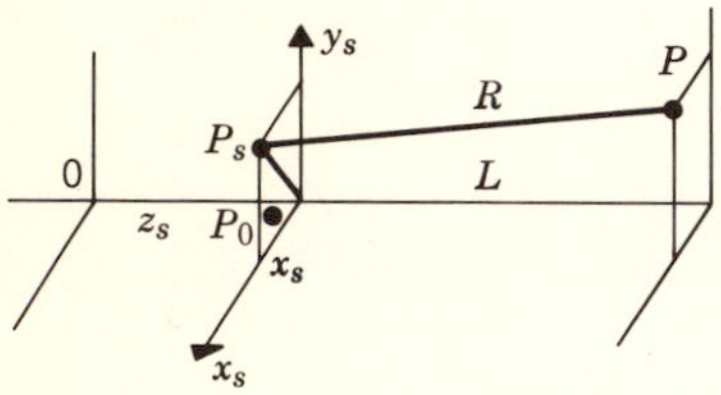

Fig. 9.3

We assume the light propagation to be from left to right in the general direction of the $+z$ axis, and consider the electric field in a convenient transverse plane to be a *source of secondary Huygens wavelets.* Figure 9.3 gives an example.

The field at a point P is then given approximately by the Fresnel–Kirchoff diffraction integral (without the inclination factor) as

$$\tilde{E}(P) = \frac{i}{\lambda} \int\!\!\int_{-\infty}^{\infty} \tilde{E}(x_s, y_s, z_s)\, \frac{e^{-ikR}}{R}\, dx_s\, dy_s$$

$$R = |\overline{P_s P}| \tag{9.1}$$

If there is a real point source of strength A at $P_0(x_0, y_0, z_s)$, the field

*We should point out the the polarization properties of light that result from the vector nature of the electromagnetic field cannot be neglected if accurate calculations are to be done with large aperture systems.

at P will be given by

$$\tilde{E}(P) = \frac{Ae^{-ikR_0}}{R_0}, \qquad R_0 = |\overline{P_0P}| \tag{9.2}$$

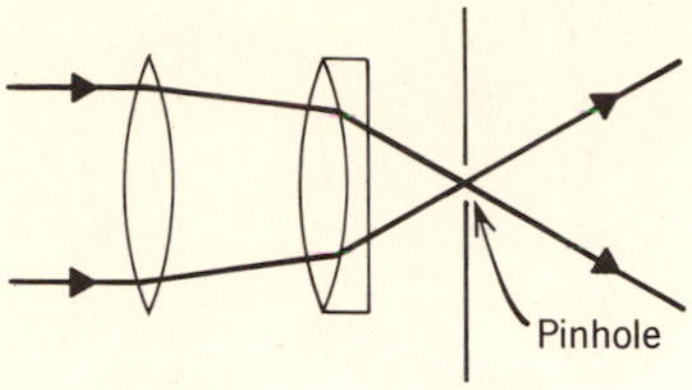

Fig. 9.4. Method of obtaining a "point source" with a uniphase laser beam. The pinhole removes unwanted stray light that often accompanies such beams.

This result can be included in Eq. (9.1) if we allow our initial field distribution to be a delta function:

$$\tilde{E}(x_s, y_s, z_s) = \frac{A\lambda\delta(x_s - x_0)\delta(y_s - y_0)}{i} \tag{9.3}$$

A strongly peaked delta-function type field will be obtained when a laser beam is focused to a diffraction-limited* spot with a microscope objective, as shown in Fig. 9.4.

2. The Fresnel Propagation Transformation. To use Eq. (9.1) we make an approximation for the distance R which is equivalent to those made in the theory of Fresnel diffraction. With reference to the situation in Fig. 9.5, we have

$$\tilde{E}(P_2) = \tilde{E}(x_2, y_2, z_0 + L) = \frac{i}{\lambda} \int\int_{-\infty}^{\infty} \tilde{E}(x_1, y_1, z_0) \frac{e^{-ikR_{12}}}{R_{12}} \, dx_1 \, dy_1$$

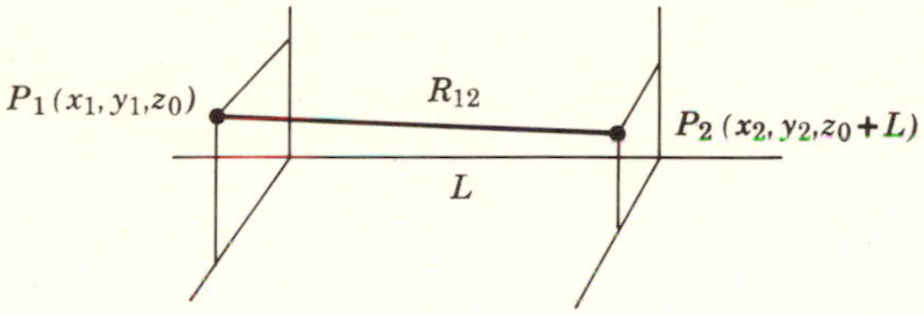

Fig. 9.5

The parabolic approximation for R_{12} is

$$R_{12} = \sqrt{L^2 + (x_2 - x_1)^2 + (y_2 - y_1)^2} \cong L + \frac{(x_2 - x_1)^2 + (y_2 - y_1)^2}{2L}.$$

This yields

$$\tilde{E}(x_2, y_2, z_0 + L) = \frac{ie^{-ikL}}{\lambda L} \int\int_{-\infty}^{\infty} \exp\left\{-i \frac{k}{2L} [(x_2 - x_1)^2 + (y_2 - y_1)^2]\right\} \times \tilde{E}(x_1, y_1, z_0) \, dx_1 \, dy_1 \tag{9.4}$$

This integral transforms the field at the plane $z = z_0$ to that at the plane $z = z_0 + L$. We call this the *Fresnel transformation.*

If we propagate beyond this plane to the plane $z = z_0 + L + M$, we could apply another Fresnel transformation and obtain

$$\tilde{E}(x_3, y_3, z_0 + L + M) = \frac{ie^{-ikM}}{\lambda M} \int\int_{-\infty}^{\infty} \exp\left\{-i \frac{k}{2M} [(x_3 - x_2)^2 + (y_3 - y_2)^2]\right\} \times \tilde{E}(x_2, y_2, z_0 + L) \, dx_2 \, dy_2 \tag{9.5}$$

*A diffraction-limited image is one as perfect as diffraction by the aperture stop will allow.

with $\tilde{E}(x_2, y_2, z_0+L)$ given by Eq. (9.4). On the other hand, we might expect the left-hand side of Eq. (9.5) to be given directly by a Fresnel transformation from the plane $z = z_0$:

$$\tilde{E}(x_3, y_3, z_0+L+M) = \frac{ie^{-ik(L+M)}}{\lambda(L+M)} \iint_{-\infty}^{\infty} \exp\left\{-\frac{ik}{2(L+M)}[(x_3-x_1)^2 + (y_3-y_1)^2]\right\} \tilde{E}(x_1, y_1, z_0)\, dx_1\, dy_1 \tag{9.5'}$$

This equation will be compatible with Eqs. (9.4) and (9.5) provided that the following equation holds:

$$\frac{(i)^2 e^{-ik(L+M)}}{\lambda^2 LM} \iint_{-\infty}^{\infty} \exp\left\{-i\frac{k}{2L}[(x_2-x_1)^2 + (y_2-y_1)^2]\right\} \times \exp\left\{-i\frac{k}{2M}[(x_3-x_2)^2 + (y_3-y_2)^2]\right\} dx_2 dy_2$$
$$= \frac{ie^{-ik(L+M)}}{\lambda(L+M)} \exp\left\{-\frac{ik}{2(L+M)}[(x_3-x_1)^2 + (y_3-y_1)^2\right\} \tag{9.6}$$

The proof of Eq. (9.6) goes along the lines of our discussion of Fresnel diffraction in Section 9.2A. For the exponents we can write, in analogy with Eq. (8.14*a*),

$$\frac{(x_2-x_1)^2}{2L} + \frac{(x_3-x_2)^2}{2M} = \frac{(x_3-x_1)^2}{2(L+M)} + \frac{(x_2-x_m)^2}{2\rho}$$

with

$$\frac{1}{\rho} = \frac{1}{L} + \frac{1}{M}, \qquad x_m = \frac{Mx_1 + Lx_3}{M+L}$$

Similar expressions hold for the y's. The left side of Eq. (9.6) becomes

$$\frac{(i)^2 e^{-ik(L+M)}}{\lambda^2 LM} \exp\left\{-i\frac{k}{2(L+M)}[(x_3-x_1)^2 + (y_3-y_1)^2]\right\}$$
$$\times \iint_{-\infty}^{\infty} \exp\left\{-\frac{\pi i}{\lambda\rho}[(x_2-x_m)^2 + (y_2-y_m)^2]\right\} dx_2\, dy_2$$

The integral here is the same as that in Eq. (8.30*c*), which equals $\lambda\rho/i$. Since $\rho = LM/(L+M)$, we see that Eq. (9.6) indeed holds. *Thus, two successive Fresnel transformations are equivalent to a third overall Fresnel transformation.* This conclusion also holds for the more complicated exact transformations that result if the simplifying parabolic approximation is not made.

Equations (9.4), (9.5), (9.5′), and (9.6) are valid for both positive and negative values of L and M. The significance of a Fresnel transformation over a distance $L < 0$ is the following. It tells us what field

$$\tilde{E}(x_2, y_2, z_0 + L) = \tilde{E}(x_2, y_2, z_0 - |L|)$$

we need to have in the plane $z = z_0 - |L|$ in order to obtain the field $\tilde{E}(x_1, y_1, z_0)$ farther on in the plane $z = z_0$.

Another mathematical property of the Fresnel transform results from the physical requirement that a transformation corresponding to a zero value of L must give us the original field distribution. Thus as $L \to 0$ the right hand side of Eq. (9.4) must tend toward $\tilde{E}(x_2, y_2, z_0)$. This means that the expression

$$\frac{ie^{-ikL}}{\lambda L} \exp\left\{-\frac{ik}{2L}\left[(x_2 - x_1)^2 + (y_2 - y_1)^2\right]\right\} \tag{9.7}$$

must behave like a two-dimensional delta function $\delta(x_1 - x_2) \cdot \delta(y_1 - y_2)$ as $L \to 0$. It does, but we omit the proof.

3. ***Focusing Action of a Lens.*** A lens will have two effects on the light passing through it. It will limit the spatial extent of the electric field because of its finite aperture, and it will produce a focusing action. The aperturing effect will be discussed later along with other apertures. We neglect light losses that result from reflections from the lens surfaces. We also assume that the refractive index is the same on both sides of the lens.

The focusing action can be described by a position-dependent phase shift, that is, the transmitted electric field will be given in terms of the incident field by

$$\tilde{E}_t(x, y) = \tilde{E}_i(x, y)e^{-i\phi(x, y)} \tag{9.8}$$

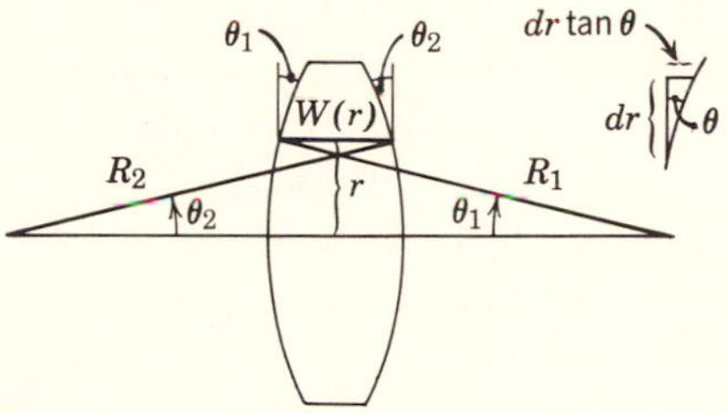

Fig. 9.6

where (x, y) locates the region of the lens under consideration. If the thickness of the lens cannot be neglected, then the incident field in Eq. (9.8) should be that at the first principal plane and the transmitted field that at the second principal plane.

The phase can be determined to within an irrelevant additive constant by the following argument. The result is independent of the thickness of the lens, but it is easier to derive if we assume a thin lens. Then a paraxial ray will pass through the lens at an essentially constant distance r from the axis, as shown in Fig. 9.6. The phase shift for such a ray relative to that for the axial ray is given by

$$\phi(r) - \phi(0) = 2\pi(n-1)\frac{[w(r) - w(0)]}{\lambda}$$

where λ is the vacuum wavelength, and $w(r) - w(0)$ is the difference in path length through the lens for the two rays. From the figure we see that

$$\frac{dw}{dr} = -\tan\theta_1 - \tan\theta_2 \approx -r\left(\frac{1}{R_1} - \frac{1}{R_2}\right)$$

We use the sign convention of Chapter 3, so that for the lens in Fig. 9.6 R_2 is negative. An integration gives

$$w(r) - w(0) = \frac{-r^2}{2}\left(\frac{1}{R_1} - \frac{1}{R_2}\right)$$

Hence

$$\phi(r) - \phi(0) = \frac{-kr^2}{2}(n-1)\left(\frac{1}{R_1} + \frac{1}{R_2}\right) = \frac{-kr^2}{2f}$$

where f is the focal length of the lens as given by the lens-makers equation (3.30).

The desired expression to describe the focusing action of the lens is then

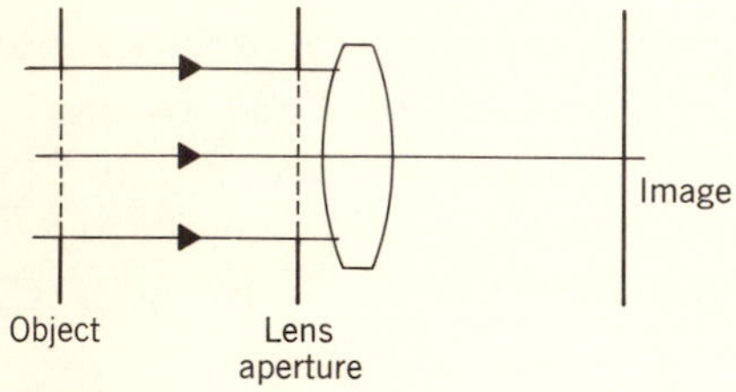

Fig. 9.7

$$\tilde{E}_t(x, y) = \tilde{E}_i(x, y)e^{(ik/2f)(x^2+y^2)}e^{-i\phi(0)} \tag{9.9}$$

Note the parabolic dependence on x^2 and y^2. The $e^{-i\phi(0)}$ factor can be ignored, since it is a constant phase factor.

*4. **Objects and Apertures.*** Our "object" will be any essentially two-dimensional, partially transmitting film or screen that modulates the incident light falling on it (Fig. 9.7). It will be described by a transmission function $\tau_0(x, y)$. The object could be a simple rectangular or circular opening or something much more complicated, like a diffraction grating or a photographic negative. We shall also want to treat systems having both an object or a generalized aperture in a general position (O) and a finite aperture generally at or near the lens (A). If the aperture is explicitly at the lens, we shall call it a lens aperture (LA). It will be possible to describe the behavior in the limit of a very large aperture (The infinite aperture case—IA) for which diffraction effects are neglected, and then to correct the results for the effects of diffraction by a finite aperture.

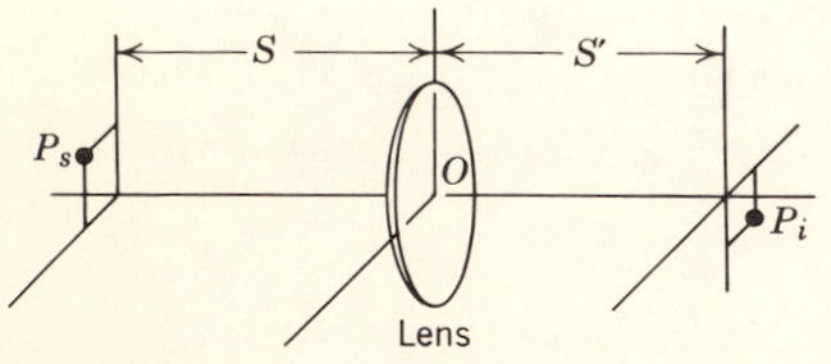

Fig. 9.8

B. GENERAL TRANSFORMATION BETWEEN CONJUGATE PLANES

*1. **Derivation.*** We shall now derive the transformation that starts with the electric field $E(x_s, y_s, -S)$ in one plane to the left of the lens (source plane) and gives the field in the conjugate or image plane at $z = S'$ (Fig. 9.8). We shall assume that there is nothing except the lens between these two planes. The lens, assumed thin for simplicity, will be at the origin. The result of this derivation, if all apertures are neglected, is given in Eq. (9.15), (9.15′), or (9.15″). A simple interpretation of the result follows Eq. (9.15′).

The field just before the lens will be given in terms of the field in the source plane by the Fresnel transformation (Eq. 9.4):

$$\tilde{E}(x',y',0)=\frac{ie^{-ikS}}{\lambda S}\iint_{-\infty}^{\infty}\tilde{E}(x_s,y_s,-S)\times\exp\left\{-\frac{ik}{2S}\left[(x'-x_s)^2+(y'-y_s)^2\right]\right\}dx_s\,dy_s$$

(Here S is positive if P_s is to the left of O.) The overall aperturing and focusing effects of the lens will be described by the combined transmission function $\{\tau_{\mathrm{LA}}(x',y')\exp[(ik/2f)(x'^2+y'^2)]\}$. In the ILA (infinite lens aperture) limit we simply set $\tau_{\mathrm{LA}}(x',y')\equiv 1$.

At an arbitrary plane a distance z to the right of the lens the electric field can be obtained via another Fresnel transformation:

$$\begin{aligned}\tilde{E}(x,y,z)&=\frac{ie^{-ikz}}{\lambda z}\iint_{-\infty}^{\infty}\tau_{\mathrm{LA}}(x',y')\\&\quad\times\exp\left[\frac{ik}{2f}(x'^2+y'^2)\right]\tilde{E}(x',y',0)\\&\quad\times\exp\left\{\frac{ik}{2z}\left[(x-x')^2+(y-y')^2\right]\right\}dx'dy'\\&=\frac{(i)^2e^{-ik(S+z)}}{\lambda^2Sz}\iiiint_{-\infty}^{\infty}\tilde{E}(x_s,y_s,-S)\tau_{\mathrm{LA}}(x',y')\\&\quad\times\exp\left\{-\frac{ik}{2S}\left[(x_s-x')^2+(y_s-y')^2\right]\right\}\\&\quad\times\exp\left[\frac{ik}{2f}(x'^2+y'^2)\right]\\&\quad\times\exp\left\{\frac{ik}{2z}\left[(x-x')^2+(y-y')^2\right]\right\}dx_s\,dy_s\,dx'\,dy'\end{aligned}\tag{9.10}$$

The phase $\Phi=\Phi_x+\Phi_y$ of the exponent in the last integrand is the sum of x- and y-dependent terms having the same form, for example,

$$\Phi_x=\frac{-ik}{2}\left[x'^2\left(\frac{1}{S}+\frac{1}{z}-\frac{1}{f}\right)+\frac{x_s^2}{S}+\frac{x^2}{z}-2x'\left(\frac{x_s}{S}+\frac{x}{z}\right)\right]\tag{9.11}$$

The formidable-appearing equations (9.10) and (9.11) can be simplified, particularly in the *image plane*, where $z=S'$. In the image plane the x'^2 term in Φ_x drops out, since

$$\frac{1}{S}+\frac{1}{S'}-\frac{1}{f}=0$$

and (9.10) becomes

$$\tilde{E}(x, y, S') = \frac{(i)^2 e^{-ik(S+S')}}{\lambda^2 SS} \iint_{-\infty}^{\infty} \tilde{E}(x_s, y_s, -S)$$
$$\times \exp\left[-\frac{ik}{2}\left(\frac{x_s^2+y_s^2}{S}+\frac{x^2+y^2}{S'}\right)\right]$$
$$\times\left(\iint_{-\infty}^{\infty} \tau_{\mathrm{LA}}(x', y') \exp\left\{\frac{2\pi i}{\lambda}\left[\left(\frac{x_s}{S}+\frac{x}{S'}\right)x'\right.\right.\right.$$
$$\left.\left.\left.+\left(\frac{y_s}{S}+\frac{y}{S'}\right)y'\right]\right\} dx'\, dy'\right) dx_s\, dy_s \qquad (9.12a)$$

The factor in large parentheses in this equation can be written

$$\{\cdots\} = \hat{\tau}_{\mathrm{LA}}(u, v)$$

using notation consistent with Eq. (7.34),

$$u = -\frac{1}{\lambda}\left(\frac{x_s}{S}+\frac{x}{S'}\right), \qquad v = -\frac{1}{\lambda}\left(\frac{y_s}{S}+\frac{y}{S'}\right) \qquad (9.12b)$$

where $\hat{\tau}_{\mathrm{LA}}$ is the Fourier transform of the transmission function of the lens aperture:

$$\hat{\tau}_{\mathrm{LA}}(u, v) = \iint_{-\infty}^{\infty} \tau_{\mathrm{LA}}(x', y') e^{-2\pi i(ux'+vy')}\, dx'\, dy' \qquad (9.12c)$$

If the object is a point source, the object field distribution $\tilde{E}(x_s, y_s, -S)$ is delta function, and the integral in Eq. (9.12*a*) is trivial.

In the image plane we observe the Fraunhofer diffraction pattern of the lens aperture, and the Fourier transform is characteristic of Fraunhofer diffraction. This occurs because the quadratic terms involving x'^2 and y'^2 in the phase vanish in the image plane.

Outside the image plane, where $z \neq S'$, the quadratic terms do not vanish, and Fresnel diffraction will occur as described by a Fresnel transformation. Fortunately this general case can be treated by first finding the field in the image plane and then applying a Fresnel transformation over the distance $z-S'$.

2. The Limiting Form of the Transformation in the Case of Infinite Aperture. In the ILA case the Fourier transform in Eq. (9.12) becomes a two-dimensional delta function. To see this, assume an explicit form for the lens aperture, namely, a $2x_L$-by-$2y_L$ rectangle, and then let x_L and $y_L \to \infty$. For such an aperture we have (Eq. [7.40′]).

$$\hat{\tau}_{\mathrm{LA}}(u, v) = \hat{\tau}_1(u)\hat{\tau}_2(v) = \left[\frac{\sin 2\pi u x_L}{\pi u}\right]\left[\frac{\sin 2\pi v y_L}{\pi v}\right] \tag{9.12c}$$

Dirichlet's formula from Appendix A, Eq. (A5),

$$\lim_{x_L \to \infty} \int_{-a}^{+b} f(u+u') \frac{\sin 2\pi u x_L}{\pi u}\, du = \int_{-a}^{+b} f(u+u')\, \delta(u)\, du = f(u') \qquad (a, b > 0) \tag{9.13}$$

says that each of the factors in (9.12*c*) has the limiting delta function property when used in an integral. Thus we may write that the ILA limit of $\tau_{\mathrm{LA}}(u, v)$ is

$$\lim_{\substack{x_L \to \infty \\ y_L \to \infty}} \hat{\tau}_{\mathrm{LA}}(u, v) = \delta(u)\, \delta(v) \tag{9.14}$$

We may then apply Eq. (9.14) and do some manipulating to match the arguments of the delta functions with the variables of integration. The property required is $\delta(ax) = \delta(x)/|a|$.

We obtain the following simple expression for the case where all apertures between object and image are neglected:

$$\tilde{E}_{\mathrm{IA}}(x, y, S') = \frac{e^{-ik(S+S')}}{m} \exp\left[-\frac{ik}{2}\left(\frac{x^2+y^2}{m^2 S} + \frac{x^2+y^2}{S'}\right)\right] \tilde{E}\left(\frac{x}{m}, \frac{y}{m}, -S\right) \qquad \left(m = \frac{-S'}{S}\right) \tag{9.15}$$

3. Interpretation. In the parabolic approximation we can use Fig. 9.9 and write

$$L = \left[\left(\frac{x}{m}\right)^2 + S^2\right]^{1/2} \cong S + \frac{x^2}{2m^2 S}$$

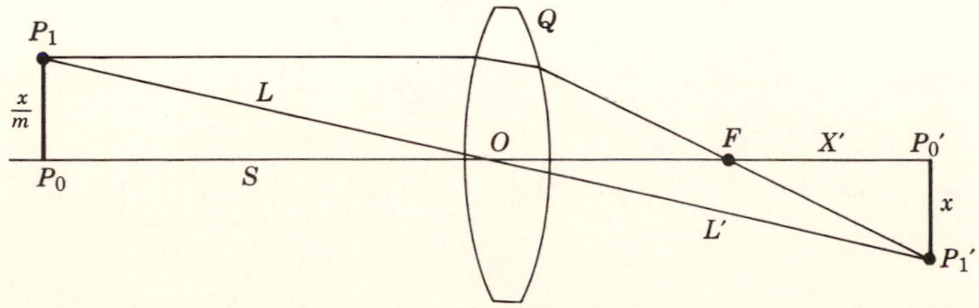

Fig. 9.9

and

$$L' = \sqrt{x^2 + S'^2} \cong S' + \frac{x^2}{2S'}$$

so that

$$\widehat{P_1OP'_1} = L + L' \cong S + S' + \frac{x^2}{2m^2S} + \frac{x^2}{2S'} \tag{9.16}$$

Thus the combined exponent in (9.15) is $e^{-2\pi i(L+L')/\lambda}$, giving

$$\tilde{E}_{\mathrm{IA}}(x, y, S') = \frac{e^{-2\pi i(L+L')/\lambda}}{m} \tilde{E}\left(\frac{x}{m}, \frac{y}{m}, -S\right) \tag{9.15$'$}$$

It is now quite easy to interpret all the factors in Eq. (9.15′): With no apertures the field at (x, y, S') is proportional to the field at the conjugate point $(x/m, y/m, -S)$. There is also a phase shift $-2\pi(L+L')/\lambda$ to account for the distance traveled by the light and a factor $1/m$ to guarantee that the total flux at $z = S'$ equals that at $z = -S$.

Referring again to Fig. 9.9, we consider the ray P_1Q parallel to the axis and can write

$$|\widehat{P_1OP'_1}| = |\widehat{P_1QP'_1}| = |\widehat{P_1QF}| + |FP'_1|$$

Since in the parabolic approximation we have

$$|FP'_1| = X' + \frac{x^2}{2X'}$$

we may write

$$|\widehat{P_1OP'_1}| = |P_1QF| + |FP'_0| + \frac{x^2}{2X'} = |P_0OF| + |FP'_0| + \frac{x^2}{2X'} = |P_0OP'_0| + \frac{x^2}{2X'} = S + S' + \frac{x^2}{2X'}$$

Thus from (9.16) we have

$$\frac{x^2+y^2}{2m^2S} + \frac{x^2+y^2}{2S'} = \frac{x^2+y^2}{2X'} \tag{9.17}$$

This result can also be obtained algebraically. Another way to write Eq. (9.15) is then

$$\tilde{E}_{\mathrm{IA}}(x, y, S') = \frac{e^{-ik(S+S')}}{m} e^{-(ik/2X')(x^2+y^2)} \tilde{E}\left(\frac{x}{m}, \frac{y}{m}, -S\right) \tag{9.15$''$}$$

References for Section 9.1 will be found at the end of this chapter.

PROBLEMS

9.1.1 In a uniphase laser beam there is one transverse plane where the field is approximately given by the Gaussian function:

$$\tilde{E}(x, y, 0) = e^{-r^2/r_0^2} = e^{-(kr^2/2R)}, \qquad r^2 = x^2 + y^2, \qquad r_0 = r_0 = \left(\frac{\lambda R}{\pi}\right)^{1/2} \tag{1}$$

Such a beam is often called a Gaussian beam.* Here r_0 is called the radius parameter of the laser beam, and R is called the diffraction length.

a. Show, by applying a Fresnel transformation, that the field in the plane $z = L$ is given by

$$\tilde{E}(x', y', L) = \frac{e^{-ikL}}{1 - iL/R} \exp\left[-\frac{kr'^2}{2(R - iL)}\right] \tag{2}$$

and that

$$|\tilde{E}(x', y', L)| = \frac{1}{\sqrt{1 + L^2/R^2}} \exp\left[-\frac{k}{2}\frac{r'^2 R}{(R^2 + L^2)}\right] \tag{3}$$

Then as far as the magnitude of $\tilde{E}$ is concerned, a radius parameter r_L may be defined so that

$$|\tilde{E}| = \frac{r_0}{r_L} e^{-r'^2/r_L{}^2}, \qquad r_L = r_0\left(1 + \frac{L^2}{R^2}\right)^{1/2} \tag{4}$$

b. Now let $M = L_1 + L_2$. The field at $z = M$ may be obtained by applying Eq. (2) with $L = M$, or by applying Eq. (2) twice, once with $L = L_1$ and once with $L = L_2$ and R replaced by $R - iL_1$. Show that both methods give the same result.

9.1.2 Suppose that the field at $z = 0$ from a laser is $E(x, y, 0) = e^{-r^2/r_0{}^2}$ as in Problem 9.1.1 and a lens of focal length f is placed at $z = L$.

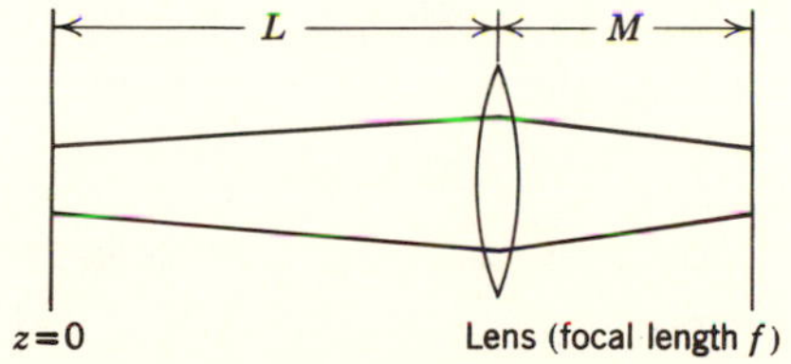

a. Find the field at $z = M + L$.

Answer:

$$\tilde{E}(x, y, M + L) = Ae^{-(kr^2/2S)}$$

where

$$S = \frac{f(L + M) - ML + iR(f - M)}{R + i(f - L)}$$

and where

$$A = (\text{phase factor}) \times \frac{Rif}{f(L + M) - ML + iR(f - M)}$$

(Hint: Apply the results of Problem 9.1.1 to get the field just to the left of the lens. Apply Problem 9.1.1 suitably modified to go from just to the right of the lens to $z = L + M$.)

*H. Kogelnick, "On the Propagation of Gaussian Beams of Light through Lenslike Media," *Applied Optics*, **4**, 1562 (1965).

b. Show that $|E(x, y, L+M)|$ is Gaussian with a radius parameter that for fixed f is a minimum when $L = M = f$. Show that the value of this minimum radius parameter is $r_0 f/R = \lambda f/(\pi r_0)$.

9.1.3 Show that for $L = M = f$ in Problem 9.1.2 the field $E = e^{-r^2/r_0^2}$ will reproduce itself apart from a constant phase factor in the plane $z = 2f$ provided that

$$r_0 = \left(\frac{f\lambda}{\pi}\right)^{1/2}$$

The diffraction length R then equals the focal length f. What is the radius parameter at the lens?

A generalization of the above results may also be made. Assume $L = M \neq f$. The field at $z = 2L$ will reproduce that at $z = 0$ provided that

$$R^2 = 2fL - L^2$$

$$r_0^4 = \frac{\lambda^2}{\pi^2}(2fL - L^2)$$

$$r_L^4 = \frac{\lambda^2}{\pi^2}\frac{f^2 L}{2f - L} \qquad (r_L \text{ is defined in Problem 9.1.1})$$

Prove these equations. These results will be derived somewhat differently in a discussion of laser cavities in Appendix C.

9.1.4 Suppose that we have two coherent optical systems using light of the same wavelength. System 2 has an electric field in the plane $z = 0$ that is the same as that for system 1, but with a scale change, that is,

$$\tilde{E}_2(x, y, 0) = \tilde{E}_1(mx, my, 0)$$

Show that, except for a constant phase factor, a Fresnel transformation gives

$$\tilde{E}_2(x', y', L) = \tilde{E}_1(mx', my', m^2 L)$$

9.2 THE GENERALIZED ABBE THEORY OF IMAGE FORMATION AND SPATIAL FILTERING

The Abbe theory of image formation with coherent light in the form usually discussed considers partially transmitting objects. The incident light beam is considered to be scattered into a superposition of plane waves in passing through the object. Each of these waves is brought to a focal spot in the back focal plane of the lens. The Huygens wavelets emerging from these spots then interfere in the image plane to produce the image.

Abbe's explanation of image formation with coherent light represents a substantial departure from the usual geometrical explanation of imaging, and it was not accepted at first. It was verified by numerous experiments, many performed with microscopes, many by Abbe himself.

A. THE ABBE THEORY OF IMAGE FORMATION

The basic ingredients for a discussion of the diffraction theory of image formation with coherent light are now at our disposal. We start with a few words about how the object is illuminated.

1. ***Condensers***. Consider a simple condenser such as is used in optical information processing. Two general types are shown in Fig. 9.10. In Fig. 9.10*a* the source, say a lamp filament, is

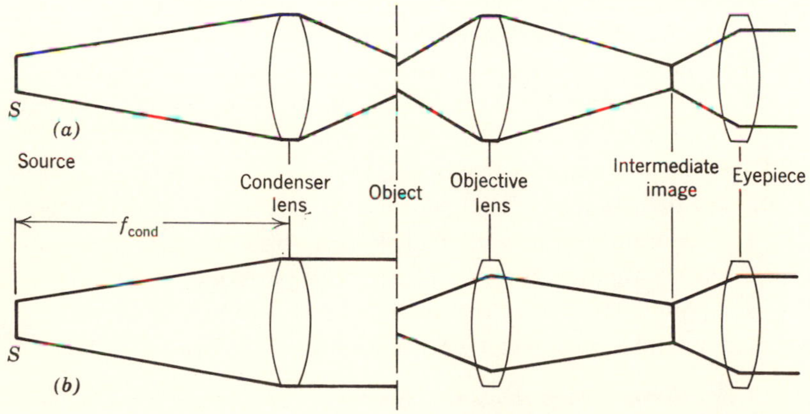

Fig. 9.10

focused directly onto the plane of the object. This gives nearly incoherent illumination and is not of primary interest here. When a very small source is defocused by the condenser to produce a parallel beam as in Fig. 9.10*b*, we obtain transverse coherence over the entire field of view if the source and object are sufficiently small. The simplest system of coherent illumination is with such a parallel beam. Such a beam can also be readily obtained with a gas laser. The effect of the condenser used this way is to put the source point P_s at infinity, as shown in Fig. 9.11*a*.

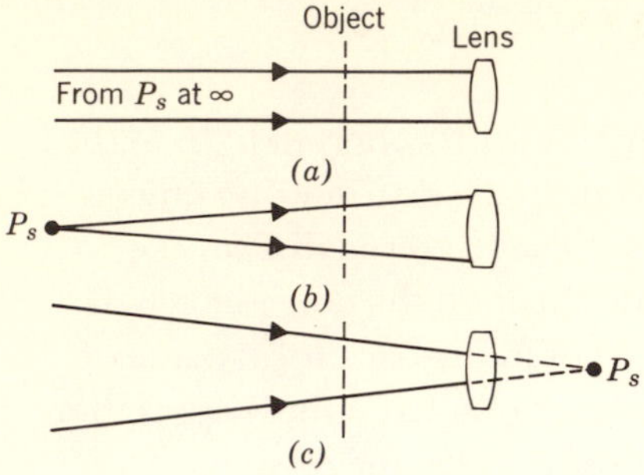

Fig. 9.11

The situation can be varied by changing the focus of the condenser lens to illuminate with a diverging beam from a real source as in Fig. 9.11*b*, or with a converging beam and hence a virtual source as in Fig. 9.11*c*.

2. The Two Essential Ideas of the Abbe Theory. Let the source point be at infinity as in Fig. 9.12. The conjugate plane is then the second focal plane of the lens; it is here that the interesting Fourier property of the image-forming process is revealed.

The two essential ideas of the Abbe theory now may be stated with reference to Fig. 9.12.

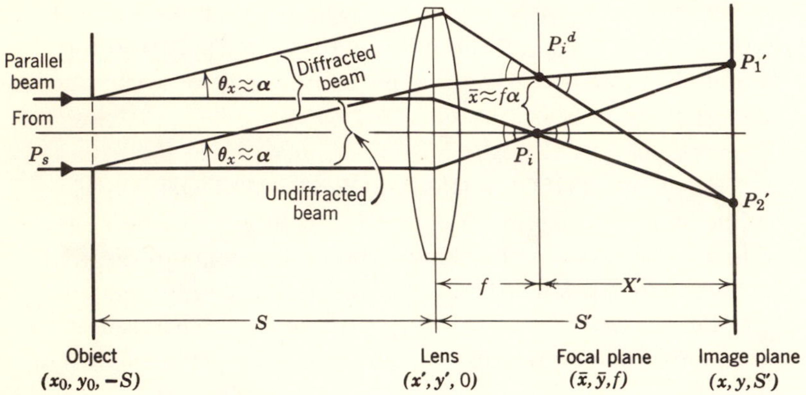

Fig. 9.12

1. With no object present, the incident parallel light beam will be focused at P_i, the image of P_s. With a partially transmitting object in place, the resulting diffracted beam may be described as a superposition of several parallel beams propagating in various directions. Each beam will be focused onto a spot P_i^d in the focal plane. The lens aperture broadens these spots into Airy disks or similar diffraction patterns, as discussed below.*

2. The optical signal will diverge from these spots in the focal plane and ultimately form an image of the original object in the image plane. This image results from interference among the Huygens wavelets emerging from the various spots.

3. Mathematical Statement of the First Part of the Abbe Theory

(a) *The process of diffraction into a superposition of parallel beams.* Consider a spot in the focal plane having coordinates $(\bar{x}, \bar{y}, f)$ with respect to an origin in the center of the second principal plane of the lens. Geometrical optics says that this spot will be the focal point for all rays incident onto the lens having direction cosines (α, β, γ) with α, β given by

$$\alpha = \frac{\bar{x}}{\sqrt{f^2 + \bar{x}^2 + \bar{y}^2}} \approx \frac{\bar{x}}{f}; \qquad \beta = \frac{\bar{y}}{\sqrt{f^2 + \bar{y}^2 + \bar{x}^2}} \approx \frac{\bar{y}}{f} \tag{9.18}$$

*Lens aberrations are neglected in the present discussion.

Wave optics says that the spatial dependence of plane waves propagating in this direction (α, β, γ) will be described by a complex electric field proportional to

$$e^{-ik(x\alpha + y\beta + z\gamma)}$$

For a fixed value of z we may write the x and y dependence in the form

$$e^{-ik(x\alpha+y\beta)} = e^{-2\pi i(\alpha x/\lambda + \beta y/\lambda)} = e^{-2\pi i(ux+vy)}$$

We can now turn the argument around. If we have monochromatic light with wave number $k = 2\pi/\lambda$ propagating from left to right so that the field distribution in a fixed transverse plane is proportional to $e^{-ik(x\alpha+y\beta)}$, we may say that this represents the value of a plane three-dimensional wave proportional to $e^{-ik(x\alpha+y\beta+z\gamma)}$ on this plane. The connection between the wave-optical and geometrical-optical pictures lies in the fact that this wave would be brought to a point focus in the focal plane at $(\bar{x}, \bar{y}, f)$ with $\bar{x}, \bar{y}$ given by Eqs. (9.18).

To obtain the desired decomposition of the light transmitted by the object into plane waves, we express the electric field in a plane just to the right of the object as a two-dimensional Fourier integral

$$\tilde{E}(x_0, y_0, -S) = \int\int_{-\infty}^{\infty} \hat{E}(-u, -v; -S)\, e^{-2\pi i(ux_0+vy_0)}\, du\, dv \qquad (9.19)$$

The third variable in $\hat{E}$, which appears after the semicolon, specifies the plane $z = \text{const}$ over which we calculate the two-dimensional Fourier transform. The two-dimensional plane waves appearing in Eq. (9.19) are said to possess the *spatial frequencies* u, v. Each separate component

$$\hat{E}(-u, -v; -S)\, e^{-2\pi i(ux_0+vy_0)} = \hat{E}(-u, -v; -S)\, e^{-ik(x_0\alpha+y_0\beta)}$$

can be interpreted as the value of the plane wave

$$\hat{E}(-u, -v; -S)\, e^{-ik[x_0\alpha+y_0\beta+(S+z)\gamma]}$$

in the plane $z = -S$.

This three-dimensional plane wave must satisfy the Helmholtz equation

$$(\nabla^2 + k^2)E = 0$$

This yields the condition

$$\alpha^2 + \beta^2 + \gamma^2 = 1.$$

(b) *Focusing of the parallel diffracted beams*. The plane wave will be brought to a focus at $(\bar{x}, \bar{y}, f)$ with $\bar{x} = \alpha f = \lambda uf$ and $\bar{y} = \beta f = \lambda vf$

(Eq. 8.68).* Indeed, we expect to be able to write the field at $(\bar{x}, \bar{y}, f)$ as proportional to this Fourier component:

$$\tilde{E}_{\mathrm{IA}}(\bar{x}, \bar{y}, f) \propto \hat{E}\left(\frac{-\bar{x}}{\lambda f}; \frac{-\bar{y}}{\lambda f}; -S\right)$$

The full expression can be obtained by setting $z = f$ and $\tau_{LA} = 1$ in Eq. (9.10) and performing the integrations over x', y', x_s, and y_s. The result is

$$\tilde{E}_{\mathrm{IA}}(\bar{x}, \bar{y}, f) = \frac{ie^{-ik(S+f)}}{\lambda f} e^{(ik/2X')(x^2+y^2)} \hat{E}\left(\frac{-\bar{x}}{\lambda f}, \frac{-\bar{y}}{\lambda f}; -S\right) \tag{9.20}$$

where $X' = S' - f$, as shown in Fig. 9.12.

This equation represents the mathematical statement of the first half of the Abbe theory, namely that the field distribution in the focal plane is proportional to the Fourier transform of the field at the object.

4. Mathematical Statement of the Second Part of the Abbe Theory

The second half of the Abbe theory says that the Huygens wavelets from the spots in the focal plane combine to give the correct field in the image plane. Mathematically, this superposition of Huygens wavelets is described by a Fresnel transformation from the plane $z = f$ to the plane $z = S'$. The following equation should therefore be valid:

$$\tilde{E}_{\mathrm{IA}}(x, y, S') = \frac{ie^{-ikX'}}{\lambda X'} \iint_{-\infty}^{\infty} \tilde{E}_{\mathrm{IA}}(\bar{x}, \bar{y}, f) \exp\left\{\frac{-ik}{2X'}[(x-\bar{x})^2 + (y-\bar{y})^2]\right\} \times d\bar{x}\, d\bar{y}$$

$$(X' = S' - f) \tag{9.21}$$

where $\tilde{E}_{\mathrm{IA}}(\bar{x}, \bar{y}, f)$ is given by Eq. (9.20) and where $\tilde{E}_{\mathrm{IA}}(x, y, S')$ is given by (9.15″). Equation (9.21) may be verified by expressing $\tilde{E}_{IA}(\bar{x}, \bar{y}, f)$ in the form given by (9.20) and then manipulating the exponent of the resulting integrand on the right-hand side of (9.21) into a convenient form for taking the inverse Fourier transform.

We now explicitly introduced the mathematical consequence of Fig. 9.12. The incident light beam is parallel to the axis and will be a constant E_0 across the plane $z = -S$. If the object has a trans-

*This statement holds true if $\gamma = \sqrt{1-\alpha^2-\beta^2} = \sqrt{1-u^2\lambda^2-v^2\lambda^2}$ is real, i.e., if $(u^2+v^2) < 1/\lambda^2$. A sufficient condition is that $\hat{E}(-u, -v; -S) \approx 0$ for spatial frequencies u, v larger in magnitude than $1/\lambda$. This condition is consistent with the assumptions behind the Fresnel-Kirchoff diffraction theory, for the only way to obtain a field containing high spatial frequencies greater than $1/\lambda$ is by passing light through an aperture having structure on a scale smaller than λ. Such an aperture cannot be treated by ordinary diffraction theory.

mission function $\tau_0(x_0, y_0)$, just to the right the field will be

$$\tilde{E}(x_0, y_0, -S) = E_0\tau_0(x_0, y_0) \tag{9.22a}$$

Then, by Eq. (9.15″) the conjugate image field is

$$\boxed{\tilde{E}_{IA}(x, y, S') = e^{-ik(S+S')} \frac{e^{-(ik/2X')(x^2+y^2)}}{m} E_0\tau_0\left(\frac{x}{m}, \frac{y}{m}\right)} \tag{9.22b}$$

and the field in the focal plane is

$$\boxed{\tilde{E}_{IA}(\bar{x}, \bar{y}, f) = \frac{ie^{-ik(S+f)}}{\lambda f} e^{(ik/2X')(\bar{x}^2+\bar{y}^2)} E_0\hat{\tau}_0\left(\frac{-\bar{x}}{\lambda f}, \frac{-\bar{y}}{\lambda f}\right)} \tag{9.22c}$$

5. ***Some Implications.*** The phase factor

$$e^{(ik/2X')(\bar{x}^2+\bar{y}^2)}$$

in Eq. (9.22*c*) contains all the important information about the location of the object and is essential if a Fresnel transformation applied to (9.22*c*) is to yield (9.22*b*). The distribution of flux density in the focal plane will be proportional to $|\tilde{E}_A(x, y, f)|^2$, which no longer contains information about the location of the object. *The pattern of flux density is independent of object position* (Fig. 9.13). This result can be readily verified in the laboratory.

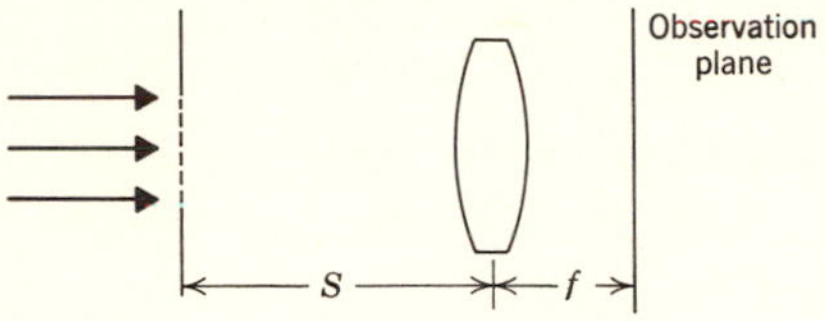

Fig. 9.13. For a large lens, the distribution of flux density in the second focal plane is independent of the object distance S.

One method of recording the phase in the field at $z = f$ is to make a hologram by adding a reference beam to the field there and recording the resulting interference pattern, as described above in Section 8.3.

Note that there is a further simplification when the object is at the first focal plane of the lens, for then $S = f$, $S' = \infty$, and $X' = \infty$. Our phase factor is then unity, and the field in the second focal plane is given by the simple expression

$$\tilde{E}_{IA}(\bar{x}, \bar{y}, f) = \frac{ie^{-ik2f}}{\lambda f} E_0\hat{\tau}_0\left(\frac{-\bar{x}}{\lambda f}, \frac{-\bar{y}}{\lambda f}\right) \tag{9.23}$$

6. ***Statement of the Abbe Theory with Nonparallel Incident Light.*** As mentioned in our discussion of condensers, the point source P_s need not be at infinity. When it is brought in a finite distance $D > f$ from the lens, its image will be found at P_i a distance $D' > f$ to the right of the lens. With no object at $z = -S$, the image plane at $z = S'$ will be essentially uniformly illuminated with light diverging from P_i. Now put in a semitransparent object. If the transmission function $\tau_0(x_0, y_0)$ has a Fourier component at spatial frequency $u = \alpha/\lambda$, a diffracted beam is produced that for small α can be obtained from the original beam by rotation through the angle α, as sketched in Fig. 9.14. This diffracted beam will be focused at

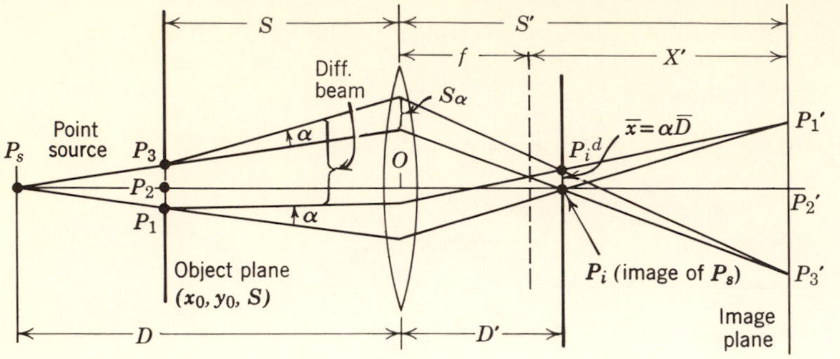

Fig. 9.14. Modification of Fig. 9.12 when the point source P_s moves in from infinity.

P_i^d a distance $\bar{x} = \alpha\bar{D}$ from the axis, where $\bar{D}$ can be written

$$\bar{D} = D' + S\left(1 - \frac{D'}{f}\right) = \frac{(S' - D')S}{S'} = \frac{(S' - D')}{-m}$$

$$\left(m = -\frac{S'}{S}\right)$$

The exact expression for the field in the plane $z = D'$ is given by Eq. (9.28) below. Secondary wavelets from points such as P_i^d then combine with those from P_i to produce the image of the original object in the conjugate image plane.

7. *Mathematical Results for Nonparallel Incident Light.* There is now a spherical wave $A\,e^{-ikR}/R$ leaving the source P_s. In the parabolic approximation the incident field at the object plane is

$$A\frac{e^{-ik(D-S)}}{(D-S)}\exp\left[-\frac{ik}{2(D-S)}(x_0^2 + y_0^2)\right]$$

After modulation by the object the field is

$$\tilde{E}(x_0, y_0, -S) = E_0 \exp\left[-\frac{ik}{2(D-S)}(x_0^2 + y_0^2)\right]\tau_0(x_0, y_0) \tag{9.24}$$

where

$$E_0 = A_0\frac{e^{-ik(D-S)}}{(D-S)} \tag{9.25}$$

represents the incident field at the object on the axis at P_2.

We first find the field in the image plane. Equation (9.15″) gives

$$\tilde{E}_{\text{IA}}(x, y, S') = \frac{E_0}{m}e^{-ik(S+S')}\exp\left[-i\frac{k}{2(D-S)m^2}(x^2 + y^2)\right]$$
$$\times \exp\left[-i\frac{k}{2X'}(x^2 + y^2)\right]\tau_0\left(\frac{x}{m}, \frac{y}{m}\right). \tag{9.26}$$

Consider the point P_1' a distance $\sqrt{x^2 + y^2}$ from the axis in Fig. 9.14. The phase shift in (9.26) described by the factor

$$\exp\left\{-\frac{ik}{2}\left[\frac{x^2 + y^2}{(D-S)m^2} + \frac{x^2 + y^2}{X'}\right]\right\}$$

should agree with the phase shift caused by propagation from P_s to P_1' minus that caused by propagation from P_s to P_2', namely,

$$-ik(|P_iP_1'|-|P_iP_2'|) = -\frac{ik}{2(S'-D')}|P_1'P_2'|^2 = \frac{-ik}{2(S'-D')}(x^2+y^2)$$

(Here we have used the fact that all optical path lengths from P_s to the conjugate point P_i are equal.) This will agree with the phase in (9.26), provided that

$$\frac{1}{X'}+\frac{1}{m^2(D-S)}=\frac{1}{S'-D'}$$

Some algebraic manipulations, omitted here, will verify this relation. Then Eq. (9.26) can be written

$$\boxed{\tilde{E}_{IA}(x,y,S') = \frac{E_0}{m}e^{-ik(S+S')}\exp\left[-i\frac{k(x^2+y^2)}{2(S'-D')}\right]\tau_0\left(\frac{x}{m},\frac{y}{m}\right)} \tag{9.27}$$

This gives the field in the image plane in terms of the scaled transmission function of the object.

Equation (9.27) is a generalization of (9.22*b*) to the case $D' \neq f$. It is interesting to note that the field it yields is exactly what would be obtained just behind an object at $z = S'$ (that is, in "image space") that has a transmission function $\tau_0(x/m, y/m)$. It would be impossible to distinguish between the two cases by any observation made farther along the z axis. This statement holds only if the lens aperture is neglected.

The various diffracted beams are brought to a focus in the plane $z = D'$. The field in this plane is given by

$$\boxed{\begin{aligned}\tilde{E}_{IA}(x,y,D') &= \frac{-ime^{-ik(S+D')}}{\lambda(S'-D')}E_0\exp\left[\frac{ik(x^2+y^2)}{2(S'-D')}\right]\\ &\times\hat{\tau}_0\left[\frac{m\bar{x}}{\lambda(S'-D')},\frac{m\bar{y}}{\lambda(S'-D')}\right]\end{aligned}} \tag{9.28}$$

Equation (9.28) is a generalization of Eq. (9.22*c*) to the case $D' \neq f$, and of course it reduces to the latter equation when $D' = f$, because then $S'-D' = X'$, and $m = -X'/f$, so that $m/(S'-D') = -1/f$.

When the object is in the first focal plane of the lens, we have $S \to f$, $S' \to \infty$, $m = -S'/S \to -\infty$, and $m/(S'-D') \to -1/S = -1/f$. Then Eq. (9.28) becomes

$$\tilde{E}_{IA}(\bar{x},\bar{y},D') = \frac{ie^{-ik(f+D')}}{\lambda f}E_0\hat{\tau}_0\left(\frac{-\bar{x}}{\lambda f},\frac{-\bar{y}}{\lambda f}\right) \tag{9.28$'$}$$

which is very similar to Eq. (9.23).

8. Periodic Transmission Function (Diffraction Grating). Suppose that the transmission function $\tau_0(x, y)$ depends only on x and has period l. Then

$$\tau_0(x, y) = \tau(x) = \tau(x+l) \tag{9.29a}$$

Of course real objects will be finite in extent, but this case is useful for illustration of the Abbe theory; indeed, Abbe himself discussed it in some detail. According to Eqs. (7.89) we can expand $\tau(x)$ in a Fourier series

$$\tau(x) = \sum_{n=-\infty}^{\infty} \tau_n e^{2\pi inx/l} \tag{9.29b}$$

with

$$\tau_n = \frac{1}{l} \int_{-l/2}^{+l/2} \tau(z) e^{-2\pi inz/l}\, dz \tag{9.29c}$$

Equations (7.88′) and (7.89*b*) show that the fourier transform of $\tau(x)$ may be written

$$\hat{\tau}(u) = \sum_{n=-\infty} \delta\left(u - \frac{n}{l}\right) \tau_n \tag{9.29d}$$

This is the limiting behavior of $\tau(u)$ as N, the number of elements in the grating, goes to infinity. For finite N the delta functions have a finite width $\Delta u = 1/(Nl)$.

In the limit where the object has infinite spatial extent in the $\pm y$ direction, the v-dependent part of the two-dimensional Fourier transform is simply $\delta(v)$. We then obtain

$$\hat{\tau}_0(u, v) = \delta(v)\hat{\tau}(u) = \sum_{-\infty}^{\infty} \delta(v)\,\delta\left(u - \frac{n}{l}\right) \tau_n \tag{9.29e}$$

Equation (9.28) then tells us that the diffracted field in the plane $z = D'$ is a sum of delta functions:

$$\tilde{E}_{\mathrm{IA}}(\bar{x}, \bar{y}, D') = -ie^{-ik(S+D')} \frac{\lambda(S' - D')}{m} \sum_{-\infty}^{+\infty} \exp\left[\frac{ikx_n^2}{2(S' - D')}\right] \times \delta(\bar{y})\,\delta(\bar{x} - x_{\hat{n}})\,\tau_{\hat{n}} \tag{9.29f}$$

where

$$x_n = nx_1 = \frac{n\lambda(S' - D')}{ml} \tag{9.29g}$$

This result is illustrated schematically in Fig. 9.15. The nth-order Fourier component is focused at P_n. For simplicity only three orders $n = 0, \pm 1$ are shown.

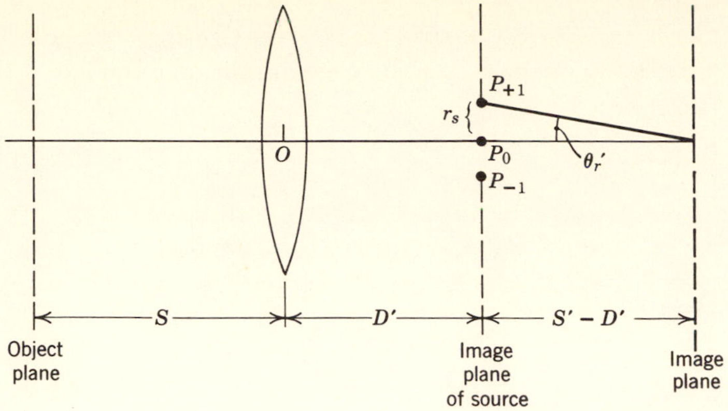

Fig. 9.15

Example (a): *Sinusoidal transmission grating*. Consider a diffraction grating with a transmission function of the form

$$\tau(x) = \frac{a}{2}\left[1 + b\sin\left(\frac{2\pi x}{l} + \phi\right)\right]$$

$$= \frac{a}{2} + \frac{ab}{4i}e^{i\phi}e^{2\pi i x/l} - \frac{ab}{4i}e^{-i\phi}e^{-2\pi i x/l} \qquad (9.30a)$$

The quantity b is called the *modulation*. It is positive and less than or equal to one. There are exactly three Fourier coefficients given by

$$\tau_0 = \frac{a}{2}, \qquad \tau_{+1} = \frac{ab}{4i}e^{i\phi}, \qquad \tau_{-1} = -\frac{ab}{4i}e^{-i\phi} \qquad (9.30b)$$

This yields three diffraction spots that correspond to the three terms in the expression

$$E_{IA}(\bar{x}, \bar{y}, D') = -ie^{-ik(S+D')}\frac{\lambda(S'-D')}{m}\delta(y)\left\{\frac{a}{2}\delta(x) + \exp\left[\frac{ikx_1^2}{2(S'-D')}\right]\right.$$
$$\left.\times\left[\frac{ab}{4i}e^{i\phi}\delta(\bar{x}-x_1) - \frac{ab}{4i}e^{-i\phi}\delta(\bar{x}+x_1)\right]\right\},$$
$$\left[x_1 = \frac{\lambda(S'-D')}{ml}\right] \qquad (9.30c)$$

Example (b): *Square transmission grating*. Let $\tau(x)$ have the form shown in Fig. 9.16*a*. The Fourier coefficients may be obtained from Eq. (9.29*c*):

$$\tau_n = \frac{1}{l}\int_{-l/2}^{-l/2}\tau(z)e^{-2\pi i n z/l}dz \equiv \frac{1}{l}\int_0^{l/2}e^{-2\pi i n z/l}\,dz = \frac{1}{2} \qquad \text{for} \quad n = 0$$
$$= \frac{1}{\pi i n}, \qquad n \text{ odd} \qquad (9.31a)$$
$$= 0, \qquad n \text{ even}, \quad \neq 0$$

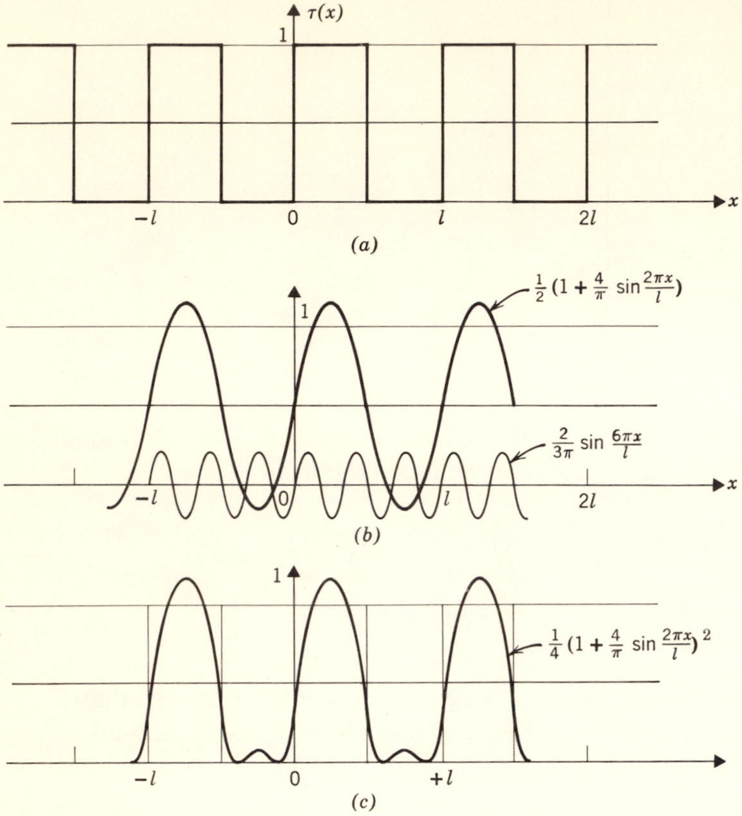

Fig. 9.16

When we combine the exponentials of order $n = \pm(2p+1)$, we obtain

$$\tau(x) = \frac{1}{2}\left[1 + \frac{4}{\pi}\sum_{p=0}^{\infty}\frac{1}{(2p+1)}\sin\frac{2\pi(2p+1)x}{l}\right] \tag{9.31b}$$

Note that if we truncate the series in (9.31*b*) after $p = 0$, we obtain (9.30*a*) with $a = 1$, $b = 4/\pi$, $\phi = 0$. The sum of the first two terms in (9.31*b*) is sketched in Fig. 9.16*b*, where it can be compared with the complete function $\tau(x)$. The next term

$$\frac{2}{3\pi}\sin\frac{6\pi x}{l}$$

is also sketched in.

In general, the more terms we take in the series, the more closely it approximates the square wave $\tau(x)$. The term

$$\frac{2}{\pi n}\sin\frac{2\pi n x}{l} \qquad (n \text{ odd})$$

yields the $\pm\, n$th order diffracted spots P_{+n} and P_{-n} in the plane $z = D'$. In a real optical system there must be a limit to how far we may go off axis; hence, beyond a certain order the higher diffraction spots will be blocked. This alters the shape of the final image.

For instance, if only P_0, P_1, and P_{-1} pass through, the result will be the same as if we only had these three orders in the first place, that is, if we had a transmission in the object plane given by

$$\tau(x) = \frac{1}{2} + \frac{2}{\pi} \sin \frac{2\pi x}{l}$$

as plotted in Fig. 9.16*b*. The flux density in the image plane at $z = S'$ will be proportional to

$$\left(\frac{1}{2} + \frac{2}{\pi} \sin \frac{2\pi x}{l'}\right)^2$$

where $l' = ml = -(S'/S)l$ is the magnified grating constant. This function referred to the object plane is plotted in Fig. 9.16*c*.

We now turn to a more complete discussion of the processing or filtering that occurs at the image when certain Fourier components are blocked or otherwise altered.

B. SPATIAL FILTERING

1. ***Introduction.*** Consider again the situation in Fig. 9.14. The field at the plane $z = D'$ is proportional to the Fourier transform of the original transmission function $\tau_0(x_0, y_0)$. As ordinarily written, the independent variables (u, v) of this transform

$$\hat{\tau}_0(u, v) = \iint_{-\infty}^{\infty} \tau_0(x_0, y_0) e^{-2\pi i(ux_0 + vy_0)}\, dx_0\, dy_0$$

have the dimensions of reciprocal length and are called spatial frequencies. The connection between the spatial frequencies u, v and the coordinates $\bar{x}$, $\bar{y}$ in the plane $z = D'$ appears in Eq. (9.28) and is given by

$$u = \frac{m\bar{x}}{\lambda(S' - D')} = \frac{(f - S')\bar{x}}{\lambda f(S' - D')}$$

$$v = \frac{m\bar{y}}{\lambda(S' - D')} = \frac{(f - S')\bar{y}}{\lambda f(S' - D')} \tag{9.32}$$

The phase factor

$$\exp\left[\frac{ik(\bar{x}^2 + \bar{y}^2)}{2(S' - D')}\right]$$

in Eq. (9.28) combines with that of a Fresnel transformation from the plane $z = D'$ to the plane $z = S'$ in such a way that the inverse Fourier transform is taken, and Eq. (9.27) results.

The details of this argument will now be given. The Fresnel transformation is

$$\tilde{E}_{IA}(x, y, S') = \frac{ie^{-ik(S'-D')}}{\lambda(S'-D')} \iint\limits_{-\infty}^{\infty} \tilde{E}_{IA}(\bar{x}, \bar{y}, D') \exp\left\{-\frac{ik}{2(S'-D')} \times [(\bar{x}-x)^2 + (\bar{y}-y)^2]\right\} d\bar{x}\, d\bar{y}$$

with $\tilde{E}_{IA}(\bar{x}, \bar{y}, D')$ given by (9.28). The total resulting exponent is

$$-ik(S'-D') - ik(S+D') - \frac{ik}{2(S'-D')}[(\bar{x}-x)^2 - \bar{x}^2 + (\bar{y}-y)^2 - \bar{y}^2]$$

$$= -ik(S+S') - \frac{ik}{2(S'-D')}(x^2+y^2) + \frac{2\pi i}{\lambda(S'-D')}(x\bar{x} + y\bar{y})$$

This gives

$$\tilde{E}_{IA}(x, y, S') = E_0 e^{-ik(S+S')} \frac{e^{-[ik/2(S'-D')](x^2+y^2)}m}{\lambda^2(S'-D')^2}$$

$$\times \iint\limits_{-\infty}^{\infty} \hat{\tau}_0\left[\frac{m\bar{x}}{\lambda(S'-D')}, \frac{m\bar{y}}{\lambda(S'-D')}\right] \exp\left\{2\pi i\left[\frac{(x\bar{x}+y\bar{y})}{\lambda(S'-D)}\right]\right\} d\bar{x}\, d\bar{y}$$

If we use Eqs. (9.32), we obtain

$$\tilde{E}_{IA}(x, y, S')$$

$$= \frac{E_0 e^{-ik(S+S')} e^{-[ik/2(S'-D')](x^2+y^2)}}{m} \iint\limits_{-\infty}^{\infty} \hat{\tau}_0(u, v) e^{2\pi i(ux/m + vy/m)}\, du\, dv \qquad (9.33)$$

The double integral gives simply $\tau_0(x/m, y/m)$. Hence (9.33) gives (9.27).

We now insert a "filter" in the form of a thin, partially transmitting sheet at the plane $z = D'$ where the spatial frequencies appear. If this sheet has a transmission function $\tau_F(\bar{x}, \bar{y})$, the field just beyond it is obtained by multiplying Eq. (9.28) by τ_F. The Fourier transform $\tau_0(u, v)$ is then replaced by the product

$$\hat{\tau}_0(u, v) \cdot \tau_F\left[\frac{\lambda(S'-D')u}{m}, \frac{\lambda(S'-D')v}{m}\right]$$

This has the effect of "filtering" each Fourier component by the factor

$$\tau_F\left[\frac{\lambda(S'-D')u}{m}, \frac{\lambda(S'-D')v}{m}\right]$$

This process is called spatial filtering.

The function

$$\tau_F\left[\frac{\lambda(S'-D')u}{m}, \frac{\lambda(S'-D')v}{m}\right]$$

must also be inserted into the integrand of Eq. (9.33). Many important cases of spatial filtering may be treated without having to compute the integral. We shall now discuss four of them.

2. Spatial Filtering by a Finite Aperture—Resolving Power. Suppose that the aperture stop of the system is at $z = D'$. If it has a radius r_s, we have

$$\begin{aligned} \tau_F(\bar{x}, \bar{y}) &= 1 \qquad \text{if } \bar{x}^2 + \bar{y}^2 < r_s^2 \\ &= 0 \qquad \text{otherwise} \end{aligned}$$

This has the effect of blocking all spatial frequencies u, v, that satisfy

$$u^2 + v^2 > w^2, \qquad \text{with} \quad w = \frac{|m|r_s}{\lambda(S'-D')} \tag{9.34}$$

The integral in Eq. (9.33) then must be taken over the limited domain $\sqrt{u^2+v^2} < w$. The resulting expression can be reworked into a form that will be derived below when we treat the effect of apertures.

We discussed above how a one-dimensional periodic object yields diffraction spots in the plane $z = D'$. The maximum order $n_{\max}$ passed by our circular aperture will be the largest integer that satisfies

$$\frac{n}{l} \leqslant w = \frac{|m|r_s}{\lambda(S'-D')} \tag{9.35}$$

This has the effect of truncating the Fourier series (9.29*b*) at $\pm n_{\max}$.

The resolving power of a coherently illuminated optical system is quite difficult to define. An early criterion was developed by Abbe as an outgrowth of his diffraction theory of the image formation of a periodic object. He argued that the zero- and first-order diffraction spots are necessary to preserve a semblance of the original periodic signal. Figure 9.16 shows that this is indeed the case. The limit is reached, as shown in Fig. 9.15, when P_{+1} and P_{-1} are just inside the aperture. This happens when the image grating has spacing

$$l' = ml = \frac{\lambda(S'-D')}{r_s} \approx \frac{\lambda}{\theta'_m} \tag{9.36}$$

Equation (9.36) represents one expression for the limiting resolution of our system; a grating with a smaller spacing would not be resolved at all. We can show that for large angle systems obeying the sine condition we should replace θ_m' by $\sin\theta_m'$. If the result is transformed to object space, where the index is n, the minimum resolvable value of the grating spacing is

$$l_{\min} = \frac{\lambda}{n \sin \theta_m} \tag{9.37a}$$

This expression is often taken as the definition of the resolution of a microscope under coherent illumination at normal incidence (P_0 is on axis), but it has a clear significance only for a periodic object. Note that $l_{\min}$ in (9.37*a*) is 1.6 times greater than the value

$$x_{S,\min} = \frac{0.61\,\lambda}{n \sin \theta_m} \tag{9.37b}$$

given for incoherent illumination in Eq. (7.66).

If we are willing to suffer a loss of contrast and other sometimes undesirable effects, resolution can be improved by oblique illumination with coherent light. Suppose that the zero-order diffracted spot P_0 in Fig. 9.15 is no longer on axis, but is moved down until it is just inside the lower rim of the aperture where P_{-1} used to be. Then if P_{+1} is still just inside the upper rim, we have doubled the limiting spatial frequency and therefore halved the resolution limit to

$$l_{\min} = \frac{0.50\,\lambda}{n \sin \theta_m} \tag{9.37c}$$

which is 22 per cent better than the value for incoherent illumination.

With this oblique arrangement all diffraction spots of negative order are blocked. This will tend to reduce contrast, unless the zero-order spot is artificially attenuated. The asymmetry in the missing orders may produce other effects, but we shall not pursue them here. For more details the reader is referred to the book by Martin cited in the list of references.

3. Spatial Filtering of a Laser Beam. Spatial filtering may be used to smooth the output beam from a gas laser. Except for inevitable spreading because of diffraction (see Problem 9.1.1), the beam may be rendered parallel, and when operated in a uniphase mode the radial dependence of the field in the beam will be approximately Gaussian. We assume that it is exactly Gaussian and write for this ideal case

$$\tilde{E}_{\text{ideal}}(x_s, y_s, -f) = Be^{-(x_s^2 + y_s^2)/\rho^2} \tag{9.38}$$

We have chosen to evaluate the field in the plane $z = -f$. Although B and ρ are constants that may be complex, we treat them as real for simplicity. Here ρ is a measure of the radius of the beam.

Since there is no partially transmitting object in the plane $z=-f$, we may simply replace the product $E_0\tau_0(x_s, y_s)$ in the above equations by E_{ideal}. The one-dimensional Fourier transform of $\exp(-x_s^2/\rho^2)$ was calculated in Chapter 6 (compare Eqs. 6.43 and 6.45 with $\nu_0 = 0$) and was found to be

$$\sqrt{\pi}\rho e^{-\pi^2u^2\rho^2}$$

Thus

$$\hat{E}_{\text{ideal}}(u, v; -f) = B\pi\rho^2\, e^{-\pi^2\rho^2(u^2+v^2)} \tag{9.39a}$$

If we neglect the effects of a finite lens aperture, we would have a field distribution in the plane $z=f$ given by Eq. (9.33), or

$$\tilde{E}_{\text{ideal}}(\bar{x}, \bar{y}, f) = \frac{ie^{-ik(S+f)}}{\lambda f} B\pi\rho^2 \exp\left[\frac{-\pi^2\rho^2(\bar{x}^2+\bar{y}^2)}{(\lambda f)^2}\right] \tag{9.39b}$$

Equation (9.39*b*) says that the field distribution in the focal plane is also Gaussian. This field distribution is limited to values of $\bar{r} = \sqrt{\bar{x}^2+\bar{y}^2}$ not much larger than $\lambda f/\rho$. The situation is indicated schematically in Fig. 9.17. A second lens L' can then be used to give a nearly parallel beam having a "radius parameter" $\rho' = f'\rho/f$.

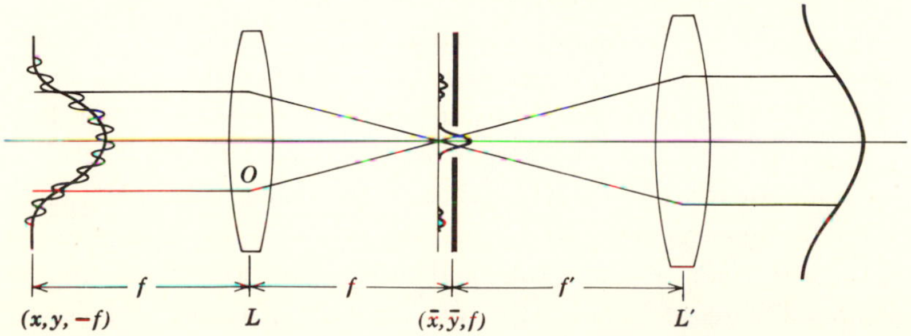

Fig. 9.17

In actual cases there are departures from the ideal situation just described because of inhomogeneities in the glass used for windows and mirrors within the laser and because of dust. The actual field distribution has a "rough" part owing to these effects superposed on the smooth ideal distribution and may be written

$$\tilde{E}_{\text{actual}} = \tilde{E}_{\text{ideal}} + \Delta\tilde{E}$$

As shown in Fig. 9.18*a*, $\Delta\tilde{E}$ varies rapidly and randomly over distances small compared with ρ. Let $l_0 \ll \rho$ be a characteristic wavelength for a fluctuation in $\Delta\tilde{E}$. Whereas the Fourier transform $\hat{E}_{\text{ideal}}$ contains spatial frequencies from zero to about $1/\rho$, as given by Eq. (9.39*a*), the transform $\Delta\hat{E}$ will contain much higher frequencies centered about $1/l_0$. The resultant distribution of irradiance is sketched in Fig. 9.18*b*. We readily see that we can remove

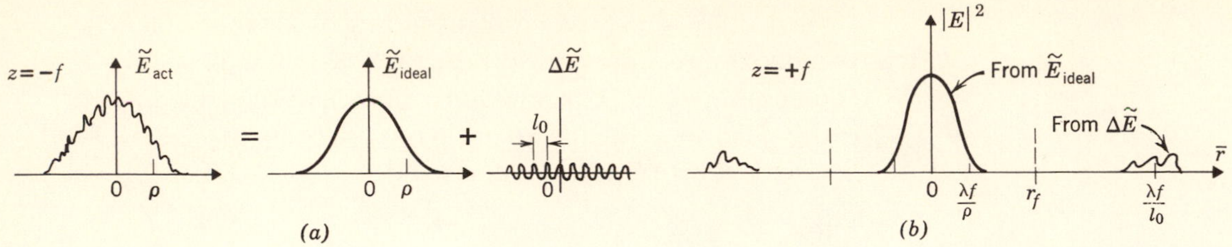

Fig. 9.18

the Fourier components of the noise $\Delta\hat{E}$ if we insert a pinhole in the second focal plane of L having a radius r_f that lies between $\lambda f/\rho$ and $\lambda f/l_0$. The field distribution in the second focal plane of L' will then be a smooth Gaussian corresponding to $\tilde{E}_{\text{ideal}}$ alone.

Note that we have not used the explicit Gaussian form for $\tilde{E}_{\text{ideal}}$ – we require only that $\tilde{E}_{\text{ideal}}(r)$ be smooth and have a radius parameter ρ, so that the Fourier transform is limited to spatial frequencies about equal to $1/\rho$ or less.

Photographs and photometric scans of the intensity distributions of laser beams with and without spatial filtering are presented in Fig. 9.19.

2. Spatial Filtering to Remove Unwanted Ordered Structure. Unwanted structure or spurious structure, such as scratches that have regularities, in a transparency can often be removed by spatial filtering. The transparency is used as an object in the set-up of Fig. 9.10 or 9.12. The unwanted structure must have a Fourier transform narrowly limited as a function of the spatial frequencies u, v. Then the region in the filtering plane $z = f$ or $z = D'$ corresponding to these frequencies can be blocked and the spurious structure removed from the final image. For instance, consider a series of parallel lines, not necessarily equally spaced, oriented at an angle θ from the horizontal, as shown in Fig. 9.20*a*. Each line will yield a Fraunhofer diffraction pattern quite like that from a long, narrow slit. This will be centered on the axis of the system and will be elongated in the direction θ from the vertical, as shown in Fig. 9.20*b*. To block these Fourier components we need a diaphragm in the shape of two narrow wedges which do not quite meet at P_i, as shown in Fig. 9.20*c*.

Actual results of spatial filtering of this type are shown in the photographs of Plate 3. The top photograph is a picture of the surface of the moon as telemetered to earth by an unmanned lunar orbiter. The "grating" of equally spaced lines is an artifact of the telemetering process. Figure 9.21 shows the diffraction pattern in the filtering plane when a negative transparency of the moon photograph is illuminated as in Fig. 9.10. The equally spaced large dots represent diffraction maxima resulting from the

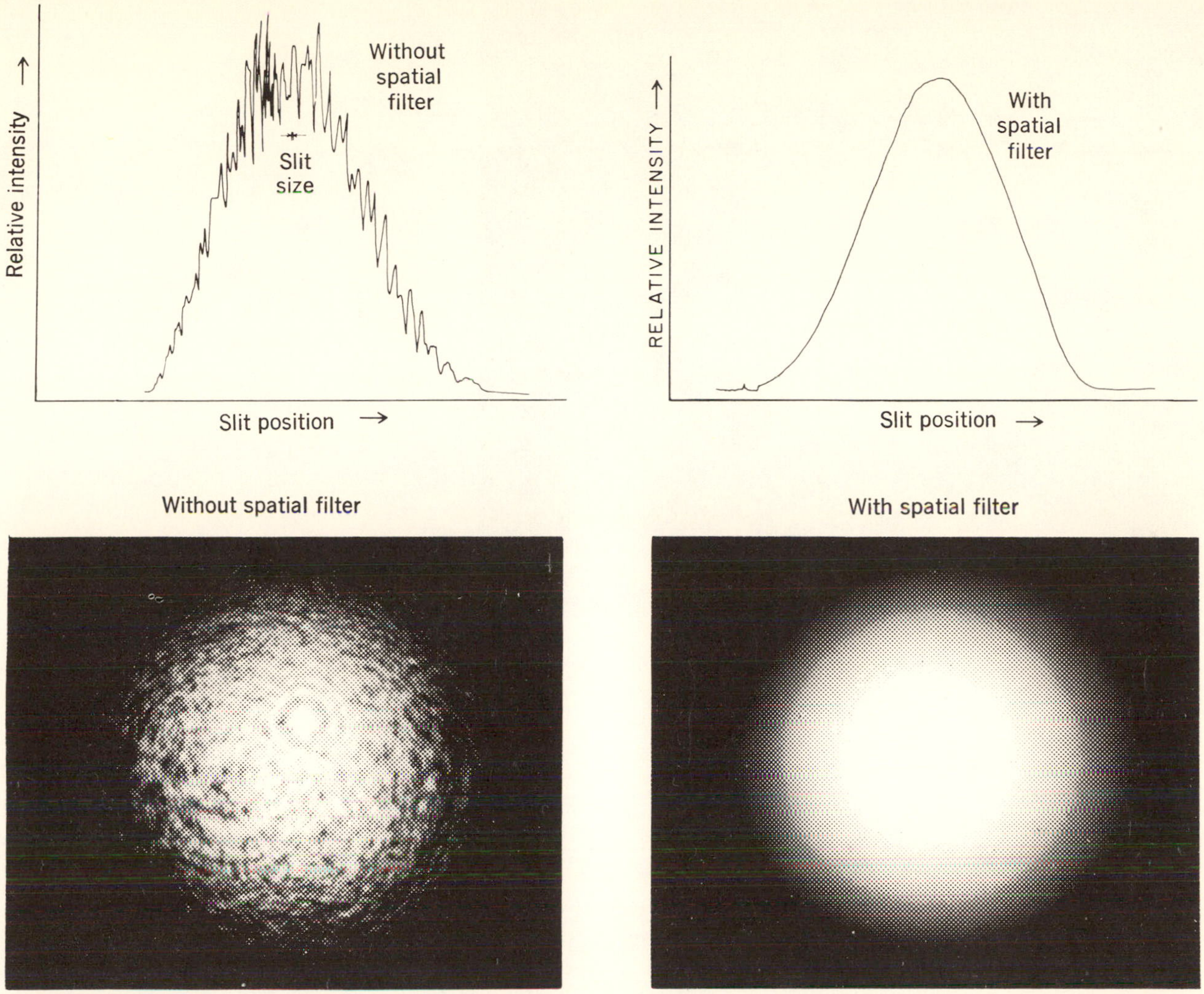

Fig. 9.19. The effect of spatial filtering on a uniphase laser beam. The plots show the results of scanning across the beam with a narrow slit in front of a photometer. (Illustrations courtesy of Spectra-Physics, Inc.)

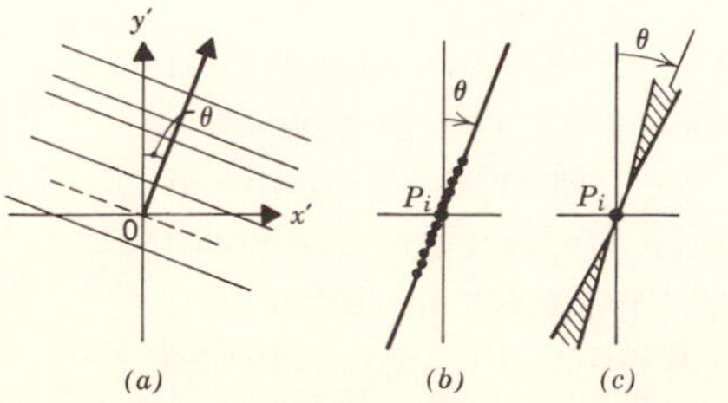

Fig. 9.20. (*a*) Object. (*b*) Diffraction pattern from (*a*) in filtering plane. (*c*) Diaphragm.

"grating" in the moon photograph and can be blocked in the filtering plane by the diaphragm of Fig. 9.20*c* with $\theta = \pi/2$. A positive photograph can be made in the image plane that is the original minus the unwanted grating (Plate 3).

5. ***The Phase Contrast Microscope.*** The phase contrast microscope was invented by Frits Zernike in 1935. He explained its operation in terms of the Abbe theory of image formation. A modern explanation can be readily cast in terms of spatial filtering.

Many biological and other microscope specimens are almost completely transparent, so that they do not modulate the amplitude of a light beam passing through them; only the phase is modulated. The transmission function of a microscope slide containing such

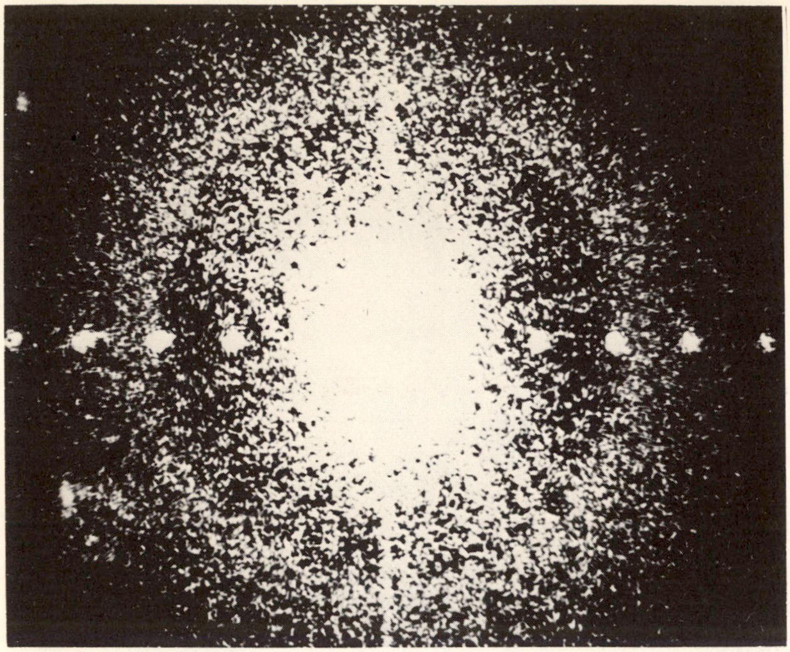

Fig. 9.21. Fraunhofer diffraction pattern (two-dimensional distribution of flux density versus spatial frequencies) for the unfiltered moon photograph at the top of Plate 3. (Courtesy of Conduction Corp.)

specimens has the form

$$\tau_0(x_0, y_0) = e^{i\phi(x_0, y_0)}$$

where

$$\phi = \frac{-2\pi}{\lambda}[n(x_0, y_0) - 1]\, d(x_0, y_0) \tag{9.40}$$

Here n and d represent the local values of the refractive index and thickness of the specimen. Since the human eye responds to amplitude, not phase, these samples will appear invisible with an ordinary microscope. The problem is to convert somehow the *phase contrast* represented by Eq. (9.40) into amplitude contrast.

Let the microscope be illuminated coherently with light from an effective point source at P_s, as discussed in connection with Figs. 9.10 and 9.11. Figure 9.12 will then apply to the objective lens of the microscope. The image it forms at $z = S'$ then acts as an object for the eyepiece, not shown. If the object has a transmission function given by Eq. (9.40), the flux density in the image plane will be constant, and the object will be invisible. The object can be rendered visible if a spatial filter in the form of a *phase plate* is inserted in the plane $z = D'$ where the image of the source is formed.

The analysis is simplified if we assume that $\phi(x_0, y_0)$ is much less than unity, as is quite often the case. Then we may write, to first order in ϕ,

$$\tau_0(x_0, y_0) \simeq 1 + i\phi(x_0, y_0) \tag{9.40'}$$

The first term (1) in (9.40′) yields the undiffracted spot at P_i; the second term ($i\phi$) yields all the diffracted spots such as $P_i{}^d$. Each of these spots is actually an Airy disk with rings, but this is not relevant to our argument.

Without spatial filtering the electric field in the image plane is given by Eq. (9.26) with

$$\tau_0\left(\frac{x}{m}, \frac{y}{m}\right) \cong 1 + i\phi\left(\frac{x}{m}, \frac{y}{m}\right) \tag{9.41a}$$

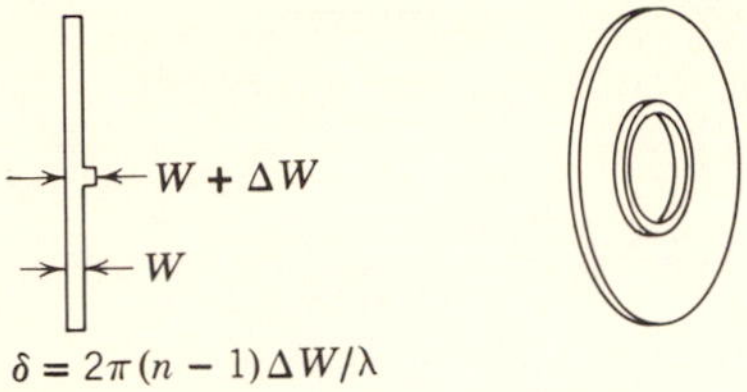

(*a*) (*b*)

Fig. 9.22

and

$$|\tau_0|^2 \cong 1$$

The phase plate is made so that it changes the phase of the spot at P_i by an angle $-\delta$ with respect to that of all the diffracted spots $P_i{}^d$. One way of doing this is shown in Fig. 9.22*a*. The first term in Eq. (9.41*a*) is then changed from unity to $e^{-i\delta}$.

Suppose that we have a quarter-wave phase plate so that $\delta = \pi/2$. Then

$$e^{-i\delta} = e^{-i\pi/2} = -i$$

and (9.41*a*) becomes

$$\tau_0\left(\frac{x}{m}, \frac{y}{m}\right) = i\left[1 - \phi\left(\frac{x}{m}, \frac{y}{m}\right)\right] \tag{9.41b}$$

Now to first order in ϕ the irradiance will be proportional to

$$|\tau_0|^2 = 1 - 2\phi\left(\frac{x}{m}, \frac{y}{m}\right) \tag{9.41c}$$

We see that the phase contrast has indeed been converted to amplitude contrast. The object is now visible.

The visibility of the object may be improved if some attenuation $\tau < 1$ is introduced along with the phase change $-\delta$ at P_i. Then for a general value of δ we have

$$\tau_0\left(\frac{x}{m}, \frac{y}{m}\right) \cong \tau e^{-i\delta} + i\phi\left(\frac{x}{m}, \frac{y}{m}\right)$$

$$= e^{-i\delta}\left[\tau + e^{i[(\pi/2)-\delta]}\,\phi\left(\frac{x}{m}, \frac{y}{m}\right)\right] \tag{9.41d}$$

and

$$|\tau_0|^2 \approx \tau^2 + \phi^2 + 2\tau\phi\cos\left(\frac{\pi}{2} - \delta\right) \tag{9.41e}$$

This will yield maximum contrast when $\delta = \pm\pi/2$ and $\tau \approx |\phi|$.

Practical phase contrast microscopes often use a condenser in such a way that the effective source has the shape of a thin ring rather than a point. The raised part of the phase plate must match the image of this source in the plane $z = D'$ and will then be ring-shaped, as shown in Fig. 9.22*b*. Photomicrographs are shown in Plate 4 that clearly reveal the improvement in visual contrast when the phase contrast method is used.

6. Digression on Interference Contrast Microscopy. There are other methods of converting phase contrast into amplitude contrast that do not involve spatial filtering. They depend on interference of two light beams and, as we already know, interference methods are capable of revealing phase differences.

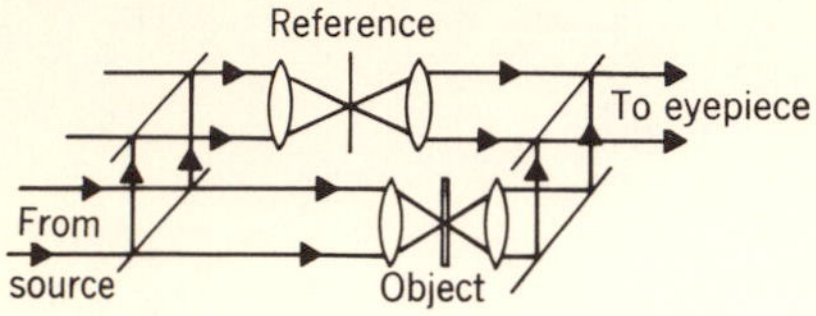

Fig. 9.23

In one method the beams are split, sent on symmetrical paths, one of which passes through the object, and recombined before reaching the eyepiece, as shown schematically in Fig. 9.23. This modification of the classical two-beam interference arrangement gives an intensity distribution that depends on the phase shift introduced by the object.

Another method, due to Nomarski, produces interference by the method of *wave shearing* and uses two Wollaston or similar prisms (Chapter 10). In Fig. 9.24 we indicate very schematically the behavior of wavefronts, assumed nearly plane for simplicity, at

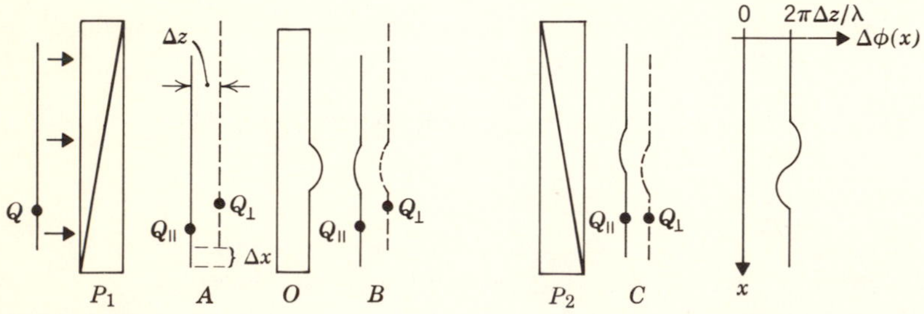

Fig. 9.24

several places in the system. The first prism P_1 acting on the initial beam polarized at 45° to the plane of the paper produces two beams of different polarization (parallel and perpendicular to the plane of the paper) traveling together through the same optical system, but with a lateral displacement (shear) in addition to a possible longitudinal retardation in phase. Surfaces of constant phase for the two beams are indicated by the solid and dashed lines at *A*. A point Q on the original wavefront separates and propagates as $Q_{\parallel}$ and $Q_{\perp}$. After passing through the phase object O, the wavefronts are distorted as shown at *B*. The second prism P_2 is oriented so that it has the reverse effect of P_1. It removes the shear produced by P_1 and produces the wavefronts at *C*. Note that $Q_{\parallel}$ and $Q_{\perp}$ no longer have a lateral separation. The total electric field is the sum of the two fields represented by these wavefronts at C. It is the sum of two sinusoidal signals having mutual coherence and a phase difference $\Delta\phi(x)$ as shown. The resulting intensity is proportional to $[1+\cos\Delta\phi(x)]$ and hence is no longer a constant independent of x. If Δx is small, the system may be adjusted so that essentially the derivative of the phase function $d\phi/dx$ is what is observed. Plate 4 gives one example of how well the Nomarski method works.

References for Section 9.2 will be found at the end of this chapter.

PROBLEMS

9.2.1 Derive Eq. (9.20).

9.2.2 a. A holographic method for recording the Fourier transform of the transmission of a thin object is shown in the figure. Work out the formula for the flux density at the recording plane under the usual assumption that the reference beam flux density is greater than that from the object.

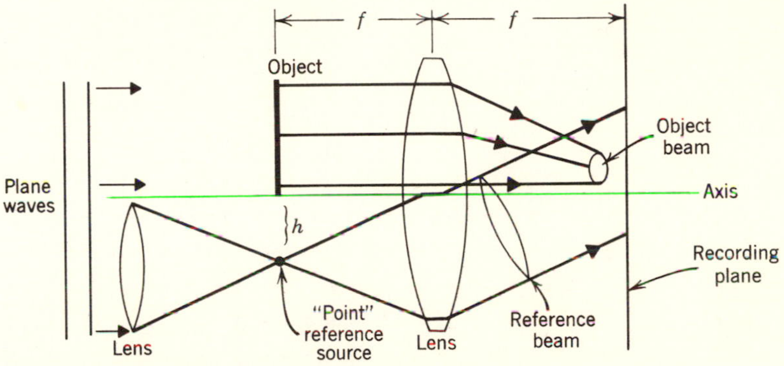

b. Show that the arrangement shown in the second figure will work as well as that in (a) by calculating the flux density in the recording plane. What is the role of the large lens? (Calculate the flux density with it removed.)

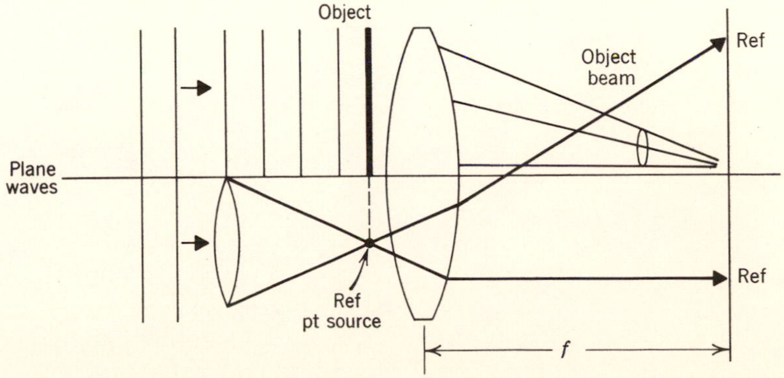

9.2.3 Consider the situation shown in Fig. 9.12. The object consists of a phase grating periodic with period l with transmission function for $0 \leqslant x \leqslant l$ given by

$$\tau(x') = 1, \qquad 0 \leqslant x' \leqslant \frac{l}{2}$$

$$\tau(x') = e^{i\phi}, \qquad \frac{l}{2} < x' < l$$

a, Write down an expression for the field in the plane $z = S'$. ("Infinite aperture case.")

b. A filter in the plane $z = f$ blocks all orders but $n = 0, \pm 1$ and lets these three through unattenuated. What is the field at $z = S'$? Neglect effects owing to finite aperture or finite object height and width.

9.2.4 Consider an object made of window screen, periodic in both x and y with period l having a transmission function as shown over one periodicity square. A long, narrow slit is used as a spatial filter in the plane $z = D'$ (Fig. 9.14). Describe the image when the slit makes the following angles with the x axis: 0, 45°, 90°.

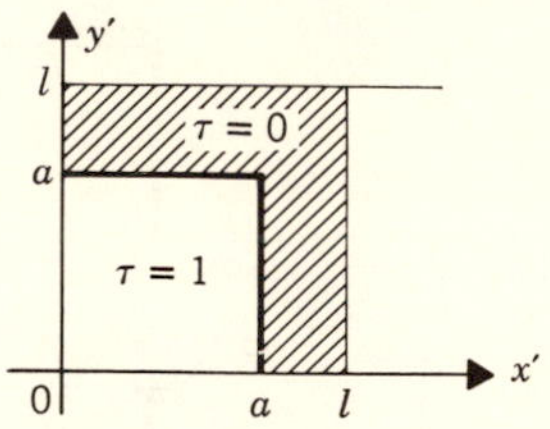

9.2.5 Work out the theory behind Fig. 9.20. Assume each scratch to be an infinitely long line with a width w and a transmission function $\tau = \frac{1}{2}$. Let the lines be at perpendicular distances $s_1, s_2, \ldots.$ from the origin in the object plane. Calculate the field distribution in the plane $z = z_i$. What happens if the lines are not quite parallel but vary in θ by $\pm\Delta\theta$ ($\Delta\theta \ll 1$)?

9.3 THE EFFECTS OF DIFFRACTION BY APERTURES AND THE ROLE OF ABERRATIONS IN OPTICAL IMAGES

In this section we cover two phenomena neglected so far, namely diffraction by the aperture stop or other apertures and the role of aberrations in changing the predictions of the diffraction theory of image formation.

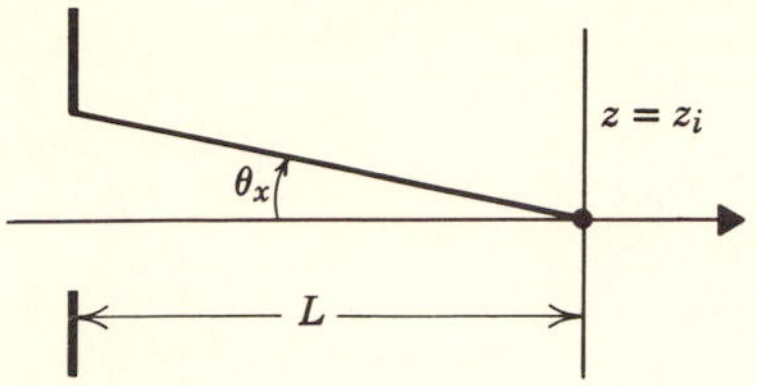

Fig. 9.25

A. THE EFFECT OF AN APERTURE STOP

1. Alteration of the Electric Field by an Aperture. We now study how the above results are changed when diffraction by the aperture stop is taken into account. Suppose that the electric field in the plane $z = z_i$ is given by $\tilde{E}_{\mathrm{IA}}(x_1, y_1, z_i)$. We move to the left a distance L as in Fig. 9.25 and assume that there are no intervening apertures. The field on the plane $z = z_i - L$ must then be given by the inverse Fresnel transformation:*

$$\tilde{E}_{\mathrm{IA}}(x_2, y_2, z_i - L) = \frac{-i}{\lambda L} e^{ikL} \iint_{-\infty}^{\infty} \tilde{E}_{\mathrm{IA}}(x_1, y_1, z_i) \times \exp\left\{\frac{ik}{2L}[(x_2 - x_1)^2 + (y_2 - y_1)^2]\right\} dx_1\, dy_1 \tag{9.42a}$$

Without changing anything else, we now insert a thin stop or other aperture in the plane $z = z_i - L$ having a transmission function $\tau_A(x_2, y_2)$. The field just in front of the aperture will be given by Eq. (9.42a) and that just behind it by

$$\tilde{E}_A(x_2, y_2, z_i - L) = \tilde{E}_{\mathrm{IA}}(x_2, y_2, z_i - L)\tau_A(x_2, y_2) \tag{9.42b}$$

What then is the field $\tilde{E}_A(x, y, z_i)$ in the plane $z = z_i$? The answer is given by a forward Fresnel transformation on Eq. (9.42b):

$$\begin{aligned}\tilde{E}_A(x, y, z_i) &= \frac{i}{\lambda L} e^{-ikL} \iint_{-\infty}^{\infty} \tilde{E}_A(x_2, y_2, z_i - L) \\ &\quad \times \exp\left\{-\frac{ik}{2L}[(x - x_2)^2 + (y - y_2)^2]\right\} dx_2\, dy_2 \\ &= \frac{1}{(\lambda L)^2} \iint_{-\infty}^{\infty} \tilde{E}_{\mathrm{IA}}(x_1, y_1, z_i) \left(\iint_{-\infty}^{\infty} \tau_A(x_2, y_2) \right. \\ &\quad \times \exp\left\{i\frac{k}{2L}[(x_2 - x_1)^2 - (x - x_2)^2 \right. \\ &\quad \left.\left. + (y_2 - y_1)^2 + (y - y_2)^2]\right\} \times dx_2\, dy_2 \right) dx_1\, dy_1\end{aligned}$$

*Recall the significance of an inverse Fresnel transform: $\tilde{E}_{\mathrm{IA}}(x_2, y_2, z_i - L)$ as given in (9.42a) will freely propagate and give the field $\tilde{E}_{\mathrm{IA}}(x, y, z_i)$ in the plane $z = z_i$.

The exponent in the integral in large parentheses may be written

$$\frac{ik}{2L}[(x_1^2-x^2)+(y_1^2-y^2)]-\frac{2\pi i}{\lambda L}[(x_1-x)x_2-(y_1-y)y_2)]$$

The integral in large parentheses then becomes

$$(\cdots)=\exp\left[i\frac{k}{2L}(x_1^2+y_1^2-x^2-y^2)\right]\iint_{-\infty}^{+\infty}\tau_A(x_2,y_2)$$

$$\times\exp\left\{-\frac{2\pi i}{\lambda L}[(x_1-x)x_2+(y_1-y)y_2]\right\}dx_2\,dy_2$$

$$=\exp\left[i\frac{k}{2L}(x_1^2+y_1^2-x^2-y^2)\right]\hat{\tau}_A\left(\frac{x_1-x}{\lambda L},\frac{y_1-y}{\lambda L}\right) \qquad (9.42c)$$

Here $\hat{\tau}_A(u, v)$ is the Fourier transform $\tau_A(x, y)$.
This gives

$$\tilde{E}_A(x,y,z_i)=\iint_{-\infty}^{\infty}\tilde{E}_{\text{IA}}(x_1,y_1,z_i)\exp\left[i\frac{k}{2L}(x_1^2+y_1^2-x^2-y^2)\right]$$

$$\times\hat{\tau}_A\left(\frac{x_1-x}{\lambda L},\frac{y_1-y}{\lambda L}\right)\frac{dx_1\,dy_1}{(\lambda L)^2} \qquad (9.43)$$

2. *Interpretation.* Equation (9.43) tells how the presence of the aperture at $z=z_i-L$ changes the field $\tilde{E}_{\text{IA}}(x_1, y_1, z_i)$ to the field $\tilde{E}_A(x, y, z_i)$. Each point x_1, y_1 is the center of a diffraction pattern that results from the aperture

$$\hat{\tau}_A\left(\frac{x_1-x}{\lambda L},\frac{y_1-y}{\lambda L}\right)\exp\left[i\frac{k}{2L}(x_1^2+y_1^2-x^2-y^2)\right]$$

having strength $\tilde{E}_{\text{IA}}(x_1, y_1, z_1)\,dx_1\,dy_1/(\lambda L)^2$ proportional to the no-aperture field. The sum of all such patterns gives the resulting field when the aperture is present.

Note from the integral in Eq. (9.42*c*) that the aperture coordinates appear only in the form x_2/L, y_2/L in τ_A. We may define angles

$$\theta_x\approx\frac{x_2}{L}, \qquad \theta_y\approx\frac{y_2}{L}$$

These approximations are consistent with our use of the parabolic approximations. Then the integral in Eq. (9.42*c*) may be written

$$L^2\iint_{-\infty}^{\infty}T_A(\theta_x,\theta_y)\exp\left\{-\frac{2\pi i}{\lambda}[(x_1-x)\theta_x+(y_1-y)\theta_y]\right\}d\theta_x\,d\theta_y$$

$$=L^2\hat{T}_A\left(\frac{x_1-x}{\lambda},\frac{y_1-y}{\lambda}\right)$$

where we define

$$T_A(\theta_x, \theta_y) \equiv \tau_A(L\theta_x, L\theta_y) \tag{9.44}$$

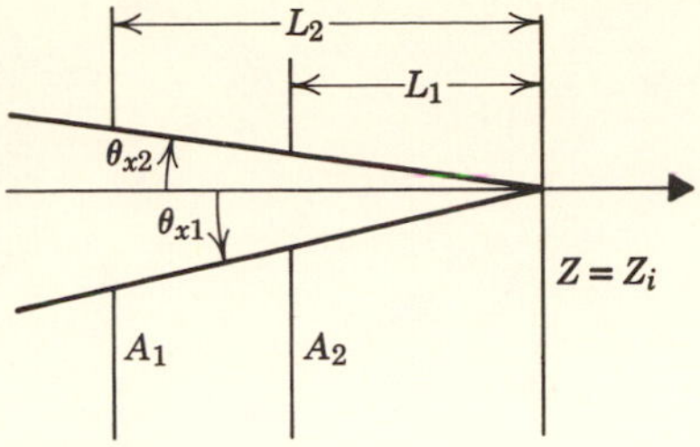

Fig. 9.26

Then an alternate form of Eq. (9.43) is

$$\tilde{E}_A(x, y, z_i) = \iint_{-\infty}^{\infty} \tilde{E}_{IA}(x_1, y_1, z_i) \exp\left[\frac{ik}{2L}(x_1^2 + y_1^2 - x^2 - y^2)\right] \times \hat{T}_A\left(\frac{x_1 - x}{\lambda}, \frac{y_1 - y}{\lambda}\right)\frac{dx_1\, dy_1}{\lambda^2} \tag{9.43$'$}$$

In Fig. 9.26 we illustrate a one-dimensional case of two apertures A_1 and A_2 that have the same limiting angles θ_{x1} and θ_{x2}. For them we would have

$$\begin{aligned} T_A &= 1 \qquad \text{if} \quad -\theta_{x2} \leq \theta_x \leq \theta_{x1} \\ &= 0 \qquad \text{otherwise} \end{aligned}$$

and the x-dependent part of $\hat{T}_A$ would be

$$\hat{T}_{Ax} = \frac{\lambda}{2\pi i(x - x_1)}\left\{\exp\left[\frac{2\pi i}{\lambda}(x - x_1)\theta_{x1}\right] - \exp\left[-\frac{2\pi i}{\lambda}(x - x_1)\theta_{x2}\right]\right\}$$

We can distinguish between apertures A_1 and A_2 only through the phase factor

$$\exp\left[i\frac{k}{2L}(x_1^2 + y_1^2 - x^2 - y^2)\right] \qquad (L = L_1 \text{ or } L_2)$$

appearing in Eqs. (9.43) and (9.43′). This justifies the statements made in Chapter 7 in connection with Fig. 7.24.

We remark that as the aperture size approaches infinity, we may invoke the delta-function limit of Eq. (9.14), from which it is easy to show that the limiting form of $\tilde{E}_A(x, y, z_i)$ is $\tilde{E}_{IA}(x, y, z_i)$.

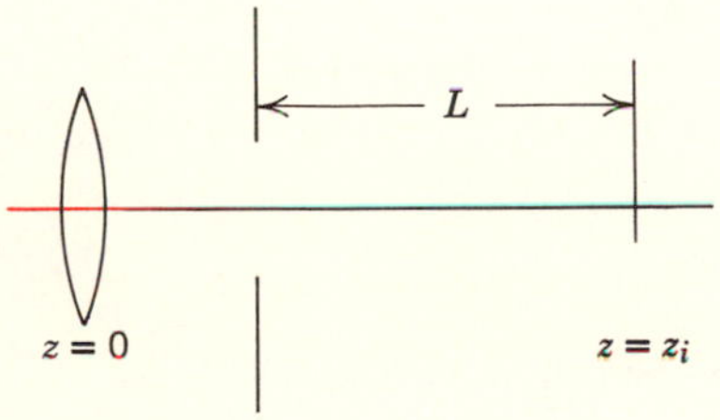

Fig. 9.27

The above results may now be applied to show how a stop or other aperture placed anywhere between a lens and a plane of interest modifies the field that would otherwise be produced there (Fig. 9.27). Note that the stop acts as its own exit pupil. Fields that we might want to modify by such an aperture are given by Eq. (9.27) ($z_i = S'$) and Eq. (9.28) ($z_i = D'$).

3. ***Aperture to the Left of the Lens.*** Instead of placing the aperture between the lens and the plane $z = z_i$, we now place it to the left of the lens. In so doing we assume that there is no important aperture to the right of the one under consideration, and thus neglect the aperturing role of the lens rim itself.

As shown in Fig. 9.28, let the plane conjugate to the plane of observation be at $z = -z_0$, where z_0 obeys the lens equation

$$\frac{1}{z_0} + \frac{1}{z_i} = \frac{1}{f}$$

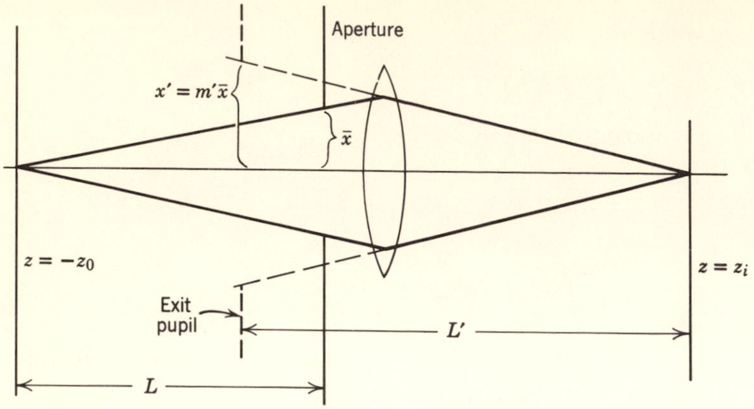

Fig. 9.28

Let the aperture be a distance L to the right of the conjugate plane, namely, at $z = -z_0 + L$. The exit pupil is the conjugate image of the aperture in image space and will be located at $z = z_i - L'$, where

$$\frac{1}{z_0 - L} + \frac{1}{z_i - L'} = \frac{1}{f}$$

To an observer near the plane $z = z_i$, the exit pupil appears to limit the wavefronts passing through the system, and a detailed diffraction calculation with proper use of Fresnel propagation transformations will substantiate this notion. We can then prove that the effect of the aperture may be described entirely in image space as an effect of the exit pupil. If m' represents the magnification between the aperture and the pupil, then a typical coordinate x' in the pupil will be given by $x' = m'\bar{x}$, where $\bar{x}$ is the coordinate of the conjugate point in the aperture, as shown in Fig. 9.28. In image space the pupil behaves as it if has the transmission function

$$\tau_P(x', y') = \tau_A\left(\frac{x'}{m'}, \frac{y'}{m'}\right) \tag{9.45}$$

If, as usual $\tilde{E}_{IA}(x, y, z_i)$ represents the field if the aperture is very large, the field with a finite aperture can be obtained by an inverse Fresnel transformation to the plane $z = z_i - L'$, multiplication by τ_P, and then a forward Fresnel transformation back to the plane $z = z_i$. The result is

$$\tilde{E}_A(x, y, z_i) = \iint_{-\infty}^{\infty} \tilde{E}_{IA}(x_1, y_1, z_i) \exp\left[\frac{ik}{2L'}(x_1^2 + y_1^2 - x^2 - y^2)\right] \times \hat{\tau}_P\left(\frac{x_1 - x}{\lambda L'}, \frac{y_1 - y}{\lambda L'}\right) \frac{dx_1\, dy_1}{(\lambda L')^2} \tag{9.46}$$

This equation is the same formally as Eq. (9.43), which describes the effect of an aperture to the right of the lens, where it is its own exit pupil so that $\tau_P = \tau_A$. It is not necessary to distinguish between the two cases if we work entirely in image space. We thus adopt a common notation: $\tau_P(x', y')$ will be the transmission function for the exit pupil, and L will be the distance of the pupil to the left of the plane of observation. With this notation Eqs. (9.43) and (9.46) may both be written

$$\boxed{\begin{aligned}\tilde{E}_A(x, y, z_i) = \iint_{-\infty}^{\infty} \tilde{E}_{\mathrm{IA}}(x_1, y_1, z_i) \exp \frac{ik}{2L}(x_1^2+y_1^2-x^2-y^2)\Big] \\ \times \hat{\tau}_P\left(\frac{x_1-x}{\lambda L}, \frac{y_1-y}{\lambda L}\right)\frac{dx_1\,dy_1}{(\lambda L)^2}\end{aligned}} \tag{9.47}$$

Our remark in connection with Fig. 9.26 about the dependence of the function τ_A on the angles θ_x and θ_y now holds for the function τ_P—the magnitude of the field $\tilde{E}_A(x, y, z_i)$ depends only on these angles; the location of the exit pupil occurs only in the phase factor. This establishes the result quoted without proof in connection with Fig. 7.24.

4. Point Source. Suppose that we have a point source that, according to Gaussian geometrical optics, would yield a point image at $x = x_i$, $y = y_i$. The large aperture field, that is, the field when diffraction is neglected, may then be represented by a delta function as in Eq. (9.3):

$$\tilde{E}_{\mathrm{IA}}(x_1, y_1, z_i) = -\frac{A}{i}\lambda\delta(x_1-x_i)\delta(y_1-y_i)$$

The minus sign is chosen so that Eq. (9.42*a*) gives an *incoming* field

$$\tilde{E}_{\mathrm{IA}}(x_2, y_2, z_i-L) \cong \frac{A\,e^{ikR}}{R}$$

with

$$R \cong L+\frac{(x_2-x_i)^2+(y_2-y_i)^2}{2L}$$

The field with an aperture is then given by

$$\begin{aligned}\tilde{E}_A(x, y, z_i) = -\frac{A\lambda}{i}\exp\left[\frac{ik}{2L}(x_i^2+y_i^2-x^2-y^2)\right] \\ \times\hat{\tau}_P\left(\frac{x_i-x}{\lambda L}, \frac{y_i-y}{\lambda L}\right)\frac{1}{(\lambda L)^2}\end{aligned} \tag{9.48}$$

A *simple pupil* will be defined as one with a transmission function that is either unity or zero. Let Σ_P represent the open surface of the pupil. Then

$$\hat{\tau}_P(u, v) = \iint_{\Sigma_P}^{\infty} e^{-2\pi i(ux'+vy')}\,dx'\,dy'$$

The *triangle inequality*

$$\left|\iint f(x', y')\, dx'\, dy'\right| \leq \iint \left|f(x', y')\right| dx'\, dy'$$

may be applied to the above integral to obtain

$$|\hat{\tau}_P(u, v)| \leq \iint_{\Sigma_P} 1 \cdot dx'\, dy' = \sigma_P = \hat{\tau}_P(0, 0)$$

where σ_P is the area of the opening Σ_P. Thus the field having the maximum is

$$\tilde{E}_A(x_i, y_i, z_i) = i\lambda A\hat{\tau}_P(0, 0)\frac{}{(\lambda L)^2} = \frac{iA}{\lambda}\frac{\sigma}{L^2} = \frac{iA}{\lambda}\Omega \tag{9.49}$$

where $\Omega = \sigma/L^2$ is the solid angle subtended by the exit pupil at the image. This result agrees with Eq. (7.20).

By gathering together the results in Eqs. (7.34), (7.35), (7.36*a*,*b*), (7.40′), (7.50), and (7.51), we obtain the following expressions for the Fourier transform in two important special cases of pupils centered on the z axis.

(a) *Rectangle* $2x_0$ *and* $2y_0$

$$\hat{\tau}_P\left(\frac{x_1 - x}{\lambda L}, \frac{y_1 - y}{\lambda L}\right)\frac{1}{(\lambda L)^2} = \frac{2y_0 \cdot 2x_0}{(\lambda L)^2}\frac{\sin t_x}{t_x}\frac{\sin t_y}{t_y},$$

$$\text{where } t_x = \frac{2\pi(x_1 - x)}{\lambda}\left(\frac{x_0}{L}\right) \text{ and } t_y = \frac{2\pi(y_1 - y)}{\lambda}\left(\frac{y_0}{L}\right) \tag{9.50a}$$

(b) *Circle of radius* r_0

$$\hat{\tau}_P\left(\frac{x_1 - x}{\lambda L}, \frac{y_1 - y}{\lambda L}\right)\frac{1}{(\lambda L)^2} = \frac{1}{\Delta r\lambda}\left(\frac{r_0}{L}\right)J_1\left[\frac{2\pi\Delta r}{\lambda}\left(\frac{r_0}{L}\right)\right] \tag{9.50b}$$

$$= \frac{\pi r_0^2}{(\lambda L)^2}\left[\frac{2J_1(t)}{t}\right], \qquad t = \frac{2\pi\Delta r}{\lambda}\left(\frac{r_0}{L}\right)$$

$$\left[\Delta r = \sqrt{(x_1 - x)^2 + (y_1 - y)^2}\right]$$

When the center of the exit pupil is not on the axis, but at the point (x_c, y_c), we must multiply Eqs. (9.50*a*,*b*) by $e^{-i\Delta}$ with

$$\Delta = \frac{2\pi}{\lambda L}[(x_1 - x)x_c + (y_1 - y)y_c] \tag{9.50c}$$

In Fig. 9.29 we sketch roughly what the field distributión looks like with a circular aperture for a point source conjugate to the point $x = x_1$, $y = y_1$. The phase factor $\exp i(k/2L)(x_1^2 + y_1^2 - x^2 - y^2)$ in Eq. (9.48) has been neglected. The discussion following Eq. (9.43) makes it clear that each point (x_1, y_1) in the image plane gives rise to a pattern like that in Fig. 9.29.

5. ***Behavior near Focus with a Circular Aperture.*** Equation (9.47) is difficult to evaluate in most specific situations. A few interesting

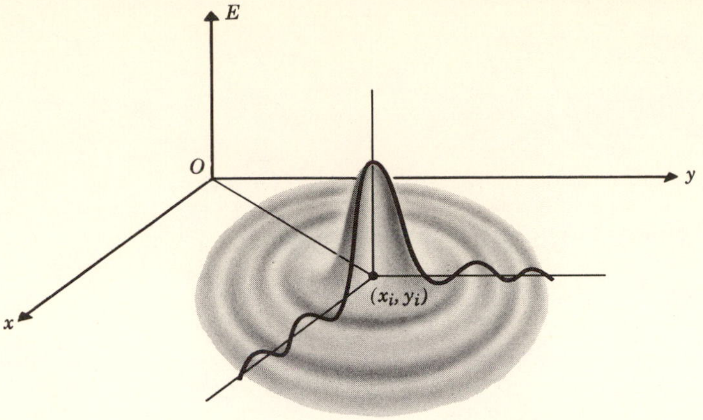

Fig. 9.29

cases have been worked out, however. One involves a calculation of the intensity and phase distribution near the geometrical focus of a coherent beam limited by diffraction from a circular aperture.

Part of such a calculation will now be given. The geometry is shown in Fig. 9.30. The origin O is at the geometrical focus, P is the observation point, P' is a point in the exit pupil (of radius r_0).

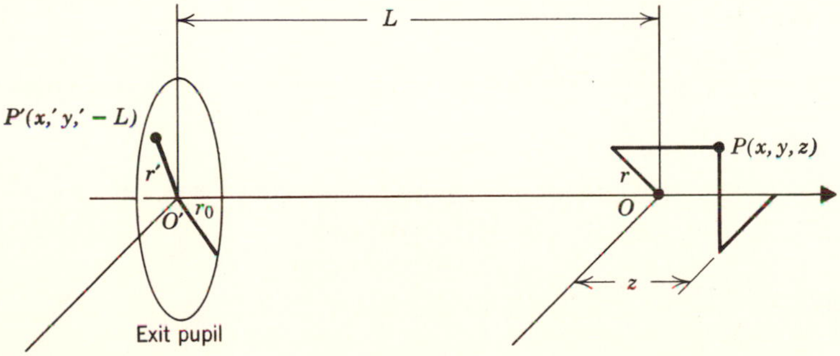

Fig. 9.30

Rather than use (9.47), we find it easier to start with the field at the pupil corresponding to a wave converging towards O.

$$\tilde{E}(x', y', -L) = \frac{A}{|P'O|} e^{ik|P'O|} \approx \frac{A}{L} e^{ikL} \exp\left[i \frac{k}{2L} (x'^2 + y'^2)\right]$$

and apply a Fresnel transformation. The result is

$$\tilde{E}_A(x, y, z) = \frac{iA}{\lambda L(L+z)} e^{-ikz} \iint_{\Sigma_p} \exp\left[i \frac{k}{2L} (x'^2 + y'^2)\right]$$

$$\times \exp\left\{-i \frac{k}{2(L+z)} [(x-x')^2 + (y-y')^2]\right\} dx' \, dy' \tag{9.51a}$$

If we assume that $|z| \ll L$, we may write the exponent in the integrand in the form

$$\frac{-ikr^2}{2L}+\frac{ikr'^2z}{2L^2}+\frac{ikrr'\cos\theta}{L}$$

where

$$r^2=x^2+y^2,\ r'^2=x'^2+y'^2,\ rr'\cos\theta=xx'=\overrightarrow{OP}\cdot\overrightarrow{OP'}$$

It is conventional to introduce dimensionless variables

$$\xi=kz\left(\frac{r_0}{L}\right)^2,\qquad \eta=kr\left(\frac{r_0}{L}\right)$$
$$\rho=\frac{r'}{r_0} \tag{9.51b}$$

Then $dx'dy'=\rho\,d\rho\,d\theta r_0^2$, and (9.51$a$) becomes

$$\tilde{E}_A(P)=\frac{iAr_0^2}{\lambda L^2}e^{-ikz}e^{-i(k/2L)r^2}\int_0^1 e^{i(\xi\rho^2/2)}\rho\,d\rho\int_0^{2\pi}e^{i\eta\rho\cos\theta}d\theta$$

According to the equation just below Eq. (7.54′), the angular integral here yields $2\pi J_0(\eta\rho)$, where J_0 is the zero-order Bessel Function. We obtain

$$\tilde{E}_A(P)=\frac{iA}{\lambda}\frac{\pi r_0^2}{L^2}e^{-ikz}e^{-i(kr^2/2L)}2\int_0^1 e^{i(\xi\rho^2/2)}J_0(\eta\rho)\rho\,d\rho \tag{9.51c}$$

It is easily seen that the integral gives $\frac{1}{2}$ when $\xi=\eta=0$, for then $J_0(0)=1$. Then we can write $\tilde{E}_A(P)$ in terms of

$$\tilde{E}_A(O)=\frac{iA}{\lambda}\frac{\pi r_0^2}{L^2}=\frac{iA}{\lambda}\Omega$$

as

$$\tilde{E}_A(P)=\tilde{E}_A(O)e^{-ikz}e^{-i(k/2L)r^2}2\int_0^1 e^{(\xi\rho^2/2)}J_0(\eta\rho)\,\rho\,d\rho \tag{9.51d}$$

In general, the integral in Eq. (9.51d) must be approximately evaluated using special procedures, but there are certain familiar and important special cases. For instance, in the plane of the point image where $z=\xi=0$ we use Eq. (7.55),

$$tJ_1(t)=\int_0^t t'J_0(t')\,dt' \tag{9.51e}$$

$$\tilde{E}_A(P)=\tilde{E}_A(O)e^{-i(k/2L)r^2}\left[\frac{2J_1(\eta)}{\eta}\right]$$

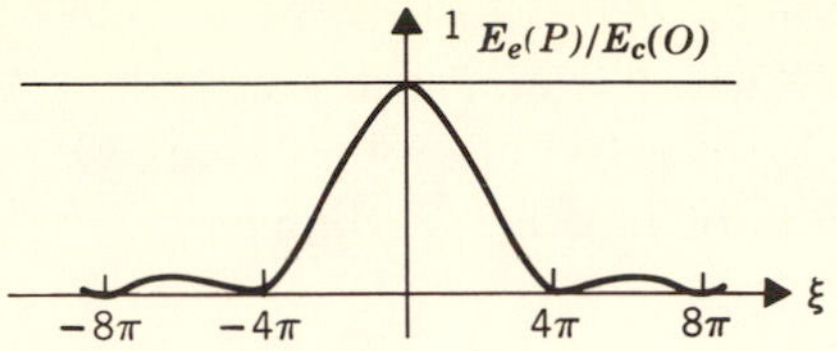

Fig. 9.31. The flux density along the axis near the geometrical point image for the case of diffraction by a circular exit pupil of radius r_0 a distance L away. Plotted as a function of the dimensionless variable $\xi = kz(r_0/L)^2$. Here z is the axial distance from the geometrical image.

Similarly, along the axis where $r = \eta = 0$, we have

$$\tilde{E}_A(P) = \tilde{E}_A(O)e^{-ikz}\int_0^1 e^{i\xi\rho^2/2}\,d\rho^2$$

$$= \tilde{E}_A(O)e^{-ikz}e^{i\xi/4}\left[\frac{\sin\xi/4}{\xi/4}\right] \tag{9.51f}$$

Thus the flux density along the axis is

$$E_e(P) = E_e(O)\left[\frac{\sin\xi/4}{\xi/4}\right]^2 \tag{9.52}$$

This familiar function is plotted in Fig. 9.31.

Contours of constant irradiance and of constant phase have been calculated by Wolf and by Linfoot and Wolf (see References). Results for the irradiance contours are shown in Fig. 9.32. The

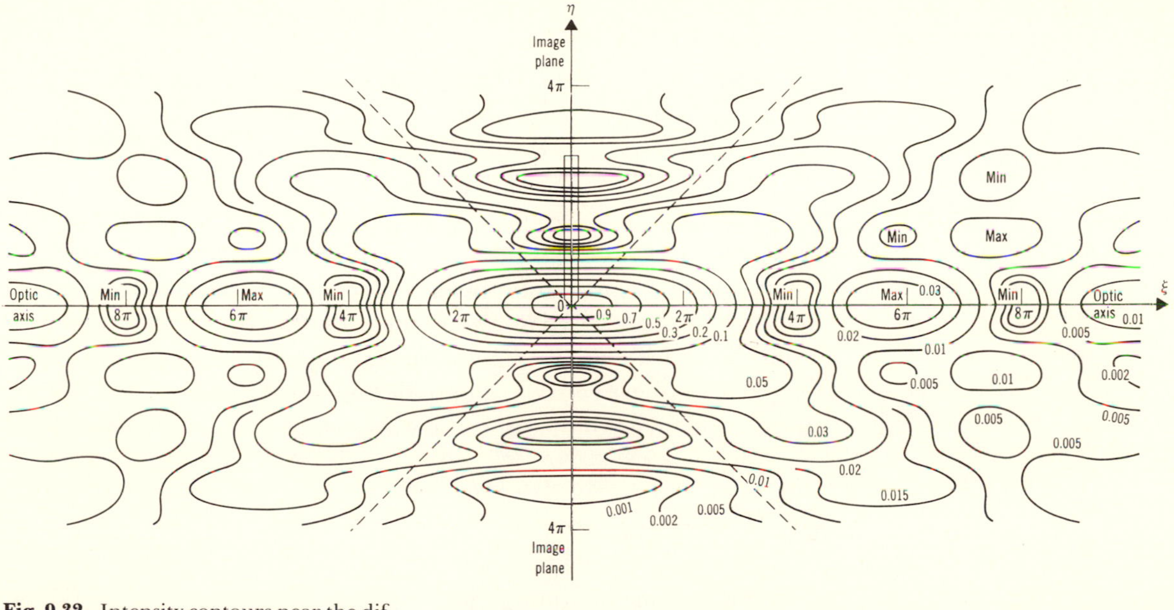

Fig. 9.32. Intensity contours near the diffraction-limited point image for a circular exit pupil of radius r_0 a distance L from the geometrical point focus. Plotted as a function of the reduced coordinates $\xi = kz(r_0/L)^2$ and $\eta = kr(r_0/L)$. r is the distance of the observation point from the axis, and z is the distance along the axis from the image. The dashed lines limit the geometrical shadow. [After E. H. Linfoot and E. Wolf, *Proc. Phys. Soc.* **B69**, 823 (1956). Used with permission of the Institute of Physics and the Physical Society.]

dashed lines $\eta = \pm\xi$ correspond to the limits of the geometrical shadow of the aperture. The irradiance distribution in the image plane is the Airy disk plus rings.

*6. **The Effect of Aberrations.*** We have been discussing the effects of diffraction in what would otherwise be Gaussian image-forming systems. Diffraction can be thought of as resulting from the limitation of the size of the wave surfaces emerging through a finite exit pupil. With a point source these wave surfaces are considered to be perfect spheres (or rather paraboloids, since the parabolic approximation has been applied). In our discussion of aberrations in Chapter 4 we compared the emerging wavefronts actually obtained from a point object with a reference sphere centered about the Gaussian point image P_i (in our present notation). This is illustrated in Fig. 9.33, which is similar to Fig. 4.24. Here Σ is the emerging wavefront, Σ_s the reference sphere centered about P', and Q is a point in the exit pupil having coordinates

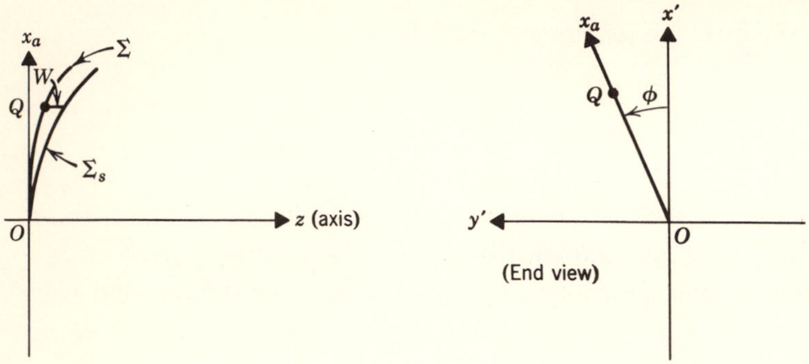

Fig. 9.33

(x', y') with respect to an origin in the center of the pupil. The so-called aberration function $W(Q, P_i) = W(x', y'; P_i)$ was introduced in Chapter 4; it equals the optical path difference between the two surfaces Σ and Σ_s at Q (see Eq. 4.46). In general, W depends on the transverse coordinates of both Q and P_i. In the case shown W is positive.

Our above discussions of the diffraction theory of image formation have been based on the assumption that $W = 0$, that is, that Σ and Σ_s coincide. If W is small compared with $|OQ|$, we can imagine the perfect sphere converted into the true wave surface Σ_s by shifting the phase of the spherical waves at the point Q by the amount

$$\Delta = \frac{2\pi}{\lambda} W$$

This can be done by adding the factor

$$e^{-i\Delta} = \exp\left[-\frac{2\pi i}{\lambda} W(x', y'; P_i)\right] \tag{9.53a}$$

to the transmission function $\tau_P(x', y')$ of the pupil. We then obtain an effective *pupil function*

$$\tau_P(x', y'; P_i) = \tau_P(x', y')e^{-ikW(x', y'; P_i)} \tag{9.53b}$$

This describes the combined effect of diffraction and aberrations on the wavefront at the pupil headed toward the point P_i. The Fourier transform $\hat{\tau}_P(u, v; P_i)$ of this pupil function must then be used in an equation like (9.48), and the field in the plane of the

Gaussian geometrical point image P_i becomes

$$\boxed{\tilde{E}_A(x, y, z_i) = \frac{A\lambda}{i} \exp\left[\frac{ik}{2L}(x_i^2 + y_i^2 - x^2 - y^2)\right] \hat{\tau}_P(u, v; P_i) \frac{1}{(\lambda L)^2}} \tag{9.54a}$$

with

$$u = \frac{x_i - x}{\lambda L}, \qquad v = \frac{y_i - y}{\lambda L}$$

and

$$\hat{\tau}_P(u, v; P_i) = \iint_{-\infty}^{\infty} \tau_P(x', y'; P_i) e^{-2\pi i(ux' + vy')} \, dx' \, dy' \tag{9.54b}$$

Because of the dependence of the aberration function W on P_i, $\hat{\tau}_P(u, v; P_i)$ will not in general be a function merely of the differences $x_i - x$, $y_i - y$. For a simple pupil having an open portion Σ_p we may write

$$\boxed{\hat{\tau}_P(u, v; P_i) = \iint_{\Sigma_p} e^{-ikW(x', y'; P_i)} e^{-2\pi i(ux' + vy')} \, dx' \, dy'} \tag{9.54c}$$

B. THE POINT SPREAD FUNCTION AND THE OPTICAL TRANSFER FUNCTION

How do we characterize a lens so that its performance may be predicted in a variety of situations? It is ordinarily not sufficient to know merely the ultimate resolution for small-scale detail. Distortion and loss of contrast on a larger scale may also be important. In recent years several quantitative methods of lens evaluation have been developed for use with incoherently illuminated objects. Several are based on common assumptions and are closely related. The optical transfer function to be discussed here is perhaps the most often employed. It bears a close relationship to other frequency response functions widely used in science and engineering.

Like other frequency response functions, the optical transfer function gives the response of the system to a sinusoidally varying input. For a general linear system such as a linear electronic amplifier, a sinusoidal input will yield a sinusoidal output, with in general a frequency-dependent amplitude and phase shift. If a few reasonable assumptions are made, an optical system with incoherent illumination may be regarded as linear, and the above statement will then apply. A linear electronic amplifier processes signals that are functions of time; the frequencies of the individual sinusoidal components are temporal frequencies. An optical system processes signals that are functions of two spatial variables; the corresponding frequencies are spatial frequencies.

The reader should be cautioned against carrying the useful analogy between optical systems and linear electronic amplifiers too far. Optical systems connected together in series will not generally yield a system having a response function that is the product of the individual response function, as is true in the electronic case.

1. Incoherent Objects. A perfectly incoherent source does not exist. Even light from a blackbody has a coherence length at its surface equal to a mean wavelength $\bar{\lambda}$. An indirectly illuminated diffuse surface will have a coherence length not much larger than $\bar{\lambda}$ if the illuminating light covers a wide angle. If an image is formed, the coherence length increases, since the light from each point in the object forms a focal spot that is finite, because of diffraction and aberrations. The resulting coherence length has to be at least as large as the size of this focal spot, which is ordinarily much larger than $\bar{\lambda}$ times the magnification. In this case we can neglect the very small coherence length of the source itself.

This is the assumption we shall make for self-luminous (or thermal) objects or for diffuse objects that are indirectly illuminated with wide-angle beams. We neglect all mutual coherence between light from two separate points in the object. The flux density is then obtained by summing the individual contributions from all points in the object.

2. Point Spread Function. Consider light from a point object that is the Gaussian conjugate point to P_i. Let Σ_p be the open part of the exit pupil a distance L away, as shown in Fig. 9.34. This is the situation assumed for Eq. (9.54*a*).

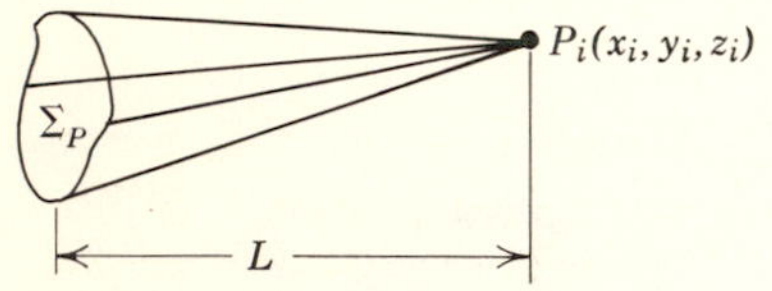

Fig. 9.34

The field at the exit pupil will have magnitude A/L, and the flux density there will be $E_e = C(A/L)^2$, where C is a constant given by Eq. (7.38′). The total flux passing through the pupil, assumed "simple," will be

$$\Phi = \iint_{\Sigma_p} E_e \, d\sigma = C\left(\frac{A}{L}\right)^2 \sigma_p$$

where σ_p is the area of the exit pupil. This will be unchanged by diffraction and by aberrations.

For unit total flux we must have an amplitude A that obeys

$$A^2 = \frac{L^2}{C\sigma_p} \tag{9.55}$$

According to Gaussian geometrical optics, this flux always stays inside the cone with apex at P_i and base Σ_p. The field at P_i would be infinite.

In the presence of aberrations and diffraction, the field in the Gaussian image plane $z = z_i$ is given by Eq. (9.54*a*), and the flux density, instead of being a delta function at P_i, will be given by

$$\frac{A^2\lambda^2}{(\lambda L)^4} C|\hat{\tau}_P(u, v; P_i)|^2 = \frac{1}{\sigma_p(\lambda L)^2} |\hat{\tau}_P(u, v; P_i)|^2 \tag{9.56}$$

$$u = \frac{x_i - x}{\lambda L}, \qquad v = \frac{y_i - y}{\lambda L}$$

In this result we have used (9.55) for A^2. Thus the integral of (9.56) over x and y must give a total flux of unity (Problem 9.3.4). This equation shows how the Gaussian point image at (x_i, y_i) spreads because of diffraction and aberrations.

There are two types of dependence on (x_i, y_i) in Eq. (9.56). There is a rapid dependence occurring through u and v that gives the size of the focal spot. There is also a slow dependence through P_i that comes ultimately from the P_i dependence of the aberration function $W(x', y', P_i)$. We expect the aberrations to change slowly throughout the image field and to remain essentially constant as P_i is moved a distance equal to the size of the focal spot. In fact, there is generally a much larger area than that of the focal spot over which W and hence the aberrations may be considered constant. It is called an *isoplanatism patch*, and its actual size depends on the strictness of our criteria. Under these circumstances x_i and y_i enter into $\hat{\tau}(u, v; P_i)$ through P_i essentially as parameters. Then the *point spread function*, defined by

$$O(x - x_i, y - y_i; P_i) = \frac{1}{\lambda^2 L^2 \sigma_p} \left| \hat{\tau}_P\left(\frac{x_i - x}{\lambda L}, \frac{y_i - y}{\lambda L}; P_i\right)\right|^2 \qquad 9.57a)$$

may be regarded primarily as a function of the difference variables $x - x_i$, $y - y_i$ with a weak additional dependence on P_i. The normalization of O is unity:

$$\iint_{-\infty}^{\infty} O(x - x_i, y - y_i; P_i)\, dx\, dy = \iint_{-\infty}^{\infty} O(x - x_i, y - y_i; P_i)\, dx_i\, dy_i = 1 \tag{9.57b}$$

The spread function for the ideal, diffraction-limited case where there are no aberrations will be denoted by $O_0(x - x_i, y - y_i)$. Using Eqs. (9.50*a*, *b*) for rectangular and circular pupils, we obtain the following.

(a) *Rectangle*, $2x_0$ *by* $2y_0$

$$O_0(x - x_i, y - y_i) = \frac{(2x_0 \cdot 2y_0)}{\lambda^2 L^2} \left[\frac{\sin t_x}{t_x} \frac{\sin t_y}{t_y}\right]^2 \tag{9.58a}$$

with

$$t_x = \frac{2\pi(x_i - x)x_0}{\lambda L}, \qquad t_y = \frac{2\pi(y_i - y)y_0}{\lambda L}$$

(b) *Circle of radius* r_0

$$O_0(x-x_i, y-y_i) = \frac{\pi r_0^2}{\lambda^2 L^2}\left[\frac{2J_1(t)}{t}\right]^2 \tag{9.58b}$$

with

$$t = \frac{2\pi}{\lambda}\left(\frac{r_0}{L}\right)\sqrt{(x_i-x)^2+(y_i-y)^2}$$

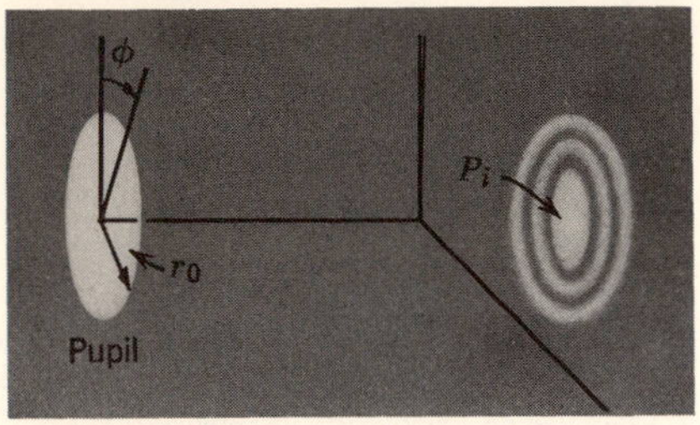

Fig. 9.35

These two results share a general property for any simple aperture, namely, that

$$O_0(O, O) = \frac{\sigma_p}{(\lambda L)^2} \tag{9.58c}$$

The distribution of flux density in the image plane described by Eq. 9.58*b*) is shown schematically in Fig. 9.35.

With no aberrations the spread function is the Airy disk with rings centered at P_i. Calculations have shown that for a small amount of spherical aberration, where the maximum value of the aberration function W is a small fraction of a wavelength, the main effect is to transfer energy from the disk to the rings. With off-axis aberrations, such as coma, the function O no longer has circular symmetry; this makes the rings nonuniform when W is less than λ. Photographs of images from a point object are shown in Plate 1 when primary coma is present. It is not necessary to show images for distortion and curvature of field; they are Airy patterns with their centers displaced laterally and longitudinally, respectively. With other aberrations also, the size of the spread function can usually be reduced if the observation plane is shifted somewhat from $z = z_i$. For a small amount of primary spherical aberration, the spread is minimized halfway between paraxial and marginal foci. This is apparent from Fig. 9.38*b* in the somewhat different context of the optical transfer function.

First a word about so-called focus errors, which are also related to curvature of field. According to Eq. (4.67), in order to obtain the best focus one should shift the observation plane along the axis a distance Δz if there is a term in W proportional to

$$\frac{nr'^2\Delta z}{2L^2}$$

where n is the index in image space. This can be written

$$W = m\lambda\left(\frac{r'}{r_0}\right)^2$$

where λ is the wavelength in image space and where $m = (\Delta z/2\lambda)(r_0/L)^2$ gives the number of wavelengths of "focus error" at the edge ($r' = r_0$) of the exit pupil.

For general calculations with aberrations it is convenient to express W in multiples of λ as a function of the reduced radius r'/r_0 and azimuthal angle ϕ at the exit pupil (see Figs. 9.33, 9.35). Thus, $W = \frac{1}{2}\lambda(r'/r_0)^4$ would correspond to $\frac{1}{2}$ wavelength of primary spherical aberration, $W = m\lambda(r'/r_0)^3 \cos\phi$ to m wavelengths of primary coma. In the latter case the dimensionless coefficient m is proportional to h', the displacement of P_i from the axis, as shown in Section 4.2C. With a combination of astigmatism and curvature of field we have

$$W = m\lambda\left(\frac{r'}{r_0}\right)^2 [\cos^2\phi - \mu]$$

where m is proportional to h'^2. If the observation plane is the tangential image plane, then $\mu = 1$; in the sagittal image plane we have $\mu = 0$; halfway between, we have $\mu = \frac{1}{2}$.

When the aberration function $W(x', y'; P_i)$ is small compared with λ, one can take the first three terms in a series expansion of e^{-ikW} in Eq. (9.54*c*) and derive the following approximate expression for the value of the spread function at the Gaussian image point (where $u = 0$, $v = 0$)

$$O(0, 0; P_i) = \frac{\sigma_p}{\lambda^2 L^2}[1 - k^2(\langle W^2\rangle - \langle W\rangle^2)] \tag{9.59a}$$

where the moments $\langle W^n\rangle$ of W are defined by

$$\langle W^n\rangle = \frac{1}{\sigma_p}\iint_{\Sigma_p} [W(x', y'; P_i)]^n\, dx'\, dy' \tag{9.59b}$$

Equation (9.59*a*) describes how aberrations reduce the intensity of what would otherwise be the center of the Airy disk. Some applications of these results will be made in the problems.

3. ***Distribution of Sources.*** Unit flux from a point source conjugate to (x_i, y_i) spreads into a focal spot described by the function $O(x - x_i, y - y_i)$. Now consider a continuous object described by a function E_{eg} such that a small region of the source conjugate to area $d\sigma$ about (x_i, y_i) radiates flux

$$d\Phi = E_{eg}(x_i, y_i; z_i)\, d\sigma$$

through the exit pupil. The *Gaussian flux density* E_{eg} represents the prediction of Gaussian geometrical optics for the flux density at $z = z_i$. This will spread by diffraction and aberrations and yield the flux density

$$O(x - x_i, y - y_i; P_i)\, d\Phi$$

in the plane $z = z_i$. The total flux density at the point (x, y, z_i) is then obtained by integration:

$$E_e(x, y, z_i) = \iint_{-\infty}^{\infty} O(x - x_i, y - y_i; P_i) E_{eg}(x_1, y_i, z_i)\, dx_i\, dy_i \qquad (9.60)$$

Equation (9.60) represents E_e in terms of a linear integral operator operating on E_{eg}. Linearity is represented by the following property. If E_{eg} is expressed as the linear superposition of two terms,

$$E_{eg} = aE_{eg}^{(1)} + bE_{eg}^{(2)}$$

then E_e can be written in the same form,

$$E_e = aE_e^{(1)} + bE_e^{(2)}$$

with $E_e^{(1)}$ and $E_e^{(2)}$ given by integrals like (9.60) involving $E_{eg}^{(1)}$ and $E_{eg}^{(2)}$, respectively. It is convenient to represent E_{eg} as a superposition of simple functions like sines and cosines. Then E_e will be given as the same superposition of the transformed sines and cosines. It happens that these functions are a particularly happy choice, since transformed sine and cosine functions are still sines and cosines if they are restricted to a single isoplanatism patch, as we are about to prove.

4. The Response to a Sinusoidal Incoherent Object. We consider a single isoplanatism patch in the plane $z = z_i$. An arbitrary positive function E_{eg} describing the Gaussian flux density there can be expressed as a two-dimensional Fourier integral over spatial frequencies u_1, u_2. We consider just one of the Fourier components and add to it a constant term to keep E_{eg} nonnegative:

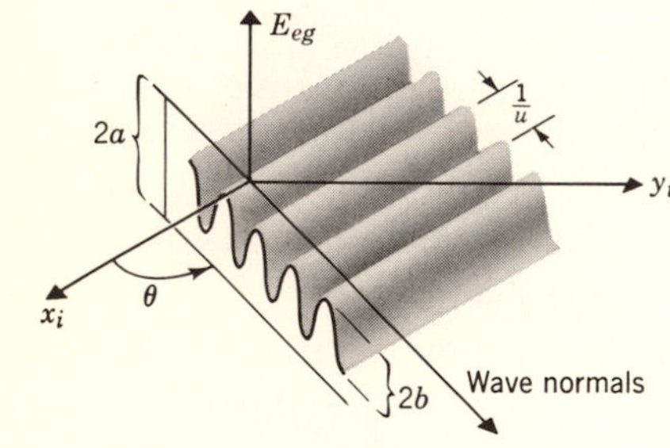

Fig. 9.36

$$\begin{aligned} E_{eg}(x_i, y_i, z_i) &= a\{1 + b \cos [2\pi(x_i u_1 + y_i u_2 + \phi]\} \\ &= \operatorname{Re} a\{1 + be^{-i\phi} \exp [-2\pi i(x_i u_1 + y_i u_2)]\} \qquad (9.61a) \end{aligned}$$

Here b is the *modulation* and is less than or equal to unity. The sign of $i = \sqrt{-1}$ is arbitrary as far as (9.61*a*) is concerned. We anticipate the appearance of the Fourier transform in Eq. (9.61*c*). This function is indicated schematically in two dimensions in Fig. 9.36. Polar coordinates for the spatial frequencies are $u = \sqrt{u_1^2 + u_2^2}$ and $\theta = \tan^{-1} u_2/u_1$. Here θ gives the direction of the wave normals, and $1/u$ gives the crest-to-crest distance in this direction.

The distribution of flux density at the object is the Gaussian conjugate to Eq. (9.61*a*) and is given by

$$E_e(x_0, y_0, z_0) = \text{const} \cdot E_{eg}(mx_0, my_0, z_i) \qquad (9.61b)$$

where $m = -z_i/z_0$ is the lateral magnification. To obtain the actual flux density in the image plane we apply Eq. (9.60) to (9.61*a*) and obtain

$$E_e(x, y, z_i) = \mathrm{Re}\left[a + abe^{-i\phi}\iint\limits_{-\infty}^{\infty} O(x - x_i, y - y_i; P_i)e^{-2\pi i(x_i u_1 + y_i u_2)}\, dx_i dy_i\right]$$

$$= \mathrm{Re}\{a + abe^{-i\phi}e^{-2\pi i(u_1 x + u_2 y)}$$

$$\times \iint\limits_{-\infty}^{\infty} O(x - x_i, y - y_i; P_i)e^{-2\pi i[u_1(x_i - x) + u_2(y_i - y)]}\, dx_i dy_i\}$$

$$= \mathrm{Re}\,[a + abe^{-i\phi}e^{-2\pi i(u_1 x + u_2 y)}\hat{O}(u_1, u_2; P_i)] \tag{9.61c}$$

Since we stay on one isoplanatism patch, we have treated P_i in Eq. (9.61*c*) as a parameter, not a variable to be integrated over. The function $\hat{O}$ appearing in Eq. (9.61*c*) as a Fourier transform is called the *optical transfer function* (OTF). If we write it in the form

$$\hat{O} = |\hat{O}|e^{i\psi} \tag{9.61d}$$

we obtain

$$E_e = a\{1 + b|\hat{O}(u_1, u_2; P_i)| \cos 2\pi\,[(xu_1 + yu_2) + \phi - \psi]\} \tag{9.61e}$$

We have just shown that the image of an incoherent sinusoidal object is sinusoidal with the same frequencies (u_1, u_2) as that of the geometrical Gaussian image. The amount of modulation b will be changed by the factor $|\hat{O}|$, and the phase shifted by ψ. Because of its role in Eq. (9.61*e*), $|\hat{O}|$ is called the *modulation transfer function* (MTF). The symbol ψ is sometimes called the *phase transfer function* (PTF).

5. Relationship between the Optical Transfer Function and the Pupil Function. The pupil function (9.53*b*) is

$$\tau_P(x', y'; P_i) = \tau_P(x', y')e^{-ikW(x', y'; P_i)}$$

where for a simple pupil $\tau_P(x', y')$ is defined to be unity on Σ_p and zero off Σ_p. The geometrical theory of aberrations yields expressions for W and hence for the pupil function. We then want the OTF $\hat{O}$ in terms of the pupil function. The result of a short calculation (See Problem 9.3.5) gives

$$\hat{O}(u_1, u_2; P_i) = \iint\limits_{-\infty}^{\infty} \tau_P(x', y'; P_i)\tau_P{*}(x' - \lambda L u_1, y' - \lambda L u_2, P_i)\,\frac{dx'\,dy'}{\sigma_p} \tag{9.62a}$$

This formula expresses $\hat{O}$ as the two-dimensional normalized autocorrelation function of the complex pupil function. Examples of autocorrelation functions were encountered in Chapter 6. There we correlated mostly real functions. A complex correlation takes the form of (9.62*a*) with one factor having the complex conjugate.

We say (9.62*a*) is normalized because at zero spatial frequency we have

$$\hat{O}_0(0, 0; P_i) = \iint_{-\infty}^{\infty} |\tau_p(x', y'; P_i)|^2 \frac{dx'\,dy'}{\sigma_p} = \iint_{-\infty}^{\infty} \tau_p(x' - y')^2 \frac{dx'\,dy'}{\sigma_p} = 1$$

To obtain this result we have made use of the fact that

$$\begin{aligned} \tau_P{}^2(x', y') &= 1 \qquad \text{on } \Sigma_p \\ &= 0 \qquad \text{off } \Sigma_p \end{aligned}$$

and the fact that σ_P is the area of Σ_p and have thus assumed a simple pupil. For such a pupil we may write (9.62*a*) as

$$\hat{O}(u_1, u_2; P_1) = \iint_{\Sigma_{pc}} \exp\{-ik[W(x', y'; P_i) - W(x' - \lambda L u_1, y' - \lambda L u_2; P_i)]\} \frac{dx'\,dy'}{\sigma_p} \tag{9.62b}$$

where Σ_{pc} represents the intersection of Σ_p with itself when its center is displaced to $(\lambda L u_1, \lambda L u_2)$.

Note that for the diffraction-limited case, that is, where there are no aberrations and $W = 0$, Eq. (9.62*b*) gives $\hat{O}$ as simply the ratio of the area of this intersection to the area of Σ_p.

Another important general property of the OTF follows from the so-called triangle inequality, true for an arbitrary complex function of one or more variables:

$$\left| \int_a^b f(x)\,dx \right| \leqslant \int_a^b \left| f(x) \right| dx$$

This gives us an inequality for the modulation transfer function $|\hat{O}|$:

$$|\hat{O}(u_1, u_2; P_i)| \leqslant \iint_{-\infty}^{\infty} |\tau_p(x', y'; P_i)|\,|\tau_p(x' - \lambda L u_1, y' - \lambda L u_2; P_i)| \times \frac{dx'\,dy'}{\sigma_p} \tag{9.63a}$$

But $|\tau_P(x', y'; P_i)| = \tau_P(x', y')$, which is the value of $\tau_P(x', y'; P_i)$ when there are no aberrations. Thus, if we let $\hat{O}_0(u_1, u_2)$ denote the OTF for the diffraction-limited no-aberration case, we may write

$$\begin{aligned} |\hat{O}(u_1, u_2; P_i)| &\leqslant \iint_{-\infty}^{\infty} \tau_p(x', y')\tau_p(x' - \lambda L u_1, y' - \lambda L u_2) \frac{dx'\,dy'}{\sigma_p} \\ &= \hat{O}_0(u_1, u_2) \end{aligned} \tag{9.63b}$$

Thus the MTF of a system is never greater than the OTF of an ideal aberration-free system with the same aperture.

6. The Diffraction-Limited OTF with a Circular Aperture. Besides acting as an upper bound on the MTF for real systems with aberrations, the diffraction-limited OTF is of interest in its own right, because in some cases, such as "diffraction-limited" telescope objectives (on or near axis), it is closely approached. From the definition of $\tau_P(x', y')$ we recognize that the product $\tau_P(x', y')\,\tau_P(x' - \lambda L u_1, y' - \lambda L u_2)$ gives unity if and only if the points (x', y') and $(x' - \lambda L u_1, y' - \lambda L u_2)$ are both in the pupil opening. The integral in Eq. (9.63*b*) then represents the area of the intersection of the pupil with itself displaced from the origin $x' = y' = 0$ to the point $(x' = \lambda L u_1, y' = \lambda L u_2)$. This is shown shaded in Fig. 9.37*a*. Thus $\hat{O}_0(u_1, u_2)$ represents this shaded area divided by the area of the

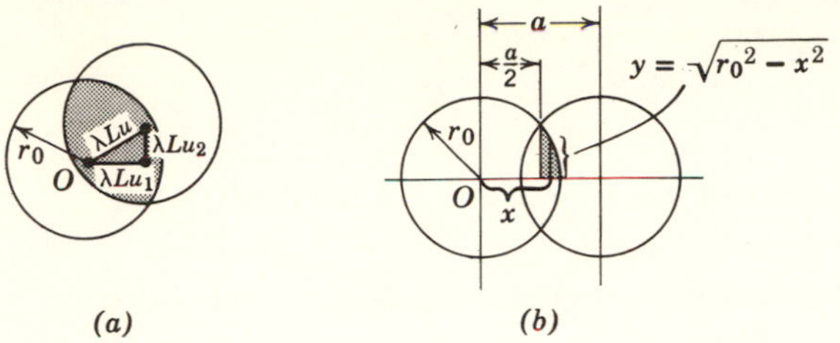

Fig. 9.37

pupil $\sigma_P = \pi r_0^2$. It is clear that the result can depend only on the distance $a = \lambda L u = \lambda L \sqrt{u_1^2 + u_2^2}$ that separates the centers of the two circles. The desired area is four times that shaded in Fig. 9.37*b*, and the OTF is given by

$$\hat{O}_0(u_1, u_2) = \frac{4}{\pi r_0^2} \int_{a/2}^{r_0} y\,dx = \frac{4}{\pi r_0^2} \int_{a/2}^{r_0} \sqrt{r_0^2 - x^2}\,dx$$

$$= \frac{2}{\pi} \left\{ \cos^{-1} \frac{a}{2r_0} - \frac{a}{2r_0} \left[1 - \left(\frac{a}{2r_0} \right)^2 \right]^{1/2} \right\}, \quad \text{if } 0 \leq \frac{a}{2r_0} \leq 1$$

$$= 0 \quad \text{if } \frac{a}{2r_0} \geq 1. \tag{9.64a}$$

The maximum spatial frequency that is transmitted by the system corresponds to $a = 2r_0$ and is

$$u_m = \frac{2r_0}{\lambda L} \tag{9.64b}$$

$$\hat{O}_0(u_1, u_2) = \frac{2}{\pi} \left\{ \cos^{-1} \frac{u}{u_m} - \frac{u}{u_m} \left[1 - \left(\frac{u}{u_m} \right)^2 \right]^{1/2} \right\}, \qquad u = \sqrt{u_1^2 + u_2^2} \tag{9.64c}$$

This function is plotted as the dotted line in Figs. 9.39*a*, *b*.

We should emphasize that $\hat{O}_0$ describes the loss of contrast in an ideal diffraction limited system. For instance, when $\hat{O}_0 = \frac{1}{4}$, we would have the situation shown in Fig. 9.38, if the object were a fully modulated sinusoidal test pattern. If the spatial frequency

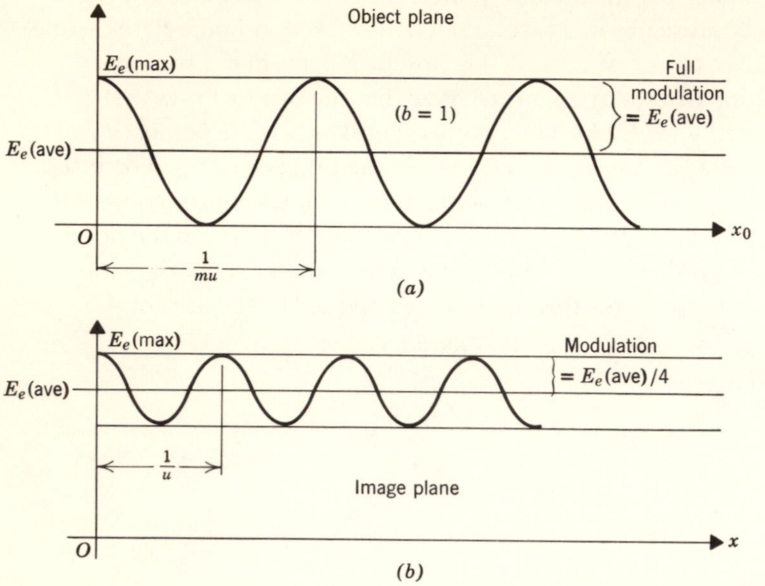

Fig. 9.38. Object flux density (a) and image flux density (b) when OTF = $\frac{1}{4}$.

in image space is u, it is mu in object space, where $m = -z_i/z_0$ is the lateral magnification.

7. ***Some Remarks about OTF's for Real Systems.*** There has been an extensive literature on the measurement of the OTF's for real lenses and the calculation of OTF's for lenses with aberrations. We cannot devote much space here to the many details necessary for a good understanding. Some calculated results are shown in Figs. 9.39a, b and 9.41.

A circular aperture of radius r_0 is assumed in all cases.

Figure 9.39 covers two cases where the aberration function does not depend on P_i. The spread function is then a function only of $\sqrt{(x-x_i)^2+(y-y_i)^2}$, and the OTF is then a function only of $u = \sqrt{u_1^2+u_2^2}$ and is real. For m wavelengths of focusing error the aberration function is

$$W = m\left(\frac{r'}{r_0}\right)^2 \lambda \tag{9.65a}$$

The resulting OTF curves are shown in Fig. 9.39a for $m\pi = 0, 1, 2, 3, 4$. For m wavelengths of primary spherical aberration we have

$$W = m\left(\frac{r'}{r_0}\right)^4 \lambda$$

This is for an observation plane at the paraxial focus. If we move

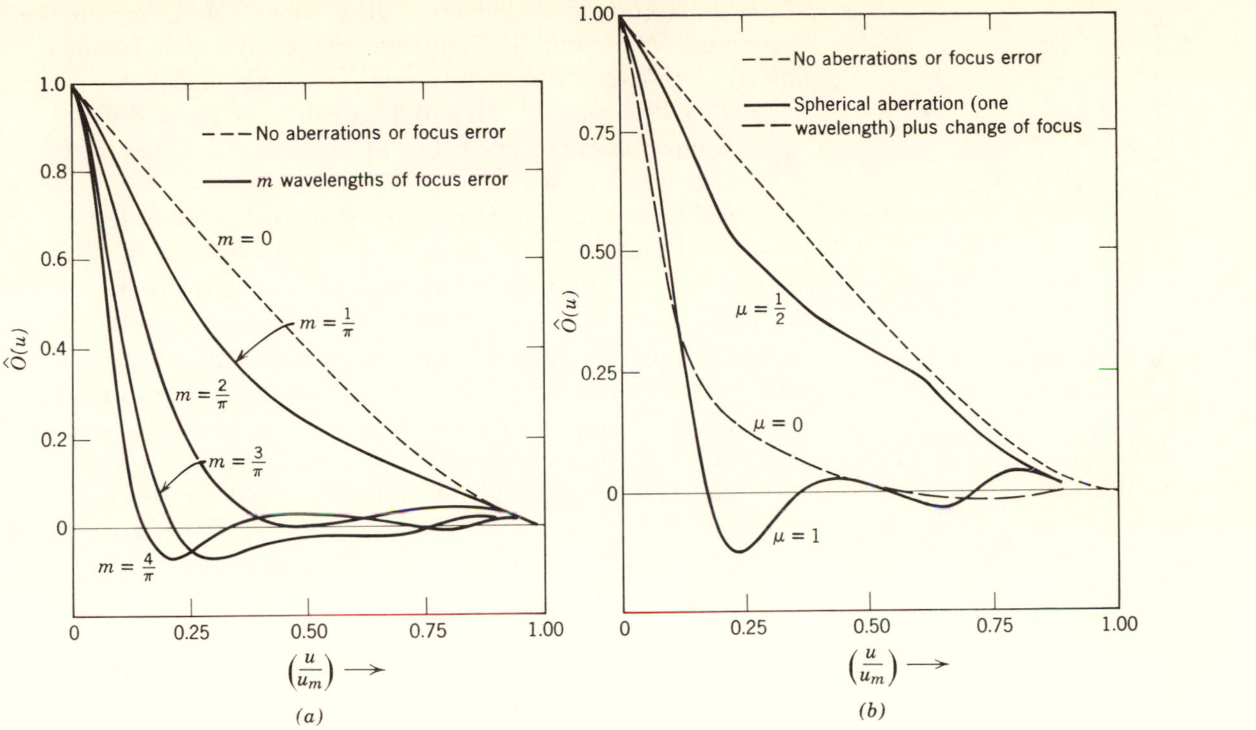

Fig. 9.39. (*a*) The optical transfer function for a system with focus error. [After H. H. Hopkins, *Proc. Roy. Soc.* (London) **A231**, 91 (1955). Used with permission of The Royal Society.] (*b*) The optical transfer function for one wavelength of spherical aberration. $\mu = 0$ corresponds to the paraxial focus, $\mu = 1$ to the marginal focus. [After G. Black and E. H. Linfoot, *Proc. Roy. Soc.* (London) **A239**, 522 (1957). Used with permission of the Royal Society.]

the observation plane along the axis, we must add a term like (9.65*a*). This gives

$$W = m\lambda \left[\left(\frac{r'}{r_0} \right)^4 - 2\mu \left(\frac{r'}{r_0} \right)^2 \right] \tag{9.65b}$$

The parameter μ is zero at the paraxial focus and unity at the marginal focus. Curves for the OTF for $m = 1$ are shown in Fig. 9.39*b* for $\mu = 0, \frac{1}{2}, 1$. When m is small, the best performance is obtained for $\mu = \frac{1}{2}$, halfway between paraxial and marginal foci. Note that some of the curves in Fig. 9.39 have the OTF negative. This can be described by a sudden jump in the phase ψ from 0 to π. It corresponds to a reversal of contrast (dark regions in the object appear light) and leads to spurious resolution, since for objects containing spatial frequencies where the OTF is negative we have *apparent* resolution, whereas for objects containing lower spatial frequencies near where the OTF is passing through zero there is little or no resolution.

Off-axis aberrations will introduce a dependence on the off-axis distance OP_i into the spread function and the OTF. The spread function will no longer have rotational symmetry about P_i, but it will by symmetric about the line OP_i. Because of this, it is convenient to define a rotated coordinate system with the $\bar{x}$ axis along this line. The OTF will then depend both on the magnitude u of

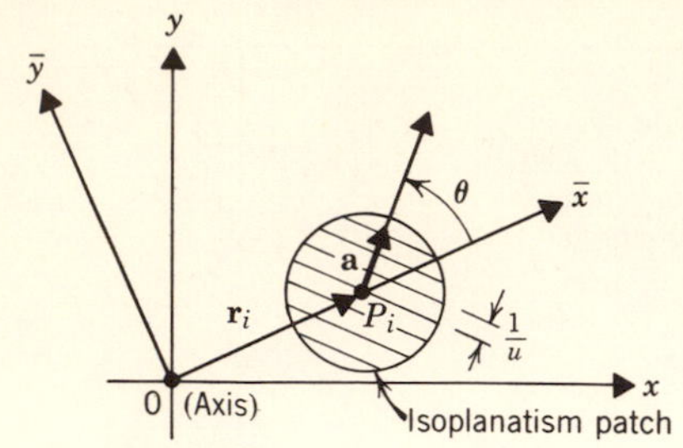

Fig. 9.40

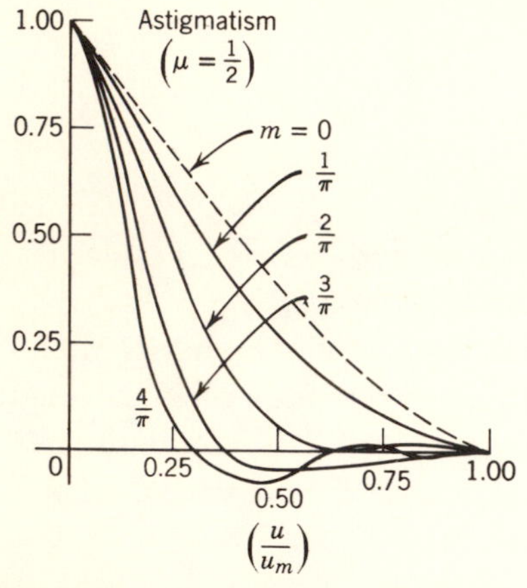

Fig. 9.41. The optical transfer function at the circle of least confusion for *m* wavelengths of primary astigmatism. [After M. De, *Proc. Roy. Soc.* (London) **A233**, 91 (1955). Used with permission of The Royal Society.]

the spatial frequency and on the angle θ between the normal to the wave crests in the sinusoidal pattern and the x axis. This is shown in Fig. 9.40. When $\theta = \pi/2$, the crests in the sinusoidal pattern are parallel to the symmetry axis and there can be no phase shift ψ, apart from a sudden reversal of contrast. For other values of θ, ψ will generally be nonzero.

Actually for primary astigmatism, the phase ψ is either zero or π as with the other cases discussed so far. The reason is that the point spread function, while no longer possessing rotational symmetry about P_i, still has inversion symmetry about P_i, that is, is the same at $\mathbf{r}_i + \mathbf{a}$ and at $\mathbf{r}_1 - \mathbf{a}$ in Fig. 9.40. Such a function will have no sine Fourier transform, only a cosine Fourier transform (see Appendix A) and an exponential transform that is purely real. Halfway between tangential and sagittal image planes the aberration function for m wavelengths of primary astigmatism is given by

$$W = m\lambda \left(\cos^2\phi - \frac{1}{2}\right)\left(\frac{r'}{r_0}\right)^2 \tag{9.65c}$$

This is where geometrical optics gives the circle of least confusion. The point spread function has four-fold symmetry about P_i. This means that the OTF for angle θ is the same as that for angle $\pm(\pi/2) \pm \theta$. Some OTF curves are shown in Fig. 9.41 for $\theta = 0$ (or $\theta = \pi$) and for $\theta = \pm\frac{1}{2}\pi$ (or $\theta = \pm\frac{3}{2}\pi$).

For real lenses the small-aberration results discussed above are often inadequate, and other methods based on ray tracing or other digital computational techniques must be employed. For a discussion, see the book by Smith cited in the references. Problem 9.3.16 develops one experimental method of measuring the OTF; for a detailed discussion, see the article by Murata cited in the references.

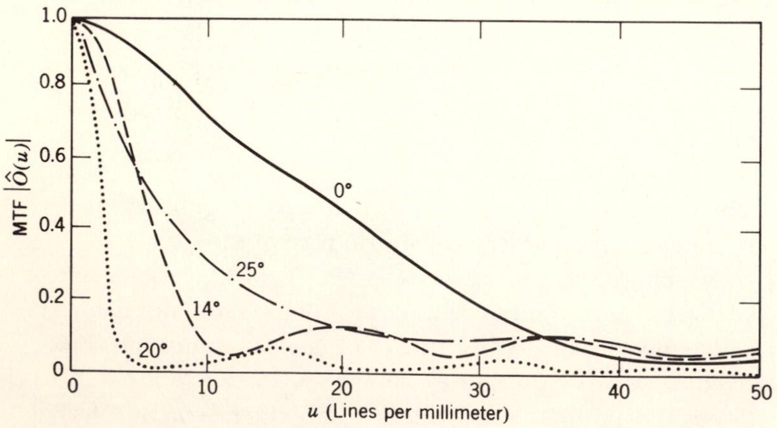

Fig. 9.42. The modulation transfer function for a Cooke triplet camera lens of 100 mm focal length. [After Kingslake in *Applied Optics and Optical Engineering*, ed. by R. Kingslake, Vol. III (Academic Press, New York, 1965), p. 19.]

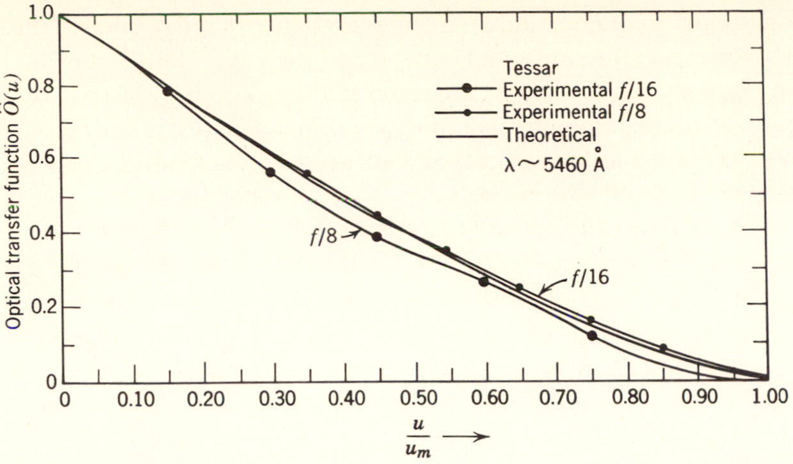

Fig. 9.43. The optical transfer function for a Tessar lens at two different F numbers. Note that u_m is different in the two cases. [After R. R. Shannon in *Applied Optics and Optical Engineering*, ed. by R. Kingslake, Vol. III, (Academic Press, New York, 1965) p. 220.]

As an example we show the MTF for a Cooke triplet camera lens in Fig. 9.42 for various values of the off-axis angle. As another example plots are given in Fig. 9.43 of the on-axis MTF for a Tessar lens at two different F numbers (F number = focal length/lens diameter). This shows that the theoretical diffraction-limited behavior is closely approached for well-designed lenses used at small apertures.

PROBLEMS

9.3.1 Consider diffraction by the shaded aperture (*a*) having a real transmission function $h(x)$ plotted in (*b*). The pattern is observed in the Gaussian image plane where a point image would otherwise be formed. Show that along the x axis in the observation plane the field is the same if the shaded aperture is replaced by a template with transmission $\tau = 0$ or 1 as limited by the curves $y = \pm bh(x)$ shown in (*c*).

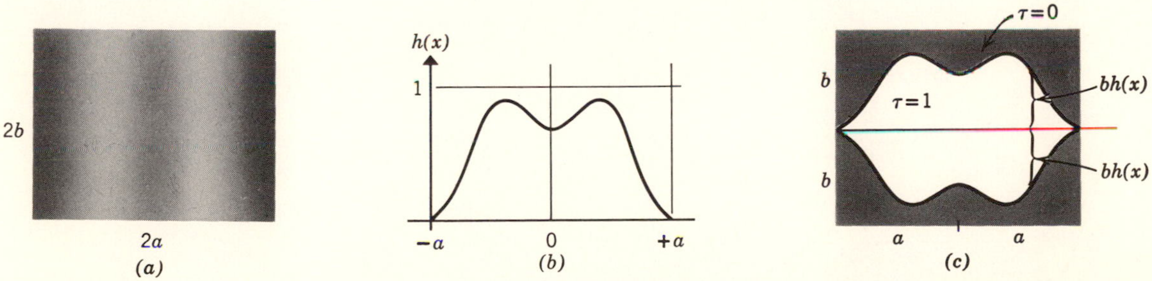

Consider the original object to be an incoherent line source, such as would be obtained from an incoherently illuminated slit. Suppose that the Gaussian image then is an incoherent line coinciding with the y axis. Show that the flux density in the Gaussian image plane with either aperture (*a*) or aperture (*c*) depends only on x for y not too large, that it is the same with either aperture, and that it is proportional to the flux density along the x axis with a single point source, as in the first part of this problem.

9.3.2 The diffraction pattern produced by a point object depends on the transmission function of the aperture stop, or the equivalent transmission function of the exit pupil. By using a shaded aperture one can reduce unwanted features in the diffraction pattern such as large amounts of energy in the tails of the pattern, at the cost of an increase in the size of the main maximum. This process is known as *apodization*. The calculations called for in this problem will bring out some features of the apodization process. For simplicity we restrict ourselves to the case where only one independent variable, x, is of interest. The results of Problem 9.3.1 are then relevant.

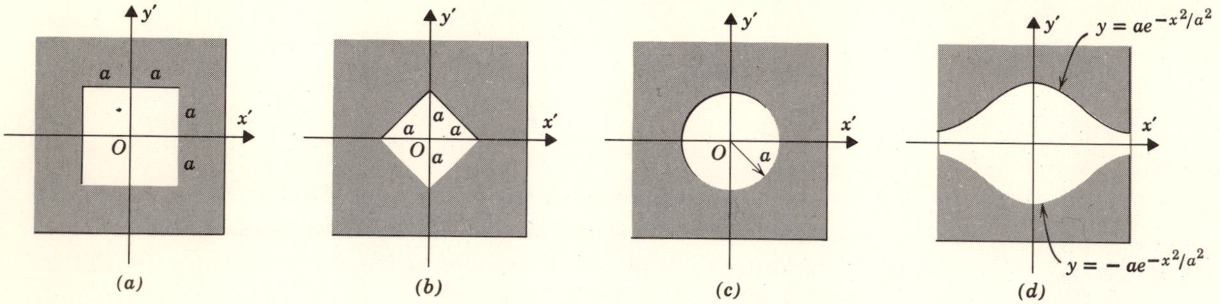

Consider apertures having the shapes shown above. Calculate the flux densities along the line Ox in the plane of a Gaussian point image. In all cases but (*c*), estimate how far in the x direction one has to move from the point image for the flux density to reach and remain 10^{-6} times its peak value.

9.3.3 Prove Parseval's theorem,

$$\int_{-\infty}^{\infty} |f(x)|^2\, dx = \int_{-\infty}^{\infty} |\hat{f}(u)|^2\, du$$

by direct calculation using definition of the Fourier transform

$$\hat{f}(u) = \int_{-\infty}^{\infty} f(x)\, e^{-2\pi iux}\, dx$$

Prove Parseval's theorem in two-dimensions,

$$\iint_{-\infty}^{\infty} |f(x, y)|^2\, dx\, dy = \iint_{-\infty}^{\infty} |\hat{f}(u, v)|^2\, du\, dv$$

9.3.4 Using Parseval's theorem (previous problem), prove directly that the integral of (9.56) over x and y gives unity. What is the result when the pupil is not simple, that is, when τ_P takes on values other than 0 or 1?

9.3.5 The *convolution* of two functions $f_1(x)$ and $f_2(x)$ denoted by $f_1 * f_2$ is defined by the integral

$$f(x) = f_1 * f_2 = \int_{-\infty}^{\infty} f_1(x')\, f_2(x - x')\, dx'$$

For functions of two variables the definition is

$$f(x, y) = f_1 * f_2 = \iint_{-\infty}^{\infty} f_1(x', y') f_2(x-x', y-y')\, dx'\, dy'$$

Let $\hat{f}_1(u)$ be the Fourier transform (F.T.) of $f_1(x)$, $\hat{f}_2(u)$ be the F.T. of $f_2(x)$, and $\hat{f}(u)$ be the F.T. of $f(x)$. Show that

$$\hat{f}(u) = \hat{f}_1(u)\hat{f}_2(u)$$

that is, that the *convolution theorem* is valid:

$$\text{F.T.}(f * g) = [\text{F.T.}(f)] \cdot [\text{F.T.}(g)]$$

Show that this result also holds in two dimensions.

The convolution theorem may be used to prove Eq. (9.62*a*). Except for constants, Eq. (9.57*a*) says that the OTF $\hat{O}$ is the F.T. of

$$\hat{\tau}_P \hat{\tau}_P^* = \text{F.T.}(\hat{\tau}_P * \hat{\tau}_P^*) \qquad (1)$$

We have used the convolution theorem here. When the F.T. is taken of both sides of Eq. (1), and when the exact constants and arguments of τ_P and $\hat{\tau}_P$ are used, Eq. (9.62*a*) results. Supply these details.

9.3.6 Show that the Fourier transform of the product of two functions is equal to the convolution of the transforms of the two functions. By this we mean the following. Let $F(x) = f_1(x)f_2(x)$ and let $\hat{F}(u)$ be the F.T. of $F(x)$. Show that

$$\hat{F}(u) = \hat{f}_1 * \hat{f}_2 = \int_{-\infty}^{\infty} \hat{f}_1(u')\hat{f}_2(u-u')\, du'$$

9.3.7 Using Problem 9.3.6, show that

$$\int_{-\infty}^{\infty} f_1(x)f_2(x)\, dx = \int_{-\infty}^{\infty} \hat{f}_1(u)\hat{f}_2(-u)\, du$$

Now let $f_2(x) = f_3(x)^*$. Show that $\hat{f}_2(-u) = \hat{f}_3(u)^*$. This gives

$$\int_{-\infty}^{\infty} f_1(x)f_3(x)^*\, dx = \int_{-\infty}^{\infty} \hat{f}_1(u)\hat{f}_3(u)^*\, du$$

This is a generalization of Parseval's theorem.

9.3.8 Consider an aperture consisting of two superimposed apertures so that the pupil transmission function is

$$\tau_P(x', y') = \tau_1(x', y')\tau_2(x', y')$$

When an arrangement like that in Fig. 9.28 is used, the field in the plane $z = z_i$ will be given by Eq. (9.47).

a. Apply the result of Problem 9.3.6 to express $\tilde{E}_A$ in terms of the Fourier transforms of τ_1 and τ_2.
b. Obtain the same result as that in (a) by first using τ_1 in (9.47) to obtain a field $\tilde{E}_1$. Then use $\tilde{E}_1$ inside the integral with τ_2 to obtain $\tilde{E}_A$.
   As a special case, consider the situation where $\tau_1 = \tau_2 = 0$ or 1 only.

Then $\tau_P = \tau_1 = \tau_2$. The results of part (a) or (b) then imply that $\tau_P = \tau_1 * \tau_1 = \tau_1$. Prove this by direct calculation of the convolution of τ_1 with itself, using the definition of τ_1.

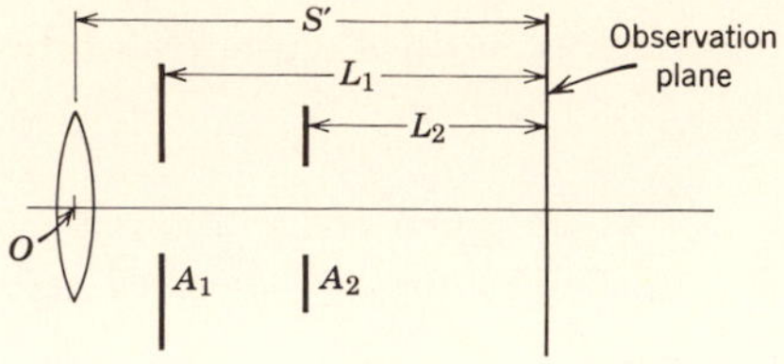

9.3.9 Assume that the lens in the above system produces a field distribution $\tilde{E}_{IA}(x, y, S')$ in the plane $z = S'$ if the lens aperture and all other apertures are neglected. What then is the field in the plane $z = S'$ with both apertures A_1 (transmission function $T_1[x'/L_1, y'/L_1]$) and A_2 (transmission function $T_2[x'/L_2, y'/L_2]$) in place? (Neglect the lens aperture.)
Hint. Use Eq. (9.43′) twice.

This result also holds also if A_1 and/or A_2 are images in image space of physical apertures occurring farther to the left in the system. When $L_1 \neq L_2$, your result should account for Fresnel diffraction resulting from propagation from A_1 to A_2 by the appearance of a factor

$$\exp\left[\frac{ik}{2}(x_1^2 + y_1^2)\left(\frac{1}{L_1} - \frac{1}{L_2}\right)\right]$$

Even for simple apertures where $T_1 = T_2 = 0$ or 1, this factor keeps the combined effect of A_1 and A_2 from being that of A_1 or A_2 acting alone.

9.3.10 a. Prove that for any simple aperture Eq. (9.58*c*) is true:

$$O_0(0, 0) = \frac{\sigma_p}{(\lambda L)^2}$$

b. For a nonsimple aperture, show that the above result generalizes to

$$O_0(0, 0) = \frac{1}{(\lambda L)^2} \iint_{-\infty}^{\infty} |t_P(x', y')|^2 \, dx' \, dy'$$

9.3.11 Derive Eqs. (9.59*a*, *b*).

9.3.12 Because of Eq. (9.58*c*) we may rewrite Eq. (9.59*a*) to give

$$O(0, 0; P_i) = O_0(0, 0)\,[1 - D]$$

$$D = \frac{4\pi^2}{\lambda^2}(\langle W^2 \rangle - \langle W \rangle^2)$$

where D represents the fractional decrease in $O(0, 0; P_i)$ because of a small amount of aberration. Calculate D for the following aberrations for a spherical exit pupil of radius r_0 ($\rho = r'/r_0$).

a. m wavelengths of focus error (curvature of field):

$$W = m\lambda\rho^2 \qquad \left[\text{Answer: } D = \frac{\pi^2}{3} m^2\right]$$

b. m wavelengths of primary spherical aberration

$$W = m\lambda\rho^4 \qquad \left[\text{Answer: } D = \frac{16\pi^2}{45} m^2\right]$$

c. Spherical aberration plus change in focal plane

$$W = m\lambda\rho^2\,[\rho^2 - 2\mu]$$

Show that D is minimized when $\mu = \frac{1}{2}$. (Halfway between paraxial and marginal foci.)

d. Astigmatism and curvature of field

$$W = m\lambda\rho^2\,[\cos^2\phi - \mu]$$

Show that D is minimized when $\mu = \frac{1}{2}$. (Halfway between paraxial and sagittal foci, at the circle of least confusion.)

e. Coma

$$W = m\lambda\rho^3\cos\phi \qquad \left[\text{Answer: } D = \frac{\pi^2}{2} m^2\right]$$

9.3.13 Despite the presence of aberrations, the initial slope of the MTF is the same as that for a diffraction-limited system. Show that for a spherical exit pupil of radius the modulation transfer function obeys

$$\frac{d}{du}|\hat{O}(u)|_{u=0} = \frac{d}{du}\hat{O}_0(u)|_{u=0} = \frac{-4}{\pi u_m}, \qquad u_m = \frac{2r_0}{\lambda L}$$

9.3.14 Show that for a rectangular aperture $2x_0$ by $2y_0$ the diffraction-limited OTF is

$$\hat{O}_0(u_1, u_2) = \left(1 - \left|\frac{u_1}{u_{1m}}\right|\right)\left(1 - \left|\frac{u_2}{u_{2m}}\right|\right) \qquad \text{for } |u_1| \leq u_{1m}, \qquad |u_2| \leq u_{2m}$$
$$= 0 \qquad \text{otherwise}$$

where

$$u_{1m} = \frac{2x_0}{\lambda L}, \qquad u_{2m} = \frac{2y_0}{\lambda L}$$

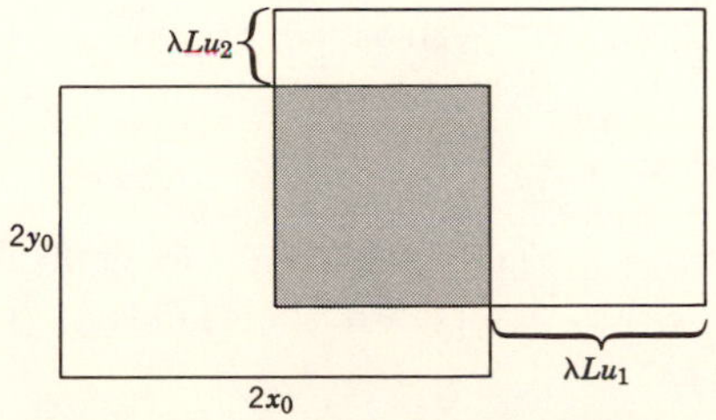

9.3.15 Show that with focusing error where $W = [n(x'/x_0)^2 + m(y'/y_0)^2]$ the optical transfer function for a $2x_0$ by $2y_0$ rectangular pupil is given by multiplying the result of 9.3.14 by $(\sin t_x/t_x) \cdot (\sin t_y/t_y)$,

$$t_x = 8\pi n \frac{u_1}{u_{1m}}\left|1 - \frac{u_1}{u_{1m}}\right|, \qquad t_y = 8\pi m \frac{u_2}{u_{2m}}\left|1 - \frac{u_2}{u_{2m}}\right|$$

9.3.16 This problem discusses the theoretical basis for one method of measuring the OTF.

In an isoplanatism patch centered at P_i in image space there is an incoherent source in the form of a fine line that gives a Gaussian image as shown in (*a*). Take a coordinate system (ξ, η) centered at P_i with axes

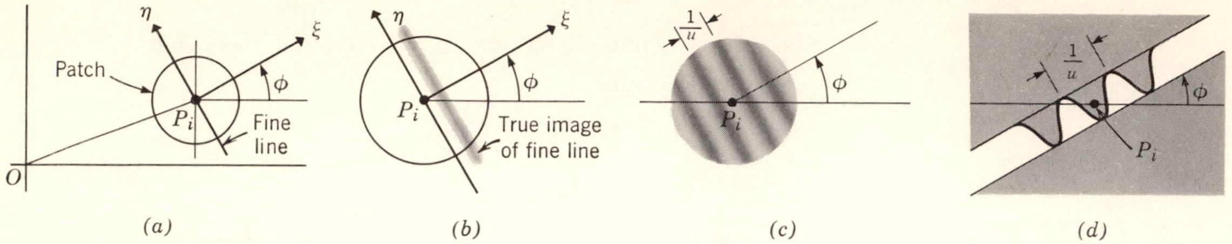

parallel to and perpendicular to the fine line. In Fig. (*b*) the true image of the line is represented. Define the "fine-line response function" F to be the flux density in this image as a function of ξ, normalized to give unity when integrated over ξ. (The integral should really be limited to the isoplanatism patch, but we neglect edge effects here and in what follows.)

a. Show that the fine-line response function is given by

$$F(\xi, \phi; P_i) = \int_{-\infty}^{\infty} O(\xi \cos \phi - \eta \sin \phi, \eta \cos \phi + \xi \sin \phi; P_i)\, d\eta$$

b. Show that as a function of the coordinates $u = \sqrt{u_1^2 + u_2^2}$ and $\phi = \arctan u_2/u_1$ the OTF is given by the Fourier transform of the fine-line response function:

$$\hat{O}(u, \phi; P_i) = |O(u, \phi; P_i)|e^{i\psi} = \int_{-\infty}^{\infty} F(\xi, \phi; P_i) e^{-2\pi i u \xi} d\xi$$

c. A shaded aperture is placed in the image plane as shown in Fig. (*c*). It has a transmission function proportional to

$$[1 + \cos(2\pi u \xi + \beta)]$$

and can be moved in the ξ direction. Show that the transmitted flux density is proportional to

$$[1 + |\hat{O}(u, \phi; P_i)| \cos(2\pi u \xi + \beta + \psi)]$$

d. The shaded aperture at the image plane is replaced by the sinusoidal template as shown in Fig. (*d*). Show that the result of part (c) still holds.

9.3.17 It is not true that the OTF for the combined two-lens system shown in the figure is equal to the product of the individual OTF's for L_1 and

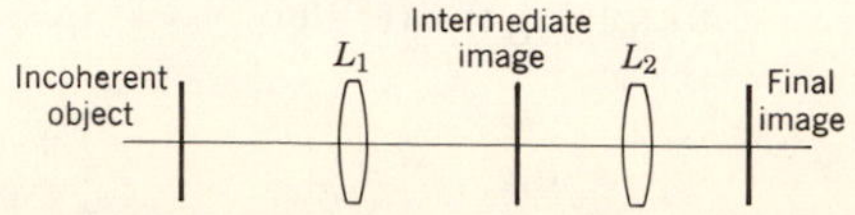

L_2. Prove this by assuming it to be true and then deriving a property of the point spread functions that can be shown to be false. (Hints. See Prob. 9.3.5. If the light at the intermediate image were completely incoherent, then the OTF's would multiply. But the light is partially coherent. Why?)

REFERENCES FOR CHAPTER 9

Many optics textbooks have good discussions of the Abbe theory of image formation, but they omit a complete discussion of the phases in the focal plane and elsewhere.

L. C. Martin's *The Theory of the Microscope*, Chapters IV–VIII, gives detailed coverage of most of the topics of this section.

The methods of spatial filtering and processing of coherent optical images have received much attention recently from electrical engineers. See for example *Introduction to Fourier Optics* by J. W. Goodman. or *Optical and Electro-Optical Information Processing*, edited by J. T. Tippett et al.; especially Chapter 4, "Use of the Fourier Transformable Properties of Lenses for Signal Spectrum Analysis," by K. Preston, Jr. or G. W. Stroke, *An Introduction to Coherent Optics and Holography*.

A detailed exposition of the diffraction theory of aberrations will be found in Sections 9.4 and 9.5 of *Principles of Optics* by M. Born and E. Wolf. In the latter section the optical transfer function is referred to as the "frequency response function."

The optical transfer function and other methods of lens evaluation are discussed from the practical point of view in Chapter 11 of *Modern Optical Engineering* by W. J. Smith. A discussion from a similar practical point of view will be found in the article "The Testing of Complete Objectives" by R. R. Shannon in Vol. III of *Applied Optics and Optical Engineering*, edited by R. Kingslake.

A short but clear discussion of the optical transfer function is found in Chapter 14 of *Geometrical and Physical Optics* by R. S. Longhurst. In Chapters 5 and 6 of *Introduction to Statistical Optics* by E. L. O'Neill will be found a detailed discussion of the OTF, including many references. A review article, "Instruments for the Measuring of Optical Transfer Functions," by K. Murata will be found in *Progress in Optics*, edited by E. Wolf, Vol. V (1966).

Some important original works are the following:

Lord Rayleigh, "On the Theory of Optical Images, with Special Reference to the Microscope," *Phil. Mag.*, (5) **42**, 167 (1896).

P. M. Duffieux, *L'Integrale de Fourier et ses applications à l'optique*, Besançon: Privately printed, 1946.

H. H. Hopkins, "On the Diffraction Theory of Optical Images," *Proc. Roy. Soc.* (London), **A217**, 408 (1953); "The Frequency Response of Optical Systems," *Proc. Phys. Soc.* (London), **B69**, 452 (1956).

In this chapter and the next we discuss important optical phenomena that can be understood only in terms of the transverse nature of electromagnetic waves. Since the field vectors are perpendicular to the propagation direction, as will be shown in Chapter 11, there is a degree of freedom involving their orientation not available in the case of scalar waves. The polarization properties of light are manifestations of this degree of freedom.

Most of the material in Chapter 10 is descriptive, rather than mathematical, with the exception of parts of Section 10.1 and of Section 10.3. The latter section, which uses the correlation function techniques of Chapter 6 to describe coherent light, can be omitted without adversely affecting the reader's understanding of the rest of Chapter 10.

CHAPTER 10

THE TRANSVERSE NATURE OF LIGHT–POLARIZATION

CHAPTER 10 CONTENTS

10.1 INTRODUCTION TO POLARIZED LIGHT

In a uniform isotropic medium, light waves and electromagnetic waves generally may be described as a superposition of plane waves. These consist of sinusoidally varying electric and magnetic fields aligned perpendicular to one another and to the direction of propagation. For light it is sufficient to specify the behavior of the electric field $\mathbf{E}(\mathbf{r}, t)$,* for then the magnetic induction $\mathbf{B}(\mathbf{r}, t)$ can be determined from $\mathbf{E}$ via Maxwell's equations, as shown in Chapter 11. For now, we limit ourselves to a discussion of $\mathbf{E}$.

A. THE ELEMENTARY TYPES OF POLARIZED LIGHT

Since the electric field is a vector quantity, we must specify both its magnitude and its direction. Consider a plane wave with wave vector $\mathbf{k}$. The electric field $\mathbf{E}$ may point along any direction in the plane perpendicular to $\mathbf{k}$, and $\mathbf{B}$ is along the vector $\mathbf{k} \times \mathbf{E}$. Let $\mathbf{k}$ be along the z-axis and have magnitude $k = \omega/v$, where $v = c/n$ is the phase velocity. The monochromatic wave can then be written in the form

$$\mathbf{E}(x, y, z, t) = \text{Re}\{\mathbf{E}_0 e^{i(\omega t - kz)}\} \tag{10.1}$$

where $\mathbf{E}_0$ is a constant vector in the xy plane that may be complex. This means that we can write

$$\mathbf{E}_0 = (\tilde{E}_x, \tilde{E}_y, 0) \tag{10.2}$$

where $\tilde{E}_x$ and $\tilde{E}_y$ may be complex:

$$\begin{aligned} E_x &= A_x e^{i\phi_x} \\ E_y &= A_y e^{i\phi_y} \end{aligned} \tag{10.3}$$

with positive amplitudes A_x, A_y and phase angles ϕ_x and ϕ_y. Then the three components of the real, physical, electric field will be given by

$$\begin{aligned} E_x(x, y, z, t) &= A_x \cos(\omega t - kz + \phi_x) \\ E_y(x, y, z, t) &= A_y \cos(\omega t - kz + \phi_y) \\ E_z(x, y, z, t) &= 0 \end{aligned} \tag{10.4}$$

1. Classification of Types. The classification of the types of behavior implied in Eq. (10.4) will depend on the relative phase

$$\phi = \phi_y - \phi_x \tag{10.5}$$

and on the relative sizes of A_x and A_y.

*In older books the direction of polarization is that of $\mathbf{B}$.

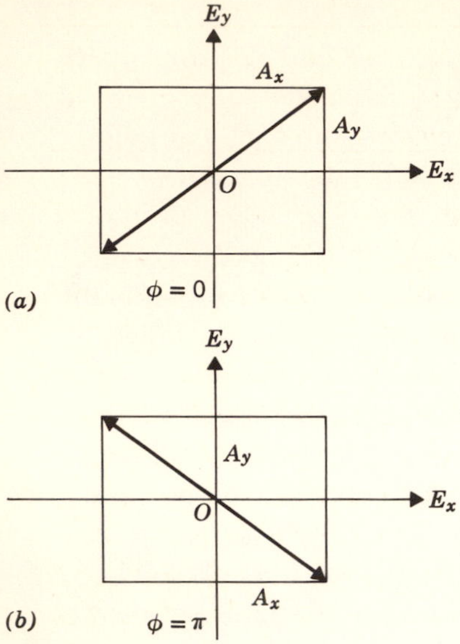

Fig. 10.1. Plane polarized light.

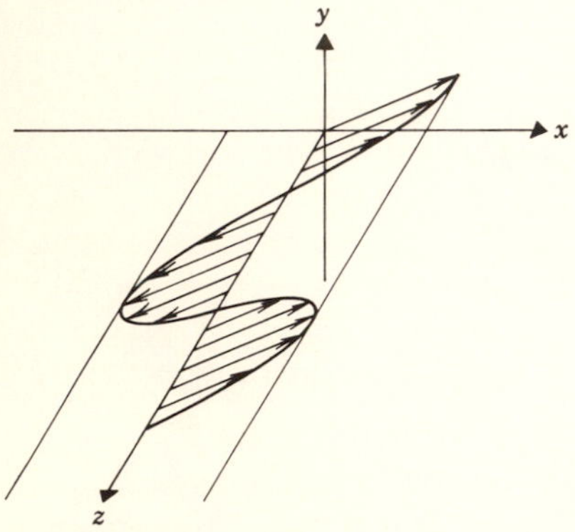

Fig. 10.2. ($\phi = 0$).

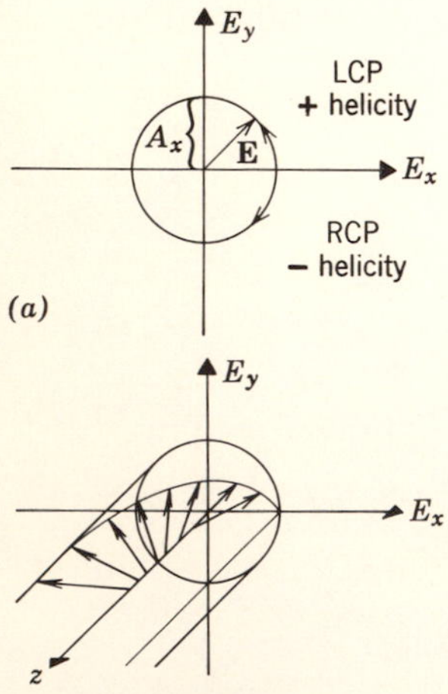

(b) RCP wave advancing in $+z$ direction

Fig. 10.3. Circularly polarized light.

Plane polarized light results when

$$\phi = 0 \qquad \text{or} \qquad \phi = \pi$$

for then E_y is proportional to E_x:

$$E_y = \frac{A_y}{A_x} E_x \qquad \text{for} \quad \phi = 0$$

$$E_y = -\frac{A_y}{A_x} E_x \qquad \text{for} \quad \phi = \pi$$

as can be seen from Eq. (10.4). At a fixed position z, the electric field undergoes simple harmonic motion along a line (Fig. 10.1). At a given value of t the electric field has a sinusoidal z dependence and is found in the plane containing the z axis and the line $y = +(A_y/A_x)x$ for $\phi = 0$, or $y = -(A_y/A_x)x$ for $\phi = \pi$ (Fig. 10.2).

Circularly polarized light results when

$$A_x = A_y \qquad \text{and} \qquad \phi = \pm\frac{\pi}{2} \tag{10.6}$$

Then

$$E_x = A_x \cos(\omega t - kz + \phi_x)$$

$$E_y = A_x \cos\left(\omega t - kz + \phi_x \pm \frac{\pi}{2}\right)$$

$$= \mp A_x \sin(\omega t - kz + \phi_x) \quad \left(\text{minus sign when } \phi = +\frac{\pi}{2}\right) \tag{10.7}$$

When $\phi = +\pi/2$, E_y leads E_x by 90°; that is, E_y reaches its maximum value a quarter of a cycle before E_x does. The electric field describes a *clockwise circle* in the xy plane as seen head-on. Using the language of modern particle physics, we would say that this light has "negative helicity." It is conventional in optics, however, to call this case where $\phi = +\pi/2$, *right circularly polarized* (RCP).

The other case of *left circularly polarized* (LCP) light or light of positive helicity occurs when $\phi = -\pi/2$. Then E_y lags E_x by a quarter of a cycle, and the electric field describes a counterclockwise circle in the xy plane (Fig. 10.3*a*).

We also obtain circular polarization when $\phi = \pm 3\pi/2$, for then the second of Eqs. (10.7) becomes

$$E_t = \pm A_x \sin(\omega t - kz + \phi_x)$$

In this case the minus sign gives clockwise rotation and right circular polarization, the plus sign counterclockwise rotation and left circular polarization. We can summarize in a table.

	Rotation with t	Rotation with z	Sense of polarization	Helicity
$+\pi/2$ or $-3\pi/2$	Clockwise	Counterclockwise	RCP	$-$
$-\pi/2$ or $+3\pi/2$	Counterclockwise	Clockwise	LCP	$+$

For a fixed value of t, the electric field describes a spiral on the surface of a circular cylinder of radius A_x with its axis along z (Fig. 10.3*b*). Since an increase in z amounts to a *retardation* in phase, RCP light will spiral counterclockwise along the z axis for fixed t and hence look like a *right-handed* screw. This explains the conventional optical terminology. As t increases this screw rotates clockwise, that is, backward.

When A_x no longer equals A_y, but when we still have $\phi = \pm\pi/2$, then the curve traced out by the E vector in the xy plane becomes an ellipse:

$$E_x = A_x \cos(\omega t - kz + \phi_x)$$
$$E_y = \pm A_y \sin(\omega t - kz + \phi_x) \qquad (10.8)$$

We can transform this into a familiar form:

$$\frac{E_x^2}{A_x^2} + \frac{E_y^2}{A_y^2} = \cos^2(\omega t - kz + \phi_x) + \sin^2(\omega t - kz - \phi_x) = 1 \qquad (10.9)$$

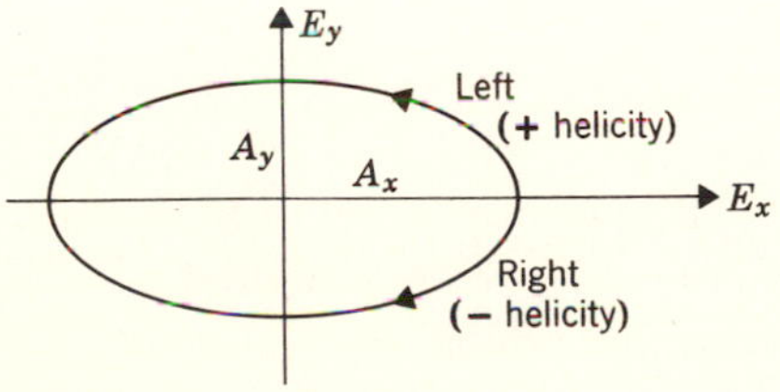

Fig. 10.4

Thus the semiaxes of the ellipse are given by A_x and A_y, and it is oriented along the x and y axes. In this case the electric field traverses an elliptical path in the xy plane, the sense of traversal being clockwise (or right-handed) for $\phi = \pm\pi/2$ and counterclockwise (or left-handed) for $\phi = -\pi/2$ (Fig. 10.4). This is a special case of *elliptically polarized light.*

The general case occurs when $\phi \neq \pi/2, \pm\pi$. We put $\phi_y = \phi_x + \phi$ and set

$$\theta = (\omega t - kz) \qquad (10.10)$$

to obtain

$$E_x = A_x \cos(\theta + \phi_x)$$
$$E_y = A_y \cos(\theta + \phi_y) \qquad (10.11a)$$

We may regard Eqs. (10.11*a*) as representing **E** as a superposition of x and y linearily polarized components or, using unit vectors $\hat{x}, \hat{y}$, as

$$\mathbf{E} = E_x\hat{x} + E_y\hat{y} \qquad (10.11b)$$

We call this a linearly polarized basis.

The ellipse described by Eqs. (10.11) is inscribed inside a rectangle having dimensions $2A_x$ by $2A_y$, as shown below in Fig. 10.7. The eccentricity of the ellipse depends on the relative phase difference $\phi = \phi_y - \phi_x$, reaching a minimum when $\phi = \pm\pi/2$ (Fig. 10.4) and a maximum when $\phi = \pm\pi$ (Fig. 10.1). The ellipses shown in Fig. 10.8 correspond to intermediate cases.

2. Elliptically Polarized Light as a Superposition of Circularly Polarized Components. A convenient method of describing the general case covered by Eq. (10.4) is as a superposition of right and left circularly polarized components rather than, as in (10.11*b*), a superposition of two linear polarized components vibrating along *x* and *y* axes. It is useful to use a complex notation to describe the circularly polarized components.

According to Eq. (10.7) for LCP (positive helicity) we can write

$$E_x = A_+ \cos(\theta + \phi_+) = \operatorname{Re} A_+ e^{+i\phi_+} e^{i\theta}$$
$$E_y = A_+ \sin(\theta + \phi_+) = \operatorname{Im} A_+ e^{i\phi_+} e^{i\theta}$$

where $A_+ > 0$ and ϕ_+ are constants. The form in these equations suggests that we make E_x and E_y into a two-dimensional vector or complex number

$$\tilde{E} = E_x + iE_y = A_+ e^{i\phi_+} e^{i\phi} \tag{10.12}$$

In a similar way for RCP (negative helicity) we may write

$$E_x = A_- \cos(\theta - \phi_-) = \operatorname{Re} A_- e^{i\phi_-} e^{-i\theta}$$
$$E_y = -A_- \sin(\theta - \phi_-) = \operatorname{Im} A_- e^{i\phi_-} e^{-i\theta}$$

or

$$\tilde{E} = E_x + iE_y = A_- e^{i\phi_+} e^{-i\theta} \tag{10.13}$$

Then a general superposition of light of plus and minus helicity may be written

$$\tilde{\mathrm{E}} = A_+ e^{i\phi_+} e^{i\theta} + A_- e^{i\phi_-} e^{-i\theta}$$

or introducing complex coefficients $a_\pm$,

$$\tilde{E} = a_+ e^{i\theta} + a_- e^{-i\theta} \tag{10.14}$$

Vector diagrams for the two terms in (10.14) are indicated in Fig. 10.5. They are similar, except that for positive helicity the vector $a_+ = A_+ e^{i\phi_+}$ rotates counterclockwise with increasing θ, whereas the opposite is true for negative helicity.

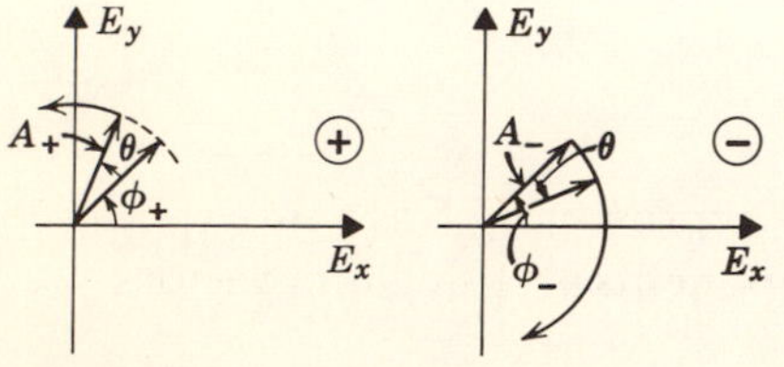

Fig. 10.5

We should mention here an important property of circularly polarized light beams, namely, that they have an *intrinsic angular momentum* or *spin*. We can show that the *z*-component of the angular momentum per unit volume is

$$L_z = \pm\frac{u}{\omega} \tag{10.15}$$

for ± helicity, respectively, where u is the electromagnetic energy density per unit volume in the circularly polarized beam,* and ω is the angular frequency.

Returning now to our general superposition (10.14), we can write (10.11) in the same form. First we obtain

$$E_x = A_x \cos(\theta + \phi_x) = \tfrac{1}{2}A_x e^{i\phi_x} e^{i\theta} + \tfrac{1}{2}A_x e^{-i\phi_x} e^{-i\theta} = \tfrac{1}{2}a_x e^{i\theta} + \tfrac{1}{2}a_x^* e^{-i\theta}$$

$$E_y = A_y \cos(\theta + \phi_y) = \tfrac{1}{2}A_y e^{i\phi_y} e^{i\theta} + \tfrac{1}{2}A_y e^{-i\phi_y} e^{-i\theta} = \tfrac{1}{2}a_y e^{i\theta} + \tfrac{1}{2}a_y^* e^{-i\theta}$$

with

$$a_x = A_x e^{i\phi_x}, \qquad a_y = A_y e^{i\phi_y}$$

Then we can write

$$\tilde{E} = E_x + iE_y = \tfrac{1}{2}(a_x + ia_y)e^{i\theta} + \tfrac{1}{2}(a_x^* + ia_y^*)e^{-i\theta} \tag{10.16}$$

A comparison of Eqs. (10.14) and (10.16) yields

$$a_+ = \tfrac{1}{2}(a_x + ia_y) \tag{10.17a}$$

$$a_- = \tfrac{1}{2}(a_x^* + ia_y^*) \tag{10.17b}$$

Now take the complex conjugate of Eq. (10.17*b*):

$$a_-^* = \tfrac{1}{2}(a_x - ia_y) \tag{10.17c}$$

Equations (10.17*a*, *c*) can then be inverted to yield

$$a_x = a_+ + a_-^*$$
$$a_y = -i(a_+ - a_-^*) \tag{10.18}$$

Equations (10.17) and (10.18) relate the two different sets of complex coefficients. They correspond to equivalent ways of describing the same light field.

3. Interpretation. The advantage of using Eq. (10.14) instead of (10.11) lies in the greater ease of interpretation when circularly polarized components are used as a basis. We introduce the sum and difference

$$\beta = \phi_- + \phi_+, \qquad \alpha = \phi_- - \phi_+ \tag{10.19a}$$

Then we may write Eq. (10.14) as

$$\tilde{E} = E_x + iE_y = e^{i\beta/2}\left[A_+ e^{i(\theta - \alpha/2)} + A_- e^{-i(\theta - \alpha/2)}\right] \tag{10.19b}$$

The behavior of the two terms in square brackets in Eq. (10.20) is indicated in Fig. 10.6. The vectors are shown when $\theta = \alpha/2$, at which time they are parallel. Then we have

$$\tilde{E} = e^{i\beta/2}(A_+ + A_-)$$

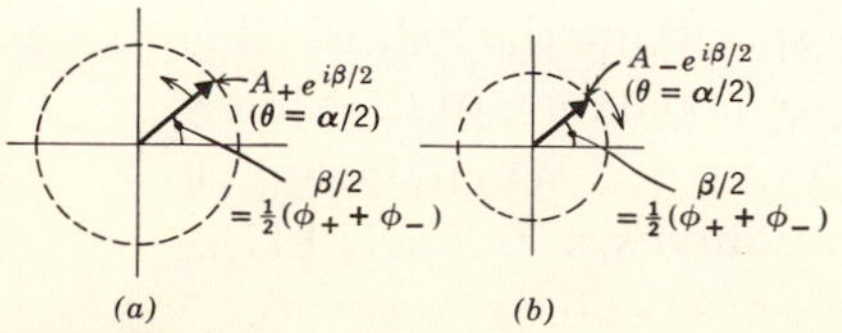

Fig. 10.6

*See, for example, J. D. Jackson, *Classical Electrodynamics*, p. 201.

This corresponds to point A in Fig. 10.7. When $\theta = \alpha/2 + \pi/2$, the vector of Fig. 10.6*a* has turned counterclockwise $\pi/2$ radians and that of Fig. 10.6*b* has turned clockwise $\pi/2$ radians. If A_- is less than A_+ and if $(\pi/2) > (\beta/2) > 0$, the resulting $\tilde{E}$ vector will be

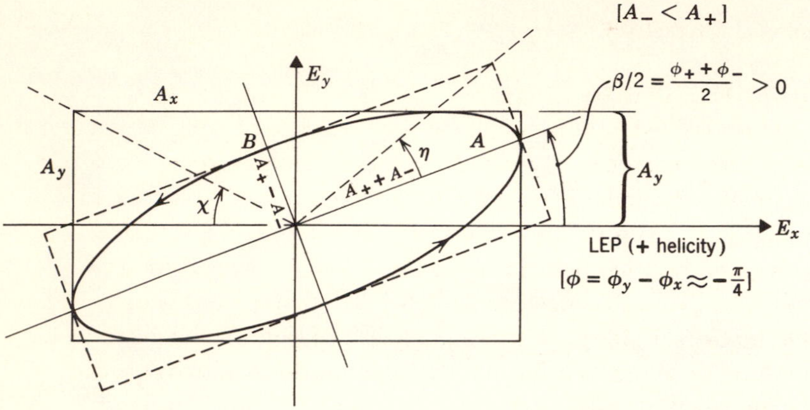

Fig. 10.7

in the second quadrant at the point B in Fig. 10.7. We see that the $\tilde{E}$ vector traces an ellipse with semimajor axis of length $(A_+ + A_-)$ at inclination $\beta/2$ and semiminor axis of length $(A_+ - A_-)$ and angle $\beta/2 + \pi/2$. The light in this case has positive helicity and is left elliptically polarized (LEP).

Other possibilities are illustrated in Fig. 10.8. When $A_+ = A_-$, we obtain linearly polarized light with E along the line at angle $\beta/2$.

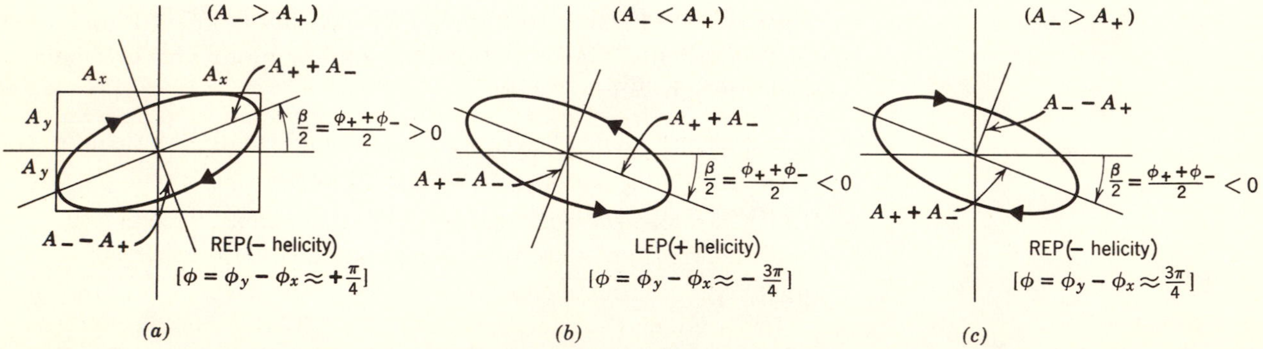

Fig. 10.8

With reference to Fig. 10.7, note that the elliptically polarized light is completely characterized by specifying the helicity, the angle $\beta/2$, and the angle η, which gives axial ratio of the ellipse via the equation, $\tan\eta = |A_+ - A_-|/(A_+ + A_-)$. When the light is described in terms of its x and y components, as in Eq. (10.11), the appropriate parameters are the phase difference $\phi = \phi_y - \phi_x$ and the ratio $A_y/Ax \equiv \tan\chi$.

The two sets of parameters are related by the following equations, which will not be derived here:

$$\tan\phi = \pm\frac{\tan 2\eta}{\sin\beta} \qquad (+\text{ for REP}, -\text{ for LEP}) \qquad (10.20a)$$

$$\cos 2\chi = \cos\beta \cos 2\eta \qquad (10.20b)$$

B. UNPOLARIZED LIGHT

Unpolarized light is not an elementary state of polarization and is in a sense more complicated than the examples we have already discussed. In general an individual microscopic source of light (such as an atom, or molecule) emits light that in a given direction has a well-defined polarization state. It is the superposition of many individual contributions to form the total macroscopic electric field that can produce unpolarized light. In Problem 10.1.2 you will be asked to show that the superposition of two *coherent* beams of elliptically polarized light will give another beam of elliptically polarized light, generally with a different ellipse. The nature of the resulting ellipse depends on the phase difference between the light in the two primary beams. If this phase difference varies, this ellipse may alter in orientation and shape. If the two beams have a finite coherence time, the phase difference between them will change randomly, with a root mean square change of about one radian in a time equal to the coherence time of the light. Thus, for two *incoherent* beams, where the measurement time is longer than the coherence time, the ellipse will change in a rapid, random way, and only its average properties will be measurable.

When many such incoherent beams are present, we can obtain unpolarized light if two conditions are satisfied. For the *xy* linearly polarized basis of Eq. (10.11) we require E_x and E_y to be incoherent and the amplitudes to have equal time averages: $\langle A_x^2\rangle = \langle A_y^2\rangle$. Since the total intensity is given by $I \propto \langle A^2\rangle = \langle A_x^2\rangle + \langle A_y^2\rangle$, each component carries half the total intensity. Since the coordinate system is arbitrary, a polarizer will always pass half the intensity, independently of its orientation.

In the circularly polarized basis of Eq. (10.14) a similar statement can be made, namely, that $E_+ = a_+e^{i\theta}$ and $E_- = a_-\,e^{-i\theta}$ must be incoherent and $\langle |a_+|^2\rangle = \langle |a_-|^2\rangle$. A more powerful method of discussing unpolarized light and mixtures of unpolarized and polarized light (partially polarized light) will be developed in Section 10.3.

One further point should be emphasized here, however. If we form a superposition of several incoherent beams, each in the same definite fully polarized state, the resulting beam will also be fully polarized in that state. The proof is quickly given. We use the linearly polarized basis. Suppose that each component beam has the same value of $\phi = \phi_y - \phi_x$ and the same value of the ratio $r = A_y/A_x$. Then we have for the jth beam

$$E_{xj} = A_{xj}\, e^{i\phi_{xj}}\, e^{i\theta}, \qquad E_{yj} = re^{i\phi}E_{xj}$$

The resultant is that $\mathbf{E}_{\text{tot}} = (E_{x\text{tot}}, E_{y\text{tot}})$ has components

$$E_{x\,\text{tot}} = \Sigma_j A_{xj}\, e^{i\phi_{xj}}\, e^{i\theta}, E_{y\,\text{tot}} = \Sigma_j E_{yj} = re^{i\phi}\Sigma_j E_{xj} = re^{i\phi}E_{x\,\text{tot}} \qquad (10.21)$$

Even if the individual phases ϕ_{xj} and amplitudes A_{xj} fluctuate, the light will be in a pure polarization state described by r and ϕ.

PROBLEMS

10.1.1 Describe semiquantitatively with a sketch the state of polarization of light waves obeying the following equations:

a. $E_x = A\cos(\omega t - kz)$
   $E_y = 2A\cos(\omega t - kz)$

b. $E_x = A\cos\left(\omega t - kz + \dfrac{\pi}{4}\right)$
   $E_y = A\cos(\omega t - kz)$

c. $E_x = A\cos\left(\omega t - kz - \dfrac{\pi}{4}\right)$
   $E_y = \left(\dfrac{A}{2}\right)\cos(\omega t - kz)$

10.1.2 Show that, in general, the superposition of two mutually coherent quasimonochromatic beams of elliptically polarized light will yield a beam of elliptically polarized light.

10.1.3 Show that Eqs. (10.17) and (10.18) give

$$A_+^2 + A_-^2 = \tfrac{1}{2}(A_x^2 + A_y^2)$$

10.1.4 a. Show that circularly polarized light can be represented as the coherent superposition of two plane polarized waves vibrating at right angles.

b. Show that plane polarized light can be represented as the coherent superposition of right circularly polarized light and left circularly polarized light.

c. Show further that changing the sum of the phases of the RCP and LCP light in (b) changes the plane of vibration of the resultant plane polarized light.

10.1.5 Find numerical values for A_+, A_-, α and β, sketch the ellipse, and indicate its helicity for light described by

$$E_x = A_x\cos\theta$$
$$E_y = A_y\cos(\theta + \phi)$$

with

a. $\phi = \dfrac{\pi}{2}$, $\quad A_y = \dfrac{1}{2}A_x$

b. $\phi = \dfrac{\pi}{4}$, $\quad A_y = A_x$

c. $\phi = \dfrac{\pi}{4}$, $\quad A_y = \dfrac{1}{2}A_x$

d. $\phi = \dfrac{5\pi}{4}$, $\quad A_y = A_x$

e. $\phi = \dfrac{-\pi}{6}$, $\quad A_y = A_x$

10.2 AN ELEMENTARY DISCUSSION OF METHODS OF PRODUCING POLARIZED LIGHT AND SHIFTING PHASE

A. WAYS OF PRODUCING POLARIZED LIGHT

We now discuss briefly some of the methods commonly employed to produce polarized light. Most, but not all, of the simple methods yield plane polarized light.

*1. **Optical Sources that Emit Light that is Already Polarized.*** One such source is the continuous *gas laser*, about which we shall say more later. Such a device emits a very monochromatic narrow light beam that is often plane polarized. The laser emits light in such a polarized "mode" because of the use of windows set at *Brewster's angle* (see below).

Another source of polarized light, at least in principle, is provided when certain sources of narrow spectral lines are placed in a magnetic field. A splitting of the lines known as the *Zeeman effect* then occurs; each component is either plane polarized or circularly polarized. This effect is difficult to produce and is not presently of practical importance as a source of polarized light.

*2. **Polarization by Reflection and Transmission.*** Light reflected from a smooth interface becomes partially polarized if the original angle of incidence is nonzero. This effect becomes particularly strong and hence useful at a dielectric interface when the angle of incidence is at or near the so-called *Brewster's angle.* The details will be discussed in Chapter 11.

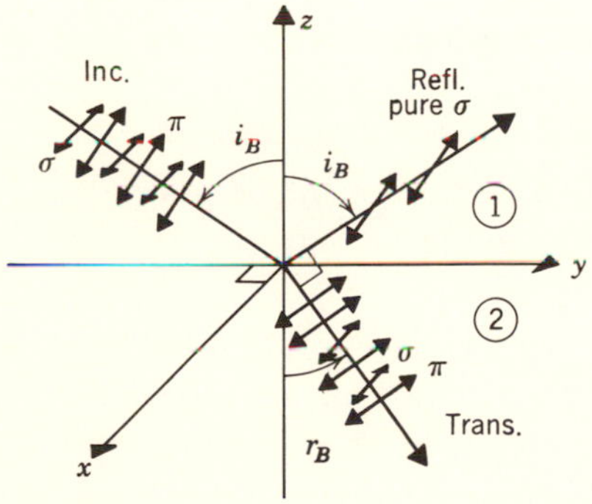

Fig. 10.9

In Fig. 10.9 such an interface is shown in the *xy* plane. When the reflected and transmitted rays meet at an angle of 90°, it is impossible for light to be reflected if its electric field lies in the plane of polarization. This "π light" (π for parallel to the plane of incidence) will be transmitted, however. Light with its electric field vibrating perpendicular to the plane of incidence may be reflected. This is the so-called σ light (σ for senkrecht, which is German for perpendicular).

At Brewster's angle of incidence i_B we must have

$$i_B + r_B = \frac{\pi}{2} \tag{10.22}$$

Since Snell's law says that

$$n_{21} = \frac{n_2}{n_1} = \frac{\sin i_B}{\sin r_B}$$

and since $\sin r_B = \cos i_B$ when $i_B + r_B = \pi/2$, we must have

$$\tan i_B = n_{21} \tag{10.23}$$

at Brewster's angle.

We shall see later how this polarization by reflection follows as a consequence of the Maxwell equations of electromagnetic theory, but for now a simple explanation will be given in terms of Huygens' principle. The transmitted and reflected wavefronts can be found as the envelope of secondary Huygens wavelets emitted at the interface. In the usual application of this principle this is a fictitious mathematical reradiation, but in the present case the atoms at the surface perform an actual physical reradiation.

Electrical oscillations are induced in these atoms along the direction of the electric field *in the second medium*. These oscillating atoms then reradiate. A property of these oscillators in isotropic media is their inability to radiate light that travels in the direction along which they are oscillating, which in this case is along the direction of the electric field in the second medium. For σ light the electric field is always perpendicular to the plane of incidence and hence to the direction of the reflected and transmitted rays. The induced atomic oscillators are able to oscillate in these directions. For π light the induced vibrations must be *perpendicular* to the direction of the transmitted ray and they must be *parallel* to the direction of the reflected ray. Thus there can be no reflection in this case.

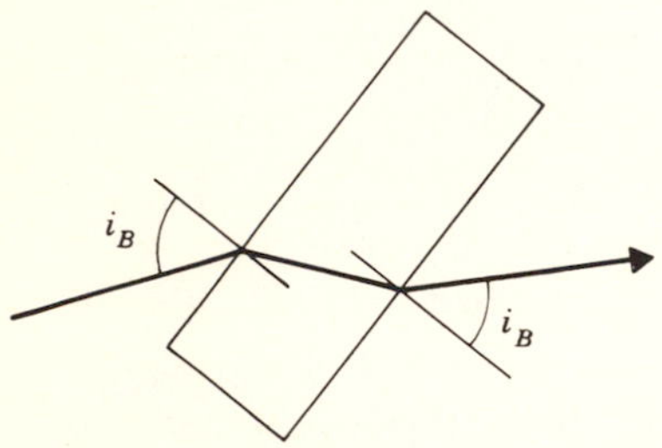

Fig. 10.10

If the light incident at Brewster's angle is unpolarized, the reflected light is pure σ light, and the transmitted light is partially polarized with more π than σ light present. If, as in Fig. 10.10, the medium is in the shape of a parallel slab with a lower interface that goes back to the first medium, we can show (Problem 10.2.1) that Brewster's condition is satisfied there also. Thus the light reflected there will be pure σ. This process may be repeated.

A pile of such parallel plates or slabs will then be an efficient polarizer for collimated light. The reflected beam will be pure σ polarized; the transmitted beam can be made virtually pure π polarized by using enough slabs.

If the angle of incidence is not exactly equal to i_B, the reflected light will have a small component of π polarization, and the transmitted light will have correspondingly less π light.

Such a "pile of plates" can be used as a polarizer, either in reflection or in transmission, especially with well-collimated beams and where other more convenient polarizers are not available. Indeed, this type of polarizer was the first to be used in the nineteenth century.

3. ***Polarization by Scattering.*** Light scattering can occur whenever there is an irregular distribution of small scattering centers. Examples are provided by dust particles in the air, small compared with the wavelength, the particles in a colloidal suspension, or even the gas molecules in the air. The basic mechanism for the production of polarized light from unpolarized light is similar to that just discussed. If the particles are small enough and simple enough, oscillations will be induced in them by the electric field of the incident light. There will be reradiation of light in all directions save that of the original electric field.

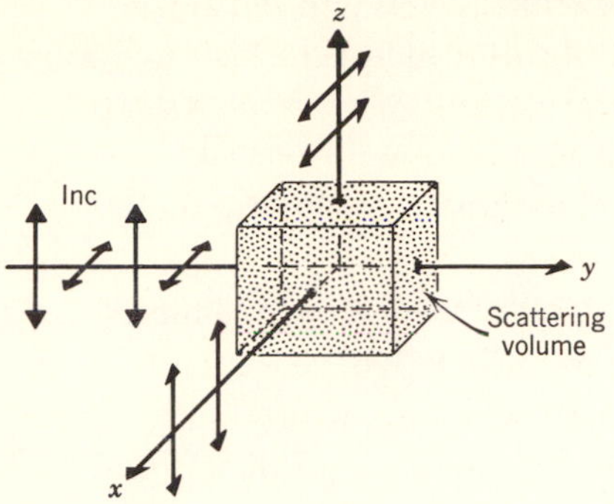

Fig. 10.11

This is shown in Fig. 10.11. The light scattered at a direction 90° to the incident propagation direction will be plane polarized with the electric field in a direction perpendicular to the scattering plane, that is, the plane determined by the incident and scattered propagation directions. This mechanism of polarization by scattering is active in the earth's atmosphere. The blue color of the sky is due to preferential scattering of short wavelength blue light over long wavelength red light. The blue light from the sky is partially polarized with the electric field vector vibrating at right angles to the plane containing the incident and scattered light rays. At 90° viewing, as shown in Fig. 10.11, the polarization is complete.

*4. **Polarization by Dichroism.*** There is a class of anisotropic optical media that exhibit a phenomenon known as *dichroism*, which means that light propagating in a given direction will be *absorbed* differently depending on the orientation of the electric field. The *absorption constant* (to be discussed in more detail later) thus depends on the direction of vibration. In the extreme case, light vibrating in a particular direction will be completely absorbed, and light vibrating in the perpendicular direction will be essentially completely transmitted. The latter direction is called the *direction of easy passage*, or *pass direction*. A tourmaline crystal is an example of a dichroic polarizer, but the best-known example is a polaroid sheet, which consists of oriented dichroic molecules embedded in plastic.

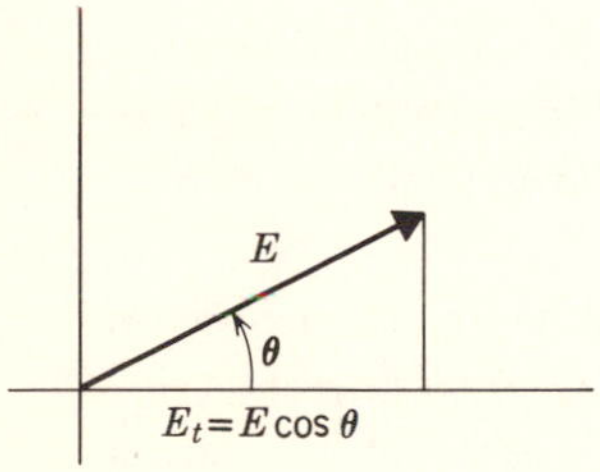

Fig. 10.12

In calculating the effect of a polarizer on an incident light beam, we should resolve the incident electric field along axes parallel and perpendicular to the pass direction. If θ is the angle between the E field and the pass direction, the component passed is $E_p = E \cos\theta$, and the transmitted intensity I_t is given in terms of the incident intensity I_i by the law of Malus,

$$I_t = I_i \cos^2\theta \tag{10.24}$$

*5. **Polarization by Double Refraction.*** *Double refraction* occurs in optically anisotropic media, that is, media having optical properties dependent on direction. The phenomenon is also referred to as *birefringence*. A detailed discussion of this subject can become quite complicated and will be postponed for now. The "double" or "bi" in the name of this effect refers to the *two* different directions of propagation that a given incident ray can take in such media, depending on the direction of polarization.

We discuss here the simplest kind of birefringent medium, a *uniaxial medium.* If diffraction effects are neglected, propagation of waves in such media can be described by a generalization of Huygens' principle. The spherical secondary wavelets must be replaced by wave surfaces of greater complexity, consisting of a spherical wave plus a wave that is an ellipsoid of revolution. The ellipsoid is tangent to the sphere at two points that lie on a line through the center of the elliptical waves. This line is called the optic axis (O.A.). It forms the axis for the ellipsoid of revolution. See Fig. 10.13.

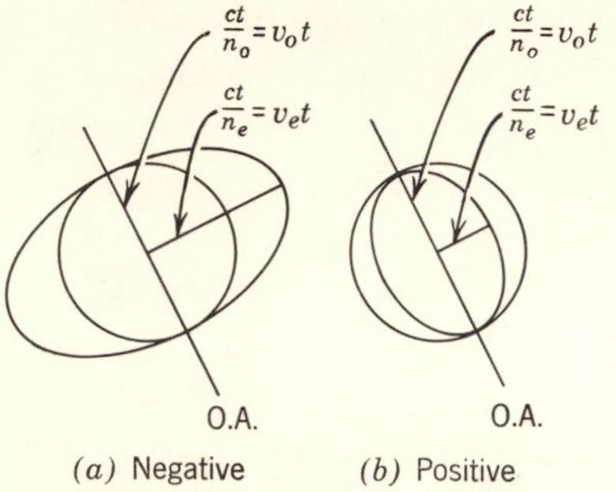

Fig. 10.13. Huygens' wavelets in uniaxial media.

If we pick a coordinate system having its z axis along this optic axis, the equation for this ellipsoid may be written

$$\frac{(x^2+y^2)}{v_e^2}+\frac{z^2}{v_o^2}=t^2 \tag{10.25}$$

The spherical wavelet spreads with a velocity $v_o = c/n_o$, and so at time t its radius will be $(c/n_o)t$. The ellipsoidal wavelet spreads with a velocity ranging between c/n_o along the optic axis to a velocity $v_e = c/n_e$ in a plane perpendicular to the O.A. For "positive" crystals n_e is greater than n_o and for "negative" crystals n_o is greater. Examples are furnished by *calcite* (negative) and *crystalline quartz* (positive). (Quartz also exhibits a phenomenon known as *optical activity,* which we ignore for the moment. This is connected with the fact that the sphere and the ellipsoid never quite touch along the optic axis.)

The spherical wave is referred to as *ordinary* (o wave), and its refractive index n_o as the *ordinary index.* The elliptical wave is the *extraordinary wave,* (e wave) and the extreme value of its refractive index n_e is called the *extraordinary index.*

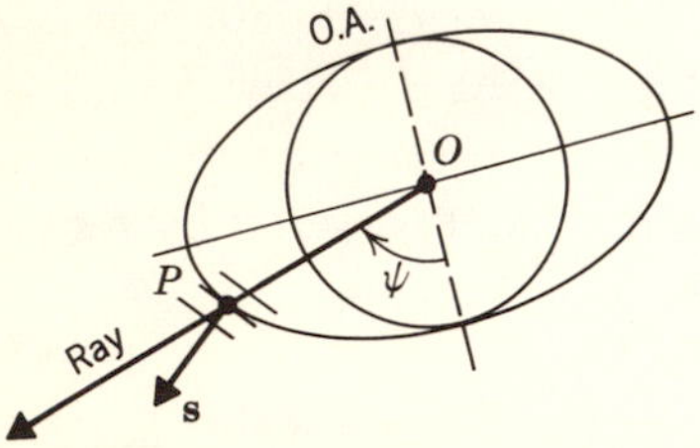

Fig. 10.14

The notion of a light ray must be redefined for the extraordinary wave. It represents the direction of energy flow and is given by a line from the origin to the point in question on the ellipsoid, such as OP in Fig. 10.14. The point P will move outward with the "ray velocity" v_r, intermediate between v_o and v_e, which depends on the angle ψ of OP with respect to the O.A . Using $|OP| = v_r t$, we can readily derive

$$\frac{1}{v_r^2}=\frac{\sin^2\psi}{v_e^2}+\frac{\cos^2\psi}{v_0^2} \tag{10.26}$$

The surfaces of constant phase (wavefronts) for the e wave are tangent to the e wave surface at P. The wave normal **s** is not parallel to the ray direction OP. Since the distance $|OP|$ is proportional to the ray velocity, these wave surfaces are also called *ray velocity surfaces.* These surfaces are useful for qualitative discussions of double refraction, but for quantitative work other surfaces, the *index surfaces*, to be introduced in Section 11.4, are more convenient.

The extraordinary ray traveling in the direction OP will be *plane polarized* with the electric field vibrating in the plane containing the direction of propagation and the optic axis. This plane is called the *principal section*. A novel aspect of the extraordinary wave is that except when $\theta = 0$ or $\pi/2$ the electric displacement vector **D** is *not* perpendicular to the propagation direction OP. The electric field **E** is perpendicular to OP, however, as are the **B** and **H** fields. These matters will be discussed in Chapter 11.

The ordinary spherical wave is always plane polarized perpendicular to the principal section, normal to both the O.A. and the direction of propagation, that is, normal to the plane of the paper in Fig. 10.14.

A disturbance of *arbitrary* pure state of polarization will, in general, propagate in a complicated way in a birefringent medium. It can be resolved, however, into plane polarized components, which become the *e* and *o* rays just discussed. The state of polarization of these components does not change, although that of the original disturbance can change quite markedly when it emerges from the birefringent medium. For example, it can change from plane to circular polarization.

Wavefronts associated with extraordinary rays will *not* be normal to the rays. They can be determined by Huygens' principle. An example is shown in Fig. 10.15. Suppose that the line AA' represents the location of the primary wave surface. Each point then

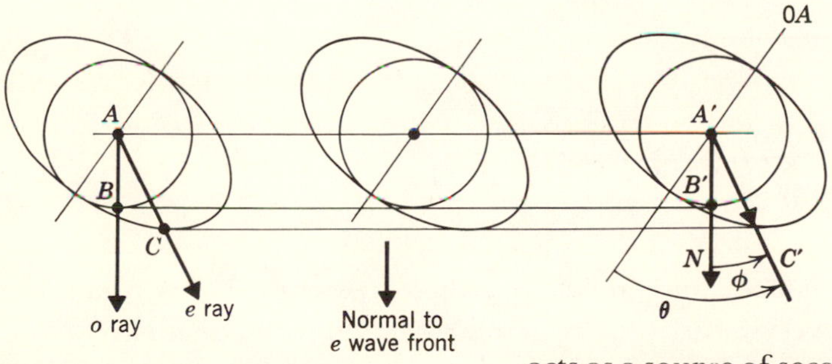

Fig. 10.15

acts as a source of secondary wavelets. The envelope of the spherical wavelets BB' then locates the ordinary wave surface at a later time. The envelope CC' locates the extraordinary wave surface, which is moving in the direction AC and making an angle ϕ with the normal to the wave front $A'N$. The *wave velocity* v_w is the velocity in the normal direction $A'N$ and is given by

$$v_w = v_r \cos \phi \tag{10.27}$$

where v_r is given by Eq. (10.26).

The angular separation between the *e* ray and the *o* ray can be used to isolate one or the other of them and hence produce plane polarized light. One of the common ways of doing this is employed in the *Nicol prism,* which is made from a natural uniaxial birefringent crystal called *calcite* or *Iceland spar*. This prism is often employed as a high-precision polarizer.

This crystal cleaves naturally into a rhomb with parallel sides and edges. The direction of the optic axis makes equal angles with the edges at the most obtuse corner, where all the edges meet at an angle of 102°, as shown in Fig. 10.16.

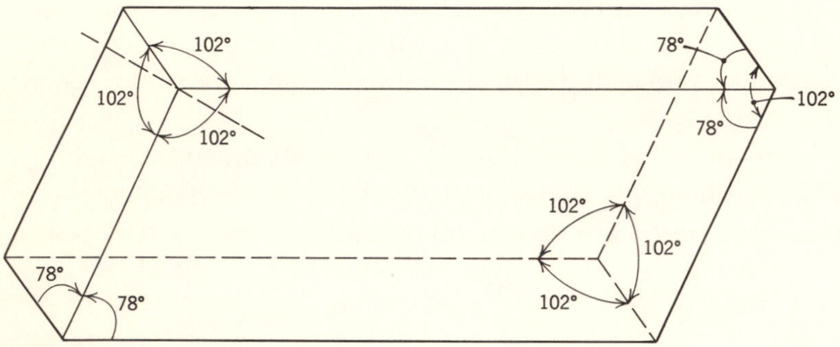

Fig. 10.16. A calcite rhomb showing natural cleavage faces.

The Nicol prism is cut from a calcite crystal as shown in Fig. 10.17. We start with a principal section of a cleaved rhomb containing the optic axis, as shown dotted in the figure. The ends are then cut to a more acute angle of 68° . A long diagonal

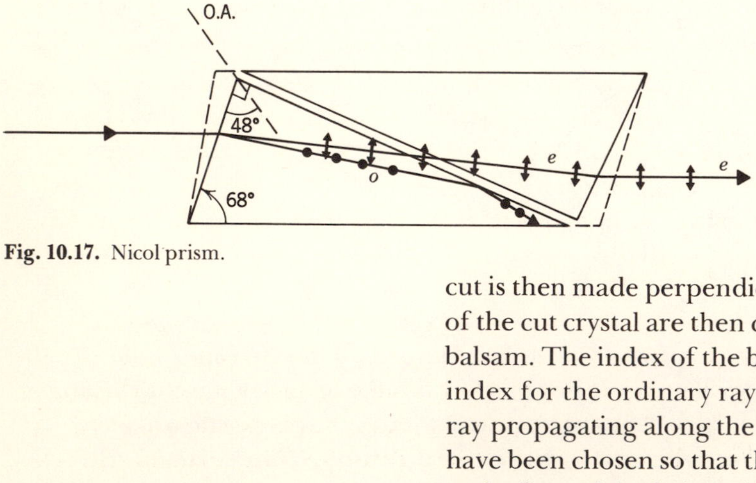

Fig. 10.17. Nicol prism.

cut is then made perpendicular to these new ends. The two halves of the cut crystal are then cemented together with Canadian balsam. The index of the balsam is 1.55, which lies between the index for the ordinary ray of 1.658 and that of the extraordinary ray propagating along the axis of the prism of 1.486. The angles have been chosen so that the *o* ray is totally internally reflected at the first calcite-balsam interface. This ray is then absorbed by a black coating on the sides of the prism.

6. Other Double Refracting Polarizers. The Nicol prism has several disadvantages, among which are the parallel displacement suffered by the transmitted beam and the relatively small angular spread of the beam that may be transmitted. The Glan-Thompson prism shown in Fig. 10.18 has neither of these disadvantages, although it makes less efficient use of calcite. It consists of two sections of calcite, both having ends cut perpendicular to the light beam with parallel optic axes as shown. The index of the cement is such that at the angle of the cut the ordinary wave suffers total internal reflection, whereas the extraordinary wave is transmitted with little loss.

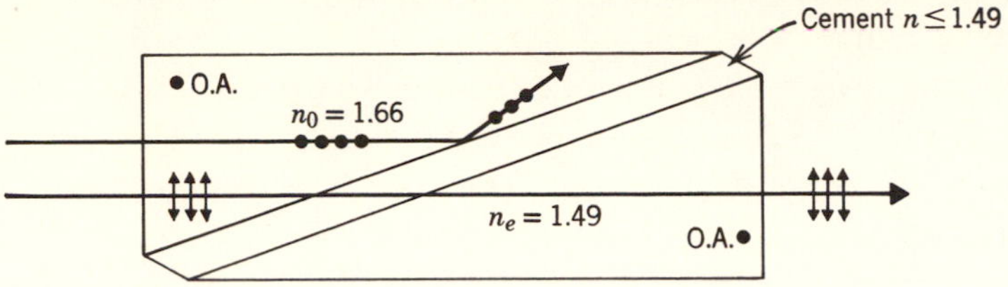

Fig. 10.18. Glan-Thompson polarizer.

In Fig. 10.19 two types of beam-splitting polarizing prism are shown. With unpolarized incident light they produce two transmitted beams with mutually perpendicular polarization directions and different directions of propagation. These prisms are usually made of either quartz or calcite and have their optic axes oriented as shown. In the first part of the Rochon prism both waves have the ordinary index n_o. In the second part the *o* wave vibrating in the plane of the paper continues to have index n_o and is not deviated. The *e* wave vibrating perpendicular to the plane of the paper has index n_e and is deviated up or down according to whether $n_e < n_o$ or $n_e > n_o$.

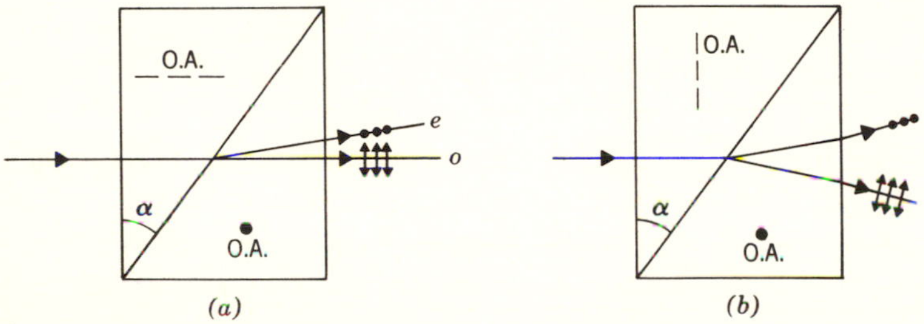

Fig. 10.19. Rochon prism (*a*) and Wollaston prism (*b*). The optic axis for the right-hand parts of both is perpendicular to the plane of the paper.

In the Wollaston prism the *o* wave in the first part vibrating perpendicular to the plane of the paper becomes the *e* wave in the second part, and the *e* wave in the first part, vibrating in the plane of the paper, becomes the *o* wave in the second part. If $n_e < n_o$, the rays will be both deviated as shown. If $n_e > n_o$, the polarizations of the two emerging beams will be reversed.

B. OPTICAL PHASE-SHIFTING DEVICES

A simple method of optical phase shifting employs a parallel slab of birefringent material. If the optic axis lies parallel to the

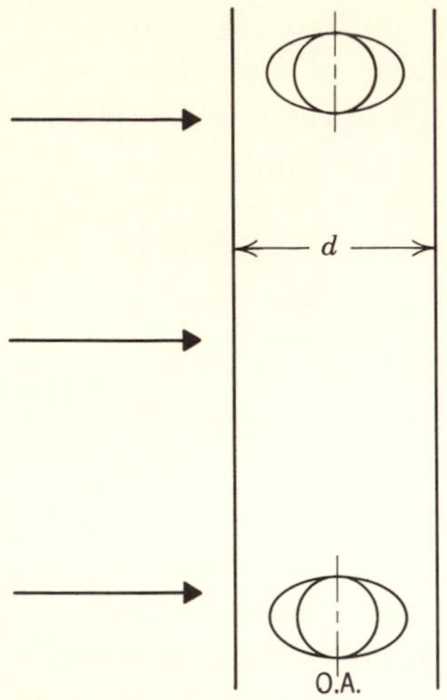

Fig. 10.20

surface and if light falls on the slab at normal incidence as shown in Fig. 10.20, both *o* and *e* rays will continue to propagate in the same direction, but with different velocities. The phase change produced by propagation will then be different for the two rays.

The phase decrease in a distance *d* suffered by the *o* ray will be given by

$$\phi_o = \frac{2\pi}{\lambda} n_o d$$

and by the *e* ray

$$\phi_e = \frac{2\pi}{\lambda} n_e d$$

giving a phase difference

$$\Delta\phi = \phi_e - \phi_o = \frac{2\pi}{\lambda}(n_e - n_o)d \tag{10.28}$$

Such a phase shifter is often referred to as a *compensator*.

1. Quarter-Wave Plate. A very useful device called a *quarter-wave* plate is produced when $|\Delta\phi| = \pi/2$, that is, when

$$|n_e - n_o|d = \frac{\lambda}{4} \tag{10.29}$$

Then the *e* ray is retarded or advanced by a quarter of a cycle with respect to the *o* ray; retardation occurs if $n_e > n_o$ (positive unaxial crystal like quartz). The direction of polarization that is advanced is along the *fast axis*; the retarded direction is the *slow axis*. The *e* ray is polarized along the slow axis for the above example of quartz. For negative calcite the *e* ray is along the fast axis. Note that $\Delta\phi$ depends on λ. This limits the practical wavelength range for precision work (see Problem 10.2.6).

A device called a Fresnel rhomb produces a $\pi/2$ relative shift by means of total internal reflection and is much less wavelength-dependent. The light beam undergoes a parallel displacement, however. See Problem 11.3.7.

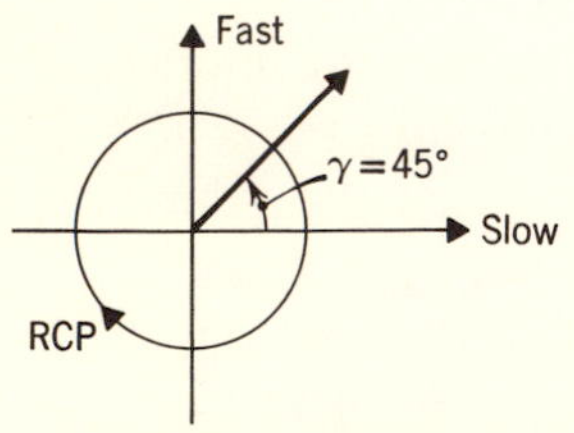

Fig. 10.21

A quarter-wave plate will convert plane polarized light to elliptically polarized light as shown in Fig. 10.21. Here the plane of polarization makes an angle of 45° with the fast axis. In this special case the $\lambda/4$ plate produces right circularly polarized light. In the general case where $\gamma \neq 45°$ we would have

Incident light
$$\begin{aligned} E_x &= A\cos\gamma\cos\omega t \\ E_y &= A\sin\gamma\cos\omega t \end{aligned} \tag{10.30}$$

After transmission, there will be a common phase shift ϕ, plus a relative phase shift of $-\pi/2$:

Transmitted light

$$E_x = A\cos\gamma\cos(\omega t - \phi)$$
$$E_y = A\sin\gamma\cos\left(\omega t - \phi - \frac{\pi}{2}\right) \tag{10.31}$$
$$= A\sin\gamma\sin(\omega t - \phi)$$

This will give LEP light. The axes of the ellipse coincide with the x and y axes and have a length ratio given by $\tan\gamma$.

Elliptically polarized light will be converted to plane polarized light when run through a quarter-wave plate, provided the axes of the quarter-wave plate and the ellipse coincide. The sequence from Eq. (10.31) to Eq. (10.30) is then essentially inverted.

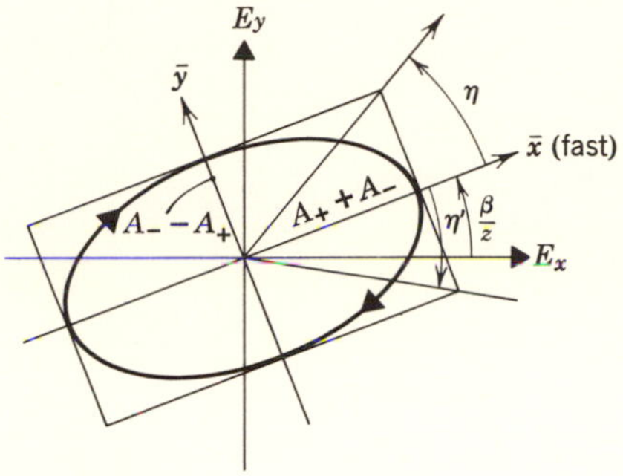

Fig. 10.22

Consider the elliptically polarized light appropriate to Fig. 10.22. Suppose that the light is REP. If the fast axis of the $\lambda/4$ plate makes an angle $\beta/2$ with the x axis, that is, if it coincides with the major axis of the ellipse ($\bar{x}$ axis), it will speed up $\bar{E}_x$, the component of **E** along the $\bar{x}$ axis, by a quarter of a cycle with respect to $\bar{E}_y$. In the original REP light $\bar{E}_x$ was lagging $\bar{E}_y$ by a quarter of a cycle. Thus it is now in phase, and the result is plane polarized light vibrating at an angle η to the $\bar{x}$ axis, that is, at an angle $\beta/2 + \eta$ to the x axis. This polarized light and its orientation can be readily detected by use of a polarizer. The polarizer gives the angle $\beta/2 + \eta$ to the x axis. The fast axis of the $\lambda/4$ plate gives the angle $\beta/2$. Subtraction gives η, which yields the axial ratio of the ellipse:

$$\frac{A_- - A_+}{A_+ + A_-} = \tan\eta \tag{10.32}$$

If the light were LEP instead of REP, the resulting plane polarized light would vibrate at the angle $\eta' = -\eta < 0$ with respect to the $\bar{x}$ axis and hence at the angle $\beta/2 + \eta' = \beta/2 + -\eta$ with respect to the x axis. Equation (10.32) will now hold with η replaced by η' and will yield

$$\frac{A_- - A_+}{A_+ + A_-} = \tan\eta' = -\tan\eta < 0$$

Hence we have $A_+ > A_-$, as is necessary for LEP light. This difference can be used to decide between REP and LEP light.

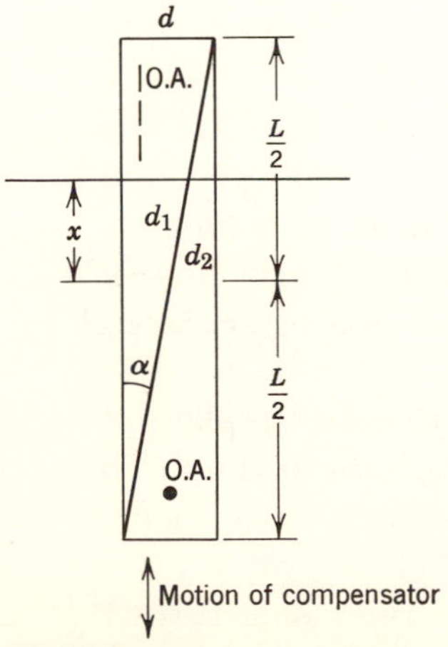

Fig. 10.23. Babinet compensator.

2. Babinet Compensator. The Babinet compensator shown in Fig. 10.23 is similar to a Wollaston prism, but the angle $\alpha \approx d/L$ is sufficiently small that the angular deviation suffered by the two emerging beams can be neglected. The main effect then is the production of a relative phase difference, which depends on the height of passage of the beam through the device. In the first part of the compensator, light vibrating perpendicular to the plane of the paper is the *o* wave and is retarded in phase an angle

$$\Delta_1 = \frac{2\pi(n_o - n_e)d_1}{\lambda}$$

relative to the wave vibrating in the plane of the paper. In the second part of the compensator the roles are reversed, and the relative phase shift is

$$\Delta_2 = \frac{2\pi(n_e - n_o)d_2}{\lambda}$$

The total phase shift is then

$$\Delta = \Delta_1 + \Delta_2 = \frac{2\pi(n_o - n_e)(d_1 - d_2)}{\lambda}$$

Since

$$d_1 - d_2 = 2d_1 - d = \frac{2xd}{L}$$

this gives

$$\Delta(x) = \frac{4\pi(n_o - n_e)xd}{\lambda L} \tag{10.33}$$

As the Babinet compensator is moved vertically across a narrow beam, the phase difference will vary linearly with position x. This device can thus be used to produce a known phase shift $\Delta(x)$.

3. ***Optical Activity.*** Optical activity can be found in crystalline and monocrystalline solids, liquids, and gases. In crystals it is often accompanied by double refraction, and the resulting phenomena can be quite complex. Some cubic crystals, for example, sodium chlorate and sodium bromate, show optical activity uncomplicated by double refraction.

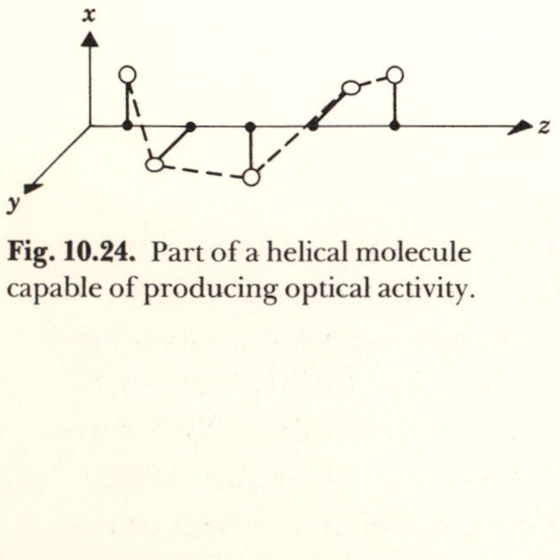

Fig. 10.24. Part of a helical molecule capable of producing optical activity.

Optical activity originates in the inherent helical or screw-like character of the molecules of the material that produces it (Fig. 10.24). In its pure form, that is, without double refraction, it results in different refractive indices for left and right circularly polarized light beams. Only such beams will propagate through the medium without change.

A plane polarized beam will be modified; in fact its plane of vibration will rotate as it propagates through the medium. This is one of the clearest manifestations of optical activity. A brief description will now be given.

Let n_+ and n_- be refractive indices for + and − helicity (LCP and RCP, respectively). Let $n_0 = \frac{1}{2}(n_+ + n_-)$ be the mean index and

$k_0 = 2\pi n_0/\lambda$ the mean magnitude of the wave vector. The *specific rotary power* is defined to be

$$\rho = \frac{\pi(n_+ - n_-)}{\lambda} \tag{10.34}$$

Then the wave vectors for left and right circularly polarized light have magnitudes

$$\begin{aligned} k_+ &= k_0 + \rho \\ k_- &= k_0 - \rho \end{aligned} \tag{10.35}$$

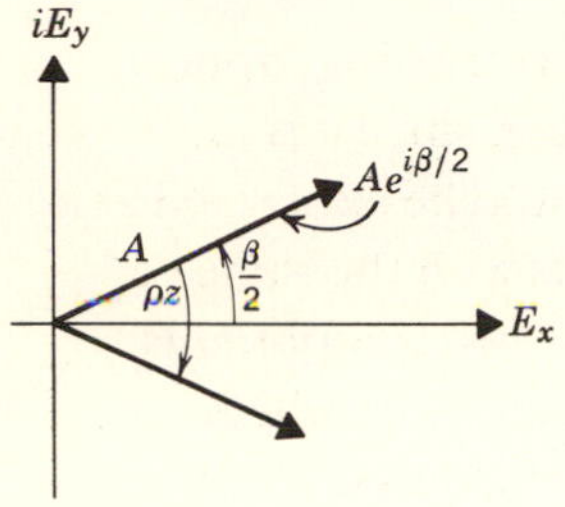

Fig. 10.25

It is convenient to use the complex notation of Eqs. (10.19) and Fig. 10.6. Let z be the direction of propagation and let $\tilde{E} = E_x + iE_y$. Suppose that in the plane $z = 0$ there is plane polarized light with amplitude A vibrating in the direction $\beta/2$, as shown in Fig. 10.25. Then

$$\tilde{E} = e^{i\beta/2} A \cos \omega t$$

or

$$E_x = A \cos \omega t \cos \frac{\beta}{2}$$

$$E_y = A \cos \omega t \sin \frac{\beta}{2}$$

This can also be written as a superposition of circularly polarized components

$$\tilde{E}_+ = e^{i\beta/2} \tfrac{1}{2} A e^{i\omega t}$$

with

$$\tilde{E}_- = e^{i\beta/2} \tfrac{1}{2} A e^{-i\omega t}$$

For finite z we must make the replacement

$$\omega t \rightarrow \omega t - k_{\pm} z = \theta_0 \mp \rho z$$

for the two components, where $\theta_0 = \omega t - k_0 z$. This gives

$$\tilde{E}_+ = e^{i\beta/z} \tfrac{1}{2} A e^{i(\theta_0 - \rho z)}$$

$$\tilde{E}_- = e^{i\beta/z} \tfrac{1}{2} A e^{-i(\theta_0 + \rho z)}$$

or

$$\tilde{E} = E_x + iE_y = e^{i\beta/z} \tfrac{1}{2} A [e^{i(\theta_0 - \rho z)} + e^{-i(\theta_0 + \rho z)}]$$

$$= e^{i(\beta/2 - \rho z)} \tfrac{1}{2} A (e^{i\theta_0} + e^{-i\theta_0}) = e^{i(\beta/2 - \rho z)} A \cos \theta_0$$

Hence

$$\theta_0 = \omega t - k_0 z \tag{10.36}$$

$$E_x = A \cos \theta_0 \cos \left(\frac{\beta}{2} - \rho z\right)$$

$$E_y = A \cos \theta_0 \sin \left(\frac{\beta}{2} - \rho z\right) \tag{10.37}$$

These equations describe plane polarized light vibrating at angle $\beta/2 - \rho z$. Thus the parameter ρ gives the angle of rotation per unit distance of propagation.

Crystalline quartz is an example of a doubly refracting, optically active crystal. If it were merely doubly refracting, since it is uniaxial, light propagating along the optic axis would be expressed as a superposition of linearly polarized components with the same refractive index, the ordinary index. Instead, the two components that propagate unaltered are circularly polarized with indices n_+ and $n_- \neq n_+$. The specific rotation at $\lambda = 589.3$ nm is $\rho = \pm 3.8$ rad/cm. Two types of quartz exist: positive or right-handed quartz for which ρ is positive, and negative or left-handed quartz for which ρ is negative.

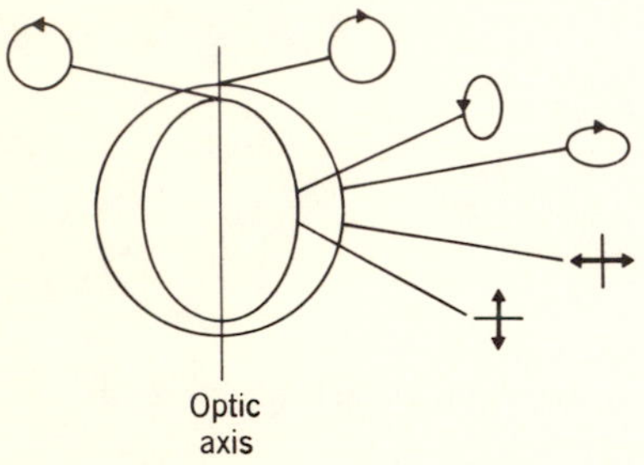

Fig. 10.26. Wavefronts and polarization states for the waves that propagate without change in crystalline quartz. This is schematic and not quantitatively correct.

For propagation in quartz not along the optic axis the two states of polarization that propagate unchanged correspond to elliptically polarized light (Fig. 10.26). The ellipses for the two components are oriented at right angles. They become circles for propagation along the optic axis and degenerate into straight lines for propagation perpendicular to the optic axis.

A demonstration of optical activity in quartz is easily given using the polarized light beam from a powerful continuous gas laser.* Figure 10.27 shows schematically the rotation of the plane of polarization of light propagating along the optic axis. This can be

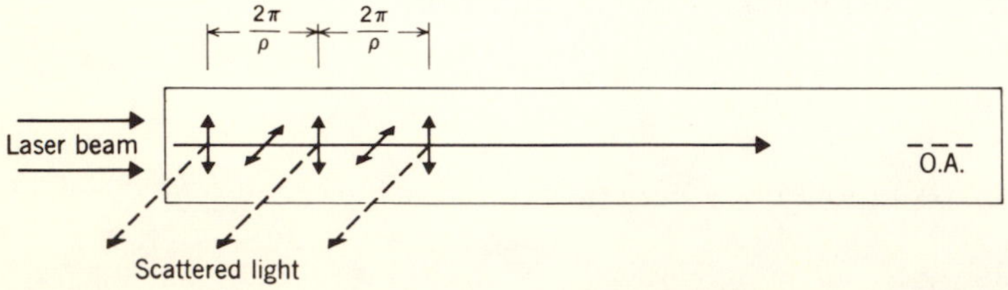

Fig. 10.27

seen directly by means of scattering of the intense beam by imperfections in the crystal. These act like dust particles in a gas. As discussed in connection with Fig. 10.11, the scattering will be most intense in directions perpendicular to the E vector of the

*C. A. Arguello, D. L. Rousseau, S. P. S. Porto, L. E. Cheesman, and J. F. Scott, *Applied Optics*, **7**, 1913 (1968).

beam and zero in directions parallel to the E vector. When we look at right angles to the laser beam, the intensity of the scattered light varies periodically with maxima when E is perpendicular to the direction of observation. These occur at equally spaced distances of $2\pi/\rho = 5{\cdot}5$ mm for $\lambda = 488$ nm.

REFERENCES FOR SECTION 10.2

R. W. Ditchburn, *Light.*
J. M. Stone, *Radiation and Optics.*
R. S. Longhurst, *Geometrical and Physical Optics.*

PROBLEMS

10.2.1 Prove that for a plane parallel slab if Brewster's condition is satisfied at the top interface it will be satisfied also for internal incidence onto the second interface.

10.2.2 Derive Eq. (10.26) from (10.15).

10.2.3 The indices of refraction for calcite and quartz for the sodium yellow line are

calcite: $n_0 = 1.658$, $n_e = 1.486$
quartz: $n_0 = 1.544$, $n_e = 1.553$

Calculate the thickness of a quarter-wave plate made from these materials and say whether the optic axis is a fast or slow axis of the plate.

10.2.4 Let the incident beam have polarization parameters $r = A_-/A_+ \neq 1$ and $\alpha = \phi_- - \phi_+$. Show that there are four values of δ for which extinction can be obtained with the polarizer and find the resulting values of δ and ψ.

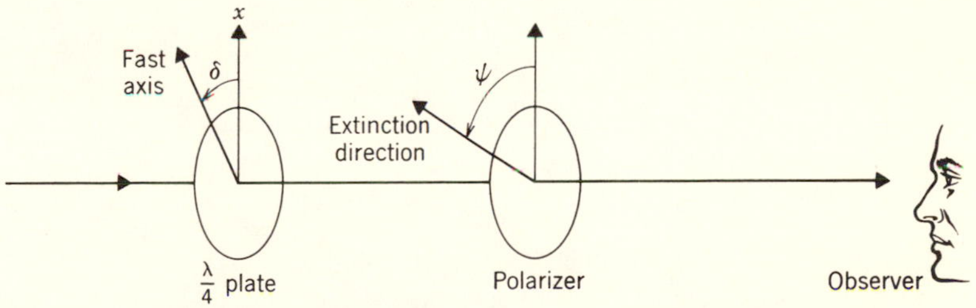

10.2.5 Discuss the action of a "half-wave plate" (for which $|n_e - n_o|d = \lambda/2$) on various types of polarized light.

10.2.6 Consider a quarter-wave plate designed for wavelength λ_0. Discuss its behavior for wavelengths λ near λ_0, that is, for small $\epsilon = (\lambda_0/\lambda - 1)$. Imagine the use of this plate in the arrangement of Problem 10.2.4. Show that for $\epsilon = \pm 0.1$ and for circularly polarized incident light the ratio of maximum to minimum transmitted intensity, as the angle ψ is varied, is 400.

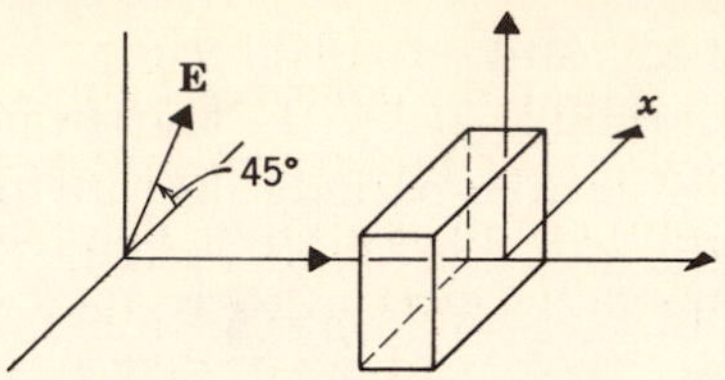

10.2.7 A uniform monochromatic light beam is incident on a Babinet compensator. It is linearly polarized at a 45° angle with respect to the long dimension of the compensator. Show that the polarization state of the emerging light depends on the coordinate x and ranges from linearly polarized through elliptically polarized to circularly polarized.
What happens if the E field of the incident light is circularly polarized? linearly polarized along the x direction?

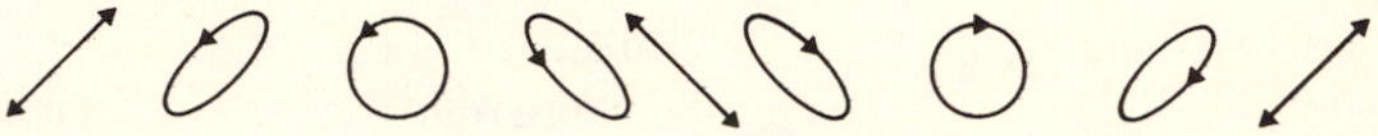

There will be additional problems on the propagation of light in anisotropic media in Chapter 11.

10.3 PARTIALLY POLARIZED LIGHT

A partially polarized light beam is a mixute of a beam in a pure polarization state and an unpolarized beam. This decomposition is unique. It depends on the mutual coherence between the electric fields vibrating along two perpendicular directions and can be determined experimentally by a method discussed in Section 10.3 B.6.

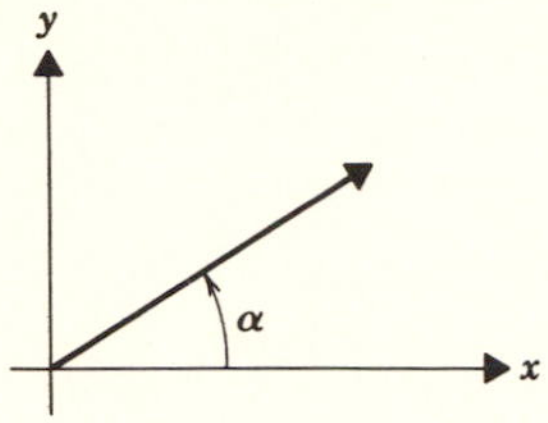

A. DESCRIPTION OF ELEMENTARY MEASUREMENTS

1. ***Intensity Transmitted by a Polarizer.*** Consider a polarizer having its direction of easy passage, or pass direction, inclined at angle α to the x axis (see Fig. 10.28. If E_x and E_y represent the real Cartesian components of the electric field, the projection at angle α is given by $E(\alpha) = E_x \cos\alpha + E_y \sin\alpha$. The transmitted-intensity will be proportional to the average

$$\langle E(\alpha)^2 \rangle = \langle E_x{}^2 \rangle \cos^2\alpha + \langle E_y{}^2 \rangle \sin^2\alpha + 2\langle E_x E_y \rangle \sin\alpha\cos\alpha \qquad (10.38)$$

If the correlation function $\langle E_x E_y \rangle$ is zero, the intensities associated with x and y components add. This can occur through either of two ways, or a combination of both: (1) E_x and E_y can have no mutual coherence, or (2) E_x and E_y can be correlated but have a relative phase difference of $\pi/2$ radians. If $\langle E_x E_y \rangle \neq 0$, there must be mutual coherence between the components, but Eq. (10.38) does not suffice to determine it completely. Additional phase shifts must be introduced artificially, and then the light passed through a rotatable polarizer, for the light to be uniquely characterized.

2. ***The Analytic Signal.*** If we want to specify the polarization of a light beam in more detail, we must do more than pass it through a polarizer; we must shift the relative phase of the two field components. As Eq. (10.28) shows, phase shifters (also called compensators) are usually wavelength dependent devices. Thus we are now forced to deal with light of essentially one frequency—quasimonochromatic light—although the requirements are usually not very strict. For such light the x component of E at a particular point in space takes the form

$$E_x(t) = A_x \cos(\omega t + \phi_x) = \mathrm{Re}\,\tilde{E}_x(t)$$

where

$$\tilde{E}_x(t) = A_x e^{i\phi_x} e^{i\omega t}$$

A similar expression holds for $E_y(t)$. Here A_x is the amplitude and ϕ_x the phase of the component field. Both are functions of time, but both are assumed to be slowly varying with respect to $\cos\omega t$. The complex field $\tilde{E}_x(t)$ is called the *analytic signal*. We shall use it here in our description of partially polarized light because of the ease with which phase shifts may be treated.

The flux density or irradiance associated with E_x is proportional to

$$\langle E_x{}^2\rangle = \tfrac{1}{2}\langle A_x{}^2\rangle = \tfrac{1}{2}\langle \tilde{E}_x\tilde{E}_x^*\rangle$$

The analytic signal passing through the polarizer in Fig. 10.28 is given by the sum of the projections of x and y components of **E** along α;

$$\tilde{E}(\alpha) = \tilde{E}_x \cos\alpha + \tilde{E}_y \sin\alpha$$

and the flux density is proportional to

$$\tfrac{1}{2} I(\alpha) = \tfrac{1}{2}\langle \tilde{E}(\alpha)\tilde{E}(\alpha)^*\rangle = \tfrac{1}{2}\langle \tilde{E}_x\tilde{E}_x^*\rangle \cos^2\alpha + \tfrac{1}{2}\langle \tilde{E}_y\tilde{E}_y^*\rangle \sin^2\alpha$$
$$+ \tfrac{1}{2}[\langle \tilde{E}_x\tilde{E}_y^*\rangle + \langle \tilde{E}_x^*\tilde{E}_y\rangle] \sin\alpha\cos\alpha \qquad (10.39)$$

3. ***Compensator plus Polarizer.*** Suppose that a compensator is is inserted before the polarizer so that the phase of the y component is shifted from ϕ_y to $\phi_y - \delta + \epsilon$, and the phase of the x component is shifted to $\phi_x + \epsilon$. Then $\tilde{E}_x$ is replaced by

$$A_x e^{i\phi_x} e^{i(\omega t+\epsilon)} = e^{i\epsilon}\tilde{E}_x$$

and $\tilde{E}_y$ is replaced by

$$A_y e^{i(\phi_y - \delta)}\, e^{i(\omega t+\epsilon)} = e^{-i(\delta-\epsilon)}\tilde{E}_y$$

and the signal passed by the polarizer set at angle α becomes

$$\tilde{E}(\alpha, \delta) = [\tilde{E}_x \cos\alpha + e^{-i\delta}\,\tilde{E}_y \sin\alpha]e^{i\epsilon}$$

with a flux density proportional to

$$\tfrac{1}{2} I(\alpha, \delta) = \tfrac{1}{2}\,\{\langle \tilde{E}_x\tilde{E}_x^*\rangle \cos^2\alpha + \langle \tilde{E}_y\tilde{E}_y^*\rangle \sin^2\alpha + [\langle \tilde{E}_x\tilde{E}_y^*\rangle e^{i\delta}$$
$$+ \langle \tilde{E}_x^*\tilde{E}_y\rangle\, e^{-i\delta}] \sin\alpha\cos\alpha\} \qquad (10.40)$$

For convenience we shall refer to $I(\alpha, \delta)$ as the "intensity," although the units and usage are technically incorrect.

It is clear from Eq. (10.40) that $I(\alpha, \delta)$ is sensitive only to the relative phase shift $(-\delta)$ between E_y and E_x produced by the compensator, and insensitive to any common phase shift ϵ.

B. THE COHERENCY MATRIX

There are four averages appearing in Eq. (10.40) that determine the function $I(\alpha, \delta)$. They are

$$J_{xx} = \langle \tilde{E}_x\tilde{E}_x^*\rangle, \qquad J_{xy} = \langle \tilde{E}_x\tilde{E}_y^*\rangle = J_{yx}^*$$
$$J_{yx} = \langle \tilde{E}_x^*\tilde{E}_y\rangle, \qquad J_{yy} = \langle \tilde{E}_y\tilde{E}_y^*\rangle \qquad (10.41)$$

and they form the elements of a two-by-two matrix called the coherency matrix:

$$\mathbf{J} = \begin{pmatrix} J_{xx} & J_{xy} \\ J_{yx} & J_{yy} \end{pmatrix} \tag{10.42}$$

This matrix is *Hermitian*. For such a matrix the components satisfy

$$J_{xx}^* = J_{xx}, \qquad J_{xy}^* = J_{yx}, \qquad J_{yy}^* = J_{yy} \tag{10.43}$$

The intensity $I(\alpha, \delta)$ can then be written

$$I(\alpha, \delta) = J_{xx} \cos^2 \alpha + J_{yy} \sin^2 \alpha + \sin \alpha \cos \alpha \, (J_{xy} e^{i\delta} + J_{yx} e^{-i\delta}) \tag{10.44}$$

Writing $J_{xy} = |J_{xy}| e^{-i\phi}$, $J_{yx} = |J_{xy}| e^{i\phi}$, we obtain an alternate form of (10.44):

$$I(\alpha, \delta) = J_{xx} \cos^2 \alpha + J_{yy} \sin^2 \alpha + 2 |J_{xy}| \sin \alpha \cos \alpha \cos (\delta - \phi) \tag{10.44'}$$

The average over α of Eq. (10.44) is

$$\langle I(\alpha, \delta) \rangle_\alpha = \tfrac{1}{2} I_0 = \tfrac{1}{2}(J_{xx} + J_{yy})$$

Here I_0 represents the sum of intensities measured in two perpendicular pass directions, at angle α and at angle $\alpha \pm \pi/2$

$$I_0 = J_{xx} + J_{yy} = I(\alpha, \delta) + I\left(\alpha \pm \frac{\pi}{2}, \delta\right) \tag{10.45}$$

and is the total intensity in the beam. Note that $J_{xx} = I(0, \delta)$, $J_{yy} = I(\pi/2, \delta)$.

The sum of the diagonal elements $J_{xx} + J_{yy}$ of a matrix is called its "trace" and is denoted by Tr $\mathbf{J}$. Thus

$$I_0 = J_{xx} + J_{yy} = \operatorname{Tr} \mathbf{J} \tag{10.45'}$$

The nature of the polarized light in the beam is independent of I_0 and depends on the ratios of the matrix elements J_{ij}. For this reason we define a *reduced* coherency matrix

$$\mathbf{j} = \frac{1}{I_0} \mathbf{J} \tag{10.46a}$$

having elements $j_{xx} = J_{xx}/I_0$, and so on. Then

$$\operatorname{Tr} \mathbf{j} = 1 \tag{10.46b}$$

1. Determination of the Elements of the Coherency Matrix. Since we can write

$$J_{xy} = J_{yx}^* = |J_{xy}| \, e^{-i\phi}$$

there are four independent real quantities that determine $\mathbf{J}$: $J_{xx}, J_{yy}, |J_{xy}|$, and ϕ. They can be determined from four intensity measurements using a polarizer and a compensator. There is no unique way of doing this, but we give here an example of a set of four such measurements.

a. Polarizer along the x axis. Then we measure $I(0, \delta) = J_{xx}$.
b. Polarizer along the y axis. Then we measure $I(\pi/2, \delta) = J_{yy}$.
c. Polarizer set at 45° to the x axis. This gives

$$I\left(\frac{\pi}{4}, 0\right) = \tfrac{1}{2}[J_{xx}+J_{yy}+J_{xy}+J_{yx}]$$
$$= \tfrac{1}{2}(J_{xx}+J_{yy}) + \operatorname{Re} J_{xy}$$

d. Quarter-wave plate with fast axis along x ($\delta = \pi/2$), followed by a polarizer at 45° to the x axis. This gives $I(\pi/4, \pi/2) = \tfrac{1}{2}[J_{xx}+J_{yy}+i(J_{xy}-J_{yx})] = \tfrac{1}{2}(J_{xx}+J_{yy}) - \operatorname{Im} J_{xy}$.

These four equations may be solved for J_{xx}, J_{yy}, $\operatorname{Re} J_{xy} = |J_{xy}| \cos \phi$, and $\operatorname{Im} J_{xy} = |J_{xy}| \sin \phi$.

2. The Coherency Matrix for Certain Types of Polarized Light.

(a) *Plane polarized.* Suppose that the plane of vibration of the light makes an angle α_0 with the x axis (Fig. 10.29). Then the components of $\tilde{E}$ are

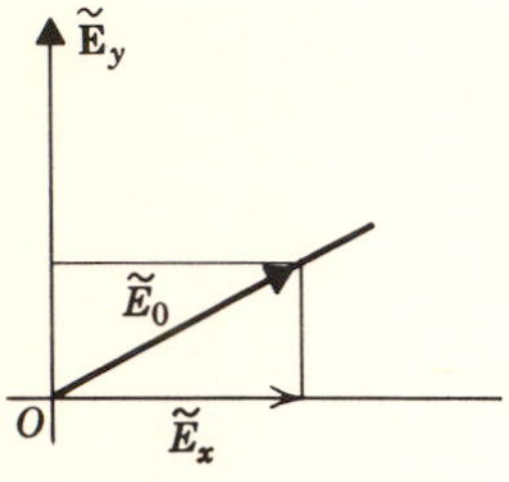

Fig. 10.29

$$\tilde{E}_x = \tilde{E}_0 \cos \alpha_0$$

$$\tilde{E}_y = \tilde{E}_0 \sin \alpha_0$$

When the easy passage direction α of the polarizer equals $\pm\alpha_0$, we observe a maximum transmitted intensity given by

$$I_0 = \langle \tilde{E}_0 \tilde{E}_0^* \rangle$$

where I_0 is the total intensity in the beam. When α equals $\pm\alpha_0 \pm \pi/2$, I takes its minimum value of zero.

For a general value of α, Eq. (10.44) holds with

$$J_{xx} = I_0 \cos^2 \alpha_0, \qquad J_{xy} = J_{yx} = I_0 \sin \alpha_0 \cos \alpha_0$$
$$Jyy = I_0 \sin^2 \alpha_0,$$

The coherency matrix thus takes the form

$$\mathbf{J} = I_0 \begin{pmatrix} \cos^2 \alpha_0 & \sin \alpha_0 \cos \alpha_0 \\ \sin \alpha_0 \cos \alpha_0 & \sin^2 \alpha_0 \end{pmatrix} \tag{10.47}$$

For zero compensating phase shift δ (10.44) can then be rearranged to give

$$I(\alpha, 0) = I_0 \cos^2 (\alpha - \alpha_0)$$

in agreement with the law of Malus.

Whenever it is possible to obtain extinction with just a polarizer, we must have pure plane polarized light with plane of vibration of the E vector perpendicular to the pass direction of the polarizer when it is set to produce extinction.

(b) *Unpolarized light.* Unpolarized light is characterized by a constant transmitted intensity independent of the polarizer angle α or the phase shift δ. Thus

$$I(\alpha, \delta) = \text{const} = \frac{I_0}{2}$$

This can only hold if

$$J_{xy} = J_{yx} = 0, \qquad J_{xx} = J_{yy} = \frac{I_0}{2}$$

Here I_0 represents the total intensity measured when there is no polarizer in the beam. The coherence matrix then takes the form

$$\mathbf{J} = \frac{I_0}{2}\begin{pmatrix} 1 & 0 \\ 0 & 1 \end{pmatrix} \tag{10.48}$$

that is, it is a multiple of the unit matrix

$$\begin{pmatrix} 1 & 0 \\ 0 & 1 \end{pmatrix}$$

(c) *Circularly polarized light.* When the light is circularly polarized, we have $A_x = A_y$ and

$$\phi_y - \phi_x = \frac{-\pi}{2} \qquad \text{for LCP (+ helicity)}$$

$$\phi_y - \phi_x = \frac{\pi}{2} \qquad \text{for RCP (− helicity)}$$

Then

$$J_{xx} = J_{yy} = \langle A_x^2 \rangle$$

$$J_{xy} = \langle A_x^2 \rangle e^{i(\phi_y - \phi_x)} = \langle A_x^2 \rangle e^{\pi i/2} = i\langle A_x^2 \rangle = -J_{yx} \qquad \text{for LCP}$$

$$J_{xy} = -J_{yx} = -i\langle A_x^2 \rangle \qquad \text{for RCP} \tag{10.49}$$

When circularly polarized light passes through a quarter-wave plate with fast axis along *x* or *y* axes, we obtain linearly polarized light at angle $\alpha_0 = \pm\pi/4$ and having a coherency matrix (for $\alpha_0 = \pi/4$)

$$\mathbf{J} = \begin{pmatrix} J_{xx} & J_{xx} \\ J_{xx} & J_{xx} \end{pmatrix}$$

The intensity is maximum when $\alpha = \alpha_0$ and equals $I_0 = 2J_{xy}$. Thus we may write

$$J_{xy} = \langle A_x^2 \rangle = \tfrac{1}{2} I_0$$

and the coherency matrix of the original light takes the form

$$\mathbf{J} = \tfrac{1}{2} I_0 \begin{pmatrix} 1 & i \\ -i & 1 \end{pmatrix} \qquad \text{for LCP}$$

$$\mathbf{J} = \tfrac{1}{2} I_0 \begin{pmatrix} 1 & -i \\ i & 1 \end{pmatrix} \qquad \text{for RCP} \tag{10.50}$$

(d) *Elliptically polarized light.* For general elliptically polarized light we have

$$\tilde{E}_x = A_x e^{i\phi_x} e^{i\omega t}$$

$$\tilde{E}_y = A_y e^{i\phi_y} e^{i\omega t}$$

where the ratio A_y/A_x and the phase difference $\phi = \phi_y - \phi_x$ must be constant. We introduce the total intensity from Eq. (10.45):

$$I_0 = J_{xx} + J_{yy} = \langle A_x^2 \rangle + \langle A_y^2 \rangle$$

Then we define a positive angle γ [χ in Eq. (10.20)] by

$$\gamma = \tan^{-1} \frac{A_y}{A_x}$$

The coherence matrix then may be written as

$$\mathbf{J} = I_0 \begin{pmatrix} \cos^2 \gamma & e^{-i\phi} \sin \gamma \cos \gamma \\ e^{i\phi} \sin \gamma \cos \gamma & \sin^2 \gamma \end{pmatrix} \tag{10.51}$$

This includes Eqs. (10.47) and (10.50) as special cases.

Note that for all pure states of polarization

$$\det \mathbf{J} = J_{xx} J_{yy} - J_{xy} J_{yx} = 0 \tag{10.52}$$

If light described by Eq. (10.51) passes through a compensator so that the phase of $\tilde{E}_y$ is retarded by an angle $\delta = -\phi$ with respect to that of $\tilde{E}_x$, then the coherency matrix becomes

$$\mathbf{J}' = I_0 \begin{pmatrix} \cos^2 \gamma & \sin \gamma \cos \gamma \\ \sin \gamma \cos \gamma & \sin^2 \gamma \end{pmatrix}$$

This has the same form as Eq. (10.47) and thus represents linearly polarized light polarized in direction γ.

3. ***Addition of Two Mutually Incoherent Light Beams.*** Two light beams propagating in the same direction will be mutually incoherent if their analytic signals are uncorrelated, that is, if

$$\langle \tilde{E}_x^{(1)} \tilde{E}_x^{(2)*} \rangle = \langle \tilde{E}_y^{(1)} \tilde{E}_y^{(2)*} \rangle = \langle \tilde{E}_x^{(1)} \tilde{E}_y^{(2)*} \rangle = \langle \tilde{E}_y^{(1)} \tilde{E}_x^{(2)*} \rangle = 0$$

The coherency matrix J for the combined analytic signal with components

$$\begin{pmatrix} \tilde{E}_x \\ \tilde{E}_y \end{pmatrix} = \begin{pmatrix} \tilde{E}_x^{(1)} + \tilde{E}_x^{(2)} \\ \tilde{E}_y^{(1)} + \tilde{E}_y^{(2)} \end{pmatrix}$$

is then simply the sum of the individual coherency matrices,

$$\begin{pmatrix} \langle \tilde{E}_x \tilde{E}_x^* \rangle & \langle \tilde{E}_x \tilde{E}_y^* \rangle \\ \langle \tilde{E}_x^* \tilde{E}_y \rangle & \langle \tilde{E}_y \tilde{E}_y^* \rangle \end{pmatrix} = \begin{pmatrix} \langle \tilde{E}_x^{(1)} \tilde{E}_x^{(1)*} \rangle & \langle \tilde{E}_x^{(1)} \tilde{E}_y^{(1)*} \rangle \\ \langle \tilde{E}_x^{(1)*} \tilde{E}_y^{(1)} \rangle & \langle \tilde{E}_y^{(1)} \tilde{E}_y^{(1)*} \rangle \end{pmatrix}$$

$$+ \begin{pmatrix} \langle \tilde{E}_x^{(2)} \tilde{E}_x^{(2)*} \rangle & \langle \tilde{E}_x^{(2)} \tilde{E}_y^{(2)*} \rangle \\ \langle \tilde{E}_x^{(2)*} \tilde{E}_y^{(2)} \rangle & \langle \tilde{E}_y^{(2)} \tilde{E}_y^{(2)*} \rangle \end{pmatrix}$$

or

$$\mathbf{J} = \mathbf{J}^{(1)} + \mathbf{J}^{(2)}$$

The nature of the polarized beam is determined uniquely by the reduced coherency matrix $\mathbf{j} = \mathbf{J}/I_0$, according to Eqs. (10.46). If the reduced matrices for the component beams are the same,

$$\mathbf{j}^{(1)} = \frac{\mathbf{J}^{(1)}}{I_1} = \mathbf{j}^{(2)} = \frac{\mathbf{J}^{(2)}}{I_2} = \mathbf{j}$$

then the combined beam has that same reduced matrix

$$\mathbf{J} = I_1\mathbf{j} + J_2\mathbf{j} = (I_1 + I_2)\mathbf{j}$$

This says that the combination of two incoherent beams in the same state of polarization yields a beam having the same polarization as its components.

The notion of superposition of incoherent beams is useful for describing unpolarized light. For instance, unpolarized light can be described as a superposition of two equal incoherent plane polarized components with mutually perpendicular planes of vibration. Let one plane have an inclination angle α. Then one component coherency matrix is

$$\mathbf{J}^{(1)} = \mathbf{J}(\alpha_0) = \frac{I_0}{2}\begin{pmatrix} \cos^2\alpha_0 & \sin\alpha_0\cos\alpha_0 \\ \sin\alpha_0\cos\alpha_0 & \sin^2\alpha_0 \end{pmatrix} \tag{10.53a}$$

The other component must then have inclination angle $\alpha_0 \pm \pi/2$ and coherency matrix

$$\mathbf{J}^{(2)} = \mathbf{J}\left(\alpha_0 \pm \frac{\pi}{2}\right) = \frac{I_0}{2}\begin{pmatrix} \sin^2\alpha_0 & -\sin\alpha_0\cos\alpha_0 \\ -\sin\alpha_0\cos\alpha_0 & \cos^2\alpha_0 \end{pmatrix} \tag{10.53b}$$

Then

$$\mathbf{J} = \mathbf{J}^{(1)} + \mathbf{J}^{(2)} = \frac{I_0}{2}\begin{pmatrix} 1 & 0 \\ 0 & 1 \end{pmatrix}$$

Unpolarized light may also be described as a superposition of equal incoherent right and left circularly polarized beams:

$$\mathbf{J}^{(1)} = \mathbf{J}\,(\text{left}) = \tfrac{1}{4}I_0\begin{pmatrix} 1 & i \\ -i & 1 \end{pmatrix} \tag{10.54a}$$

$$\mathbf{J}^{(2)} = \mathbf{J}\,(\text{right}) = \tfrac{1}{4}I_0\begin{pmatrix} 1 & -i \\ i & 1 \end{pmatrix} \tag{10.54b}$$

$$\mathbf{J} = \mathbf{J}^{(1)} + \mathbf{J}^{(2)} = \tfrac{1}{2}I_0\begin{pmatrix} 1 & 0 \\ 0 & 1 \end{pmatrix}$$

Two incoherent beams, each in a pure elliptically polarized state, may be combined to yield unpolarized light. The notation of Eq.

(10.51) is convenient for this purpose. Let the coherency matrices for the two beams be

$$\mathbf{J}^{(1)} = I_1 \begin{pmatrix} \cos^2 \delta_1 & e^{-i\phi_1} \sin \delta_1 \cos \delta_1 \\ e^{i\phi_1} \sin \delta_1 \cos \delta_1 & \sin^2 \delta_1 \end{pmatrix};$$

$$\mathbf{J}^{(2)} = I_2 \begin{pmatrix} \cos^2 \delta_2 & e^{-i\phi_2} \sin \delta_2 \cos \delta_2 \\ e^{i\phi_2} \sin \delta_2 \cos \delta_2 & \sin^2 \delta_2 \end{pmatrix} \tag{10.55}$$

Then let

$$\delta_2 = (\pi/2) - \delta_1, \qquad \phi_2 = \pi + \phi_1, \qquad I_1 = I_2 = \frac{I_0}{2}$$

to obtain

$$\mathbf{J} = \mathbf{J}^{(1)} + \mathbf{J}^{(2)} = \frac{I_0}{2} \begin{pmatrix} 1 & 0 \\ 0 & 1 \end{pmatrix}$$

This last example includes the first two as special cases—when $\phi_1 = 0$, $\delta_1 = \alpha_0$, the first example is recovered, and when $\phi_1 = -\pi/2$, $\delta_1 = \pi/4$, the second is recovered. The two different components in the three cases are shown in Fig. 10.30. Note that the major axes of the two ellipses are 90° apart and that the ellipses have the same size and shape with opposite helicity.

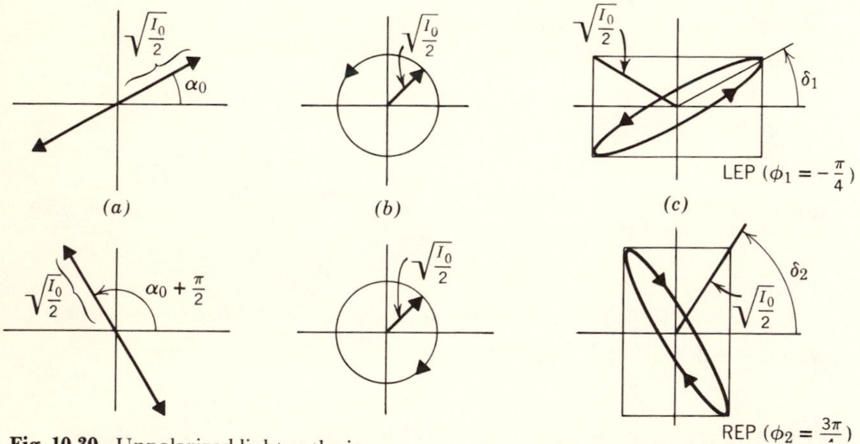

Fig. 10.30. Unpolarized light as the incoherent superposition of two components: (*a*) plane, (*b*) circular, (*c*) elliptical.

4. Partially Polarized Light as a Superposition of Unpolarized Light and Elliptically Polarized Light. We now show that partially polarized light can be expressed as a superposition of unpolarized light and of pure polarized light (elliptically polarized, in general), and that this decomposition is unique. To do this we show that an arbitrary coherency matrix **J** may be written in the form

$$\mathbf{J} = \mathbf{J}^{(1)} + \mathbf{J}^{(2)}$$

or

$$\begin{pmatrix} J_{xx} & J_{xy} \\ J_{xy}^* & J_{yy} \end{pmatrix} = \begin{pmatrix} A & 0 \\ 0 & A \end{pmatrix} + \begin{pmatrix} B & D \\ D^* & C \end{pmatrix} \tag{10.56}$$

The second matrix $\mathbf{J}^{(2)}$ should represent a pure state of polarization. Hence by Eq. (10.50) its determinant must vanish:

$$BC - DD^* = 0 \tag{10.57a}$$

Another requirement on $\mathbf{J}^{(2)}$ is that it give nonnegative values of intensity when used in Eq. (10.39) or (10.40). In particular, the requirement $I(0) \geqslant 0$ gives

$$B \geqslant 0 \tag{10.57b}$$

and the requirement $I(\pi/2) \geqslant 0$ gives

$$C \geqslant 0 \tag{10.57c}$$

The matrix Eq. (10.56) gives four ordinary equations

$$\begin{aligned} J_{xx} &= A + B \\ J_{xy} &= D \\ J^*_{xy} &= D^* \\ J_{yy} &= A + C \end{aligned}$$

Thus $B = J_{xx} - A$ and $C = J_{yy} - A$, and (10.57a) becomes

$$(J_{xx} - A)(J_{yy} - A) - J^2_{xy} = 0 \tag{10.58}$$

or

$$A^2 - A(J_{xx} + J_{yy}) + J_{xx}J_{yy} - J_{xy}J_{yx} = 0$$

or

$$A^2 - A \operatorname{Tr} \mathbf{J} + \det \mathbf{J} = 0 \tag{10.58'}$$

with

$$\operatorname{Tr} \mathbf{J} = J_{xx} + J_{yy}; \qquad \det \mathbf{J} = J_{xx}J_{yy} - J_{xy}J_{yx}$$

Equation (10.58) also represents the so-called *characteristic equation* for the matrix $\mathbf{J}$. Such an equation is usually obtained in the form

$$\det(\mathbf{J} - A\mathbf{I}) = 0$$

where $\mathbf{I}$ is the unit matrix

$$\mathbf{I} = \begin{pmatrix} 1 & 0 \\ 0 & 1 \end{pmatrix}$$

or

$$\det \begin{pmatrix} J_{xx} - A & J_{xy} \\ J_{yx} & J_{yy} - A \end{pmatrix} = 0$$

There are two solutions of Eq. (10.58′) given by

$$A_{1,2} = \frac{\operatorname{Tr} \mathbf{J} \pm \sqrt{(\operatorname{Tr} \mathbf{J})^2 - 4 \det \mathbf{J}}}{2} \tag{10.59}$$

The argument of the radical is

$$\begin{aligned}&(J_{xx}+J_{yy})^2-4(J_{xx}J_{yy}-|J_{xy}|^2)\\&=J_{xx}^2+J_{yy}^2-2J_{xx}J_{yy}+4|J_{xy}|^2\\&=(J_{xx}-J_{yy})^2+4|J_{xy}|^2\geqslant 0\end{aligned}\tag{10.60}$$

so that the radical is real.

Both solutions $A_{1,2}$ are nonnegative. This obvious for

$$A_1=\frac{\mathrm{Tr}\,\mathbf{J}+\sqrt{(\mathrm{Tr}\,\mathbf{J})^2-4\det\mathbf{J}}}{2}$$

To see it for

$$A_2=\frac{\mathrm{Tr}\,\mathbf{J}-\sqrt{(\mathrm{Tr}\,\mathbf{J})^2-4\det\mathbf{J}}}{2}$$

we need only note that

$$\sqrt{(\mathrm{Tr}\,\mathbf{J})^2-4\det\mathbf{J}}\leqslant\mathrm{Tr}\,\mathbf{J}$$

so that

$$-\sqrt{(\mathrm{Tr}\,\mathbf{J})^2-4\det\mathbf{J}}\geqslant-\mathrm{Tr}\,\mathbf{J}$$

and that

$$A_2=\frac{\mathrm{Tr}\,\mathbf{J}-\sqrt{(\mathrm{Tr}\,\mathbf{J})^2-4\det\mathbf{J}}}{2}\geqslant\frac{\mathrm{Tr}\,\mathbf{J}-\mathrm{Tr}\,\mathbf{J}}{2}=0$$

We also have from (10.60)

$$\sqrt{(\mathrm{Tr}\,\mathbf{J})^2-4\det\mathbf{J}}\geqslant|J_{xx}-J_{yy}|$$

so that

$$\begin{aligned}&A_1\geqslant J_{xx}\qquad\text{if}\quad J_{xx}\geqslant J_{yy}\\&\text{or}\\&A_1\geqslant J_{yy}\qquad\text{if}\quad J_{yy}\geqslant J_{xx}\end{aligned}\tag{10.61}$$

This violates one or the other of the inequalities (10.57*b*, *c*):

$$B=J_{xx}-A\geqslant 0$$
$$C=J_{yy}-A\geqslant 0$$

Thus our physically meaningful solution must be

$$A=A_2=\frac{\mathrm{Tr}\,\mathbf{J}-\sqrt{(\mathrm{Tr}\,\mathbf{J})^2-4\det\mathbf{J}}}{2}\tag{10.62a}$$

Then

$$B=J_{xx}-A_2=\frac{J_{xx}-J_{yy}+\sqrt{(\mathrm{Tr}\,\mathbf{J})^2-4\det\mathbf{J}}}{2}\tag{10.62b}$$

and

$$C=J_{yy}-A_2=\frac{J_{yy}-J_{xx}+\sqrt{(\mathrm{Tr}\,\mathbf{J})^2-4\det\mathbf{J}}}{2}\tag{10.62c}$$

The total intensity is given by Eq. (10.45) as

$$I_{\text{tot}} = \text{Tr}\,\mathbf{J} = A_1 + A_2 \tag{10.63a}$$

The intensity of the polarized part is

$$I_{\text{pol}} = \text{Tr}\,\mathbf{J}^{(2)} = B + C = (A_1 - A_2) \tag{10.63b}$$

and the intensity of the unpolarized part is

$$I_{\text{un}} = I_{\text{tot}} - I_{\text{pol}} = 2A_2 \tag{10.63c}$$

*5. **The Degree of Polarization.*** The ratio of intensity of the polarized part to that of the total beam is defined to be the *degree of polarization P*:

$$P = \frac{I_{\text{pol}}}{I_{\text{tot}}} = \left[1 - \frac{4\det\mathbf{J}}{(\text{Tr}\,\mathbf{J})^2}\right]^{1/2} \tag{10.64}$$

Here P can also be written

$$P = \frac{A_1 - A_2}{A_1 + A_2} \tag{10.64$'$}$$

The matrix $\mathbf{J}^{(1)}$ represents an unpolarized beam:

$$\mathbf{J}^{(1)} = \begin{pmatrix} A_2 & 0 \\ 0 & A_2 \end{pmatrix} = \tfrac{1}{2} I_{\text{un}} \begin{pmatrix} 1 & 0 \\ 0 & 1 \end{pmatrix} \tag{10.65a}$$

The matrix $\mathbf{J}^{(2)}$ represents a pure elliptically polarized beam and may be written

$$\mathbf{J}^{(2)} = \begin{pmatrix} B & J_{xy} \\ J^*_{xy} & C \end{pmatrix} = \begin{pmatrix} B & |J_{xy}|e^{-i\phi} \\ |J_{xy}|e^{i\phi} & C \end{pmatrix}$$

$$= I_{\text{pol}} \begin{pmatrix} \cos^2\alpha_0 & e^{-i\phi}\sin\alpha_0\cos\alpha_0 \\ e^{i\phi}\sin\alpha_0\cos\alpha_0 & \sin^2\alpha_0 \end{pmatrix} \tag{10.65b}$$

with

$$\alpha_0 = \tan^{-1}\left(\frac{C}{B}\right)^{1/2}$$

If the beam is passed through a compensator with phase shift $\delta = -\phi$, $\mathbf{J}^{(2)}$ is changed into

$$\mathbf{J}^{(2)\prime} = \begin{pmatrix} B & |J_{xy}| \\ |J_{xy}| & C \end{pmatrix} = I_{\text{pol}} \begin{pmatrix} \cos^2\alpha_0 & \sin\alpha_0\cos\alpha_0 \\ \sin\alpha_0\cos\alpha_0 & \sin^2\alpha_0 \end{pmatrix}$$

Here $\mathbf{J}^{(1)}$ is left unchanged. Then if the beam is passed through a polarizer with pass angle α the measured intensity, $I(\alpha, -\phi)$ in our original notation (Eq. 10.44), will be given by

$$\begin{aligned} I(\alpha, -\phi) &= \tfrac{1}{2} I_{\text{unpol}} + I_{\text{pol}}\cos^2(\alpha - \alpha_0) \\ &= I_{\text{tot}}\left\{\tfrac{1}{2} + P\left[\cos^2(\alpha - \alpha_0) - \tfrac{1}{2}\right]\right\} \end{aligned} \tag{10.66}$$

This has a maximum value of

$$I_{\text{max}} = \tfrac{1}{2}I_{\text{tot}}(1+P) = A_1$$

and a minimum value of

$$I_{\text{min}} = \tfrac{1}{2}I_{\text{tot}}(1-P) = A_2 \tag{10.67}$$

Thus the degree of polarization can be written

$$P = \frac{I_{\text{max}} - I_{\text{min}}}{I_{\text{max}} + I_{\text{min}}} \tag{10.64$''$}$$

When δ is not equal to $-\phi$, the maximum and minimum values of $I(\alpha, \delta)$ will be respectively smaller and larger than the values given above.

6. Experimental Characterization of a Partially Polarized Light Beam. A general partially polarized light beam is a mixture of elliptically polarized light and unpolarized light. It will be uniquely specified by five parameters, the total intensity I_{tot}, the degree of polarization P, the orientation angle $\beta/2$ of the major axis of the ellipse, the ratio $\tan \eta$ of the lengths of minor to major axes, and the sense of rotation or helicity of the light. (See Fig. 10.22.)

We need a polarizer and a quarter-wave plate to determine these parameters. The polarizer is first placed alone in the beam and rotated while the transmitted intensity is observed. The two pass directions giving maximum intensity correspond to E vibrating along the major axis of the ellipse and hence correspond to the angles $\beta/2$ and $\beta/2+\pi$. The pass directions giving minimum intensity are at $\beta/2 \pm \pi/2$. So far the light could lie between the following two extremes: (1) pure elliptically polarized light with $\tan \eta = \sqrt{I_{\text{min}}/I_{\text{max}}}$, and (2) a mixture of unpolarized light and plane polarized light vibrating along the direction $\beta/2$.

It is convenient to consider a new set of axes along $\bar{x}$ and $\bar{y}$ as shown in Fig. 10.22. Place a quarter-wave plate in the incident beam so that its fast axis is along the $\bar{x}$ axis. Any elliptically polarized light with major axis along $\bar{x}$ will then be converted to plane polarized light vibrating at an angle of $\pm\eta$ with respect to the $\bar{x}$ axis, where the plus sign refers to right elliptically polarized light and the minus sign to left elliptically polarized light. If the original light is a mixture of unpolarized light and plane polarized light vibrating along $\bar{x}$, the quarter-wave plate will have no effect.

For the mixed case there will be an unpolarized component present after the light has passed through the quarter-wave plate. The angle η can be determined by using a polarizer after the quarter-wave plate. With the pass direction of the polarizer along

η or $\eta + \pi$ (with respect to the $\bar{x}$ axis) the intensity will have a maximum value of

$$I_{\text{max}} = I_{\text{pol}} + \tfrac{1}{2} I_{\text{unpol}}$$

With the pass direction along $\eta + \pi/2$, the intensity will have a minimum value of

$$I_{\text{min}} = \tfrac{1}{2} I_{\text{unpol}}$$

The degree of polarization is then

$$P = \frac{I_{\text{max}} - I_{\text{min}}}{I_{\text{max}} + I_{\text{min}}}$$

and the total intensity is

$$I_{\text{tot}} = I_{\text{max}} + I_{\text{min}}$$

REFERENCES FOR SECTION 10.3

E. L. O'Neill, *Introduction to Statistical Optics*, Chapter 9.
M. Born and E. Wolf, *Principles of Optics*, Section 10.8.

PROBLEMS

10.3.1 Partially polarized light can be expressed in a unique way as the incoherent superposition of two elliptically polarized beams. The ellipses have the same ellipticity, opposite helicity, and major axes oriented at right angles.

a. Show that we can write

$$\mathbf{J} = A_1 \begin{pmatrix} \cos^2\alpha_0 & e^{-i\phi} \sin \alpha_0 \cos \alpha_0 \\ e^{i\phi} \sin \alpha_0 \cos \alpha_0 & \sin^2\alpha_0 \end{pmatrix} + A_2 \begin{pmatrix} \sin^2\alpha_0 & -e^{-i\phi} \sin \alpha_0 \cos \alpha_0 \\ -e^{i\phi} \sin \alpha_0 \cos \alpha_0 & \cos^2\alpha_0 \end{pmatrix}$$

where the phase ϕ is defined by $J_{xy} = |J_{xy}|e^{-i\phi}$, where $A_{1,2}$ are defined in Eq. (10.59) and B and C in Eqs. (10.62*b*, *c*). Then $\alpha_0 = \tan^{-1} \sqrt{C/B}$.

b. Then justify the statements made at the beginning of this problem.

10.3.2 Two calcite plates may be used to isolate spatially the *e* ray from the *o* ray, as shown in the figure. The thicknesses of the two plates are the same, and their optical axes have opposite tilt angles. The *e* ray has electric vibrations in the plane of the paper along the *y* direction, and the *o* ray has vibrations perpendicular to the plane of the paper along the *x* direction. The second plate brings the two beams together at *Q*. The symbol *S* is a transparent plate used as a phase shifter to make the total optical path lengths for *o* and *e* rays nearly the same, in particular to make the optical path difference less than the longitudinal coherence length for the light. Note that because of its larger wave surface and greater velocity, it is the *e* ray that must be retarded to match the optical path lengths. Let Δ be the net resulting phase difference $\phi_y - \phi_x = \phi_e - \phi_o$.

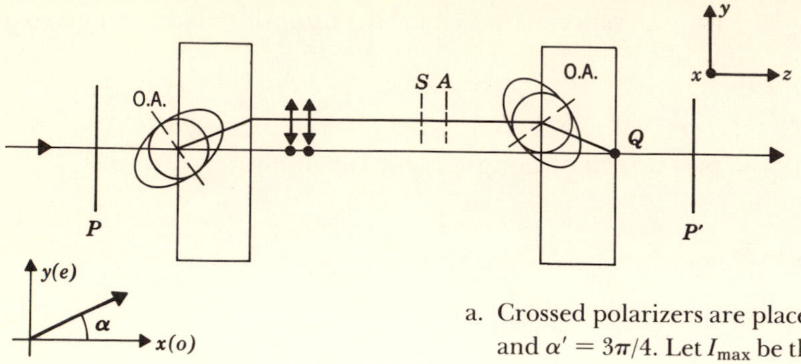

a. Crossed polarizers are placed at P and P' having pass angles $\alpha = \pi/4$ and $\alpha' = 3\pi/4$. Let $I_{\max}$ be the maximum transmitted intensity as a function of Δ. Then Δ is adjusted to give zero transmitted intensity. One of the two beams in the region between the plates is blocked out. Light of intensity $I_{\max}/4$ is now transmitted! Explain.

b. Let partially polarized light with a general coherency matrix

$$\mathbf{J} = \begin{pmatrix} J_{xx} & J_{xy} \\ J_{yx} & J_{yy} \end{pmatrix}$$

be incident from the left. In addition to the phase shifter S, an absorber A is now inserted in the e beam. It produces no phase shift but changes the amplitude by the factor $\tau_0 < 1$. What is the resulting coherency matrix $\mathbf{J}'$ at Q for general values of Δ and τ_0?

10.3.3 Consider unpolarized laser light with the following properties. (a) Coherence time τ long enough to be readily accessible to direct measurement, say 10^{-3} sec. (b) Total intensity which is constant on a time scale much less than the coherence time (but much longer than the period $2\pi/\omega$). Discuss qualitatively the nature of the intensity fluctuations observed when this light is passed through a polarizer.

10.3.4 Determine the coherency matrix $\mathbf{J}$ for partially polarized light having the parameters defined in Section B.7. Write the components of $\mathbf{J}$ first with respect to the $\overline{xy}$ coordinate system (Fig. 10.22). Then rotate coordinates and calculate the elements of $\mathbf{J}$ with respect to the xy system.

10.3.5 A diffraction grating is covered with two polaroid sheets with mutually perpendicular pass directions oriented at 45° to the grating

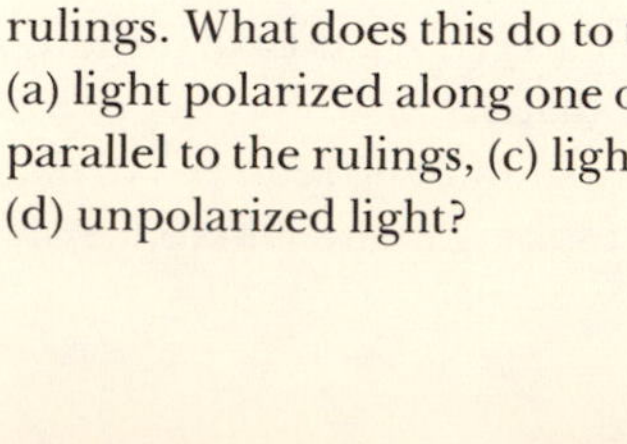

rulings. What does this do to the ultimate resolution of the grating for (a) light polarized along one of the pass directions, (b) light polarized parallel to the rulings, (c) light polarized perpendicular to the rulings, (d) unpolarized light?

We turn now to a description of light as electromagnetic waves. The equations of electricity and magnetism, Maxwell's equations, describe these waves in vacuum. The equations may be extended to treat interactions of electromagnetic waves with matter, provided we introduce auxiliary quantities such as macroscopic fields and electric and magnetic polarizations and assume certain "material relations" between these quantities. What results is a phenomenological theory *of the interaction of light with matter.*

The material relations are given in terms of quantities such as the dielectric constant and magnetic susceptibility. These depend on the microscopic properties of the type of matter under consideration. The theory of, say, the dielectric constant of a gas, liquid, or solid is a proper branch of the physics and chemistry of the respective gas, liquid, or solid. As such, it is beyond the scope of this book. Nevertheless, we do attempt to present the very beginnings of such theories with one or two examples, using mathematical models on which one can base a microscopic theory of the dielectric constant.

CHAPTER 11

THE ELECTROMAGNETIC NATURE OF LIGHT

CHAPTER 11 CONTENTS

11.1 ELECTROMAGNETIC THEORY APPLIED TO FREE SPACE AND APPLIED PHENOMENOLOGICALLY TO MATTER

A. ELECTROMAGNETIC WAVES IN FREE SPACE

In this section it is shown that the equations of electromagnetic theory, Maxwell's equations, when applied to a region free from charges and currents, lead to the three-dimensional wave equation. The transverse nature of waves will be brought out and energy-related quantities derived. The treatment is brief; standard texts on electromagnetic theory should be consulted for detailed discussions.

In free space the four Maxwell equations are

$$\begin{aligned} \nabla \cdot \mathbf{E} &= 0 \\ \nabla \cdot \mathbf{B} &= 0 \\ \nabla \times \mathbf{B} &= \frac{1}{c}\frac{\partial \mathbf{E}}{\partial t} \\ \nabla \times \mathbf{E} &= -\frac{1}{c}\frac{\partial \mathbf{B}}{\partial t} \end{aligned} \tag{11.1}$$

We use c.g.s. Gaussian units. The symbol c is the velocity of light in vacuum. The electric field E is in statvolts/cm or dynes/statcoulomb. The magnetic induction B is in gauss.

1. Derivation of the Wave Equation. Take the curl of both sides of the curl $\mathbf{E}$ equation. Note that

$$\nabla \times (\nabla \times \mathbf{E}) = \nabla(\nabla \cdot \mathbf{E}) - \nabla^2\mathbf{E} = -\nabla^2\mathbf{E}$$

The last equality holds because $\nabla \cdot \mathbf{E} = 0$. Then

$$-\nabla^2\mathbf{E} = -\frac{1}{c}\frac{\partial}{\partial t}(\nabla \times \mathbf{B})$$

Then use the curl $\mathbf{B}$ equation from (11.1),

$$-\nabla^2\mathbf{E} = -\frac{1}{c^2}\frac{\partial^2\mathbf{E}}{\partial t^2}$$

or

$$\nabla^2\mathbf{E} - \frac{1}{c^2}\frac{\partial^2\mathbf{E}}{\partial t^2} = 0 \tag{11.2a}$$

In a similar way, take the curl of the curl $\mathbf{B}$ equation, use

$$\nabla \times (\nabla \times \mathbf{B}) = \nabla(\nabla \cdot \mathbf{B}) - \nabla^2\mathbf{B} = -\nabla^2\mathbf{B}$$

and then make use of the curl $\mathbf{E}$ equation to obtain

$$\nabla^2\mathbf{B} - \frac{1}{c^2}\frac{\partial^2\mathbf{B}}{\partial t^2} = 0 \tag{11.3}$$

Thus, each component of **B** and **E** obeys the usual three-dimensional wave equation, for example,

$$\nabla^2 E_x - \frac{1}{c^2}\frac{\partial^2}{\partial t^2}E_x = 0 \tag{11.2b}$$

These equations and their solutions are not independent of one another because Eqs. (11.1) must still apply. When wavelike solutions of Eqs. (11.2*a, b*) or (11.3) are obtained, the connections between **B** and **E** contained in (11.1) yield the transverse polarization properties of electromagnetic waves, as we now show with some examples.

2. ***Plane Polarized, Sinusoidal Plane Waves.*** We know already that

$$\cos(\omega t - \mathbf{k}\cdot\mathbf{r} + \phi) = \operatorname{Re} e^{i(\omega t - \mathbf{k}\cdot\mathbf{r}+\phi)}$$

obeys the wave equation (11.2*b*) if $\omega = c|\mathbf{k}|$ and ϕ = constant. Therefore the following two equations will automatically be solutions of Eqs. (11.2*a*) and (11.3):

$$\mathbf{E} = \operatorname{Re}\mathbf{E}_0 e^{i(\omega t-\mathbf{k}\cdot\mathbf{r}+\phi)} \tag{11.4a}$$

$$\mathbf{B} = \operatorname{Re}\mathbf{B}_0 e^{i(\omega t-\mathbf{k}\cdot\mathbf{r}+\phi)} \tag{11.4b}$$

where $\mathbf{E}_0$ and $\mathbf{B}_0$ are constant vectors. *Hereafter when there is no chance for misunderstanding, we shall drop the "Re" in equations such as (11.4).* It shall be understood that the physical electric and magnetic fields will be given by the real part of the complex fields appearing in our equations.

We now show that $\mathbf{E}_0$ and $\mathbf{B}_0$ must be perpendicular to the direction of propagation **k**. From Eq. (11.1)

$$\nabla\cdot\mathbf{E} = 0$$

Using Eq. (11.4*a*), we calculate

$$\begin{aligned}\nabla\cdot\mathbf{E} &= \sum_j \frac{\partial}{\partial x_j}E_{0j}\exp\left[i\left(\omega t+\phi-\sum_l k_l r_l\right)\right] = -i\sum_j k_j E_{0j}\\ &\quad\times\exp\left[i(\omega t+\phi-\mathbf{k}\cdot\mathbf{r})\right]\\ &= -i\mathbf{k}\cdot\mathbf{E}\end{aligned} \tag{11.5a}$$

Thus, $\mathbf{k}\cdot\mathbf{E} = 0$, which implies that **E** must be perpendicular to the direction of propagation represented by **k**.

In a similar fashion it can be shown that

$$\mathbf{k}\cdot\mathbf{B} = 0$$

Thus, **B** must also be perpendicular to **k**. Therefore, **E** and **B** are both *transverse* vibrations.

The next task is to relate the amplitude vectors $\mathbf{E}_0$ and $\mathbf{B}_0$. We can readily obtain the relations

$$\nabla \times \mathbf{E} = -i\mathbf{k} \times \mathbf{E} \tag{11.5b}$$

$$\frac{\partial \mathbf{E}}{\partial t} = i\omega \mathbf{E} \tag{11.5c}$$

with similar equations for $\mathbf{B}$. Indeed, whenever the entire space-time dependence of a wave is described by the factor $\exp i(\omega t - \mathbf{k}\cdot\mathbf{r})$, then the space-time derivatives may be replaced as follows:

$$\frac{\partial}{\partial t} \to i\omega$$

$$\nabla \to -i\mathbf{k}$$

The last of Eqs. (11.1),

$$c\nabla \times \mathbf{E} = -\frac{\partial \mathbf{B}}{\partial t}$$

then becomes

$$-ic\mathbf{k} \times \mathbf{E} = -i\omega \mathbf{B}$$

or

$$\mathbf{B} = \frac{c}{\omega}\,\mathbf{k} \times \mathbf{E} = \frac{(\mathbf{k} \times \mathbf{E})}{k}$$

Thus

$$\mathbf{B}_0 = \frac{(\mathbf{k} \times \mathbf{E}_0)}{k} \tag{11.6}$$

Equation (11.6) says two important things: $\mathbf{B}$ is perpendicular to $\mathbf{E}$, and $\mathbf{B}$ is in phase with $\mathbf{E}$. The magnitudes of $\mathbf{B}$ and $\mathbf{E}$ are also the same for free space (in the Gaussian system of units).

The three vectors $\mathbf{E}_0$, $\mathbf{B}_0$, and $\mathbf{k}$ form a right-handed Cartesian coordinate system. The resulting field pattern for a fixed value of time is sketched in Fig. 11.1. This pattern moves in the $+\mathbf{k}$ direction with velocity c.

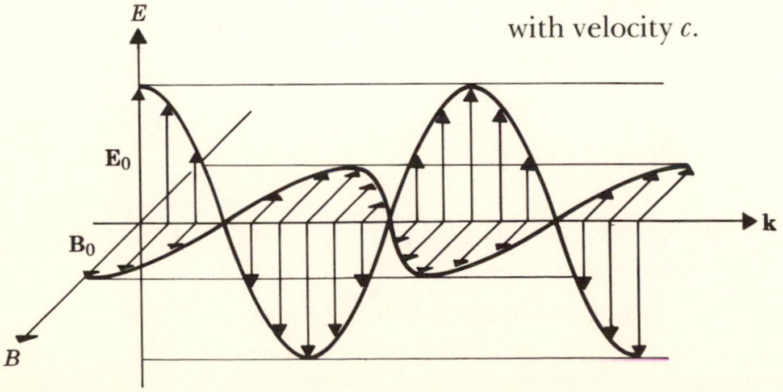

Fig. 11.1

Note that $\mathbf{B}_0$ is uniquely determined once $\mathbf{E}_0$ and $\mathbf{k}$ are known, since $\mathbf{B}_0 = (\mathbf{k}\times\mathbf{E}_0)/k$. Note also that $\mathbf{E}_0$ can take any direction in the transverse plane perpendicular to $\mathbf{k}$. Two different waves of the type just discussed having the same ω and $\mathbf{k}$ may be superimposed, yielding, in general, elliptically polarized light. As discussed in Chapter 10, the nature of the ellipse will depend on the ratio of amplitudes and value of the phase difference for these two waves.

*3. **Energy Density and Energy Flow.*** Electromagnetic theory gives for the energy density (in ergs/cm³) associated with electric and magnetic fields the expression

$$U = \frac{1}{8\pi}\left(|\mathbf{E}|^2 + |\mathbf{B}|^2\right) \tag{11.7}$$

The *Poynting vector* is defined by

$$\mathbf{S} = \frac{c}{4\pi}\left(\mathbf{E}\times\mathbf{B}\right) \tag{11.8}$$

This important vector gives the energy flow per unit time per unit area in a direction $\mathbf{n}$ by means of the scalar product $\mathbf{S}\cdot\mathbf{n}$.

A quadratic expression in the fields such as $\mathbf{S}$ or U must be calculated with care when complex notation is used for the fields. One safe method of calculation is to multiply the appropriate real fields. For plane polarized plane waves in vacuum we have the following expressions for the real, physical fields:

$$\mathbf{E} = \mathbf{E}_0 \cos\theta$$

$$\mathbf{B} = \mathbf{B}_0\cos\theta = \frac{\mathbf{k}\times\mathbf{E}_0}{k}\cos\theta, \qquad \theta = (\omega t - \mathbf{k}\cdot\mathbf{r} + \phi)$$

Hence, $\mathbf{S}$ is given by

$$\mathbf{S} = \frac{c}{4\pi}\mathbf{E}_0\times\frac{(\mathbf{k}\times\mathbf{E}_0)}{k}\cos^2\theta = \frac{c}{4\pi}|\mathbf{E}_0|^2\frac{\mathbf{k}}{k}\cos^2\theta \tag{11.9}$$

Since the time average over many cycles of $\cos^2(\omega t - \mathbf{k}\cdot\mathbf{r}+\phi)$ is $\frac{1}{2}$, the time average of $\mathbf{S}$ is given by

$$\langle\mathbf{S}\rangle = \frac{c}{8\pi}|\mathbf{E}_0|^2\frac{\mathbf{k}}{k} \tag{11.10}$$

The time-averaged energy flow is along the wave vector $\mathbf{k}$, in the direction of propagation of the wave.

For a wave in vacuum the energy density is given by

$$U = \frac{1}{8\pi}\left[|\mathbf{E}|^2 + |\mathbf{B}|^2\right] = \frac{1}{4\pi}|\mathbf{E}_0|^2\cos^2\theta$$

with a time average

$$\langle U \rangle = \frac{1}{8\pi} |\mathbf{E}_0|^2 \tag{11.11}$$

Note that the magnitude of $\langle \mathbf{S} \rangle$ obeys

$$S = |\langle \mathbf{S} \rangle| = \langle U \rangle c \tag{11.12}$$

This is a general result discussed in Chapter 1: energy flux = (energy density) × (propagation velocity).

In the language of Chapter 4, S represents electromagnetic flux density or irradiance. Suppose that $\mathbf{E}_0$ has a magnitude of 1 statvolt/cm, which equals 300 volts/cm or $3 \cdot 10^4$ volts/m in the M.K.S. system of units. Then S is

$$S = \frac{c}{8\pi} = \frac{3 \times 10^{10}}{8\pi} = 1.195 \times 10^9 \text{ erg/sec-cm}^2 = 119.5 \text{ watt/cm}^2$$

B. MAXWELL'S EQUATIONS IN THE PRESENCE OF MATTER

When charges and currents are present, Maxwell's equations become

$$\begin{aligned} \nabla \cdot \mathbf{E} &= 4\pi\rho \\ \nabla \cdot \mathbf{B} &= 0 \\ \nabla \times \mathbf{E} &= -\frac{1}{c}\frac{\partial \mathbf{B}}{\partial t} \\ \nabla \times \mathbf{B} &= \frac{4\pi}{c}\mathbf{j} + \frac{1}{c}\frac{\partial \mathbf{E}}{\partial t} \end{aligned} \tag{11.13}$$

Here ρ is the total charge density, and $\mathbf{j}$ is the total current density. In a general situation, parts of ρ and $\mathbf{j}$ come from "free" charges and currents and parts come from the response of matter to the fields $\mathbf{B}$ and $\mathbf{E}$. But $\mathbf{B}$ and $\mathbf{E}$ in turn depend on these same currents and charges, giving us a coupled system.

The charges and currents associated with the atoms and molecules of matter have sharp discontinuities on an atomic scale, that is, over distances of about one Angstrom (10^{-8} cm). The resulting $\mathbf{E}(\mathbf{r})$ and $\mathbf{B}(\mathbf{r})$ fields change very rapidly over similar distances, as the field point $\mathbf{r}$ moves about, say inside an atom or molecule. These then are *microscopic fields*. Macroscopic measuring devices will ordinarily not be able to observe the sharp discontinuities in these microscopic fields.

1. Macroscopic Fields. It is very useful to introduce the *macroscopic* fields $\overline{\mathbf{E}}$ and $\overline{\mathbf{B}}$ obtained by averaging the microscopic fields over an element of volume ΔV that is macroscopically small but large enough to contain very many atoms or molecules. The sharp spatial fluctuations in the microscopic fields will be smoothed over,

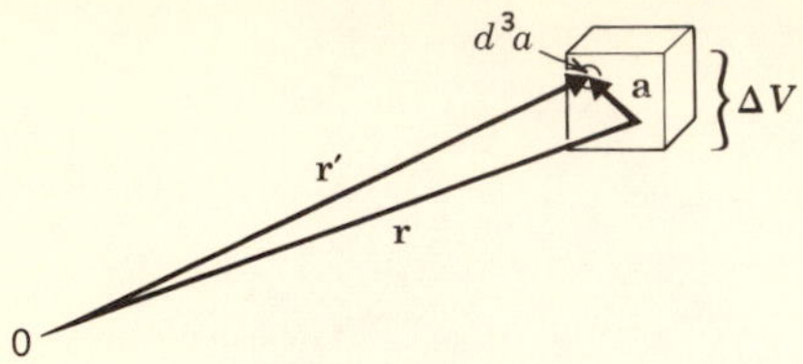

Fig. 11.2

leaving well-behaved functions for $\bar{\mathbf{E}}$ and $\bar{\mathbf{B}}$. As shown in Fig. 11.2, we let **a** be a small vector that generates the volume ΔV:

$$\Delta V = \iiint d^3a$$

Then the macroscopic field $\bar{\mathbf{E}}(r)$ is given by

$$\bar{\mathbf{E}}(\mathbf{r}) = \frac{1}{\Delta V}\iiint \mathbf{E}(\mathbf{r}+\mathbf{a})\, d^3a \tag{11.14}$$

A similar expression holds for $\mathbf{B}(\mathbf{r})$. We also define total *macroscopic* charge and current densities by the equations

$$\bar{\rho}_{\text{tot}}(\mathbf{r}) \equiv \frac{1}{\Delta V}\iiint \rho(\mathbf{r}+\mathbf{a})\, d^3a$$

$$\bar{\mathbf{j}}_{\text{tot}}(\mathbf{r}) \equiv \frac{1}{\Delta V}\iiint \mathbf{j}(\mathbf{r}+\mathbf{a})\, d^3a \tag{11.15}$$

It is then a straightforward matter to show that *when the microscopic Maxwell equations are averaged over the volume element ΔV, the corresponding set of equations is obtained for the macroscopic quantities*

$$\nabla\cdot\bar{\mathbf{E}} = 4\pi\bar{\rho}_{\text{tot}}$$

$$\nabla\cdot\bar{\mathbf{B}} = 0$$

$$\nabla\times\bar{\mathbf{E}} = -\frac{1}{c}\frac{\partial\bar{\mathbf{B}}}{\partial t}$$

$$\nabla\times\bar{\mathbf{B}} = \frac{4\pi}{c}\bar{\mathbf{j}}_{\text{tot}} + \frac{1}{c}\frac{\partial\bar{\mathbf{E}}}{\partial t} \tag{11.16}$$

The magnetic induction and electric field dealt with in the laboratory are these average fields $\bar{\mathbf{E}}$ and $\bar{\mathbf{B}}$. The microscopic fields are not usually accessible to measurement.

2. Bound Charges and Currents. Macroscopic regions of matter can contain a net charge or current density. For instance, a pith ball can be charged electrostatically, or a metallic conductor can be carrying a macroscopic current. These are called *free charges or currents*. They are to be distinguished from *bound* charges and currents. The argument for bound charges is summarized as follows.

A given molecule may have a net charge; if so, it contributes to the free charge density ρ_f. Whether or not the molecule has a net charge, it may be polarized; that is, there may be a relative separation of positive charges (atomic nuclei) and negative charges (electrons) within it. Let $\mathbf{r}_{+j}$ be the position vector of the jth positive charge $(+q_j)$ with respect to, say, the center of mass of the molecule, and let $\mathbf{r}_{-i}$ be the position vector of the ith negative charge $(-q_i)$.

Then the *electric dipole moment* of the molecule is defined to be

$$\mathbf{p} = \sum_j q_j \mathbf{r}_{+j} - \sum_i q_i \mathbf{r}_{-i} \tag{11.17}$$

Generally we distinguish between *permanent* electric dipoles, as is the case with an NaCl molecule which is composed of two oppositely charged ions, and *induced* dipoles resulting from charge displaced in an external field. These two cases are illustrated in Fig. 11.3.

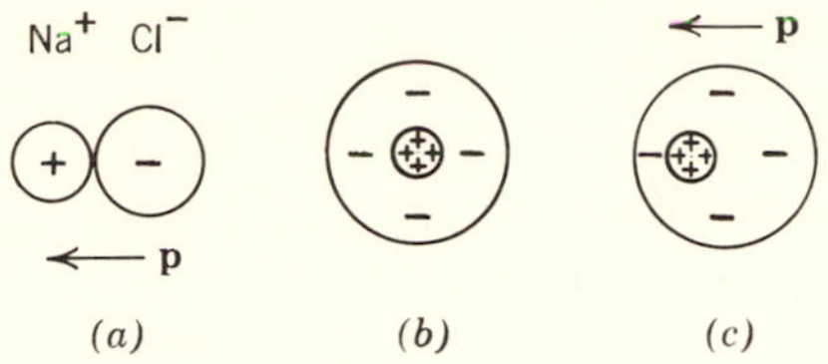

Fig. 11.3. (*a*) Permanent dipole. (*b*) Unpolarized atom ($\mathbf{p} = 0$). (*c*) Polarized atom (induced dipole).

The *polarization vector* **P** is defined to be the average dipole moment per unit volume:

$$\mathbf{P} = \frac{1}{\Delta V} \sum \mathbf{p} \tag{11.18}$$

where the sum is over all dipoles in the macroscopically small but microscopically large volume ΔV.

"Bound charge" is defined to be macroscopic charge in matter resulting from the presence of macroscopic polarization **P**. When **P** is constant, each macroscopic volume element contains the same number of positive and negative bound charges, and the bound charge density ρ_b is zero. When $\mathbf{P}(\mathbf{r})$ is spatially varying, a nonzero bound charge density can appear. This is shown schematically in Fig. 11.4*a*. The mean dipole moment $\langle \mathbf{p} \rangle$ increases gradually from

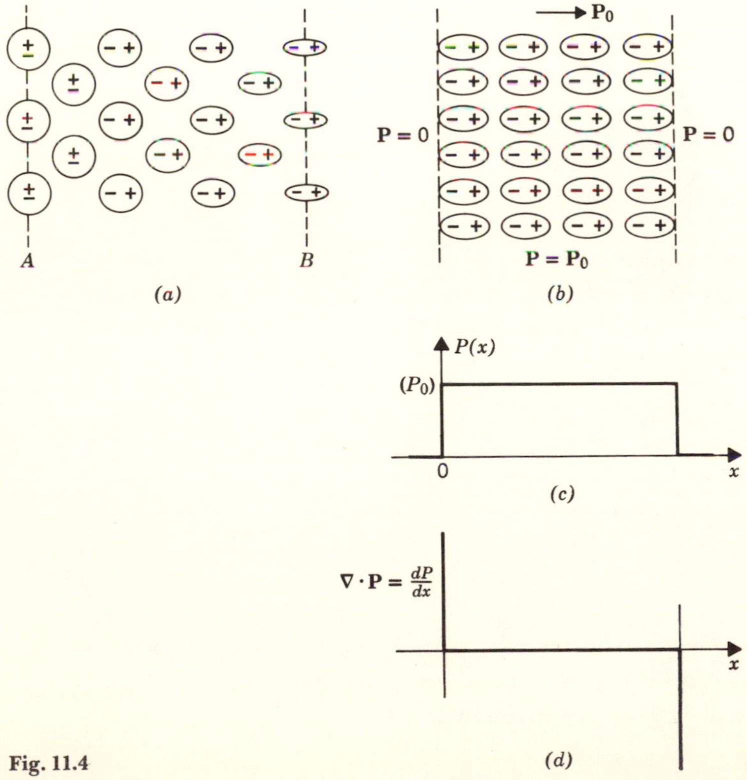

Fig. 11.4

zero in plane A to a net value in the $+x$ direction in plane B. We see that the B plane tends to cut through some dipoles, whereas the A plane does not. There is thus a net negative (bound) charge in the volume between planes A and B.

The general formula for the bound charge density is

$$\rho_b = -\nabla \cdot \mathbf{P} \tag{11.19}$$

We shall not derive Eq. (11.19) here, but instead shall make it plausible by reference to Fig. 11.4. Part (*b*) of the figure shows the individual molecular dipoles lined up producing a uniform polarization $\mathbf{P}_0$ inside a dielectric slab. The function $P(x)$ gives the magnitude of $\mathbf{P}$ and is sketched in part (*c*). Then $\nabla \cdot \mathbf{P} = dP/dx$ is sketched in part (*d*). The latter consists of two x-dependent delta functions as shown. From part (*b*) we see that there is a net positive surface charge density on the right surface and a negative surface charge on the left. In this example the x-dependent delta functions give the bound surface charge density.

Bound current results from motion of the charges bound in matter. The resulting current density is

$$\mathbf{j}_b = \frac{1}{\Delta V} \sum \, (q_{+i}\mathbf{r}_{+i} - q_{-i}\mathbf{r}_{+i}) = \dot{\mathbf{P}} \tag{11.20}$$

(The dot denotes differentiation with respect to time.)

In magnetic material, account must also be taken of the magnetization $\mathbf{M}$, which is defined to be the average magnetic dipole moment per unit volume:

$$\mathbf{M} = \frac{1}{\Delta V} \sum \mathbf{m} \tag{11.21}$$

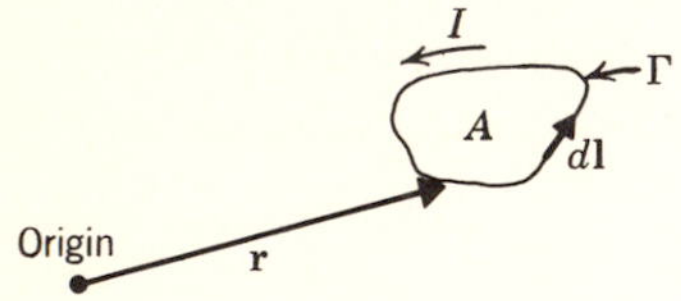

Fig. 11.5

The individual atomic or molecular magnetic moments $\mathbf{m}$ appearing in Eq. (11.21) are associated with current loops, as illustrated in Fig. 11.5. For current I flowing around a path Γ the magnetic moment is given by

$$\mathbf{m} = \frac{I}{2c} \oint_{\Gamma} \mathbf{r} \times d\mathbf{l} \tag{11.22a}$$

and has magnitude

$$|\mathbf{m}| = \frac{IA}{c} \tag{11.22b}$$

where A is the area of the loop enclosed by Γ. We can then show that a nonuniform $\mathbf{M}$ contributes to the bound current density $\mathbf{j}_b$ so that the total $\mathbf{j}_b$ is given by

$$\mathbf{j}_b = \dot{\mathbf{P}} + c\, \nabla \times \mathbf{M} \tag{11.20$'$}$$

Free charges and currents are then defined by the differences

$$\rho_f = \bar{\rho}_{\text{tot}} - \rho_b$$
$$\mathbf{j}_f = \bar{\mathbf{j}}_{\text{tot}} - \mathbf{j}_b \tag{11.23}$$

*3. **Maxwell's Equations in Matter.*** The first of Eqs. (11.16) now becomes

$$\nabla \cdot \bar{\mathbf{E}} = 4\pi(\rho_f + \rho_b) = 4\pi\rho_f - 4\pi\, \nabla \cdot \mathbf{P} \tag{11.24}$$

The last of Eqs. (11.16) is changed also:

$$\nabla \times \bar{\mathbf{B}} = \frac{1}{c}\frac{\partial \mathbf{E}}{\partial t} + \frac{4\pi}{c}(\mathbf{j}_f + \mathbf{j}_b) = \frac{1}{c}\frac{\partial \bar{\mathbf{E}}}{\partial t} + \frac{4\pi}{c}\mathbf{j}_f + \frac{4\pi\dot{\mathbf{P}}}{c} + 4\pi\nabla \times \mathbf{M} \tag{11.25}$$

We introduce the electric displacement **D**:

$$\mathbf{D} \equiv \mathbf{E} + 4\pi\mathbf{P} \tag{11.26}$$

We have dropped the bar over **E** in Eq. (11.26). *From now on it is understood that* **E** *and* **B** *refer to macroscopic fields.* We also introduce the *magnetic field* **H**:

$$\mathbf{H} \equiv \mathbf{B} - 4\pi\mathbf{M} \tag{11.27}$$

Equation (11.24) then may be written

$$\nabla \cdot \mathbf{D} = 4\pi\rho_f \tag{11.28}$$

and Eq. (11.25) may be written

$$\nabla \times \mathbf{H} = \frac{4\pi\mathbf{j}_f}{c} + \frac{1}{c}\frac{\partial \mathbf{D}}{\partial t} \tag{11.29}$$

Along with the middle two of Eq. (11.16), which remain unchanged, Eqs. (11.28) and (11.29) give the *four Maxwell equations in the presence of matter*:

$$\nabla \cdot \mathbf{D} = 4\pi\rho_f$$
$$\nabla \cdot \mathbf{B} = 0$$
$$\nabla \times \mathbf{E} = -\frac{1}{c}\frac{\partial \mathbf{B}}{\partial t}$$
$$\nabla \times \mathbf{H} = \frac{4\pi\mathbf{j}_f}{c} + \frac{1}{c}\frac{\partial \mathbf{D}}{\partial t} \tag{11.30}$$

*4. **Energy Relations.*** In the presence of matter it becomes necessary to redefine the energy density as

$$U = \frac{1}{8\pi}(\mathbf{D} \cdot \mathbf{E} + \mathbf{B} \cdot \mathbf{H}) \tag{11.31}$$

The Poynting vector is now

$$\mathbf{S} = \frac{c}{4\pi}(\mathbf{E} \times \mathbf{H}) \tag{11.32}$$

5. Linear Material Relations. To proceed, we need some relationship between **B** and **H** and between **E** and **D**. All four vectors may depend on position and time. It is often a good approximation to assume that the relationships are *local* and *linear*. The most general local, linear form is given by these equations for the Cartesian components at the same values of **r**:

$$D_i = \sum_{j=1}^{3} \epsilon_{ij} E_j \tag{11.33a}$$

$$B_i = \sum_{j=1}^{3} \mu_{ij} H_j \tag{11.33b}$$

Here $\{\epsilon_{ij}\}$ form the elements of the dielectric tensor $\boldsymbol{\epsilon}$ and $\{\mu_{ij}\}$ form the elements of the permeability tensor $\boldsymbol{\mu}$. Often we also have a linear relationship between the current density and the electric field,

$$j_{f,i} = \sum_{j=1}^{3} \sigma_{ij} E_j \tag{11.33c}$$

This is a generalization of Ohm's law. The $\{\sigma_{ij}\}$ form the elements of the *electrical conductivity* tensor $\boldsymbol{\sigma}$.

The quantities ϵ_{ij}, μ_{ij}, σ_{ij} may be determined directly by experiment. It is then the task of the physicist to make mathematical models of the material medium and calculate these same quantities. In the next section (11.2) we shall give examples of such models.

6. Isotropic Media. If the linear medium is also isotropic, the three tensors become multiples of the unit tensor $\mathbf{I}$:

$$\boldsymbol{\epsilon} = \epsilon\mathbf{I}, \qquad \boldsymbol{\mu} = \mu\mathbf{I}, \qquad \boldsymbol{\sigma} = \sigma\mathbf{I}$$

Thus we write $\epsilon_{ij} = \epsilon\delta_{ij}$, and so on, where $\delta_{ij} = 1$ if $i = j$ and $\delta_{ij} = 0$ if $i \neq j$. Here ϵ, μ, and σ are called the *dielectric constant* the *permeability*, and the *conductivity*, respectively. Then Eqs. (11.33) become

$$\begin{aligned} \mathbf{D} &= \epsilon\mathbf{E} \\ \mathbf{B} &= \mu\mathbf{H} \\ \mathbf{j} &= \sigma\mathbf{E} \end{aligned} \tag{11.34}$$

The energy density becomes

$$U = \frac{1}{8\pi}\left(\epsilon|\mathbf{E}|^2 + \frac{1}{\mu}|\mathbf{B}|^2\right) \tag{11.31$'$}$$

and the Poynting vector may be written

$$\mathbf{S} = \frac{c}{4\pi\mu}(\mathbf{E}\times\mathbf{B}) \tag{11.32'}$$

7. Plane Waves in Linear Isotropic Nonconducting Media. When there are no free charges or currents, we can derive the following wave equation from the last two of Eqs. (11.30):

$$\nabla^2\mathbf{E} - \frac{1}{v^2}\frac{\partial^2\mathbf{E}}{\partial t^2} = 0; \qquad \nabla^2\mathbf{B} - \frac{1}{v^2}\frac{\partial^2\mathbf{B}}{\partial t^2} = 0 \tag{11.35}$$

with

$$v = \frac{c}{n}, \tag{11.36}$$

where $n = \sqrt{\epsilon\mu}$ is the index of refraction. These same equations hold for **D** and **H**. They differ from the free-space equations by the substitution of the velocity $v = c/n$ for c.

Plane wave solutions to (11.35) exist in the form

$$\mathbf{E} = \mathbf{E}_0 e^{i(\omega t - \mathbf{k}\cdot\mathbf{r})}, \qquad \mathbf{B} = \mathbf{B}_0 e^{i(\omega t - \mathbf{k}\cdot\mathbf{r})}$$

$$\mathbf{D} = \epsilon\mathbf{E}, \qquad \mathbf{H} = \frac{\mathbf{B}}{\mu}$$

with

$$\omega = vk, \qquad k = |\mathbf{k}| \tag{11.37}$$

From the third Maxwell equation we then find

$$\mathbf{B} = \frac{c}{\omega}\mathbf{k}\times\mathbf{E} = n\mathbf{k}\times\frac{\mathbf{E}}{k}, \qquad [|\mathbf{B}| = n|\mathbf{E}|] \tag{11.38}$$

It then follows that

$$\epsilon|\mathbf{E}|^2 = \frac{|\mathbf{B}|^2}{\mu}$$

and the mean energy density becomes

$$\langle U\rangle = \frac{1}{8\pi}\epsilon|\mathbf{E}_0|^2 \tag{11.39}$$

The mean energy flux may be written

$$S = \langle\mathbf{S}\rangle = \frac{nc}{\mu}\frac{|\mathbf{E}_0|^2}{8\pi} \tag{11.40}$$

Equation (11.40) contains a very important result for light; when $\mu \approx 1$, as is often the case (nonmagnetic media), the energy flux is proportional to the product $n|\mathbf{E}_0|^2$, and not simply to $|\mathbf{E}_0|^2$.

8. Complex Refractive Index—Absorption. Often the index of refraction is complex. When this is the case we write it as $\tilde{n}$ and separate real and imaginary parts as follows:

$$\tilde{n} = n_r - in_i \tag{11.41}$$

The significance of n_r and n_i will now be discussed.

The imaginary part n_i of the complex index of refraction is often called the extinction coefficient.* The terminology refers to the extinction of an electromagnetic wave as it propagates through the medium. To exhibit this extinction, note that the solutions to Maxwell's equations in a medium lead to plane waves of the form, say for propagation in the $+z$ direction,

$$\mathbf{E} = \mathbf{E}_0 e^{(\omega t - kz)}$$

with

$$\tilde{k} = \frac{\omega \tilde{n}}{c} = \frac{\omega n_r}{c} - \frac{i\omega n_i}{c}$$

The real, physical field takes the form

$$\mathbf{E}(\text{real}) = \text{Re}\,\{\mathbf{E}_0 e^{i\omega(t - n_r z/c)} e^{-\omega n_i z/c}\} = \mathbf{E}_0 e^{-\omega n_i z/c} \cos \omega \left(t - \frac{n_r z}{c}\right) \tag{11.42}$$

There is an exponential decay of the signal strength with z; the energy flux density goes like

$$I = I_0 e^{-2\omega n_i z/c} = I_0 e^{-Kz} \tag{11.43}$$

with $I_0 = \text{constant} = I(z = 0)$ and with

$$K = \frac{2\omega n_i}{c} = \frac{4\pi n_i}{\lambda} \qquad (\lambda = \text{vacuum wavelength}) \tag{11.44}$$

Here K is called the absorption coefficient. The flux density is $(1/e)$ times its initial value when $z = 1/K$. This value of z is sometimes called the penetration depth. Note that when $n_i = 1, K = 4\pi/\lambda$, and the penetration depth of the light is $\lambda/4\pi$, one order of magnitude less than λ. In many cases the absorption coefficient can be determined directly from measured values of I/I_0. Then n_i is obtained from $\lambda K/(4\pi)$. If there is not too much absorption of light, the real part n_r of the complex index of refraction may be measured in the usual way an ordinary refractive index is measured.

*Some authors write $\tilde{n} = n_r(1 - i\kappa)$, κ is called the attenuation index. Often n_r is written simply as n. Others use a time dependence $e^{-i\omega t}$, which changes the sign of i in all our equations, for example, the index of refraction would be written $\tilde{n} = n_r + in_i$, and $n_i(\omega)$ would be the same function of ω used here. Equation (11.43) will hold regardless of the sign of the exponential time dependence.

11.2 ELEMENTARY MICROSCOPIC THEORY OF THE INTERACTION OF LIGHT WITH MATTER

Phenomenologically the presence of matter results in a dielectric constant $\epsilon \neq 1$ (and sometimes in a $\mu \neq 1$). How then is ϵ determined from the microscopic properties of matter? More basically, can we see directly how the presence of atoms and molecules modifies the propagation law for light waves? These are some of the questions to be discussed in this section. To simplify, the discussion will be limited to nonmagnetic media where $\mu = 1$ and $\mathbf{B} = \mathbf{H}$. Magnetic media may be described by theories somewhat analogous to the ones to be developed below.

A key role is played by the polarization $\mathbf{P}$ since it yields the bound charge $\rho_b = -\nabla \cdot \mathbf{P}$ and the bound current density $\mathbf{j}_b = \dot{\mathbf{P}}$. If we assume that the "free" currents and charges, $\mathbf{j}_f$ and ρ_f, are known, then once $\mathbf{P}$ is known, all the sources of $\mathbf{E}$ and $\mathbf{B}$ are known, and $\mathbf{E}$ and $\mathbf{B}$ may be determined (in principle at least) by integration of Maxwell's equations or of equations derived from them.

How is $\mathbf{P}$ determined? In a linear approximation and in a local theory (where $\mathbf{P}(\mathbf{r})$ depends only on the field $\mathbf{E}(\mathbf{r})$ at the same point $\mathbf{r}$), we may write phenomenologically

$$\mathbf{P}(\mathbf{r}) = \chi \mathbf{E}(\mathbf{r}) \tag{11.45}$$

For simplicity we assume spatial isotropy here, so that χ is a scalar, called the *electric susceptibility*. In terms of it we readily derive

$$\epsilon = 1 + 4\pi\chi \tag{11.46}$$

with an obvious generalization if χ is a tensor. Here ϵ and χ depend on the medium and may be a function of $\mathbf{r}$.

A. MODEL CALCULATION FOR A DILUTE NONPOLAR GAS

Consider a medium composed of molecules. The polarization $\mathbf{P}$ is the average dipole moment per unit volume. We assume that in a field-free environment the molecules have no net dipole moment; any dipoles must therefore be induced as in Fig. 11.3*b*. Such a medium is called *nonpolar*.

The induced dipole moment on the *i*th molecule will be given in a linear approximation by

$$\mathbf{p}_i = \boldsymbol{\alpha} \cdot \mathbf{E}_m(\mathbf{r}_i) \tag{11.47}$$

Here $\mathbf{E}_m(\mathbf{r}_i)$ is the local microscopic electric field at the center of mass $\mathbf{r}_i$ of the molecule. We hereby neglect variations in the external field over the spatial extent of the molecule; this is a good assumption for visible light and for molecules of ordinary size. For polymers or crystals, for which, technically speaking, the molecules have macroscopic dimensions, the argument must be applied to a small subunit, or cell, that is repeated throughout the structure. The tensor $\boldsymbol{\alpha}$ in Eq. (11.47) is called the *polarizability tensor*. In the c.g.s. system of units it has the dimensions of volume.

The microscopic local field $\mathbf{E}_m(\mathbf{r}_i)$ at $\mathbf{r}_i$ depends not only on $\mathbf{r}_i$ but also on the positions $\mathbf{r}_j$ of all the other molecules in the system and on the orientation and magnitude of the dipoles induced on these molecules. If we average the microscopic field $\mathbf{E}_m$ over all the molecules in a macroscopically small but microscopically large volume element at $\mathbf{r}$, we obtain the average local field $\mathbf{E}_l(\mathbf{r})$ acting on a molecule at $\mathbf{r}$. This local field should be contrasted with the ordinary macroscopic field $\mathbf{E}(\mathbf{r})$ which is the average of the microscopic field over *all* positions in a small volume element at $\mathbf{r}$, not merely over positions occupied by molecules.

Two tasks must now be performed if the theory is to be carried farther: (1) a model for a molecule must be constructed and $\boldsymbol{\alpha}$ calculated, and (2) the local field $\mathbf{E}_l$ must be related to the macroscopic field $\mathbf{E}$.

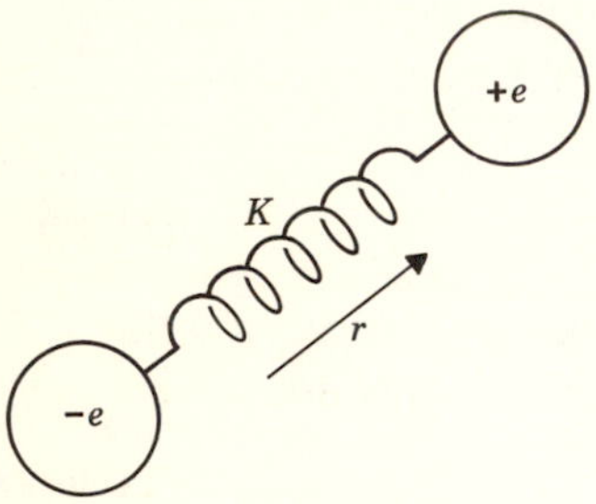

Fig. 11.6

1. Simple Molecular Model. A simple model of a nonpolar molecule is provided by a harmonic oscillator (Fig. 11.6). Let a charge e of mass m be attached to another charge $-e$ of essentially infinite mass (for simplicity) by a harmonic spring with spring constant K. The equilibrium separation $\mathbf{r}$ of the two charges is zero if no external forces are applied.

For frequencies corresponding to the visible region of the spectrum, the major contribution to α and hence ultimately to $\boldsymbol{\epsilon}$ comes from oscillating electrons. Then e would be minus the electronic charge and m would be the mass of an electron. In the infrared region of the spectrum the oscillating charges can be ions or parts of molecules having much large values of m. Fortunately the present derivation can apply to a variety of oscillating systems.

The equation of motion for the position vector of the movable charge will be written

$$m\ddot{\mathbf{r}} = -K\mathbf{r} - \frac{m\dot{\mathbf{r}}}{\tau} + \mathbf{F} \tag{11.48a}$$

Here the external force $\mathbf{F}$ is given by

$$\mathbf{F} = e\mathbf{E}_m \tag{11.48b}$$

where $\mathbf{E}_m$ is the microscopic electric field at the molecule. We have included a damping or frictional force $-m\dot{\mathbf{r}}/\tau$ in (11.48*a*) proportional to the velocity to describe irreversible losses because of collisions with other molecules or to radiation damping. This term may not always yield the correct quantitative behavior resulting from collisions, but it does give the main qualitative effects. The constant τ can be interpreted as a damping or collision time—it is a measure of the mean time of duration of an oscillation.

The natural resonance frequency ω_0 of the undamped oscillator is given by

$$\omega_0 = \sqrt{K/m} \tag{11.48c}$$

Equations (11.48*a*, *b*) refer to real physical quantities $\mathbf{r}$(real) and $\mathbf{E}_m$(real). We assume a single angular frequency ω and write

$$\mathbf{E}_m(\text{real}) = \text{Re}[\mathbf{E}_m e^{i\omega t}] \qquad \text{and} \qquad \mathbf{r}(\text{real}) = \text{Re}[\mathbf{r}e^{i\omega t}]$$

where now $\mathbf{E}_m$ and $\mathbf{r}$ are complex, time-independent amplitudes. Equations (11.48*a*, *b*, *c*) are combined, and use is made of

$$\dot{\mathbf{r}} = i\omega\mathbf{r} \qquad \text{and} \qquad \ddot{\mathbf{r}} = -\omega^2\mathbf{r}$$

to obtain

$$\mathbf{r} = \frac{e/m}{\omega_0^{\,2} - \omega^2 + i\omega/\tau}\mathbf{E}_m \tag{11.49a}$$

The induced dipole moment is then $\mathbf{p} = e\mathbf{r}$.

Equation (11.47) then becomes simply $\mathbf{p} = \alpha\mathbf{E}_m$ with

$$\alpha = \frac{e^2/m}{\omega_0^{\,2} - \omega^2 + i\omega/\tau} \tag{11.49b}$$

The polarization is then the average dipole moment per unit volume or

$$\mathbf{P}(\mathbf{r}) = Ne\langle\mathbf{r}\rangle = \frac{Ne^2/m}{\omega_0^{\,2} - \omega^2 + i\omega/\tau}\langle\mathbf{E}_m\rangle = \frac{Ne^2/m}{\omega_0^{\,2} - \omega^2 + i\omega/\tau}\mathbf{E}_l(\mathbf{r}) \tag{11.50}$$

Here N is the mean number of molecules per unit volume, and $\mathbf{E}_l(\mathbf{r})$ is, as explained above, the average field $\langle\mathbf{E}_m\rangle$ acting on a molecule at $\mathbf{r}$. Thus

$$\mathbf{P}(\mathbf{r}) = N\alpha\mathbf{E}_l(\mathbf{r}) \tag{11.51}$$

Since $i = \sqrt{-1}$ appears in the denominator of Eqs. (11.49*b*) and (11.50), α will be a complex function of ω. Write it in the form

$$\alpha = \alpha_r - i\,\alpha_i = \frac{(e^2/m)(\omega_0^{\,2} - \omega^2)}{(\omega_0^{\,2} - \omega^2)^2 + \omega^2/\tau^2} - i\,\frac{(e^2/m)(\omega/\tau)}{(\omega_0^{\,2} - \omega^2)^2 + \omega^2/\tau^2}$$

The real, physical electric field is given by

$$\mathbf{E}_l(\text{real}) = \text{Re}\mathbf{E}_l\, e^{i\omega t} = \mathbf{E}_l \cos\omega t$$

Similarly the real, physical polarization will be

$$\mathbf{P}(\text{real}) = \text{Re}\{N\alpha\mathbf{E}_l\, e^{i\omega t}\} = N\mathbf{E}_l\alpha_r \cos\omega t + N\mathbf{E}_l\,\alpha_i \sin\omega t$$

The first term is oscillating in phase with the local field, the second term is $\pi/2$ radians out of phase.

The existence of a nonzero imaginary part for α implies that the dielectric constant ϵ has an imaginary part and, since $\mathbf{D} = \epsilon\mathbf{E}$, there must be a part of $\mathbf{D}$ that is out of phase with $\mathbf{E}$. Similarly there will be an imaginary part to the index of refraction n, and since the magnitude of the wave vector $\mathbf{k}$ is given by $k = n\omega/c$, k will therefore have an imaginary part. As described by Eqs. (11.42)–(11.44), this leads to exponential attenuation of a light beam.

2. Very Dilute Gas. Having determined the polarizability, we now must decide what to use for the local field $\mathbf{E}_l$. For a very dilute system the molecules have such a small effect on the fields that $\mathbf{E}_l(\mathbf{r})$ can be approximated well enough by the macroscopic field $\mathbf{E}(\mathbf{r})$. We shall later make a better approximation for $\mathbf{E}_l$ and shall therefore be in a position to assess the applicability of our present assumption that $\mathbf{E}_l = \mathbf{E}$.

With $\mathbf{E}_l = \mathbf{E}$, the susceptibility becomes

$$\chi = N\alpha$$

and the dielectric constant is given by

$$\epsilon = 1 + 4\pi\chi = 1 + \frac{4\pi Ne^2/m}{\omega_0^2 - \omega^2 + i\omega/\tau} \tag{11.52a}$$

If we separate real and imaginary parts via $\epsilon = \epsilon_r - i\epsilon_i$, we find

$$\epsilon_r - 1 = \frac{(4\pi Ne^2/m)(\omega_0^2 - \omega^2)}{(\omega_0^2 - \omega^2)^2 + \omega^2/\tau^2} \tag{11.52b}$$

$$\epsilon_i = \frac{(4\pi Ne^2/m)(\omega/\tau)}{(\omega_0^2 - \omega^2)^2 + (\omega^2/\tau^2)} \tag{11.52c}$$

The low density assumption allowing us to set $\mathbf{E}_l = \mathbf{E}$ will be shown later to hold provided that

$$\frac{4\pi}{3}|\chi| \ll 1$$

This inequality will be used to simplify the calculation of the complex index of refraction $\tilde{n} = \epsilon^{1/2}$. Use of the approximation $\sqrt{1+\delta} \approx 1 + \frac{1}{2}\delta$ for small δ leads to

$$\tilde{n} = \sqrt{1 + 4\pi\chi} \approx 1 + 2\pi\chi = 1 + \frac{1}{2}(\epsilon - 1) \tag{11.53a}$$

$$= 1 + \frac{(2\pi Ne^2/m)(\omega_0^2 - \omega^2 - i\omega/\tau)}{(\omega_0^2 - \omega^2)^2 + \omega^2/\tau^2}$$

If we split $\tilde{n}$ into real and imaginary parts by writing $\tilde{n} = n_r - in_i$, we obtain the expressions

$$n_r - 1 = \frac{1}{2}(\epsilon_r - 1) = \frac{(2\pi Ne^2/m)(\omega_0^2 - \omega^2)}{(\omega_0^2 - \omega^2)^2 + \omega^2/\tau^2} \tag{11.53b}$$

$$n_i = \frac{1}{2}\epsilon_i = \frac{(2\pi Ne^2/m)(\omega/\tau)}{(\omega_0^2 - \omega^2)^2 + \omega^2/\tau^2} \tag{11.53c}$$

In Eqs. (11.53*b*, *c*) and in many equations to follow we give the corresponding formulas for ϵ_r, ϵ_i. They will be useful later when we discuss the case of high molecular densities.

The constant

$$\omega_p = \left(\frac{4\pi Ne^2}{m}\right)^{1/2} \tag{11.54}$$

will appear often in the equations that follow. Here ω_p has the dimensions of (angular) frequency and is called the (angular) *plasma frequency*. It also has a physical significance of its own — ω_p gives the frequency of the oscillation that results if one part of a gas of free particles of charge e, mass m, and number density N is displaced with respect to the rest of the gas and with respect to a fixed oppositely charged background charge distribution. In terms of ω_p, Eqs. (11.53*b*, *c*) become

$$n_r - 1 = \frac{1}{2}(\epsilon_r - 1) = \frac{\omega_p{}^2(\omega_0{}^2 - \omega^2)/2}{(\omega_0{}^2 - \omega^2)^2 + \omega^2/\tau^2} \tag{11.55a}$$

$$n_i = \frac{1}{2}\epsilon_i = \frac{\omega\,\omega_p{}^2/(2\tau)}{(\omega_0{}^2 - \omega^2)^2 + \omega^2/\tau^2} \tag{11.55b}$$

According to Eq. (11.43), the intensity (or flux density or other energy-related quantity in the beam) decays exponentially,

$$I = I_0 e^{-2\omega n_i z/c}$$

The energy lost from the light beam goes into the irreversible processes responsible for the collision or damping time τ. For instance, if collisions between molecules provide the main mechanism that limits τ, then these collisions provide a means of converting electromagnetic energy into energy associated with the translational motion of the molecules, that is, into heat.

3. ***Some Results.*** Plots of n_r and n_i versus ω are shown in Fig. 11.7 for the case $\omega_0\tau \gg 1$, which is the usual one for not too dense a gas. In this limit, for ω close to the resonance frequency ω_0, we may replace ω by ω_0 everywhere in Eqs. (11.55*a*, *b*) except in the expression $(\omega^2 - \omega_0{}^2)$ which we can approximate as follows:

$$\omega^2 - \omega_0{}^2 = (\omega - \omega_0)(\omega + \omega_0) \approx 2\omega_0(\omega - \omega_0) \tag{11.56}$$

Thus, *near resonance* the real and imaginary parts of the index of refraction are

$$n_r = 1 + \frac{\omega_p{}^2(\omega_0 - \omega)/(4\omega_0)}{(\omega_0 - \omega)^2 + 1/(2\tau)^2} \tag{11.57a}$$

$$= 1 + \frac{1}{2}(\epsilon_r - 1)$$

$$n_i = \frac{\omega_p{}^2/(8\tau\omega_0)}{(\omega_0 - \omega)^2 + 1/(2\tau)^2} \tag{11.57b}$$

$$= \frac{\epsilon_i}{2}$$

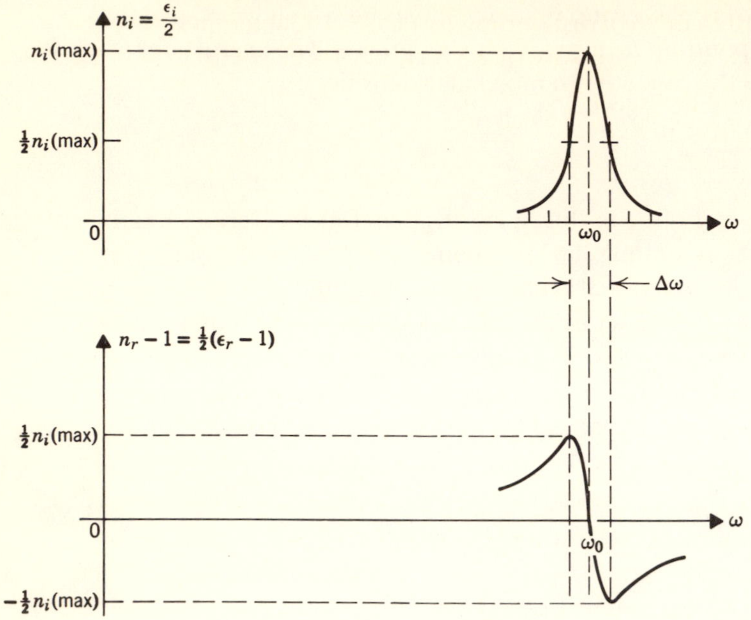

Fig. 11.7

At resonance ($\omega = \omega_0$), n_i takes on its maximum value

$$n_i(\text{max}) = \frac{\epsilon_i(\text{max})}{2} = \frac{\omega_p{}^2\tau}{2\omega_0} = \frac{2\pi Ne^2\tau}{m\omega_0} \tag{11.58}$$

and both n_r and ϵ_r equal 1. Above resonance, n_r is less than 1, below resonance it is greater than 1.

n_i in Eq. (11.57*b*) is expressed as a *Lorentzian* function of ω. When $\omega = \omega_0 \pm 1/2\tau$, n_i is reduced to half its maximum value. The full width at half maximum is thus

$$\Delta\omega = 2 \cdot \frac{1}{2\tau} = \frac{1}{\tau} \tag{11.59}$$

Since we have assumed $\omega_0\tau \gg 1$, the width is very much less than the resonance frequency.

At the half-maximum points of n_i the real part of the index n_r takes on its maximum and minimum values of

$$n_r(\text{max}) = 1 + \frac{\omega_p{}^2\tau}{4\omega_0} = 1 + \frac{n_i(\text{max})}{2} \qquad \text{when} \quad \omega = \omega_0 - \frac{\Delta\omega}{2}$$

and (11.60)

$$n_r(\text{min}) = 1 - \frac{\omega_p{}^2\tau}{4\omega_0} = 1 - \frac{n_i(\text{max})}{2} \qquad \text{when} \quad \omega = \omega_0 + \frac{\Delta\omega}{2}$$

4. Numerical Example. As a numerical example, take

$\omega_0 = 3 \times 10^{15}$ rad/sec (wavelength of light at resonance is 628 nm = 6280 A)

$\Delta\omega = 1/\tau = 10^{11}$/sec

($\omega_0\tau = 3 \times 10^4 \gg 1$)

$N = 10^{17}/\text{cm}^3$ (corresponding to a gas at room temperature at a pressure of 2.5 torr = 2.5 mm of Hg).

$$\left.\begin{aligned} e &= 4.8 \times 10^{-20} \text{ e.s.u.} \\ m &= 0.91 \times 10^{-27} \text{ g} \end{aligned}\right\} \text{electron}$$

Then

$$\omega_p{}^2 = 3.18 \times 10^{26} \text{ sec}^{-2}$$

and

$$n_i(\max) = \frac{3.18 \times 10^{26} \cdot 10^{-11}}{6 \times 10^{15}} = 0.53$$

This is too large a value to be consistent with the assumption $|4\pi\chi|/3 \ll 1$. If we want to make this assumption for the above example, we must stay away from the peak in n_i, that is, we must be certain that ω is far enough from ω_0 so that $n_i \ll 1$ and $|n_r - 1| \ll 1$.

Equation (11.57) represents a slightly simplified version of Eq. (11.55) near resonance. Away from resonance the second term ω^2/τ^2 in the denominators of the latter equations is small and can be neglected, in comparison with $(\omega_0{}^2 - \omega^2)$, which will be comparable with $\omega_0{}^2$. Thus, away from resonance we can write

$$n_r - 1 = \frac{1}{2}(\epsilon_r - 1) = \frac{\omega_p{}^2/2}{\omega_0{}^2 - \omega^2}$$

$$n_i = \frac{1}{2}\epsilon_i = \frac{(\omega_p{}^2/2)(\omega/\tau)}{(\omega_0{}^2 - \omega^2)^2} = (n_r - 1)\left(\frac{\omega^2}{\omega_0{}^2 - \omega^2}\right)\frac{1}{\omega\tau} \tag{11.61}$$

Since $\omega_0\tau \gg 1$ and since $(\omega_0{}^2 - \omega^2)$ is not too close to zero, the factor $|\omega^2/(\omega_0{}^2 - \omega^2)|(1/\omega\tau)$ is $\ll 1$. This means that $n_i \ll |n_r - 1|$, so that *away from resonance n can be considered to be purely real*. For the above numerical example the off-resonance value of n will be given by $n = 1 + 1.77 \cdot 10^{-5}(1 - \omega^2/\omega_0{}^2)^{-1}$.

5. Remarks. Note that above resonance n is less than unity, and below resonance it is greater than unity. As the frequency tends to zero the index tends to the static limit

$$n(0) = 1 + \frac{\omega_p{}^2}{2\omega_0{}^2} \qquad (= 1 + 1.77 \cdot 10^{-5} \text{ for our example}) \tag{11.62}$$

Note that when damping or collisions are neglected so that τ is allowed to go to infinity, the complex quantities α, χ, ϵ, and n all go to infinity at resonance. A finite value of τ keeps these quantities finite at $\omega = \omega_0$, but they still have a large absolute value there. Specifically, the real parts of α and χ are zero at resonance, and the real parts of ϵ and $\tilde{n}$ are unity there, whereas the imaginary parts of all four functions become very large. We say that these functions have a "near singularity" at $\omega^2 = \omega_0{}^2$.

In the case of higher number densities the functions ϵ, χ, and $\tilde{n}$ still have near singularities, but not at $\omega_0{}^2$, where the polarizability still has its near singularity. Then ϵ and χ keep the same functional form they had at low densities. In particular, the equations for ϵ_r and ϵ_i look exactly like the ones written above, with ω_0^2 replaced by a new frequency parameter $\bar{\omega}^2$. The real and imaginary parts of the index of refraction become more complicated at higher densities—so complicated that we often choose to give our results in terms of the dielectric constant. Of course, when ϵ_i is very small, so is n_i, and we may write simply

$$n = n_r = \sqrt{\epsilon_r}$$

The discussion of high-density dielectrics will be postponed until Section C after the following discussion of wave propagation in conducting media.

B. CONDUCTING MEDIA

In a conducting medium such as a metal or an ionized gas there is a cloud of (nearly) free electrons moving through a background of relatively stationary positive charge. The molecular model of Section A above may be modified to describe this new case by setting the force constant K and hence the resonance frequency ω_0 equal to zero. We can argue (with some success, it turns out) that for a free electron gas the local field acting on a given electron is more likely to be simply the macroscopic field, even at high densities, than for a gas or liquid composed of molecules. Molecules have a well-defined volume which they occupy to the exclusion of other molecules. This exclusion affects the distribution of other molecules about a given molecule—it is not the same as the distribution of molecules about an arbitrary point in space. Thus the average electric field acting on a given molecule is likely to be different from the average field at an arbitrary point in space, especially at high molecular densities. Electrons, on the other hand, occupy no such well-defined volume, and the average field seen by one of them is much more likely to be well approximated by the average field at an arbitrary point in space.

The classical calculation of the dielectric constant and the index of refraction of the electron gas proceeds by setting $\omega_0 = 0$ and $\mathbf{E}_l = \mathbf{E}$. Then

$$\epsilon = 1 + \frac{4\pi Ne^2/m}{i\omega/\tau - \omega^2} = 1 - \frac{\omega_p^{\,2}}{\omega^2 - i\omega/\tau} \tag{11.63}$$

where again the plasma frequency

$$\omega_p = \left(\frac{4\pi Ne^2}{m}\right)^{1/2}$$

has been introduced.

An alternate formula for the frequency-dependent dielectric constant can be given in terms of the low frequency or dc electrical conductivity σ, defined by the equation

$$\mathbf{J} = \sigma \mathbf{E}$$

In this equation the field $\mathbf{E}$ is assumed to be oscillating at a frequency much, much less than the plasma frequency. In the present case the current density $\mathbf{J}$ is given by

$$\mathbf{J} = \dot{\mathbf{P}} = \frac{\epsilon - 1}{4\pi}\dot{\mathbf{E}} = \frac{\epsilon - 1}{4\pi}(i\omega)\mathbf{E}$$

By analogy with the low frequency case, we write the proportionality factor between $\mathbf{J}$ and $\mathbf{E}$ as a frequency-dependent conductivity $\sigma(\omega)$. Then we have

$$\sigma(\omega) = \frac{\epsilon - 1}{4\pi}(i\omega) = \frac{(Ne^2/m)(i\omega)}{i\omega/\tau - \omega^2} = \frac{Ne^2\tau}{m(1 + i\omega\tau)}$$

The low frequency limit of $\sigma(\omega)$, denoted simply by σ, is then

$$\sigma = \frac{Ne^2\tau}{m}$$

This formula may be used to estimate τ for real metals or gas discharges from the value of σ. The frequency-dependent complex dielectric constant may therefore be written

$$\epsilon = 1 + \frac{4\pi\sigma}{i\omega(1 + i\omega\tau)} \tag{11.64}$$

For the numerical example discussed above, N was $10^{17}/\text{cm}^3$ and $\omega_p \approx 1.8 \times 10^{13}/\text{sec}$. For the ionized gas layers in the ionosphere, N is about $10^5/\text{cm}^3$, and hence $\omega_p \sim 2 \cdot 10^7$, which corresponds to a circular frequency of 3 megacycles. For a metal, N is about $10^{22}/\text{cm}^3$, and hence $\omega_p \sim 6 \times 10^{15}$ /sec.

As a first approximation at high frequencies where $\omega\tau \gg 1$ we can often go to the limit $\tau \to \infty$. The assumption is then equivalent to assuming that the electron gas or plasma is *collisionless*.* Then

$$\epsilon = 1 - \frac{{\omega_p}^2}{\omega^2} \tag{11.65}$$

A major consequence of this result is that the index of refraction $n = \sqrt{\epsilon}$ is purely imaginary for $\omega < \omega_p$ and purely real for $\omega > \omega_p$. In the former case we have

$$\tilde{n} = -i\, n_i, \qquad n_i = \frac{\sqrt{{\omega_p}^2 - \omega^2}}{\omega} \tag{11.66}$$

*The term "plasma" is often used in discussions of the high-frequency behavior of an electron gas to emphasize the collective behavior of the motion of the electrons, which is often describable as motion of a continuous, charged fluid, the plasma.

A medium with a purely imaginary index of refraction cannot propagate electromagnetic waves. We shall see below that the waves are completely reflected at the boundary of the medium.

In spite of the fact that there is no propagation of waves, there is penetration of fields into the conducting medium. Equation (11.43) is applicable to this case and gives for the physical electric field

$$\mathbf{E}(\text{real}) = \mathbf{E}_0 e^{-\omega n_i z/c} \cos \omega t = \mathbf{E}_0 e^{-z/\delta} \cos \omega t \tag{11.67a}$$

Here

$$\delta = \frac{c}{\omega n_i} = \frac{\lambda}{2\pi n_i} = \frac{c}{\sqrt{\omega_p{}^2 - \omega^2}} \tag{11.67b}$$

is called the skin depth.

When $\omega > \omega_p$ (and $\tau \to \infty$), then the dielectric constant is positive and $\tilde{n}$ is purely real, $n = n_r$,

$$n_r = \frac{\sqrt{\omega^2 - \omega_p{}^2}}{\omega} \tag{11.68}$$

Electromagnetic waves then propagate with no losses with phase velocity

$$v = \frac{\omega}{k} = \frac{c}{n_r}$$

For a metal ω_p corresponds to frequencies in the near ultraviolet region of the spectrum. We thus expect visible and infrared light to be strongly reflected, since for such light $\omega < \omega_p$, but ultraviolet light for which $\omega > \omega_p$ should be transmitted. This is indeed the case; for instance, sodium metal is relatively transparent for wavelengths below 2100 A.* For the ionosphere where the plasma frequency of about 3 megacycles is in the radio frequency region, we expect higher frequencies to be transmitted and lower frequencies to be reflected. This is, roughly, what actually happens, but the real situation is much more complicated, in part because of the presence of the earth's magnetic field.

In the more general case of finite damping the dielectric constant ϵ from Eq. (11.64*a*) when separated into real and imaginary parts via $\epsilon = \epsilon_r - i\epsilon_i$ gives

$$\begin{aligned} \epsilon_r &= 1 - \frac{\omega_p{}^2}{\omega^2 + 1/\tau^2} = \frac{\omega^2 + (1/\tau^2) - \omega_p{}^2}{\omega^2 + 1/\tau^2} \\ \epsilon_i &= \left(\frac{\omega_p{}^2}{\omega^2 + 1/\tau^2}\right)\frac{1}{\omega\tau} \end{aligned} \tag{11.69}$$

Since

$$\epsilon = \epsilon_r - i\epsilon_i = \tilde{n}^2 = (n_r - in_i)^2 = n_r{}^2 - n_i{}^2 - 2in_r n_i$$

*R. W. Wood, *Phys. Rev.*, **44**, 353 (1933); J. C. Sutherland, E. T. Arakawa, and R. N. Hamm, *J. Opt. Soc. Am.*, **57**, 645 (1967).

we find

$$n_r^2 - n_i^2 = \epsilon_r$$
$$2n_r n_i = \epsilon_i \tag{11.70}$$

The real and imaginary parts of the dielectric constant and of the index of refraction are plotted in Fig. 11.8 for $\omega_p = 6 \cdot 10^{15}$ and $1/\tau = 3 \cdot 10^{13}$, typical values for a simple metal like sodium. The critical frequency ω_c at the onset of transparency is now defined to be the frequency at which ϵ_r is zero. Thus

$$\omega_c^2 = \omega_p^2 - \frac{1}{\tau^2} \tag{11.71}$$

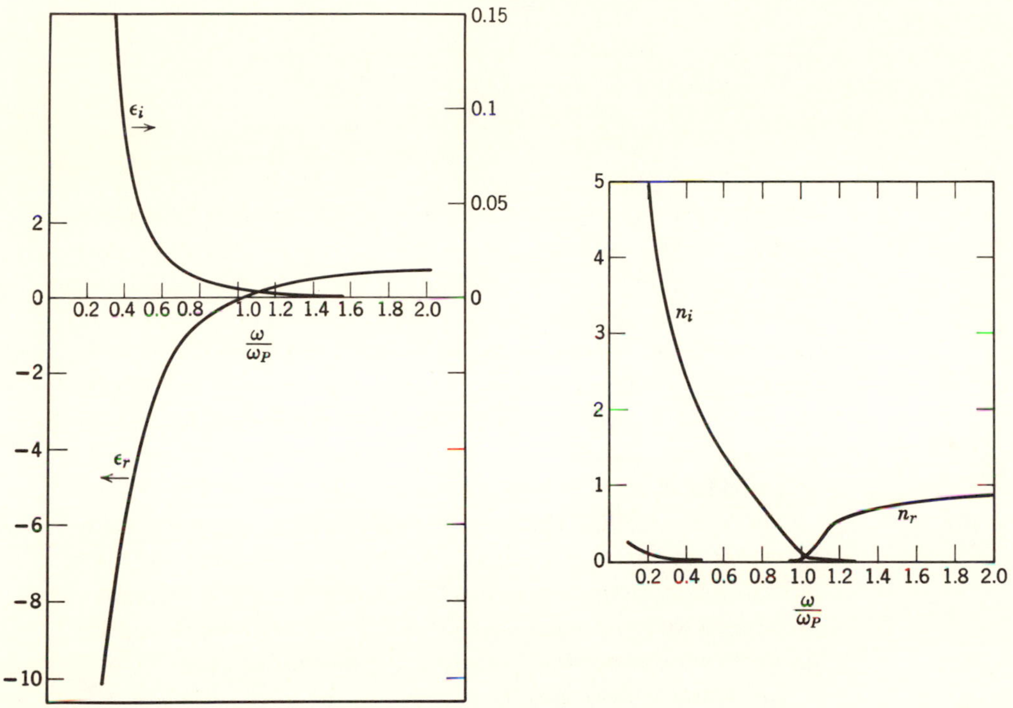

Fig. 11.8. Real and imaginary parts of ϵ and $\tilde{n}$ for a metal.

C. DENSER DIELECTRICS

Return now to the model of a dielectric medium as a collection of molecules, and consider higher densities so that the local field cannot be expected to be the same as the macroscopic field. In many cases it is a good approximation to write

$$\mathbf{E}_l = \mathbf{E} + \frac{4\pi}{3}\mathbf{P} \tag{11.72}$$

The term

$$\frac{4\pi \mathbf{P}}{3}$$

is called the Lorentz local field correction.

*1. **Justification for Lorentz Local Field.*** Some of the arguments that lead to the Lorentz local field will now be given with the help of Fig. 11.9. The molecules are represented by circles, their centers by dots, and random positions in the medium by ×'s. Each molecule has an induced dipole moment, which is not shown. Two extremes of molecular arrangements are depicted – a completely ordered crystalline array in (*a*) and a disordered fluid or amorphous solid in (*b*).

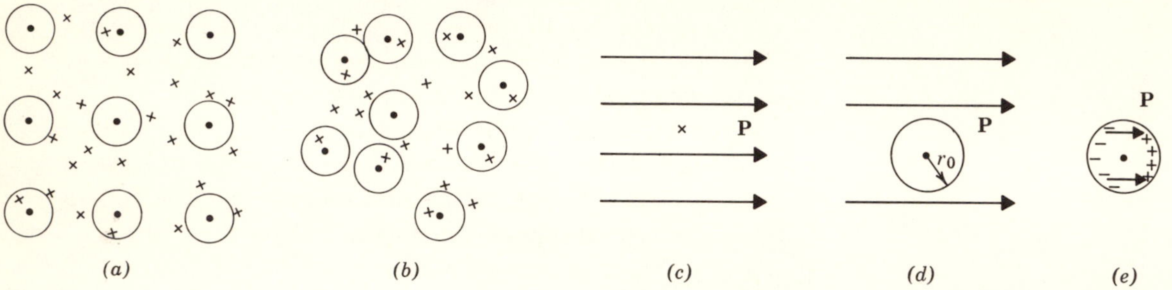

Fig. 11.9

The microscopic field $\mathbf{E}_m(\mathbf{r})$ is the external field from sources outside the medium plus the field from the dipoles. The macroscopic field $\mathbf{E}$ is the average of $\mathbf{E}_m$ as $\mathbf{r}$ ranges over random positions in a limited volume, that is, over the ×'s. This will be the same as the field inside a dielectric continuum having macroscopic polarization $\mathbf{P}$, as indicated schematically in Fig. 11.9*c*. The actual value of $\mathbf{E}$ will depend on the value of the external field, on the position (and time) dependence of $\mathbf{P}$, and on the shape of the medium; fortunately it is not necessary to produce an explicit formula for $\mathbf{E}$.

The local field $\mathbf{E}_l$ is the average of $\mathbf{E}_m(\mathbf{r})$ as $\mathbf{r}$ ranges over all the black dots in Fig. 11.9. In this case the "self-field" from the molecule at the dot in question must be excluded. Since each molecule occupies a well-defined volume, the spacing between dots is at least twice the molecular radius r_m; the nearest molecules contributing to the field must be at least this distance $2r_m$ away. Furthermore, the distribution of molecules about a given molecule will have inversion symmetry and be isotropic for a fluid or amorphous solid composed of spherical molecules. For certain simple crystal structures, such as cubic structures (but not for all crystal structures), the symmetry is high enough to be also considered isotropic in this context. These two features of the molecular distribution about a given molecule – a minimum separation $2r_m$ and an isotropic distribution – will be preserved during the averaging process that defines $\mathbf{E}_l$. The result is that the dipoles become essentially smeared into a dielectric continuum with polarization $\mathbf{P}$. The minimum separation is preserved by having a cavity in this continuum of molecular dimensions centered about $\mathbf{r}$, and the isotropic distribution can be assured by making the cavity spherical. Then $\mathbf{E}_l$ is the field at the center of this cavity, shown in Fig. 11.9*d*.

Two valid approximations greatly simplify the calculation of $\mathbf{E}_l$. The cavity has molecular dimensions and hence is much smaller than the wavelength of visible light. This means that **P** can be treated as a constant in the region of the cavity. This also means that the time for light to propagate across the region of the cavity is very much less than the period of oscillation of the polarization field, so that retardation effects resulting from the finite velocity of electromagnetic waves may also be neglected.

If the cavity were to be filled with a uniformly polarized sphere having polarization **P**, the field at its center would then be the macroscopic field **E**. Let $\mathbf{E}_s$ be the field at the center of an isolated uniformly polarized sphere. Then we must have

$$\mathbf{E}_l + \mathbf{E}_s = \mathbf{E}$$

or

$$\mathbf{E}_l = \mathbf{E} - \mathbf{E}_s$$

The field $\mathbf{E}_s$ (Fig. 11.9*e*) is readily shown to be

$$\mathbf{E}_s = -\frac{4\pi\mathbf{P}}{3}$$

This gives Eq. (11.72),

$$\mathbf{E}_l = \mathbf{E} + \frac{4\pi\mathbf{P}}{3}$$

2. Calculation of χ and ϵ. Equation (11.72) leads to

$$\mathbf{P} = N\alpha\mathbf{E}_l = N\alpha\mathbf{E} + \frac{4\pi N\alpha}{3}\mathbf{P}$$

The electric susceptibility is then

$$\chi = \frac{P}{E} = \frac{N\alpha}{1 - 4\pi N\alpha/3} \tag{11.73a}$$

and the dielectric constant is given by

$$\epsilon = 1 + 4\pi\chi = 1 + \frac{4\pi N\alpha}{1 - 4\pi N\alpha/3} \tag{11.73b}$$

The earlier low-density results were

$$\chi = N\alpha$$

$$\epsilon = 1 + 4\pi N\alpha$$

They hold when the second term in the denominator in Eq. (11.73*a*, *b*) can be neglected, or when

$$\left|\frac{4\pi N\alpha}{3}\right| \ll 1$$

Equation (11.73*b*) may be rearranged to give

$$\frac{\epsilon-1}{\epsilon+2}=\frac{4\pi N\alpha}{3} \tag{11.73c}$$

This is known as the Clausius-Mosotti-Lorentz-Lorenz equation and named after the four people that derived various versions of it. There are other ways of writing it. For instance, we can put

$$N = N_A n_m$$

where N_A is Avogadro's number and n_m is the number of moles per unit volume. Then

$$\frac{\epsilon-1}{\epsilon+2}=\frac{4\pi}{3}\alpha_m n_m$$

where

$$\alpha_m = N_A \cdot \alpha$$

is the polarizability per mole (molar polarizability).

For simplicity we shall consistently use Eq. (11.73*c*), and for brevity we shall call it the Clausius–Mosotti equation. It says that the quantity

$$\frac{1}{N}\frac{(\epsilon-1)}{(\epsilon+2)}$$

should be a constant independent of the number density N, the constant being

$$\frac{4\pi\alpha}{3}$$

If the Clausius–Mosotti formula is taken seriously, we should be able to calculate α from the refractive index in the gas phase and then predict ϵ in the liquid phase at much higher density. This often works very well (within a few per cent) for nonpolar molecules in the transparent regions of the optical spectrum. In the absorptive regions the molecular polarizability can change at high densities because of changes in the damping constant τ resulting from short-range interactions between molecules—essentially τ becomes much shorter.

3. Generalizations. Consider a medium that is a homogeneous mixture of more than one type of molecule, and let the jth type have number density N_j and molecular polarizability α_j. The polarization is then written

$$\mathbf{P} = \sum_j N_j \alpha_j \mathbf{E}_l$$

The use of the Lorentz local field with this formula leads to a modified Clausius–Mosotti equation that says that the quotient $(\epsilon - 1)/(\epsilon + 2)$ is additive:

$$\frac{\epsilon - 1}{\epsilon + 2} = \frac{4\pi}{3} \sum_j N_j \alpha_j \tag{11.74a}$$

This formula holds very well for mixtures of nonpolar gases at not too high densities.

Still another change may be made. Suppose that a given molecule contains several types of oscillator. Each such oscillator contributes to the molecular polarizability a term

$$\alpha_j = \frac{e^2/m}{\omega_j^2 - \omega^2 + i\omega/\tau_j}$$

The total polarizability is

$$\alpha = \sum_j f_j \alpha_j \tag{11.74b}$$

The constant f_j is called the oscillator strength of the jth oscillator. It is introduced in an *ad hoc* fashion in the classical theory of dispersion. A quantum-mechanical calculation of the polarizability leads to exactly the same result with explicit expressions for the f_j in terms of "matrix elements" for intramolecular dipole transitions. Each transition is identified with a classical oscillator, and the transition energy is given by Planck's constant h times the oscillator frequency.

If Eq. (11.74*b*) is used with the Lorentz local field, we obtain

$$\frac{\epsilon - 1}{\epsilon + 2} = \frac{4\pi}{3} N \sum_j f_j \alpha_j \tag{11.74c}$$

which is formally the same as (11.74*a*) for a mixture of different molecules if Nf_j is replaced by N_j.

4. Significance of Higher-Density Expressions. Consider for simplicity the case of a single oscillator so that

$$\alpha(\omega) = \frac{e^2/m}{\omega_0 \;\; - \omega^2 + i\omega/\tau}$$

We say that α has a "near singularity" at $\omega^2 = \omega_0^2$ because its absolute value is maximum there. Specifically, at resonance we have

$$\alpha(\omega_0) = \frac{e^2\tau}{im\omega_0}$$

and in absolute value this is $\omega_0\tau$ times larger than the zero-frequency value of

$$\alpha(0) = \frac{e^2}{m\omega_0^2}$$

In the low-density limit when it is a good approximation to write

$$\epsilon = 1 + \frac{4\pi N\alpha}{3}$$

this near singularity of α at $\omega^2 = \omega_0{}^2$ is shared by the complex dielectric constant and the complex index of refraction. The near singularity appears as a sharply peaked function $n_i(\omega) = \frac{1}{2}\epsilon_i(\omega)$ having a maximum at resonance and a characteristic dispersive wiggle for $n_r(\omega) - 1 = \frac{1}{2}[\epsilon_r(\omega) - 1]$, that sends it through zero at resonance. Optical measurements on such a dilute dielectric medium would yield a value for ω_0.

In dense media where the Lorentz local field correction applies, this is no longer the case. For a single oscillation, Eqs. (11.73*c*) and (11.49*b*) give

$$\epsilon = 1 + \frac{\omega_p{}^2}{(\omega_0{}^2 - \omega_p{}^2/3) - \omega \ + i\omega/\tau} \tag{11.75}$$

where

$$\omega_p = \left(\frac{4\pi Ne^2}{m}\right)^{1/2}$$

represents the plasma frequency that the oscillating charges would have if they were released from their restoring springs.

Now ϵ has a near singularity at

$$\bar{\omega}^2 \equiv \omega_0{}^2 - \frac{\omega_p{}^2}{3}$$

not at

$$\omega^2 = \omega_0{}^2$$

The resulting peak in

$$\epsilon_i = \frac{\omega_p{}^2\omega/\tau}{(\bar{\omega}^2 - \omega^2)^2 + \omega^2/\tau^2} \tag{11.76a}$$

and dispersive behavior of

$$\epsilon_r - 1 = \frac{\omega_p{}^2(\bar{\omega}^2 - \omega^2)}{(\bar{\omega}^2 - \omega^2)^2 + \omega^2/\tau^2} \tag{11.76b}$$

now occur at $\omega^2 = \bar{\omega}^2$. Any optical measurements would suggest that $\bar{\omega}$, not ω_0, is the resonance frequency of the system; indeed this is precisely the case—the resonance *has* shifted to ω.

The resonance frequency of isolated molecules thus shifts downward when the molecules are assembled into dense matter. The shift results from the smoothed-out effect of long-range dipole-dipole interactions as introduced via the Lorentz local field term $4\pi\mathbf{P}/3$. That the shift should be downward can be understood from Fig. 11.10. If the molecules are vibrating together, a given molecule sees an electric field, from its neighbors that tends to increase its dipole moment, that is, stretch its spring. This weakens the restoring effect of the spring, lowers the effective spring constant, and gives a lower resonance frequency for the coupled system.

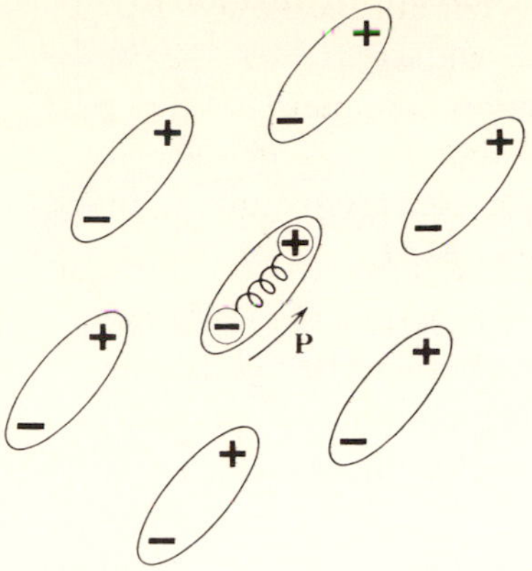

Fig. 11.10

In quantum mechanics a shift in the frequency of an oscillator corresponds to a shift in the energy of the transition that corresponds to it. Such energy shifts are common in dense many-particle systems and go by the name of "renormalization."

5. Some High-Density Results. The high-density expressions (11.75), (11.76*a*), and (11.76*b*) are formally the same as the low-density expressions (11.52), (11.55), and (11.57) for ϵ_i and ϵ_r, with ω_0 replaced by $\bar{\omega}$, but it is no longer true that

$$n_i = \tfrac{1}{2}\epsilon_i \qquad \text{and} \qquad (n_r - 1) = \tfrac{1}{2}(\epsilon_r - 1)$$

Instead we must use Eqs. (11.70):

$$n_r^2 - n_i^2 = \epsilon_r$$

$$2n_r n_i = \epsilon_i$$

The resulting expressions for n_r and n_i are complicated and will not be given here, in analytical form. Typical behavior near resonance is shown in Fig. 11.11

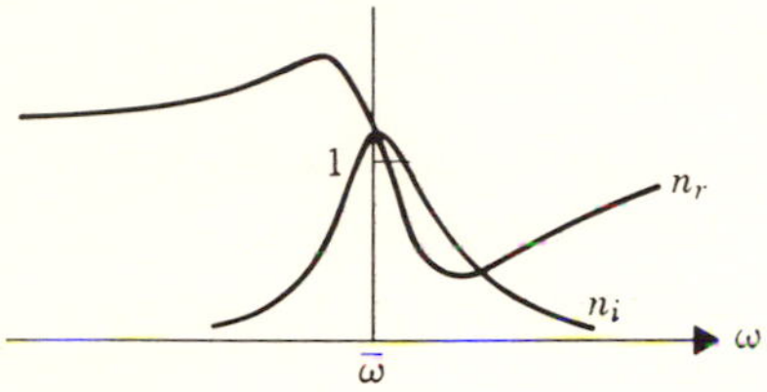

Fig. 11.11

All earlier low-density results for ϵ_i and $\epsilon_r - 1$ may now be written with $\bar{\omega}$ in place of ω_0. For instance, near resonance*

$$\epsilon_r - 1 = \frac{\omega_p^2(\bar{\omega}-\omega)/(2\bar{\omega})}{(\bar{\omega}-\omega)^2 + 1/(2\tau)^2} \tag{11.77}$$

$$\epsilon_i = \frac{\omega_p^2/(4\tau\bar{\omega})}{(\bar{\omega}-\omega)^2 + 1/(2\tau)^2}$$

Here ϵ_i is described by a Lorentzian function having a maximum

$$\epsilon_i\ (\text{max}) = \frac{\omega_p^2\tau}{\bar{\omega}}$$

and a width

$$\Delta\omega = \frac{1}{\tau} \tag{11.78}$$

The imaginary part ϵ_i, as approximated by Eq. (11.77), has integrated area

$$A = \int_0^\infty \epsilon_i(\omega)\, d\omega = \frac{\pi\omega_p^2}{\bar{\omega}} = \frac{4\pi^2 N e^2}{m\bar{\omega}} \tag{11.79}$$

*This means $|\omega - \bar{\omega}| \ll \bar{\omega}$

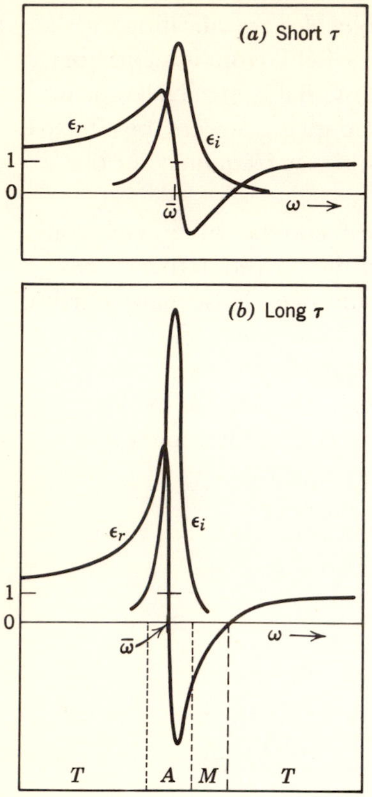

Fig. 11.12. Real and imaginary parts of the dielectric constant for two values of the lifetime τ. In part (*b*) T denotes the transparent regions, A the region of absorption, whose boundaries are ill-defined, and M the region of "metallic" reflection.

independent of τ. This expression is accurate when $\Delta\omega \ll \bar{\omega}$, for then all the contribution to ϵ_i comes when $|\omega - \bar{\omega}| \ll \bar{\omega}$, that is, when Eq. (11.77) is a good approximation. As τ increases, the peak for ϵ_i rises and narrows, but keeps constant area. This is seen by comparing Figs. 11.12*a* and 11.12*b*. The maximum and minimum of ϵ_r still occur at the half-maximum points of ϵ_i and may be readily shown to be

$$\begin{aligned} \epsilon_r(\max) &= 1 + \frac{\omega_p{}^2\tau}{2\bar{\omega}} = 1 + \frac{1}{2}\,\epsilon_i(\max) \qquad \text{when} \quad \omega = \bar{\omega} - \frac{\Delta\omega}{2} \\ \epsilon_i(\min) &= 1 - \frac{\omega_p{}^2\tau}{2\bar{\omega}} = 1 - \frac{1}{2}\,\epsilon_i(\min) \qquad \text{when} \quad \omega = \bar{\omega} + \frac{\Delta\omega}{2} \end{aligned} \tag{11.80}$$

Except for the region $\omega - (\Delta\omega/2) \leqslant \omega \leqslant +(\Delta\omega/2)$, ϵ_r increases with increasing frequency. The real part of the index of refraction, n_r, has similar behavior, although the frequency range is not quite the same. (See Fig. 11.11.) This region where n_r is not increasing with ω, called the region of anomalous dispersion, coincides with the region of greatest absorption where n_i is large. For the usual or "normal" dispersion, n increases with increasing frequency (decreasing wavelength), so that for visible light, n is greater in the blue than in the red.

6. The Significance of a Negative Value of $\boldsymbol{\epsilon_r}$. Note from Fig. 11.12*b* that when ϵ_i has a strong, narrow resonance peak, ϵ_r necessarily is negative over a frequency range just above resonance. The significance of this behavior is most apparent when τ is very long. Then $\Delta\omega$ is very small, and except for ω very close to $\bar{\omega}$ (within a few multiples of $\Delta\omega$), we may neglect the imaginary part ϵ_i of ϵ and write

$$\epsilon \cong \epsilon_r = 1 + \frac{\omega_p{}^2}{\bar{\omega}^2 - \omega^2} = \frac{\bar{\omega}^2 + \omega_p{}^2 - \omega^2}{\bar{\omega} \ \ - \omega^2} \tag{11.81}$$

Unlike Eq. (11.77), this expression is valid for all values of ω not too close to $\bar{\omega}$. According to Eq. (11.81), ϵ is negative for

$$\bar{\omega}^2 < \omega^2 < \bar{\omega}^2 + \omega_p{}^2$$

To be consistent with our neglect of ϵ_i, in the left-hand inequality, we should keep ω^2 greater than $(\bar{\omega} + N\Delta\omega)^2 \approx \bar{\omega}^2 + 2N\bar{\omega}\Delta\omega$, where N is of the order of 5 or 10. Thus we require

$$\bar{\omega}^2 + 2N\bar{\omega}\Delta\omega < \omega^2 < \bar{\omega}^2 + \omega_p{}^2 \tag{11.82a}$$

as the condition for both a negative value of ϵ_r and a very small value of ϵ_i. If (11.82*a*) is to hold for a nonzero range of ω, it is necessary that

$$\omega_p{}^2 > 2N\bar{\omega}\Delta\omega = \frac{2N\bar{\omega}}{\tau}$$

In this case we find

$$\epsilon_i(\max) = \frac{\omega_p{}^2\tau}{\bar{\omega}} > 2N \tag{11.82b}$$

The greater the maximum value of ϵ_i exceeds this inequality, the greater the range of ω over which (11.82*a*) is satisfied.

Thus, under the appropriate conditions stated above we have a dielectric constant given by

$$\epsilon = -\left(\frac{\bar{\omega}^2 + \omega_p{}^2 - \omega^2}{\omega^2 - \bar{\omega}^2}\right) < 0 \tag{11.83a}$$

The index of refraction $\tilde{n} = \sqrt{\epsilon}$ then is purely imaginary:

$$\tilde{n} = n_r - in_i$$

$$n_r = 0, \qquad n_i = \left(\frac{\omega^2 + \omega_p{}^2 - \omega^2}{\omega^2 - \omega^2}\right)^{1/2} \tag{11.83b}$$

As discussed in connection with Eq. (11.43), a nonzero value of n_i leads to an exponential attenuation of flux density or intensity within the medium of the form

$$I = I_0 e^{-2\omega n_i z/c}$$

A medium with a purely imaginary refractive index $\tilde{n} = -in_i$ will *totally reflect* a light beam incident onto it from outside. Reflection of light will be discussed in depth in Section 11.3, but for *normal incidence* we may use a result from Chapter 1, namely, Eq. (1.28), for the amplitude reflection coefficient

$$\rho = \frac{v_2 - v_1}{v_2 + v_1} = \frac{n_1 - n_2}{n_1 + n_2} = \frac{1 + in_i}{1 - in_i} \tag{11.84a}$$

Here we have used $v_{1,2} = c/n_{1,2}$ and have assumed $n_1 = 1$, $n_2 = -in_i$. The energy reflection coefficient or "reflectivity" is given by

$$R = |\rho|^2 = \rho \cdot \rho^* = \left(\frac{1 + in_i}{1 - in_i}\right)\left(\frac{1 - in_i}{1 + in_i}\right) = 1 \tag{11.84b}$$

All the incident flux is seen to be reflected. The exponential penetration of the fields into medium 2 is accompanied by no energy loss. This phenomenon will be discussed in more detail in Section 11.3. It is very similar to the reflection off good metals of light of frequency less than the plasma frequency, for there also n_i is of order unity and n_r is much less than unity (Fig. 11.8).

The reflection coefficient R deviates from unity when ϵ_i is not zero. For a small $\epsilon_i \neq 0$, there is a small nonzero real part n_r to $\tilde{n}$. The reflectivity for normal incidence is then rigorously given by

$$R = \left|\frac{1 - n_r + in_i}{1 + n_r - in_i}\right|^2 = \frac{(1 - n_r)^2 + n_i^2}{(1 + n_r)^2 + n_i^2} < 1 \tag{11.84c}$$

Some implications of this result will be explored in the problems.

7. *Several Widely Separated Resonance Frequencies.* Suppose that the molecules when isolated have several widely spaced resonance frequencies $\omega_1^2 \ll \omega_2^2 \ll \omega_3^2$, and so on. Then the contribution of the jth oscillator to the molecular polarizability

$$\alpha_j(\omega) = \frac{f_j e^2/m_j}{\omega_j^2 - \omega^2 + i\omega/\tau_j}$$

is nearly constant near the resonance frequency ω_l of the lth oscillator. For oscillators j at lower frequencies than $\omega_l(\omega_j^2 \ll \omega_l^2)$ we have for ω near ω_l

$$\alpha_j(\omega) \approx \frac{-e^2 f_j}{m_j\omega^2}$$

and for oscillators at higher frequencies $(\omega_j^2 \gg \omega_l^2)$

$$\alpha_j \approx \frac{e^2 f_j}{m_j\omega_j^2}$$

The contribution of one of the oscillators, say the lth, to ϵ may be described as follows using Eq. (11.74*a*). Define ϵ_{0l} by

$$\frac{\epsilon_{0l}-1}{\epsilon_{0l}+2} = \frac{4\pi}{3}\sum_{j\neq l} N_j\alpha_j \tag{11.85a}$$

Here ϵ_{0l} is the dielectric constant if the lth oscillator is not present. The dielectric constant ϵ in the presence of the lth oscillator is given by

$$\frac{\epsilon-1}{\epsilon+2} = \frac{4\pi}{3}\sum_{j\neq l} N_j\alpha_j + \frac{4\pi}{3}N_l\alpha_l = \frac{\epsilon_{0l}-1}{\epsilon_{01}+1} + \frac{4\pi}{3}N_l\alpha_l \tag{11.85b}$$

with

$$\alpha_l = \frac{f_l e^2/m_l}{\omega_l^2 - \omega^2 + i\omega/\tau_l}$$

Define

$$\Delta\epsilon = \epsilon - \epsilon_{0l}$$

Then it is easy to show that

$$\Delta\epsilon = \frac{f_l\omega_{pl}^2\,[(\epsilon_{0l}+2)/3]^2}{\omega_l^2 - (\omega_{pl}^2/3)\,[(\epsilon_{0l}+2)/3]\,f_l - \omega^2 + i\omega/\tau_l} \tag{11.85c}$$

with

$$\omega_{pl}^2 = \frac{4\pi N_l e^2}{m_l}$$

Here $\Delta\epsilon$ shows a resonance at

$$\bar{\omega}_l^2 = \omega_l^2 - \frac{\omega_{pl}^2}{3}\left(\frac{\epsilon_{0l}+2}{3}\right) f_l \tag{11.85d}$$

This is shifted down from ω_l^2 because of interactions among dipoles. Roughly speaking, the dipoles resonating at $\bar{\omega}_l$ interact within a medium having a dielectric constant; hence the presence of ϵ_{0l} in Eq. (11.85*d*).

Far below resonance $\Delta\epsilon$ tends to a constant

$$\Delta\epsilon_l \equiv \frac{f_l\omega_{pl}^2}{\bar{\omega}_l^2}\left(\frac{\epsilon_{0l}+2}{3}\right)^2 \tag{11.85e}$$

and far above it $\Delta\epsilon$ tends to zero. In both cases these limits are essentially attained before the next isolated resonance frequency ω_{l-1} or ω_{l+1} is reached, because of the assumed inequalities $\omega_{l-1}^2 \ll \omega_l^2 \ll \omega_{l+1}^2$. As the resonance is transversed from above, the dielectric constant increases by $\Delta\epsilon_l$. To the extent that the N_l's are approximately the same, so are the plasma frequencies ω_{pl}^2. The dielectric constant shift $\Delta\epsilon_l$ is then approximately proportional to $(\bar{\omega}_l)^{-2}$, according to Eq. (11.85*e*). This will be quite small for large resonance frequencies.

The resulting behavior of ϵ is shown semiquantitatively in Fig. 11.13. A logarithmic scale is used as an abscissa so that the condition $\bar{\omega}_1^2 \ll \bar{\omega}_2^2 \ll \bar{\omega}_3^2$ can be satisfied and all interesting behavior shown on one plot. Only the real part ϵ_r is plotted; the imaginary part ϵ_i will be nonzero essentially only in the regions of rapid change in ϵ_r—the regions of anomalous dispersion.

The type of behavior shown in Fig. 11.13 is exhibited by many real nonconducting systems. The lowest resonance frequency

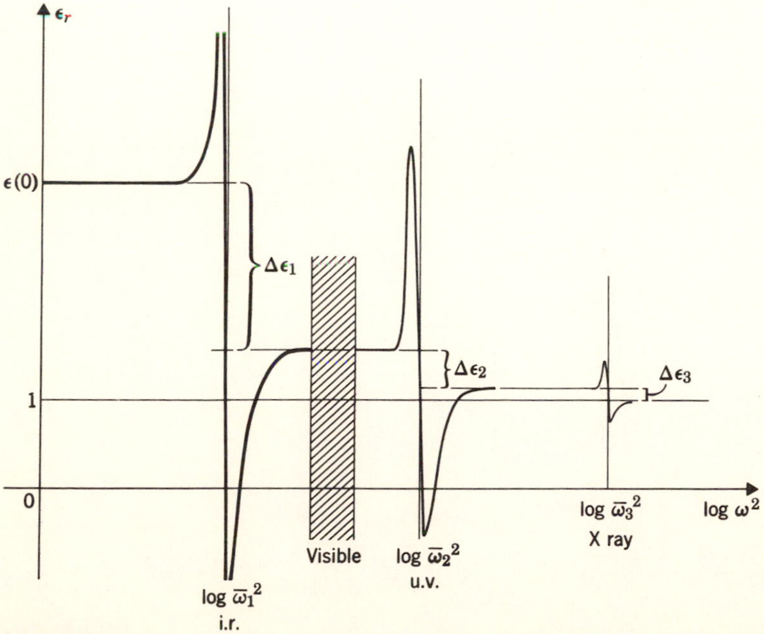

Fig. 11.13

$\bar{\omega}_1$ might typically be about $2 \cdot 10^{14}$ rad/sec ($\lambda = 10\ \mu$) and would correspond to the infrared-active internal vibrations of one part of the molecule with respect to another. It is often called the fundamental infrared absorption frequency. Several such vibrations often occur, but we show only one. For diatomic molecules such as O_2 and N_2 and for monatomic molecules such as A or He, there is no infrared-active mode; thus the $\bar{\omega}_1$ resonance is missing.

An example of the behavior of ϵ_r in the vicinity of the fundamental infrared absorption frequency for a real substance is shown in Fig. 11.14. The curves give ϵ_r for an LiF crystal at various temperatures as a function of the "wavenumber" or reciprocal wavelength $1/\lambda = \omega/(2\pi c)$.* Note the region where ϵ_r is negative, for example, between 320 and 670 cm^{-1} for the 7.5°K curve. Over this

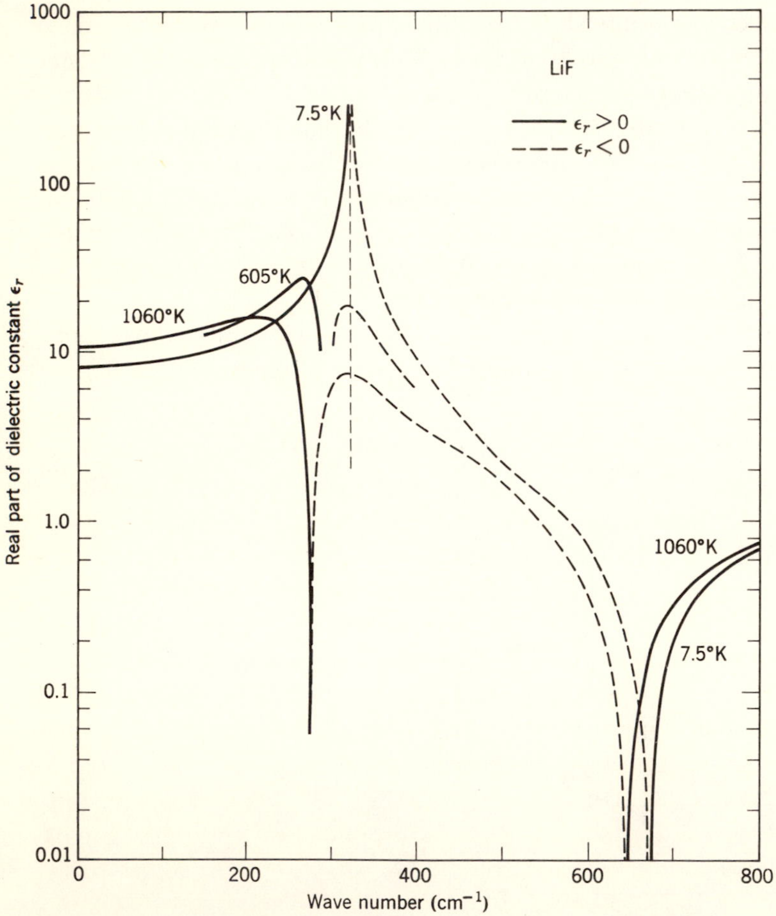

Fig. 11.14. Real part of the complex dielectric constant of LiF as a function of wave number and temperature. In the dashed regions ϵ_r is negative. (After Jasperse *et al.*)

*J. R. Jasperse, A. Kahan, J. N. Plendl, and S. S. Mitra, *Phys. Rev.*, **146**, 526 (1966).

range the reflectivity is high, as shown in Fig. 11.15. The points represent the actual experimental data; the solid curves are calculated from the simple damped oscillator theory discussed here using equations like Eq. (11.85*c*). Two oscillators were assumed in

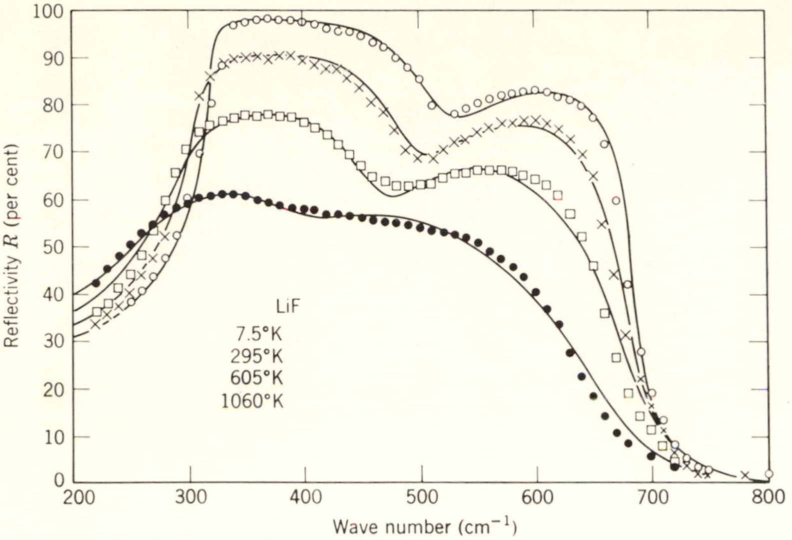

Fig. 11.15. Infrared reflectivity of LiF as a function of wave number and temperature. The various symbols are experimental points, and the solid lines are theoretical curves calculated from a two-oscillator model. (After Jasperse *et al.*)

these calculations; at low temperatures their frequencies (expressed in wave numbers) were 320 and 520 cm^{-1}. The lower-frequency oscillator was the stronger by a factor of 60.

The two oscillators give the two peaks in the curves for ϵ_i of Fig. 11.16. The presence of the weak high-frequency oscillator makes ϵ_i much larger near 500 cm^{-1} than otherwise would be the case. The result is that the real part of the refractive index n_r is not small there, as can be seen from Fig. 11.17. According to Eq. (11.84*c*) these relatively large values of n_r near 500 cm^{-1} lead to a reflectivity R noticeably less than unity. This explains the dip in the reflectivity curves of Fig. 11.15. For completeness, curves for the imaginary part of the index of refraction are shown in Fig. 11.18.

To fit the experimental reflectivity data, it was necessary to change the frequencies and lifetimes of the two oscillators, as well as their strengths with changing temperature. The resulting shifts in peak position and width are apparent from Fig. 11.16.

The fundamental infrared oscillator for a LiF crystal—the one at 320 cm^{-1} at low frequencies—corresponds to a relative motion of Li^+ and F^- ions. These ions bear charges close to $\pm e$, respectively, where e is the magnitude of the electronic charge. The oscillator strength f is within 20 per cent of unity. In the expression for the plasma frequency

$$\omega_p{}^2 = \frac{4\pi N e^2}{m}$$

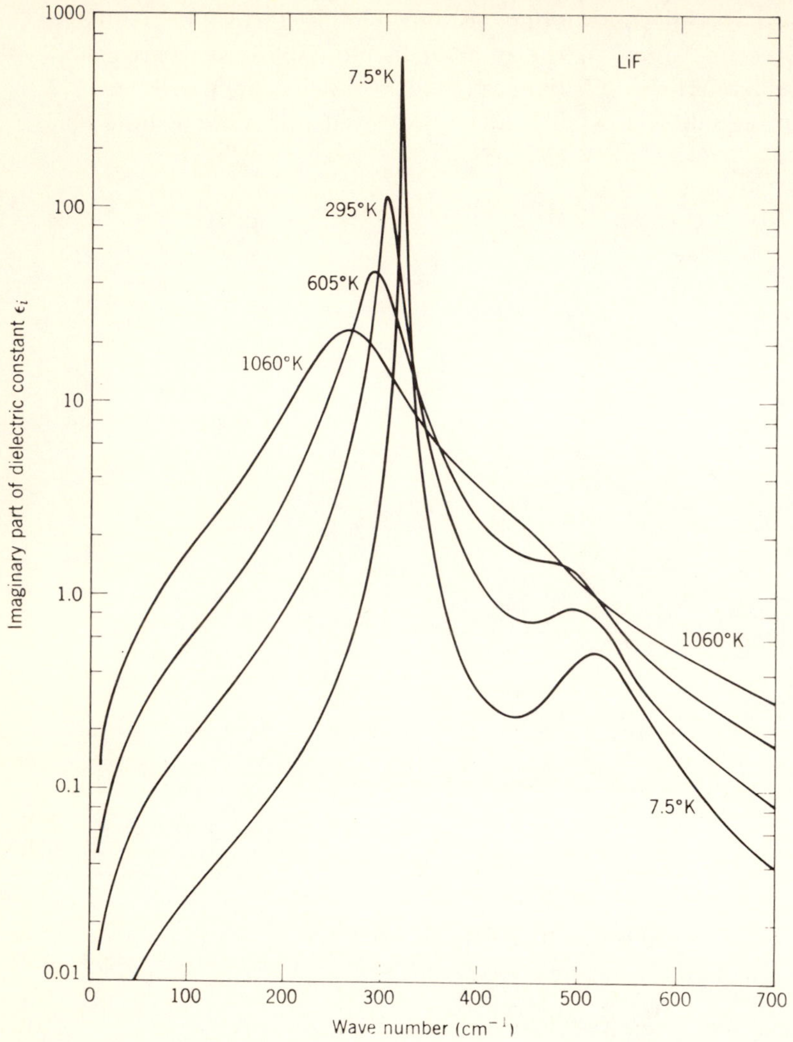

Fig. 11.16. Imaginary part of the complex dielectric constant of LiF as a function of wave number and temperature. (After Jasperse *et al.*)

the mass m should be the reduced mass of an LiF molecule

$$m = \frac{m(L_i)m(F)}{m(Li) + m(F)}$$

and the number density N should be the number of molecules per unit volume. The weaker, higher-frequency oscillator is believed to be a combination of two oscillators coupled by means of non-harmonic forces.

The next highest fundamental frequency region of Fig. 11.13, that of ω_2 is typically in the ultraviolet at about 10^{16} rad/sec ($\lambda = 200$ nm $= 2000$ A). It corresponds to an excitation in the outer electron shell of the molecule. In real materials there can be more than one such excitation. In solids ω_2 becomes a frequency band where ϵ_i is large and shows structure and where ϵ_r also shows structure, but

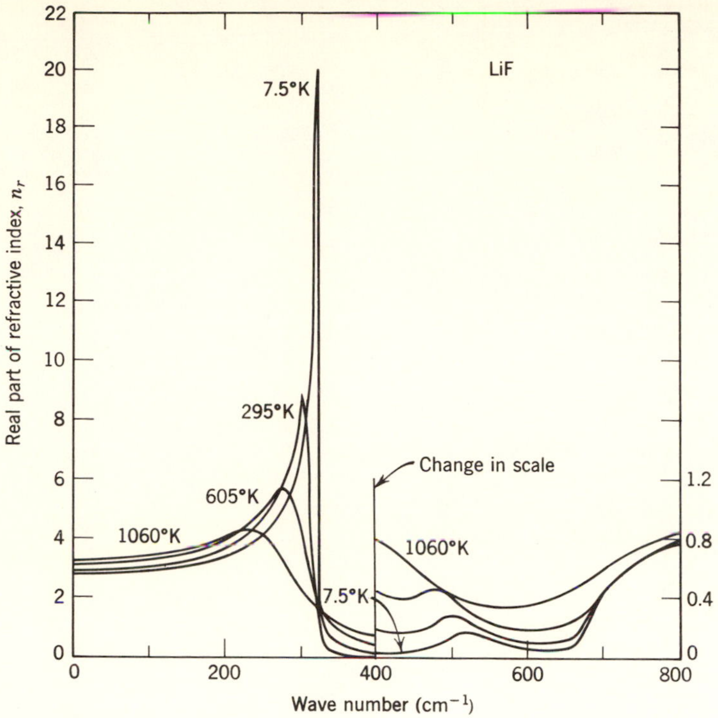

Fig. 11.17. Refractive index (real part) of LiF as a function of wave number and temperature. Note change in vertical scale. (After Jasperse *et al.*)

outside this band the behavior is more or less as indicated in Fig. 11.13. The electronic excitation frequencies of the ω_2 type can often extend into the visible region (where the medium becomes colored) or even into the infrared region.

Finally ω_3 will have an order of magnitude of about 10^{18} rad/sec ($\lambda = 2$ m$\mu = 20$ A) in the X ray region, corresponding to electronic excitations of one or more inner electron shells of the atoms in the molecule. Then $\Delta\epsilon$ will be of order 10^{-6}.

In this section we have painted a rough picture of the dispersion properties of most transparent materials. The frequency dependence of ϵ_r or n_r is intimately connected with the existence of regions of large absorption. These regions are associated with various characteristic excitation processes. Measurement of the optical properties of a medium provide one important method of studying these processes.

D. WAVE PACKETS AND GROUP VELOCITY

When light propagates through a transparent medium that has a resonance frequency ω less than the frequency ω of the light, the dielectric constant $\epsilon = n^2$ is less than unity. The phase velocity $v = c/n$ will be greater than the velocity of light in vacuum c. According to the principles of special relativity, no signal can propagate at a velocity greater than c. At this point we must ask

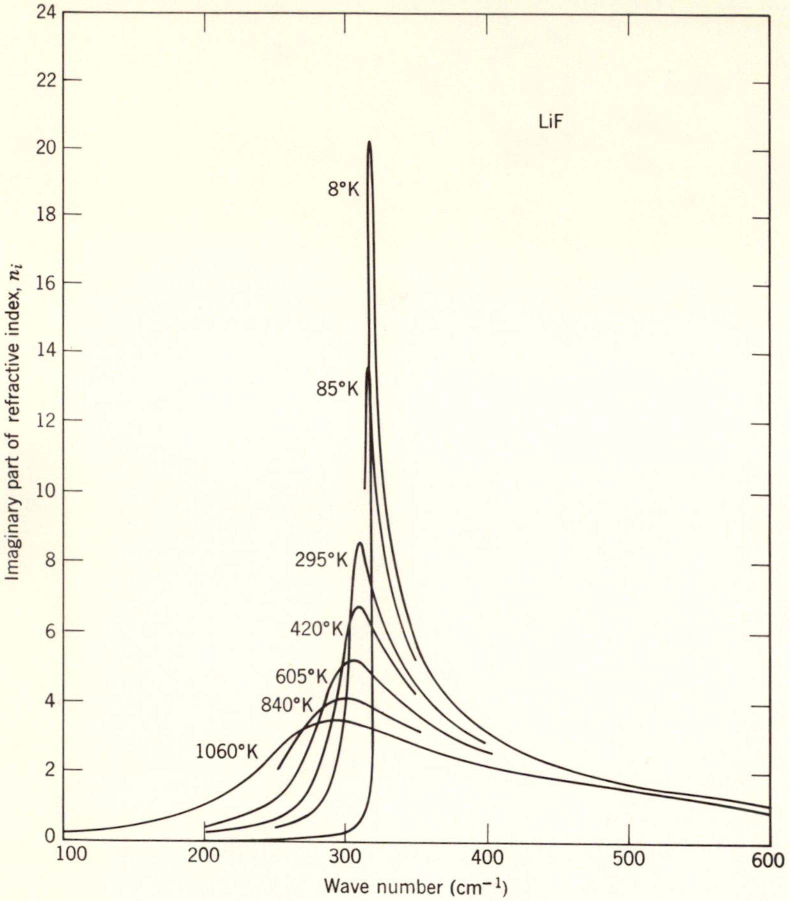

Fig. 11.18. Imaginary part of the refractive index (extinction coefficient) for LiF as a function of wave number and temperature. (After Jasperse *et al.*)

ourselves about the mathematical nature of a light signal, as opposed to an idealized sinusoidal wave.

In order to carry information, there must be imposed upon an ideal purely monochromatic sinusoidal light wave some form of modulation or change in amplitude, phase, or both. Because of the modulation, the light signal can no longer be a pure sinusoidal wave, but must be considered to be a superposition of sinusoidal components as described by a Fourier integral. Each component travels with the phase velocity $v = c/n$, but the resulting signal can exhibit somewhat different behavior.

For simplicity, assume plane waves propagating in the $+x$ direction. Then we may follow the discussion of Appendix B and write, by analogy with Eqs. (7.1) and (7.2),

$$E(x, t) = \int_{-\infty}^{\infty} \hat{E}(x, \nu) e^{2\pi i\nu t} d\nu = Re \int_{0}^{\infty} 2\hat{E}(x, \nu) e^{2\pi i\nu t} d\nu \qquad (11.86a)$$

The Fourier coefficient $\hat{E}(x, \nu)$ obeys the Helmholtz equation

$$\nabla^2 \hat{E}(x, \nu) + k(\nu)^2 \hat{E}(x, \nu) = 0 \qquad (11.86b)$$

with

$$k(\nu) = \frac{2\pi\nu n(\nu)}{c}$$

The solution we seek is

$$\hat{E}(x, \nu) = \hat{E}(0, \nu)e^{-ik(\nu)x}$$

Thus

$$E(x, t) = \text{Re} \int_0^\infty 2\hat{E}(0, \nu)e^{i[2\pi\nu t - k(\nu)x]}\, d\nu \tag{11.86c}$$

Thus far we have assumed nothing specific about the spectral nature of the light. We now assume that the light is quasimonochromatic with its frequency distribution centered about $\nu = \nu_0$. We therefore write

$$\hat{E}(0, \nu) = \hat{g}(\nu - \nu_0)$$

where $|\hat{g}(\nu - \nu_0)|$ is nonzero only for $|\nu - \nu_0| \ll \nu_0$.

We next expand the exponent about $\nu = \nu_0$:

$$2\pi\nu t - k(\nu)x = 2\pi\nu_0 t - k(\nu_0)x + \left[2\pi t - \left(\frac{dk}{d\nu}\right)_{\nu 0} x\right](\nu - \nu_0) + \cdots$$

By using only the terms shown in this expansion, we obtain

$$E(x, t) \cong 2\text{Re}\left\{e^{i[2\pi\nu_0 t - k(\nu_0)x]} \int_0^\infty \hat{g}(\nu - v_0)e^{i[2\pi t - (dk/d\nu)_{\nu 0}x](\nu - \nu_0)}\, d\nu\right\} \tag{11.87}$$

Because $\hat{g}(\nu - \nu_0)$ is sharply peaked about $\nu = \nu_0$, we may replace the lower limit on the integral by $-\infty$. The integral then gives

$$\int_{-\infty}^{\infty} \hat{g}(\mu)e^{2\pi i\mu(t - x/v_g)}\, d\mu = g\left(t - \frac{x}{v_g}\right) \tag{11.88}$$

where g is the inverse Fourier transform of $\hat{g}(\mu)$, and where the group velocity v_g is defined by

$$v_g{}^{-1} = \frac{1}{2\pi}\frac{dk}{d\nu} = \frac{dk}{d\omega} = \frac{d}{d\omega}\left[\frac{\omega n(\omega)}{c}\right] \tag{11.89a}$$

This can also be written

$$v_g = \frac{d\omega}{dk} \tag{11.89b}$$

with k considered to be the independent variable. The phase velocity v at frequency ν_0 is $v = 2\pi\nu_0/k(\nu_0) = \omega_0/k(\omega_0)$.

Equation (11.87) then becomes

$$E(x, t) = \text{Re}\left\{2e^{2\pi i\nu_0(t - x/v)}\, g\left(t - \frac{x}{v_g}\right)\right\} \tag{11.90}$$

The analytic signal, which is the complex function in curly brackets in Eq. (11.90), consists of a pure rapidly oscillating exponential with frequency ν_0:

$$e^{2\pi i\nu_0(t-x/v)}$$

propagating in the $+x$ direction with the phase velocity v multiplied by a complex "envelope function" $g(t-x/v_g)$. The latter propagates in the $+x$ direction with velocity v_g. Since the Fourier transform $\hat{g}(\mu)$ is nonzero only for $|\mu| \ll \nu_0$, the function $g(t')$ contains oscillatory components having frequencies much less than ν_0. Thus it must be slowly varying on a time scale comparable with the period ν_0^{-1} of the rapidly oscillating exponential. If we write $g(t')$ in the form

$$g(t') = A(t')e^{i\phi(t')}$$

where the phase angle ϕ is real and where the amplitude A is real and positive, we obtain

$$E(x, t) = 2A\left(t-\frac{x}{v_g}\right)\cos\left[2\pi\nu_0\left(t-\frac{x}{v}\right)+\phi\left(t-\frac{x}{v_g}\right)\right] \tag{11.91}$$

If the phase ϕ in Eq. (11.91) is constant, the light signal has pure amplitude modulation

$$E(x, t) = 2A\left(t-\frac{x}{v_g}\right)\cos\left[2\pi\nu_0\left(t-\frac{x}{v}\right)+\phi\right] \tag{11.92}$$

This case is relatively easy to draw and is indicated in Fig. 11.19. A finite modulated signal of the type illustrated is often called a wave packet.

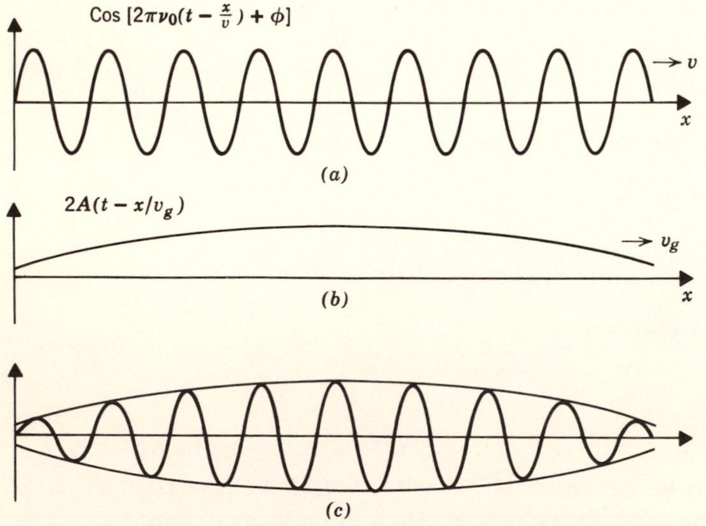

Fig. 11.19. A wave packet. (*a*) Sinusoidal wave. (*b*) Envelope. (*c*) Combined optical signal.

Except in the regions of anomalous dispersion, the derivative $dn/d\omega$ is positive. As can be seen in Fig. 11.11 or Fig. 11.17, the imaginary part n_i is large when $dn/d\omega$ is negative, causing the light to be absorbed after traveling less than a wavelength, and the above derivation does not apply.

Thus, in the nonanomalous, transparent regions we can say that $v_g < v$. It is more difficult to show that v_g is less than c, that is, that

$$\frac{c}{v_g} = n + \omega \frac{dn}{d\omega} > 1 \tag{11.93a}$$

The inequality is obvious when $n > 1$ (and $dn/d\omega > 0$). For the case of n less than unity, which means $v > c$, we write the inequality as follows:

$$n^2 + \omega^2 \frac{dn^2}{d\omega^2} > n$$

This inequality will be satisfied if the more stringent inequality

$$n^2 + \omega^2 \frac{dn^2}{d\omega^2} > 1$$

is satisfied. This can be rewritten

$$(n^2 - 1) + \omega^2 \frac{d(n^2 - 1)}{d\omega^2} > 0 \tag{11.93b}$$

The inequality (11.93*b*) and hence (11.93*a*) can be shown to hold in any specific case. For instance, suppose that the situation described by Eq. (11.85*c*) has an index of refraction away from resonance that obeys an equation of the form

$$n^2 = \epsilon_{0l} + \frac{A}{\omega_l^2 - \omega^2} \tag{11.94}$$

where $\epsilon_{0l} > 1$ and where A is the positive constant

$$A = f_l \omega_{pl}^2 \left(\frac{\epsilon_{0l} + 2}{3}\right)^2$$

A direct calculation gives the result

$$n^2 - 1 + \omega^2 \frac{d(n^2 - 1)}{d\omega^2} = \epsilon_{0l} - 1 + \frac{A}{(\omega_l^2 - \omega^2)} + \frac{A\omega^2}{(\omega_l^2 - \omega^2)^2} = (\epsilon_{0l} - 1) + \frac{A\omega_l^2}{(\omega_l^2 - \omega^2)^2} > 0$$

In this way we can convince ourselves that in the transparent regions the group velocity is less than the velocity of light.*

*For a discussion of the propagation of a wave packet when absorption cannot be neglected, see J. A. Stratton, *Electromagnetic Theory*, pp. 301–309.

REFERENCES FOR SECTION 11.2

Intermediate level

R. W. Ditchburn, *Light*, Chapter XV.
G. R. Fowles, *Introduction to Modern Optics*, Chapter 5.
R. S. Longhurst, *Geometrical and Physical Optics*, Chapter 20.
H. R. Von Hippel, *Dielectrics and Waves*, Pages 93–104.
M. Garbuny, *Optical Physics*, Chapters 4 and 5. (Some advanced material.)
R. W. Pohl, *Optik und Atomphysik*, Chapter XIII (In German.)

Treatments of Local Field Effects

J. G. Kirkwood, "On the Theory of Dielectric Polarization," *J. Chem. Phys.*, **4**, 592 (1936).
R. H. Cole, "Theories of Dielectric Polarization and Relaxation," in *Progress in Dielectrics*, J. B. Birks, ed., Vol. 3 (Heywood, London, 1961), pp. 47.

Optical Properties of Solids

F. C. Brown, *The Physics of Solids*. Chapters 7, 8, 9.
C. Kittel, *Introduction to Solid State Physics*, Third Edition. Chapter 12.
F. Seitz, *Modern Theory of Solids*, Chapter XVII.
F. Stern, "Elementary Theory of the Optical Properties of Solids," in *Solid State Physics*, F. Seitz and D. Turnbull, ed., Vol. 15 (Academic Press, New York, 1963), pp. 300.

PROBLEMS

11.2.1 A normalized Lorentzian function may be written

$$L(x;\gamma)=\frac{\gamma/\pi}{x^2+\gamma^2}$$

Its maximum value of $1/(\pi\gamma)$ occurs at $x=0$. The half-width at half maximum is γ. The area under the peak is unity:

$$\int_{-\infty}^{\infty} L(x;\gamma)\,dx=1$$

a. Prove the above statements.

b. Consider a dilute system of N oscillators per unit volume embedded in a dielectric medium. The oscillators have charge e, mass m, natural frequency ω_0, oscillator strength f, and damping time τ, where $\omega_0\tau \gg 1$. The medium itself has no absorption near ω_0, so that its index of refraction n may be treated as a real constant number for ω near ω_0. Show that for ω near ω_0 the absorption coefficient is well approximated by

$$K(\omega)=\frac{2\pi^2Ne^2}{mc}\frac{(n+2)^2}{9n}fL\left[(\omega-\omega_0);\frac{1}{2\tau}\right]$$

c. Change variables to $1/\lambda=\omega/2\pi c$. Show that

$$\int K\left(\frac{1}{\lambda}\right)d\left(\frac{1}{\lambda}\right)=f\frac{(n+2)^2}{9n}\frac{\pi}{\lambda_p{}^2}$$

where $\lambda_p = 2\pi c/\omega_p$ represents the vacuum wavelength of light at the plasma frequency $\omega_p = \sqrt{4\pi Ne^2/m}$. The integral is to be taken over the wave number range about the resonance peak where K is nonzero. (This is one form of "Smakula's equation." See the book by Seitz cited in the list of references for other versions.)

11.2.2 There is a class of electronic impurity centers in crystals called color centers. The absorption constant produced by one such center, the "*F* center" in an NaCl crystal is shown in the figure. Treat this center as a harmonic oscillator with the mass of an electron imbedded in the crystal (of refractive index $n = 1.55$). Assume the experimental line shape to be Lorentzian, and from it estimate the lifetime of the oscillator and its oscillator strength. (Hint: Use the results of the previous problem.) Calculate a value for the maximum change in the real part of the index of refraction caused by the presence of the F centers in this sample. (Answer: $f = 0.85$.)

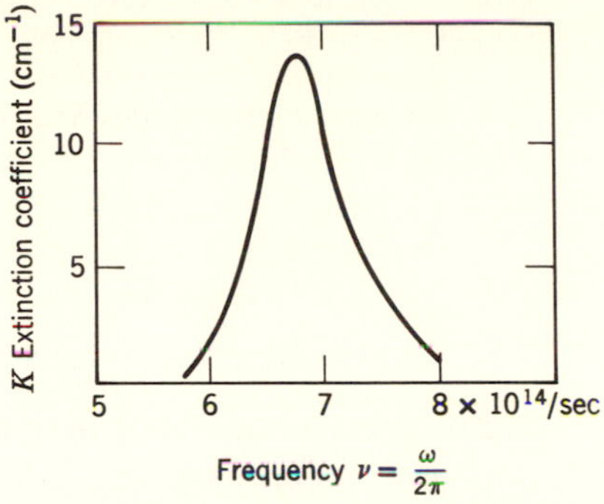

11.2.3 For the model of a metal assumed for Fig. 11.8, the plasma frequency $\omega_p = 6 \cdot 10^{15}$/sec corresponds to a wavelength $\lambda_p = 314$ nm. Calculate numerical values for the penetration depth $1/K$ (Eq. 11.44) for light at normal incidence at these wavelengths: 349, 314, and 286 nm. Answer: 57.8, 500, and 5060 nm, respectively.)

11.2.4 Consider an oscillator O with strength f, mass m, isolated resonance frequency ω_0, (long) damping time τ, and concentration N imbedded in a medium with index of refraction $n_\infty(\omega)$; that is, the dielectric constant would be $[n_\infty(\omega)]^2$ if the oscillators in questions were not present. Assume that n_∞ may be treated as a constant for all frequencies from the vicinity of ω_0 to zero. This means that the oscillators that contribute to n_∞ have their resonances at frequencies much higher than ω_0.

a. Write the analogues of Eq. (11.77) for ϵ_i and ϵ_r and Eq. (11.79)

$$\int_0^{(\text{above } \omega_0)} \epsilon_i(\omega)\, d\omega.$$

b. Let ω_t be the frequency of the oscillator O in the medium at which ϵ_i attains its maximum value (resonance frequency in the medium). Assume N and τ large enough so that ϵ_r passes through zero near resonance. Let ω_l be the frequency at which ϵ_r becomes positive again. Assume that $\omega_l - \omega_t \gg 1/\tau$. Show that

$$\omega_l^2 - \omega_t^2 = f\frac{4\pi Ne^2}{m}\frac{(n_\infty^2+2)^2}{n_\infty^2}$$

and that

$$\frac{\omega_l^2}{\omega_t^2} = \frac{\epsilon(\omega=0)}{n_\infty^2}$$

This last equation is known as the Lyddane-Sachs-Teller relation (LST relation).

c. From the 7.5°K curves of Figs. 11.14 and 11.16, pick off values of $\epsilon(\omega = 0)$, ω_t, and ω_l. Then estimate n_∞ from the LST relation. Compare with the accepted values of $n = 1.44$ at 200 nm and 1.38 at 1500 nm.

11.2.5 Consider a substance in the "metallic" reflection region where ϵ_r is negative and ϵ_i quite small.

a. Show that to first order in ϵ_i, the reflectivity R at normal incidence obeys

$$1 - R = \frac{2\epsilon_i/n_i}{1+n_i^2}$$

b. Calculate $(1-R)$ for a metal with $\omega_p\tau = 100$ when $\omega/\omega_p = 0.8$ and 0.95.

11.2.6 A useful approximate equation to describe the index of refraction of transparent media in the visible is Cauchy's equation

$$n = A + \frac{B}{\lambda^2}$$

Show that this approximation may be derived from equations like (11.76a, *b*) under the assumption that the frequency of the light is much less than the frequencies ω_j of all the important oscillators.

11.2.7 Consider a generalization of Eq. (11.94) where ϵ_{0l} is given by an equation of the form

$$\epsilon_{0l} = 1 + \frac{B}{\omega_0^2 - \omega^2}$$

Show that the inequality (11.93*b*) still holds.

11.3 REFLECTION AND TRANSMISSION OF LIGHT AT AN INTERFACE

In this section we derive and discuss the coefficients that describe the reflection and transmission of light at an interface between two media having different optical properties. The treatment is the analogue for electromagnetic waves of the theory of reflection and transmission of one-dimensional waves on a string treated in Section 1.1D.

A. DERIVATION OF BASIC EQUATIONS

We assume that there is an abrupt interface between two otherwise isotropic and homogeneous optical media, which may have quite general optical properties such as complex indices of refraction. The four relevant Maxwell's equations, given in Eq. (11.30), are rewritten here for convenience:

$$\nabla \cdot \mathbf{D} = 4\pi\rho_f \tag{11.30a}$$

$$\nabla \cdot \mathbf{B} = 0 \tag{11.30b}$$

$$\nabla \cdot \mathbf{E} = -\frac{1}{c}\frac{\partial \mathbf{B}}{\partial t} \tag{11.30c}$$

$$\nabla \cdot \mathbf{H} = \frac{1}{c}\frac{\partial \mathbf{D}}{\partial t} + \frac{4\pi \mathbf{j}_f}{c} \tag{11.30d}$$

Our first task is to determine boundary conditions that match the values of **B**, **H**, **E**, and **D** on both sides of the interface.

*1. **Boundary Conditions.*** When Eqs. (11.30) are integrated over suitable regions containing the boundary between the two optical media, we may obtain the desired boundary conditions obeyed by the fields on the two sides. Consider the "pill box" shown in Fig. 11.20*a*. The height h is assumed infinitesimally small. The area σ is finite but arbitrary. The volume $V = \sigma h$ is infinitesimally small. We assume the free charge density ρ_f to be everywhere finite, so that the integral over the volume of the pillbox

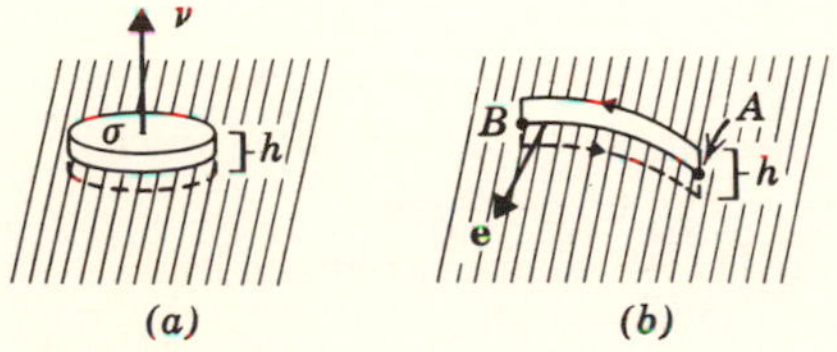

Fig. 11.20. Aids in determining boundary conditions.

$$\int_V \rho_f \, dV$$

is infinitesimally small. Thus, if S denotes the entire surface of the pill box, we obtain

$$0 = 4\pi \int_V \rho_f \, dV = \int_V \nabla \cdot \mathbf{D} \, dV = \int_S D_n \, d\sigma$$

The last equality followed by use of the divergence theorem. where D_n is the outward normal component of D. Because h is infinitesimal, there is no finite contribution to the integral from the sides. What remains is

$$0 = \int_{S_2} D_n \, d\sigma + \int_{S_1} D_n \, d\sigma = \int_{\sigma} (\mathbf{D}_2 - \mathbf{D}_1) \cdot \boldsymbol{\nu} \, d\sigma$$

where $\mathbf{D}_1$ and $\mathbf{D}_2$ represent the displacement fields just below and above the interface. Because the surface σ is arbitrary, we may conclude that

$$\mathbf{D}_2 \cdot \boldsymbol{\nu} = \mathbf{D}_1 \cdot \boldsymbol{\nu} \tag{11.95a}$$

This equation says that the normal component of **D** is continuous across the interface. The same derivation applies to **B**. Hence

$$\mathbf{B}_2 \cdot \boldsymbol{\nu} = \mathbf{B}_1 \cdot \boldsymbol{\nu} \tag{11.95b}$$

The normal component of **B** is also continuous.

Consider now the "ribbon" shown in Fig. 11.20*b* and take the scalar product of Eq. (11.30*d*) with a unit vector **e** normal to the ribbon (and lying in the interface). Then integrate the result over the area S_Γ. The quantity $\partial \mathbf{D}/\partial t$, which for a sinusoidal oscillation is given by $i\omega\mathbf{D}$, will remain finite at the surface, since we assume that there are no free charges there. We also assume that the free current density $\mathbf{j}_f$ remains finite at the surface. We then obtain

$$0 = \frac{1}{c} \int_{S_\Gamma} \left(\frac{\partial \mathbf{D}}{\partial t} + 4\pi \mathbf{j}_f \right) \cdot \mathbf{e} \, d\sigma = \int_{S_\Gamma} (\nabla \times \mathbf{H}) \cdot \mathbf{e} \, d\sigma$$

An application of Stokes theorem gives, since $S_\Gamma \approx 0$,

$$0 = \oint_{\Gamma} \mathbf{H} \cdot d\mathbf{l}$$

The line integral is to be taken around the boundary curve Γ. The curve Γ is arbitrary. Thus we conclude that

$$(\mathbf{H}_2 - \mathbf{H}_1) \cdot d\mathbf{l} = 0 \tag{11.95c}$$

where $d\mathbf{l}$ is any small vector lying in the interface between the two media. Equation (11.95*c*) is equivalent to the statement that $\mathbf{H}_{\tan}$, the tangential component of **H**, is continuous across the interface.

An analogous derivation yields

$$(\mathbf{E}_2 - \mathbf{E}_1) \cdot d\mathbf{l} = 0 \tag{11.95d}$$

Thus, $\mathbf{E}_{\tan}$, the tangential component of **E**, must also be continuous across the interface.

We shall use the required continuity of $\mathbf{E}_{\tan}$ and $\mathbf{H}_{\tan}$ to derive the reflection and transmission coefficients; the continuity of D_n and B_n may be shown to follow.

2. Notation, Phase-Matching Condition. Let the interface lie in the xy plane and let the plane of incidence be the zy plane as shown in Fig. 11.21. The fields **E** and **H** may be resolved into their components in the plane of incidence and their x components normal to the plane of incidence. Two independent cases then arise:

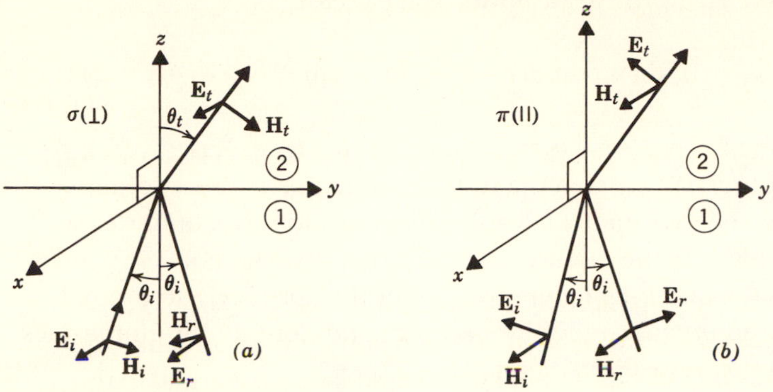

Fig. 11.21. (*a*) E perpendicular to the plane of incidence. (*b*) E parallel to the plane of incidence.

the case where **E** is perpendicular to the plane of incidence (σ case) with **H** in the plane of incidence, and the case where **E** is in the plane of incidence (π case) with **H** perpendicular to it.

We use a complex notation and assume time dependence $e^{i\omega t}$ and a spatial dependence $e^{-i\mathbf{k}_i\cdot\mathbf{r}}$ for the incident wave, $e^{-i\mathbf{k}_r\cdot\mathbf{r}}$ for the reflected wave, and $e^{-i\mathbf{k}_t\cdot\mathbf{r}}$ for the transmitted wave. The wave vectors will have components

$$\mathbf{k}_i = (0, k_{iy}, k_{iz}) = n_1\left(\frac{\omega}{c}\right)(0, \sin\theta_i, \cos\theta_i) \tag{11.96a}$$

$$\mathbf{k}_r = (k_{rx}, k_{ry}, k_{rz}) \qquad \text{with} \quad k_{rx}^2 + k_{ry}^2 + k_{rz}^2 = n_1{}^2\left(\frac{\omega}{c}\right)^2 \tag{11.96b}$$

$$\mathbf{k}_t = (k_{tx}, k_{ty}, k_{tz}) \qquad \text{with} \quad k_{tx}^2 + k_{ty}^2 + k_{tz}^2 = n_2{}^2\left(\frac{\omega}{c}\right)^2 \tag{11.96c}$$

where n_1 and n_2 are the refractive indices in the two media.

The continuity requirements on $\mathbf{E}_{\tan}$ and $\mathbf{H}_{\tan}$ across the interface at $z = 0$ then take the form

$$\mathbf{V}_{i,\tan}\, e^{-iyk_{iy}} + \mathbf{V}_{r,\tan}\, e^{-i(xk_{rx} + yk_{ry})} = \mathbf{V}_{t,\tan}\, e^{-i(xk_{tx} + yk_{ty})}$$

where **V** stands for either **E** or **H**. This equation must hold for all x and y and will be valid if and only if the phases are everywhere the same. One consequence is the condition

$$k_{rx} = k_{tx} = 0 \tag{11.97a}$$

This says that the reflected and transmitted wave vectors must lie in the plane of incidence (yz plane). Other requirements are

$$k_{ry} = k_{iy} \tag{11.97b}$$

and

$$k_{ty} = k_{iy} \tag{11.97c}$$

Using the notation of Fig. 11.21, for the reflected wave vector we may write

$$\mathbf{k}_r = n_1 \left(\frac{\omega}{c}\right) (0, \sin\theta_i, -\cos\theta_i) \tag{11.97d}$$

and similarly for the transmitted wave vector,

$$\mathbf{k}_t = n_2 \left(\frac{\omega}{c}\right) (0, \sin\theta_r, \cos\theta_r) = \left(\frac{\omega}{c}\right) (0, n_1 \sin\theta_i, n_2 \cos\theta_r) \tag{11.97e}$$

Thus Snell's law, $n_2 \sin\theta_t = n_1 \sin\theta_i$, follows from Eqs. (11.96*a*, *c*) and (11.97*c*). We see that the full electromagnetic theory of light does indeed explain the laws of reflection and refraction; we have only to identify the rays with directions of propagation.

With the phase-matching conditions thus properly accounted for, the boundary conditions become conditions on the amplitudes **E**, **H** of the waves $\mathbf{E}e^{i(\omega t - \mathbf{k}\cdot\mathbf{r})}$ and $\mathbf{H}e^{i(\omega t - \mathbf{k}\cdot\mathbf{r})}$:

$$\sigma \begin{cases} E_{ix} + E_{rx} = E_{tx} & (11.98a) \\ H_{iy} + H_{ry} = H_{ty} & (11.98b) \end{cases}$$

$$\pi \begin{cases} H_{ix} + H_{rx} = H_{ty} & (11.98c) \\ E_{iy} + E_{ry} = E_{ty} & (11.98d) \end{cases}$$

3. ***σ Case—E Perpendicular to the Plane of Incidence.*** The boundary conditions (11.98) pair off as shown. For the σ case, **E** has only an x component. Then **H** must lie in the yz plane, and Eqs. (11.98*a*,*b*) apply. Equation (11.38) gives **H** in terms of **E**:

$$\mathbf{H} = \mu^{-1}\mathbf{B} = \mu^{-1}\left(\frac{c}{\omega}\right)(\mathbf{k} \times \mathbf{E}) \qquad \text{with} \quad |\mathbf{k}| = \frac{n\omega}{c} \tag{11.98e}$$

If we assume that both media are nonmagnetic, then the μ's are so close to unity that we make no practical error by setting $\mu_1 = 1$ and $\mu_2 = 1$. Then the required continuity of $\mathbf{H}_{\tan}$ implies that $(\mathbf{k} \times \mathbf{E})_{\tan}$ be continuous. Equation (11.98*b*) then becomes

$$(\mathbf{k}_i \times \mathbf{E}_i)_y + (\mathbf{k}_r \times \mathbf{E}_r)_y = (\mathbf{k}_t \times \mathbf{E}_t)_y$$

Now the **E**'s have only x components, and we may write

$$E_{ix} \equiv E_i, \qquad E_{rx} \equiv E_r, \qquad E_{tx} \equiv E_t$$

The **k**'s have only y and z components; thus, $(\mathbf{k} \times \mathbf{E})_y = k_z E_x$. Using these specific relationships, we obtain from Eq. (11.98*a*)

$$E_i + E_r = E_t \tag{11.99a}$$

and from Eq. (11.98*b*)

$$k_{iz}E_i + k_{rz}E_r = k_{tz}E_t$$

But $k_{rz} = -k_{iz}$. Thus

$$E_i - E_r = \frac{k_{tz}}{k_{iz}} E_t \tag{11.99b}$$

Equations (11.99*a*, *b*) are readily solved:

$$E_t = \frac{2E_i}{1 + k_{tz}/k_{iz}} \equiv \tau_\sigma E_i \tag{11.100a}$$

$$E_r = \frac{1 - k_{tz}/k_{iz}}{1 + k_{tz}/k_{iz}} E_i \equiv \tau_\sigma E_i \tag{11.100b}$$

From Eqs. (11.97*d*,*e*) we have

$$a \equiv \frac{k_{tz}}{k_{iz}} = \frac{n_2 \cos \theta_t}{n_1 \cos \theta_i} = \frac{\tan \theta_i}{\tan \theta_t} \tag{11.101}$$

the last equality following from Snell's law. Using Eq. (11.101) and some trigonometric identities, we find for the *transmission coefficient*

$$\tau_\sigma = \frac{2}{1 + a} = \frac{2n_1 \cos \theta_i}{n_1 \cos \theta_i + n_2 \cos \theta_t} = \frac{2 \sin \theta_t \cos \theta_i}{\sin (\theta_i + \theta_t)} \tag{11.102a}$$

and similarly for the *reflection coefficient*

$$\rho_\sigma = \frac{1 - a}{1 + a} = \frac{n_1 \cos \theta_i - n_2 \cos \theta_t}{n_1 \cos \theta_i + n_2 \cos \theta_t} = -\frac{\sin (\theta_i - \theta_t)}{\sin (\theta_i + \theta_t)} \tag{11.102b}$$

The last equalities for τ_σ and ρ_σ in Eqs. (11.102) are known as the *Fresnel relations* for the σ case.

At this point it is useful to compare the above results with those that hold when the roles of incident and transmitted waves are interchanged. Let us therefore consider light incident onto the interface from medium 2 at angle $\theta_i' = \theta_t$. The angle of refraction is now $\theta_t' = \theta_i$. The ratio a becomes

$$a' = \frac{k_{tz}'}{k_{iz}'} = \frac{n_1 \cos \theta_t'}{n_2 \cos \theta_i'} = \frac{n_1 \cos \theta_i}{n_2 \cos \theta_t} = \frac{1}{a}$$

The reflection coefficient is

$$\frac{E_r'}{E_i'} = \rho_\sigma' = \frac{1 - a'}{1 + a'} = \frac{1 - a^{-1}}{1 + a^{-1}} = \frac{a - 1}{a + 1} = -\rho_\sigma \tag{11.103a}$$

The minus sign in the relation

$$\rho_\sigma' = -\rho_\sigma$$

corresponds to the well-known 180° phase difference when reflection coefficients corresponding to two directions are compared.

The transmission coefficient is

$$\frac{E_t'}{E_i'} = \tau_\sigma' = \frac{2}{1+a'} = \frac{2}{1+a^{-1}} = \frac{2a}{1+a} = a\tau_\sigma$$

or

$$\tau_\sigma \tau_\sigma' = a{\tau_\sigma}^2 = \frac{4a}{(1+a)^2}$$

This should be compared with

$$1-{\rho_\sigma}^2 = 1-\frac{(1-a)^2}{(1+a)^2} = \frac{(1+a)^2-(1-a)^2}{(1+a)^2} = \frac{4a}{(1+a)^2}$$

Thus

$$\tau_\sigma \tau_\sigma' = 1-{\rho_\sigma}^2 \tag{11.103b}$$

Equations (11.103), which also hold for the π case, were used several times in Chapter 5.

4. π Case—E in the Plane of Incidence. In this case **H** has only an x component. The resulting directions of $\mathbf{E}_i$, $\mathbf{E}_r$, and $\mathbf{E}_t$ are as shown in Fig. 11.21*b*. This figure determines the sign convention that we adopt for the scalar quantities

$$H_{ix} \equiv H_i = n_i E_i, \qquad H_{rx} \equiv H_r = n_1 E_r$$

and

$$H_{tx} \equiv H_t = n_2 E_t$$

Here we have used the result from Eq. (11.98*e*) that $|\mathbf{H}| = n|\mathbf{E}|$ when $\mu = 1$.

Now **E** may be determined from **H** via Eq. (11.30*e*), which for $e^{i(\omega t - \mathbf{k}\cdot\mathbf{r})}$ space-time dependence gives

$$\mathbf{D} = \epsilon\mathbf{E} = \left(\frac{c}{\omega}\right)\mathbf{k}\times\mathbf{H} \tag{11.104}$$

For $\mu = 1$, we have $\epsilon = n^2$. Then the continuity of $\mathbf{E}_{\tan}$ implies the continuity $n^{-2}(\mathbf{k}\times\mathbf{H})_{\tan}$. With only an x component for **H** and no x component for **k**, our requirement for E_y continuous is that $n^{-2}k_z H_x$ be continuous. Thus, continuity of $\mathbf{H}_{\tan}$ gives

$$n_1 E_i + n_1 E_r = n_2 E_t \tag{11.105a}$$

and continuity of $\mathbf{E}_{\tan}$ gives

$$\frac{k_{iz} n_1 E_i}{{n_1}^2} + \frac{k_{rz} n_1 E_r}{{n_1}^2} = \frac{k_{tz} n_2 E_t}{{n_2}^2}$$

or, since $k_{rz} = -k_{iz}$,

$$n_1(E_i - E_r) = \left(\frac{n_1}{n_2}\right)^2 \frac{k_{tz}}{k_{iz}} n_2 E_t \tag{11.105b}$$

The solutions to Eqs. (11.105) are

$$E_t = \tau_\pi E_i, \qquad E_r = \rho_\pi E_i$$

with

$$\tau_\pi = \frac{2n_1/n_2}{1+b}, \qquad \rho_\pi = \frac{1-b}{1+b} \tag{11.106}$$

and

$$b = \left(\frac{n_1}{n_2}\right)^2 \frac{k_{tz}}{k_{iz}} = \left(\frac{n_1}{n_2}\right)^2 a$$

Alternate versions may be derived, namely,

$$\tau_\pi = \frac{2n_1 \cos\theta_i}{n_2 \cos\theta_i + n_1 \cos\theta_t} = \frac{2\sin\theta_t \cos\theta_i}{\sin(\theta_i+\theta_t)\cos(\theta_i-\theta_t)} \tag{11.107a}$$

$$\rho_\pi = \frac{n_2 \cos\theta_i - n_1 \cos\theta_t}{n_2 \cos\theta_i + n_2 \cos\theta_t} = \frac{\tan(\theta_i-\theta_t)}{\tan(\theta_i+\theta_t)} \tag{11.107b}$$

The last equalities in these two equations express the Fresnel relations for the π case.

For the inverse case of incidence onto the interface from medium 2 at angle $\theta_i' = \theta_t$, and so on, we find that we must replace b by $b' = b^{-1}$ and n_1/n_2 by n_2/n_1. This gives

$$\rho_\pi' = \frac{1-b'}{1+b'} = -\rho_\pi \tag{11.108a}$$

and

$$\tau_\pi' = \frac{2n_2/n_1}{1+b'} = \frac{2bn_2/n_1}{1+b}$$

so that

$$\tau_\pi \tau_\pi' = 1 - \rho_\pi^2 \tag{11.108b}$$

*5. **Energy Reflection and Transmission Coefficients.*** The average Poynting vector $\langle \mathbf{S} \rangle$ gives the time-average power per unit area; for a plane wave this is a maximum for a plane normal to the direction of propagation. From Eq. (11.40) we have

$$\langle \mathbf{S} \rangle = \frac{S\mathbf{k}}{|k|} = \frac{nc}{\mu} \frac{|\mathbf{E}|^2}{8\pi} \frac{\mathbf{k}}{|k|}$$

The average power crossing unit area of a plane surface parallel to the interface is given by the magnitude of the z-component of $\langle \mathbf{S} \rangle$, which is $S_z = S\cos\theta$. Thus we have for the three waves

Incident $\qquad S_{iz} = \dfrac{cn_1}{8\pi\mu_1} |\mathbf{E}_i|^2 \cos\theta_i$

Reflected $\qquad S_{rz} = \dfrac{cn_1}{8\pi\mu_1} |\mathbf{E}_r|^2 \cos\theta_i$

Transmitted $\qquad S_{tz} = \dfrac{cn_2}{8\pi\mu_2} |\mathbf{E}_t|^2 \cos\theta_t$

The reflectivity R is defined by the ratio

$$R = \frac{S_{rz}}{S_{iz}} = \frac{|\mathbf{E}_r|^2}{|\mathbf{E}_i|^2} = |\rho|^2 \tag{11.109a}$$

and the transmissivity is defined by the ratio

$$T = \frac{S_{tz}}{S_{iz}} = \frac{|\mathbf{E}_t|^2 n_2 \cos\theta_t}{|\mathbf{E}_i|^2 n_1 \cos\theta_i} = |\tau|^2 \frac{n_2 \cos\theta_t}{n_1 \cos\theta_i} \tag{11.109b}$$

We can readily show that for both σ and π cases we have

$$T + R = 1 \tag{11.109c}$$

which must follow from conservation of energy, since the power per unit area of the interface in the incident beam equals the power in the reflected beam plus the power in the transmitted beam.

B. DISCUSSION OF THE DIELECTRIC CASE

When both media are dielectrics, the indices n_1 and n_2 are real. At normal incidence the distinction between π and σ cases is lost. We have $a = b^{-1} = n_2/n_1$ and hence

$$\rho_\sigma = \frac{n_1 - n_2}{n_1 + n_2} = -\rho_\pi \tag{11.110a}$$

$$\tau_\sigma = \frac{2n_1}{n_1 + n_2} = \tau_\pi \tag{11.110b}$$

The change in sign in Eq. (11.110*a*) is an artifact of the sign convention adopted for the π case (Fig. 11.21*b*).

For a simple numerical example, take $n_2/n_1 = 1.5$. Then

$$\rho_\sigma = -\rho_\pi = -0.2$$

and the reflectivity is

$$R = R_\sigma = R_\pi = \rho_\sigma^2 = 0.04$$

The transmission coefficient is

$$\tau_\sigma = \tau_\pi = 0.8$$

and the transmissivity is

$$T = T_\sigma = T_\pi = n_2\tau^2 = 0.96 = 1 - R$$

The next cases to be discussed involve a nonzero angle of incidence.

*1. **Oblique Incidence onto a Denser Medium.*** In this case medium 2 has greater optical density and hence $n_1 < n_2$. The parameter a given by

$$a = \frac{k_{tz}}{k_{iz}} = \frac{\cos\theta_t}{\cos\theta_i} = \frac{\sqrt{n_2^2 - n_1^2 \sin^2\theta_i}}{n_1 \cos\theta_i}$$

is real for all values of θ_i and positive because of our sign convention for the components of **k**. Furthermore, from Snell's law — $\sin\theta_t = (n_1/n_2)\sin\theta_i$ — we find that for all values of θ_i between zero and $\pi/2$ the inequality $\theta_t < \theta_i$ is obeyed. Thus

$$\rho_\sigma = -\frac{\sin(\theta_i - \theta_t)}{\sin(\theta_i + \theta_t)}$$

is always negative. At grazing incidence when $\theta_i \to \pi/2$ we have $a \to +\infty$ and hence $\rho_\sigma \to -1$. The resulting behavior of ρ_σ and $R_\sigma = \rho_\sigma{}^2$ as a function of θ_i is shown in Fig. 11.22 for the case $n_2/n_1 = 1.5$.

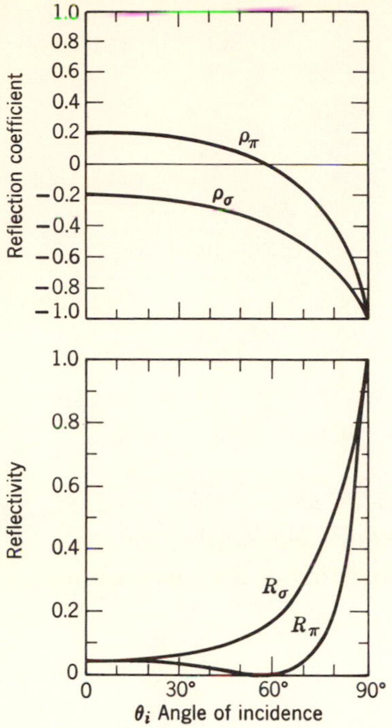

Fig. 11.22. Reflection coefficients and reflectivities for a dielectric interface with $n_2/n_1 = 1.5$.

For the π case, we must consider the parameter $b = (n_1/n_2)^2 a$ and a reflection coefficient given by

$$\rho_\pi = \frac{1-b}{1+b} = \frac{\tan(\theta_i - \theta_t)}{\tan(\theta_i + \theta_t)}$$

The limiting value at grazing incidence is $\rho_\pi(\pi/2) = -1$. At normal incidence ρ_π is positive:

$$\rho_\pi(0) = \frac{n_2 - n_1}{n_2 + n_1} \text{ [= 0.2 for our numerical example]}$$

As θ_i ranges from 0 to $\pi/2$, ρ_π decreases monotonically from 0.2 to -1. It equals zero when

$$\theta_i + \theta_t = \frac{\pi}{2} \tag{11.111a}$$

because then $\tan(\theta_i + \theta_t)$ becomes infinite. When θ_i satisfies Eq. (11.111a), it equals Brewster's angle θ_B (see Eq. 10.22). When Snell's law is used, Eq. (11.111a) gives for $\theta_i = \theta_B$

$$\tan\theta_B = \frac{n_2}{n_1} \tag{11.111b}$$

For our numerical example of $n_2/n_1 = 1.5$, $\theta_B = 56.3°$. The corresponding angle of refraction is

$$\theta_{r,B} = 33.7°$$

Whereas ρ_π is zero at Brewster's angle, the other component has a reflection coefficient

$$\rho_\sigma(\theta_B) = -\sin(\theta_B - \theta_{v,B}) = -\sin 22.6° = -0.384$$

and the reflectivity $R(\theta_B)$ is $0.147 \approx 15\%$.

2. *Oblique Incidence onto a Less Dense Medium — $\theta < \theta_c$*. In this case we have $n_1 > n_2$. Then by Snell's law — $\sin\theta_t = (n_1/n_2)\sin\theta_i$ — we have $\theta_t > \theta_i$ until $\theta_t = \pi/2$. This defines the *critical angle of incidence* θ_c:

$$\sin\theta_c = \frac{n_2}{n_1}$$

For $n_1/n_2 = 1.5$, $\theta_c = 41.8°$.

At angles of incidence greater than the critical angle, we have $\sin\theta_t > 1$; this means that the angle θ_t becomes imaginary. In this subsection we shall assume that the angle of incidence is less than the critical angle. Then θ_t is real, as is

$$a = \frac{k_{tz}}{k_{iz}} = \frac{\cos\theta_t}{\cos\theta_i} = \frac{\sqrt{n_2^2 - n_1^2\sin^2\theta_i}}{n_1\cos\theta_i}$$

At normal incidence we now have a positive value for

$$\rho_\sigma(0) = -\rho_\pi(0) = \frac{n_1 - n_2}{n_1 + n_2} \quad \left[= +0.2 \qquad \text{for} \quad \frac{n_1}{n_2} = 1.5\right]$$

At the critical angle we have $a = b = 0$, so that

$$\rho_\sigma(\theta_c) = \rho_\pi(\theta_c) = 1$$

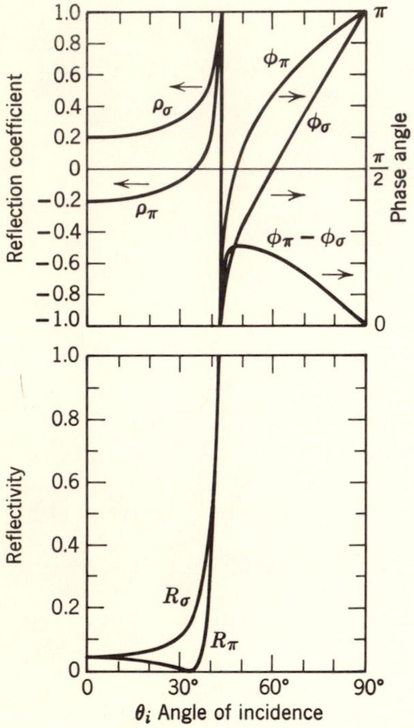

Fig. 11.23. Reflection coefficients, phase shifts, and reflectivities for a dielectric interface with $n_1/n_2 = 1.5$.

The coefficient ρ_π now increases monotonically from $-0{\cdot}2$ to $+1$ as θ_i increases from 0 to θ_c. Now ρ_π passes through zero at the new Brewster's angle given by

$$\tan\theta_B = \frac{n_2}{n_1}$$

For $n_1/n_2 = 1.5$, $\theta_B = 33.7°$. The resulting behavior of ρ_σ, ρ_π, R_σ, and R_π is shown in Fig. 11.23.

3. ***Total Internal Reflection.*** For light incident at an angle greater than the critical angle, the parameter a becomes purely imaginary, and we therefore introduce a new real parameter γ so that

$$a = \frac{k_{tz}}{k_{iz}} = -i\gamma \equiv \frac{-i\sqrt{n_1^2\sin^2\theta_i - n_2^2}}{n_1\cos\theta_i} \tag{11.112}$$

The choice of the minus sign in Eq. (11.112) is dictated by the need to have a decreasing exponential dependence in the z direction for the transmitted wave. The full complex transmitted field is then given by

$$E_t(y, z, t) = E_t e^{i(\omega t - \mathbf{k}_t \cdot \mathbf{r})} = E_t e^{i(\omega t - y k_{iy})} e^{-z\gamma k_{iz}} \tag{11.113a}$$

with $k_{iy} = n_1\omega\sin\theta_i/c$, $k_{iz} = n_1\omega\cos\theta_i/c = 2\pi n_1\cos\theta_i/\lambda_0$ and with λ_0 = vacuum wavelength. The intensity or flux density in the second medium is then proportional to

$$|E_t|^2 e^{-2\gamma k_{iz} z} = |E_t|^2 e^{-2z/\delta} \tag{11.113b}$$

with a "skin depth" given by

$$\delta = \frac{1}{k_{iz}\gamma} = \frac{\lambda_0\cos\theta_i}{2\pi\sqrt{n_1^2\sin^2\theta_i - n_2^2}} \tag{11.113c}$$

The necessity of having a real exponential z dependence in Eq. (11.113a) can be seen directly from the wave equation, which yields the condition

$$\frac{n_2{}^2\omega^2}{c^2} = |\mathbf{k}_t|^2 = k_{ty}^2 + k_{tz}^2 = k_{iy}^2 + k_{tz}^2 = \frac{n_1{}^2\omega^2 \sin^2\theta_i}{c^2} + k_{tz}^2 \qquad (11.113d)$$

The third equality made use of the phase-matching condition $k_{ty} = k_{iy}$. Thus, when $\theta_i > \theta_c$, Eq. (11.113d) is satisfied only by an imaginary value of k_{tz}. A wave such as (11.113a) that has a real exponential decay in one spatial direction is called an inhomogeneous plane wave.

The reflection coefficients become

$$\rho_\sigma = \frac{1+i\gamma}{1-i\gamma}, \qquad \rho_\pi = \frac{1+i(n_1/n_2)^2\gamma}{1-i(n_1/n_2)^2\gamma}$$

The reflectivities are unity:

$$R_\sigma = |\rho_\sigma|^2 = \left|\frac{1+i\gamma}{1-i\gamma}\right|^2 = \frac{1+\gamma^2}{1+\gamma^2} = 1, \qquad R_\pi = |\rho_\pi|^2 = 1$$

The light is thus seen to be totally reflected when incident on the interface from the more dense medium at an angle of incidence greater than the critical angle. This is called total internal reflection.

Since $R_\sigma = R_\pi = 1$, the transmissivities T_σ and T_π are zero, and there is no flow of energy in the second medium in a direction normal to the interface. This can also be shown by an explicit calculation of $\langle S_z \rangle$, the time average of the z component of the Poynting vector. We can show however, that Eq. (11.113a) gives expressions for **H** and for **S** such that there is a nonzero component of **S** in the y direction parallel to the interface. (See Problems.)

The coefficients ρ_π, ρ_σ are complex and impose a phase change upon the electric field, although the amplitude remains unchanged. It is convenient to introduce the angles ϕ_σ and ϕ_π shown in Fig. 11.24 by the equations

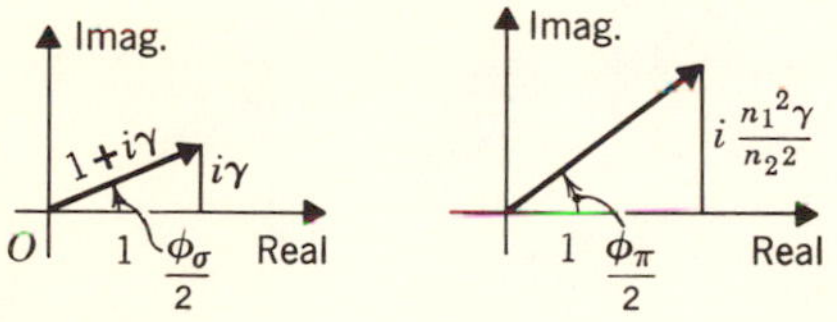

Fig. 11.24

$$\frac{\phi_\sigma}{2} = \tan^{-1}\gamma \qquad \frac{\phi_\pi}{2} = \tan^{-1}\left(\frac{n_1}{n_2}\right)^2\gamma \qquad (11.114a)$$

It is then possible to write $1+i\gamma = \sqrt{1+\gamma^2}e^{i\phi_\sigma/2}$, $1-i\gamma = \sqrt{1+\gamma^2}e^{-i\phi_\sigma/2}$, and so on, and obtain

$$\begin{aligned} \rho_\sigma &= e^{i\phi_\sigma} \\ \rho_\pi &= e^{i\phi_\pi} \end{aligned} \qquad (11.114b)$$

with

$$\tan\frac{\phi_\sigma}{2} = \gamma, \quad \tan\frac{\phi_\pi}{2} = \frac{n_1{}^2\gamma}{n_2{}^2}$$

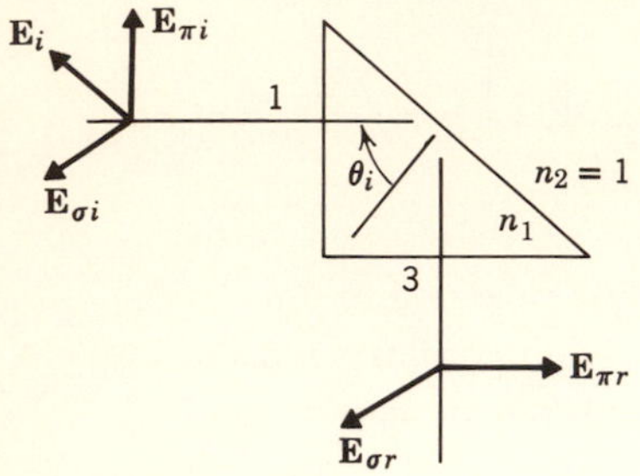

Fig. 11.25

The phase angles ϕ_σ, ϕ_π are plotted in the right half of the top of Fig. 11.23 for $\theta_c \leq \theta_i \leq \pi/2$. The existence of a phase difference $\phi = \phi_\pi - \phi_\sigma$ produces a change in the state of polarization of any light beam that is at least partially polarized and has both σ and π components. For example, consider the simple situation depicted in Fig. 11.25. Assume plane polarized incident light and resolve it into σ and π components. Then if reflection losses caused by surfaces 1 and 3 are neglected, the resulting reflected field is

$$\mathbf{E}_r = \mathbf{E}_{\pi r} + \mathbf{E}_{\sigma r}$$

with

$$\mathbf{E}_{\pi r} = e^{i\phi_\pi}\,\mathbf{E}_{\pi i} \qquad \mathbf{E}_{\sigma r} = e^{i\phi_\sigma}\,\mathbf{E}_{\sigma i}$$

The real, physical reflected electric fields may be written in the form of Eq. (10.11*a*):

$$E_{\pi r} = A_\pi \cos\,(\theta + \phi_\pi)$$

$$E_{\sigma r} = A_\sigma \cos\,(\theta + \phi_\sigma)$$

with $\theta = \omega t - kz$ and with z a coordinate along the final propagation direction.

For a numerical example, assume the refractive index of the prism to be 1.5 and let $\theta_i = 45°$. Then we find

$$\gamma = 0.333$$

$$\phi_\sigma = 36.8°, \qquad \phi_\pi = 73.8°$$

$$\phi = \phi_\sigma - \phi_\pi = -37°$$

Let us determine the resulting state of polarization of the reflected light if the incident light is plane polarized at 45° to the plane of incidence. Then the two amplitudes are equal:

$$A_\pi = A_\sigma = A$$

The reflected fields are

$$E_x = E_{\pi r} = A \cos\,(\theta + \phi_\pi)$$

$$E_y = E_{\sigma r} = A \cos\,(\theta - \phi + \phi_\pi)$$

These equations describe elliptically polarized light, which can be most conveniently characterized by the method of Eqs. (10.20):

$$\tan\phi = \pm\frac{\tan 2\eta}{\sin\beta} \tag{10.20a}$$

(+ for REP, − for LEP)

$$\cos 2\chi = \cos\beta\cos 2\eta \tag{10.20b}$$

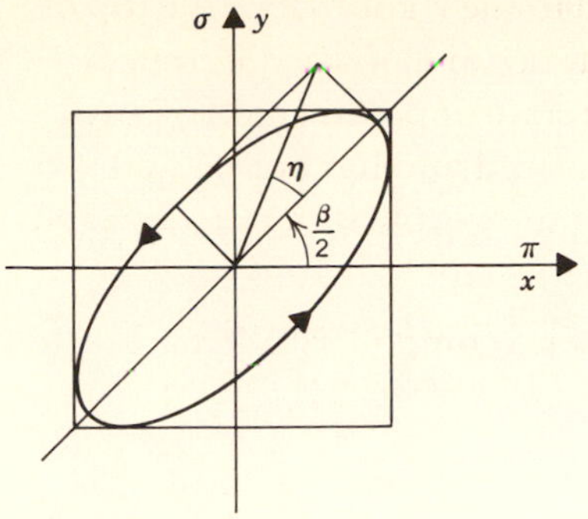

Fig. 11.26

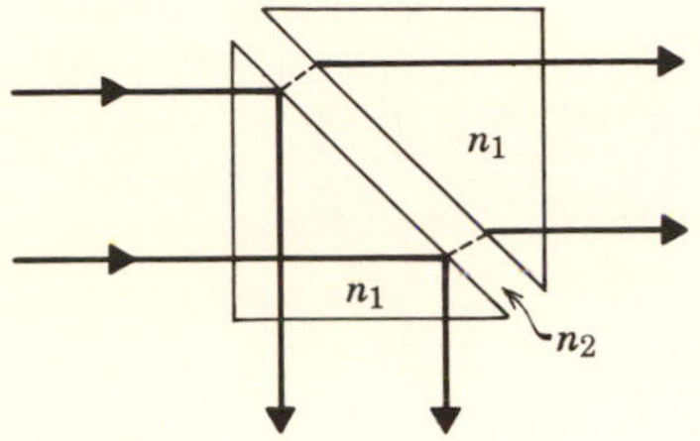

Fig. 11.27

Here $\tan \chi = A_y/A_x = 1$. Then $2\chi = \pi/2$. Equation (10.20*b*) then gives $\beta/2 = \pi/4$. Equation (10.20*a*) yields

$$\tan \phi = -\tan 2\eta$$

which means the light is LEP, and that $\eta = (-\phi)/2 = 18.5°$. Thus the axial ratio of the ellipse (Fig. 11.26) is

$$\tan \eta = \tan 18.5 = 0.335$$

*4. **Frustrated Total Internal Reflection.*** The exponential decay $E \sim e^{-z/\delta}$ of the fields in the second medium can be detected if a second interface is placed at a certain depth z in that medium. Often the third medium has the same refractive index as the first, as in the example shown in Fig. 11.27. If d is somewhat larger than the skin depth δ, we can neglect multiple reflection effects – the first multiple reflected beam will travel a distance $3d$ and hence will be attenuated by the factor $e^{-3d/\delta}$, which we assume is much less than $e^{-d/\delta}$.

Just inside the second interface the transmitted field for, say, the π component will be given by

$$E_{t\pi} = E_{i\pi}\tau_\pi\tau'_\pi e^{(-d/\delta)}e^{i(\omega t - yk_{iy})} \qquad (11.115a)$$

where $E_{i\pi}$ is the incident electric field just outside the first interface. There is a similar expression for the σ component. Our previous results for the τ's hold with the parameter a replaced by $i\gamma$. In particular, we have

$$\tau_\pi\tau'_\pi = 1 - \rho_\pi{}^2 = 1 - e^{2i\phi_\pi}$$

The coefficient for energy transmission across the gap is then given by

$$T_\pi = 2e^{-2d/\delta}(1 - \cos 2\phi_\pi) \qquad (d \gg \delta) \qquad (11.115b)$$

and, since there are no dissipative, loss-producing mechanisms, by the principle of energy conservation, the beam reflected by the $n_1 - n_2 - n_1$ "sandwich" would be described by a reflectivity

$$R_\pi = 1 - T_\pi$$

When d is not greater than a few times the skin depth, these results must be modified to take multiple reflections into account or, equivalently, to take the proper boundary conditions into account. (See Eqs. (11.123*a*, *b*) below.)

The phenomenon of frustrated total reflection is exactly analogous to the quantum-mechanical phenomenon of penetrating or tunneling of a plane wave through a one-dimensional rectangular barrier.

*5. **Attenuated Total Reflection (ATR).*** An interesting modification of the phenomenon of total internal reflection occurs when the second medium is somewhat "lossy," that is, when its index n_2 is not entirely a real quantity. The exponential penetration of light into the second medium is now accompanied by an irreversible loss of energy. The resulting reflectivity will be less than unity, and we say that the otherwise totally internally reflected beam has been attenuated.

Consider the second medium to have a complex refractive index

$$\tilde{n}_2 = n_{2r} - i n_{2i}$$

so that

$$\tilde{n}_2{}^2 = n_{2r}^2 - n_{2i}^2 - 2 i n_{2r} n_{2i}$$

The parameter γ is now complex and is given by

$$\tilde{\gamma} = \frac{\sqrt{n_1{}^2 \sin^2 \theta_i - n_{2r}^2 + n_{2i}^2 + 2 i n_{2r} n_{2i}}}{n_1 \cos \theta_i} \tag{11.116a}$$

The discussion is simplified by assuming two inequalities:

a. $n_{2i} \ll n_{2r}$

and

b. $n_1{}^2 \sin^2 \theta_i - n_{2r}^2 \gg 2 n_{2i} n_{2r}$

Then we write

$$\tilde{\gamma} = \gamma_r + i\gamma_i$$

with

$$\gamma_r \approx \frac{\sqrt{n_1{}^2 \sin^2 \theta_i - n_{2r}^2}}{n_1 \cos \theta_i} \tag{11.116b}$$

and

$$\gamma_i = \left[\frac{\gamma_r n_{2r}}{n_1{}^2 \sin^2 \theta_i - n_{2r}^2}\right] n_{2i} \tag{11.116c}$$

and find

$$\gamma_i \ll \gamma_r$$

This gives us for the reflection coefficient

$$\rho_\sigma = \frac{1 + i\tilde{\gamma}}{1 - i\tilde{\gamma}} = \frac{(1 - \gamma_i) + i\gamma_r}{(1 + \gamma_i) - i\gamma_r}$$

and for the reflectivity

$$R_\sigma = |\rho_\sigma|^2 = \frac{(1 - \gamma_i)^2 + \gamma_r{}^2}{(1 + \gamma_i)^2 + \gamma_r{}^2} \approx \frac{1 + \gamma_r{}^2 - 2\gamma_i}{1 + \gamma_r{}^2 + 2\gamma_i} \approx 1 - \frac{4\gamma_i}{1 + \gamma_r{}^2}$$

The reduction in reflectivity from the case of total internal reflection is

$$1 - R_\sigma = \frac{4\gamma_i}{1 + \gamma_r{}^2} \tag{11.117a}$$

The π case can be treated by replacing $\tilde{\gamma}$ by $(n_1/n_2)^2\tilde{\gamma}$ with the result

$$1 - R_\pi = \frac{4\gamma_i(n_1/n_2)^2}{1+\gamma_r^2(n_1/n_2)^4} \tag{11.117b}$$

Often the imaginary part n_{2i} owes its origin to the presence of a relatively dilute concentration of absorbing atoms or molecules in the second medium. In an ordinary transmission measurement, the absorption constant would be

$$K(\omega) = \frac{2\omega n_{2i}}{c} = \frac{4\pi n_{2i}}{\lambda_0}$$

For $n_{2i} \ll 1$, the effect of these atoms or molecules on n_{2r} would be very slight, so that for a fixed angle of incidence θ_i and a limited range of wavelength, we could treat γ_r and the coefficient of n_{2i} in Eq. (11.116*c*) as constants. Hence Eqs. (11.116*b*, *c*) and (11.117*a*, *b*) may be used to give n_{2i} from measurements of $(1-R)$.

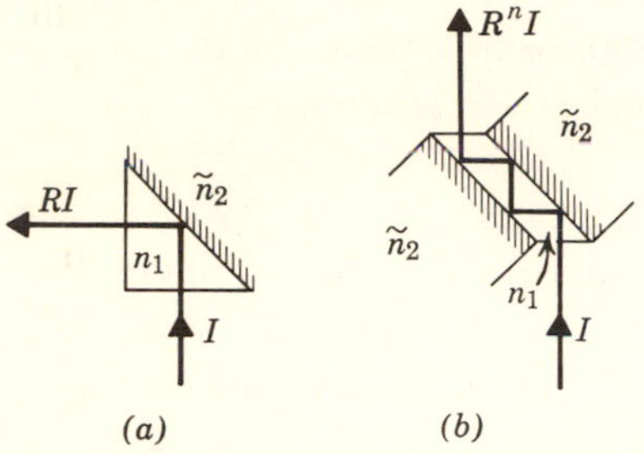

Fig. 11.28. (*a*) Single reflection. (*b*) Multiple reflection.

Experiments using the attenuated total reflection technique are often performed as indicated in Fig. 11.28. The medium with index n_1 is used as a tool to study the medium with index $\tilde{n}_2$. The former medium must be highly transparent in the desired wavelength range and of high optical density. The interface between the two media should be of good optical quality, and the two surfaces should be in good optical contact if one wants the expressions derived here to apply quantitatively; otherwise each interface is actually a complicated n_2-air-n_1 sandwich. The multiple reflection technique of Fig. 11.28*b* is used to amplify the effects of a small reduction in reflectivity. If $(1-R) \ll 1$, then after n reflections, we have

$$R^n = [1+(R-1)]^n \approx 1 + n(R-1)$$

and the reduction from the case of total internal reflection is

$$1 - R^n \cong n(1-R)$$

giving an n-fold amplification.

As an example, consider the case where $n_1 = 2.2$ and $n_{2r} = 1.5$. Then $\theta_c = 43.1°$. Let the angle of incidence be $\theta_i = 50°$. We then find the following:

$$\gamma_r = 0.54, \qquad \gamma_i = 1.40 n_{2i}$$

$$1 - R_\sigma = 4.3 n_{2i}$$

$$1 - R_\pi = 5.2 n_{2i}$$

If n_{2i} is 10^{-2}, then

$$1 - R_\sigma = 0.043$$

$$1 - R_\pi = 0.052$$

In a transmission measurement the absorption constant would be

$$K = \frac{4\pi n_{2i}}{\lambda_0} = \frac{0.125}{\lambda_0} = 1.25 \times 10^3\ \text{cm}^{-1} \qquad \text{for} \quad \lambda_0 = 1\ \mu\text{m} = 10^{-4}\ \text{cm}$$

To measure this in absorption we need a thin sample. Suppose that we can measure an attenuation by a factor of 100. Then we set $e^{-K\Delta z} = 10^{-2}$ and find that the resulting value of Δz would be $3.7 \cdot 10^{-3}$ cm $= 37\ \mu$m. In some cases it may be impractical to prepare samples this thin. The ATR technique would yield the same information with thick samples provided that surface preparation was of sufficiently high quality, and even if it were not, the ATR technique could be used to locate conveniently the positions of maxima in n_{2i} as a function of wavelength.

The ATR technique has been used in practice mainly in the infrared spectral region with relatively strongly absorbing samples. In the longer-wavelength infrared regions the requirements on surface quality are not as stringent as they are in the visible.

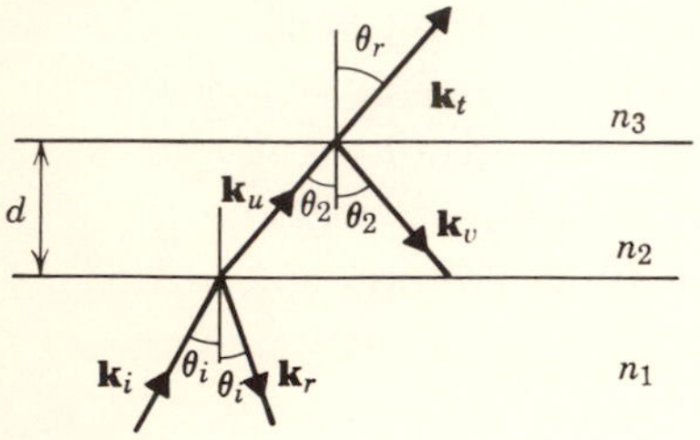

Fig. 11.29

C. PLANE PARALLEL DIELECTRIC LAYER

Consider a plane electromagnetic wave incident on the plane parallel dielectric slab shown in Fig. 11.29. We wish to determine the resulting reflected and transmitted waves. This type of problem was treated by the multiple reflection technique in Section 5.6. We now wish to treat it using the electromagnetic theory of light and the proper boundary conditions. The results are the same with both methods.

1. ***Basic Equations.*** We follow as closely as possible the notation and geometry of Fig. 11.21. The n_1-n_2 interface is located in the plane $z = 0$, the n_2-n_3 interface in the plane $z = d$. As before, let $\mathbf{k}_i$, $\mathbf{k}_t$, and $\mathbf{k}_r$ denote the incident, transmitted, and refracted wave vectors, all lying in the yz plane. The phase-matching condition

$$k_{iy} = k_{ty} = k_{ry}$$

still holds. In the second medium there will be both an oblique upward moving wave E_u, with wave vector

$$\mathbf{k}_u = (0, k_{iy}, k_{uz}) \tag{11.118a}$$

and an oblique downward moving wave E_v with wave vector

$$\mathbf{k}_v = (0, k_{iy}, -k_{uz}) \tag{11.118b}$$

These two waves can be thought of as arising from multiple internal reflections interfering with each other and with an original transmitted wave. The amplitudes E_u and E_v will be determined from boundary conditions. Since the magnitude of $\mathbf{k}_u$ is $n_2\omega/c$, and since its y component is k_{iy}, the z component is given by

$$k_{uz} = \frac{n_2\omega \cos\theta_2}{c} = \left[\left(\frac{n_2{}^2\omega^2}{c^2}\right) - k_{iy}^2\right]^{1/2} \tag{11.118c}$$

This can be written in the form

$$k_{uz} = a_{12}k_{iz}, \qquad a_{12} = \frac{\sqrt{n_2^2 - n_1^2 \sin^2 \theta_i}}{n_1 \cos \theta_i} \tag{11.118d}$$

To keep the notation consistent, it is useful to write a similar expression for the second interface,

$$k_{tz} = a_{23}k_{uz}, \qquad a_{23} = \frac{\sqrt{n_3^2 - n_2^2 \sin^2 \theta_2}}{n_2 \cos \theta_2} \tag{11.118e}$$

Once again we separate the electric fields into components parallel and perpendicular to the plane of incidence, and discuss two cases.

2. *σ Case—E Perpendicular to the Plane of Incidence.* The boundary conditions require the continuity of E_x and of the product $k_z E_x$. Denoting E_{ix} by E_i, and so on, we find, for the boundary conditions at $z = 0$,

$$E_i + E_r = E_u + E_v \tag{11.119a}$$

$$E_i - E_r = a_{12}(E_u - E_v) \tag{11.119b}$$

These are simply the generalizations of Eqs. (11.99) for the case of a nonzero E_v.

We have a choice of phase for the transmitted wave E_t. We choose to write the wave in medium 3 in the form

$$E_t(y, z, t) = E_t \exp \{i[\omega t - yk_{iy} - k_{tz}(z-d)]\}$$

The two waves in the second medium have the form

$$E_u \exp \{i[\omega t - yk_{iy} - zk_{2z}]\} \qquad \text{and} \qquad E_v \exp \{i[\omega t - yk_{iy} + zk_{2z}]\}$$

Then the boundary conditions at $z = d$ are

$$E_u e^{-i\phi_2} + E_v e^{i\phi_2} = E_t \tag{11.119c}$$

$$E_u e^{-i\phi_2} - E_v e^{i\phi_2} = a_{23}E_t \tag{11.119d}$$

where $\phi_2 = dk_{tz}$ represents the phase shift resulting from propagation through the dielectric layer.*

One must solve Eqs. (11.119*c*, *d*) for E_u and E_v in terms of E_t, insert the results in Eqs. (11.119*a*, *b*), and solve the resulting equations for $\rho = E_r/E_i$ and $\tau = E_t/E_i$. The result for ρ is

$$\rho = \frac{\rho_{12} + \rho_{23}e^{-i\Delta}}{1 + \rho_{12}\rho_{23}e^{-i\Delta}} \tag{11.120a}$$

with

$$\Delta = 2\phi_2 = 2dk_{2z} = \frac{n_2\omega d \cos \theta_2}{c} = \frac{4\pi n_2 d \cos \theta_2}{\lambda_0}\ ^* \tag{11.120b}$$

*This will be well defined only if the light has a coherence length greater than $2n_2 d \cos \theta_2$. For light with a much smaller coherence length, the transmission and reflection phenomena at the two surfaces are independent. (See Problems).

and with

$$\rho_{12} = \frac{1 - a_{12}}{1 + a_{12}} \tag{11.120c}$$

which is the expression for the reflection coefficient for a single n_1-n_2 interface. In addition we have

$$\rho_{23} = \frac{1 - a_{23}}{1 + a_{23}} \tag{11.120d}$$

which is the expression for ρ_σ for a single n_2-n_3 interface.

For τ the result is

$$\tau = \frac{\tau_{12}\tau_{23}e^{-i\Delta/2}}{1 + \rho_{12}\rho_{23}e^{-i\Delta}} \tag{11.120e}$$

with

$$\tau_{12} = \frac{2}{1 + a_{12}}, \qquad \tau_{23} = \frac{2}{1 + a_{23}} \tag{11.120f, g}$$

and τ's for the individual n_1-n_2 and n_2-n_3 interfaces.

To compare with earlier results, note that Eq. (11.120*a*) is identical with Eq. (5.72) and that Eq. (11.120*e*) agrees with the multiple reflection of Eq. (5.74) when $n_3 = n_1$.

3. ***π Case—E Parallel to the Plane of Incidence.*** For this geometry we may write as the generalization of Eqs. (11.105) the following boundary conditions at $z = 0$:

$$n_1E_i + n_1E_r = n_2E_u + n_2E_v \tag{11.121a}$$

$$n_1E_i - n_1E_r = b_{12}(n_2E_u - n_2E_v) \tag{11.121b}$$

with

$$b_{12} = \left(\frac{n_1}{n_2}\right)^2 a_{12}$$

At $z = d$ the boundary conditions are

$$n_2E_ue^{-i\phi_2} + n_2E_ve^{i\phi_2} = n_3E_t \tag{11.121c}$$

$$n_2E_ue^{-i\phi_2} + n_2E_ve^{i\phi_2} = b_{23}n_3E_t \tag{11.121d}$$

with

$$b_{23} = \left(\frac{n_2}{n_3}\right)^2 a_{23}$$

These equations for the quantities n_1E_i, and so on, have the same form as Eqs. (11.119) for the fields alone, E_i, and so on. The result for $\rho = E_r/E_i$ is the same as Eq. (11.120*a*) but with new expressions for ρ_{12} and ρ_{23} appropriate for the π case:

$$\rho_{12} = \frac{1 - b_{12}}{1 + b_{12}}, \qquad \rho_{23} = \frac{1 - b_{23}}{1 + b_{23}} \tag{11.122a, b}$$

For $\tau = E_t/E_i$, Eq. (11.120*e*) still holds, but with the π-case results for τ_{12} and τ_{23}:

$$\tau_{12} = \frac{2(n_1/n_2)}{1+b_{12}}, \qquad \tau_{23} = \frac{2(n_2/n_3)}{1+b_{23}} \tag{11.122c, d}$$

4. Generalizations. The formal expressions given above are valid for a dielectric slab when k_{uz} or k_{tz}, or both are imaginary. The former situation occurs when $n_1 > n_3$ and $\sin\theta_i > n_3/n_1$; the latter case occurs when $n_1 > n_3$ and $\sin\theta_1 > n_3/n_1$. Then we need only write $a_{12} = -i\gamma_{12}$, and so on, wherever appropriate, and use the expressions already derived.

For example, consider the case of frustrated total reflection at the n_1-n_2 interface. Thus assume that $\sin\theta_i > n_2/n_1$, but that $\sin\theta_i < n_3/n_1$. Then a_{12}, a_{23}, and k_{uz} are imaginary, but k_{tz} is real. We write $a_{12} = -i\gamma_{12}$ and find that $e^{-i\Delta/2} = e^{-d/\delta}$, where δ is the skin depth of Eq. (11.113*b*). Equation (11.120*e*) then gives for the overall transmission coefficient the exact result

$$\tau_0 = \frac{\tau_{12}\tau_{23}e^{-d/\delta}}{1+\rho_{12}\rho_{23}e^{-2d/\delta}} \tag{11.123a}$$

When $n_3 = n_1$, we have $\tau_{12} \equiv \tau$, $\tau_{21} = \tau'$, $\rho_{12} = \rho_{\pi,\sigma}e^{i\phi_{\pi,\sigma}}$, $\rho_{21} = -\rho_{12}$.

Then

$$\tau_0 = \frac{\tau\tau' e^{-d/\delta}}{1+\rho\rho' e^{-2d/\delta}} = \frac{(1-\rho^2)e^{-d/\delta}}{1-\rho^2 e^{-2d/\delta}} = \frac{(1-e^{2i\phi_{\pi,\sigma}})e^{-d/\delta}}{1-e^{2i\phi_{\pi,\sigma}}e^{-2d/\delta}} \tag{11.123b}$$

Here $\phi_{\pi,\sigma}$ represent the appropriate phase ϕ_π or ϕ_σ of ρ in the π and σ cases. Equation (11.123*b*) is the exact version of Eq. (11.115*a*).

The results for the overall transmission and reflection coefficients also hold when either n_2 or n_3, or both, are complex. The quantities a_{12}, b_{12}, a_{23}, and b_{23} are then complex, as are ρ and τ, but the formal expressions remain unchanged.

D. NONTRANSPARENT MEDIA

This section discusses some of the optical properties of nontransparent media. Such materials have a large value of n_i, the imaginary part of the complex index of refraction. Light incident at all angles is then strongly attenuated. It is useful to discuss two limiting cases that have sizable values for n_i, although it is true that most real systems represent an intermediate case.

In one limit the dielectric constant is real, but negative. We saw examples of this in Figs. 11.8 and 11.14–11.18. The phenomenon is very similar to total internal reflection. The index of refraction $\tilde{n}_2$ is purely imaginary, as are the parameters a and b. The reflectance R_σ or R_π is unity. The transmissions coefficients τ_σ and τ_π

have magnitudes of order unity. The fields **E** and **H** just inside the surface are relatively large. They do not propagate as a wave, however. The components of **E** and **H** that contribute to the normal or z component of the Poynting vector **S** are $\pi/2$ radians out of phase, so that the time average of S_z then behaves like $\langle \cos \omega t \sin \omega t \rangle = 0$.

In the other limiting case the dielectric constant is complex with positive real part, but with a sizable imaginary part. The refractive index $\tilde{n}_2$ is large in absolute value compared with n_1; $|a|$ is large compared with unity and $|b|$ is small. Then R_σ and R_π are close to unity. The transmissions coefficients t_σ and t_π are both small in absolute value because of the large mismatch $|\tilde{n}_2| \gg n_1$, and the fields **E**, **H** just inside the interfaces are quite small. The power lost to dissipative absorption processes in the medium is then small, not because the processes are weak, but because the fields are weak.

We shall now work out an example of the second limiting case and then discuss the more general situation.

1. ***Low-Frequency Behavior of a Metal.*** In metals the electrons can be classified as bound or free. The former behave quite like those in a dielectric and contribute a term, which we shall call ϵ_0, to the dielectric constant. For frequencies far below that of any resonances resulting from these bound electrons, which usually occur in the mid to far ultraviolet and X-ray regions, ϵ_0 will be a real positive constant. The free electrons form an electron gas having a plasma frequency

$$\omega_p{}^2 = \frac{4\pi N e^2}{m}$$

and a damping or collision time τ, which we shall assume is a constant, independent of frequency. The extreme low frequency (or dc) electrical conductivity σ is related to τ by the equation

$$4\pi\sigma = \omega_p{}^2\tau \tag{11.124}$$

A slight generalization of Eq. (11.64*b*) gives for the complex dielectric constant as a function of frequency

$$\epsilon_2 = \epsilon_0 + \frac{4\pi\sigma}{i\omega(1+i\omega\sigma)} = \epsilon_0 - \frac{\omega_p{}^2\tau}{\omega^2\tau - i\omega} \tag{11.125}$$

We assume that $\omega_p\tau \gg 1$ so that, for frequencies ω of order of the plasma frequency, $\omega\tau \gg 1$. Then we have

$$\epsilon_2 \approx \epsilon_0 - \frac{\omega_p{}^2}{\omega^2} \qquad \text{(high frequency, } \omega\tau \gg 1\text{)}$$

$$\epsilon_2 \approx \epsilon_0 - \frac{4\pi\sigma i}{\omega} = \epsilon_0 - \frac{\omega_p{}^2\tau i}{\omega} \qquad \text{(low frequency, } \omega\tau \ll 1\text{)}$$

The second case is the one we wish to consider here. In the example used for Fig. 11.8, we assumed $1/\tau = 3 \cdot 10^{13}$ rad/sec. Then the requirement of low frequency would be $\omega \ll 3 \cdot 10^{13}$ rad/sec. The requirement on the vacuum wavelength would be $\lambda = 2\pi c/\omega \gg$ 63 μm = 0.063 mm. This will put the radiation into the far, far infrared or microwave region of the spectrum. The imaginary part of ϵ_2 dominates, since

$$\frac{\omega_p^2\tau}{\omega} = \frac{(\omega_p\tau)^2}{\omega\tau} \ggg 1$$

Then we have approximately

$$\tilde{n}_2^2 = \epsilon_2 = -\frac{\omega_p^2\tau i}{\omega} = \frac{\omega_p^2\tau e^{-\pi i/2}}{\omega}$$

This gives $\tilde{n}_2 = n_{2r} - in_{2i} = |\tilde{n}_2|e^{-i\alpha}$, with $|\tilde{n}_2| = \omega_p\sqrt{\tau/\omega}$ and $\alpha = \pi/4$. Thus

$$n_{2r} = n_{2i} = \frac{|\tilde{n}_2|}{\sqrt{2}} = \omega_p\left(\frac{\tau}{2\omega}\right)^{1/2} \gg 1 \tag{11.126}$$

Note that the skin depth $\delta = 2\pi\lambda/n_2$ is very much less than λ.

The parameter a is

$$a = \frac{\sqrt{\tilde{n}_2^2 - n_1^2\sin^2\theta_i}}{n_1\cos\theta_i} \approx \frac{\tilde{n}_2}{n_1\cos\theta_i} \qquad \text{with} \quad |a| \gg 1$$

and the parameter b is

$$b = \frac{n_1^2}{\tilde{n}_2^2}a \approx \frac{n_1}{\tilde{n}_2\cos\theta_i} \qquad \text{with} \quad |b| \ll 1$$

unless θ_i is very close to $\pi/2$. The reflection coefficients are

$$\rho_\sigma = \frac{1-a}{1+a} = \frac{-(1-1/a)}{(1+1/a)} \approx -(1-2/a) = -\left(1-\frac{2n_1\cos\theta_i}{\tilde{n}_2}\right)$$

$$= -\left[1-\frac{n_1\cos\theta_i(1+i)}{n_{2_i}}\right] \tag{11.127a}$$

which is very close to -1, and

$$\rho_\pi = \frac{1-b}{1+b} \approx 1-2b = 1-\frac{2n_1}{\tilde{n}_2\cos\theta_i} = 1-\frac{n_1(1+i)}{n_{2i}\cos\theta} \tag{11.127b}$$

which is very close to $+1$. The expressions for the reflectivity are

$$R_\sigma \approx 1-\frac{2n_1\cos\theta_i}{n_{2i}} \qquad \text{(good for all } \theta_i\text{)} \tag{11.127c}$$

$$R_\pi \approx 1-\frac{2n_1}{n_{2i}\cos\theta_i} \qquad \left(\text{good for all } \theta_i \text{ except close to } \frac{\pi}{2}\right) \tag{11.127d}$$

At normal incidence,

$$R_\sigma = R_\pi = R = 1-\frac{2n_1}{n_{2i}}.$$

Hence

$$\frac{n_{2i}}{n_1} = \frac{2}{1-R} \tag{11.127e}$$

Again we stress that this is a low-frequency result.

Note that, except for normal incidence, R_π is less than R_σ. Actually R_π has a minimum for θ_i near $\pi/2$, where $n_{2i}\cos\theta_i$ is of order unity. Then the "small-b" approximation is no good, and more exact expressions must be used. This will be done below.

In regions of strong and rapidly varying optical absorption, that is, large and variable n_{2i}, the reflectivity R, say at normal incidence, also tends to be large (close to 1) and rapidly varying. Often peaks in R are very close to the corresponding peaks in n_{2i}. When ϵ_{2r} is negative the situation is more complicated, as the LiF case shows (Figs. 11.15 and 11.18), for then R can be large over a wide range, whereas the peak in n_{2i} can be narrow. The latter peak does coincide, however, with the initial sharp rise in R at low frequencies. Thus, if one is careful, the shape of the reflectivity curves may be used for information about n_{2i}. This can be done quantitatively using the method of *dispersion relations*,* provided that $R = |\rho|^2$ is known accurately over a wide frequency range. The phase ϕ of the reflection coefficient $\rho = \sqrt{R}e^{i\phi}$ can then be determined from the equation

$$\phi(\omega) = \frac{\omega}{\pi}\int_0^\infty \frac{lnR(\omega')\,d\omega'}{\omega^2-\omega'^2}$$

and the optical constants n_{2r}, n_{ri} may be obtained by inverting the equation

$$\rho = \sqrt{R}\,e^{i\phi} = \frac{n_1 - n_{2r} + in_{2i}}{n_1 + n_{2r} - in_{2i}}$$

2. Additional Methods in Reflection Measurements. In the example just discussed, all the useful information is contained in the small difference between R and unity, and this is not easy to measure with accuracy. In addition we often want to exploit the information contained in the phases of the reflection coefficients without using the complicated method of dispersion relations. Consider in detail the geometry of a reflection experiment shown in Fig. 11.30. The incident and reflected fields $\mathbf{E}_i$ and $\mathbf{E}_r$ are resolved into their π and σ components in a manner consistent with the sign convention of Fig. 11.21.

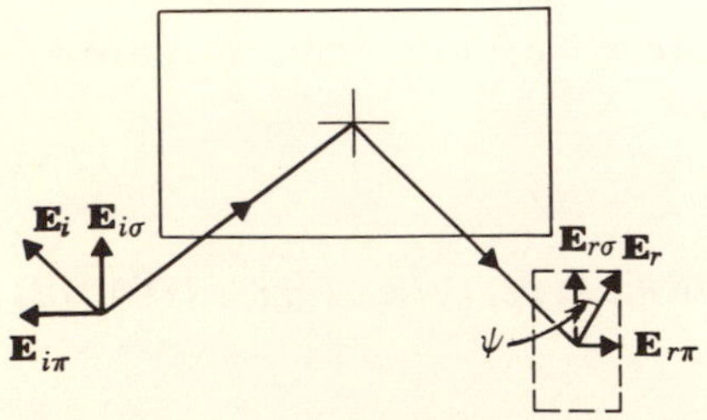

Fig. 11.30

Let the incident light be plane polarized at 45° with respect to the plane of incidence, that is, let $E_{i\sigma} = E_{i\pi}$. The ratio of

*See, for example, J. S. Toll, *Phys. Rev.*, **104**, 1760 (1956).

reflected electric field components is then simply

$$\frac{E_{r\pi}}{E_{r\sigma}} = \frac{\rho_\pi E_{i\pi}}{\rho_\sigma E_{i\sigma}} = \frac{\rho_\pi}{\rho_\sigma}$$

For a dielectric at less than the critical angle the ratio ρ_π/ρ_σ is real, and we may define a real angle, called the *azimuth angle*, by

$$\tan\psi = \frac{E_{r\pi}}{E_{r\sigma}} = \frac{\rho_\pi}{\rho_\sigma} \tag{11.128a}$$

When ρ_π and ρ_σ are complex, we define ψ by the equation

$$\tan\psi = \left|\frac{E_{r\pi}}{E_{r\sigma}}\right| = \left|\frac{\rho_\pi}{\rho_\sigma}\right| \tag{11.128b}$$

The reflected light is then elliptically polarized, the ellipse being inscribed within the rectangle shown dotted in Fig. 11.30.

For a dielectric we have sketched the behavior of the azimuth ψ as a function of θ_i in Fig. 11.31. Note that ψ is zero at Brewster's angle.

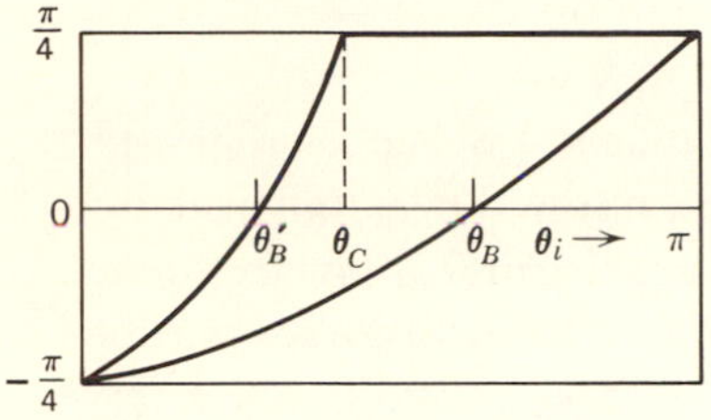

Fig. 11.31. Azimuth angle versus angle of incidence for $n_1 > n_2$ (left hand curve) and for $n_1 < n_2$ (right hand curve).

For nontransparent media with complex values for n_2 and ρ_π/ρ_σ, the azimuth ψ is never zero. But it does take a minimum value near $\bar{\psi}$, called the "principal azimuth." The precise definition of $\bar{\psi}$ is that it corresponds to a 90° phase shift between ρ_π and ρ_σ. This means that

$$\frac{\rho_\pi}{\rho_\sigma} = i\tan\bar{\psi} \tag{11.129}$$

The relative phase is as written, that is, $E_{r\pi}$ is $\pi/2$ radians ahead of $E_{r\sigma}$. This makes the light LEP. The principal azimuth is found at a value $\bar{\theta}_i$ called the principal angle of incidence.

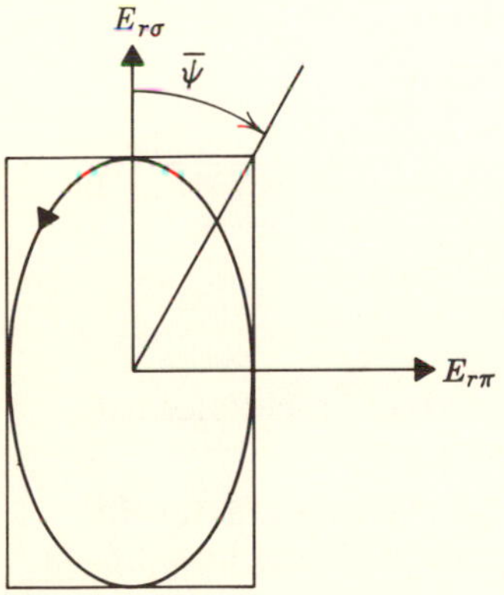

Fig. 11.32

At the principal angle the axes of the ellipse are horizontal and vertical, as shown in Fig. 11.32; the helicity is as shown. Hence a quarter-wave plate placed with its fast axis in the σ direction will yield plane polarized light with electric field vibrating at an angle ψ with respect to the σ axis. This can be completely extinguished by a polarizer set with its pass direction at $\bar{\psi} \pm \pi/2$. At any other angle of incidence $\theta_i \neq \bar{\theta}_i$ the quarter-wave plate with fast axis along the σ axis will give elliptically polarized light that cannot be extinguished by the polarizer. Thus, $\bar{\psi}$ and $\bar{\theta}_i$ may be determined.

3. Determination of n_r and n_i from $\bar{\psi}$ and $\bar{\theta}_i$. The experimental values for $\bar{\psi}$ and $\bar{\theta}_i$ may be used to determine n_r and n_i. Consider first a nontransparent second medium for which n_{2r} and n_{2i} are much larger than n_1. The following approximations are then valid:

$$a \approx \frac{\tilde{n}_2}{n_1\cos\theta_i}, \qquad b \approx \frac{n_1}{\tilde{n}_2\cos\theta_i}$$

$$\rho_\sigma = \frac{n_1\cos\theta_i - \tilde{n}_2}{n_1\cos\theta_i + \tilde{n}_2}, \qquad \rho_\pi = \frac{\tilde{n}_2\cos\theta_i - n_1}{\tilde{n}_2\cos\theta_i + n_1}$$

With these assumptions we can readily show that $R_\pi = |\rho_\pi|^2$ is a minimum when θ_i satisfies the equation

$$n_1 = |\tilde{n}_2| \cos \theta_i \tag{11.130}$$

Note that $\cos \theta_i \ll 1$. If we write

$$\tilde{n}_2 = n_{2r} - in_{2i} = |\tilde{n}_2| e^{-i\alpha}$$

we find that for this value of θ_i that makes R_π a minimum

$$\rho_\pi = \frac{e^{-i\alpha} - 1}{e^{-i\alpha} + 1} = -\frac{e^{i\alpha/2} - e^{-i\alpha/2}}{e^{i\alpha/2} + e^{i\alpha/2}} = -i \tan \alpha/2$$

$$\rho_\sigma = \frac{\cos^2\theta_i - e^{-i\alpha}}{\cos^2\theta_i + e^{-i\alpha}} \approx -1$$

Thus, $\rho_\pi/\rho_\sigma = i \tan \alpha/2$, which means that $\psi = \bar{\psi}$. We see that equation (11.130) not only gives the minimum for R_π, but also the principal angle $\bar{\theta}_i$, for this limiting case. We thus can write

$$\frac{\alpha}{2} = \bar{\psi} \tag{11.131a}$$

For the complex index of refraction we obtain from Eqs. (11.130) and (11.131*a*)

$$\tilde{n}_2 = |\tilde{n}_2| e^{-i\alpha} \cong \frac{n_1 e^{-2i\bar{\psi}}}{\cos \bar{\theta}_i} \tag{11.131b}$$

This is the equivalent of the two real equations

$$\frac{n_{2i}}{n_{2r}} = \tan 2\bar{\psi} \tag{11.131c}$$

$$n_{2r} = \frac{n_1 \cos 2\bar{\psi}}{\cos \bar{\theta}_i} \tag{11.131d}$$

These equations hold for large principal angles of incidence for which $\cos \theta_i$ is small.

4. Exact Treatment. Exact versions of Eqs. (11.131) will now be derived. The derivation is not hindered if we also consider general values of the angle of incidence. The ratio

$$\eta = \frac{\rho_\pi}{\rho_\sigma}$$

then will have a general phase angle. We write it in the form

$$\eta = \frac{\rho_\pi}{\rho_\sigma} = \tan \psi \, e^{i\phi} \tag{11.132}$$

Using the expressions for ρ_π and ρ_σ, we find

$$\eta = \left(\frac{1-b}{1+b}\right)\left(\frac{1+a}{1-a}\right)$$

Note that

$$\frac{1+\eta}{1-\eta}=\frac{1-ab}{b-a}=\frac{\tan^2\theta_i}{a}=\frac{n_1\tan\theta_i\sin\theta_i}{\sqrt{\tilde{n}_2{}^2-n_1{}^2\sin^2\theta_i}}$$

Defining

$$Q=\left(\frac{1+\eta}{1-\eta}\right)^2$$

we find that $\tilde{n}_2$ is given by the expression

$$\tilde{n}_2=n_1\tan\theta_i\left[1-\frac{(Q-1)}{Q}\sin^2\theta_i\right]^{1/2}=n_1\tan\theta_i\left[1-\frac{4\eta}{(1+\eta)^2}\sin^2\theta_i\right]^{1/2} \tag{11.133a}$$

$$=n_1\tan\theta_i\left[1-\frac{4\tan\psi e^{i\phi}\sin^2\theta_i}{(1+\tan\psi e^{i\phi})^2}\right]^{1/2}$$

Since $\cos\theta_i$ is often a small quantity, it is useful to write this result in the form

$$\tilde{n}_2=\frac{n_1}{\cos\theta_i}\left\{(1-\cos^2\theta_i)\left[\frac{(1-e^{i\phi}\tan\psi)^2+4e^{i\phi}\tan\psi\cos^2\theta_i}{(1+e^{i\phi}\tan\psi)^2}\right]\right\}^{1/2} \tag{11.133b}$$

Equations (11.133*a*, *b*) may be separated into real and imaginary parts and used to give n_{2r} and n_{2i} from experimental values of ϕ and ψ at a given value of the angle of incidence θ_i. The angles ϕ and ψ are best determined in an instrument called an ellipsometer using a compensator and a polarizer. The compensator is oriented so that its axis coincides with the plane of incidence (π direction). It then introduces an extra variable phase shift Δ between the reflected π and σ components. The compensator is adjusted so that $\Delta=-\phi$, at which point plane polarized light is obtained vibrating at the angle ψ with respect to the σ axis. The polarizer can then be used to determine ψ.

Instead of a compensator, we can use a rotatable quarter-wave plate followed by a polarizer to determine the axial ratio of the ellipse and its orientation. These are determined from the parameters $\beta/2$ and η in Figs. 10.7 and 10.20.

At the principal angle $\bar{\theta}_i$ where $\phi=\pi/2$ we need only a fixed quarter-wave plate (with fast axis in the σ direction) and a polarizer. Equations (11.133*a*, *b*) give

$$\tilde{n}_2=n_1\tan\bar{\theta}_i\left[1-\frac{4i\tan\bar{\psi}\sin^2\bar{\theta}_i}{(1+i\tan\bar{\psi})^2}\right]^{1/2} \tag{11.134a}$$

$$=\frac{n_1}{\cos\theta_i}\left\{(1-\cos^2\bar{\theta}_i)\left[\frac{(1-i\tan\bar{\psi})^2+4i\tan\bar{\psi}\cos^2\bar{\theta}_i}{(1+i\tan\bar{\psi})^2}\right]\right\}^{1/2} \tag{11.134b}$$

Equation (11.134*b*) gives the earlier approximate result, Eq. (11.131*b*) when $|\tilde{n}_2|$ is large, for then $\cos\theta_i$ is small and can be neglected in the term in curly brackets. We then need only to note that

$$\frac{1-i\tan\bar{\psi}}{1+i\tan\bar{\psi}}=\frac{\cos\bar{\psi}-i\sin\bar{\psi}}{\cos\bar{\psi}+i\sin\bar{\psi}}=e^{-2i\bar{\psi}}$$

to obtain the desired agreement.

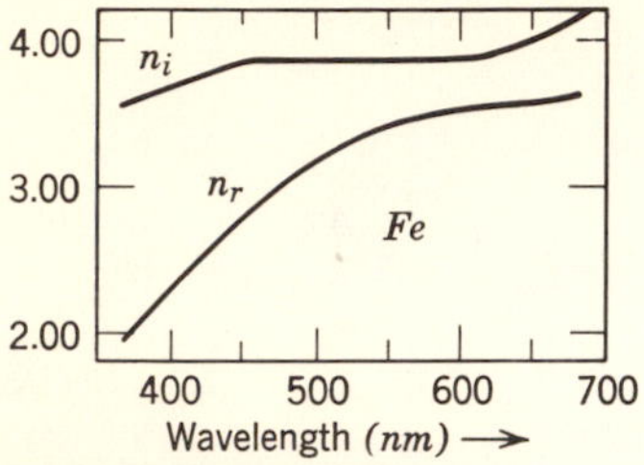

Fig. 11.33. Real and imaginary parts of the complex refractive index of iron. (After Yolken and Kruger.)

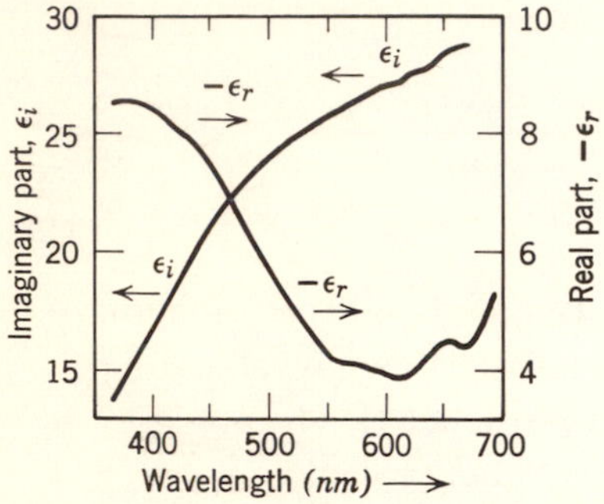

Fig. 11.34. Real and imaginary parts of the complex dielectric constant of iron. (After Yolken and Kruger.)

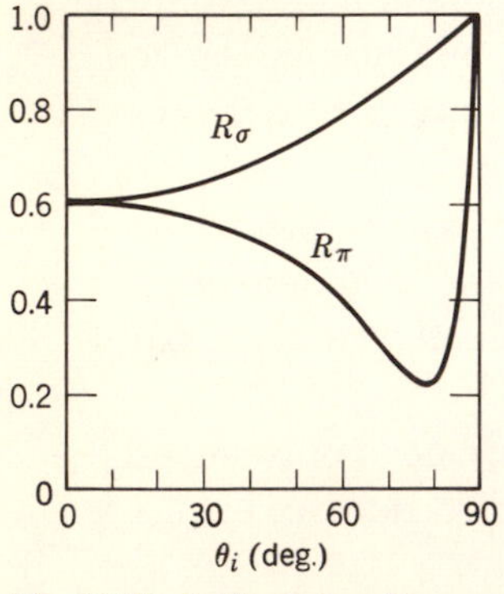

Fig. 11.35. Reflectivity of iron versus angle of incidence at a wavelength of 510 nm.

Many of the data in the literature on the optical constants of metals are not completely indicative of true metallic behavior, because the surfaces studied were of insufficient purity and optical quality. An oxide layer or an oil film on the surface can appreciably alter the optical constants. We give experimental data in Fig. 11.33 on the optical constants of a high-quality iron crystal, measured using the technique discussed below Eq. (11.113).* The crystal was 99.99 per cent pure, and the measurements were done in a vacuum of $5\cdot 10^{-10}$ torr. The surface was polished, lightly etched, and then heat-treated in hydrogen to remove oxygen contamination. If these precautions are not observed and an ordinary "clean surface" measured, the results may differ by as much as 30 per cent from those given here.

Calculated values for the real and imaginary parts of the dielectric constant of iron are plotted in Fig. 11.34. Note that ϵ_i is large and that ϵ_r is negative. Thus, iron represents a real mixture of the two limiting cases discussed earlier. The conduction electrons are quite heavily damped in iron; in fact, at least some of them may be partially bound and not form a very good "free electron gas."

In Fig. 11.35 R_σ and R_π are shown as functions of the angle of incidence at $\lambda = 510$ nm (5100 A) where $n_r = 3.19$, $n_i = 3.86$.

REFERENCES FOR SECTION 11.3

Treatments of the reflection and transmission of light will be found in all books on physical optics and on electromagnetic theory. Special attention is called to the following.

M. Born and E. Wolf, *Principles of Optics*, Sections 1.5 (1.6), 13.1–13.3 (13.4).
P. Drude, *Theory of Optics*.
R. W. Wood, *Physical Optics*.
M. Garbuny, *Optical Physics*, Chapter 5 (with many examples of the optical properties of real systems).

References on the attenuated total reflection method
N. J. Harrick, *Internal Reflection Spectroscopy*.
W. W. Wendlandt, ed., *Modern Aspects of Reflectance Spectroscopy*.

*H. T. Yolken and J. Kruger, *J. Optical Soc. Am.*, **55**, 842 (1965).

PROBLEMS

11.3.1 Show that the equations used to determine the reflection and transmission coefficients satisfy the requirements of continuity of the normal components of **B** and **D**.

11.3.2 Derive the Fresnel relations, Eqs. (11.102*a*, *b*), (11.107*a*, *b*).

11.3.3 Prove energy conservation, Eq. (11.109*c*), for π and σ cases.

11.3.4 Show by direct calculation of $\mathbf{E} \times \mathbf{H}$ that for total internal reflection the normal or z component of the Poynting vector **S** has zero time average in the second medium.

11.3.5 Consider total internal reflection at a dielectric interface. Consider the inhomogeneous plane wave in the second medium. What are the surfaces of constant amplitude? The surfaces of constant phase? The phase velocity?

11.3.6 Consider total internal reflection of an incident light beam plane polarized at 45° to the plane of incidence. With reference to Eqs. (11.114), show that the relative phase shift $\phi = \phi_\pi - \phi_\sigma$ obeys the equation

$$\tan\frac{\phi}{2} = \gamma\frac{\cos^2\theta_i}{\sin^2\theta_i}, \qquad \gamma = \frac{\sqrt{n_1^2 \sin^2\theta_i - n_2^2}}{n_1 \cos\theta_i}$$

Show that the relative phase shift is a maximum when

$$\cos\theta_i = \sqrt{\frac{n_1^2 - n_2^2}{n_1^2 + n_2^2}}$$

11.3.7 With references to the previous problem, consider the prism shown, for which two internal reflections occur. This doubles the relative phase shift, producing a net shift of 2ϕ. When $2\phi = \pi/2$, the device is called a *Fresnel rhomb*. It then functions like a quarter-wave plate. With $n_2 = 1$, $n_1 = n$, find the minimum value of n and the corresponding angle of incidence that can be used in such a rhomb.

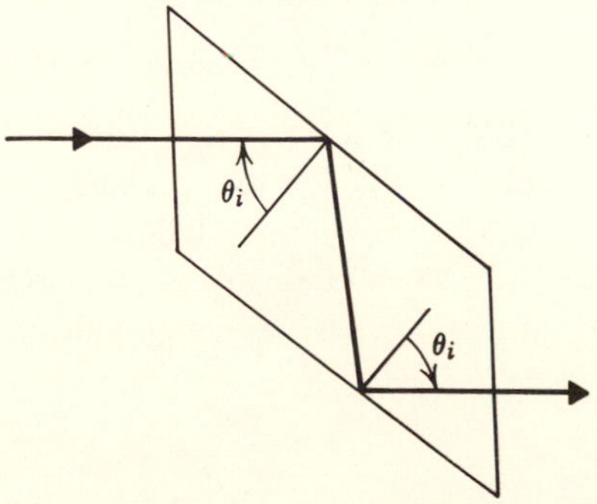

11.3.8 Show that for a plane parallel dielectric slab used with parallel coherent light, the sum of the reflected and transmitted intensities equals the incident intensity.

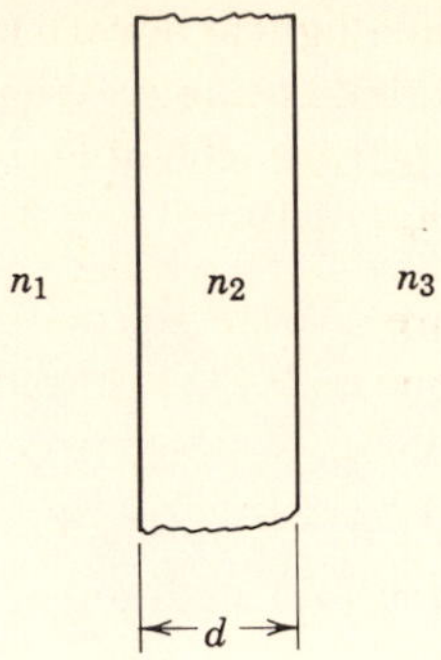

11.3.9 Consider reflection and transmission of a light beam by a dielectric slab under conditions where the coherence length is much less than $2n_2\, d \cos\theta_2$. This problem may be treated by summing the various intensities, possibly after multiple reflection.

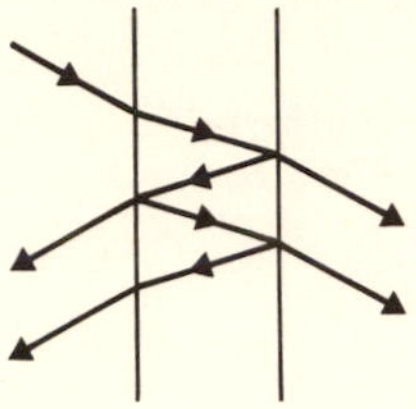

a. Show that the transmitted intensity is

$$I_t = \frac{|\tau_{12}|^2|\tau_{23}|^2}{1-|\rho_{23}|^2|\rho_{12}|^2} I_i$$

where I_i is the incident intensity, and that the reflected intensity is

$$I_r = \left[|\rho_{12}|^2 + \frac{|\tau_{12}|^2|\tau_{21}|^2|\rho_{23}|^2}{1-|\rho_{23}|^2|\rho_{12}|^2}\right] I_i$$

b. For a n_1-n_2-n_1 dielectric slab, show that these expressions reduce to

$$I_t = \frac{1-R_{12}}{1+R_{12}} I_i \qquad I_r = \frac{2R_{12}}{1+R_{12}} I_i$$

where R_{12} is the reflectivity for a single n_1-n_2 interface.

11.3.10 Show that the entire π polarized component of light incident at Brewster's angle onto a plane parallel dielectric slab is transmitted for light with coherence lengths both long and short with respect to $2n_2d \cos\theta_2$. What is the overall transmissivity for the σ component at Brewster's angle for the coherent and incoherent cases?

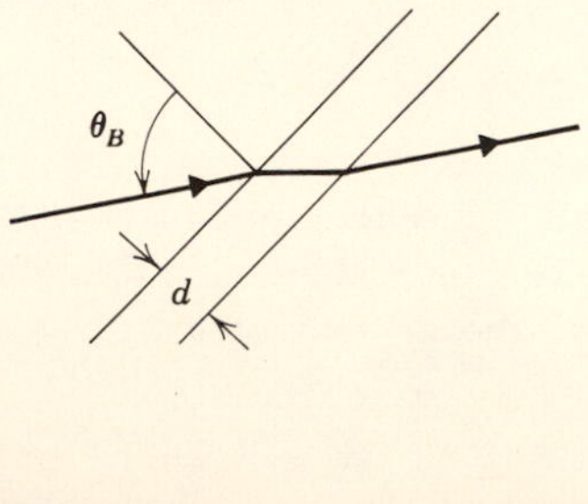

11.3.11 Consider light at normal incidence onto a metal surface, the incident electric field being E_i. What is the electric field just inside the surface for (a) the metal of Fig. 11.8 at $\omega = 0.6\,\omega_p$, and for (b) iron at 510 nm where $n_r = 3.19$ and $n_i = 3.86$?

11.3.12 You are given a pure sample with a good polished surface of an unknown opaque material with a high reflectivity at a certain wavelength and at normal incidence. How could you tell whether the large value of R was due to a large value of n_i or to a large value of n_r?

11.3.13 Show that for a thin n_1-n_2-n_3 layer the expressions for ρ and τ go over to those for a n_1-n_3 interface as $d \to 0$.

11.4 AN INTRODUCTION TO CRYSTAL OPTICS

Many crystals have symmetry sufficiently low that their optical properties are anisotropic. Then the propagation of electromagnetic radiation is direction-dependent, and double refraction will occur. The description of these effects belongs to the field of crystal optics. Other subjects such as electrooptics and magneto-optics are traditionally included in complete discussions of crystal optics, and one might add the recently developed subject of non-linear optics. In the treatment below, only the first mentioned class of phenomena will be discussed, namely, the propagation of electromagnetic radiation in anisotropic media.

A. ELECTROMAGNETIC WAVES IN AN ANISTROPIC DIELECTRIC MEDIUM

The four Maxwell equations in the absence of free charges or currents are

$$\nabla \cdot \mathbf{D} = 0$$

$$\nabla \cdot \mathbf{B} = 0$$

$$\nabla \times \mathbf{E} = -\frac{1}{c}\frac{\partial \mathbf{B}}{\partial t}$$

$$\nabla \times \mathbf{H} = \frac{1}{c}\frac{\partial \mathbf{D}}{\partial t}$$

The medium will be assumed nonmagnetic so that $\mu = 1$ and $\mathbf{B} = \mathbf{H}$. For a single monochromatic component, the real physical fields will be the real parts of complex fields (analytic signals) having $e^{i\omega t}$ time dependence. Then we find that $\partial \mathbf{H}/\partial t = i\omega \mathbf{H}$, and so on, and we obtain

$$\nabla \times \mathbf{E} = -\frac{i\omega}{c}\mathbf{H}$$

$$\nabla \times \mathbf{H} = \frac{i\omega}{c}\mathbf{D}$$

or

$$\nabla \times (\nabla \times \mathbf{E}) = \frac{\omega^2}{c^2}\mathbf{D}$$

1. The Relationship between D and E. A solution will be sought of the plane-wave type, for which the spatial dependence of **D**, **E**, and **H** is given by the factor $e^{-i\omega n \mathbf{s}\cdot\mathbf{r}/c}$, where **s** is a unit vector normal to the plane surfaces of constant phase, and n = index of refraction. In other words, the combined space-time dependence of the analytic signals corresponding to **D**, **H**, and **E** is of the form $e^{i(\omega t - \mathbf{k}\cdot\mathbf{r})}$ with $\mathbf{k} = \omega n \mathbf{s}/c$. The magnitude of the *phase velocity* c/n will then be the speed at which the surfaces of constant phase propagate. This will also be denoted as the *wave velocity* $\mathbf{v}_w = \mathbf{s}c/n$.

We then find

$$\nabla \times (\nabla \times \mathbf{E}) = -\frac{\omega^2 n^2}{c^2} \mathbf{s} \times (\mathbf{s} \times \mathbf{E}) = -\frac{\omega^2 n^2}{c^2} [\mathbf{s}(\mathbf{s} \cdot \mathbf{E}) - \mathbf{E}]$$

since $\mathbf{s} \cdot \mathbf{s} = 1$. The last equality made use of the vector identity

$$\mathbf{A} \times (\mathbf{B} \times \mathbf{C}) = \mathbf{B}(\mathbf{A} \cdot \mathbf{C}) - \mathbf{C}(\mathbf{A} \cdot \mathbf{B})$$

We then find the following expression for **D** in terms of **E**:

$$\mathbf{D} = n^2[\mathbf{E} - \mathbf{s}(\mathbf{s} \cdot \mathbf{E})] \tag{11.135}$$

This is perpendicular to **s**.

Equation (11.135) has been derived from Maxwell's equations using only the assumption of $e^{i\omega(t - n\mathbf{s}\cdot\mathbf{r}/c)}$ space-time dependence. The properties of the dielectric medium are introduced by the additional equation

$$\mathbf{D} = \boldsymbol{\epsilon}\mathbf{E} \tag{11.136}$$

or in the component form

$$D_i = \sum_{j=1}^{3} \epsilon_{ij} E_j \tag{11.137}$$

The coefficients ϵ_{ij} form the elements of the dielectric tensor $\boldsymbol{\epsilon}$. In a nonabsorbing medium with no optical activity it can be shown that $\boldsymbol{\epsilon}$ is real and symmetric. Thus

$$\epsilon_{ij} = \epsilon_{ij}^* \tag{11.138}$$

and

$$\epsilon_{ij} = \epsilon_{ji}$$

For such a tensor we can always find an orthogonal Cartesian coordinate system that diagonalizes the tensor, so that with respect to such a coordinate system only the diagonal components $(i = j)$ of $\boldsymbol{\epsilon}$ appear. The new axes are called *principal axes*. With respect to them $\boldsymbol{\epsilon}$ takes the form

$$\boldsymbol{\epsilon} = \begin{pmatrix} \epsilon_x & 0 & 0 \\ 0 & \epsilon_y & 0 \\ 0 & 0 & \epsilon_z \end{pmatrix} \tag{11.139}$$

The quantities ϵ_x, ϵ_y, and ϵ_z are called *principal values* of $\boldsymbol{\epsilon}$. They are all nonnegative. When we combine Eqs. (11.135) and (11.136), we find that for principal axes

$$n^2 \left[E_i - s_i \sum_j s_j E_j\right] - \epsilon_i E_i = 0 \tag{11.140a}$$

Another version of this equation is

$$\sum_{j=1} M_{ij}E_j = 0 \tag{11.140b}$$

with

$$\mathbf{M} = \begin{pmatrix} n^2(1-s_x^2)-\epsilon_x & -n^2s_xs_y & -n^2s_xs_z \\ -n^2s_xs_y & n^2(1-s_y^2)-\epsilon_y & -n^2s_ys_z \\ -n^2s_xs_z & -n^2s_ys_z & n^2(1-s_z^2)-\epsilon_z \end{pmatrix} \tag{11.140c}$$

For a given unit vector $\mathbf{s} = (s_x, s_y, s_z)$, Eqs. (11.140) have a solution if and only if the *characteristic equation*

$$\det(M_{ij}) = 0 \tag{11.141}$$

is satisfied. An equivalent equation may be derived from Eq. (11.140*a*). It may be written as three equations of the form

$$(n^2-\epsilon_x)E_x = n^2s_x(\mathbf{s}\cdot\mathbf{E})$$

Or

$$E_x = \frac{n^2}{n^2-\epsilon_x}s_x(\mathbf{s}\cdot\mathbf{E})$$

and so on. Or

$$s_xE_x = \frac{n^2s_x^2}{n^2-n_x^2}(\mathbf{s}\cdot\mathbf{E}) \tag{11.142}$$

and so on, where $n_x^2 \equiv \epsilon_x$. When the three equations of the form of (11.142) are summed, we find

$$(\mathbf{s}\cdot\mathbf{E}) = \left[\frac{n^2s_x^2}{n^2-n_x^2}+\frac{n^2s_y^2}{n^2-n_y^2}+\frac{n^2s_z^2}{n^2-n_z^2}\right](\mathbf{s}\cdot\mathbf{E})$$

When $(\mathbf{s}\cdot\mathbf{E}) \neq 0$, it may be canceled from both sides, leaving *Fresnel's equation*

$$\frac{1}{n^2} = \frac{s_x^2}{n^2-n_x^2}+\frac{s_y^2}{n^2-n_y^2}+\frac{s_z^2}{n^2-n_z^2} \tag{11.143}$$

Equations (11.141) and (11.143) both appear to be cubic in n^2, but in both cases when they are written in terms of a polynomial in n^2, the coefficients of n^6 vanish identically, leaving a quadratic equation in n^2, which can be shown to have two positive roots, n_1^2 and n_2^2.

Thus, for a given propagation direction **s**, there are in general two values of the refractive index, n_1 and n_2. If these are substituted one at a time into the matrix (M_{ij}), Eq. (11.140*b*) may be solved for the components of **E**, to within an undetermined multiplicative constant. Let the two solutions for **E** corresponding to n_1 and n_2 be denoted by $\mathbf{E}_1$ and $\mathbf{E}_2$; respectively. The displacement vector is then

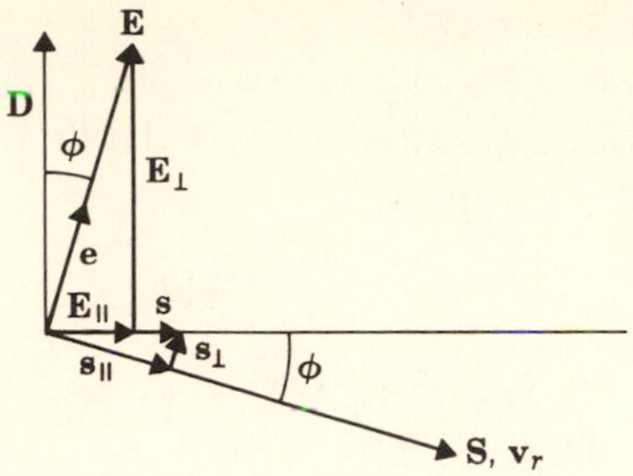

Fig. 11.36

given by Eq. (11.135). If **E** is resolved into components parallel and perpendicular to **s** as shown in Fig. 11.36,

$$\mathbf{E} = \mathbf{E}_{\parallel} + \mathbf{E}_{\perp}$$

Then $\mathbf{E}_{\parallel} = (\mathbf{E} \cdot \mathbf{s})\mathbf{s}$ and $\mathbf{E}_{\perp} = \mathbf{E} - (\mathbf{E} \cdot \mathbf{s})\mathbf{s}$, and **D** is given by the simple relation

$$\mathbf{D} = n^2 \mathbf{E}_{\perp} \tag{11.144a}$$

*Here **D**, **E**, and **s** are seen to be coplanar.*

Actually there are two equations of the form of (11.144*a*), one for each solution of Eq. (11.140*a*):

$$\mathbf{D}_1 = n_1{}^2 \mathbf{E}_{1\perp} \tag{11.144b}$$

$$\mathbf{D}_2 = n_2{}^2 \mathbf{E}_{2\perp} \tag{11.144c}$$

It will now be shown that they are perpendicular,

$$(\mathbf{D}_1 \cdot \mathbf{D}_2) = (\mathbf{E}_{1\perp} \cdot \mathbf{E}_{2\perp}) = 0 \tag{11.144d}$$

Consider the scalar product

$$\begin{aligned}(\mathbf{E}_2 \cdot \mathbf{D}_1) &= \sum_i E_{2i} D_i = \sum_{ij} E_{2i} \epsilon_{ij} E_{1j} = \sum_{ij} E_{1j} \epsilon_{ji} E_{2i} = \sum_j E_{1j} D_{2j} \\ &= (\mathbf{E}_1 \cdot \mathbf{D}_2)\end{aligned}$$

The third equality made use of the symmetry of the dielectric tensor. Use of Eqs. (11.144*b*,*c*) gives

$$n_1{}^2 (\mathbf{E}_2 \cdot \mathbf{E}_{1\perp}) = n_2{}^2 (\mathbf{E}_1 \cdot \mathbf{E}_{2\perp})$$

Set $\mathbf{E}_2 = \mathbf{E}_{2\perp} + \mathbf{E}_{2\parallel}$, where $\mathbf{E}_{2\parallel}$ is parallel to **s** and hence perpendicular to $\mathbf{E}_{1\perp}$. Then $(\mathbf{E}_2 \cdot \mathbf{E}_{1\perp}) = (\mathbf{E}_{2\perp} \cdot \mathbf{E}_{1\perp})$. Similarly $(\mathbf{E}_1 \cdot \mathbf{E}_{2\perp}) = (\mathbf{E}_{1\perp} \cdot \mathbf{E}_{2\perp})$. Thus

$$(n_1{}^2 - n_2{}^2)(\mathbf{E}_{2\perp} \cdot \mathbf{E}_{1\perp}) = 0$$

When $n_1{}^2 \neq n_2{}^2$, this gives (11.144*d*). The case $n_1 = n_2$ occurs in isotropic media and along certain directions (wave optic axes) in anisotropic media. The directions of $\mathbf{D}_1$ and $\mathbf{D}_2$ are then arbitrary as long as they are perpendicular to **s**, and it is convenient to choose them to be mutually perpendicular in this case also.

2. *The Poynting Vector and Ray Velocity.* The magnetic field **H** will be given by

$$\mathbf{H} = n\mathbf{s} \times \mathbf{E}$$

The Poynting vector

$$\mathbf{S} = \frac{c}{4\pi} \mathbf{E} \times \mathbf{H}$$

is perpendicular to both **E** and **H**, and since **H** is normal to the plane of **D**, **E**, and **s**, **S** will lie in that plane. It will, in general, not be parallel to the wave normal **s**; in fact it makes the same angle ϕ with **s** as does **E** with **D**. This is illustrated in Fig. 11.36. An explicit expression for **S** may be obtained by writing $\mathbf{E} = E\mathbf{e}$, where **e** is a unit vector in the direction of **E**. Then **S** is given by

$$\mathbf{S} = \frac{c}{4\pi} nE^2[\mathbf{s} - \mathbf{e}(\mathbf{s}\cdot\mathbf{e})] \tag{11.145a}$$

Now $\mathbf{e}(\mathbf{s}\cdot\mathbf{e})$ is the component $\mathbf{s}_\perp$ of **s** along **e**, and $\mathbf{s} - \mathbf{e}(\mathbf{s}\cdot\mathbf{e})$ is the component $\mathbf{s}_\parallel$ of **s** along **S**. Thus, (11.45*a*) may also be written

$$\mathbf{S} = \frac{c}{4\pi} nE^2\mathbf{s}_\parallel \tag{11.145b}$$

The Poynting vector **S** gives the direction and rate of energy flow. Its direction defines the ray direction. We have just seen that, in general, for anisotropic media this is not the direction of the wave normal **s**. It can be shown that when dispersion is present, that is, when the components of $\boldsymbol{\epsilon}$ depend on frequency, the group velocity vector for a wave packet will be along **S**.*

The wave velocity $\mathbf{v}_w$ is the velocity of propagation of the surfaces of constant phase $\omega t - \omega n(\mathbf{s}\cdot\mathbf{r})/c = \text{const}$ and is given by

$$\mathbf{v}_w = \frac{\mathbf{s}c}{n} \tag{11.146}$$

The ray velocity $\mathbf{v}_r$ is along **S** and has a magnitude according to Eq. (10.27),

$$v_r = \frac{v_w}{\cos\phi} = \frac{c}{n\cos\phi} \tag{11.147a}$$

A unit vector along **S** may be constructed from the parallel vector

$$\mathbf{s}_\parallel = \mathbf{s} - \mathbf{e}(\mathbf{s}\cdot\mathbf{e})$$

which has magnitude $\cos\phi$. Thus the desired unit vector is $\mathbf{s}_\parallel/\cos\phi$, and $\mathbf{v}_r$ may be written

$$\mathbf{v}_r = \frac{c}{n\cos^2\phi}\mathbf{s}_\parallel = \frac{c}{n\cos^2\phi}[\mathbf{s} - \mathbf{e}(\mathbf{s}\cdot\mathbf{e})] \tag{11.147b}$$

Another version is obtained by noting that

$$|\mathbf{s}_\parallel|^2 = \cos^2\phi = 1 - (\mathbf{s}\cdot\mathbf{e})^2$$

Hence

$$\mathbf{v}_r = \frac{c[\mathbf{s} - \mathbf{e}(\mathbf{s}\cdot\mathbf{e})]}{n[1 - (\mathbf{s}\cdot\mathbf{e})^2]} \tag{11.147c}$$

Yet another version may be easily derived:

$$\mathbf{v}_r = \frac{cn[\mathbf{s} - \mathbf{e}(\mathbf{s}\cdot\mathbf{e})]}{\sum_{ij} e_i\epsilon_{ij}e_j} = \frac{cn\,\mathbf{s}_\parallel}{\sum_{ij} e_i\epsilon_{ij}e_j} \tag{11.147d}$$

*D. A. Kleinman, *Phys. Rev.*, **174**, 1027 (1968) (Appendix A).

B. INDEX SURFACES AND THEIR USES

There are several constructions useful in solving problems of wave propagation in optically anisotropic crystals. One of the most useful is that of the *index surfaces* or *normal surfaces*, as they are sometimes called. We first have to find the two indices n_1, n_2 associated with the wave normal **s** for various directions of **s**. (A construction to help do this will be discussed below under the heading "Poinsot ellipsoid.") Starting at a suitable origin O, we procede along **s** distances equal to n_1 and n_2 and locate two points, as shown in Fig. 11.37. The loci of all such points form the two index surfaces. Examples of such surfaces are given in Figs. 11.43 and 11.45 below.

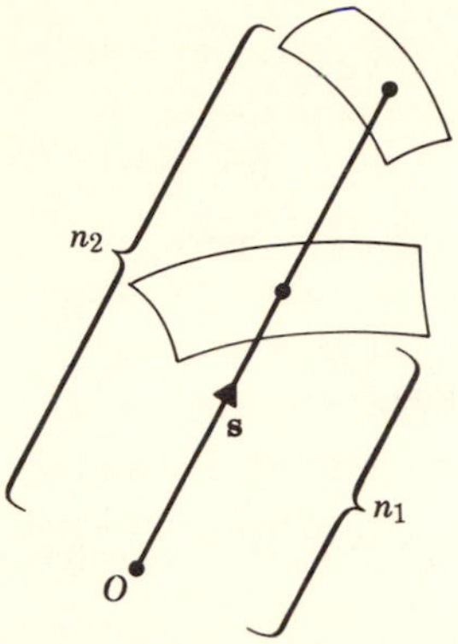

Fig. 11.37. Portions of index surfaces.

*1. **Finding the Ray Direction from the Index Surface.*** A given index surface $n(\mathbf{s})$ may be used to determine the directions of **E**, **H**, **D**, and **S** for the electromagnetic wave with normal **s** and index $n(\mathbf{s})$. It is convenient to introduce the vector

$$\boldsymbol{\eta} \equiv n\mathbf{s} \tag{11.148}$$

Then the Poynting vector **S** may be written

$$\mathbf{S} = \frac{c}{4\pi} E^2[\boldsymbol{\eta} - \mathbf{e}(\boldsymbol{\eta}\cdot\mathbf{e})] \tag{11.149}$$

We now show that the direction of the Poynting vector **S**, that is, the ray direction associated with the vector $\boldsymbol{\eta}$, is normal to the index surface at the point $\boldsymbol{\eta}$. To do so we consider a first-order change $d\mathbf{s}$ in **s**. This will change the index n, so that the change in $\boldsymbol{\eta}$ is

$$d\boldsymbol{\eta} = n d\mathbf{s} + \mathbf{s} dn$$

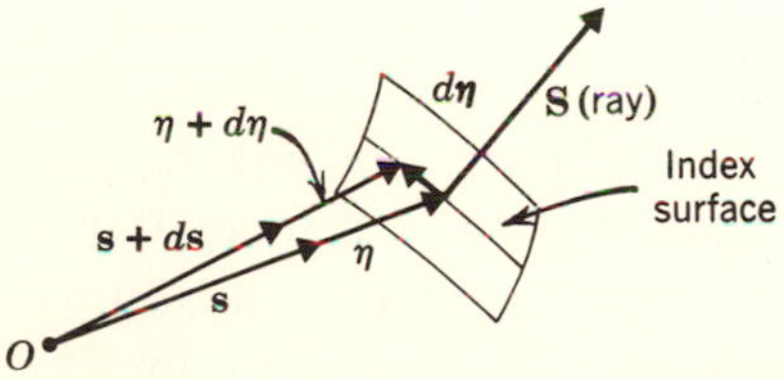

Fig. 11.38

Since $\boldsymbol{\eta} + d\boldsymbol{\eta}$ is on the index surface (Fig. 11.38), as is $\boldsymbol{\eta}$, the vector $d\boldsymbol{\eta}$ must be tangent to the index surface. To show that the Poynting vector **S** is along the normal to the surface, it is sufficient to show that $\mathbf{S}\cdot d\boldsymbol{\eta} = 0$ for all $d\boldsymbol{\eta}$.

Equation (11.135) may be written using $\boldsymbol{\eta} = n\mathbf{s}$ and $\mathbf{E} = E\mathbf{e}$ as

$$\mathbf{D} = E[(\boldsymbol{\eta}\cdot\boldsymbol{\eta})\mathbf{e} - \boldsymbol{\eta}(\boldsymbol{\eta}\cdot\mathbf{e})]$$

Then form the scalar product

$$\mathbf{E}\cdot\mathbf{D} = E^2[(\boldsymbol{\eta}\cdot\boldsymbol{\eta}) - (\boldsymbol{\eta}\cdot\mathbf{e})^2]$$

We ask how this changes in first order when $\boldsymbol{\eta} \to \boldsymbol{\eta} + \delta\boldsymbol{\eta}$ and $\mathbf{e} \to \mathbf{e} + \delta\mathbf{e}$, but when E is held constant. The answer is

$$\delta(\mathbf{E}\cdot\mathbf{D}) = E^2\{2\delta\boldsymbol{\eta}\cdot[\boldsymbol{\eta} - \mathbf{e}(\boldsymbol{\eta}\cdot\mathbf{e})] - 2(\boldsymbol{\eta}\cdot\delta\mathbf{e})(\boldsymbol{\eta}\cdot\mathbf{e})\} \tag{11.150a}$$

The term involving $\delta\boldsymbol{\eta}$ can be eliminated by using the fact that

$$D_i = \sum_j \epsilon_{ij} E_j$$

so that

$$(\mathbf{E}\cdot\mathbf{D}) = \sum_i E_i D_i = E^2 \sum_{ij} e_i \epsilon_{ij} e_j$$

Then

$$\delta(\mathbf{E}\cdot\mathbf{D}) = E^2 \sum_{ij} \delta e_i \epsilon_{ij} e_j + E^2 \sum_{ij} e_i \epsilon_{ij} \delta e_j$$

Now use the symmetry of $\boldsymbol{\epsilon}$, $\epsilon_{ij} = \epsilon_{ji}$, to obtain

$$\begin{aligned}\delta(\mathbf{E}\cdot\mathbf{D}) &= 2E^2 \sum_{ij} \delta e_i \epsilon_{ij} e_j = 2E \sum_i \delta e_i \sum_j \epsilon_{ij} E e_j \\ &= 2E(\delta\mathbf{e}\cdot\mathbf{D}) = 2E^2\,\delta\mathbf{e}\cdot[(\boldsymbol{\eta}\cdot\boldsymbol{\eta})\mathbf{e}-\boldsymbol{\eta}(\boldsymbol{\eta}\cdot\mathbf{e})] \\ &= -2E^2\,(\delta\mathbf{e}\cdot\boldsymbol{\eta})(\boldsymbol{\eta}\cdot\mathbf{e}) \qquad (11.150b)\end{aligned}$$

The last equality followed from the fact that **e** is a unit vector so that

$$2\mathbf{e}\cdot\delta\mathbf{e} = \delta(\mathbf{e}\cdot\mathbf{e}) = \delta(1) = 0$$

A comparison of Eq. (11.150*a*) with Eq. (11.150*b*) gives

$$2E^2\delta\boldsymbol{\eta}\cdot[\boldsymbol{\eta}-\mathbf{e}(\boldsymbol{\eta}\cdot\mathbf{e})] = 0$$

Since, according to Eq. (11.149), **S** is proportional to the vector in square brackets, this gives the desired result

$$(\delta\boldsymbol{\eta}\cdot\mathbf{S}) = 0 \qquad (11.151)$$

What we have shown is the following. With a known index surface $n(\mathbf{s})$, a given **s** yields n and hence $\boldsymbol{\eta}$. The normal to the surface then gives the direction of **S**. The plane of **s** and **S** is also the plane of **D** and **E**. As illustrated in Fig. 11.36, it is then possible to locate in this plane the direction of **D** (perpendicular to **s**) and **E** (perpendicular to **S**).

*2. **Use of Index Surfaces for Phase Matching at an Interface.*** Consider an interface between two nonabsorbing nonoptically active media. Both will be anisotropic, in general. Let a plane wave be incident in medium 1 with wave normal direction $\mathbf{s}_i$ and index n_i. Then there will generally be two transmitted waves in medium 2 and two reflected waves in medium 1. The three rays in medium 1 will have, in general, three different insides n_i, n_{ra}, and n_{rb}, and three different directions of linear polarization. (Fig. 11.39).

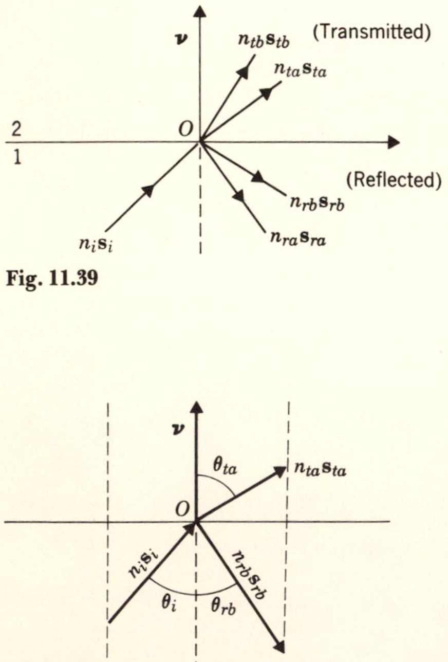

Fig. 11.39

We shall not attempt to solve the entire reflection-transmission problem, which is to determine the two transmitted and the two reflected fields, if the incident electric field is known. (This can be done by requiring that the tangential components of **E** and **H** be continuous across the interface.) Instead, we ask only for the directions **s** of the four wave normals and the corresponding values of the indices n. Let $\boldsymbol{\nu}$ be the normal to the interface. With the origin O as shown in Fig. 11.39 or Fig. 11.40, the plane interface obeys the equation $\boldsymbol{\nu}\cdot\mathbf{r} = 0$. The incident wave will be of the form

$$\mathbf{E} = \mathbf{E}_i e^{i\omega t} e^{-i\omega n_i \mathbf{s}_i\cdot\mathbf{r}/c}$$

with two reflected waves, one of which is

$$\mathbf{E}_{rb} e^{i\omega t} e^{-i\omega n_{rb}\mathbf{s}_{rb}\cdot\mathbf{r}/c}$$

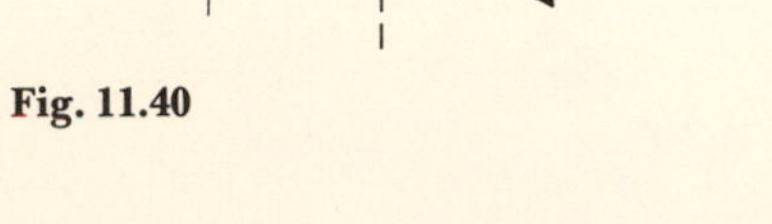

Fig. 11.40

and two transmitted waves, one of which is

$$\mathbf{E}_{ta}e^{i\omega t}e^{-i\omega n_{ta}\mathbf{s}_{ta}\cdot\mathbf{r}/c}$$

Imposition of the boundary conditions on $\mathbf{E}_{\tan}$ and $\mathbf{H}_{\tan}$ will lead, among other things, to the condition that the phases in the exponents of each term must match at the interface. Thus for all $\mathbf{r}$ such that $(\boldsymbol{\nu}\cdot\mathbf{r})=0$, we must have

$$n_{ta}(\mathbf{s}_{ta}\cdot\mathbf{r}) = n_{tb}(\mathbf{s}_{tb}\cdot\mathbf{r}) = n_{ra}(\mathbf{s}_{ra}\cdot\mathbf{r}) = n_{rb}(\mathbf{s}_{rb}\cdot\mathbf{r}) = n_i(\mathbf{s}_i\cdot\mathbf{r}) \tag{11.152a}$$

This leads to four equations of the form

$$n_{ra}\mathbf{s}_{ra,p} = n_i\mathbf{s}_{i,p} \tag{11.152b}$$

where the extra subscript p means the component *in* the plane of the interface. Then the vectors

$$n_{ra}\mathbf{s}_{ra} - n_i\mathbf{s}_i$$

(and so on) must be parallel to the normal $\boldsymbol{\nu}$, since the components perpendicular to $\boldsymbol{\nu}$ are zero by Eq. (11.152*b*). This leads to four equations of the form

$$n_{ra}\mathbf{s}_{ra} = n_i\mathbf{s}_i + \text{const}\ \boldsymbol{\nu} \tag{11.152c}$$

Thus all the wave normal vectors $\mathbf{s}_{ra}$, $\mathbf{s}_{rb}$, $\mathbf{s}_{ta}$, and $\mathbf{s}_{tb}$ shown in Fig. 11.39 are in the plane determined by $\mathbf{s}_i$ and $\boldsymbol{\nu}$, that is, the *plane of incidence*. Since the components of the vectors $n\mathbf{s}$ in the plane of the interface are the same, we obtain a generalized Snell's law for anisotropic media:

$$\begin{aligned} n_{ta}\sin\theta_{ta} &= n_i\sin\theta_i \\ n_{tb}\sin\theta_{tb} &= n_i\sin\theta_i \\ n_{ra}\sin\theta_{ra} &= n_i\sin\theta_i \\ n_{rb}\sin\theta_{rb} &= n_i\sin\theta_i \end{aligned} \tag{11.153}$$

The simple law of reflection $\theta_{ra}=\theta_i$, $\theta_{rb}=\theta_i$ will not generally hold, since the n's are different for the reflected waves.

Total internal reflection can occur in two ways for light incident onto a doubly refracting medium. They are illustrated in Fig. 11.41.

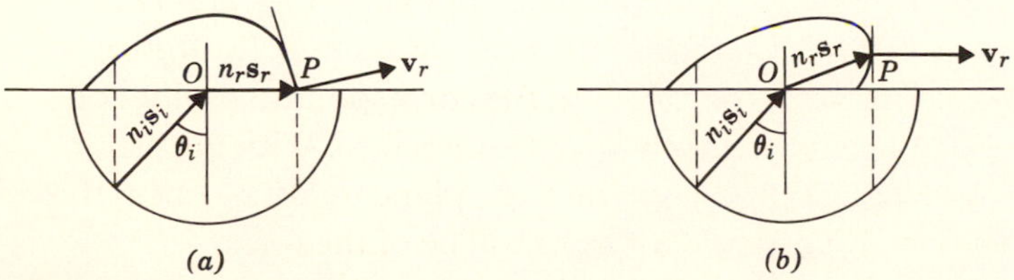

Fig. 11.41. Two types of total internal reflection. The subscripts i and r refer to incident and refracted waves. $\mathbf{v}_r$ is the refracted ray velocity.

The upper medium is anistropic, and one of the index surfaces is drawn. For case (*a*) we increase the angle of incidence θ_i until $n_i \sin\theta_i = |OP|$. The refracted wave normal $\mathbf{s}_r$ is then along the interface. The refracted ray velocity $\mathbf{v}_r$ for this case is not yet grazing the interface, but nevertheless for larger values of θ_i than that shown the wave is totally reflected, because Eqs. (11.153) can no longer be satisfied by a real angle θ_r.

In case (*b*), as θ_i is increased a point is reached for which $\mathbf{v}_r$ is parallel to the interface, whereas $\mathbf{s}_r$ is not yet grazing. The Poynting vector is then parallel to OP, and for any larger value of θ_i beyond that for which $n_i \sin\theta = |OP|$ there can be no energy flow into the upper medium for this branch of the index surface. This situation also corresponds to total reflection.

3. Index Surfaces and Fields in Uniaxial Media. In a uniaxial medium the dielectric tensor, when written in diagonal form as in Eq. (11.139), has two equal elements, and if the axes are chosen so that the third element is ϵ_2, it takes the form

$$\boldsymbol{\epsilon} = \begin{pmatrix} n_0^2 & 0 & 0 \\ 0 & n_0^2 & 0 \\ 0 & 0 & n_e^2 \end{pmatrix}. \tag{11.154}$$

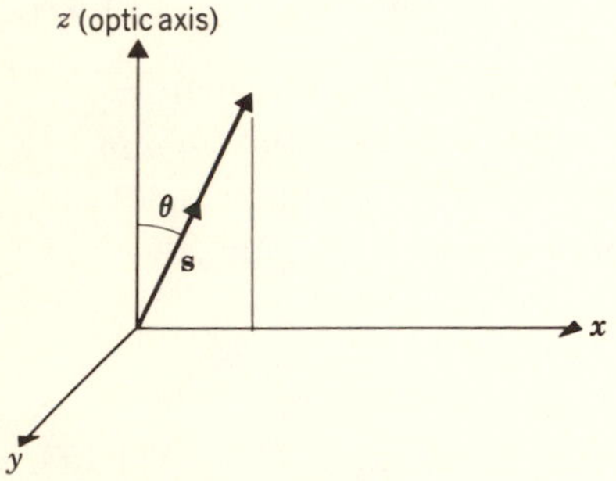

Fig. 11.42

Thus, $\epsilon_x = \epsilon_y = n_0^2$, $\epsilon_z = n_e^2$. The optical properties are now invariant with respect to rotations about the z axis, which is the optic axis. (See Fig. 11.42). Without loss of generality we can assume that the wave normal $\mathbf{s}$ is in the xz plane. Hence, $\mathbf{s} = (\sin\theta, 0, \cos\theta)$, and Eqs. (11.140*a*) when written out in detail give

$$\begin{aligned} (n^2 - n_0^2)E_x - n^2 \sin\theta(\sin\theta\, E_x + \cos\theta\, E_z) &= 0 \\ (n^2 - n_0^2)E_y &= 0 \\ (n^2 - n_e^2)E_z - n^2 \cos\theta\, (\sin\theta\, E_x + \cos\theta\, E_z) &= 0 \end{aligned} \tag{11.155}$$

One solution of (11.155) corresponds to the *ordinary wave* $\mathbf{E}_0$ and is obtained by setting

$$E_{0x} = E_{0z} = 0, \quad E_{0y} \neq 0, \quad n^2 = n_0^2 \tag{11.156}$$

Here $\mathbf{E}_0$ is perpendicular to both the wave normal $\mathbf{s}$ and to the optic axis. The corresponding $\mathbf{D}$ field has components

$$D_{0y} = n_0^2 E_{0y}, \qquad D_{0x} = D_{0z} = 0$$

Hence

$$\mathbf{D}_0 = n_0^2 \mathbf{E}_0 \tag{11.157}$$

The wave and ray velocities are

$$\mathbf{v}_w = \mathbf{v}_r = \frac{c\mathbf{s}}{n_0} \tag{11.158}$$

The solution for the extraordinary wave is obtained by setting $E_y = 0$ and by solving the two remaining equations, which can be written

$$(n_0^2 - n^2 \cos^2\theta)E_x + n^2 \sin\theta \cos\theta\, E_z = 0 \tag{11.159a}$$

$$n^2 \sin\theta \cos\theta\, E_x + (n_e^2 - n^2 \sin^2\theta)E_z = 0 \tag{11.159b}$$

For a solution to exist, the determinant of the coefficients of E_x and E_y must equal zero:

$$n_0^2 n_e^2 + n^4 \sin^2\theta \cos^2\theta - n^4 \sin^2\theta \cos^2\theta - n^2(n_0^2 \sin^2\theta + n_e^2 \cos^2\theta) = 0$$

The second and third terms cancel, giving a first-order equation for n^2, which can be written

$$\frac{1}{n^2} = \frac{\sin^2\theta}{n_e^2} + \frac{\cos^2\theta}{n_0^2} \tag{11.160a}$$

This then represents the equation for the *e*-wave index surface. For propagation along the optic axis $(\theta = 0, \pi)$ the index n is n_0; for propagation in the x,y plane $(\theta = \pi/2)$ the index n is n_e.

In terms of the vector $\boldsymbol{\eta} = n\mathbf{s} = n(\sin\theta, 0, \cos\theta)$, Eq. (11.160*a*) may be written as the equation of an ellipse

$$\frac{\eta_x^2}{n_e^2} + \frac{\eta_z^2}{n_0^2} = 1 \tag{11.160b}$$

Thus, for a uniaxial crystal the index surfaces consist of a sphere and an ellipsoid of revolution. They are tangent along the optic axis. Examples are shown in Fig. 11.43.

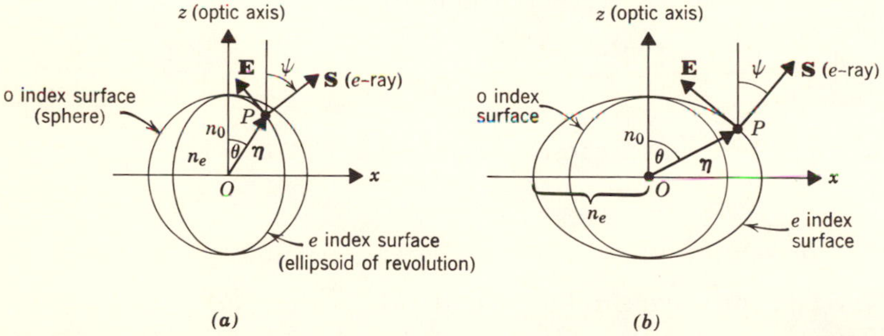

Fig. 11.43. Index surfaces for a uniaxial crystal. (*a*) Negative uniaxial crystal. $(n_e < n_0)$. (*b*) Positive uniaxial crystal $(n_e > n_0)$.

4. Relation to Wave Surfaces. The index surfaces have a kind of inverse relationship to the ray velocity surfaces of Fig. 10.14 or Fig. 11.44. For the latter the vector *OP* is along a ray direction with a length equal to the ray velocity, and the normal to the surface gives the direction of the wave normal **s**.

The **E** vector for the *e* wave lies in the plane of **s** and **S** and is perpendicular to **S**. Hence it must be tangent to the extraordinary

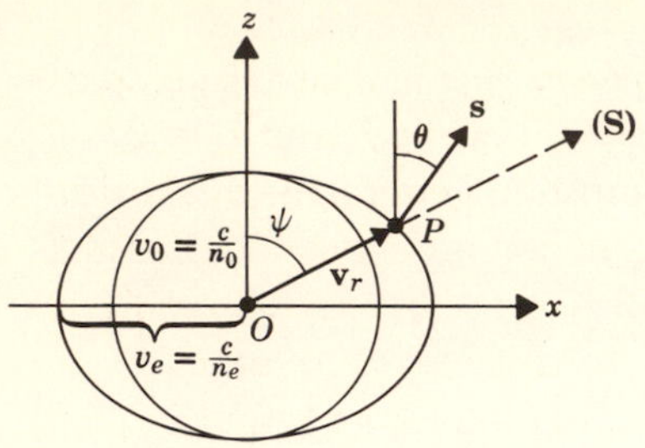

(a)

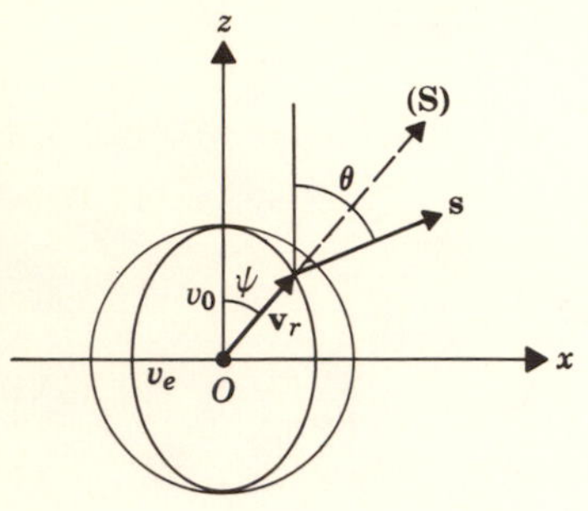

(b)

Fig. 11.44. Ray velocity surfaces corresponding to Fig. 11.43. (Wave surfaces at time $t = 1$). (*a*) Negative uniaxial crystal ($v_0 < v_e$). (*b*) Positive uniaxial crystal ($v_o > v_e$).

index surface at P. The tangent can be obtained by differentiating Eq. (11.160*b*), which gives

$$\frac{d\eta_z}{d\eta_x} = -\frac{n_0^2}{n_e^2}\frac{\eta_x}{\eta_z} = -\frac{n_0^2}{n_e^2}\tan\theta \tag{11.161}$$

This must also equal E_z/E_x, as can be shown directly by use of Eq. (11.159*a*) or (11.159*b*) with (11.160*b*).

The extraordinary ray direction is at right angles to the **E** field. Hence

$$\tan\psi = \frac{v_{rx}}{v_{rz}} = \frac{S_x}{S_z} = -\frac{d\eta_z}{d\eta_x} = \frac{n_0^2}{n_e^2}\tan\theta = \frac{n_0^2}{n_e^2}\frac{s_x}{s_z} \tag{11.162}$$

The extraordinary ray velocity surface (the *e*-wave surface at time $t = 1$) is an ellipse, as can be seen from Eq. (10.26) with $v_{rx} = v_r \sin\psi$, $v_{ry} = v_r \cos\psi$:

$$\frac{v_{rx}^2}{v_e^2} + \frac{v_{rz}^2}{v_0^2} = 1 \tag{11.163a}$$

With $v_e = c/n_e$, $v_0 = c/n_0$, this can also be written

$$n_e^2 v_{rx}^2 + n_0^2 v_{rz}^2 = c^2 \tag{11.163b}$$

Equation (11.163*b*) may be readily derived from some of our above results. For the *e* wave and for the *zx* coordinate system in use, we can write, using Eq. (11.154),

$$\Sigma_{ij} e_i \epsilon_{ij} e_j = n_0^2 e_x^2 + n_e^2 e_z^2$$

The ray velocity $\mathbf{v}_r$ is perpendicular to **e**; hence

$$e_x^2 = \left(\frac{v_{rz}}{v_r}\right)^2 \qquad e_z^2 = \left(\frac{v_{rx}}{v_r}\right)^2$$

When these results are inserted into Eq. (11.147*d*), we find

$$\mathbf{v}_r = \frac{v_r^2 cn\, \mathbf{s}_\parallel}{n_0^2 v_{rz}^2 + n_e^2 v_{rx}^2}$$

Now take the scalar product of both sides with $\mathbf{s}_\parallel$ and use the fact that $\mathbf{v}_r \cdot \mathbf{s}_\parallel = v_r \cos\phi$ and $|\mathbf{s}_\parallel|^2 = \cos^2\phi$ to obtain

$$v_r \cos\phi = \frac{v_r^2 cn \cos^2\phi}{n_0^2 v_{rz}^2 + n_e^2 v_{rx}^2}$$

Use of Eq. (11.147*a*) then yields Eq. (11.163*b*). These results enable us to convert all quantities related to the index surface to those related to the ray velocity surface for the extraordinary wave in uniaxial media.

C. BIAXIAL CRYSTALS

*1. **Some Properties of the Index Surfaces.*** For biaxial anisotropic crystals, all three principal values $\epsilon_x = n_x^2$, $\epsilon_y = n_y^2$, and $\epsilon_z = n_z^2$ of $\boldsymbol{\epsilon}$ are different, and Eqs. (11.140*b*, *c*) must, in general, be dealt with without further simplification. The treatment here shall be limited to the case where the wave normal **s** lies in a plane perpendicular to one of the principal axes of the dielectric tensor. Assume for convenience that $n_x < n_y < n_z$, and suppose that **s** lies in the *yz* plane. Then $\mathbf{s} = (0, s_y, s_z)$, and Eqs. (11.140*b*, *c*) give

$$(n^2 - n_x^2)E_x = 0 \tag{11.164a}$$

$$(n_y^2 - n^2 s_z^2)E_y - n^2 s_y s_z E_z = 0 \tag{11.164b}$$

$$-n^2 s_y s_z E_y + (n_z^2 - n^2 s_y^2)E_z = 0 \tag{11.164c}$$

One solution is simply

$$E_y = 0, \qquad E_z = 0, \qquad E_x \neq 0, \qquad n = n_x \tag{11.165a}$$

The **D** field has components

$$D_y = D_z = 0 \qquad \text{and} \qquad D_x = n_x^2 E_x \tag{11.165b}$$

$$\mathbf{D} = n_x^2 \mathbf{E}$$

The ray direction for this solution is along **s**. The index surface for this mode intersects the *zy* plane in a circle of radius n_x, as shown in Fig. 11.45, and the **D** and **E** directions are normal to this plane.

Another solution of Eqs. (11.164) will have $E_x = 0$, $E_y \neq 0$, $E_z \neq 0$. Equations (11.164*b*, *c*) have the same form as Eqs. (11.159), and the solution for n will have the same form as (11.160*a*), namely,

$$\frac{1}{n^2} = \frac{s_y^2}{n_z^2} + \frac{s_z^2}{n_y^2} \tag{11.166a}$$

Thus, for this mode when **s** is along the *y* axis, $n = n_z$, and when **s** is along the *z* axis, $n = n_y$. The **D** vectors will be in the *yz* plane perpendicular to **s**, and the **E** vectors will be in the *yz* plane tangent to the ellipse formed by the intersection of the wave surface for this mode with that plane. The ellipse has semimajor axis n_z and semiminor axis n_y.

Similar results are obtained for **s** perpendicular to the *y* and *x* axes. In the former case one of the index surfaces intersects the *xz* plane in a circle of radius n_y. The other index surface intersects the plane in the ellipse

$$\frac{1}{n^2} = \frac{s_z^2}{n_x^2} + \frac{s_x^2}{n_z^2} \tag{11.166b}$$

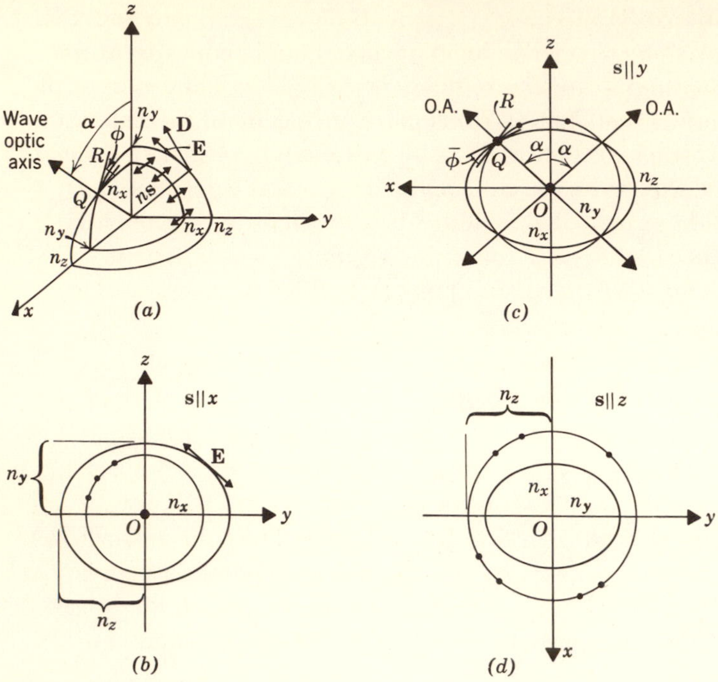

Fig. 11.45. Sections of the index surfaces for a biaxial crystal with $n_x < n_y < n_z$. The directions of **D** and **E** for the circles are out of the plane of the paper as indicated by the dots.

Since $n_x < n_y < n_z$, this ellipse will intersect the circle of radius n_y at a point Q. For **s** along OQ, the two indices of refraction are both equal to n_y.

The direction OQ is one of the two *wave optic axes*. It makes an angle α with the $+z$ axis that can be determined as follows. Let a bar over **s** indicate that it is along an optic axis. Then

$$\bar{s}_z = \cos\alpha, \qquad \bar{s}_x = \sin\alpha, \qquad \text{and} \quad n = n_y$$

Equation (11.166*b*) then gives

$$\frac{1}{n_y^2} = \frac{\cos^2\alpha}{n_x^2} + \frac{\sin^2\alpha}{n_z^2} = \cos^2\alpha\left(\frac{1}{n_x^2} - \frac{1}{n_z^2}\right) + \frac{1}{n_z^2}$$

or

$$\cos^2\alpha = \frac{n_y^{-2} - n_z^{-2}}{n_x^{-2} - n_z^{-2}} \qquad (11.167a)$$

From this result we readily obtain

$$\sin^2\alpha = \frac{n_x^{-2} - n_y^{-2}}{n_x^{-2} - n_z^{-2}} \qquad (11.167b)$$

$$\tan^2\alpha = \frac{n_x^{-2} - n_y^{-2}}{n_y^{-2} - n_z^{-2}} \qquad (11.167c)$$

If α is less than 45°, the crystal is said to be a *positive biaxial crystal*. If α is greater than 45°, the crystal is said to be *negative*. The other wave optic axis is also in the xz plane. It makes an angle α with the $-z$ axis. When α is zero or 90°, the two optic axes coincide, and the crystal becomes a positive or negative uniaxial crystal, respectively.

Figure 11.45*a* enables us to visualize one octant of the index surfaces for a biaxial medium. The full surfaces may be obtained by reflection in the xy, yz, and zx planes. Note that there are two interpenetrating sheets that meet at the optic axes. Figures 11.45 *b*, *c*, *d* show the intersections of the full surfaces with each coordinate plane. Note that these intersections consist of a circle and an ellipse. The circle corresponds to light with **E** and **D** perpendicular to the plane, the ellipse to light with **E** and **D** in the plane.

*2. **Internal Conical Refraction.*** When a narrow unpolarized light beam is incident normally onto a plane parallel biaxial crystalline plate cut so that the surface is normal to an optic axis, we find that the light rays inside the crystal form a hollow cone, and the emerging light forms a hollow cylinder. The cone and cylinder are tangent to the optic axis. The light at each point on the cylinder is found to be linearly polarized. Two extreme polarization cases are indicated schematically in Fig. 11.46. This phenomenon is known as *internal conical refraction.*

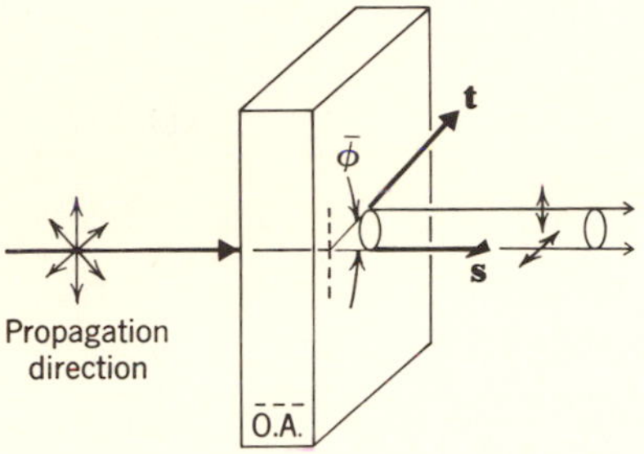

Fig. 11.46. Internal conical refraction.

To explain this effect we must first realize that when the wave normal **s** is along an optic axis, any direction of **D** perpendicular to it will give rise to an allowed linearly polarized optical mode, that is, to a solution of Eqs. (11.140). This will be discussed in more detail below with the aid of the Poinsot ellipsoid. Figure 11.47 shows this situation for **s** along the optic axis. The plane perpendicular to the optic axis is labelled the "*D* plane." A typical direction for **D** might be along the line QA, making an angle γ with the y axis. The corresponding direction of **E** would be along QB in the "*E* plane." This latter plane intersects the D plane along the y axis and makes an angle $\bar{\phi}$ with it. The points Q, R, and L are in the xz plane of Fig. 11.45. When **D** is along QR, **E** is along QL, and the ray direction is along **t**. This is one extreme case. When **D** is along the y axis, so is **E**, and the ray direction is along the optic axis. Then **D** along the direction QA would correspond to an intermediate case, and the ray direction is as shown in Fig. 11.47. As the angle γ varies, the ray direction produces the conical pattern shown.

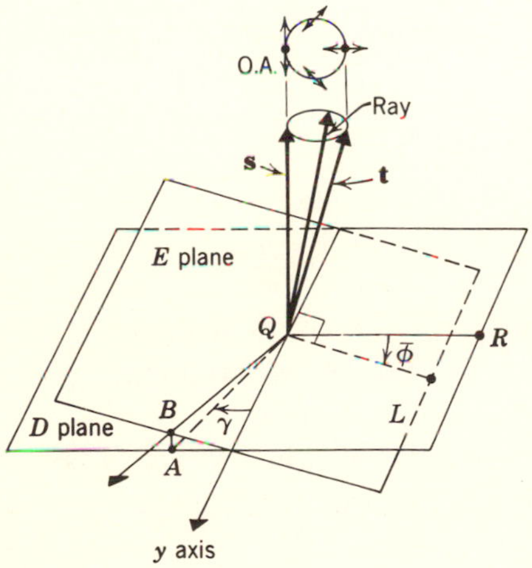

Fig. 11.47

An unpolarized light beam incident along the optic axis can be considered in a somewhat oversimplified way to be a superposition of many incident linearly polarized beams with all planes of vibration of **D** equally represented. The corresponding directions of **E** and the Poynting vector **S** then differ in the way just described, so that all ray directions on the cone are represented. Each ray is plane polarized, because it has a well-defined direction for **D**. When the light rays emerge from the biaxial plate, the tangential component of **s** must remain zero, so that they must all propagate with both wave normals and ray velocities normal to the plate, thus producing the hollow cylinder of rays.

3. ***The Poinsot Ellipsoid.*** In a coordinate system aligned along the principal axes of the dielectric tensor we can write

$$E_x = \frac{D_x}{n_x^2}, \qquad E_y = \frac{D_y}{n_y^2}, \qquad E_z = \frac{D_z}{n_z^2} \tag{11.168}$$

Hence the scalar product $W = (\mathbf{D} \cdot \mathbf{E})$ is given by

$$W = (\mathbf{D} \cdot \mathbf{E}) = \frac{D_x^2}{n_x^2} + \frac{D_y^2}{n_y^2} + \frac{D_z^2}{n_z^2} \tag{11.169}$$

With W considered to be a function of **D**, the equation $W = 1$ gives the equation of an ellipsoid

$$\frac{D_x^2}{n_x^2} + \frac{D_y^2}{n_y^2} + \frac{D_z^2}{n_z^2} = 1 \tag{11.170}$$

called the *Poinsot ellipsoid* or sometimes the *index ellipsoid*. The former nomenclature will be used here to avoid confusion with the index surfaces, which are quite different entities. The Poinsot ellipsoid is sketched in Fig. 11.48. Let the center be at O, and let P be a point on the ellipsoid. Then if $\mathbf{D} = \mathbf{OP} = (D_x, D_y, D_z)$, it will satisfy the equation $W(\mathbf{D}) = 1$.

The normal $\boldsymbol{\nu}$ to the ellipsoid at P is along the gradient of the function $W(\mathbf{D})$ with respect to the components of **D**, that is, we have

$$\boldsymbol{\nu} \propto \nabla_D W = \left(\frac{\partial W}{\partial D_x}, \frac{\partial W}{\partial D_y}, \frac{\partial W}{\partial D_z}\right) = 2\left(\frac{D_x}{n_x^2}, \frac{D_y}{n_y^2}, \frac{D_z}{n_z^2}\right) = 2\mathbf{E}$$

where **E** is the electric field associated with the vector $\mathbf{D} = \mathbf{OP}$. Thus the electric field is along the normal to the index surface, whereas the **D** field is along **OP**.

The relative magnitudes of **E** and **D** may also be determined from the Poinsot ellipsoid. Construct the tangent plane at P and let **OQ** be parallel to the normal, that is, parallel to **E**. As usual, let the angle between **D** and **E** be denoted by ϕ. Then with $\mathbf{D} = \mathbf{OP}$ on the ellipsoid and **E** given by Eq. (11.168), we can write Eq. (11.169) as

$$W = (\mathbf{E} \cdot \mathbf{D}) = |\mathbf{E}|\, |\mathbf{OP}| \cos\phi = 1$$

The magnitude of the electric field is then

$$|E| = \frac{1}{|\mathbf{OP}| \cos\phi} = \frac{1}{|OQ|} \tag{11.171a}$$

We emphasize that this value of **E** corresponds to the special case of **D** on the ellipsoid, that is, $|\mathbf{D}| = |\mathbf{OP}|$. If **D** is parallel to **OP** but has magnitude $|\mathbf{D}| \neq |\mathbf{OP}|$, we would have to change (11.171*a*) to

$$|\mathbf{E}| = \frac{|\mathbf{D}|}{|OP|\,|OQ|} \tag{11.171b}$$

As is clear from Fig. 11.36, the plane of **D** and **E** is also the plane of **s** and **S**, with $\mathbf{s} \perp \mathbf{D}$, $\mathbf{S} \perp \mathbf{E}$. As shown in Fig. 11.48, the ray direction **S** will be along QP in the tangent plane at P. Thus the construction of Fig. 11.48 determines the directions of **E**, **s**, and **S**, once the direction of **D** is given. But as we have seen above, it is often required to find the directions of **D** and **E**, given the direction of the wave normal **s**. Previously we employed the index surface to do this, but a geometrical solution can also be found using the Poinsot ellipsoid. This is illustrated in Fig. 11.49.

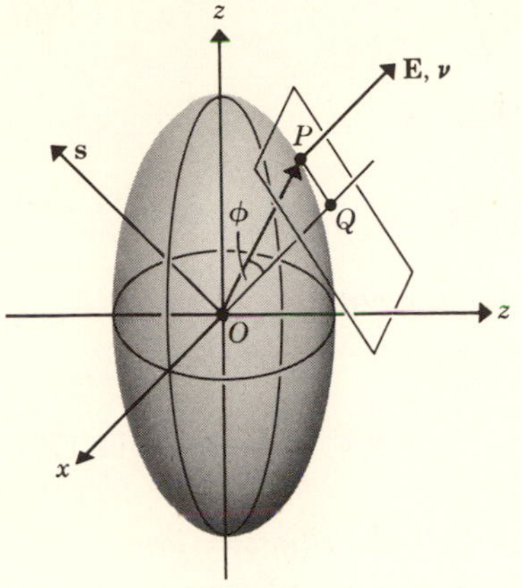

(*a*)

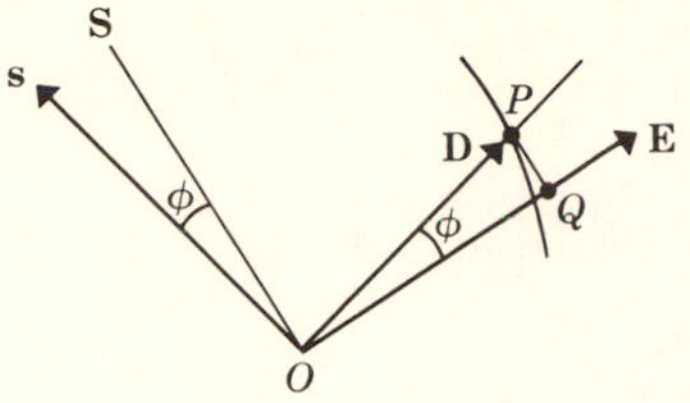

(*b*)

Fig. 11.48. The Poinsot ellipsoid. (*a*) Three-dimensional view. (*b*) Section in the plane OPQ. PQ is a section of the tangent plane at P.

First assume that for a given **s**, Eqs. (11.140*a*) have been solved for n_1^2, $\mathbf{E}_1$, n_2^2, and $\mathbf{E}_2$. Recall

$$\mathbf{D}_1 = n_1^2\,\mathbf{E}_{1\perp} \tag{11.144b}$$

$$\mathbf{D}_2 = n_2^2\,\mathbf{E}_{2\perp} \tag{11.144c}$$

$$(\mathbf{D}_1 \cdot \mathbf{D}_2) = 0 \tag{11.144d}$$

Choose a Cartesian coordinate system with z' axis along **s**, and x' and y' axes along the directions $\mathbf{OP}_1$, $\mathbf{OP}_2$ of $\mathbf{D}_1$ and $\mathbf{D}_2$, respectively. Now let $\mathbf{D}_1$ be on the Poinsot ellipsoid, that is, let $|\mathbf{D}_1| = |OP_1|$. Then $|\mathbf{E}_{1\perp}| = |OP_1|/n_1^2$ and

$$W = (D_1 \cdot E_1) = |\mathbf{D}_1|\,|\mathbf{E}_{1\perp}| = \frac{|OP_1|^2}{n_1^2}$$

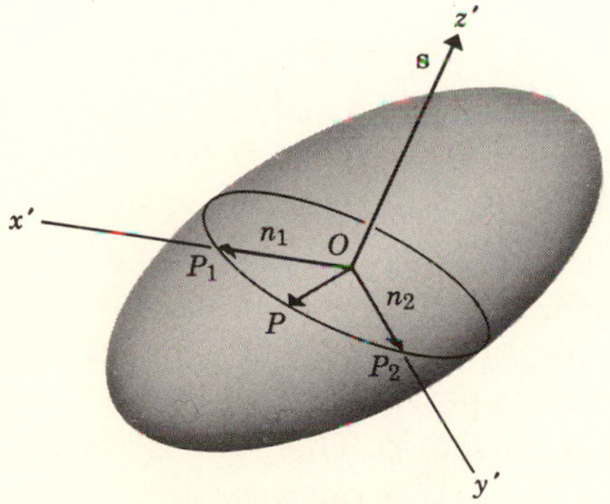

Fig. 11.49. The Poinsot ellipsoid intersects the plane normal to **s** in an ellipse whose semi-axes, along OP_1 and OP_2, have lengths n_1 and n_2 and give the allowed directions of **D**.

Since $\mathbf{D}_1$ is on the Poinsot ellipsoid, we must have $W = 1$, or

$$|\mathbf{OP}_1| = n_1$$

Similarly (11.172)

$$|\mathbf{OP}_2| = n_2$$

We now seek an equation to describe a more general **D** vector **OP** that is both on the Poinsot ellipsoid and perpendicular to **s**. Therefore, P must lie on the closed curve generated by the intersection of the ellipsoid with the plane normal to **s**. We let $\mathbf{d}_1$ and $\mathbf{d}_2$ be unit vectors along $\mathbf{OP}_1$ and $\mathbf{OP}_2$ and write for arbitrary **D** in the $x'y'$ plane

$$\mathbf{D} = D_1\mathbf{d}_1 + D_2\mathbf{d}_2 = \mathbf{D}_1 + \mathbf{D}_2$$

The corresponding **E** vector has components perpendicular to **s** given by

$$\mathbf{E}_\perp = \mathbf{E}_{1\perp} + \mathbf{E}_{2\perp}$$

where, according to Eqs. (11.144*b*, *c*)

$$\mathbf{E}_{1\perp} = \frac{\mathbf{D}_1}{{n_1}^2}$$

$$\mathbf{E}_{2\perp} = \frac{\mathbf{D}_2}{{n_2}^2}$$

Thus

$$\mathbf{E}_\perp = \frac{D_1\mathbf{d}_1}{{n_1}^2} + \frac{D_2\mathbf{d}_2}{{n_2}^2}$$

The requirement that $\mathbf{D} = \mathbf{OP}$ be on the Poinsot ellipsoid is

$$1 = W = (\mathbf{E}\cdot\mathbf{D}) = (\mathbf{E}_\perp\cdot\mathbf{D}) = \frac{{D_1}^2}{{n_1}^2} + \frac{{D_2}^2}{{n_2}^2}$$

This is the equation of an ellipse with semiaxes of length n_1 and n_2 along $\mathbf{OP}_1$ and $\mathbf{OP}_2$, respectively.

The prescription for finding $\mathbf{D}_1$, n_1, $\mathbf{D}_2$, and n_2 from the Poinsot ellipsoid for a given wave normal **s** is therefore as follows. The section of the ellipsoid cut by the plane normal to **s** will be an ellipse. The mutually perpendicular semiaxes of this ellipse give the directions of vibration for the **D** fields in the two linearly polarized waves that will propagate with wave normal **s**. The lengths of these semiaxes give the corresponding values of n. The **E** field is along the normal to the ellipse, as shown in Fig. 11.48, and has a magnitude given by Eq. (11.171*b*).

This prescription is unique except when **s** is along an optic axis of the crystal. As is clear from Fig. 11.45, the two indices of refraction n_1 and n_2 must both be equal to n_y. This means that the plane normal to the optic axis must cut the Poinsot ellipsoid in a circle of radius n_y. This is shown in Fig. 11.50. Since the semiaxes of a circle are not defined, all directions in the normal plane are possible directions for **D**.

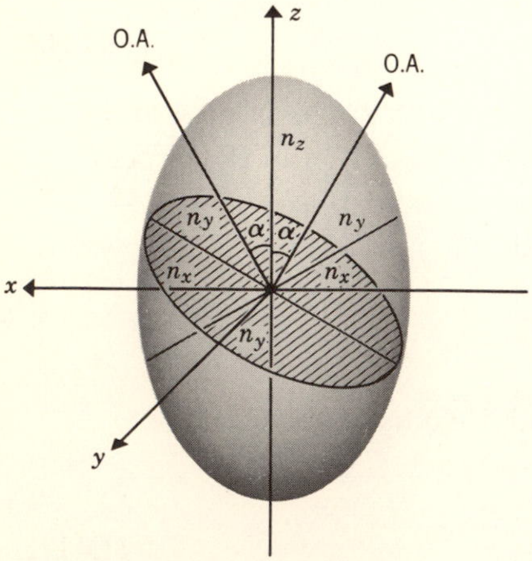

Fig. 11.50. The Poinsot ellipsoid and the optic axes.

REFERENCES FOR SECTION 11.4

F. A. Jenkins and H. A. White, *Fundamentals of Optics*, Chapter 16.
J. M. Stone, *Radiation and Optics*, Chapter 17.
R. S. Longhurst, *Geometrical and Physical Optics*, Chapter 22.
R. W. Ditchburn, *Light*, Chapter 16.
A. Yariv, *Quantum Electronics*, Chapter 18. (For applications to nonlinear optics.)
M. Born and E. Wolf, *Principles of Optics*, Chapter 14.
G. N. Ramachandran and S. Ramaseshan, "Crystal Optics," *Encyclopedia of Physics* (*Handbuch der Physik*), Vol. 25, Part 1.

PROBLEMS

11.4.1 Prove that the values of n^2 that solve Eqs. (11.140 *a*, *b*) must be positive (Hint. Eqs. (11.140 *a*, *b*) are equivalent to

$$\mathbf{D} = \boldsymbol{\epsilon}\mathbf{E} = n^2\mathbf{E}_\perp$$

The tensor $\boldsymbol{\epsilon}$ is *positive definite*, which means that

$$\sum_{i,j} \epsilon_{ij} E_i E_j > 0$$

unless $E_i = 0$ for all i.)

11.4.2 Show that the two linearly polarized waves for a given wave normal **s** have electric fields $\mathbf{E}_1$ and $\mathbf{E}_2$ that are in general *not* orthogonal.

11.4.3 Describe in detail how an index surface $n(\mathbf{s})$ allows us to find all quantities **D**, **E**, **H**, and **S**, given **s** and the magnitude of **E**.

11.4.4 Prove Eq. (11.147*d*).

11.4.5 Derive Eq. (11.166*a*) from Fresnel's equation (11.143) with $s_x = 0$.

Problems 11.4.6. through 11.4.8. refer to anisotropic crystals cut into prisms to be used in a spectrometer such as that of Fig. 7.29 or a similar instrument with nearly parallel light. The plane of incidence is perpendicular to the apex AC in the accompanying illustration.

11.4.6 Show that when a uniaxial crystal is made into a prism with its apex AC parallel to the optic axis, both ordinary and extraordinary rays obey the ordinary law of refraction with indices n_0 and n_e, respectively.

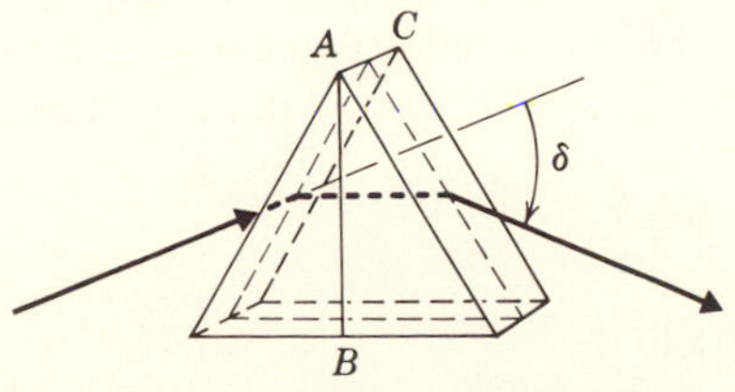

11.4.7 Consider a prism made from a uniaxial crystal with its optic axis parallel to AB. Show that light with **E** polarized perpendicular to the plane of incidence is refracted in the usual way with index n_0. Show that light with **E** vibrating in the plane of incidence behaves in a more complicated fashion unless the deviation angle δ is a minimum δ_m, at which point the simple law of refraction holds with index n_e given by

$$n_e = \frac{\sin \frac{1}{2}(\alpha + \delta_m)}{\sin \frac{1}{2}\alpha}$$

(See Problem 1.3.6.)

11.4.8 If a biaxial crystal is made into a prism as shown above, with the apex *AC* parallel to the *j*th principal axis of the dielectric tensor, show that light polarized parallel to *AC* will obey the simple form of Snell's law with index n_j. If, in addition, another principal axis, say the *k*th, is oriented parallel to *AB*, then light polarized in the plane of incidence will have an angle δ_m of minimum deviation satisfying

$$n_k = \frac{\sin \frac{1}{2}(\alpha + \delta_m)}{\sin \frac{1}{2}\alpha}$$

Show this.

11.4.9 An unpolarized yellow light beam from a sodium lamp is incident normally onto a natural cleavage face of a calcite rhomb, such as the one in Fig. 10.16. It emerges from the opposite face as two parallel separated beams with perpendicular polarizations. The situation in a principal

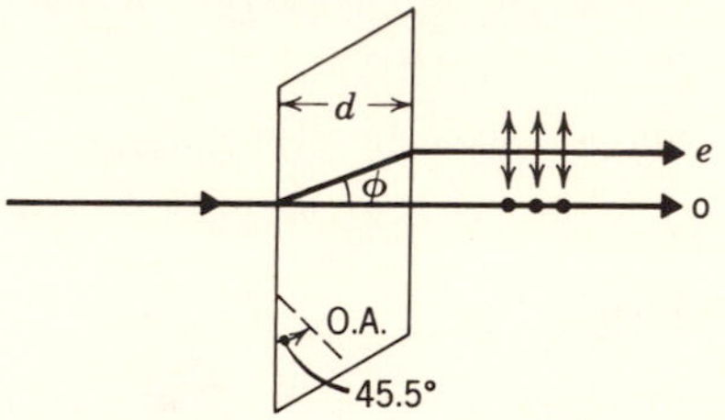

section (that is, a plane containing the optic axis) is illustrated here. For the sodium yellow line the indices of calcite are $n_0 = 1.658$, $n_e = 1.486$. Show that for this geometry the extraordinary ray has index $n' = 1.57$ and that $\phi = 6.3°$. Show that the emerging rays have been retarded by a relative phase angle

$$\Delta_0 - \Delta_e = \frac{4\pi d}{\lambda}(n_0 - n')$$

11.4.10 Show that in a uniaxial crystal the largest angle $\phi = \psi - \theta$ between the ray direction and the direction of the wave normal **s** occurs when **s** makes an angle with the optic axis of

$$\theta = \tan^{-1}\left(\frac{n_e}{n_0}\right)$$

and that

$$\tan \phi_{\max} = \frac{n_0^2 - n_e^2}{2n_e n_0}$$

11.4.11 Show that it is possible to make a quarter-wave plate to use near normal incidence from a thin slice of a biaxial crystal. Thin sheets of mica oriented almost exactly parallel to the *yz* plane may be prepared easily by cleaving. What thickness should such a sheet have to make a quarter-wave plate for $\lambda = 500$ nm if the principal indices are $n_x = 1.5601$, $n_y = 1.5936$, and $n_z = 1.5877$?

11.4.12 Consider the calcite prisms in the figure used with unpolarized light at normal incidence as shown. Use $n_0 = 1.668$ and $n_e = 1.491$ and calculate the range of α for which only one polarized component is critically reflected at the second surface. Which one is it? Are prisms (*a*) and (*b*) equivalent in this regard? Are they equivalent for the dashed rays?

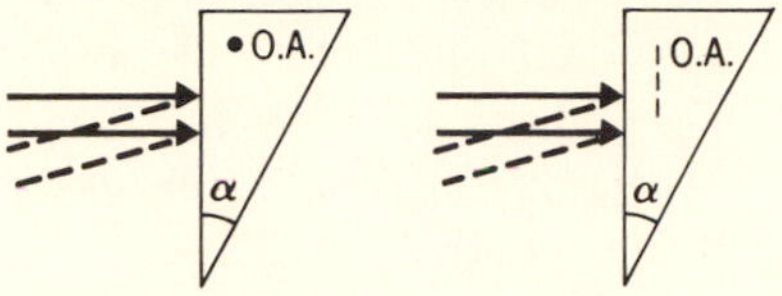

(*a*) (*b*)

11.4.13 A parallel incident unpolarized light beam is incident normally onto one face of a turquoise crystal, which is biaxial and has principal indices of $n_x = 1.520$, $n_y = 1.523$, and $n_z = 1.530$. With $\alpha = 30°$, calculate the angles of deviation for the following ways of cutting the prism: $AB//x$ and $BC//y$, with $AB//y$ and $BC//z$, and with $AB//z$ and $BC//x$. With each of the above orientations, find the range of α for which only one of the two mutually perpendicularly polarized rays is critically reflected.

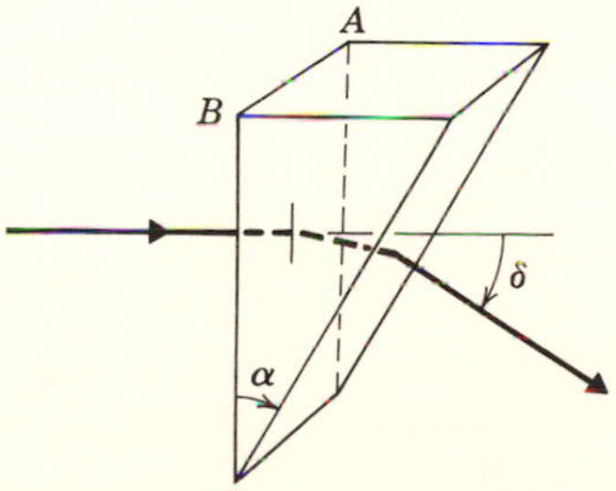

11.4.14 Consider Rochon and Wollaston prisms such as those of Fig. 10.19 made of quartz for which $n_0 = 1.54467$ and $n_e = 1.55379$. With unpolarized light incident normally, calculate the angles made by both ordinary and extraordinary rays with the normal to the second surface after they have emerged from it ($\alpha = 35°$).

11.4.15 Consider an air-spaced polarizing prism made of calcite. It

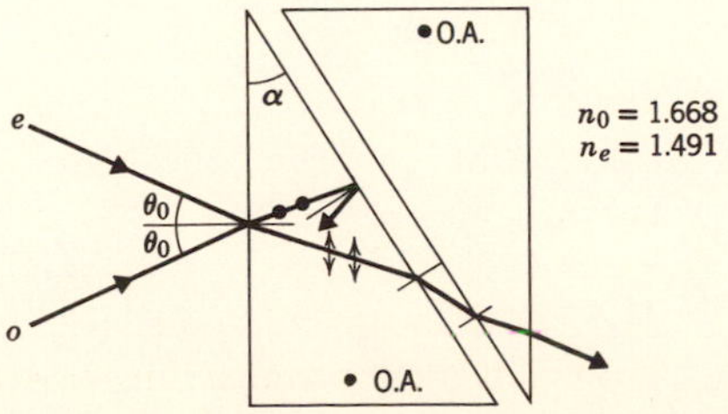

works by totally reflecting the ordinary ray at the air gap, while passing the extraordinary ray. There is only a finite range of the angle of incidence θ for which both of these effects occur. The prism is designed so that if the *o* ray is incident below the normal at an angle greater than θ_0, it ceases to be totally reflected, whereas if the *e* ray is incident above the normal at an angle greater than θ_0, it ceases to be transmitted and is also totally reflected. Use these two criteria to determine the maximum angular spread $2\theta_0$ and the prism angle α. (Answer: $2\theta_0 = 8.17°$ and $\alpha = 39.5°$.)

11.4.16 When the air gap of the prism of Problem 11.4.15 is replaced by linseed oil ($n = 1.494$), the angular aperture $2\theta_0$ can be considerably increased provided that the angle α is made much larger. This gives the Glan-Thompson polarizer of Fig. 10.19. With $\alpha = 76.4°$, find the maximum value of the angle of incidence for the *o* ray for which it will still be totally reflected.

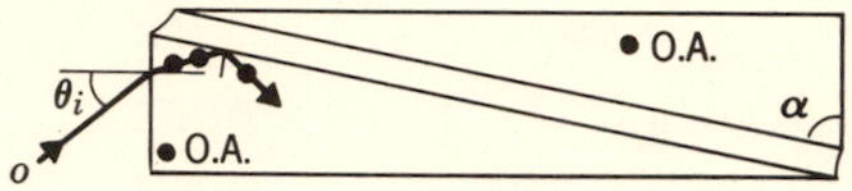

APPENDICES

CONTENTS

APPENDIX A THE DELTA FUNCTION AND THE FOURIER INTEGRAL

THE DELTA FUNCTION

1. PROPERTIES

The delta function is a singular function that can be defined as the limit of a sequence of well-behaved or regular functions. The functions in a sequence giving the delta function $\delta(x-x_0)$ should be normalized to unit area and be strongly peaked about $x=x_0$. The resulting delta function is defined by the properties

$$\delta(x-x_0)=0 \qquad \text{for} \quad x \neq x_0 \tag{i}$$

$$\int_{x_0-a}^{x_0+b} \delta(x-x_0)\,dx = 1 \qquad \text{for } a, b \text{ positive} \tag{ii}$$

$$\int_{x_0-a}^{x_0+b} f(x)\,\delta(x-x_0)\,dx = f(x) \qquad \text{for } a, b \text{ positive} \tag{iii}$$

and $f(x)$ continuous at x_0

If the limits of integration in (ii) or (iii) are changed so that the range of integration does not include $x=x_0$, then the integrals give zero.

An important additional property is

$$\delta(\alpha x)=\frac{\delta(x)}{|\alpha|}\ (\text{real } \alpha) \tag{iv}$$

This equation will be proved by showing that it satisfies (i)–(iii). It automatically satisfies Property (i). Properties (ii) and (iii) can be treated together. The left-hand side of (iii) gives

$$\int_{-a}^{+b} f(x)\,\delta(\alpha x)\,dx = \left(\frac{1}{\alpha}\right)\int_{-a/\alpha}^{+b/\alpha} f\left(\frac{y}{\alpha}\right)\delta(y)\,dy = \frac{f(0)}{|\alpha|}$$

A special case involves $\alpha=-1$. Then we have

$$\delta(-x)=\delta(x) \tag{v}$$

2. REPRESENTATIONS OF THE DELTA FUNCTION

A family of functions that tends to the delta function in a limit is said to form a *representation* of the delta function. One example is provided by the function plotted in Fig. A.1 and given in Eq. (A.5). Its properties will be discussed in some detail below. Another

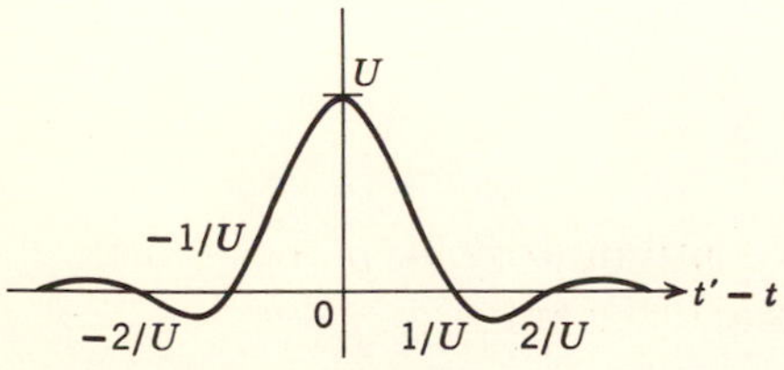

Fig. A.1. The function $\sin \pi U(t'-t)/\pi(t'-t)$ that appears in Dirichlet's formula.

good example is provided by a normalized Lorentzian function

$$L(x-x_0;\,\epsilon) = \frac{\epsilon/\pi}{(x-x_0)^2+\epsilon^2} = \frac{1}{\pi}\,\mathrm{Im}\frac{1}{x-x_0-i\epsilon}$$

which in the limit as $\epsilon \to 0$ becomes $\delta(x-x_0)$. Here Im means "imaginary part of."

Another example of a representation of the delta function is given by the Fresnel integrand of Eq. (9.5). Explicitly one can obtain the result

$$\delta(x) = \lim_{L\to 0}\left[\frac{e^{i(\pi/4-kL/2)}e^{-ikx^2/(2L)}}{\sqrt{2\pi L/k}}\right]$$

Yet another example is provided by the Fourier transform. See Eq. A.13.

THE FOURIER INTEGRAL

1. THE COMPLEX FOURIER INTEGRAL

Consider a function $f(t)$, which may be complex, with the following properties.*

1. It is "square integrable," that is, the limit of the integral

$$\int_{-T/2}^{+T/2} |f(t)|^2\,dt$$

exists as $T \to \infty$.

2. The function is continuous, and df/dt exists at all but a finite number of points in any finite range of the variable t. Then the *Fourier integral*

$$\hat{f}(u) = \int_{-\infty}^{\infty} f(t')e^{-2\pi iut'}\,dt' \tag{A.1}$$

exists and is continuous at all but a finite number of points in any finite range of the real variable u. The function $\hat{f}(u)$ is called the *Fourier transform* of $f(t)$. It is important because it furnishes an integral representation of $f(t)$ as a superposition of exponentials. Explicitly, *the integral*

$$\int_{-\infty}^{\infty} \hat{f}(u)^{2\pi iut}\,du \tag{A.2}$$

exists and equals $f(t)$ at points of continuity of f. At points of discontinuity the integral converges to the average

*These properties are sufficient for the stated relations to hold. Often weaker conditions are all that is required. These fine points will be left to the mathematicians.

$$\lim_{\epsilon\to 0} \tfrac{1}{2}[f(t+\epsilon)+f(t-\epsilon)]$$

of the right-hand and left-hand limits.

The truth of these statements rests on a formula which may be obtained by substituting Eq. (A.1) into (A.2):

$$\int_{-\infty}^{\infty} \hat{f}(u)e^{2\pi iut}\,dt = \lim_{U\to\infty} \int_{-U/2}^{+U/2} \hat{f}(u)e^{2\pi iut}\,dt$$

$$= \lim \int_{-U/2}^{+U/2} e^{2\pi iut}\,dt\left[\int_{-\infty}^{\infty} f(t')e^{-2\pi iut'}\,dt'\right]$$

We interchange the order of integration to obtain

$$\lim_{U\to\infty} \int_{-\infty}^{\infty} f(t')\,dt'\left[\int_{-U/2}^{+U/2} e^{2\pi iu(t-t')}\,du\right] = \lim_{U\to\infty} \int_{-\infty}^{+\infty} f(t')\,\frac{\sin \pi U(t'-t)}{\pi(t'-t)}\,dt'$$

$$= \lim_{\epsilon\to 0} \frac{1}{2}[f(t+\epsilon)+f(t-\epsilon)] \qquad \text{(A.3)}$$

The formula that we need in order to establish the last equality in (A.3) is called *Dirichlet's formula* and may be written in the form

$$\lim_{u\to\infty} \int_{t-a}^{t+b} f(t')\,\frac{\sin \pi U(t'-t)}{\pi(t'-t)}\,dt' = \lim_{\epsilon\to 0} \frac{1}{2}[f(t+\epsilon)+f(t-\epsilon)] \qquad \text{(A.4)}$$

where a and b are positive constants. The range of the t' integration must include the point $t' = t$, otherwise the limit gives zero.

We shall not prove Dirichlet's formula* but instead shall try to make it seem reasonable. The function

$$d(t'-t) \equiv \frac{\sin \pi U(t'-t)}{\pi(t'-t)}$$

that appears in Eq. (A.4) is plotted in Fig. A.1. It has a peak value of U at $t' = t$ and first goes through zero at $t' = t \pm 1/U$. Its integrated area is

$$\int_{-\infty}^{\infty} d(t'-t)\,dt' = 1$$

As U becomes larger, the function becomes more strongly peaked at $t' = t$ and hence more narrow. Eventually U becomes large enough so that we may ignore the variation of $f(t')$ over values of t' near t where $d(t'-t)$ is nonzero and we may set $f(t') = f(t)$. We then obtain for the integral in (A.4)

$$f(t) \int_{-\infty}^{\infty} \frac{\sin \pi U(t'-t)}{\pi(t'-t)}\,dt' = f(t)$$

This result is equivalent to the statement that $d(t'-t)$ tends to a delta function:

$$\operatorname*{Lim}_{U\to\infty}\left[\frac{\sin \pi U(t'-t)}{\pi(t'-t)}\right]=\delta(t'-t) \tag{A.5}$$

It should be clear that the key region is $|t'-t| \lesssim n/U$, where n is a small integer. If $f(t')$ remains essentially constant over the range $f(t-\Delta)$ to $f(t+\Delta)$ and if $1/U$ is somewhat less than Δ, the limiting value of the integral in Eq. (A.3) and hence in Eq. (A.2) is reached. The condition then is $U > 1/\Delta$.

2. THE REAL FOURIER INTEGRALS

The integral in Eq. (A.2) represents $f(t)$ as a superposition of exponentials $e^{2\pi i u t}$. An equivalent superposition in terms of sines and cosines is possible. Let $f(t)$ be real: $f(t')^* = f(t')$. Then take the complex conjugate of Eq. (A.1):

$$\hat{f}(u)^* = \int_{-\infty}^{\infty} f(t')e^{2\pi i u t'}\,dt' = \int_{-\infty}^{\infty} f(t')e^{-2\pi i(-u)t'}\,dt' = \hat{f}(-u) \tag{A.6}$$

Now write $\hat{f}(u)$ in terms of its absolute value and phase angle:

$$\hat{f}(u) = |\hat{f}(u)|e^{i\psi(u)} \tag{A.7a}$$

Equation (A.6) implies that

$$|\hat{f}(-u)| = |\hat{f}(u)| \tag{A.7b}$$

and

$$\psi(-u) = -\psi(u) \tag{A.7c}$$

Then we may write

$$f(t) = \operatorname{Re} f(t) = \operatorname{Re}\int_{-\infty}^{\infty} \hat{f}(u)e^{2\pi i u t}\,du = \operatorname{Re}\int_{-\infty}^{\infty} |\hat{f}(u)|e^{i[2\pi u t+\psi(u)]}\,du$$

$$= \int_{-\infty}^{\infty} |\hat{f}(u)| \cos\,[2\pi u t + \psi(u)]\,du$$

or

$$f(t) = 2\int_{0}^{\infty} |\hat{f}(u)| \cos\,[2\pi u t + \psi(u)]\,du \tag{A.8}$$

$$= 2\int_{0}^{\infty} |\hat{f}(u)| \cos\psi(u)\cos 2\pi u t\,du$$

$$-2\int_{0}^{\infty} |f(u)| \sin\psi(u)\sin 2\pi u t\,dt$$

*For a proof of a similar relation, see R. Courant, *Differential and Integral Calculus*, Vol. 1, pp. 450.

The first equality in Eq. (A.8) represents $f(t)$ as a superposition of cosine waves of different frequency u, each with an amplitude $|f(u)|$ and a phase angle $\psi(u)$. If we take real and imaginary parts of Eq. (A.1), we obtain

$$\begin{aligned} \mathrm{Re}\hat{f}(u) &= |f(u)|\cos\psi(u) = \int_{-\infty}^{\infty} f(t')\cos 2\pi ut'\,dt' \\ -\mathrm{Im} f(u) &= -|f(u)|\sin\psi(u) = \int_{-\infty}^{\infty} f(t')\sin 2\pi ut'\,dt' \end{aligned} \tag{A.9}$$

from which $|\hat{f}(u)|$ and $\psi(u)$ may be obtained. We may also write Eq. (A.8) in the form

$$f(t) = \int_0^{\infty} C_1(u)\cos 2\pi ut\,du + \int_0^{\infty} S_2(u)\sin 2\pi ut\,du \tag{A.10}$$

The functions $C_1(u) \equiv [2\,\mathrm{Re}\hat{f}(u)]$ and $S_2(u) \equiv -[2\,\mathrm{Im}\hat{f}(u)]$ are known as the cosine and sine Fourier transforms of $f(t)$.

From Eqs. (A.9) we obtain

$$\begin{aligned} C_1(u) &= 2\int_{-\infty}^{\infty} f(t')\cos 2\pi ut'\,dt' \\ S_2(u) &= 2\int_{-\infty}^{\infty} f(t')\sin 2\pi ut'\,dt' \end{aligned} \tag{A.11}$$

Note the reciprocity of Eqs. (A.1) and (A.2). If $\hat{f}(u)$ is the Fourier transform of $f(t)$, then $f(t)$ is the Fourier transform of $\hat{f}(-u) = \hat{f}(u)^*$. The equation for $\hat{f}(u)$ that is analogous to (A.3) is

$$\hat{f}(u) = \lim_{T\to\infty} \int_{-T/2}^{+T/2} \hat{f}(u')\frac{\sin \pi T(u'-u)}{\pi(u'-u)}\,du' \tag{A.12}$$

If we ask how large T must be for the limit to be reached, the answer is: Larger than $1/\Delta u$, where Δu is a measure of the range of u' about u, over which $\hat{f}(u')$ may be assumed to remain constant. Again the reason is that the function

$$\frac{\sin \pi T(u'-u)}{\pi(u'-u)}$$

will integrate to a value of unity before $\hat{f}(u')$ is able to change appreciably from its value at $u' = u$.

3. THE FOURIER TRANSFORM OF A DELTA FUNCTION AND OF A CONSTANT

Let $f(t') = \delta(t' - t_0)$. Then the Fourier transform is

$$\int_{-\infty}^{+\infty} \delta(t' - t_0) e^{-2\pi i u t'} \, dt' = e^{-2\pi i u t_0} \tag{A.13}$$

Now let $f(t') = C$, a constant. The resulting Fourier integral

$$C \int_{-\infty}^{\infty} e^{-2\pi i u t'} \, dt'$$

is indeterminate as it is written. We must understand it to be the limit of an integral over a symmetric range, say from $-T/2$ to $+T/2$; then it gives a δ function

$$\operatorname*{Lim}_{T\to\infty} C \int_{-T/2}^{+T/2} e^{-2\pi i u t'} \, dt' = C \operatorname{Lim} \frac{\sin \pi u T}{\pi u} = C\delta(u) \tag{A.14}$$

REFERENCES

Many physics books have sections on the Fourier integral, as do books on mathematical methods for physicists. See, for example, *Mathematical Methods in the Physical Sciences* by Mary L. Boas, Chapter 13, Sections 4-7.

A more rigorous treatment of delta functions and Fourier integrals will be found in M. J. Lighthill, *Introduction to Fourier Analysis and Generalized Functions*.

APPENDIX B
FOURIER ANALYSIS OF THE OPTICAL SIGNAL AND THE SPECTRAL DISTRIBUTION FUNCTION

We wish to show the equivalence of the expression (6.24) for $\gamma(\tau)$ in Section 6.1 and Eq. (6.50) for $\gamma(\tau)$ used in Section 6.2. Thus we must prove the Wiener–Khintchine theorem, Eq. (6.51),

$$\gamma(\tau) = \operatorname*{Lim}_{T\to\infty} \frac{\int_{t_0-T/2}^{t_0+T/2} E(t'-\tau)E(t')\,dt'}{\int_{t_0-T/2}^{t_0+T/2} E(t')^2\,dt'} = \int_0^\infty P(\nu)\cos 2\pi\nu\tau\,d\nu \tag{B.1}$$

A proof requires a precise definition of the spectral distribution function $P(\nu)$ in terms of $E(t')$. It is also of interest to determine how large T must be before the limit is reached.

The method of solution uses Fourier analysis, in which the electric field is explicitly represented as a superposition of sinusoidal waves. The statistical nature of the light signal adds some complications.

The signal $E(t')$ may be nonzero over an indefinitely long time span. To assure convergence of the time integrals we shall truncate it. Thus we define a function

$$\begin{aligned} E_S(t') &= E(t') \qquad \text{for} \quad |t'| \leq \frac{S}{2} \\ &= 0 \qquad\qquad \text{for} \quad |t'| > \frac{S}{2} \end{aligned} \tag{B.2}$$

We lose nothing by this if S is chosen to be much greater than the averaging time T.

We then express $E_S(t')$ and $E_S(t'-\tau)$ as Fourier integrals:

$$\begin{aligned} E_S(t') &= \int_{-\infty}^{\infty} \hat{E}_S(\nu)e^{2\pi i\nu t'}\,d\nu \\ E_S(t'-\tau) &= \int_{-\infty}^{\infty} \hat{E}_S(\nu')e^{2\pi i\nu'(t'-\tau)}\,d\nu' \end{aligned} \tag{B.3}$$

with the Fourier transform given by

$$\hat{E}_S(\nu) = \int_{-\infty}^{\infty} E_S(t')e^{-2\pi i\nu t'}\,dt' = \int_{-S/2}^{+S/2} E(t')e^{-2\pi i\nu t'}\,dt' \tag{B.4}$$

These expressions are to be inserted into the integrals in Eq. (B.1). Before we take the limit as $T\to\infty$, the numerator becomes

$$\text{Numerator} = \int_{t_0-T/2}^{t_0+T/2} dt' \left[\int_{-\infty}^{\infty} d\nu \int_{-\infty}^{\infty} d\nu' \hat{E}_S(\nu)\hat{E}_S(\nu') e^{2\pi i\nu t'} e^{2\pi i\nu'(t'-\tau)} \right]$$

$$= \iint_{-\infty}^{\infty} d\nu\, d\nu' \hat{E}_S(\nu) E_S(\nu') e^{-2\pi i\nu'\tau} \int_{t_0-T/2}^{t_0+T/2} e^{2\pi i(\nu+\nu')t'}\, dt'$$

$$= \iint_{-\infty}^{\infty} d\nu\, d\nu' \hat{E}_S(\nu)\hat{E}_S(\nu') e^{-2\pi i\nu'\tau} e^{2\pi i(\nu+\nu')t_0} \frac{\sin \pi(\nu+\nu')T}{\pi(\nu+\nu')}$$

Now let $u = \nu + \nu'$. Then we have

$$\text{Numerator} = \int_{-\infty}^{\infty} d\nu'\, e^{-2\pi i\nu'\tau} \hat{E}_S(\nu') \left[\int_{-\infty}^{\infty} \hat{E}_S(u-\nu') e^{2\pi i u t_0} \frac{\sin \pi u T}{\pi u}\, du \right] \tag{B.5}$$

At this point we invoke Dirichlet's formula (Eq. [A.4]), which says that the integral in square brackets in Eq. (B.5) tends to $\hat{E}_S(-\nu')$ as T tends to $+\infty$. This gives

$$\text{Lim (Numerator)} = \int_{-\infty}^{\infty} \hat{E}_S(\nu')\hat{E}_S(-\nu') e^{-2\pi i\nu'\tau}\, d\nu' \tag{B.6}$$

To attain the limit in Eq. (B.6) it can be shown that T must be greater than $2t_0$. Our discussion of Eq. (A.12) showed that T must also be greater than $1/\Delta\nu$, where $\hat{E}_S(\nu)$ may be assumed essentially constant for

$$-\nu' - \Delta\nu \leq \nu \leq -\nu' + \Delta\nu.$$

If we use Eq. (A.6), we find

$$\hat{E}_S(-\nu') = \hat{E}_S(\nu')^*$$

or

$$\hat{E}_S(\nu')\hat{E}_S(-\nu') = |\hat{E}_S(\nu')|^2 = |\hat{E}_S(-\nu')|^2$$

This implies that Eq. (B.6) can also be written

$$\text{Lim (Numerator)} = 2\int_{0}^{\infty} |\hat{E}_S(\nu)|^2 \cos 2\pi\nu\tau\, d\nu \tag{B.7}$$

The Fourier transform

$$\hat{E}_S(\nu) = |\hat{E}_S(\nu)| e^{i\psi_s(\nu)}$$

of the truncated function $E_S(t')$ will not tend to a well-defined limit as $S \to \infty$, because it is unbounded and because the phase ψ_s oscillates and does not converge to a limit. We can show, however, that under a wide range of assumptions about the statistical nature of the function $E(t')$, $1/S$ times the absolute value squared, $(1/S)|\hat{E}_S(\nu)|^2$, does tend to a well-defined limit $|\hat{\epsilon}(\nu)|^2$ as $S \to \infty$. On the physical grounds it is clear that this should happen, because we expect $|\hat{\epsilon}(\nu)|^2 d\nu$ to be proportional to the intensity $I(\nu)$ present in the optical signal in the frequency range from ν to $\nu + d\nu$. If we write

$$I(\nu) = \text{const}\,|\epsilon(\nu)|^2 = \text{const Lim}\left[\frac{1}{S}|\hat{E}_S(\nu)|^2\right]$$

Then Eq. (B.7) becomes for sufficiently large $T(> 1/\Delta\nu)$ and $S(\gg T)$

$$[\text{Numerator of (B.1)}] \rightarrow K \cdot S \cdot \int_0^\infty I(\nu) \cos 2\pi\nu\tau\, d\nu$$

where K is a constant. Since the denominator of Eq. (B.1) is the numerator at $\tau = 0$, this proves the desired limit:

$$\lim_{T\to\infty} \frac{\int_{t_0-T/2}^{t_0+T/2} E(t')E(t'-\tau)\,dt'}{\int_{t_0-T/2}^{t_0+T/2} E(t')^2\,dt'} = \frac{\int_0^\infty I(\nu)\cos 2\pi\nu\tau\,d\nu}{\int_0^\infty I(\nu)\,d\nu}$$

$$= \int_0^\infty P(\nu)\cos 2\pi\nu\tau\,d\nu$$

We may also state the above results in the following way. If we express the truncated function E_S by

$$E_S(t') = \int_{-\infty}^{\infty} \hat{E}_S(\nu) e^{2\pi i\nu t'}\,d\nu$$

$$= 2\int_0^\infty |\hat{E}_S(\nu)| \cos\,[2\pi\nu t' + \psi_s(\nu)]\,d\nu$$

we may identify $(1/S)\,|E_S(\nu)|^2$ as proportional to the intensity in the component wave,

$$|\hat{E}_S(\nu)| \cos\,[2\pi\nu t' + \psi_s(\nu)]$$

If there was no interference with components at other frequencies, this will make a contribution proportional to

$$|\hat{E}_S(\nu)|^2 \cos 2\pi\nu\tau$$

to the autocorrelation function, and the final result would be

$$\gamma_S(\tau) = \frac{\int_0^\infty |\hat{E}_S(\nu)|^2 \cos 2\pi\nu\tau\,d\nu}{\int_0^\infty |\hat{E}_S(\nu)|^2\,d\nu}$$

Then as $S \to \infty$ we would have

$|\hat{E}_S(\nu)|^2 \to S \cdot \text{const}\, I(\nu)$

and the autocorrelation function of $E(t')$ becomes

$$\gamma(\tau) = \frac{\int_0^\infty I(\nu) \cos 2\pi\nu\tau \, d\nu}{\int_0^\infty I(\nu)\, d\nu}$$

If we examine our argument in detail, we see that the key point comes with the use of Dirichlet's formula in Eq. (B.5). This is equivalent to the assumption of noninterference between components of different frequencies. But once we may safely use Dirichlet's formula, there will be no interference. This reason is that interference is revealed as "beats" between components of different frequency. This implies a structure in the function $\hat{E}_S(\nu)$ for frequency differences equal to the beat frequency $\delta\nu$. If the averaging time T is sufficiently large so that we average over many of these beats, they will not be measured. Thus we want $T \gg 1/\delta\nu$. This is equivalent to the condition $T \gg 1/\Delta\nu$ that we used just below Eq. (B.6).

APPENDIX C LASERS

The word "laser" is an acronym for **L**ight **A**mplification by **S**timulated **E**mission of **R**adiation. The lasers used as light sources in the optics laboratory are actually oscillators, not merely amplifiers. In electronics it is known that an amplifier with too much positive feedback may break into oscillation. Similarly a laser amplifier with enough gain will become an oscillator. Unlike the light from usual thermal sources, the light from a laser will be highly coherent. We shall describe the continuous gas laser in some detail, for it is the type most used in optics laboratories.

1. CONSTRUCTION

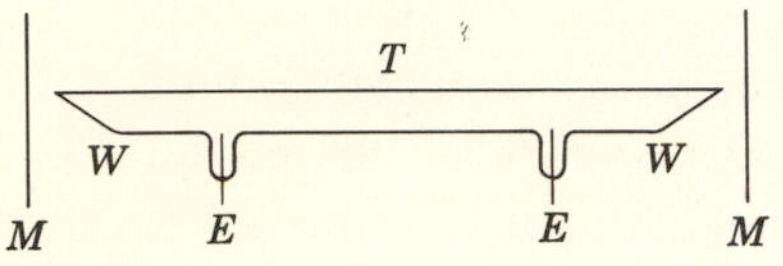

Fig. C.1

The essential features are shown in Fig. C.1. The laser tube T is set between two highly reflecting mirrors M, which form an optical cavity. The ends of the tube are usually sealed by high-quality windows W set at Brewster's angle, so that light passing along the axis of the tube polarized in the plane of the window will be transmitted essentially without loss.

The tube is filled with the desired gas. At the moment the most popular gas is a mixture of helium and neon in a ratio of about 1:5 at a total pressure of about one torr ($\approx$ 1 mm Hg), and we shall use it as an example. A discharge is established in the gas by applying a voltage across the electrodes E. For most critical applications this is a dc voltage in the range of 3–6 Kv across a ballast resistance depending on the tube length. Currents of 10 to 30 milliampers are typical.

2. ATOMIC PHYSICS

We now mention briefly some of the pertinent atomic processes involved in helium-neon laser action. The atomic species directly involved are *excited neon atoms*. The helium acts as a source of *pump power* to raise the neon atoms into the desired excited states. How this can happen is shown briefly in Fig. C.2. For helium we show only three energy levels: the ground state, whose energy is taken to be zero, the metastable $^3S_1$ state, and the metastable $^1S_0$ state.

Electrons accelerated by the electric fields applied at the electrodes collide with helium atoms and raise them to excited states, some of which decay into the $^1S_0$ state. Once there, they cannot decay to the ground state by emission of light; since this is forbidden, they need to lose energy by radiationless means.

Some energy levels of neon are shown in Fig. C.2 also. There is an important series of $(2p)^5 5s$ levels that lie just below the $^1S_0$ level of helium. By "resonant transfer," an excited helium atom passing near a neon atom in its ground state can excite the neon atom into a $(2p)^5 5s$ state and thereby fall into the ground state.

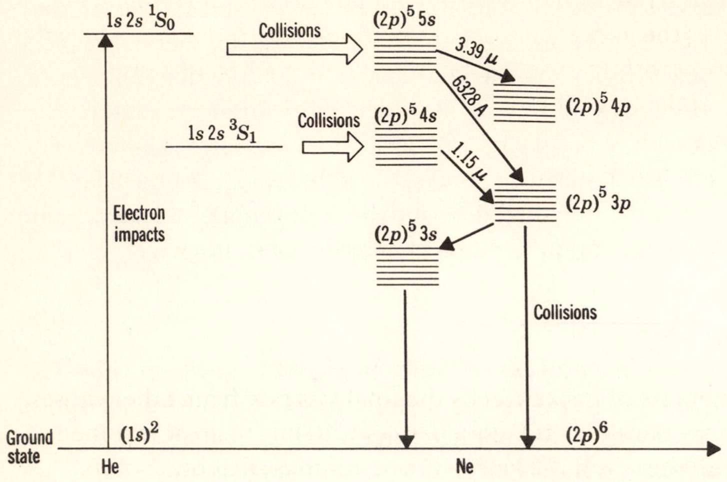

Fig. C.2

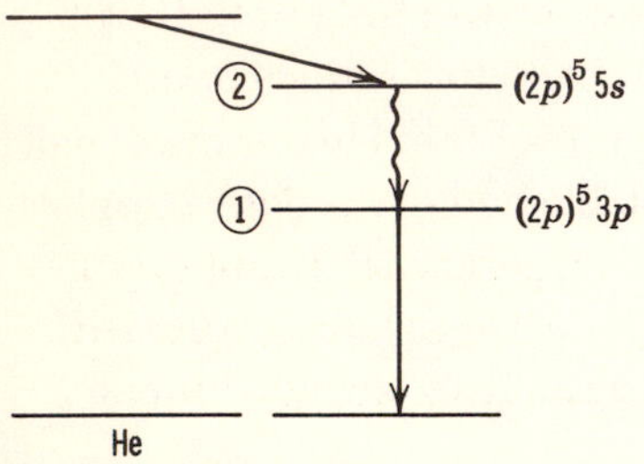

Fig. C.3

The $(2p)^5 5s$ neon states then make transitions to the $(2p)^5 3p$ states. One of these gives rise to the 6328 A line. From the $(2p)^5 3p$ states, the neon atoms can pass to the ground states in two steps by emitting radiation, or directly by collisions with the walls.

Other laser transitions are possible at 3.39 μ and 1.15 μ, as shown in Fig. C.2. The levels important for the 6328 A transition are shown in Fig. C.3. The 2-1 transition is the laser transition.

Laser action requires a *population inversion* between states 1 and 2. By this we mean that the number N_2 of atoms in state 2 must be greater than the number N_1 in state 1. In thermal equilibrium, of course, we have $N_1 > N_2$. Inversion is caused by the pumping action of the helium metastable atoms. It is enhanced by any mechanism which depletes state 1.

In the ordinary case where $N_1 > N_2$ any light present having the circular frequency ω_{12} given by

$$\omega_{12} = \frac{E_2 - E_1}{\hbar}$$

$$\hbar = \frac{h}{2\pi} \tag{C.1}$$

$h =$ Planck's constant

will tend to be absorbed by atoms in state 1 and raised to state 2. As a result, energy will be lost by the radiation field, and gained by the atoms at a rate

$$\left(\frac{dU}{dt}\right)_{\text{loss}} = N_1 B_{12} U(\omega_{12}) \tag{C.2}$$

where $U(\omega)$ represents the spectral energy density at frequency ω and B_{12} is the Einstein B coefficient.

With atoms also present in state 2, *stimulated emission* will occur under the action of the radiation already present at frequency ω_{12}. It induces the excited atoms to emit even more radiation. The latter is coherent with the stimulating light. A net rate of increase of energy in the radiation field results. It is given by:

$$\left(\frac{dU}{dt}\right)_{\text{gain}} = N_2 B_{12} U(\omega_{12}) \tag{C.3}$$

The *net gain* rate by the field is therefore

$$\left(\frac{dU}{dt}\right)_{\text{net}} = (N_2 - N_1) B_{12} U(\omega_{12}) \tag{C.4}$$

If this net rate of gain exceeds the total loss rate from other causes, the energy density U at frequency ω_{12} will start to grow exponentially with time, and the laser will burst into oscillation. As this happens, the population inversion $N_2 - N_1$ is forced to decrease, since more $2 \rightarrow 1$ transitions are now taking place. The gain rate of Eq. (C.4) is thus brought down until it just equals the loss rate from other causes. Then a stable operating condition is reached.

3. CAVITY MODES

Mechanisms for loss of energy from the laser radiation field include scattering, reflection, and diffraction. The last two occur because of the nature of the cavity formed by the mirrors. The reason we need a cavity at all is that a very long optical path is necessary to produce enough amplification by stimulated emission. A large effective length is provided by multiple reflections off the mirrors. The mirrors also reduce diffraction losses.

Their function will now be considered in detail. Neglect the effect of the discharge tube for the moment. We can imagine a certain distribution of the electric field just in front of one of the mirrors, and then by the Kirchoff-Fresnel version of Huygens' principle calculate the field just in front of the other mirror. This process of calculation can then be repeated until a definite field pattern emerges which reproduces itself at a given mirror, except possibly for small diffraction and reflection losses. Such a pattern is called a *mode*. If we can excite the electric field into such a mode, it will persist in the region between the mirrors for a relatively long time, while oscillating at its characteristic frequency. These optical modes are analogous to the modes of a microwave cavity, and the oscillations they undergo are analogous to the oscillations of an LC resonant circuit.

This concept of modes can be illustrated by the standing plane waves that would occur between two infinite, parallel mirrors a distance l apart. If there are no reflection losses, the amplitude will be constant between the plates, and there will be a net change in phase in one round trip of amount

$$-2\pi\frac{2l}{\lambda}+2\delta=-\omega\frac{2l}{c}+2\delta$$

where δ is the phase change that occurs at reflection off each mirror. For the field pattern to reproduce itself after one round trip, this phase must be an integral multiple of 2π; that is, the angular frequency ω must satisfy

$$\omega\frac{2l}{c}-2\delta=2\pi p, \qquad p=\text{integer} \tag{C.5}$$

This type of cavity represents an infinite Fabry-Perot etalon, and Eq. (C.5) is the condition for peak transmission through such an etalon.

We recall from our discussion of such a system that for large energy reflection coefficients the transmission through the etalon can be very strongly peaked as a function of frequency. We see from Eq. (C.5) that from one peak to the next there will be a separation $\Delta\omega$ given by

$$\Delta\omega=\frac{2\pi c}{2l}=\frac{\pi c}{l}$$

The corresponding frequency spread is

$$\Delta\nu=\frac{c}{2l} \tag{C.6}$$

Furthermore, each peak will be approximately Lorentzian with a *half*-width at half maximum given from Eq. (5.88) by

$$W_{1/2}=\frac{2}{\sqrt{F}}$$

measured in units of the quantity

$$\frac{4\pi l}{\lambda}=\frac{2\omega l}{c}$$

Thus

$$\delta\left(\frac{2\omega l}{c}\right)=W_{1/2}, \qquad \text{or} \quad \delta\omega_{1/2}\frac{2l}{c}=\sqrt{\frac{2}{F}}$$

or

$$\delta\omega_{1/2}=\frac{c}{2l}\sqrt{\frac{2}{F}} \tag{C.7}$$

or for the frequency $\frac{1}{2}$-width

$$\delta\nu_{1/2} = \frac{c}{4\pi l}\sqrt{\frac{2}{F}} \tag{C.8}$$

Recall that the finesse F is given by

$$F = \frac{4R_1}{(1-R_1)^2} \tag{C.9}$$

The mirrors used in He–Ne gas lasers are coated with a multilayer dielectric film, specially manufactured to give values of R_1 of 0.99 or even higher. For $R_1 = 0.99$, $F \approx 4/(.01)^2$, and Eq. (C.8) becomes

$$\delta\nu_{1/2} = \frac{3.10^8}{4\pi}\frac{\sqrt{2}(.01)}{2} = 1.7 \cdot 10^5 \text{ cycles/sec} \tag{C.10}$$

for a laser cavity 1 m long.

The frequency difference between modes is, from (C.6),

$$\Delta\nu = \frac{c}{2l} = 1.5 \times 10^8 \text{ cycles/sec} \tag{C.11}$$

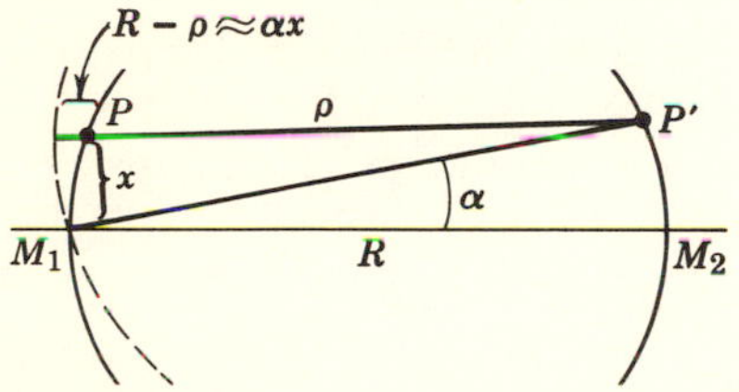

Fig. C.4

Plane mirrors are rarely used to form the optical cavity. Because of the nature of gas discharge tubes, a long cavity must constrain the electric field to a relatively small transverse dimension. A finite plane wave will spread by diffraction. Diffraction losses can be reduced by using one or more spherical mirrors. For these more general cases Eqs. (C.10) and (C.11) will still hold.

One of the simplest laser cavities to analyze is provided by the confocal configuration of Fig. C.4. Two spherical mirrors, each of radius of curvature R are located a distance R apart. Our general condition for the occurrence of a mode of the cavity is that the field distribution at M_1 reproduce after reflection off M_2. A less general condition, not obeyed by *all* modes, but obeyed by some of the most important ones, is that the field distribution at M_1 reproduce at M_2.

Let (x, y) be transverse coordinates on M_1 and (x', y') on M_2. We apply the diffraction formula of Eq. (7.10) to obtain the field on M_2 in terms of that on M_1:

$$E_2(x', y') = \frac{ik}{2\pi}\iint \frac{\epsilon^{-ik\rho}}{\rho} E_1(x, y)\, dx\, dy \tag{C.12}$$

(The obliquity factor is set equal to one.) We define the (small) angles α, β by

$$\begin{aligned} R\alpha &= x' \\ R\beta &= y' \end{aligned} \tag{C.13}$$

Then to lowest order we have

$$\rho = R - (R - \rho) \approx R - \alpha x - \beta y$$

$$= R - \frac{xx'}{R} - \frac{yy'}{R} \tag{C.14}$$

This is indicated for one dimension in Fig. C.4. The reason for the absence of the quadratic terms in x and y in Eq. (C.14) is that M_2 is a focal surface for the spherical wavefronts on M_1. At a focal point we observe Fraunhofer diffraction with only linear terms.

Equation (C.12) then becomes

$$E_2(x', y') = \frac{ik}{2\pi} \frac{e^{-ikR}}{R} \iint E_1(x, y) \exp\left[ik\left(\frac{x'x}{R} + \frac{yy'}{R}\right)\right] dx\, dy \tag{C15}$$

Now if the limits of integration in (C15) can be considered as $\pm\infty$, that is, for mirrors large compared with the size of the region where E is effectively nonzero, then $E_2(x', y')$ is proportional to the two-dimensional Fourier transform of $E_1(x, y)$.

Our condition that $E_2(x', y)$ be essentially the same as $E_1(x, y)$ says that E_1 must be its own Fourier transform. Functions which are their own Fourier transform are encountered in quantum mechanics in the theory of the harmonic oscillator. There the wave functions have the desired property. The wave functions are Hermite polynomials times a Gaussian function. Such functions will also be the solutions to our problem.

We work out the case for the lowest mode in detail. Let $E_1(x, y)$ be given by

$$E_1(x, y) = e^{-(x^2+y^2)/r_0^2} \tag{C.16}$$

where r_0 is a constant representing the distance from the axis at which E_1 is down to $1/e$ times its on-axis value. Use of (C.16) in (C.15) will give

$$E_2(x' y') = \frac{ik}{2\pi} \frac{e^{-ikR}}{R} \int e^{-(x^2/r_0^2)} e^{ik(xx'/R)}\, dx \int e^{-(y^2/r_0^2)\, ik(yy'/R)}\, dy \tag{C.17}$$

Consider the x integral. It can be evaluated by completing the square in the exponent:

$$\frac{-x^2}{r_0^2} + ik\frac{xx'}{R} = -\frac{1}{r_0^2}\left[x^2 - ik\frac{xx'r_0^2}{R} - \frac{1}{4}\left(\frac{kx'r_0^2}{R}\right)^2\right]$$

$$-\frac{k^2x'^2r_0^2}{4R^2}$$

$$= -\frac{1}{r_0^2}\left[x - \frac{ikx'r_0^2}{2R}\right]^2 - \frac{k^2x'^2r_0^2}{4R^2}$$

The x-integral then becomes

$$\int_{-\infty}^{\infty} e^{-x^2/r_0^2}\, e^{ik(xx'/R)}\, dx = e^{(-k^2x'^2r_0^2)/4R^2}$$

$$\times \int_{-\infty}^{\infty} \exp\left\{-\frac{1}{r_0^2}\left[x - ik\frac{x'r_0^2}{2R}\right]^2\right\} dx$$

$$= r\sqrt{\pi}\, e^{-k^2x'^2r_0^2/4R^2} \tag{C.18}$$

The y integration yields a similar result. Equation (C.17) then becomes

$$E_2(x'y') = \frac{ik}{2\pi}\frac{e^{-ikR}}{R} r^2\pi e^{-(k^2r_0^2/4R^2)(x'^2+y'^2)} \tag{C.19}$$

This will be the same function of x' and y' as E_1 is of x and y provided that we set

$$\frac{1}{r_0^2} = \frac{k^2r_0^2}{4R^2}$$

or

$$r_0 = \left(\frac{2R}{k}\right)^{1/2} = \left(\frac{\lambda R}{\pi}\right)^{1/2} \tag{C.20}$$

With this value of r_0, (C.19) becomes

$$E_2(x'y') = ie^{-ikR}e^{-(x'^2+y'^2)/r_0^2} \tag{C.21}$$

which reproduces $E_i(x, y)$ except for the phase factor $ie^{-ikR} = e^{[-ikR+(\pi/2)i]}$. After a complete round trip back to M_1, the phase factor becomes $e^{i(-2kR+\pi+2\delta)}$, and the condition that the mode pattern reproduce after one trip is

$$2kR - 2\delta - \pi = \frac{2\omega l}{c} - 2\delta - \pi = 2\pi p \tag{C.22}$$

This equation, which holds for the lowest confocal mode, is almost the same as Eq. (C.5), which holds for the plane parallel mirrors.

The quantity r_0 is called the radius of the mode. For $\lambda = 6328$ nm, $R = 1$ m, Eq. (C.20) gives

$$r_0 = 0.45 \text{ mm}$$

Away from the mirrors the field distribution is still Gaussian. Halfway between them the value of r_0 is reduced by a factor of $\sqrt{2}$.

The modes of other cavities using two identical spherical mirrors where $R \neq l$ can be studied by Fresnel diffraction calculations. We consider the so-called spherical cavity where $2R$ is close to l (Fig. C.5). It is equivalent to the hemispherical cavity, which puts a plane

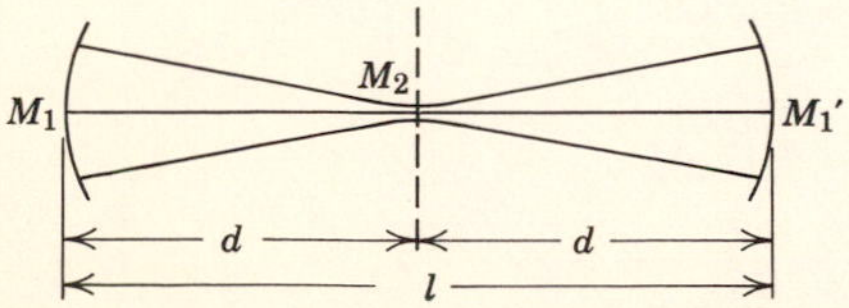

Fig. C.5

mirror M_2 halfway between the two spherical mirrors at the position shown in dotted lines in Fig. C.5. A point P on M_1 will have Cartesian coordinates $(x, y, R-\sqrt{R^2-x^2-y^2})$, and P' on M' will have coordinates $(x', y', l-\sqrt{R^2-x'^2-y'^2}+R)$ which can be approximated as follows:

$$P: \left(x, y, \frac{x^2+y^2}{2R}\right)$$
$$P': \left(x', y', l-\frac{(x'^2+y'^2)}{2R}\right) \tag{C.23}$$

Then $\rho^2 = |PP'|^2$ is given by

$$\rho^2 = (x-x')^2+(y-y')^2+\left(l-\frac{x^2+y^2+x'^2+y'^2}{2R}\right)^2$$
$$\cong (x-x')^2+(y-y')^2+l^2-l\frac{x^2+y^2+x'^2+y'^2}{R}$$

Then

$$\rho \approx l+\frac{1}{2l}\left[(x^2+x'^2)\left(1-\frac{l}{R}\right)-2xx'+(y^2+y'^2)\left(1-\frac{l}{R}\right)-2yy'\right]$$

We then use this expression for ρ in Eq. (C.12) with (C.16) for E_1. We obtain

$$E_1'(x'y') = \frac{e^{-ikl}}{l}\frac{ik}{2\pi}\exp\left[-\frac{ik}{2l}(x'^2+y'^2)\left(1-\frac{l}{R}\right)\right]$$
$$\times\iint \exp\left[-\frac{ik}{2l}(x^2+y^2)\left(1-\frac{l}{R}\right)\right]e^{(ik/l)(xx'+yy')}$$
$$\times e^{-(x^2+y^2)/r_0^2}\,dx\,dy \tag{C.24}$$

We can factor the integral into an x integral times a y integral. The argument of the exponential function in the x-integral will be

$$-x^2\left[\frac{1}{r_0{}^2}+\frac{ik}{2l}\left(1-\frac{l}{R}\right)\right]+\frac{ikxx'}{l}$$

We can complete the square in the same way we did to obtain Eq. (C.18). Equation (C.24) then becomes

$$E_1'(x', y') = \frac{e^{-ikl}}{l}\frac{ik}{2\pi}\exp\left[-\frac{ik}{2l}(x'^2+y'^2)\left(1-\frac{l}{R}\right)\right]$$
$$\times\frac{\pi}{[(1/r_0{}^2)+(ik/2l)(1-l/R)]}$$
$$\times\exp\left\{-\frac{k^2}{4l^2}\frac{1}{[(1/r_0{}^2)+(ik/2l)(1-l/R)]}(x'^2+y'^2)\right\} \tag{C.25}$$

We can make this take the form

$$E_1'(x'y') = e^{-ikl} e^{-(x'^2+y'^2)/r_0^2} e^{i\eta} \tag{C.26}$$

provided that r_0 satisfies

$$r_0^{\,4} = \frac{2lR^2}{k^2(R-l/2)} = \left(\frac{\lambda}{\pi}\right)^2 \frac{(l/2)R^2}{(R-l/2)} \tag{C.27}$$

and that η satisfies

$$\begin{aligned} \sin\eta &= \frac{2l}{kr_0^{\,2}} \\ \cos\eta &= 1 - \frac{l}{R} \end{aligned} \tag{C.28}$$

Then we have a proper cavity mode whose radius at M_1 is given by Eq. (C.27). Note that we must keep $R \neq l/2$ to avoid having r_0 become infinite.

The hemispherical cavity with a plane mirror a distance $d = l/2$ from the spherical mirror will have a mode radius r_0 given by (C.27) at the spherical mirror. The mode radius at the plane mirror is very much smaller than this. For d slightly smaller than R, the pattern at the plane mirror will be essentially the Fraunhofer diffraction pattern occurring at the center of curvature of M_1. The calculation of this pattern is the same as that leading to Eq. (C.19), and that equation still holds, but now r_0 satisfies (C.27) instead of (C.20). The radius r' of the mode at M_2 is given by

$$r'^2 = \frac{4R^2}{k^2 r_0^{\,2}}$$

or

$$\begin{aligned} r'^4 &= \frac{16R^4}{k^4 r_0^{\,4}} = \frac{16R^4(R-l/2)}{2lR^2k^2} = \frac{8R^2(R-l/2)}{2lk^2} \\ &= \frac{8R^2(R-d)}{4d(2\pi/\lambda)^2} = \left(\frac{\lambda}{\pi}\right)^2 R(R-d) \end{aligned} \tag{C.29}$$

where $d = l/2$ and where we have used the near equality of d with R. Actually one usually works with d slightly less than R.

As an example, take $d = 1$ meter, $r_0 = 2$ mm. Then from (C.27) we calculate a value of 2.5 mm for $R - d = R - l/2$. The corresponding value of r' from (C.29) is 0.1 mm.

The calculation that led to Eq. (C.29) is the same as would be needed to calculate the radius of the image formed by a lens with image distance d from the lens where the beam amplitude at the lens is given by

$$e^{-(x^2+y^2)/r_0^2}$$

The illuminance in the focal spot at M_2 will be $(kr^2/2R)^2$ times as great as that in the beam at M_1, as can be seen from Eq. (C.19). Using (C.27), this turns out to be equal to

$$\frac{E_e(2)}{E_e(1)} = \frac{d}{R-d} \tag{C.30}$$

For our above numerical example this ratio is 400, which is the square of the ratio of r' to r_0, as expected.

4. THE NATURE OF THE OUTPUT OF THE LASER

The light coming out of a laser exhibits both longitudinal and transverse coherence. Longitudinal coherence means that there is a definite correlation between the phases at a given point of the beam at different times. The coherence time τ will be related to the frequency spread $\delta\nu$ of each of the components by

$$\delta\nu\tau \approx 2\pi \tag{C.31}$$

The $\delta\nu$ here is not the $\delta\nu_{1/2}$ of Eqs. (C.9) and (C.10). The latter refer to an empty cavity. Laser action occurs only within a small frequency range near the top of each Fabry–Perot transmission maximum. This frequency spread can be thought of as caused by the noise fluctuations caused by spontaneous emission of some of the excited neon atoms. It is given by

$$\delta\nu_{\text{noise}} = \left(\frac{4\pi h\nu}{P}\right)(\delta\nu_{1/2})^2 \tag{C.32}$$

where P is the laser power, ν the frequency, h Planck's constant, $\delta\nu_{1/2}$ the half-width given by

$$\delta\nu_{1/2} = \frac{\sqrt{3}c}{4\pi lF} \tag{C.33}$$

which follows from Eq. (C.7). A typical value for $\delta\nu_{\text{noise}}$ might be 10^{-2} cycles per second!

A more important cause of $\delta\nu$ is vibrations of the mount holding the mirrors, Even if the mount is a rigid bar, it can still undergo thermal oscillations analogous to those undergone by a suspended galvanometer mirrors. In each mode of the bar there will be stored a mean energy $\langle E \rangle = k_BT$. The frequency spread from this cause can be shown to be

$$\frac{\delta\nu_{\text{thermal}}}{\nu} = \left(\frac{2k_BT}{YV}\right)^{1/2} \tag{C.34}$$

where k_B is the Boltzmann constant, Y is Young's modulus for the mirror mount material, T the temperature, and V the volume of the mount. A typical value of $\delta\nu_{\text{thermal}}$ is 2 cps. The best experimental values of $\delta\nu$ are about 20 cycles.

From the uncertainty principle $2\pi\delta\nu\tau_c \sim 1$, we obtain a coherence time τ_c of about 8 milliseconds. This represents the length of time it takes the phase of the laser light to change, on the average by 1 radian.

This value of $\delta\nu$ implies that the beat frequency between two well-stabilized but independent lasers will fluctuate by 20 cycles on a short-term basis. On a long-term basis the beat frequency will drift because of unequal thermal expansion of the two mirror mounts.

REFERENCES

B. A. Lengyel, *Introduction to Laser Physics*.
W. V. Smith and P. P. Sorokin, *The Laser*.
A. Yariv, *Quantum Electronics*.

BIBLIOGRAPHY

Boas, Mary P.: *Mathematical Methods in the Physical Sciences.* John Wiley and Sons, New York, 1966.

Born, Max, and Emil Wolf: *Principles of Optics.* Third revised edition. Pergamon Press, Oxford, 1965.

Brouwer, Willem: *Matrix Methods in Optical Instrument Design.* W. A. Benjamin, New York, 1964.

Brown, Frederick C.: *The Physics of Solids.* W. A. Benjamin, New York, 1967.

Courant, Richard: *Differential and Integral Calculus.* Vol. 1, Interscience, New York, 1965.

Cox, Arthur: *Photographic Optics*, Focal Press, London, 1966.

De Velis, John B., and George O. Reynolds: *Theory and Applications of Holography.* Addison-Wesley Publishing Co., Reading, Mass., 1967.

Ditchburn, R. W.: *Light.* Second edition. Interscience Publishers, London, 1963.

Drude, Paul: *Theory of Optics.* Longmans, Green, New York, 1902; Dover Publications, New York, 1959.

Fowles, Grant R.: *Introduction to Modern Optics.* Holt, Rinehart, and Winston, New York, 1968.

Françon, M.: *Diffraction: Coherence in Optics.* Pergamon Press, Oxford, 1966.

Françon, M.: *Modern Applications of Physical Optics.* Interscience Publishers, New York, 1963.

Françon, M.: *Optical Interferometry.* Academic Press, New York, 1966.

Garbuny, Max: *Optical Physics.* Academic Press, New York, 1965.

Goodman, Joseph W.: *Introduction to Fourier Optics.* McGraw-Hill Book Company, New York, 1968.

Hardy, A. C., and F. H. Perrin: *The Principles of Optics.* McGraw-Hill Book Co., New York, 1932.

Harrick, N. J.: *Internal Reflection Spectroscopy.* Interscience Publishers, New York, 1967.

Heavens, O. S.: *Optical Properties of Thin Solid Films.* Academic Press, New York, 1955.

Herzberger, Max: *Modern Geometrical Optics.* Interscience Publishers, New York, 1958.

Holland, L.: *Vacuum Deposition of Thin Films.* John Wiley and Sons, New York, 1956.

Jackson, John David: *Classical Electrodynamics.* John Wiley and Sons, New York, 1962.

Jenkins, Francis A., and Harvey E. White: *Fundamentals of Optics.* Third edition. McGraw-Hill Book Co., New York, 1957.

Kingslake, Rudolf (editor): *Applied Optics and Optical Engineering*, Vols. I–IV. Academic Press, New York, 1965–1967.

Kittel, Charles: *Introduction to Solid State Physics.* Third edition. John Wiley and Sons, New York, 1966.

Klauder, John R., and E. C. G. Sudarshan: *Fundamentals of Quantum Optics.* W. A. Benjamin, Inc., New York, 1968.

Kline, Morris, and Irwin W. Kay: *Electromagnetic Theory and Geometrical Optics.* Interscience Publishers, New York, 1965.

Lengyel, Bela A.: *Introduction to Laser Physics*. John Wiley and Sons, New York, 1966.

Levi, Leo: *Applied Optics*, Vol. 1. John Wiley and Sons, New York, 1968.

Lighthill, M. J.: *Introduction to Fourier Analysis and Generalized Functions*. Cambridge University Press, 1958.

Linfoot, E. H.: *Fourier Methods in Optical Image Evaluation*. Focal Press, London, 1964.

Longhurst, R. S.: *Geometrical and Physical Optics*, Second edition, Longmans, Green and Co., London; John Wiley and Sons, New York, 1967.

Martin, L. C.: *The Theory of the Microscope*, American Elsevier Publishing Co., New York, 1966.

Mertz, Lawrence: *Transformations in Optics*, John Wiley and Sons, New York, 1965.

O'Neill, E. L.: *Introduction to Statistical Optics*, Addison-Wesley Publishing Co., Reading Mass., 1963.

Pearson, J. M.: *A Theory of Waves*, Allyn and Bacon, Boston, 1966.

Pohl, R. W.: *Optik und Atomphysik*, Twelfth edition, Springer Verlag, Berlin, 1967.

Rossi, Bruno: *Optics*, Addison-Wesley Publishing Co., Reading, Mass., 1957.

Seitz, Frederick: *Modern Theory of Solids*, McGraw-Hill Book Company, New York, 1940.

Smith, Howard M.: *Principles of Holography*, Wiley-Interscience, New York, 1969.

Smith, Warren J.: *Modern Optical Engineering*, McGraw-Hill Book Co., New York, 1966.

Smith, William V., and Peter P. Sorokin: *The Laser*, McGraw-Hill Book Company, New York, 1966.

Stone, John M.: *Radiation and Optics*, McGraw-Hill Book Co., New York, 1963.

Stratton, Julius A.: *Electromagnetic Theory*, McGraw-Hill Book Company, New York, 1941.

Stroke, George W.: *An Introduction to Coherent Optics and Holography*, Second edition, Academic Press, New York, 1969.

Strong, John: *Concepts of Classical Optics*, W. H. Freeman and Co., San Francisco, 1958.

Tippett, J. T., D. A. Berkowitz, L. C. Clapp, C. J. Koester, and A . Vanderburgh, Jr. (editors): *Optical and Electro-Optical Information Processing*, Massachusetts Institute of Technology Press, Cambridge, Mass., 1964.

Troup, C. J.: *Optical Coherence Theory—Recent Developments*, Methuen and Co., London, 1967.

Von Hippel, H. R.: *Dielectrics and Waves*, John Wiley and Sons, New York, 1954.

Walsh, J. W. T.: *Photometry*, Third edition, Constable and Co., London, 1958; Dover Publishing Co., New York, 1965.

Welford, W. T.: *Geometrical Optics*, North-Holland Publishing Co., Amsterdam, 1962.

Wendlandt, W. H. (editor): *Modern Aspects of Reflectance Spectroscopy*, Plenum Press, New York, 1968.

Wolf, Emil (editor): *Progress in Optics*, North-Holland Publishing Co., Amsterdam, 1961–1966.

Wood, Robert W.: *Physical Optics* (3rd. ed.), MacMillan, New York, 1934.

Yariv, Amnon: *Quantum Electronics*, John Wiley and Sons, New York, 1967.

INDEX